ONE WEEK LOAN

The Finite Element Method for

Solid and Structural Mechanics

Sixth edition

Professor O.C. Zienkiewicz, CBE, FRS, FREng is Professor Emeritus at the Civil and Computational Engineering Centre, University of Wales Swansea and previously Director of the Institute for Numerical Methods in Engineering at the University of Wales Swansea, UK. He holds the UNESCO Chair of Numerical Methods in Engineering at the Technical University of Catalunya, Barcelona, Spain. He was the head of the Civil Engineering Department at the University of Wales Swansea between 1961 and 1989. He established that department as one of the primary centres of finite element research. In 1968 he became the Founder Editor of the *International Journal for Numerical Methods in Engineering* which still remains today the major journal in this field. The recipient of 27 honorary degrees and many medals, Professor Zienkiewicz is also a member of five academies – an honour he has received for his many contributions to the fundamental developments of the finite element method. In 1978, he became a Fellow of the Royal Society and the Royal Academy of Engineering. This was followed by his election as a foreign member to the US Academy of Engineering (1981), the Polish Academy of Science (1985), the Chinese Academy of Sciences (1998), and the National Academy of Science, Italy (Academia dei Lincei) (1999). He published the first edition of this book in 1967 and it remained the only book on the subject until 1971.

Professor R.L. Taylor has more than 40 years' experience in the modelling and simulation of structures and solid continua including two years in industry. He is Professor in the Graduate School and the Emeritus T.Y. and Margaret Lin Professor of Engineering at the University of California at Berkeley. In 1991 he was elected to membership in the US National Academy of Engineering in recognition of his educational and research contributions to the field of computational mechanics. Professor Taylor is a Fellow of the US Association of Computational Mechanics – USACM (1996) and a Fellow of the International Association of Computational Mechanics – IACM (1998). He has received numerous awards including the Berkeley Citation, the highest honour awarded by the University of California at Berkeley, the USACM John von Neumann Medal, the IACM Gauss–Newton Congress Medal and a Dr.-Ingenieur ehrenhalber awarded by the Technical University of Hannover, Germany. Professor Taylor has written several computer programs for finite element analysis of structural and non-structural systems, one of which, *FEAP*, is used world-wide in education and research environments. A personal version, *FEAPpv*, available from the publisher's website, is incorporated into the book.

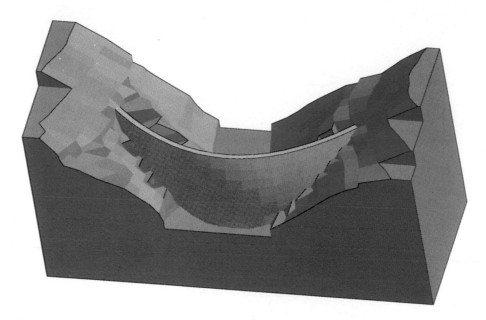

(a) Geometrical model of an arch dam

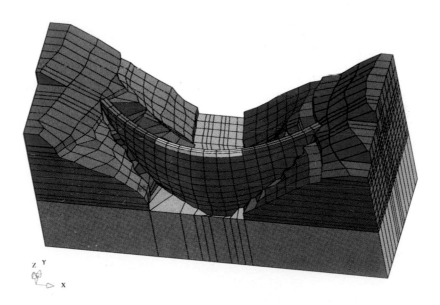

(b) Finite element discretization of an arch dam including the foundation

Plate 1 Three dimensional non-linear analysis of a dam
Courtesy of Prof. Miguel Cervera, CIMNE, Barcelona. Source: B. Suarez, M. Cervera and J. Miguel Canet, 'Safety assessment of the Suarna arch dam using non-linear damage model', *Proc. Int. Sym. New Trends and Guidelines on Dam Safety*, Barcelona, Spain, 1998. L. Berga (ed.) Balakema, Rotterdam.

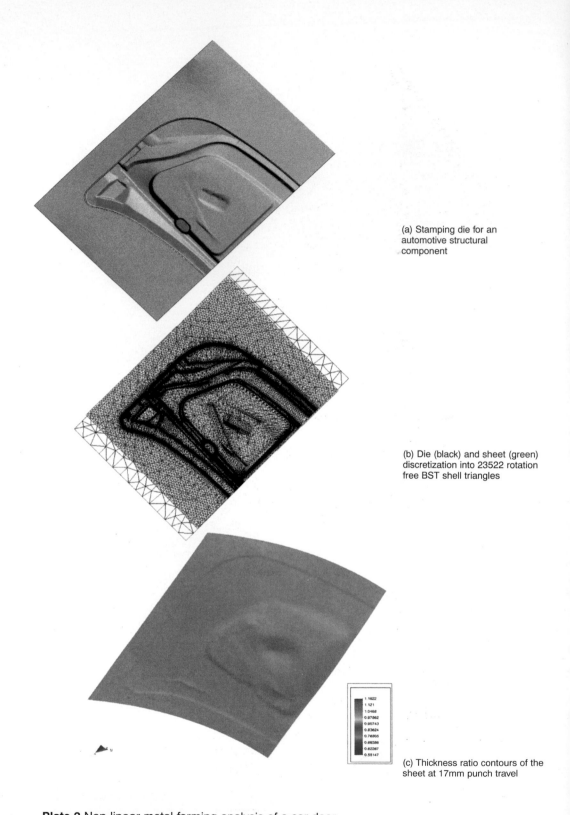

(a) Stamping die for an automotive structural component

(b) Die (black) and sheet (green) discretization into 23522 rotation free BST shell triangles

(c) Thickness ratio contours of the sheet at 17mm punch travel

Plate 2 Non-linear metal forming analysis of a car door
Courtesy of Prof. E. Oñate, CIMNE and DECAD S.A Barcelona. Source: E. Oñate, F. Zarate, J. Rojek, G. Duffet, L. Neamtui, 'Adventures in rotation free elements for sheet stamping analysis'. 4th Int. Conf. Workshop on Numerical Simulation of 3D Sheet Forming Processess (NUMISHEET' 99, Besancon, France, Sept 13-17, 1999)

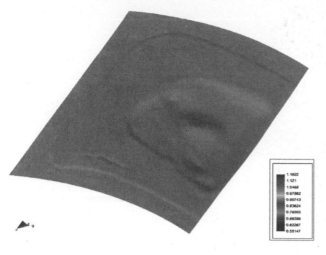

(d) Thickness ratio contours at 35mm punch travel

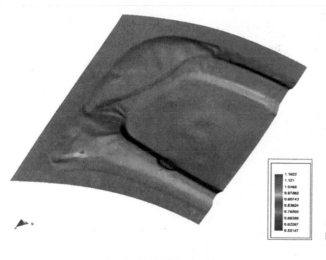

(e) Thickness ratio contours at 54mm punch travel

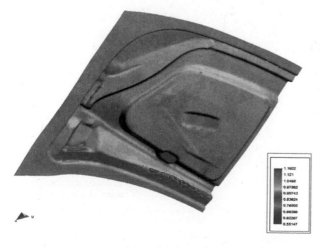

(f) Thickness ratio contours at 72mm punch travel

Plate 2 *continued* Non-linear metal forming analysis of a car door

Plate 3 Car crash analysis

Frontal crash of a Neon Car performed using LS-DYNA. Courtesy of Livermore Software Technology Corporation. Model developed by FHWA/NHTSA National Crash Analysis Center of the George Washington University

The Finite Element Method for Solid and Structural Mechanics

Sixth edition

O.C. Zienkiewicz, CBE, FRS
UNESCO Professor of Numerical Methods in Engineering
International Centre for Numerical Methods in Engineering, Barcelona
Previously Director of the Institute for Numerical Methods in Engineering
University of Wales Swansea

R.L. Taylor
Professor in the Graduate School
Department of Civil and Environmental Engineering
University of California at Berkeley
Berkeley, California

ELSEVIER
BUTTERWORTH
HEINEMANN

AMSTERDAM • BOSTON • HEIDELBERG • LONDON • NEW YORK • OXFORD
PARIS • SAN DIEGO • SAN FRANCISCO • SINGAPORE • SYDNEY • TOKYO

Butterworth-Heinemann is an imprint of Elsevier
Linacre House, Jordan Hill, Oxford OX2 8DP, UK
30 Corporate Drive, Suite 400, Burlington, MA 01803, USA

First edition published in1967 by McGraw-Hill
Fith edition published by Butterworth Heinemann 2000
Reprinted 2002
Sixth edition 2005
Reprinted 2006 (three times)

British Library Cataloguing in Publication Data
A catalogue record for this book is available from the British Library

Library of Congress Cataloging-in-Publication Data
A catalog record for this book is available from the Library of Congress

ISBN–13: 978-0-7506-6321-2
ISBN–10: 0-7506-6321-9

For information on all Butterworth-Heinemann publications
visit our website at books.elsevier.com

Printed and bound in *Great Britain*

06 07 08 09 10   10 9 8 7 6 5 4

Working together to grow
libraries in developing countries

www.elsevier.com | www.bookaid.org | www.sabre.org

ELSEVIER   BOOK AID International   Sabre Foundation

Dedication

This book is dedicated to our wives Helen and Mary Lou
and our families for their support and patience during
the preparation of this book, and also to all of our students
and colleagues who over the years have contributed to our
knowledge of the finite element method. In particular we
would like to mention Professor Eugenio Oñate and his
group at CIMNE for their help, encouragement and support
during the preparation process.

Contents

Preface

It is thirty-eight years since the *The Finite Element Method in Structural and Continuum Mechanics* was first published. This book, which was the first dealing with the finite element method, provided the basis from which many further developments occurred. The expanding research and field of application of finite elements led to the second edition in 1971, the third in 1977, the fourth as two volumes in 1989 and 1991 and the fifth as three volumes in 2000. The size of each of these editions expanded geometrically (from 272 pages in 1967 to the fifth edition of 1482 pages). This was necessary to do justice to a rapidly expanding field of professional application and research. Even so, much filtering of the contents was necessary to keep these editions within reasonable bounds.

In the present edition we retain the three volume format of the fifth edition but have decided not to pursue having three contiguous volumes – rather we treat the whole work as an assembly of three separate works. Each one is capable of being used without the others and each one appeals perhaps to a different audience. Though naturally we recommend the use of the whole ensemble to people wishing to devote much of their time and study to the finite element method.

The first volume is renamed *The Finite Element Method: Its Basis and Fundamentals*. This volume covers the topic starting from a physical approach to solve problems in linear elasticity. The volume then presents a mathematical framework from which general problems may be formulated and solved using variational and Galerkin methods. The general topic of shape functions is also presented for situations in which the approximating functions are C^0 continuous. The two- and three-dimensional problems of linear elasticity are then presented in a unified manner using higher order shape functions. This is followed by consideration of quasi-harmonic problems governed by Laplace and Poisson differential equations. The patch test is introduced and used as a means to guarantee convergence of the method. We also cover in some depth solution forms using mixed methods with special consideration given to problems in which incompressibility can occur. The solution of transient problems is presented using semi-discrete formulations and finite element in time concepts. The volume concludes with a presentation of coupled problems.

In this volume we consider more advanced problems in solid and structural mechanics while in a third volume we consider applications in fluid dynamics. It is our intent that the present volume can be used by investigators familiar with the finite element method at the level presented in the first volume or any other basic textbook on the subject. However, the volume has been prepared such that it can stand alone. The volume has been reorganized from the previous edition to cover consecutively two main subject areas. In the first part we consider non-linear problems in solid mechanics and in the second part linear and non-linear problems in structural mechanics.

In Chapters 1 to 9 we consider non-linear problems in solid mechanics. In these chapters the special problems of solving non-linear equation systems are addressed. We begin by restricting our attention to non-linear behaviour of materials while retaining the assumptions on small strain. This serves as a bridge to more advanced studies later in which geometric effects form large displacements and deformations are presented. Indeed, non-linear applications are of great importance today and of practical interest in most areas of engineering and physics. By starting our study first using a small strain approach we believe the reader can more easily comprehend the various aspects which need to be understood to master the subject matter. We cover in some detail formulations of material models for viscoelasticity, plasticity and viscoplasticity which should serve as a basis for applications to other material models. In our study of finite deformation problems we present a series of approaches which may be used to solve problems including extensions for treatment of constraints such as near incompressibility, rigid and multi-body motions and discrete element forms. The chapter on discrete element methods was prepared by Professor Nenad Bićanić of the University of Glasgow, UK.

In the second part of the volume we consider problems in structural mechanics. In this class of applications the dimension of the problem is reduced using basic kinematic assumptions. We begin the presentation in a new chapter that considers rod problems where two of the dimensions of the structure are small compared to the third. This class of problems is a combination of beam bending, axial extension and torsion. Again we begin from a small strain assumption and introduce alternative forms of approximation for the Euler–Bernoulli and the Timoshenko theory. In the former theory it is necessary now to use C^1 interpolation (i.e. continuous displacement and slope) to model the bending behaviour, whereas in the latter theory use of C^0 interpolation is permitted when special means are included to avoid 'locking' in the transverse shear response. Based upon the study of rods we then present a detailed study of problems in which only one dimension is small compared to the other two. Building on the results from rods we present a coverage for thin plates (Kirchhoff theory), thick plates (Reissner–Mindlin theory) and their corresponding forms for shells. We then consider the problem of large strains and present forms for buckling and large displacements.

The volume includes a new chapter on multi-scale effects. This is a recent area of much research and the chapter presents a summary of some notable recent results. We are indebted to Professor Bernardo Schrefler of the University of Padova, Italy, for preparing this timely contribution.

The volume concludes with a short chapter on computational methods that describes a companion computer program that can be used to solve several of the problem classes described in this volume.

We emphasize here the fact that all three of our volumes stress the importance of considering the finite element method as a unique and whole basis of approach and that it contains many of the other numerical analysis methods as special cases. Thus, imagination and knowledge should be combined by the readers in their endeavours.

The authors are particularly indebted to the International Centre of Numerical Methods in Engineering (CIMNE) in Barcelona who have allowed their pre- and post-processing code (GiD) to be accessed from the web site. This allows such difficult tasks as mesh generation and graphic output to be dealt with efficiently. The authors are also grateful to Professor Eric Kasper for his careful scrutiny of the entire text. We also acknowledge the assistance of Matt Salveson who also helped in proofreading the text.

Resources to accompany this book

Complete source code and user manual for program *FEAPpv* may be obtained at no cost from the publisher's web page: http://books.elsevier.com/companions/ or from the author's web page: http://www.ce.berkeley.edu/~rlt

OCZ and RLT

1

General problems in solid mechanics and non-linearity

1.1 Introduction

Many introductory texts on the finite element method discuss the solution for linear problems of elasticity and field equations.[1-3] In practical applications the limitation of linear elasticity, or more generally of linear behaviour, often precludes obtaining an accurate assessment of the solution because of the presence of 'non-linear' effects and/or because the geometry has a 'thin' dimension in one or more directions. In this book we describe extensions to the formulations introduced to solve linear problems to permit solutions to both classes of problems.

Non-linear behaviour of solids takes two forms: material non-linearity and geometric non-linearity. The simplest form of non-linear material behaviour is that of elasticity for which the stress is not linearly proportional to the strain. More general situations are those in which the loading and unloading response of the material is different. Typical here is the case of classical elastic–plastic behaviour.

When the deformation of a solid reaches a state for which the undeformed and deformed shapes are substantially different a state of *finite deformation* occurs. In this case it is no longer possible to write linear strain–displacement or equilibrium equations on the undeformed geometry. Even before finite deformation exists it is possible to observe *buckling* or *load bifurcations* in some solids and non-linear equilibrium effects need to be considered. The classical Euler column, where the equilibrium equation for buckling includes the effect of axial loading, is an example of this class of problem. When deformation is large the boundary conditions can also become non-linear. Examples are pressure loading that remains normal to the deformed body and also the case where the deformed boundary interacts with another body. This latter example defines a class known as *contact problems* and much research is currently performed in this area. An example of a class of problems involving non-linear effects in deformation measures, material behaviour and contact is the analysis of a rolling tyre. A typical mesh for a tyre analysis is shown in Fig. 1.1. The cross-section shown is able to model the layering of rubber and cords and the overall character of a tread. The full mesh is generated by sweeping the cross-section around the wheel axis with a variable spacing in the area which will be in contact. A formulation in which the mesh is fixed and the material rotates is commonly used to perform the analysis.[4-7]

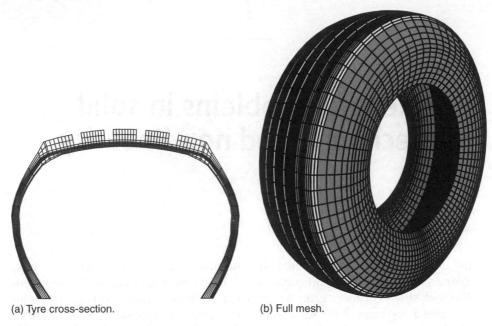

(a) Tyre cross-section.   (b) Full mesh.

Fig. 1.1 Finite element mesh for tyre analysis.

Generally the accurate solution of solid problems which have one (or more) small dimension(s) compared to the others cannot be achieved efficiently using standard two- or three-dimensional finite element formulations. Traditionally separate theories of *structural mechanics* are introduced to solve this class of problems. A *plate* is a flat structure with one thin (small) direction which is called the thickness. A *shell* is a curved structure in space with one such small thickness direction. Structures with two small dimensions are called *beams*, *frames*, or *rods*. A primary reason why use of standard two- or three-dimensional finite element formulations do not yield accurate solutions is the numerical ill-conditioning which results in their algebraic equations. In this book we combine the traditional approaches of structural mechanics with a much stronger link to the full three-dimensional theory of solids to obtain formulations which are easily solved using standard finite element approaches.

This book considers both solid and structural mechanics problems and formulations which make practical finite element solutions feasible. We divide the volume into two main parts. In the first part we consider problems in which continuum theory of solids continues to be used, whereas in the second part we focus attention on theories of structural mechanics to describe the behaviour of rods, plates and shells.

In the present chapter we review the general equations for analysis of solids in which deformations remain 'small' but material behaviour includes effects of a non-linear kind. We present the theory in both an indicial (or tensorial) form as well as in the matrix form commonly used in finite element developments. We also reformulate the equations of solids in a variational (Galerkin) form. In Chapter 2 we present a general scheme based on the Galerkin method to construct a finite element approximate solution to problems based on variational forms. In this chapter we consider both *irreducible* and *mixed* forms of finite element approximation and indicate where the mixed forms have distinct advantages. Here we also show how the linear problems of solids for steady state and transient behaviour become non-linear when the material constitutive model is represented in a non-linear form. Some discussion on the solution of transient non-linear finite element forms is included. Since the form of the inertial effects is generally unaffected by non-linearity, in the remainder of this volume we shall primarily confine our remarks to terms arising from non-linear material behaviour and finite deformation effects.

In Chapter 3 we describe various possible methods for solving non-linear algebraic equations. This is followed in Chapter 4 by consideration of material non-linear behaviour and completes the development of a general formulation from which a finite element computation can proceed.

In Chapter 5 we present a summary for the study of finite deformation of solids. Basic relations for defining deformation are presented and used to write variational (Galerkin) forms related to the undeformed configuration of the body and also to the deformed configuration. It is shown that by relating the formulation to the deformed body a result is obtained which is nearly identical to that for the small deformation problem we considered in the small deformation theory treated in the early chapters of this volume. Essential differences arise only in the constitutive equations (stress–strain laws) and the addition of a new stiffness term commonly called the *geometric* or *initial stress* stiffness. For constitutive modelling we summarize in Chapter 6 alternative forms for elastic and inelastic materials. Contact problems are discussed in Chapter 7. Here we summarize methods commonly used to model the interaction of intermittent contact between surfaces of bodies.

In Chapter 8 we show that analyses of rigid and so-called pseudo-rigid bodies[8] may be developed directly from the theory of deformable solids. This permits the inclusion in programs of options for multi-body dynamic simulations which combine deformable solids with objects modelled as rigid bodies. In Chapter 9 we discuss specialization of the finite deformation problem to address situations in which a large number of small bodies interact [multi-particle or granular bodies commonly referred to as *discrete element methods* (DEM) or *discrete deformation analysis* (DDA)].

In the second part of this book we study the behaviour of problems of *structural mechanics*. In Chapter 10 we present a summary of the behaviour of rods (beams) modelled by linear kinematic behaviour. We consider cases where deformation effects include axial, bending and transverse shearing strains (Timoshenko beam theory[9]) as well as the classical theory where transverse effects are neglected (Euler–Bernoulli theory). We then describe the solution of plate problems, considering first the problem of thin plates (Chapter 11) in which only bending deformations are included and, second, the problem in which both bending and shearing deformations are present (Chapter 12).

The problem of shell behaviour adds in-plane membrane deformations and curved surface modelling. Here we split the problem into three separate parts. The first combines simple flat elements which include bending and membrane behaviour to form a faceted approximation to the curved shell surface (Chapter 13). Next we involve the addition of shearing deformation and use of curved elements to solve axisymmetric shell problems (Chapter 14). We conclude the presentation of shells with a general form using curved isoparametric element shapes which include the effects of bending, shearing, and membrane deformations (Chapter 15). Here a very close link with the full three-dimensional analysis will be readily recognized.

In Chapter 16 we address a class of problems in which the solution in one coordinate direction is expressed as a series, for example a Fourier series. Here, for linear material behaviour, very efficient solutions can be achieved for many problems. Some extensions to non-linear behaviour are also presented.

In Chapter 17 we specialize the finite deformation theory to that which results in large displacements but small strains. This class of problems permits use of all the constitutive equations discussed for small deformation problems and can address classical problems of instability. It also permits the construction of non-linear extensions to plate and shell problems discussed in Chapters 11–15 of this volume.

We conclude the descriptions applied to solids in Chapter 18 with a presentation of multi-scale effects in solids.

In the final chapter we summarize the capabilities of a companion computer program (called *FEAPpv*) that is available at the publisher's web site. This program may be used to address the class of non-linear solid and structural mechanics problems described in this volume.

1.2 Small deformation solid mechanics problems

1.2.1 Strong form of equations – indicial notation

In this general section we shall describe how the various equations of solid mechanics* can become non-linear under certain circumstances. In particular this will occur for solid mechanics problems when non-linear stress–strain relationships are used. The chapter also presents the notation and the methodology which we shall adopt throughout this book. The reader will note how simply the transition between forms for linear and non-linear problems occurs.

The field equations for solid mechanics are given by equilibrium behaviour (balance of momentum), strain-displacement relations, constitutive equations, boundary conditions, and initial conditions.[10–15]

In the treatment given here we will use two notational forms. The first is a cartesian tensor indicial form and the second is a matrix form (see reference 1 for additional details on both approaches). In general, we shall find that both are useful to describe particular parts of formulations. For example, when we describe large strain problems the development of the so-called 'geometric' or 'initial stress' stiffness is most easily described by using an indicial form. However, in much of the remainder, we shall find that it is convenient to use a matrix form. The requirements for transformations between the two will also be indicated.

In the sequel, when we use indicial notation an index appearing once in any term is called a *free* index and a repeated index is called a *dummy* index. A dummy index may only appear twice in any term and implies summation over the range of the index.

---

* More general theories for solid mechanics problems exist that involve higher order micro-polar or couple stress effects; however, we do not consider these in this volume.

Thus if two vectors a_i and b_i each have three terms the form $a_i b_i$ implies

$$a_i b_i = a_1 b_1 + a_2 b_2 + a_3 b_3$$

Note that a dummy index may be replaced by any other index without changing the meaning, accordingly

$$a_i b_i \equiv a_j b_j$$

Coordinates and displacements

For a fixed Cartesian coordinate system we denote coordinates as x, y, z or in index form as x_1, x_2, x_3. Thus the vector of coordinates is given by

$$\mathbf{x} = x_1 \mathbf{e}_1 + x_2 \mathbf{e}_2 + x_3 \mathbf{e}_3 = x_i \, \mathbf{e}_i$$

in which $\mathbf{e}_i$ are unit base vectors of the Cartesian system and the summation convention described above is adopted.

Similarly, the displacements will be denoted as u, v, w or u_1, u_2, u_3 and the vector of displacements by

$$\mathbf{u} = u_1 \mathbf{e}_1 + u_2 \mathbf{e}_2 + u_3 \mathbf{e}_3 = u_i \, \mathbf{e}_i$$

Generally, we will denote all quantities by their components and where possible the coordinates and displacements will be denoted as x_i and u_i, respectively, in which the range of the index i is $1, 2, 3$ for three-dimensional applications (or $1, 2$ for two-dimensional problems).

Strain–displacement relations

The strains may be expressed in Cartesian tensor form as

$$\varepsilon_{ij} = \frac{1}{2} \left(\frac{\partial u_i}{\partial x_j} + \frac{\partial u_j}{\partial x_i} \right) \tag{1.1}$$

and are valid measures provided deformations are small. By a small deformation problem we mean that

$$|\varepsilon_{ij}| << 1 \quad \text{and} \quad |\omega_{ij}^2| << \|\varepsilon_{ij}\|$$

where $|\cdot|$ denotes absolute value and $\|\cdot\|$ a suitable norm. In the above ω_{ij} denotes a small rotation given by

$$\omega_{ij} = \frac{1}{2} \left(\frac{\partial u_i}{\partial x_j} - \frac{\partial u_j}{\partial x_i} \right) \tag{1.2}$$

and thus the displacement gradient may be expressed as

$$\frac{\partial u_i}{\partial x_j} = \varepsilon_{ij} + \omega_{ij} \tag{1.3}$$

Equilibrium equations – balance of momentum

The equilibrium equations (balance of linear momentum) are given in index form as

$$\sigma_{ji,j} + b_i = \rho \ddot{u}_i, \qquad i, j = 1, 2, 3 \tag{1.4}$$

where σ_{ij} are components of (Cauchy) stress, ρ is mass density, and b_i are body force components. In the above, and in the sequel, we use the convention that the partial derivatives are denoted by

$$f_{,i} = \frac{\partial f}{\partial x_i} \quad \text{and} \quad \dot{f} = \frac{\partial f}{\partial t}$$

for coordinates and time, respectively.

Similarly, moment equilibrium (balance of angular momentum) yields symmetry of stress given in indicial form as

$$\sigma_{ij} = \sigma_{ji} \tag{1.5}$$

Equations (1.4) and (1.5) hold at all points x_i in the domain of the problem Ω.

Boundary conditions

Stress boundary conditions are given by the traction condition

$$t_i = \sigma_{ji} n_j = \bar{t}_i \tag{1.6}$$

for all points which lie on the part of the boundary denoted as Γ_t. A quantity with a 'bar' denotes a specified function.

Similarly, displacement boundary conditions are given by

$$u_i = \bar{u}_i \tag{1.7}$$

and apply for all points which lie on the part of the boundary denoted as Γ_u.

Many additional forms of boundary conditions exist in non-linear problems. Conditions where the boundary of one part interacts with another part, so-called contact conditions, will be taken up in Chapter 7. Similarly, it is necessary to describe how loading behaves when deformations become large. Follower pressure loads are one example of this class and we consider this further in Sec. 5.7.

Initial conditions

Finally, for transient problems in which the inertia term $\rho \ddot{u}_i$ is important, initial conditions are required. These are given for an initial time denoted as 'zero' by

$$u_i(x_j, 0) = \bar{d}_i(x_j) \quad \text{and} \quad \dot{u}_i(x_j, 0) = \bar{v}_i(x_j) \quad \text{in } \Omega \tag{1.8}$$

It is also necessary in some problems to specify the state of stress at the initial time.

Constitutive relations

All of the above equations apply to any material provided the deformations remain small. The specific behaviour of a material is described by constitutive equations which relate the stresses to imposed strains and, often, other sources which cause deformation (e.g. temperature).

The simplest material model is that of linear elasticity where quite generally

$$\sigma_{ij} = C_{ijkl}(\varepsilon_{kl} - \varepsilon_{kl}^{(0)})$$ (1.9a)

in which C_{ijkl} are *elastic moduli* and $\varepsilon_{kl}^{(0)}$ are strains arising from sources other than displacement. For example, in thermal problems strains result from change in temperature and these may be given by

$$\varepsilon_{kl}^{(0)} = \alpha_{kl}[T - T_0]$$ (1.9b)

in which α_{kl} are coefficients of linear expansion and T is temperature with T_0 a reference temperature for which thermal strains are zero.

For linear *isotropic* materials these relations simplify to

$$\sigma_{ij} = \lambda\delta_{ij}(\varepsilon_{kk} - \varepsilon_{kk}^{(0)}) + 2\mu(\varepsilon_{ij} - \varepsilon_{ij}^{(0)})$$ (1.10a)

and

$$\varepsilon_{kl}^{(0)} = \delta_{ij}\alpha[T - T_0]$$ (1.10b)

where λ and μ are Lamé elastic parameters and α is a scalar coefficient of linear expansion.[10,11] In addition, δ_{ij} is the Kronecker delta function given by

$$\delta_{ij} = \left\{ \begin{array}{ll} 1; & \text{for } i = j \\ 0; & \text{for } i \neq j \end{array} \right.$$

Many materials are not linear nor are they elastic. The construction of appropriate constitutive models to represent experimentally observed behaviour is extremely complex. In this book we will illustrate a few classical models of behaviour and indicate how they can be included in a general solution framework. Here we only wish to indicate how a non-linear material behaviour affects our formulation. To do this we consider non-linear elastic behaviour represented by a strain–energy density function W in which stress is computed as[11]

$$\sigma_{ij} = \frac{\partial W}{\partial \varepsilon_{ij}}$$ (1.11)

Materials based on this form are called *hyperelastic*. When the strain–energy is given by the quadratic form

$$W = \tfrac{1}{2}\varepsilon_{ij}C_{ijkl}\varepsilon_{kl} - \varepsilon_{ij}C_{ijkl}\varepsilon_{kl}^{(0)}$$ (1.12)

we obtain the linear elastic model given by Eq. (1.9a). More general forms are permitted, however, including those leading to non-linear elastic behaviour.

1.2.2 Matrix notation

In this book we will often use a matrix form to write the equations. In this case we denote the coordinates as

$$\mathbf{x} = \left\{ \begin{array}{c} x \\ y \\ z \end{array} \right\} = \left\{ \begin{array}{c} x_1 \\ x_2 \\ x_3 \end{array} \right\}$$ (1.13)

and displacements as

$$\mathbf{u} = \left\{ \begin{array}{c} u \\ v \\ w \end{array} \right\} = \left\{ \begin{array}{c} u_1 \\ u_2 \\ u_3 \end{array} \right\} \tag{1.14}$$

For two-dimensional forms we often ignore the third component.

The transformation to matrix form for stresses is given in the order

$$\begin{aligned} \boldsymbol{\sigma} &= \begin{bmatrix} \sigma_{11} & \sigma_{22} & \sigma_{33} & \sigma_{12} & \sigma_{23} & \sigma_{31} \end{bmatrix}^T \\ &= \begin{bmatrix} \sigma_{xx} & \sigma_{yy} & \sigma_{zz} & \sigma_{xy} & \sigma_{yz} & \sigma_{zx} \end{bmatrix}^T \end{aligned} \tag{1.15}$$

and strains by

$$\begin{aligned} \boldsymbol{\varepsilon} &= \begin{bmatrix} \varepsilon_{11} & \varepsilon_{22} & \varepsilon_{33} & \gamma_{12} & \gamma_{23} & \gamma_{31} \end{bmatrix}^T \\ &= \begin{bmatrix} \varepsilon_{xx} & \varepsilon_{yy} & \varepsilon_{zz} & \gamma_{xy} & \gamma_{yz} & \gamma_{zx} \end{bmatrix}^T \end{aligned} \tag{1.16}$$

where symmetry of the tensors is assumed and 'engineering' shear strains are introduced as

$$\gamma_{ij} = 2\varepsilon_{ij}, \ i \neq j \tag{1.17}$$

to make writing of subsequent matrix relations in a concise manner.

The transformation to the six independent components of stress and strain is performed by using the index order given in Table 1.1. This ordering will apply to many subsequent developments also. The order is chosen to permit reduction to two-dimensional applications by merely deleting the last two entries and treating the third entry as appropriate for plane or axisymmetric applications.

The strain–displacement equations are expressed in matrix form as

$$\varepsilon = \boldsymbol{S}\mathbf{u} \tag{1.18}$$

with the three-dimensional strain operator given by

$$\boldsymbol{S}^T = \begin{bmatrix} \dfrac{\partial}{\partial x_1} & 0 & 0 & \dfrac{\partial}{\partial x_2} & 0 & \dfrac{\partial}{\partial x_3} \\ 0 & \dfrac{\partial}{\partial x_2} & 0 & \dfrac{\partial}{\partial x_1} & \dfrac{\partial}{\partial x_3} & 0 \\ 0 & 0 & \dfrac{\partial}{\partial x_3} & 0 & \dfrac{\partial}{\partial x_2} & \dfrac{\partial}{\partial x_1} \end{bmatrix}$$

Table 1.1 Index relation between tensor and matrix forms

Form	Index value					
Matrix	1	2	3	4	5	6
Tensor (1, 2, 3)	11	22	33	12 21	23 32	31 13
Cartesian (x, y, z)	xx	yy	zz	xy yx	yz zy	zx xz
Cylindrical (r, z, θ)	rr	zz	$\theta\theta$	rz zr	$z\theta$ θz	θr $r\theta$

The same operator may be used to write the equilibrium equations (1.4) as

$$\mathcal{S}^T \boldsymbol{\sigma} + \mathbf{b} = \rho \ddot{\mathbf{u}} \tag{1.19}$$

The boundary conditions for displacement and traction are given by

$$\mathbf{u} = \bar{\mathbf{u}} \quad \text{on } \Gamma_u \quad \text{and} \quad \mathbf{t} = \mathbf{G}^T \boldsymbol{\sigma} = \bar{\mathbf{t}} \quad \text{on } \Gamma_t \tag{1.20}$$

where

$$\mathbf{G}^T = \begin{bmatrix} n_1 & 0 & 0 & n_2 & 0 & n_3 \\ 0 & n_2 & 0 & n_1 & n_3 & 0 \\ 0 & 0 & n_3 & 0 & n_2 & n_1 \end{bmatrix}$$

in which $\mathbf{n} = (n_1, n_2, n_3)$ are direction cosines of the normal to the boundary Γ. We note further that the non-zero structure of $\mathcal{S}$ and $\mathbf{G}$ are the same.

For transient problems, initial conditions are denoted by

$$\mathbf{u}(\mathbf{x}, 0) = \bar{\mathbf{d}}(\mathbf{x}) \quad \text{and} \quad \dot{\mathbf{u}}(\mathbf{x}, 0) = \bar{\mathbf{v}}(\mathbf{x}) \quad \text{in } \Omega \tag{1.21}$$

The constitutive equations for a linear elastic material are given in matrix form by

$$\boldsymbol{\sigma} = \mathbf{D}(\boldsymbol{\varepsilon} - \boldsymbol{\varepsilon}_0) \tag{1.22}$$

where in Eq. (1.9a) the index pairs ij and kl for C_{ijkl} are transformed to the 6×6 matrix $\mathbf{D}$ terms using Table 1.1. For a general hyperelastic material we use

$$\boldsymbol{\sigma} = \frac{\partial W}{\partial \boldsymbol{\varepsilon}} \tag{1.23}$$

1.2.3 Two-dimensional problems

There are several classes of two-dimensional problems which may be considered. The simplest are *plane stress* in which the plane of deformation (e.g. $x_1 - x_2$) is thin and stresses $\sigma_{33} = \tau_{13} = \tau_{23} = 0$; and *plane strain* in which the plane of deformation (e.g. $x_1 - x_2$) is one for which $\varepsilon_{33} = \gamma_{13} = \gamma_{23} = 0$. Another class is called *axisymmetric* where the analysis domain is a three-dimensional body of revolution defined in cylindrical coordinates (r, θ, z) but deformations and stresses are two-dimensional functions of r, z only.

Plane stress and plane strain
For plane stress and plane strain problems which have $x_1 - x_2$ as the plane of deformation, the displacements are assumed in the form

$$\mathbf{u} = \begin{Bmatrix} u_1(x_1, x_2, t) \\ u_2(x_1, x_2, t) \end{Bmatrix} \tag{1.24}$$

and thus the strains may be defined by:[11]

$$
\varepsilon = \begin{Bmatrix} \varepsilon_{11} \\ \varepsilon_{22} \\ \varepsilon_{33} \\ \gamma_{12} \end{Bmatrix} = \mathcal{S}\mathbf{u} + \varepsilon_3 = \begin{bmatrix} \dfrac{\partial}{\partial x_1} & 0 \\ 0 & \dfrac{\partial}{\partial x_2} \\ 0 & 0 \\ \dfrac{\partial}{\partial x_2} & \dfrac{\partial}{\partial x_1} \end{bmatrix} \begin{Bmatrix} u_1 \\ u_2 \end{Bmatrix} + \begin{Bmatrix} 0 \\ 0 \\ \varepsilon_{33} \\ 0 \end{Bmatrix}
\tag{1.25}
$$

Here the ε_{33} is either zero (plane strain) or determined from the material constitution by assuming σ_{33} is zero (plane stress). The components of stress are taken in the matrix form

$$
\sigma^T = \begin{Bmatrix} \sigma_{11} & \sigma_{22} & \sigma_{33} & \tau_{12} \end{Bmatrix}
\tag{1.26}
$$

where σ_{33} is determined from material constitution (plane strain) or taken as zero (plane stress).

We note that the local 'energy' term

$$
E = \sigma^T \varepsilon
\tag{1.27}
$$

does not involve ε_{33} for either plane stress or plane strain. Indeed, it is not necessary to compute the σ_{33} (or ε_{33}) until after a problem solution is obtained.

The traction vector for plane problems is given by

$$
\mathbf{t} = \mathbf{G}^T \sigma \quad \text{where} \quad \mathbf{G}^T = \begin{bmatrix} n_1 & 0 & 0 & n_2 \\ 0 & n_2 & 0 & n_1 \end{bmatrix}
\tag{1.28}
$$

and once again we note that $\mathcal{S}$ and $\mathbf{G}$ have the same non-zero structure.

Axisymmetric problems
In an axisymmetric problem we use the cylindrical coordinate system

$$
\mathbf{x} = \begin{Bmatrix} x_1 \\ x_2 \\ x_3 \end{Bmatrix} = \begin{Bmatrix} r \\ z \\ \theta \end{Bmatrix}
\tag{1.29}
$$

This ordering permits the two-dimensional axisymmetric and plane problems to be written in a very similar manner. The body is three dimensional but defined by a surface of revolution such that properties and boundaries are independent of the θ coordinate. For this case the displacement field may be taken as

$$
\mathbf{u} = \begin{Bmatrix} u_1(x_1, x_2, t) \\ u_2(x_1, x_2, t) \\ u_3(x_1, x_2, t) \end{Bmatrix} = \begin{Bmatrix} u_r(r, z, t) \\ u_z(r, z, t) \\ u_\theta(r, z, t) \end{Bmatrix}
\tag{1.30}
$$

and, thus, also is taken as independent of θ.

The strains for the axisymmetric case are given by:[11]

$$
\varepsilon = \begin{Bmatrix} \varepsilon_{rr} \\ \varepsilon_{zz} \\ \varepsilon_{\theta\theta} \\ \gamma_{rz} \\ \gamma_{z\theta} \\ \gamma_{\theta r} \end{Bmatrix} = \begin{Bmatrix} \varepsilon_{11} \\ \varepsilon_{22} \\ \varepsilon_{33} \\ \gamma_{12} \\ \gamma_{23} \\ \gamma_{31} \end{Bmatrix} = \boldsymbol{S}\,\mathbf{u} = \begin{bmatrix} \dfrac{\partial}{\partial x_1} & 0 & 0 \\[2mm] 0 & \dfrac{\partial}{\partial x_2} & 0 \\[2mm] \dfrac{1}{x_1} & 0 & 0 \\[2mm] \dfrac{\partial}{\partial x_2} & \dfrac{\partial}{\partial x_1} & 0 \\[2mm] 0 & 0 & \dfrac{\partial}{\partial x_2} \\[2mm] 0 & 0 & \left(\dfrac{\partial}{\partial x_1} - \dfrac{1}{x_1}\right) \end{bmatrix} \begin{Bmatrix} u_1 \\ u_2 \\ u_3 \end{Bmatrix} \tag{1.31}
$$

The stresses are written in the same order as

$$
\boldsymbol{\sigma}^T = \begin{Bmatrix} \sigma_{11} & \sigma_{22} & \sigma_{33} & \tau_{12} & \tau_{23} & \tau_{31} \end{Bmatrix} \tag{1.32}
$$

Similar to the three-dimensional problem the traction is given by

$$
\mathbf{t} = \mathbf{G}^T \boldsymbol{\sigma} \quad \text{where} \quad \mathbf{G}^T = \begin{bmatrix} n_1 & 0 & 0 & n_2 & 0 & 0 \\ 0 & n_2 & 0 & n_1 & 0 & 0 \\ 0 & 0 & 0 & 0 & n_2 & n_1 \end{bmatrix} \tag{1.33}
$$

where we note that n_3 cannot exist for a complete body of revolution. Once again we note that $\boldsymbol{S}$ and $\mathbf{G}$ have the same non-zero structure.

We note that the strain–displacement relations between the u_1, u_2 and u_3 components are uncoupled. If the material constitution is also uncoupled between the first four and the last two components of strain (i.e. the first four stresses are related only to the first four strains) we may separate the axisymmetric problem into two parts: (a) a part which depends only on the first four strains which are expressed in u_1, u_2; and (b) a problem which depends only on the last two shear strains and u_3. The first problem is sometimes referred to as *torsionless* and the second as a *torsion* problem. However, when the constitution couples the effects, as in classical elastic–plastic solution of a bar which is stretched and twisted, it is necessary to consider the general case.

The torsionless axisymmetric problem is given by

$$
\varepsilon = \begin{Bmatrix} \varepsilon_{rr} \\ \varepsilon_{zz} \\ \varepsilon_{\theta\theta} \\ \gamma_{rz} \end{Bmatrix} = \begin{Bmatrix} \varepsilon_{11} \\ \varepsilon_{22} \\ \varepsilon_{33} \\ \gamma_{12} \end{Bmatrix} = \boldsymbol{S}\,\mathbf{u} = \begin{bmatrix} \dfrac{\partial}{\partial x_1} & 0 \\[2mm] 0 & \dfrac{\partial}{\partial x_2} \\[2mm] \dfrac{1}{x_1} & 0 \\[2mm] \dfrac{\partial}{\partial x_2} & \dfrac{\partial}{\partial x_1} \end{bmatrix} \begin{Bmatrix} u_1 \\ u_2 \end{Bmatrix} \tag{1.34}
$$

with stresses given by Eq. (1.26) and tractions by Eq. (1.28). Thus the only difference in these two classes of problems is the presence of the u_1/x_1 for the third strain in the axisymmetric case (of course the two differ also in the domain description of the problem as we shall point out later).

1.3 Variational forms for non-linear elasticity

For an elastic material as specified by Eq. (1.23), the above equations may be given in a variational form when no inertial effects are included. The simplest form is the *potential energy principle* where

$$\Pi_{PE} = \int_\Omega W(\boldsymbol{S}\mathbf{u})\,\mathrm{d}\Omega - \int_\Omega \mathbf{u}^T \mathbf{b}\,\mathrm{d}\Omega - \int_{\Gamma_t} \mathbf{u}^T \bar{\mathbf{t}}\,\mathrm{d}\Gamma \qquad (1.35)$$

The first variation yields the governing equation of the functional as[16]

$$\delta\Pi_{PE} = \int_\Omega \delta(\boldsymbol{S}\mathbf{u})^T \frac{\partial W}{\partial \boldsymbol{S}\mathbf{u}}\,\mathrm{d}\Omega - \int_\Omega \delta\mathbf{u}^T \mathbf{b}\,\mathrm{d}\Omega - \int_{\Gamma_t} \delta\mathbf{u}^T \bar{\mathbf{t}}\,\mathrm{d}\Gamma = 0 \qquad (1.36)$$

After integration by parts and collecting terms we obtain

$$\begin{aligned}
\delta\Pi_{PE} = &- \int_\Omega \delta\mathbf{u}^T \left(\boldsymbol{S}^T \boldsymbol{\sigma} + \mathbf{b} \right) \mathrm{d}\Omega \\
&+ \int_{\Gamma_t} \delta\mathbf{u}^T \left(\mathbf{G}^T \boldsymbol{\sigma} - \bar{\mathbf{t}} \right) \mathrm{d}\Gamma = 0
\end{aligned} \qquad (1.37)$$

where

$$\boldsymbol{\sigma} = \frac{\partial W}{\partial \boldsymbol{S}\mathbf{u}}$$

When W is given by the quadratic form (1.12) we recover the linear problem given by Eq. (1.22). In this case the form becomes the principle of minimum potential energy and the displacement field which renders W an absolute minimum is an exact solution to the problem.[11]

We note that the potential energy principle includes the strain–displacement equations and the elastic model expressed in terms of displacement-based strains. It also requires the displacement boundary condition to be stated in addition to the theorem. It is, however, the simplest variational form and only requires knowledge of the displacement field to be valid. This form is a basis for *irreducible (or displacement) methods* of approximate solution.

A general variational theorem, which includes all the equations and boundary conditions, is given by the Hu–Washizu variational theorem.[17] This theorem is given by

$$\begin{aligned}
\Pi_{HW}(\mathbf{u}, \boldsymbol{\varepsilon}, \boldsymbol{\sigma}) = &\int_\Omega \left[W(\boldsymbol{\varepsilon}) + \boldsymbol{\sigma}^T (\boldsymbol{S}\mathbf{u} - \boldsymbol{\varepsilon}) \right] \mathrm{d}\Omega \\
&- \int_\Omega \mathbf{u}^T \mathbf{b}\,\mathrm{d}\Omega - \int_{\Gamma_t} \mathbf{u}^T \bar{\mathbf{t}}\,\mathrm{d}\Gamma - \int_{\Gamma_u} \mathbf{t}^T (\mathbf{u} - \bar{\mathbf{u}})\,\mathrm{d}\Gamma
\end{aligned} \qquad (1.38)$$

in which $\mathbf{t} = \mathbf{G}^T \boldsymbol{\sigma}$. The proof that the theorem contains all the governing equations is obtained by taking the variation of Eq. (1.38) with respect to $\mathbf{u}$, $\boldsymbol{\varepsilon}$ and $\boldsymbol{\sigma}$. Accordingly, taking the variation of (1.38) and performing an integration by parts on $\delta(\boldsymbol{S}\mathbf{u})$ we obtain

$$
\begin{aligned}
\delta\Pi_{HW} = &\int_\Omega \delta\varepsilon^T \left[\frac{\partial W}{\partial \varepsilon} - \sigma\right] d\Omega \\
&+ \int_\Omega \delta\sigma^T \left[\boldsymbol{S}\mathbf{u} - \varepsilon\right] d\Omega - \int_{\Gamma_u} \delta\mathbf{t}^T \left(\mathbf{u} - \bar{\mathbf{u}}\right) d\Gamma \\
&- \int_\Omega \delta\mathbf{u}^T \left(\boldsymbol{S}^T\sigma + \mathbf{b}\right) d\Omega + \int_{\Gamma_t} \delta\mathbf{u}^T \left(\mathbf{t} - \bar{\mathbf{t}}\right) d\Gamma = 0
\end{aligned}
\tag{1.39}
$$

and it is evident that the Hu–Washizu variational theorem yields all the equations for the non-linear elastostatic problem.

We may also establish a direct link between the Hu–Washizu theorem and other variational principles. If we express the strains ε in terms of the stresses using the Laurant transformation

$$
U(\sigma) + W(\varepsilon) = \sigma^T \varepsilon
\tag{1.40}
$$

we recover the Hellinger–Reissner variational principle given by[18–20]

$$
\begin{aligned}
\Pi_{HR}(\mathbf{u}, \sigma) = &\int_\Omega \left[\sigma^T \boldsymbol{S}\mathbf{u} - U(\sigma)\right] d\Omega \\
&- \int_\Omega \mathbf{u}^T \mathbf{b} \, d\Omega - \int_{\Gamma_t} \mathbf{u}^T \bar{\mathbf{t}} \, d\Gamma - \int_{\Gamma_u} \mathbf{t}^T \left(\mathbf{u} - \bar{\mathbf{u}}\right) d\Gamma
\end{aligned}
\tag{1.41}
$$

In the linear elastic case we have, ignoring initial strain and stress effects,

$$
U(\sigma) = \tfrac{1}{2} \sigma_{ij} S_{ijkl} \sigma_{kl}
\tag{1.42}
$$

where S_{ijkl} are elastic compliances. While this form is also formally valid for general elastic problems. We shall find that in the non-linear case it is not possible to find *unique relations* for the constitutive behaviour in terms of stress forms. Thus, we shall often rely on use of the Hu–Washizu functional as the basis for a mixed formulation.

We may also establish a direct link to the minimum potential energy form and the Hu–Washizu theorem. If we satisfy the displacement boundary condition (1.20) *a priori* the integral term over Γ_u is eliminated from Eq. (1.38). Generally, in our finite element approximations based on the Hu–Washizu theorem (or variants of the theorem) we shall satisfy the displacement boundary conditions explicitly and thus avoid approximating the Γ_u term.

If we then satisfy the strain-displacement relations *a priori* then the Hu–Washizu theorem is identical with the potential energy principle. In constructing finite element approximations, the potential energy principle is a basis for developing *displacement models* (also referred to as *irreducible models*[1]) whereas the Hu–Washizu form is a basis for developing *mixed models*.[1] As we will show in Chapter 2 mixed methods have distinct advantages in constructing robust finite element formulations. However, there are also advantages in having a finite element formulation where the global problem is expressed in a displacement form. Noting how the Hu–Washizu form reduces to the potential energy principle provides a link on treating the reductions to their approximate counterparts (see Sec. 2.6).

One advantage of a variational theorem is that *symmetry conditions* are automatically obtained; however, a distinct disadvantage is that only elastic behaviour and static forms may be considered. In the next section we consider an alternative approach of *weak forms* which is valid for both elastic or inelastic material forms and directly admits the inertial effects. We shall observe that for the elastostatic problem a weak form is equivalent to the variation of a theorem.

1.4 Weak forms of governing equations

A variational (weak) form for any set of equations is a *scalar* relation and may be constructed by multiplying the equation set by an appropriate arbitrary function which has the same free indices as in the set of governing equations (which then becomes a dummy index and sums over its range), integrating over the domain of the problem and setting the result to zero.[1,17]

1.4.1 Weak form for equilibrium equation

For example, in indicial form the equilibrium equation (1.4) has the free index i, thus to construct a weak form we multiply by an arbitrary vector with index i and integrate the result over the domain Ω. Virtual work is a weak form in which the arbitrary function is a virtual displacement δu_i, accordingly using this function we obtain the form

$$\delta \Pi_{eq} = \int_{\Omega} \delta u_i \left[\rho \ddot{u}_i - \sigma_{ji,j} - b_i \right] d\Omega = 0$$

Generally stress will depend on strains which are derivatives of displacements. Thus, the above form will require computation of second derivatives of displacement to form the integrands. The need to compute second derivatives may be reduced (i.e. 'weakened') by performing an integration by parts and upon noting the symmetry of the stress we obtain

$$\delta \Pi_{eq} = \int_{\Omega} \delta u_i \, \rho \, \ddot{u}_i \, d\Omega + \int_{\Omega} \delta \varepsilon_{ij}(u_k) \, \sigma_{ij} \, d\Omega - \int_{\Omega} \delta u_i \, b_i \, d\Omega - \int_{\Gamma} \delta u_i \, t_i \, d\Omega = 0$$

$$(1.43)$$

where *virtual strains* are related to virtual displacements as

$$\delta \varepsilon_{ij}(u_k) = \tfrac{1}{2}(\delta u_{i,j} + \delta u_{j,i})$$

$$(1.44)$$

This may be further simplified by splitting the boundary into parts where traction is specified, Γ_t, and parts where displacements are specified, Γ_u. If we enforce pointwise all the displacement boundary conditions* and impose a constraint that δu_i vanishes on Γ_u, we obtain the final result

$$\delta \Pi_{eq} = \int_{\Omega} \delta u_i \, \rho \, \ddot{u}_i \, d\Omega + \int_{\Omega} \delta \varepsilon_{ij}(u_k) \, \sigma_{ij} \, d\Omega - \int_{\Omega} \delta u_i \, b_i \, d\Omega - \int_{\Gamma_t} \delta u_i \, \bar{t}_i \, d\Omega = 0$$

$$(1.45)$$

* Alternatively, we can combine this term with another from the integration by parts of the weak form of the strain–displacement equations.

or in matrix form as

$$\delta\Pi_{eq} = \int_{\Omega} \delta\mathbf{u}^T \rho\ddot{\mathbf{u}}\,d\Omega + \int_{\Omega} \delta(\boldsymbol{S}\mathbf{u})^T \boldsymbol{\sigma}\,d\Omega - \int_{\Omega} \delta\mathbf{u}^T \mathbf{b}\,d\Omega - \int_{\Gamma_t} \delta\mathbf{u}^T \bar{\mathbf{t}}\,d\Gamma = 0$$

$$(1.46)$$

The first term is the virtual work of internal inertial forces, the second the virtual work of the internal stresses and the last two the virtual work of body and traction forces, respectively.

The above weak form provides the basis from which a finite element formulation of equilibrium may be deduced for general applications. It is necessary to add appropriate expressions for the strain–displacement and constitutive equations to complete a problem formulation. Weak forms for these may be written immediately from the variation of the Hu–Washizu principle given in Eq. (1.39).

We note that the form adopted to define the matrices of stress and strain permits the internal work of stress and strain to be written as

$$\varepsilon_{ij}\,\sigma_{ij} = \boldsymbol{\varepsilon}^T \boldsymbol{\sigma} = \boldsymbol{\sigma}^T \boldsymbol{\varepsilon} \qquad (1.47)$$

Similarly, the internal virtual work per unit volume may be expressed by

$$\delta W = \delta\varepsilon_{ij}\,\sigma_{ij} = \delta\boldsymbol{\varepsilon}^T \boldsymbol{\sigma} \qquad (1.48)$$

In Chapter 4 we will discuss this in more detail and show that constructing constitutive equations in terms of six components of stress and strain must be treated appropriately in reductions from the original nine tensor components.

1.5 Concluding remarks

In this chapter we have summarized the basic steps needed to formulate a general small-strain solid mechanics problem. The formulation has been presented in a *strong form* in terms of partial differential equations and in a *weak form* in terms of integral expressions. We have also indicated how the general problem can become non-linear. In the next chapter we describe the use of the finite element method to construct approximate solutions to weak forms for non-linear transient solid mechanics problems.

References

1. O.C. Zienkiewicz, R.L. Taylor and J.Z. Zhu. *The Finite Element Method: Its Basis and Fundamentals*. Butterworth-Heinemann, Oxford, 6th edition, 2005.
2. T.J.R. Hughes. *The Finite Element Method: Linear Static and Dynamic Analysis*. Dover Publications, New York, 2000.
3. R.D. Cook, D.S. Malkus, M.E. Plesha and R.J. Witt. *Concepts and Applications of Finite Element Analysis*. John Wiley & Sons, New York, 4th edition, 2001.
4. F. de S. Lynch. A finite element method of viscoelastic stress analysis with application to rolling contact problems. *International Journal for Numerical Methods in Engineering*, 1:379–394, 1969.

5. J.T. Oden and T.L. Lin. On the general rolling contact problem for finite deformations of a viscoelastic cylinder. *Computer Methods in Applied Mechanics and Engineering*, 57:297–367, 1986.

6. P. le Tallec and C. Rahier. Numerical models of steady rolling for non-linear viscoelastic structures in finite deformation. *International Journal for Numerical Methods in Engineering*, 37:1159–1186, 1994.

7. S. Govindjee and P.A. Mahalic. Viscoelastic constitutive relations for the steady spinning of a cylinder. Technical Report UCB/SEMM Report 98/02, University of California at Berkeley, 1998.

8. H. Cohen and R.G. Muncaster. *The Theory of Pseudo-rigid Bodies*. Springer, New York, 1988.

9. S.P. Timoshenko and J.M. Gere. *Theory of Elastic Stability*. McGraw-Hill, New York, 1961.

10. S.P. Timoshenko and J.N. Goodier. *Theory of Elasticity*. McGraw-Hill, New York, 3rd edition, 1969.

11. I.S. Sokolnikoff. *The Mathematical Theory of Elasticity*. McGraw-Hill, New York, 2nd edition, 1956.

12. L.E. Malvern. *Introduction to the Mechanics of a Continuous Medium*. Prentice-Hall, Englewood Cliffs, NJ, 1969.

13. A.P. Boresi and K.P. Chong. *Elasticity in Engineering Mechanics*. Elsevier, New York, 1987.

14. P.C. Chou and N.J. Pagano. *Elasticity: Tensor, Dyadic and Engineering Approaches*. Dover Publications, Mineola, NY, 1992. Reprinted from 1967 Van Nostrand edition.

15. I.H. Shames and F.A. Cozzarelli. *Elastic and Inelastic Stress Analysis*. Taylor & Francis, Washington, DC, 1997. (Revised printing.)

16. F.B. Hildebrand. *Methods of Applied Mathematics*. Prentice-Hall (reprinted by Dover Publishers, 1992), 2nd edition, 1965.

17. K. Washizu. *Variational Methods in Elasticity and Plasticity*. Pergamon Press, New York, 3rd edition, 1982.

18. E. Hellinger. Die allgemeine Aussetze der Mechanik der Kontinua. In F. Klein and C. Muller, editors, *Encyclopedia der Mathematishen Wissnschaften*, volume 4. Tebner, Leipzig, 1914.

19. E. Reissner. On a variational theorem in elasticity. *Journal of Mathematics and Physics*, 29(2): 90–95, 1950.

20. E. Reissner. A note on variational theorems in elasticity. *International Journal of Solids and Structures*, 1:93–95, 1965.

Galerkin method of approximation – irreducible and mixed forms

2.1 Introduction

In the previous chapter we presented the basic equations for problems in non-linear solid mechanics in which strains remain small. We showed that the equations can be presented in a strong form as a set of partial differential equations or alternatively in terms of a variational principle or weak form expressed as an integral over the domain of interest. In the present chapter we use the weak form to construct approximate solutions based on the finite element method. This results in a Galerkin method for which general properties are well known.[1–4]

Although it is assumed that the reader is familiar with finite element methods for small deformation linear problems, we present a full summary of the basic steps to construct a solution for the transient problem. We emphasize the differences between linear and non-linear effects as well as the numerical procedures used to establish the final discrete form of the equations which is the form used in computer analysis. We also consider both *irreducible* and *mixed* forms of approximation. The mixed forms are introduced to overcome deficiencies arising in use of low order elements based on irreducible forms. In particular, in this chapter we consider a mixed form appropriate for use in problems in which near incompressible behaviour can occur. In the second part of this book, we consider forms for structural problems where so-called 'shear locking' can occur in bending of thin rods, plates and shells.

We conclude this chapter by applying the methods developed for the equations of solid mechanics to that for thermal analysis based on a non-linear form of the quasi-harmonic equation.

2.2 Finite element approximation – Galerkin method

The finite element approximation to a problem starts by dividing the domain of interest, Ω, into a set of subdomains (called elements), Ω_e, such that

$$\Omega \approx \hat{\Omega} = \sum_e \Omega_e \qquad (2.1a)$$

Similarly the boundary is divided into subdomains as

$$\Gamma \approx \hat{\Gamma} = \sum_e \Gamma_e = \sum_{et} \Gamma_{t_e} + \sum_{eu} \Gamma_{u_e} \qquad (2.1b)$$

where Γ_{t_e} is a boundary segment on which tractions are specified and Γ_{u_e} one where displacements are specified. We note that in general the domain of a finite element analysis is an approximation to the true domain which depends on the boundary shape of elements.

The weak form for the governing equations is written for the problem domain $\hat{\Omega}$ and also written as a sum over the element domains. Thus, the weak form given in Eq. (1.46) for the equilibrium equation becomes

$$\delta\Pi_{eq} \approx \delta\hat{\Pi}_{eq} = \sum_e \left[\int_{\Omega_e} \delta\mathbf{u}^T \rho\ddot{\mathbf{u}} \, d\Omega + \int_{\Omega_e} \delta(\boldsymbol{S}\mathbf{u})^T \boldsymbol{\sigma} \, d\Omega - \int_{\Omega_e} \delta\mathbf{u}^T \mathbf{b} \, d\Omega \right]$$

$$- \sum_{et} \left[\int_{\Gamma_{t_e}} \delta\mathbf{u}^T \bar{\mathbf{t}} \, d\Gamma \right] = 0 \qquad (2.2)$$

$$= \sum_e \delta\hat{\Pi}^e + \sum_{et} \delta\hat{\Pi}_t^e = 0$$

In the above $\delta\hat{\Pi}^e$ are terms within the domain Ω_e of each element and $\delta\hat{\Pi}_t^e$ those which belong to traction boundary surfaces Γ_{t_e}. A *Galerkin method* of solution is obtained using approximations to the dependent variables and their virtual forms.[1,2] For an irreducible (or displacement) finite element form we only need approximations for $\mathbf{u}$ and $\delta\mathbf{u}$.

In order for a variational theorem or a weak form to be split into the additive sum indicated in Eq. (2.2) the highest derivatives appearing in the functional must be at least piecewise continuous so that all the integrals exist and no contributions across interelement boundaries are present.* For a functional containing a variable with a highest derivative of order $m + 1$ the functions used to approximate the variable must have *all derivatives* up to order m continuous in the entire domain, $\hat{\Omega}$ – such functions are called C^m. For the weak forms considered for problems in solid mechanics we will encounter functionals which contain only first derivatives and thus will need only C^0 functions for the approximation. Indeed some functions in mixed forms will have no derivatives and these may be approximated by discontinuous functions in $\hat{\Omega}$. Generally, one should respect the order of approximation (i.e. C^m) where an exact solution can have discontinuous behaviour. For solid mechanics there are discontinuities in the displacement at material interfaces and at some singular load forms (e.g. point loads or line loads). Material interfaces are real; however, use of a point or line load is not and, when used, is an approximation to a physical action. Use of functions with added continuity over C^m can be beneficial where solutions are smooth. Thus there are some forms of interpolation being introduced in recent literature that have increased smoothness. In this volume, however, we will generally present only those forms which provide C^m continuity. In the first part of this volume concerning problems in solid mechanics approximations will be made with C^0 functions. In the second part concerning structural mechanics problems for rods, plates and shells we shall need C^1 functions for some formulations, whereas in others we can still use C^0 forms.

There are formulations which violate the continuity conditions leading to so-called *incompatible* approximation (e.g. see Wilson *et al.*[6]). Strang termed such approxima-

---

* It is possible to add interelement jump terms to a functional leading to a *Discontinuous Galerkin formulation* – see Cockburn *et al.*[5] for more information.

tions *a variational crime*[7] but showed convergence could still be achieved provided certain requirements were met. Most incompatible formulations that perform well have subsequently been shown to be members of a valid mixed formulation (e.g. see Simo and Rifai[8]). In the forms given for problems in solid mechanics we will use interpolations which satisfy the C^0 requirement; however, in the study of thin plates we will present some forms which violate the C^1 requirement.

In addition to the continuity requirement it is necessary for C^m functions to possess complete polynomials to order $m + 1$ to ensure that the derivatives up to order $m + 1$ can assume constant values. Both of the above requirements are covered in standard introductory texts on the finite element method (e.g. see reference 2 or 3). They remain equally valid for the study of non-linear problems – both for forms with material non-linearity as well as those with large deformations where kinematic conditions are non-linear. The patch test also remains valid in assessing the available continuity and derivatives present in any approximation (see reference 2 for a general discussion on the patch test for irreducible and mixed finite element formulations).

Displacement approximation

A finite element approximation for displacements is given by

$$\mathbf{u}(\mathbf{x}, t) \approx \hat{\mathbf{u}} = \sum_b N_b(\mathbf{x})\tilde{\mathbf{u}}_b(t) = \mathbf{N}(\mathbf{x})\tilde{\mathbf{u}}(t) \tag{2.3}$$

where N_b are element shape functions, $\tilde{\mathbf{u}}_b(t)$ are time dependent nodal displacements and the sum ranges over the number of nodes associated with an element. Alternatively, in *isoparametric form*[2] the expressions are given by (as shown in Fig. 2.1 for a four-node two-dimensional quadrilateral)

$$\mathbf{u}(\boldsymbol{\xi}, t) \approx \hat{\mathbf{u}}(\boldsymbol{\xi}, t) = \sum_b N_b(\boldsymbol{\xi})\tilde{\mathbf{u}}_b(t) = \mathbf{N}(\boldsymbol{\xi})\tilde{\mathbf{u}}(t);$$

$$\mathbf{x}(\boldsymbol{\xi}) = \sum_b N_b(\boldsymbol{\xi})\tilde{\mathbf{x}}_b = \mathbf{N}(\boldsymbol{\xi})\tilde{\mathbf{x}} \tag{2.4a}$$

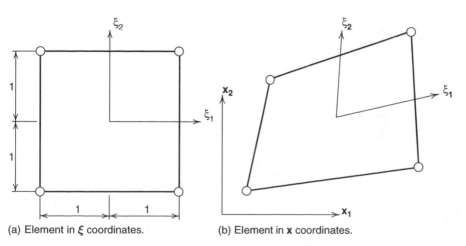

(a) Element in ξ coordinates.   (b) Element in **x** coordinates.

Fig. 2.1 Isoparametric map for 4-node two-dimensional quadrilateral.

where $\tilde{\mathbf{x}}$ represent nodal coordinate parameters and $\boldsymbol{\xi}$ are the parametric coordinates for each element.

An approximation for the virtual displacement is given by

$$\delta\mathbf{u}(\boldsymbol{\xi}) \approx \delta\hat{\mathbf{u}}(\boldsymbol{\xi}) = \sum_a N_a(\boldsymbol{\xi})\tilde{\mathbf{u}}_a = \mathbf{N}(\boldsymbol{\xi})\tilde{\mathbf{u}} \tag{2.4b}$$

A summary of procedures used to construct shape functions for some isoparametric elements is included in Appendix A.

Derivatives

The weak forms presented in Chapter 1 all include first derivative of displacements. For the isoparametric approximation given in Eq. (2.4a) we need first derivatives of the shape functions with respect to x_j. These are computed using the chain rule as:

$$\frac{\partial N_a}{\partial \xi_i} = \frac{\partial x_j}{\partial \xi_i}\frac{\partial N_a}{\partial x_j} \tag{2.5}$$

or in matrix form

$$\frac{\partial N_a}{\partial \boldsymbol{\xi}} = \mathbf{J}\frac{\partial N_a}{\partial \mathbf{x}} \tag{2.6a}$$

where

$$\frac{\partial N_a}{\partial \boldsymbol{\xi}} = \begin{Bmatrix} \dfrac{\partial N_a}{\partial \xi_1} \\ \dfrac{\partial N_a}{\partial \xi_2} \\ \dfrac{\partial N_a}{\partial \xi_3} \end{Bmatrix} ; \quad \frac{\partial N_a}{\partial \mathbf{x}} = \begin{Bmatrix} \dfrac{\partial N_a}{\partial x_1} \\ \dfrac{\partial N_a}{\partial x_2} \\ \dfrac{\partial N_a}{\partial x_3} \end{Bmatrix} ; \quad \mathbf{J} = \begin{bmatrix} \dfrac{\partial x_1}{\partial \xi_1} & \dfrac{\partial x_2}{\partial \xi_1} & \dfrac{\partial x_3}{\partial \xi_1} \\ \dfrac{\partial x_1}{\partial \xi_2} & \dfrac{\partial x_2}{\partial \xi_2} & \dfrac{\partial x_3}{\partial \xi_2} \\ \dfrac{\partial x_1}{\partial \xi_3} & \dfrac{\partial x_2}{\partial \xi_3} & \dfrac{\partial x_3}{\partial \xi_3} \end{bmatrix} \tag{2.6b}$$

in which $\mathbf{J}$ is the Jacobian transformation between $\mathbf{x}$ and $\boldsymbol{\xi}$. Using the above the shape function derivatives are given by

$$\frac{\partial N_a}{\partial \mathbf{x}} = \mathbf{J}^{-1}\frac{\partial N_a}{\partial \boldsymbol{\xi}} \tag{2.6c}$$

In two-dimensional problems only the first two coordinates are involved, thus reducing the size of $\mathbf{J}$ to a 2×2 matrix. In the sequel we will often use the notation

$$\frac{\partial N_a}{\partial x_j} = N_{a,x_j} \quad \text{and} \quad \frac{\partial N_a}{\partial \xi_i} = N_{a,\xi_i} \tag{2.7}$$

Strain–displacement equations

Using (1.18) the strain–displacement equations are given by

$$\boldsymbol{\varepsilon} = \mathcal{S}\mathbf{u} \approx \sum_b (\mathcal{S}N_b)\,\tilde{\mathbf{u}}_b = \sum_b \mathbf{B}_b\tilde{\mathbf{u}}_b = \mathbf{B}\tilde{\mathbf{u}} \tag{2.8}$$

In a general three-dimensional problem the strain matrix at each node of an element is defined by

$$\mathbf{B}_b^T = \begin{bmatrix} N_{b,x_1} & 0 & 0 & N_{b,x_2} & 0 & N_{b,x_3} \\ 0 & N_{b,x_2} & 0 & N_{b,x_1} & N_{b,x_3} & 0 \\ 0 & 0 & N_{b,x_3} & 0 & N_{b,x_2} & N_{b,x_1} \end{bmatrix} \tag{2.9a}$$

For the two-dimensional plane stress, plane strain and torsionless axisymmetric problem the strain matrix at a node is given by

$$\mathbf{B}_b^T = \begin{bmatrix} N_{b,x_1} & 0 & c\,N_b/x_1 & N_{b,x_2} \\ 0 & N_{b,x_2} & 0 & N_{b,x_1} \end{bmatrix} \tag{2.9b}$$

where $c = 0$ for plane stress and strain and $c = 1$ for the torsionless axisymmetric case. For the axisymmetric problem with torsion the strain matrix becomes

$$\mathbf{B}_b^T = \begin{bmatrix} N_{b,x_1} & 0 & N_b/x_1 & N_{b,x_2} & 0 & 0 \\ 0 & N_{b,x_2} & 0 & N_{b,x_1} & 0 & 0 \\ 0 & 0 & 0 & 0 & N_{b,x_2} & (N_{b,x_1} - N_b/x_1) \end{bmatrix} \tag{2.9c}$$

Weak form

Substituting the above forms for displacement and strains into the weak form of equilibrium given in Eq. (2.2) yields, for a single element,

$$\delta\hat{\Pi}_{eq}^e = \delta\tilde{\mathbf{u}}^T \left[\int_{\Omega_e} \mathbf{N}^T \rho \mathbf{N}\,\mathrm{d}\Omega\,\ddot{\mathbf{u}} + \int_{\Omega_e} \mathbf{B}^T \sigma\,\mathrm{d}\Omega - \int_{\Omega_e} \mathbf{N}^T \mathbf{b}\,\mathrm{d}\Omega - \int_{\Gamma_{te}} \mathbf{N}^T \bar{\mathbf{t}}\,\mathrm{d}\Gamma \right] \tag{2.10}$$

Performing the sum over all elements and noting that the virtual parameters $\delta\tilde{\mathbf{u}}$ are arbitrary we obtain a semi-discrete problem given by the set of ordinary differential equations

$$\mathbf{M}\ddot{\mathbf{u}} + \mathbf{P}(\sigma) = \mathbf{f} \tag{2.11a}$$

where

$$\mathbf{M} = \sum_e \mathbf{M}^{(e)}; \quad \mathbf{P} = \sum_e \mathbf{P}^{(e)} \quad \text{and} \quad \mathbf{f} = \sum_e \mathbf{f}^{(e)} \tag{2.11b}$$

with the element arrays specified by

$$\mathbf{M}^{(e)} = \int_{\Omega_e} \mathbf{N}^T \rho \mathbf{N}\,\mathrm{d}\Omega; \quad \mathbf{P}^{(e)}(\sigma) = \int_{\Omega_e} \mathbf{B}^T \sigma\,\mathrm{d}\Omega \quad \text{and} \quad \mathbf{f}^{(e)} = \int_{\Omega_e} \mathbf{N}^T \mathbf{b}\,\mathrm{d}\Omega + \int_{\Gamma_{te}} \mathbf{N}^T \bar{\mathbf{t}}\,\mathrm{d}\Gamma$$
$$\tag{2.11c}$$

The term $\mathbf{P}$ is often referred to as the *stress divergence* or *stress force* term.

While the form for the arrays given above is valid for all problem classes the volume element differs and is given by:

$$\begin{aligned}
\mathrm{d}\Omega &= \mathrm{d}x_1\,\mathrm{d}x_2\,\mathrm{d}x_3; && \text{General three-dimensional problems,} \\
\mathrm{d}\Omega &= \mathrm{d}x_1\,\mathrm{d}x_2; && \text{Plane strain problems,} \\
\mathrm{d}\Omega &= h_3\,\mathrm{d}x_1\,\mathrm{d}x_2; && \text{Plane stress problems,} \\
\mathrm{d}\Omega &= 2\,\pi\,x_1\,\mathrm{d}x_1\,\mathrm{d}x_2; && \text{Axisymmetric problems.}
\end{aligned}$$

In the above we assume a unit thickness in the x_3 direction for plane strain, h_3 is the thickness of a plane stress slab; and the factor $2\,\pi$ in axisymmetric problems results from the integration of $\int \mathrm{d}x_3 = \int \mathrm{d}\theta$ of the body of revolution.[*]

In the sequel we will discuss the finite element form for solids in a general context using the coordinates, x_i, displacements, u_i, etc. Unless otherwise stated, we will also assume that the forms for $\mathbf{B}$, Ω_e, $\mathrm{d}\Omega$, etc. are always replaced by that appropriate for the problem class considered (i.e. plane stress, plane strain, axisymmetric, or general three dimensions).

---

[*] Some programs, including *FEAPpv* available at the publisher's web site, omit the factor 2π in axisymmetric forms.

2.2.1 Irreducible displacement method

In the case of linear elasticity the constitutive equations are given by Eq. (1.22) and using Eq. (2.8) an *irreducible displacement method* results[2] with

$$\mathbf{P}^{(e)}(\boldsymbol{\sigma}) = \left(\int_{\Omega_e} \mathbf{B}^T \mathbf{D} \mathbf{B} \mathrm{d}\Omega \right) \tilde{\mathbf{u}} = \mathbf{K}^{(e)} \tilde{\mathbf{u}} \tag{2.12}$$

in which $\mathbf{K}^{(e)}$ is a *linear stiffness matrix*. In many situations, however, it is necessary to use non-linear or time-dependent stress–strain (constitutive) relations and in these cases we need to develop solution strategies directly from Eqs (2.11a) to (2.11c). This will be considered further in detail in later chapters for quite general constitutive behaviour. However, at this stage we simply need to note that

$$\boldsymbol{\sigma} = \boldsymbol{\sigma}(\boldsymbol{\varepsilon}) \tag{2.13}$$

and that the functional relationship can be very non-linear and occasionally non-unique. Furthermore, it will be necessary to use a *mixed approach* if constraints, such as near incompressibility, are encountered.[2] We address this latter aspect in Sec. 2.6; however, before doing so we consider the manner whereby calculation of the finite element arrays and solution of the transient equations may be computed using numerical methods.

2.3 Numerical integration – quadrature

The integrations needed to compute the finite element arrays are most conveniently performed numerically by quadrature.[2,3,9] Many forms of quadrature formulas exist; however, the most accurate for polynomial expressions is Gauss–Legendre quadrature.[10] Gauss–Legendre quadrature tables are generally tabulated over the range of coordinates $-1 < \xi < 1$ (hence our main reason for also choosing many shape function on this interval).

Gaussian quadrature integrates a function as

$$\int_{-1}^{1} f(\xi) \, \mathrm{d}\xi = \sum_{j=1}^{n} f(\xi_j) \, w_j + O\left(\frac{\mathrm{d}^{2n} f}{\mathrm{d}\xi^{2n}} \right) \tag{2.14a}$$

where ξ_j are the points where the function is evaluated and w_j is a weight. Thus, an n-point formula integrates exactly a polynomial of order $2n - 1$. Table 2.1 presents the location of points and weights for the first five members of the family.

Integrations over multi-dimensional domains may be performed by products of the one-dimensional formula. Thus, in two dimensions we use (with $\eta \equiv \xi_2$)

$$\int_{-1}^{1} \int_{-1}^{1} f(\xi, \eta) \, \mathrm{d}\xi \, \mathrm{d}\eta = \sum_{j=1}^{n} \sum_{k=1}^{n} f(\xi_j, \eta_k) \, w_j \, w_k \tag{2.14b}$$

and in three dimensions (with $\zeta = \xi_3$)

$$\int_{-1}^{1} \int_{-1}^{1} \int_{-1}^{1} f(\xi, \eta, \zeta) \, \mathrm{d}\xi \, \mathrm{d}\eta \, \mathrm{d}\zeta = \sum_{j=1}^{n} \sum_{k=1}^{n} \sum_{l=1}^{n} f(\xi_j, \eta_k, \zeta_l) \, w_j \, w_k \, w_l \tag{2.14c}$$

which are exact when polynomials in any direction are less than order $2n$.

Table 2.1 Gaussian quadrature abscissae and weights for $\int_{-1}^{1} f(\xi)\,d\xi = \sum_{j=1}^{n} f(\xi_j)\,w_j$.

Order	j	ξ_j	w_j	
$n = 1$	1	0	2	
$n = 2$	1	$+1/\sqrt{3}$	1	
	2	$-1/\sqrt{3}$	1	
$n = 3$	1	$+\sqrt{0.6}$	5/9	
	2	0	8/9	
	3	$-\sqrt{0.6}$	5/9	
$n = 4$	1	$+\sqrt{((3+a)/7)}$	$0.5 - 1/(3a)$	$a = \sqrt{4.8}$
	2	$+\sqrt{((3-a)/7)}$	$0.5 + 1/(3a)$	
	3	$-\sqrt{((3-a)/7)}$	$0.5 + 1/(3a)$	
	4	$-\sqrt{((3+a)/7)}$	$0.5 - 1/(3a)$	
$n = 5$	1	$+\sqrt{b}$	$((5c-3)d/b)$	$a = \sqrt{1120}$
	2	$+\sqrt{c}$	$((3-5b)d/c)$	$b = (70+a)/126$
	3	0	$2 - 2(w_1 + w_2)$	$c = (70-a)/126$
	4	$-\sqrt{c}$	$((3-5b)d/c)$	$d = 1/(15(c-b))$
	5	$-\sqrt{b}$	$((5c-3)d/b)$	

Volume integrals

The problem remains to transform our integrals from the element region Ω_e to the gaussian range $-1 \le \xi \le 1$. The determinant of $\mathbf{J}$ appearing in Eq. (2.6a) is used to transform the volume element from the cartesian coordinates to the natural coordinates as

$$dx_1\,dx_2\,dx_3 = \det \mathbf{J}\,d\xi_1\,d\xi_2\,d\xi_3 = j(\xi_1,\xi_2,\xi_3)\,d\xi_1\,d\xi_2\,d\xi_3 \qquad (2.15)$$

in which $\det \mathbf{J} = j$ must be positive to maintain a correct volume element.

Using the above type of transformation, integrals of finite element arrays are given by

$$\int_{\Omega_e} f(\mathbf{x})\,d\Omega = \int_{\square} \hat{f}(\xi)\,j(\xi)\,d\square \qquad (2.16)$$

where $\hat{f}$ is the function f written in terms of the parent coordinates ξ, $\square$ denotes the range of parent coordinates for the dimension of problem considered and $j(\xi)$ is the appropriate Jacobian transformation for the coordinate system considered. For the various problem classes these are given by:

Problem type	$\square$ – domain	j – Jacobian	
Three-dimensional:	$d\xi_1\,d\xi_2\,d\xi_3$	$j(\xi_1,\xi_2,\xi_3)$	
Plane strain:	$d\xi_1\,d\xi_2$	$j(\xi_1,\xi_2)$	(2.17)
Plane stress:	$d\xi_1\,d\xi_2$	$h_3\,j(\xi_1,\xi_2)$	
Axisymmetric:	$d\xi_1\,d\xi_2$	$2\pi x_1\,j(\xi_1,\xi_2)$	

where for two-dimensional problems

$$j(\xi_1,\xi_2) = \det \begin{bmatrix} \dfrac{\partial x_1}{\partial \xi_1} & \dfrac{\partial x_2}{\partial \xi_1} \\[2mm] \dfrac{\partial x_1}{\partial \xi_2} & \dfrac{\partial x_2}{\partial \xi_2} \end{bmatrix} > 0$$

The minimum number of quadrature points permissible is chosen so that

1. Elements which have constant jacobians j are exactly integrated; or
2. The resulting element stiffness matrix has full rank.

In either case, the consistency part of the patch test must also be satisfied.[2] The first criterion may be used only for linear materials in small strain (such as discussed in this chapter). The second is applicable to both linear and non-linear problems. Use of the next lower order quadrature than that satisfying the above is called *reduced quadrature* and generally should be avoided.* When some terms are integrated using 'full' order and some with 'reduced' order quadrature the method is referred to as *selective reduced integration*.[2,3]

The above form of natural coordinates $\boldsymbol{\xi}$ assumes that each finite element is a line, a quadrilateral or a hexahedron. For other shapes, such as a triangle or a tetrahedron, appropriate changes are made for the natural coordinates and integration formula used (see also Appendix A).[2,3,9]

Surface integrals

It is also necessary to compute integrals over element surfaces and this is most easily accomplished by considering $\mathrm{d}\Gamma$ as a vector oriented in the direction normal to the surface. For three-dimensional problems we form the vector product

$$\mathbf{n}\,\mathrm{d}\Gamma = \mathrm{d}\boldsymbol{\Gamma} = \frac{\partial \mathbf{x}}{\partial \xi_1} \times \frac{\partial \mathbf{x}}{\partial \xi_2}\,\mathrm{d}\xi_1\,\mathrm{d}\xi_2 \tag{2.18}$$
$$= (\mathbf{v}_1 \times \mathbf{v}_2)\,\mathrm{d}\xi_1\,\mathrm{d}\xi_2 = \mathbf{v}_n\,\mathrm{d}\xi_1\,\mathrm{d}\xi_2$$

where ξ_1 and ξ_2 are parent coordinates for the surface element and $\times$ denotes a vector cross product.

Surfaces for two-dimensional problems may be described in terms of ξ_1 only and for this case we replace $\partial \mathbf{x}/\partial \xi_2$ by $\mathbf{e}_3$ (the unit normal to the plane of deformation).

If necessary, the surface differential may be computed from

$$\mathrm{d}\Gamma = \left(\mathbf{v}_n^T \mathbf{v}_n\right)^{1/2}\,\mathrm{d}\xi_1\,\mathrm{d}\xi_2 \tag{2.19}$$

2.4 Non-linear transient and steady-state problems

To obtain a set of *algebraic equations* for transient problems we introduce a *discrete approximation in time*. We write the approximation to the solution as

$$\tilde{\mathbf{u}}(t_{n+1}) \approx \mathbf{u}_{n+1}; \quad \dot{\tilde{\mathbf{u}}}(t_{n+1}) \approx \mathbf{v}_{n+1} \quad \text{and} \quad \ddot{\tilde{\mathbf{u}}}(t_{n+1}) \approx \mathbf{a}_{n+1}$$

where the tilde on discrete variables is omitted for simplicity. Thus, the equilibrium equation (2.11a) at each discrete time t_{n+1} may be written in a *residual form* as

$$\boldsymbol{\Psi}_{n+1} = \mathbf{f}_{n+1} - \mathbf{M}\,\mathbf{a}_{n+1} - \mathbf{P}_{n+1} = \mathbf{0} \tag{2.20a}$$

where

$$\mathbf{P}_{n+1} \equiv \int_{\Omega} \mathbf{B}^T \boldsymbol{\sigma}_{n+1}\,\mathrm{d}\Omega = \mathbf{P}(\mathbf{u}_{n+1}) \tag{2.20b}$$

* Many explicit codes use reduced quadrature in combination with so-called hour-glass stabilization.[4,11]

Here we have indicated that $\mathbf{P}$ can be expressed in terms of the displacement alone. This is correct for elastic materials but with inelastic behaviour the material model will depend on the solution variables in a more general form. In Chapter 4 we will show that most constitutive models may be given in terms of *increments* of $\mathbf{u}$. Thus the above assumed form will need only a minor modification that does not significantly affect the following discussion.

For transient problems we apply the GN22 method[2] (which is identical to the Newmark procedure[12] except for the manner parameters are defined) to equations with second derivatives in time. The GN22 method relates the discrete displacements, velocities, and accelerations at t_{n+1} to those at t_n by the formulas

$$
\begin{aligned}
\mathbf{u}_{n+1} &= \mathbf{u}_n + \Delta t\, \mathbf{v}_n + \tfrac{1}{2}(1 - \beta_2)\Delta t^2 \mathbf{a}_n + \tfrac{1}{2}\beta_2 \Delta t^2 \mathbf{a}_{n+1} = \breve{\mathbf{u}}_{n+1} + \tfrac{1}{2}\beta_2 \Delta t^2 \mathbf{a}_{n+1} \\
\mathbf{v}_{n+1} &= \mathbf{v}_n + (1 - \beta_1)\Delta t\, \mathbf{a}_n + \beta_1 \Delta t\, \mathbf{a}_{n+1} = \breve{\mathbf{v}}_{n+1} + \beta_1 \Delta t\, \mathbf{a}_{n+1}
\end{aligned}
\tag{2.21}
$$

in which $\Delta t = t_{n+1} - t_n$ is a time increment and $\breve{\mathbf{u}}_{n+1}$, $\breve{\mathbf{v}}_{n+1}$ are values depending only on the solution at t_n. This one-step form is very desirable as it allows the Δt to change from one step to the next without introducing any complications (although very large changes should always be avoided). The two parameters β_1 and β_2 are selected to control accuracy and stability.

The transient problem is now obtained for each time t_{n+1} by solving the non-linear equation set (2.20a) and the pair of linear equations with *scalar* coefficients (2.21). It is possible to take as the basic unknown any one of the three variables at time t_{n+1} (i.e., $\mathbf{u}_{n+1}$, $\mathbf{v}_{n+1}$ or $\mathbf{a}_{n+1}$). Using Eq. (2.21) the non-linear equation may then be given in terms of a single unknown.

2.4.1 Explicit GN22 method

A very convenient choice is to take $\beta_2 = 0$ and select $\mathbf{a}_{n+1}$ as the primary unknown. Using Eq. (2.21₁) we immediately obtain

$$
\mathbf{u}_{n+1} = \breve{\mathbf{u}}_{n+1}
$$

and thus from Eq. (2.20a)

$$
\mathbf{M}\,\mathbf{a}_{n+1} = \mathbf{f}_{n+1} - \mathbf{P}(\breve{\mathbf{u}})
$$

may be solved directly for $\mathbf{a}_{n+1}$. The velocity $\mathbf{v}_{n+1}$ is then obtained from Eq. (2.21₂).

This leads to a so-called *explicit* scheme since only linear equations are solved.

If the $\mathbf{M}$ matrix is diagonal[2] (or lumped) the solution for $\mathbf{a}_{n+1}$ is trivial and the problem can be considered solved since

$$
\mathbf{M}^{-1} =
\begin{bmatrix}
1/M_{11} & & \\
& \ddots & \\
& & 1/M_{mm}
\end{bmatrix}
$$

where m is the total number of equations in the problem. However, explicit schemes are only *conditionally stable* with $\Delta t \le \Delta t_{crit}$, where Δt_{crit} is related to the smallest time it takes for 'wave propagation' across any element or, alternatively, the highest 'frequency' in the finite element mesh.[2] Thus a solution by an explicit scheme may require many thousands of time steps to cover a specified time interval. For many transient problems, and indeed for most static (steady state) problems, it is often more efficient to deal with *implicit* methods for which much larger time steps may be used.

2.4.2 Implicit GN22 method

In an implicit method it is convenient to use $\mathbf{u}_{n+1}$ as the basic variable and to calculate $\mathbf{v}_{n+1}$ and $\mathbf{a}_{n+1}$ using Eq. (2.21). With this form we merely set $\mathbf{v}_{n+1} = \mathbf{a}_{n+1} = \mathbf{0}$ to consider a quasi-static problem.* The equation system (2.20a) now can be written as

$$\boldsymbol{\Psi}(\mathbf{u}_{n+1}) \equiv \mathbf{f}_{n+1} - \frac{2c}{\beta_2 \Delta t^2} \mathbf{M} \left[\mathbf{u}_{n+1} - \breve{\mathbf{u}}_{n+1} \right] - \mathbf{P}_{n+1} = \mathbf{0} \tag{2.22}$$

where $c = 1$ for transient problems and $c = 0$ for quasi-static ones. The solution to this set of equations requires an iterative process when any of the terms is non-linear. We shall discuss various non-linear calculation processes in some detail in Chapter 3; however, we note here that Newton's method[†] forms the basis of most practical schemes. In this method an iteration is written as [‡]

$$\boldsymbol{\Psi}_{n+1}^{k+1} \approx \boldsymbol{\Psi}_{n+1}^{k} + d\boldsymbol{\Psi}_{n+1}^{k} = \mathbf{0} \tag{2.23a}$$

where, for $\mathbf{f}_{n+1}$ independent of deformation, the increment of Eq. (2.22) is given by

$$d\boldsymbol{\Psi}_{n+1}^{k} = - \left[\frac{2c}{\beta_2 \Delta t^2} \mathbf{M} - \frac{\partial \mathbf{P}_{n+1}}{\partial \mathbf{u}_{n+1}} \bigg|_{n+1}^{k} \right] d\mathbf{u}_{n+1}^{k} = - \mathbf{A}_{n+1}^{k} \, d\mathbf{u}_{n+1}^{k} \tag{2.23b}$$

The displacement increment is computed from

$$\mathbf{A}_{n+1}^{k} \, d\mathbf{u}_{n+1}^{k} = \boldsymbol{\Psi}_{n+1}^{k} \tag{2.24a}$$

and the solution is updated using

$$\mathbf{u}_{n+1}^{k+1} = \mathbf{u}_{n+1}^{k} + d\mathbf{u}_{n+1}^{k}$$

$$\mathbf{a}_{n+1}^{k+1} = \frac{2}{\beta_2 \Delta t^2} \left[\mathbf{u}_{n+1}^{k+1} - \breve{\mathbf{u}}_{n+1} \right] \tag{2.24b}$$

$$\mathbf{v}_{n+1}^{k+1} = \breve{\mathbf{v}}_{n+1} + \beta_1 \Delta t \, \mathbf{a}_{n+1}^{k+1}$$

An initial iterate may be taken as zero or, more appropriately, as the converged solution from the last time step. Accordingly,

$$\mathbf{u}_{n+1}^{1} = \mathbf{u}_{n} \tag{2.25a}$$

---

* A quasi-static problem may be time or load path dependent; however, inertia effects are not included.
† Often also called the Newton–Raphson method. See reference 13 for a discussion on the history of the method.
‡ Note that an italic '*d*' is used for a solution increment and an upright 'd' for a differential.

in which a quantity without the superscript k denotes a converged value. For transient problems initial velocities and accelerations are given by

$$\mathbf{a}_{n+1}^1 = \frac{2}{\beta_2\,\Delta t^2}\left[\mathbf{u}_n - \breve{\mathbf{u}}_{n+1}\right]$$

$$\mathbf{v}_{n+1}^1 = \breve{\mathbf{v}}_{n+1} + \beta_1\Delta t\,\mathbf{a}_{n+1}^1$$

(2.25b)

Iteration continues until a convergence criterion of the form

$$\|\mathbf{\Psi}_{n+1}^k\| \le \varepsilon\|\mathbf{\Psi}_{n+1}^1\|$$

(2.26)

or similar is satisfied for some small tolerance ε. A good practice when all terms of the Newton method are accurately computed is to assume the tolerance at half machine precision. Thus, if the machine can compute to about 16 digits of accuracy, selection of $\varepsilon = 10^{-8}$ is appropriate. Additional discussion on selection of appropriate convergence criteria is presented in Chapter 3. A common choice of parameters is $\beta_1 = \beta_2 = 1/2$ which is also known as the 'trapezoidal integration rule'.

The derivative of $\mathbf{P}$ appearing in Eq. (2.23b) is computed for each element from Eq. (2.11c$_2$) as

$$\left.\frac{\partial\mathbf{P}^{(e)}}{\partial\mathbf{u}_{n+1}}\right|_{n+1}^k = \int_{\Omega_e}\mathbf{B}^T\mathbf{D}_T^k\mathbf{B}\,\mathrm{d}\Omega \equiv \mathbf{K}_T^k$$

(2.27)

We note that the above relation is similar but not identical to that of linear elasticity. Here $\mathbf{D}_T^k$ is the *tangent modulus matrix* for the stress–strain relation (which may or may not be unique but generally is related to deformations in a non-linear manner) and $\mathbf{K}_T$ is the *tangent stiffness matrix*.

Various forms of non-linear elasticity have in fact been used in the present context and here we present a simple approach in which we define a *strain energy density*, W, as a function of ε

$$W = W(\varepsilon) = W(\varepsilon_{ij})$$

and we note that this definition gives us immediately

$$\sigma = \frac{\partial W}{\partial\varepsilon}$$

(2.28)

If the nature of the function W is known the tangent modulus $\mathbf{D}_T^k$ becomes

$$\mathbf{D}_T^k = \left.\frac{\partial\sigma}{\partial\varepsilon}\right|_{n+1}^k = \left.\frac{\partial^2 W}{\partial\varepsilon\partial\varepsilon}\right|_{n+1}^k$$

For problems in which path dependence is involved to compute σ_{n+1} (viz. Chapter 4) it is necessary to keep track of the *total increment* during the solution step from t_n to t_{n+1} and write

$$\mathbf{u}_{n+1}^{k+1} = \mathbf{u}_n + \Delta\mathbf{u}_{n+1}^{k+1} \qquad\text{with}\qquad \Delta\mathbf{u}_{n+1}^1 = \mathbf{0}$$

(2.29a)

The total increment can be accumulated using the solution increments as

$$\Delta\mathbf{u}_{n+1}^{k+1} = \mathbf{u}_{n+1}^{k+1} - \mathbf{u}_n = \Delta\mathbf{u}_{n+1}^k + d\mathbf{u}_{n+1}^k$$

(2.29b)

In an implicit scheme it is desirable to use the displacement from the last iteration to compute both $\mathbf{A}$ and $\mathbf{\Psi}$ – especially when inelastic material behaviour or large strains are considered.

2.4.3 Generalized mid-point implicit form

An alternative form to that just discussed satisfies the balance of momentum equation at an intermediate time between t_n and t_{n+1}. In this form we interpolate the variables as

$$\mathbf{u}_{n+\alpha} = (1 - \alpha)\,\mathbf{u}_n + \alpha\,\mathbf{u}_{n+1}$$
$$\mathbf{v}_{n+\alpha} = (1 - \alpha)\,\mathbf{v}_n + \alpha\,\mathbf{v}_{n+1} \tag{2.30}$$
$$\mathbf{a}_{n+\alpha} = (1 - \alpha)\,\mathbf{a}_n + \alpha\,\mathbf{a}_{n+1}$$

and write momentum balance as

$$\boldsymbol{\Psi}_{n+\alpha} = \mathbf{f}_{n+\alpha} - \mathbf{M}\,\mathbf{a}_{n+\alpha} - \mathbf{P}_{n+\alpha} = \mathbf{0} \tag{2.31}$$

We select the parameters in the GN22 algorithm as $\beta_1 = \beta_2 = \beta$ and, thus, obtain the simple form

$$\mathbf{u}_{n+1} = \mathbf{u}_n + \tfrac{1}{2}\,\Delta t\,(\mathbf{v}_n + \mathbf{v}_{n+1})$$
$$\frac{1}{\Delta t}\,(\mathbf{v}_{n+1} - \mathbf{v}_n) = (1 - \beta)\,\mathbf{a}_n + \beta\,\mathbf{a}_{n+1} \tag{2.32}$$

Selecting now $\alpha = \beta$ gives the momentum equation in the form

$$\boldsymbol{\Psi}_{n+\alpha} = f_{n+\alpha} - \frac{1}{\Delta t}\,\mathbf{M}\,(\mathbf{v}_{n+1} - \mathbf{v}_n) - \mathbf{P}_{n+\alpha} = \mathbf{0} \tag{2.33}$$

which is the form utilized by Simo *et al.** as part of an energy-momentum conserving method.[14,15] For a linear elastic problem it is easy to show that choosing $\alpha = \beta = 1/2$ will conserve energy during free motion (i.e. $\mathbf{f} = \mathbf{0}$). If non-linear elastic forms are used with these values, it is necessary to modify the manner by which $\sigma_{n+1/2}$ is computed to preserve the conservation property.[16] We shall address this more in Chapters 5 and 6 when we consider finite deformation forms and hyperelastic constitutive models.

The solution of the above form of the balance equation may still use $\mathbf{u}_{n+1}$ as the primary variable. The only modification is the appearance of the parameter α in terms arising from linearizations of $\mathbf{P}_{n+\alpha}$.

2.5 Boundary conditions: non-linear problems

In constructing a solution from a variational (weak) form, boundary conditions are classified into two categories: *natural* conditions that are satisfied by the variational form without special considerations; and *essential* conditions for which modifications to the solution process must be made to make the variational form valid. For example, in an irreducible displacement method the traction boundary condition is a natural form and the displacement boundary condition is an essential form and must be imposed separately.

---

* Simo *et al.* did not interpolate the inertia term and, thus, needed different parameters to obtain the conservation property.

2.5.1 Displacement (essential) condition

The specification of a boundary condition for displacement is given by Eq. (1.7). In a finite element calculation the usual procedure to specify a displacement boundary condition merely assigns the value at a node as

$$(\tilde{u}_a)_i = \bar{u}_i(\mathbf{x}_a) \tag{2.34}$$

where $(\tilde{u}_a)_i$ is the value at node a in the direction i, as shown for a two-dimensional case in Fig. 2.2. Here we note that the condition is imposed on the finite element approximation to the boundary Γ_u^h and not the true boundary Γ_u. As the mesh is refined near the boundary the two converge, generally at a rate equal to or higher than errors from other approximations.

Imposing a specified displacement condition may be implemented in several different ways. For example, consider the *linear static problem* given by

$$\begin{bmatrix} K_{11} & K_{12} \\ K_{21} & K_{22} \end{bmatrix} \begin{Bmatrix} u_1 \\ u_2 \end{Bmatrix} = \begin{Bmatrix} f_1 \\ f_2 \end{Bmatrix} \tag{2.35}$$

in which the condition $u_1 = \bar{u}_1$ is to be imposed.

1. Impose the condition by replacing the first equation (associated with δu_i in a weak form) by the boundary condition giving

$$\begin{bmatrix} 1 & 0 \\ K_{21} & K_{22} \end{bmatrix} \begin{Bmatrix} u_1 \\ u_2 \end{Bmatrix} = \begin{Bmatrix} \bar{u}_1 \\ f_2 \end{Bmatrix}$$

   which yields the desired solution.

   This method is not efficient if $\mathbf{K}$ is symmetric. However, by writing the system as

$$\begin{bmatrix} 1 & 0 \\ 0 & K_{22} \end{bmatrix} \begin{Bmatrix} u_1 \\ u_2 \end{Bmatrix} = \begin{Bmatrix} \bar{u}_1 \\ f_2 - K_{21}\bar{u}_1 \end{Bmatrix} \tag{2.36a}$$

   the problem again becomes symmetric.
2. A second approach is to perform the modification as above and eliminate all equations for which values are known. Accordingly, we then have

$$\mathbf{K}_{22}\mathbf{u}_2 = \mathbf{f}_2 - \mathbf{K}_{21}\bar{u}_1 \tag{2.36b}$$

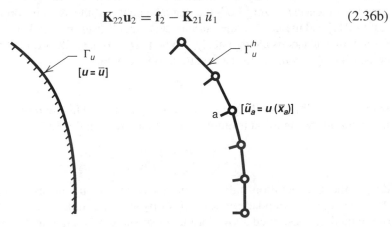

Fig. 2.2 Boundary conditions for specified displacements.

together with the known condition $u_1 = \bar{u}_1$. This approach leads to a final set of equations with a minimum number of unknowns and is the one adopted in *FEAPpv*.

3. A third method uses a 'penalty' approach[2] in which the equations are given as

$$\begin{bmatrix} k_{11} & \mathbf{K}_{12} \\ \mathbf{K}_{21} & \mathbf{K}_{22} \end{bmatrix} \begin{Bmatrix} u_1 \\ \mathbf{u}_2 \end{Bmatrix} = \begin{Bmatrix} k_{11}\,\bar{u}_1 \\ \mathbf{f}_2 \end{Bmatrix} \tag{2.36c}$$

in which $k_{11} = \alpha K_{11}$ where $\alpha \gg 1$. This method is very easy to implement but requires selection of an appropriate value of α. For simple point constraints, such as considered for $u_1 = \bar{u}_1$, a choice of $\alpha = 10^6$ to 10^8 usually is adequate.

When transient non-linear problems are encountered the imposition of displacement boundary conditions becomes slightly more involved. First, the boundary condition needs to be implemented on the incremental equations. This requires computation of initial values for the displacement, velocity and acceleration variables. As noted above there are two basic forms to consider: the *explicit* form and the *implicit* form.

Non-linear explicit problems

The explicit form is straightforward if velocity terms do not appear in the equilibrium equation. Here, for the GN22 algorithm, $\beta_2 = 0$ and the value of the displacement at t_{n+1} is obtained from Eq. (2.21$_1$) including those for the boundary Γ_u. Next we solve

$$\mathbf{M}\,\mathbf{a}_{n+1} = \mathbf{f}_{n+1} - \mathbf{P}(\sigma_{n+1}) \tag{2.37}$$

for the new acceleration, with $\mathbf{M}$ given by a diagonal form. The velocity is then computed with the known acceleration using Eq. (2.21$_2$). By employing a diagonal $\mathbf{M}$, there is no coupling of the acceleration between boundary and non-boundary nodes. If the velocity appears explicitly in the equilibrium equation an iterative strategy can be adopted where $\mathbf{v}_{n+1}^1$ is taken as $\mathbf{v}_n$ and a 'trial' value of the acceleration is computed. Computing the velocity from the trial acceleration and performing one more iteration yields results which are adequate. Here it may be necessary to devise an expression for the boundary velocity updates to maintain high accuracy in final results.

Non-linear implicit problems

In an implicit form using the GN22 algorithm both β_1 and β_2 are non-zero. If the time increment Δt is zero both the displacement and the velocity do not change [viz. Eq. (2.21)] and the new acceleration is determined from Eq. (2.37) – accounting only for any instantaneous change in $\mathbf{f}_{n+1}$. When $\Delta t > 0$, Eq. (2.24a) is used to impose the constraint $\mathbf{u}_{n+1} = \bar{\mathbf{u}}_{n+1}$. In the first iteration we obtain

$$d\mathbf{u}_{n+1}^1 = d\bar{\mathbf{u}}_{n+1}^1$$

from Eqs (2.24b$_1$) and (2.25a) such that $\mathbf{u}_{n+1}^2 \equiv \bar{\mathbf{u}}_{n+1}$. This increment of the displacement boundary condition is employed in the incremental form

$$\begin{bmatrix} \mathbf{A}_{11} & \mathbf{A}_{12} \\ \mathbf{A}_{21} & \mathbf{A}_{22} \end{bmatrix} \begin{Bmatrix} d\bar{\mathbf{u}}_1 \\ d\mathbf{u}_2 \end{Bmatrix} = \begin{Bmatrix} \psi_1 \\ \psi_2 \end{Bmatrix} \tag{2.38}$$

during the *first* iteration only. In the above the set $\bar{\mathbf{u}}_1$ are associated with known displacements of boundary nodes and set $\mathbf{u}_2$ with the 'unknown' displacements. Any of the methods described above for the linear static problem may be used to obtain the solution.

2.5.2 Traction condition

The application of a traction is a 'natural' variational boundary condition and does not affect the active nodal displacements at a boundary – it only affects the applied nodal force condition. The imposition of a non-zero traction on the boundary requires an integration over the surface of each element. Thus for a typical node a as shown in Fig. 2.3 it is necessary to evaluate the integral

$$\mathbf{f}_a = \sum_e \int_{\Gamma_t} N_a \bar{\mathbf{t}} \, \mathrm{d}\Gamma \tag{2.39}$$

where e ranges over all elements belonging to Γ_t that include node a (e.g. for the two-dimensional case shown in Fig. 2.3 this is the element above and the element below node a). Of course if $\bar{\mathbf{t}}$ is zero no evaluation of the integral is required.

Pressure loading
One important example is the application of a normal 'pressure' to a surface. Here the traction is given by

$$\bar{\mathbf{t}} = \bar{p}_n \mathbf{n}$$

where $\bar{p}_n$ is the specified normal pressure (taken as positive when in tension) and $\mathbf{n}$ is a unit outward normal to the boundary Γ_t, see Fig. 2.4. In this case Eq. (2.39) becomes

$$\mathbf{f}_a = \sum_e \int_{\Gamma_t} N_a \, \bar{p}_n \mathbf{n} \, \mathrm{d}\Gamma \tag{2.40a}$$

Using Eq. (2.18) the computation of pressure loading is given by

$$\mathbf{f}_a = \sum_e \int_\square N_a(\boldsymbol{\xi}) \, \bar{p}_n(\boldsymbol{\xi}) \, (\mathbf{v}_1 \times \mathbf{v}_2) \, \mathrm{d}\square \tag{2.40b}$$

where $\square = \mathrm{d}\xi_1 \, \mathrm{d}\xi_2$ and each element integral is performed on the natural coordinate system directly. For two-dimensional problems the surface shape functions are given

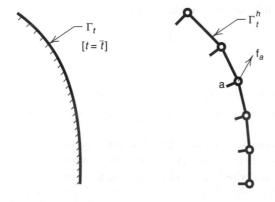

Fig. 2.3 Boundary conditions for specified traction.

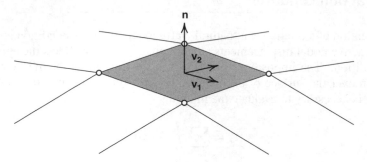

Fig. 2.4 Normal to surface.

by $N_a(\xi_1)$, $\square = d\xi_1$ and we use

$$\mathbf{v}_2 \equiv \begin{cases} \mathbf{e}_3, & \text{for plane strain} \\ h_3\,\mathbf{e}_3, & \text{for plane stress} \\ 2\pi\,x_1\,\mathbf{e}_3, & \text{for axisymmetry} \end{cases}$$

where $\mathbf{e}_3$ is the unit normal vector to the plane of deformation.

2.5.3 Mixed displacement/traction condition

The treatment of a mixed condition in which some displacement components are specified together with some traction components often requires a change in the nodal parameters. For example, a shaft with axis in the x_3 direction and radius R that rotates inside a bearing (without friction or gaps) requires $u_n = u_r(R) = 0$ and $t_\theta(R) = t_z(R) = 0$ (where the coordinate origin is placed at the centre of the shaft). In this case it is necessary to transform the degrees of freedom at each node on the boundary of the shaft such that

$$\mathbf{u}_a = \begin{Bmatrix} (\tilde{u}_1)_a \\ (\tilde{u}_2)_a \\ (\tilde{u}_3)_a \end{Bmatrix} = \begin{bmatrix} \cos\theta_a & -\sin\theta_a & 0 \\ \sin\theta_a & \cos\theta_a & 0 \\ 0 & 0 & 1 \end{bmatrix} \begin{Bmatrix} (\tilde{u}_r)_a \\ (\tilde{u}_\theta)_a \\ (\tilde{u}_z)_a \end{Bmatrix} = \mathbf{L}_a\,\tilde{\mathbf{u}}_{a'} \tag{2.41}$$

This transformation is then applied to a residual as

$$\mathbf{R}_{a'} = \mathbf{L}_a^T \mathbf{R}_a \tag{2.42a}$$

and to the mass and stiffness as

$$\mathbf{M}_{a'b'} = \mathbf{L}_a^T \mathbf{M}_{ab}\,\mathbf{L}_b;\;\; \mathbf{M}_{a'c} = \mathbf{L}_a^T \mathbf{M}_{ac};\;\; \mathbf{M}_{cb'} = \mathbf{M}_{cb}\,\mathbf{L}_b$$

$$\mathbf{K}_{a'b'} = \mathbf{L}_a^T \mathbf{K}_{ab}\,\mathbf{L}_b;\;\; \mathbf{K}_{a'c} = \mathbf{L}_a^T \mathbf{K}_{ac};\;\; \mathbf{K}_{cb'} = \mathbf{K}_{cb}\,\mathbf{L}_b \tag{2.42b}$$

where a and b belong to transformed nodes and c to a node which retains its original orientation. It is usually convenient to perform these transformations on each individual element; however, if desired they can be applied to the assembled arrays.

Once the transformation is performed, each individual displacement and traction condition may be imposed as described above.

2.6 Mixed or irreducible forms

The evaluation of the stiffness given by Eq. (2.12) was cast entirely in terms of the so-called displacement formulation which indeed is extensively used in many finite element solutions. However, on some occasions it is convenient to use *mixed finite element forms* and these are especially necessary when constraints such as (near) incompressibility arise. It has been frequently noted that certain constitutive laws, such as those of viscoelasticity and associative plasticity that we will discuss in Chapter 4, the material behaves in a nearly incompressible manner. For such problems a reformulation is necessary. On such occasions we have two choices of formulation. We can have the variables $\mathbf{u}$ and p (where p is the mean stress) as a *two-field formulation* or we can have the variables $\mathbf{u}$, p and ε_v (where ε_v is the volume change) as a *three-field formulation* (e.g. see reference 2 for more details). Here several alternatives are available and the matter of which we use may depend on the form of the constitutive equation employed. For situations where changes in volume affect only the pressure the two-field form can be easily used. However, for problems in which the response may become coupled between the *deviatoric* and *mean* components of stress and strain the three-field formulations lead to much simpler forms from which to develop a finite element model. To illustrate this point we present a general three-field mixed formulation and show in detail how such coupled effects can be easily included without any change to the previous discussion on solving non-linear problems. The development also serves as a basis for the development of an extended form which permits the treatment of finite deformation problems. This extension will be presented in Chapter 5.

Deviatoric and mean stress and strain components

The treatment of nearly incompressible materials is most easily considered by splitting the stress and strain into their deviatoric (isochoric) and mean parts. Accordingly, we define the mean stress (pressure) as

$$p = \tfrac{1}{3}[\sigma_{11} + \sigma_{22} + \sigma_{33}] = \tfrac{1}{3}\sigma_{ii} \qquad (2.43a)$$

and the deviator stress as

$$(\sigma_d)_{ij} = \sigma_{ij} - \delta_{ij}\,p \qquad (2.43b)$$

where δ_{ij} is the Kronecker delta function

$$\delta_{ij} = \left\{ \begin{array}{ll} 1; & \text{for } i = j \\ 0; & \text{for } i \neq j \end{array} \right.$$

Similarly, we define the mean strain (volume change) as

$$\varepsilon_v = [\varepsilon_{11} + \varepsilon_{22} + \varepsilon_{33}] = \varepsilon_{ii} \qquad (2.44a)$$

and the deviator strain as

$$(\varepsilon_d)_{ij} = \varepsilon_{ij} - \tfrac{1}{3}\delta_{ij}\,\varepsilon_v \qquad (2.44b)$$

Note that the placement of the $1/3$ factor appears in both, but at different locations in the expressions.

A three-field mixed method for general constitutive models

In order to develop a mixed form for use with constitutive models in which mean and deviatoric effects can be coupled we define mean and deviatoric matrix operators given by

$$\mathbf{m} = \begin{bmatrix} 1 & 1 & 1 & 0 & 0 & 0 \end{bmatrix}^T \quad \text{and} \quad \mathbf{I}_d = \mathbf{I} - \tfrac{1}{3}\mathbf{m}\mathbf{m}^T, \tag{2.45}$$

respectively, where $\mathbf{I}$ is the identity matrix.

The strains may now be expressed in a *mixed* form as

$$\varepsilon = \mathbf{I}_d(\mathcal{S}\mathbf{u}) + \tfrac{1}{3}\mathbf{m}\,\varepsilon_v \tag{2.46a}$$

where the first term is the deviatoric part and the second the mean part. Similarly, the stresses may now be expressed in a mixed form as

$$\sigma = \mathbf{I}_d\,\breve{\sigma} + \mathbf{m}\,p \tag{2.46b}$$

where $\breve{\sigma}$ is the set of stresses deduced directly from the strains, incremental strains, or strain rates, depending on the particular constitutive model form. For the present we shall denote this stress by

$$\breve{\sigma} = \sigma(\varepsilon) \tag{2.47}$$

and we note that it is not necessary to split the model into mean and deviatoric parts.

The weak form (variational Galerkin equations) for the case including transients is now given by

$$\int_\Omega \delta\mathbf{u}^T \rho\,\ddot{\mathbf{u}}\,d\Omega + \int_\Omega \delta(\mathcal{S}\mathbf{u})^T \sigma\,d\Omega = \int_\Omega \delta\mathbf{u}^T \mathbf{b}\,d\Omega + \int_{\Gamma_t} \delta\mathbf{u}^T \mathbf{t}\,d\Gamma$$

$$\int_\Omega \delta\varepsilon_v \left[\tfrac{1}{3}\mathbf{m}^T \breve{\sigma} - p \right] d\Omega = 0 \tag{2.48}$$

$$\int_\Omega \delta p \left[\mathbf{m}^T (\mathcal{S}\mathbf{u}) - \varepsilon_v \right] d\Omega = 0$$

Introducing finite element approximations to the variables as

$$\mathbf{u} \approx \hat{\mathbf{u}} = \mathbf{N}_u \tilde{\mathbf{u}}, \qquad p \approx \hat{p} = \mathbf{N}_p \tilde{\mathbf{p}} \quad \text{and} \quad \varepsilon_v \approx \hat{\varepsilon}_v = \mathbf{N}_v \tilde{\varepsilon}_v$$

and similar approximations to virtual quantities as

$$\delta\mathbf{u} \approx \delta\hat{\mathbf{u}} = \mathbf{N}_u \delta\tilde{\mathbf{u}}, \qquad \delta p \approx \delta\hat{p} = \mathbf{N}_p \delta\tilde{\mathbf{p}} \quad \text{and} \quad \delta\varepsilon_v \approx \delta\hat{\varepsilon}_v = \mathbf{N}_v \delta\tilde{\varepsilon}_v$$

the strain in an element becomes

$$\varepsilon = \mathbf{I}_d\,\mathbf{B}\,\tilde{\mathbf{u}} + \tfrac{1}{3}\mathbf{m}\,\mathbf{N}_v\,\tilde{\varepsilon}_v \tag{2.49}$$

in which $\mathbf{B}$ is the standard strain–displacement matrix given in Eq. (2.9a). Similarly, the stresses in each element may be computed by using

$$\sigma = \mathbf{I}_d\,\breve{\sigma} + \mathbf{m}\,\mathbf{N}_p\,\tilde{\mathbf{p}} \tag{2.50}$$

where again $\breve{\sigma}$ are stresses computed as in Eq. (2.47) in terms of the strains ε.

Substituting the element stress and strain expressions from Eqs (2.49) and (2.50) into Eq. (2.48) we obtain the set of finite element equations

$$\mathbf{P} + \mathbf{M}\ddot{\mathbf{u}} = \mathbf{f}$$
$$\mathbf{P}_p - \mathbf{K}_{vp}\tilde{\mathbf{p}} = \mathbf{0}$$
$$-\mathbf{K}_{vp}^T\tilde{\varepsilon}_v + \mathbf{K}_{pu}\tilde{\mathbf{u}} = \mathbf{0}$$

(2.51)

where

$$\mathbf{P} = \int_\Omega \mathbf{B}^T \boldsymbol{\sigma}\, d\Omega, \qquad\qquad \mathbf{P}_p = \frac{1}{3}\int_\Omega \mathbf{N}_v^T \mathbf{m}^T \check{\sigma}\, d\Omega$$

$$\mathbf{K}_{vp} = \int_\Omega \mathbf{N}_v^T \mathbf{N}_p\, d\Omega, \qquad\qquad \mathbf{K}_{pu} = \int_\Omega \mathbf{N}_p^T \mathbf{m}^T \mathbf{B}\, d\Omega$$

(2.52)

$$\mathbf{f} = \int_\Omega \mathbf{N}_u^T \mathbf{b}\, d\Omega + \int_{\Gamma_t} \mathbf{N}_u^T \bar{\mathbf{t}}\, d\Gamma$$

If the pressure and volumetric strain approximations are taken locally in each element and $\mathbf{N}_v = \mathbf{N}_p$ it is possible to solve the second and third equation of (2.51) in each element individually. Noting that the array $\mathbf{K}_{vp}$ is now symmetric positive definite, we may always write these as

$$\tilde{\mathbf{p}} = \mathbf{K}_{vp}^{-1}\mathbf{P}_p$$
$$\tilde{\varepsilon}_v = \mathbf{K}_{vp}^{-1}\mathbf{K}_{pu}\tilde{\mathbf{u}} = \mathbf{W}\tilde{\mathbf{u}}$$

(2.53)

The mixed strain in each element may now be computed as

$$\varepsilon = \left[\mathbf{I}_d\mathbf{B} + \frac{1}{3}\mathbf{m}\mathbf{B}_v\right]\tilde{\mathbf{u}} = \begin{bmatrix}\mathbf{I}_d & \frac{1}{3}\mathbf{m}\end{bmatrix}\begin{bmatrix}\mathbf{B}\\\mathbf{B}_v\end{bmatrix}\tilde{\mathbf{u}}$$

(2.54)

where

$$\mathbf{B}_v = \mathbf{N}_v\mathbf{W}$$

(2.55)

defines a *mixed form* of the volumetric strain–displacement equations.

From the above results it is possible to write the vector $\mathbf{P}$ in the alternative forms[17,18]

$$\mathbf{P} = \int_\Omega \mathbf{B}^T\boldsymbol{\sigma}\, d\Omega = \int_\Omega \left[\mathbf{B}^T\mathbf{I}_d + \tfrac{1}{3}\mathbf{B}_v^T\mathbf{m}^T\right]\check{\sigma}\, d\Omega = \int_\Omega \begin{bmatrix}\mathbf{B}^T & \mathbf{B}_v^T\end{bmatrix}\begin{bmatrix}\mathbf{I}_d\\\tfrac{1}{3}\mathbf{m}^T\end{bmatrix}\check{\sigma}\, d\Omega$$

(2.56)

Based on this result we observe that it is not necessary to compute the true mixed stress except when reporting final results. This is particularly important when we consider the effects of other material models in Chapter 4.

The last step in the process is the computation of the tangent for the equations. This is straightforward using forms given by Eq. (2.47) where we obtain

$$d\check{\sigma} = \check{\mathbf{D}}_T d\varepsilon$$

Use of Eq. (2.54) to express the incremental mixed strains then gives

$$\mathbf{K}_T = \int_\Omega \begin{bmatrix}\mathbf{B}^T & \mathbf{B}_v^T\end{bmatrix}\begin{bmatrix}\mathbf{I}_d\\\tfrac{1}{3}\mathbf{m}^T\end{bmatrix}\check{\mathbf{D}}_T\begin{bmatrix}\mathbf{I}_d & \tfrac{1}{3}\mathbf{m}\end{bmatrix}\begin{bmatrix}\mathbf{B}\\\mathbf{B}_v\end{bmatrix} d\Omega$$

(2.57a)

It should be noted that construction of a modified modulus term given by

$$\bar{\mathbf{D}}_T = \begin{bmatrix} \mathbf{I}_d \\ \frac{1}{3}\mathbf{m}^T \end{bmatrix} \check{\mathbf{D}}_T \begin{bmatrix} \mathbf{I}_d & \frac{1}{3}\mathbf{m} \end{bmatrix} = \begin{bmatrix} \mathbf{I}_d \check{\mathbf{D}}_T \mathbf{I}_d & \frac{1}{3}\mathbf{I}_d \check{\mathbf{D}}_T \mathbf{m} \\ \frac{1}{3}\mathbf{m}^T \check{\mathbf{D}}_T \mathbf{I}_d & \frac{1}{9}\mathbf{m}^T \check{\mathbf{D}}_T \mathbf{m} \end{bmatrix} = \begin{bmatrix} \bar{\mathbf{D}}_{11} & \bar{\mathbf{D}}_{12} \\ \bar{\mathbf{D}}_{21} & \bar{\mathbf{D}}_{22} \end{bmatrix} \quad (2.57b)$$

requires very few operations because of the sparsity and form of the arrays $\mathbf{I}_d$ and $\mathbf{m}$. Consequently, the multiplication by the coefficient matrices $\mathbf{B}$ and $\mathbf{B}_v$ in this form is far more efficient than constructing a $\bar{\mathbf{B}}$ as[2]

$$\bar{\mathbf{B}} = \mathbf{I}_d \mathbf{B} + \frac{1}{3}\mathbf{m}\,\mathbf{B}_v \quad (2.58)$$

and computing the tangent from

$$\mathbf{K}_T = \int_\Omega \bar{\mathbf{B}}^T \check{\mathbf{D}}_T \bar{\mathbf{B}}\, \mathrm{d}\Omega \quad (2.59)$$

In this form $\bar{\mathbf{B}}$ has few zero terms which accounts for the difference in effort.

Example 2.1   Linear elastic tangent
As an example we consider a linear elastic material with the constitutive equation expressed as

$$\boldsymbol{\sigma} = \left(K\,\mathbf{m}\mathbf{m}^T + 2G\,\mathbf{I}_0 \right) \boldsymbol{\varepsilon} \quad (2.60)$$

in which

$$\mathbf{I}_0 = \frac{1}{2} \begin{bmatrix} 2 & 0 & 0 & 0 & 0 & 0 \\ 0 & 2 & 0 & 0 & 0 & 0 \\ 0 & 0 & 2 & 0 & 0 & 0 \\ 0 & 0 & 0 & 1 & 0 & 0 \\ 0 & 0 & 0 & 0 & 1 & 0 \\ 0 & 0 & 0 & 0 & 0 & 1 \end{bmatrix} \quad (2.61)$$

accounts for the transformations used to define the strain.

The incremental form is identical to Eq. (2.60) with $\boldsymbol{\sigma}$, $\boldsymbol{\varepsilon}$ replaced by $d\boldsymbol{\sigma}$, $d\boldsymbol{\varepsilon}$.

The above form for the mixed element is valid for use with many different linear and non-linear constitutive models. In Chapter 4 we consider stress–strain behaviour modelled by viscoelasticity, classical plasticity, and generalized plasticity formulations. Each of these forms can lead to situations in which a nearly incompressible response is required and for many examples included in this book we shall use the above mixed formulation. Here two basic forms of finite element approximations are considered: a four-node quadrilateral or an eight-node brick isoparametric element with constant interpolation in each element for one-term approximations to N_v and N_p by unity; and a nine-node quadrilateral or a 27-node brick isoparametric element with linear interpolation for $\mathbf{N}_p$ and $\mathbf{N}_v$.* Accordingly, for the latter class of elements in two dimensions we use

$$\mathbf{N}_p = \mathbf{N}_v = \begin{bmatrix} 1 & \xi & \eta \end{bmatrix} \quad \text{or} \quad \begin{bmatrix} 1 & x & y \end{bmatrix}$$

---

* Formulations using the eight-node quadrilateral and 20-node brick serendipity elements may also be constructed; however, these elements do not fully satisfy the mixed patch test (see reference 2).

and in three dimensions

$$\mathbf{N}_p = \mathbf{N}_v = \begin{bmatrix} 1 & \xi & \eta & \zeta \end{bmatrix} \quad \text{or} \quad \begin{bmatrix} 1 & x & y & z \end{bmatrix}$$

The elements created by this process may be used to solve a wide range of problems in solid mechanics, as we shall illustrate in later chapters of this volume.

2.7 Non-linear quasi-harmonic field problems

In subsequent chapters we shall touch upon non-linear problems in the context of inelastic constitutive equations for solids, plates, and shells and in geometric effects arising from finite deformation. Non-linear effects can also be considered for various fluid mechanics situations (e.g. see reference 19). However, non-linearity occurs in many other problems and in these the techniques described in this chapter are still universally applicable. An example of such situations is the quasi-harmonic equation which is encountered in many fields of engineering. Here we consider a simple quasi-harmonic problem given by (e.g. heat conduction)

$$\rho c \dot{\phi} + \mathbf{\nabla}^T \mathbf{q} - Q(\phi) = 0 \tag{2.62}$$

with suitable boundary conditions. Such a form may be used to solve problems ranging from temperature response in solids, seepage in porous media, magnetic effects in solids, and potential fluid flow. In the above, $\mathbf{q}$ is a flux and quite generally this can be written as

$$\mathbf{q} = \mathbf{q}(\phi, \mathbf{\nabla}\phi) = -\mathbf{k}(\phi, \mathbf{\nabla}\phi)\mathbf{\nabla}\phi$$

or, after linearization,

$$d\mathbf{q} = -\mathbf{k}^0 d\phi - \mathbf{k}^1 d(\mathbf{\nabla}\phi)$$

where

$$k_i^0 = -\frac{\partial q_i}{\partial \phi} \quad \text{and} \quad k_{ij}^1 = -\frac{\partial q_i}{\partial \phi_{,j}}$$

The source term $Q(\phi)$ also can introduce non-linearity.

A discretization based on Galerkin procedures gives after integration by parts of the $\mathbf{q}$ term the problem

$$\delta\Pi = \int_\Omega \delta\phi \, \rho \, c \, \dot{\phi} \, d\Omega - \int_\Omega (\mathbf{\nabla}\,\delta\phi)^T \, \mathbf{q} \, d\Omega \\ - \int_\Omega \delta\phi \, Q(\phi) \, d\Omega - \int_{\Gamma_q} \delta\phi \, \bar{q}_n \, d\Gamma = 0 \tag{2.63}$$

and is still valid if $\mathbf{q}$ and/or Q (and indeed the boundary conditions) are dependent on ϕ or its derivatives. Introducing the interpolations

$$\phi = \mathbf{N}\tilde{\phi}(t) \quad \text{and} \quad \delta\phi = \mathbf{N}\delta\tilde{\phi} \tag{2.64}$$

a discretized form is given as

$$\mathbf{\Psi} = \mathbf{f}(\tilde{\phi}) - \mathbf{C}\dot{\tilde{\phi}} - \mathbf{P}_q(\tilde{\phi}) = \mathbf{0} \tag{2.65a}$$

where

$$\mathbf{C} = \int_\Omega \mathbf{N}^T \rho c \, \mathbf{N} \, d\Omega$$

$$\mathbf{P}_q = -\int_\Omega (\nabla \mathbf{N})^T \mathbf{q} \, d\Omega \qquad (2.65b)$$

$$\mathbf{f} = \int_\Omega \mathbf{N}^T Q(\phi) \, d\Omega - \int_{\Gamma_q} \mathbf{N}^T \bar{q}_n \, d\Gamma$$

Equation (2.65a) may be solved following similar procedures described above. For instance, just as we did with GN22, we can now use GN11 as[2]

$$\phi_{n+1} = \phi_n + (1 - \theta)\dot{\phi}_n + \theta\dot{\phi}_{n+1} \qquad (2.66)$$

Once again we have the choice of using ϕ_{n+1} or $\dot{\phi}_{n+1}$ as the primary solution variable. To this extent the process of solving transient problems follows the same lines as those described in the previous section and need not be further discussed here. We note again that the use of ϕ_{n+1} as the chosen variable will allow the solution method to be applied to static (steady state) problems in which the first term of Eq. (2.62) becomes zero.

2.8 Typical examples of transient non-linear calculations

In this section we report results of some transient problems of structural mechanics as well as field problems. As we mentioned earlier, we usually will not consider transient behaviour in later parts of this book as the solution process for transients essentially follows the path described above.

Transient heat conduction

The governing equation for this set of physical problems is discussed in the previous section, with ϕ being the temperature T now [Eq. (2.62)].

Non-linearity clearly can arise from the specific heat, c, thermal conductivity, k, and source, Q, being temperature dependent or from a radiation boundary condition

$$k\frac{\partial T}{\partial n} = -\alpha(T - T_0)^n \qquad (2.67)$$

with $n \neq 1$. Here α is a convective heat transfer coefficient and T_0 is an ambient external temperature. We shall show two examples to illustrate the above.

The first concerns the *freezing* of ground in which the latent heat of freezing is represented by varying the material properties with temperature in a narrow zone, as shown in Fig. 2.5. Further, in the transition from the fluid to the frozen state a variation in conductivity occurs. We now thus have a problem in which both matrices $\mathbf{C}$ and $\mathbf{P}$ [Eq. (2.65b)] are variable, and solution in Fig. 2.6 illustrates the progression of a freezing front which was derived by using the three-point (Lees) algorithm[20,21] with $\mathbf{C} = \mathbf{C}_n$ and $\mathbf{P} = \mathbf{P}_n$.

A computational feature of some significance arises in this problem as values of the specific heat become very high in the transition zone and in time stepping can be missed

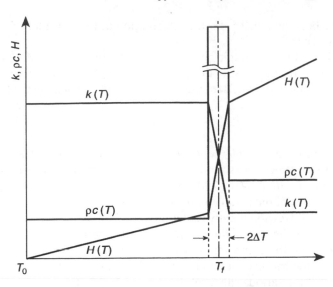

Fig. 2.5 Estimation of thermophysical properties in phase change problems. The latent heat effect is approximated by a large capacity over a small temperature interval $2\Delta T$.

if the temperature step *straddles* the freezing point. To avoid this difficulty and keep the heat balance correct the concept of enthalpy is introduced, defining

$$H = \int_0^T \rho c \, dT \qquad (2.68)$$

Now, whenever a change of temperature is considered, an appropriate value of ρc is calculated that gives the correct change of H.

The heat conduction problem involving phase change is of considerable importance in welding and casting technology. Some very useful finite element solutions of these

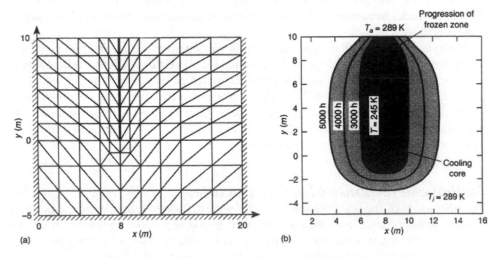

Fig. 2.6 Freezing of a moist soil (sand).

problems have been obtained.[22] Further elaboration of the procedure described above is given in reference.[23]

The second non-linear example concerns the problem of *spontaneous ignition*.[24] We will discuss the steady state case of this problem in Chapter 4 and now will be concerned only with transient behaviour. Here the heat generated depends on the temperature

$$Q = \bar{\delta}e^{T} \tag{2.69}$$

and the situation can become *physically unstable* with the computed temperature rising continuously to extreme values. In Fig. 2.7 we show a transient solution of a sphere at an initial temperature of $T = 290\,\text{K}$ immersed in a bath of 500 K. The solution is given for two values of the parameter $\bar{\delta}$ with $k = \rho c = 1$, and the non-linearity is now so severe that a full Newton iterative solution in each time increment is necessary. For the larger value of $\bar{\delta}$ the temperature increases to an 'infinite' value in a *finite time* and the time interval for the computation had to be changed continuously to account for this. The finite time for this point to be reached is known as the *induction time* and is shown in Fig. 2.7 for various values of $\bar{\delta}$.

The question of changing the time interval during the computation has not been discussed in detail, but clearly this must be done quite frequently to avoid large changes of the unknown function which will result in inaccuracies.

Structural dynamics

Here the examples concern dynamic structural transients with material and geometric non-linearity. A highly non-linear geometrical and material non-linearity generally occurs. Neglecting damping forces, Eq. (2.11a) can be explicitly solved in an efficient manner.

If the explicit computation is pursued to the point when steady state conditions are approached, that is, until $\mathbf{a} = \mathbf{v} \approx \mathbf{0}$, the solution to a static non-linear problem is obtained. This type of technique is frequently efficient as an alternative to the methods described above and in Chapter 3 has been applied successfully in the context of finite differences under the name of 'dynamic relaxation' for the solution of non-linear static problems.[25]

Two examples of explicit dynamic analysis will be given here. The first problem, illustrated in Plate 3, is a large three-dimensional problem and its solution was obtained with the use of an explicit dynamic scheme. In such a case implicit schemes would be totally inapplicable and indeed the explicit code provides a very efficient solution of the crash problem shown. It must, however, be recognized that such final solutions are not necessarily unique. As a second example Fig. 2.8 shows a typical crash analysis of a motor vehicle carried out by similar means.

Earthquake response of soil – structures

The interaction of the soil skeleton or matrix with the water contained in the pores is of extreme importance in earthquake engineering and here again solution of transient non-linear equations is necessary. As in the mixed problem which we referred to earlier, the variables include displacement, and the pore pressure in the fluid p.

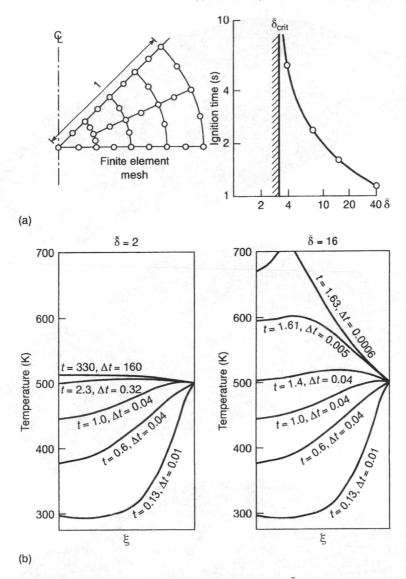

Fig. 2.7 Reactive sphere. Transient temperature behaviour for ignition ($\bar{\delta} = 16$) and non-ignition ($\bar{\delta} = 2$) cases: (a) induction time versus Frank–Kamenetskii parameter; temperature profiles; (b) temperature profiles for ignition ($\bar{\delta} = 16$) and non-ignition ($\bar{\delta} = 2$) transient behaviour of a reactive sphere.

We have in fact shown a comparison between some centrifuge results and computations elsewhere (viz. Chapter 18 of reference 2). These illustrate the development of the pore pressure arising from a particular form of the constitutive relation assumed. Many such examples and indeed the full theory are given in reference 26 and in Fig. 2.9 we show an example of comparison of calculations and a centrifuge model presented at a 1993 workshop known as VELACS.[27] This figure shows the displacements of a big retaining wall after the passage of an earthquake, which were measured in the centrifuge and also calculated.

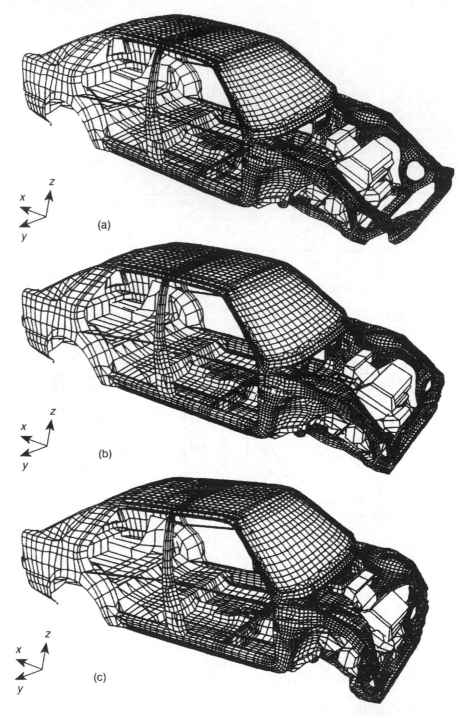

Fig. 2.8 Crash analysis: (a) mesh at $t = 0$ ms; (b) mesh at $t = 20$ ms; (c) mesh at $t = 40$ ms.

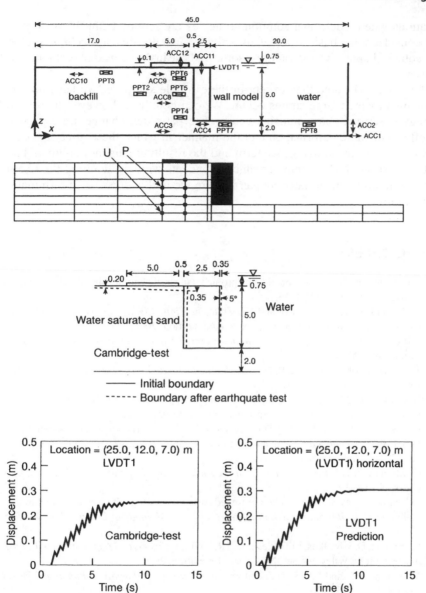

Fig. 2.9 Retaining wall subjected to earthquake excitation: comparison of experiment (centrifuge) and calculations.[26]

2.9 Concluding remarks

In this chapter we have summarized the basic steps needed to solve a general small-strain solid mechanics problem as well as the quasi-harmonic field problem. Only a standard Newton solution method has been mentioned to solve the resulting non-linear algebraic problem. For problems which include non-linear behaviour there are many situations where additional solution strategies are required. In the next chapter we will consider some basic schemes for solving such non-linear algebraic problems. In subsequent chapters we shall address some of these in the context of particular problems classes.

The reader will note that, except in the example solutions, we have not discussed problems in which large strains occur. We can note here, however, that the solution strategy described above remains valid. The parts that change are associated with the effects of finite deformation and the manner in which these affect the computing of stresses, the stress-divergence term and the resulting tangent moduli and stiffness. As these aspects involve more advanced concepts we have deferred the treatment of finite strain problems to later chapters where we will address basic formulations and applications.

References

1. B.G. Galerkin. Series solution of some problems in elastic equilibrium of rods and plates. *Vestn. Inzh. Tech.*, 19:897–908, 1915.
2. O.C. Zienkiewicz, R.L. Taylor and J.Z. Zhu. *The Finite Element Method: Its Basis and Fundamentals*. Butterworth-Heinemann, Oxford, 6th edition, 2005.
3. T.J.R. Hughes. *The Finite Element Method: Linear Static and Dynamic Analysis*. Dover Publications, New York, 2000.
4. T. Belytschko, W.K. Liu and B. Moran. *Nonlinear Finite Elements for Continua and Structures*. John Wiley & Sons, Chichester, 2000.
5. B. Cockburn, G.E. Karniadakis and Chi-Wang Shu. *Discontinuous Galerkin Methods: Theory, Computation and Applications*. Springer-Verlag, Berlin, 2000.
6. E.L. Wilson, R.L. Taylor, W.P. Doherty and J. Ghaboussi. Incompatible displacement models. In S.T. Fenves *et al.*, editor, *Numerical and Computer Methods in Structural Mechanics*, pages 43–57. Academic Press, New York, 1973.
7. G. Strang and G.J. Fix. *An Analysis of the Finite Element Method*. Prentice-Hall, Englewood Cliffs, NJ, 1973.
8. J.C. Simo and M.S. Rifai. A class of mixed assumed strain methods and the method of incompatible modes. *International Journal for Numerical Methods in Engineering*, 29:1595–1638, 1990.
9. R.D. Cook, D.S. Malkus, M.E. Plesha and R.J. Witt. *Concepts and Applications of Finite Element Analysis*. John Wiley & Sons, New York, 4th edition, 2001.
10. M. Abramowitz and I.A. Stegun, editors. *Handbook of Mathematical Functions*. Dover Publications, New York, 1965.
11. D. Kosloff and G.A. Frasier. Treatment of hour glass patterns in low order finite element codes. *International Journal for Numerical Analysis Methods in Geomechanics*, 2:57–72, 1978.
12. N. Newmark. A method of computation for structural dynamics. *J. Engineering Mechanics, ASCE*, 85:67–94, 1959.
13. N. Bićanić and K.W. Johnson. Who was 'Raphson'? *International Journal for Numerical Methods in Engineering*, 14:148–152, 1979.
14. J.C. Simo and N. Tarnow. The discrete energy-momentum method. Conserving algorithm for nonlinear elastodynamics. *Zeitschrift für Mathematik und Physik*, 43:757–793, 1992.
15. J.C. Simo and N. Tarnow. Exact energy-momentum conserving algorithms and symplectic schemes for nonlinear dynamics. *Computer Methods in Applied Mechanics and Engineering*, 100:63–116, 1992.

16. J.C. Simo and O. González. Recent results on the numerical integration of infinite dimensional hamiltonian systems. In *Recent Developments in Finite Element Analysis*. CIMNE, Barcelona, Spain, 1994.

17. T.J.R. Hughes. Generalization of selective integration procedures to anisotropic and non-linear media. *International Journal for Numerical Methods in Engineering*, 15:1413–1418, 1980.

18. J.C. Simo and T.J.R. Hughes. On the variational foundations of assumed strain methods. *J. Appl. Mech.*, 53(1):51–54, 1986.

19. O.C. Zienkiewicz, R.L. Taylor and P. Nithiarasu. *The Finite Element Method for Fluid Dynamics*. Butterworth-Heinemann, Oxford, 6th edition, 2005.

20. M. Lees. A linear three level difference scheme for quasilinear parabolic equations. *Maths. Comp.*, 20:516–622, 1966.

21. G. Comini, S. Del Guidice, R.W. Lewis and O.C. Zienkiewicz. Finite element solution of non-linear conduction problems with special reference to phase change. *International Journal for Numerical Methods in Engineering*, 8:613–624, 1974.

22. H.D. Hibbitt and P.V. Marçal. Numerical thermo-mechanical model for the welding and subsequent loading of a fabricated structure. *Computers and Structures*, 3:1145–1174, 1973.

23. K. Morgan, R.W. Lewis and O.C. Zienkiewicz. An improved algorithm for heat convection problems with phase change. *International Journal for Numerical Methods in Engineering*, 12:1191–1195, 1978.

24. C.A. Anderson and O.C. Zienkiewicz. Spontaneous ignition: finite element solutions for steady and transient conditions. *Trans ASME, J. Heat Transfer*, pages 398–404, 1974.

25. J.R.H. Otter, F. Cassel and R.E. Hobbs. Dynamic relaxation. *Proc. Inst. Civ. Eng.*, 35:633–656, 1966.

26. O.C. Zienkiewicz, A.H.C. Chan, M. Pastor, B.A. Schrefler and T. Shiomi. *Computational Geomechanics: With Special Reference to Earthquake Engineering*. John Wiley & Sons, Chichester, 1999.

27. K. Arulanandan and R.F. Scott, editors. *Proceedings of VELACS Symposium*, Rotterdam, 1993. Balkema.

3

Solution of non-linear algebraic equations

3.1 Introduction

In the solution of linear problems by a finite element method we always need to solve a set of simultaneous algebraic equations of the form

$$\mathbf{Ku} = \mathbf{f} \tag{3.1}$$

Provided the coefficient matrix is non-singular the solution to these equations is unique. In the solution of non-linear problems we will always obtain a set of algebraic equations; however, they generally will be non-linear. For example, in Chapter 2 we obtained the set (2.22) at each discrete time t_{n+1}. Here, we consider the generic problem which we indicate as

$$\boldsymbol{\Psi}_{n+1} = \boldsymbol{\Psi}(\mathbf{u}_{n+1}) = \mathbf{f}_{n+1} - \mathbf{P}(\mathbf{u}_{n+1}) \tag{3.2}$$

where $\mathbf{u}_{n+1}$ is the set of discretization parameters, $\mathbf{f}_{n+1}$ a vector which is independent of the parameters and $\mathbf{P}$ a vector dependent on the parameters. These equations may have multiple solutions [i.e. more than one set of $\mathbf{u}_{n+1}$ may satisfy Eq. (3.2)]. Thus, if a solution is achieved it may not necessarily be the solution sought. Physical insight into the nature of the problem and, usually, small-step incremental approaches from known solutions are essential to obtain realistic answers. Such increments are indeed always required if the problem is transient, if the constitutive law relating stress and strain is path dependent and/or if the load-displacement path has bifurcations or multiple branches at certain load levels.

The general problem should always starts from a nearby solution at

$$\mathbf{u} = \mathbf{u}_n, \qquad \boldsymbol{\Psi}_n = \mathbf{0}, \qquad \mathbf{f} = \mathbf{f}_n \tag{3.3}$$

and often arises from changes in the forcing function $\mathbf{f}_n$ to

$$\mathbf{f}_{n+1} = \mathbf{f}_n + \Delta \mathbf{f}_{n+1} \tag{3.4}$$

The determination of the change $\Delta \mathbf{u}_{n+1}$ such that

$$\mathbf{u}_{n+1} = \mathbf{u}_n + \Delta \mathbf{u}_{n+1} \tag{3.5}$$

will be the objective and generally the increments of $\Delta \mathbf{f}_{n+1}$ will be kept reasonably small so that path dependence can be followed. Further, such incremental procedures will

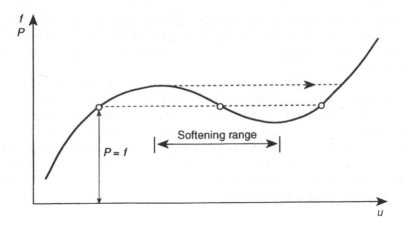

Fig. 3.1 Possibility of multiple solutions.

be useful in avoiding excessive numbers of iterations and in following the physically correct path. In Fig. 3.1 we show a typical non-uniqueness which may occur if the function Ψ_{n+1} decreases and subsequently increases as the parameter $\mathbf{u}_{n+1}$ uniformly increases. It is clear that to follow the path $\Delta\mathbf{f}_{n+1}$ will have both positive and negative signs during a complete computation process.

It is possible to obtain solutions in a single increment only in the case of mild non-linearity (and no path dependence), that is, with

$$\mathbf{f}_n = \mathbf{0}, \qquad \Delta\mathbf{f}_{n+1} = \mathbf{f}_{n+1} \tag{3.6}$$

The literature on general solution approaches and on particular applications is extensive and, in a single chapter, it is not possible to encompass fully all the variants which have been introduced. However, we shall attempt to give a comprehensive picture by outlining first the *general* solution procedures.

In later chapters we shall focus on procedures associated with rate-independent material non-linearity (plasticity), rate-dependent material non-linearity (creep and viscoplasticity), some non-linear field problems, large displacements and other special examples.

3.2 Iterative techniques

3.2.1 General remarks

The solution of the problem posed by Eqs (3.2)–(3.5) cannot be approached directly and some form of iteration will always be required. We shall concentrate here on procedures in which repeated solution of linear equations (i.e. iteration) of the form

$$\mathbf{K}^i d\mathbf{u}_{n+1}^i = \mathbf{r}_{n+1}^i \tag{3.7}$$

in which a superscript i indicates the iteration number. In these a solution increment $d\mathbf{u}_{n+1}^i$ is computed. Direct (Gaussian) elimination techniques or iterative methods can be used to solve the linear equations associated with each iteration.[1-3] However, the application of an iterative solution method may prove to be more economical, and in later chapters we shall frequently refer to such possibilities although they have not been fully explored.

Many of the iterative techniques currently used to solve non-linear problems origi-nated by intuitive application of physical reasoning. However, each of such techniques has a direct association with methods in numerical analysis, and in what follows we shall use the nomenclature generally accepted in texts on this subject.[2,4-7]

Although we state each algorithm for a set of non-linear algebraic equations, we shall illustrate each procedure by using a single scalar equation. This, though useful from a pedagogical viewpoint, is dangerous as convergence of problems with numerous degrees of freedom may depart from the simple pattern in a single equation.

3.2.2 Newton's method

Newton's method is the most rapidly convergent process for solutions of problems in which only one evaluation of $\mathbf{\Psi}$ is made in each iteration. Of course, this assumes that the initial solution is within the *zone of attraction* and, thus, divergence does not occur. Indeed, Newton's method is the only process described here in which the asymptotic rate of convergence is quadratic. The method is often called the Newton–Raphson method as it appears to have been simultaneously derived by Newton and Raphson, and an interesting history of its origins is given in reference 8.

In this iterative method we note that, to the first order, Eq. (3.2) can be approximated as

$$\mathbf{\Psi}(\mathbf{u}_{n+1}^{i+1}) \approx \mathbf{\Psi}(\mathbf{u}_{n+1}^{i}) + \left(\frac{\partial \mathbf{\Psi}}{\partial \mathbf{u}}\right)_{n+1}^{i} d\mathbf{u}_{n+1}^{i} = \mathbf{0} \tag{3.8}$$

Here the iteration counter i usually starts by assuming

$$\mathbf{u}_{n+1}^{1} = \mathbf{u}_{n} \tag{3.9}$$

in which $\mathbf{u}_{n}$ is a converged solution at a previous load level or time step. The jacobian matrix (or in structural terms the stiffness matrix) corresponding to a tangent direction is given by

$$\mathbf{K}_{\mathrm{T}} = \frac{\partial \mathbf{P}}{\partial \mathbf{u}} = -\frac{\partial \mathbf{\Psi}}{\partial \mathbf{u}} \tag{3.10}$$

Equation (3.8) gives immediately the iterative correction as

$$\mathbf{K}_{\mathrm{T}}^{i} d\mathbf{u}_{n+1}^{i} = \mathbf{\Psi}_{n+1}^{i}$$

or

$$d\mathbf{u}_{n+1}^{i} = \left(\mathbf{K}_{\mathrm{T}}^{i}\right)^{-1} \mathbf{\Psi}_{n+1}^{i} \tag{3.11}$$

A series of successive approximations gives

$$\mathbf{u}_{n+1}^{i+1} = \mathbf{u}_{n+1}^{i} + d\mathbf{u}_{n+1}^{i}$$
$$= \mathbf{u}_{n} + \Delta \mathbf{u}_{n+1}^{i} \tag{3.12}$$

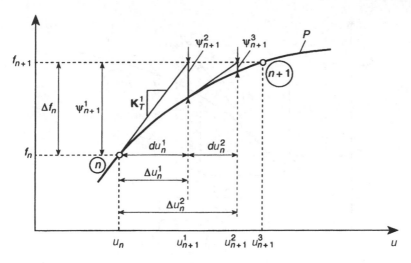

Fig. 3.2 Newton's method.

where
$$\Delta \mathbf{u}_{n+1}^{i} = \sum_{k=1}^{i} d\mathbf{u}_{n+1}^{k} \tag{3.13}$$

The process is illustrated in Fig. 3.2 and shows the very rapid convergence that can be achieved.

The need for the introduction of the total increment $\Delta \mathbf{u}_{n+1}^{i}$ is perhaps not obvious here but in fact it is essential if the solution process is path dependent, as we shall see in Chapter 4 for some non-linear constitutive equations of solids.

The Newton process, despite its rapid convergence, has some negative features:

1. a new $\mathbf{K}_{\mathrm{T}}$ matrix has to be computed at each iteration;
2. if direct solution for Eq. (3.11) is used the matrix needs to be factored at each iteration;
3. on some occasions the tangent matrix is symmetric at a solution state but unsymmetric otherwise (e.g. in some schemes for integrating large rotation parameters[9] or non-associated plasticity). In these cases an unsymmetric solver is needed in general.

Some of these drawbacks are absent in alternative procedures, although generally then a quadratic asymptotic rate of convergence is lost.

3.2.3 Modified Newton's method

This method uses essentially the same algorithm as the Newton process but replaces the variable jacobian matrix $\mathbf{K}_{\mathrm{T}}^{i}$ by a constant approximation
$$\mathbf{K}_{\mathrm{T}}^{i} \approx \bar{\mathbf{K}}_{\mathrm{T}} \tag{3.14}$$

giving in place of Eq. (3.11),
$$d\mathbf{u}_{n+1}^{i} = \bar{\mathbf{K}}_{\mathrm{T}}^{-1} \mathbf{\Psi}_{n+1}^{i} \tag{3.15}$$

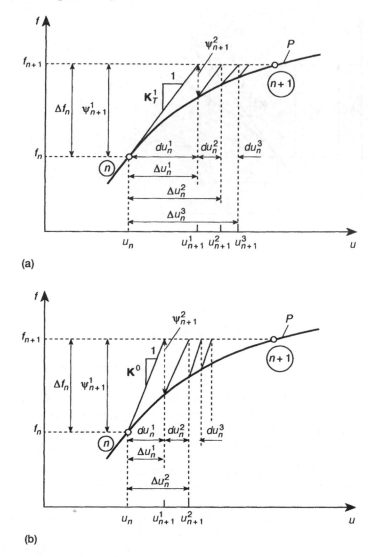

Fig. 3.3 The modified Newton method: (a) with initial tangent in increment; (b) with initial problem tangent.

Many possible choices exist here. For instance, $\bar{\mathbf{K}}_T$ can be chosen as the matrix corresponding to the first iteration $\mathbf{K}_T^1$ [as shown in Fig. 3.3(a)] or may even be one corresponding to some previous time step or load increment $\mathbf{K}^0$ [as shown in Fig. 3.3(b)]. In the context of solving problems in solid mechanics the method is also known as the *stress transfer* or *initial stress method.* Alternatively, the approximation can be chosen every few iterations as $\bar{\mathbf{K}}_T = \mathbf{K}_T^j$ where $j \leq i$.

Obviously, the procedure generally will converge at a slower rate (generally a norm of the residual $\mathbf{\Psi}$ has linear asymptotic convergence instead of the quadratic one in the full Newton method) but some of the difficulties mentioned above for the Newton process disappear. However, some new difficulties can also arise as this method fails to converge when the tangent used has opposite 'slope' to the one at the current solution (e.g. as shown by regions with different slopes in Fig. 3.1). Frequently the 'zone of attraction' for the modified process is increased and previously divergent approaches can be made to converge, albeit slowly. Many variants of this process can be used and symmetric solvers often can be employed when a symmetric form of $\bar{\mathbf{K}}_T$ is chosen.

3.2.4 Incremental-secant or quasi-Newton methods

Once the first iteration of the preceding section has been established giving

$$d\mathbf{u}_{n+1}^1 = \bar{\mathbf{K}}_T^{-1} \boldsymbol{\Psi}_{n+1}^1 \tag{3.16}$$

a secant 'slope' can be found, as shown in Fig. 3.4, such that

$$d\mathbf{u}_{n+1}^1 = \left(\mathbf{K}_s^2\right)^{-1} \left(\boldsymbol{\Psi}_{n+1}^1 - \boldsymbol{\Psi}_{n+1}^2\right) \tag{3.17}$$

This 'slope' can now be used to establish $\mathbf{u}_n^2$ by using

$$d\mathbf{u}_{n+1}^2 = \left(\mathbf{K}_s^2\right)^{-1} \boldsymbol{\Psi}_{n+1}^2 \tag{3.18}$$

Quite generally, one could write in place of Eq. (3.18) for $i > 1$, now dropping subscripts,

$$d\mathbf{u}^i = \left(\mathbf{K}_s^i\right)^{-1} \boldsymbol{\Psi}^i \tag{3.19}$$

where $(\mathbf{K}_s^i)^{-1}$ is determined so that

$$d\mathbf{u}^{i-1} = \left(\mathbf{K}_s^i\right)^{-1} \left(\boldsymbol{\Psi}^{i-1} - \boldsymbol{\Psi}^i\right) = \left(\mathbf{K}_s^i\right)^{-1} \boldsymbol{\gamma}^{i-1} \tag{3.20}$$

For the scalar system illustrated in Fig. 3.4 the determination of $\mathbf{K}_s^i$ is trivial and, as shown, the convergence is much more rapid than in the modified Newton process (generally a super-linear asymptotic convergence rate is achieved for a norm of the residual).

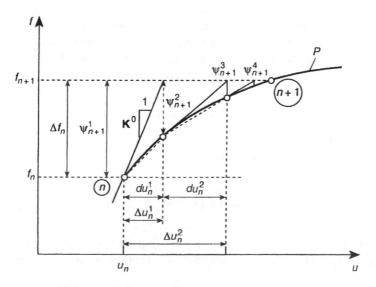

Fig. 3.4 The secant method starting from a $\mathbf{K}^0$ prediction.

For systems with more than one degree of freedom the determination of $\mathbf{K}_s^i$ or its inverse is more difficult and is not unique. Many different forms of the matrix $\mathbf{K}_s^i$ can satisfy relation (3.1) and, as expected, many alternatives are used in practice. All of these use some form of updating of a previously determined matrix or of its inverse in a manner that satisfies identically Eq. (3.20). Some such updates preserve the matrix symmetry whereas others do not. Any of the methods which begin with a symmetric tangent can avoid the difficulty of non-symmetric matrix forms that arise in the Newton process and yet achieve a faster convergence than is possible in the modified Newton procedures.

Such secant update methods appear to stem from ideas introduced first by Davidon[10] and developed later by others. Dennis and More[11] survey the field extensively, while Matthies and Strang[12] appear to be the first to use the procedures in the finite element context. Further work and assessment of the performance of various update procedures is available in references 13–16.

The BFGS update[11] (named after Broyden, Fletcher, Goldfarb and Shanno) and the DFP update[11] (Davidon, Fletcher and Powell) preserve matrix symmetry and positive definiteness and both are widely used. We summarize below a step of the BFGS update for the inverse, which can be written as

$$\left(\mathbf{K}^i\right)^{-1} = \left(\mathbf{I} + \mathbf{w}_i\mathbf{v}_i^\mathrm{T}\right)\left(\mathbf{K}^{i-1}\right)^{-1}\left(\mathbf{I} + \mathbf{v}_i\mathbf{w}_i^\mathrm{T}\right) \tag{3.21}$$

where $\mathbf{I}$ is an identity matrix and

$$\mathbf{v}_i = \left[1 - \frac{(d\mathbf{u}^{i-1})^\mathrm{T}\boldsymbol{\gamma}^{i-1}}{d(\mathbf{u}^i)^\mathrm{T}\boldsymbol{\Psi}^{i-1}}\right]\boldsymbol{\Psi}^{i-1} - \boldsymbol{\Psi}^i$$

$$\mathbf{w}_i = \frac{1}{d\mathbf{u}^{(i-1)\mathrm{T}}\boldsymbol{\gamma}^{i-1}}d\mathbf{u}^{i-1} \tag{3.22}$$

where $\boldsymbol{\gamma}$ is defined by Eq. (3.20). Some algebra will readily verify that substitution of Eqs (3.21) and (3.22) into Eq. (3.20) results in an identity. Further, the form of Eq. (3.21) guarantees preservation of the symmetry of the original matrix.

The nature of the update does not preserve any sparsity in the original matrix. For this reason it is convenient at every iteration to return to the original (sparse) matrix $\mathbf{K}_s^1$, used in the first iteration and to reapply the multiplication of Eq. (3.21) through all previous iterations. This gives the algorithm in the form

$$\mathbf{b}_1 = \prod_{j=2}^{i}\left(\mathbf{I} + \mathbf{v}_j\mathbf{w}_j^\mathrm{T}\right)\boldsymbol{\Psi}^i$$

$$\mathbf{b}_2 = \left(\mathbf{K}_s^1\right)^{-1}\mathbf{b}_1 \tag{3.23}$$

$$d\mathbf{u}^i = \prod_{j=0}^{i-2}\left(\mathbf{I} + \mathbf{w}_{i-j}\mathbf{v}_{i-j}^\mathrm{T}\right)\mathbf{b}_2$$

This necessitates the storage of the vectors $\mathbf{v}_j$ and $\mathbf{w}_j$ for all previous iterations and their successive multiplications. Further details on the operations are described well in reference 12.

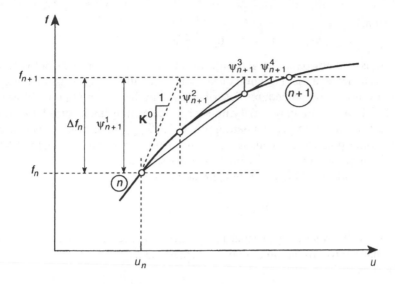

Fig. 3.5 Direct (or Picard) iteration.

When the number of iterations is large ($i > 15$) the efficiency of the update decreases as a result of incipient instability. Various procedures are open at this stage, the most effective being the recomputation and factorization of a tangent matrix at the current solution estimate and restarting the process.

Another possibility is to disregard *all* the previous updates and return to the original matrix $\mathbf{K}_s^1$. Such a procedure was first suggested by Crisfield[13,17,18] in the finite element context and is illustrated in Fig. 3.5. It is seen to be convergent at a slightly slower rate but avoids totally the stability difficulties previously encountered and reduces the storage and number of operations needed. Obviously any of the secant update methods can be used here.

The procedure of Fig. 3.5 is identical to that generally known as direct (or Picard) iteration[4] and is particularly useful in the solution of non-linear problems which can be written as

$$\mathbf{\Psi}(\mathbf{u}) \equiv \mathbf{f} - \mathbf{K}(\mathbf{u})\mathbf{u} = 0 \tag{3.24}$$

In such a case $\mathbf{u}_{n+1}^1 = \mathbf{u}_n$ is taken and the iteration proceeds as

$$\mathbf{u}_{n+1}^{i+1} = \left[\mathbf{K}(\mathbf{u}_{n+1}^i)\right]^{-1}\mathbf{f}_{n+1} \tag{3.25}$$

3.2.5 Line search procedures – acceleration of convergence

All the iterative methods of the preceding section have an identical structure described by Eqs (3.11)–(3.13) in which various approximations to the Newton matrix $\mathbf{K}_T^i$ are used. For all of these an iterative vector is determined and the new value of the unknowns found as

$$\mathbf{u}_{n+1}^{i+1} = \mathbf{u}_{n+1}^i + d\mathbf{u}_{n+1}^i \tag{3.26}$$

starting from

$$\mathbf{u}^1_{n+1} = \mathbf{u}_n$$

in which $\mathbf{u}_n$ is the known (converged) solution at the previous time step or load level. The objective is to achieve the reduction of $\mathbf{\Psi}^{i+1}_{n+1}$ to zero, although this is not always easily achieved by any of the procedures described even in the scalar example illustrated. To get a solution approximately satisfying such a scalar non-linear problem would have been in fact easier by simply evaluating the scalar $\mathbf{\Psi}^{i+1}_{n+1}$ for various values of $\mathbf{u}_{n+1}$ and by suitable interpolation arriving at the required answer. For multi-degree-of-freedom systems such an approach is obviously not possible unless some scalar norm of the residual is considered. One possible approach is to write

$$\mathbf{u}^{i+1,j}_{n+1} = \mathbf{u}^i_{n+1} + \eta_{i,j}d\mathbf{u}^i_{n+1} \tag{3.27}$$

and determine the *step size* $\eta_{i,j}$ so that a projection of the residual on the *search direction* $d\mathbf{u}^i_{n+1}$ is made zero. We could define this projection as

$$G_{i,j} \equiv \left(d\mathbf{u}^i_{n+1}\right)^{\mathrm{T}} \mathbf{\Psi}^{i+1,j}_{n+1} \tag{3.28}$$

where

$$\mathbf{\Psi}^{i+1,j}_{n+1} \equiv \mathbf{\Psi}\left(\mathbf{u}^i_{n+1} + \eta_{i,j}d\mathbf{u}^i_n\right), \qquad \eta_{i,0} = 1$$

Here, of course, other norms of the residual could be used.

This process is known as a *line search*, and $\eta_{i,j}$ can conveniently be obtained by using a *regula falsi* (or secant) procedure as illustrated in Fig. 3.6. An obvious disadvantage of a line search is the need for several evaluations of $\mathbf{\Psi}$. However, the acceleration of the overall convergence can be remarkable when applied to modified or quasi-Newton methods. Indeed, line search is also useful in the full Newton method by making the radius of attraction larger. A compromise frequently used[12] is to undertake the search only if

$$G_{i,0} > \varepsilon \left(d\mathbf{u}^i_{n+1}\right)^{\mathrm{T}} \mathbf{\Psi}^{i+1,j}_{n+1} \tag{3.29}$$

where the tolerance ε is set between 0.5 and 0.8. This means that if the iteration process directly resulted in a reduction of the residual to ε or less of its original value a line search is not used.

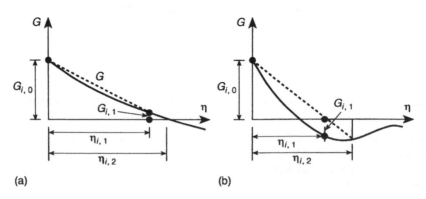

Fig. 3.6 *Regula falsi* applied to line search: (a) extrapolation; (b) interpolation.

3.2.6 'Softening' behaviour and displacement control

In applying the preceding to load control problems we have implicitly assumed that the iteration is associated with positive increments of the forcing vector, $\mathbf{f}$, in Eq. (3.4). In some structural problems this is a set of loads that can be assumed to be proportional to each other, so that one can write

$$\Delta \mathbf{f}_{n+1} = \Delta \lambda_{n+1} \mathbf{f}_0 \qquad (3.30)$$

In many problems the situation will arise that no solution exists above a certain maximum value of $\mathbf{f}$ and that the real solution is a 'softening' branch, as shown in Fig. 3.1. In such cases $\Delta \lambda_{n+1}$ will need to be negative unless the problem can be recast as one in which the forcing can be applied by displacement control. In a simple case of a single load it is easy to recast the general formulation to increments of a single prescribed displacement and much effort has gone into such solutions.[13,19–25]

In all the successful approaches of incrementation of $\Delta \lambda_{n+1}$ the original problem of Eq. (3.2) is rewritten as the solution of

$$\boldsymbol{\Psi}_{n+1} \equiv \lambda_{n+1} \mathbf{f}_0 - \mathbf{P}(\mathbf{u}_{n+1}) = 0$$

with

$$\mathbf{u}_{n+1} = \mathbf{u}_n + \Delta \mathbf{u}_{n+1} \qquad (3.31)$$

and

$$\lambda_{n+1} = \lambda_n + \Delta \lambda_{n+1}$$

being included as variables in any increment. Now an additional equation (constraint) needs to be provided to solve for the extra variable $\Delta \lambda_{n+1}$.

This additional equation can take various forms. Riks[21] assumes that in each increment

$$\Delta \mathbf{u}_{n+1}^{\mathrm{T}} \Delta \mathbf{u}_{n+1} + \Delta \lambda^2 \mathbf{f}_0^{\mathrm{T}} \mathbf{f}_0 = \Delta l^2 \qquad (3.32)$$

where Δl is a prescribed 'length' in the space of $n+1$ dimensions. Crisfield[13,26] provides a more natural control on displacements, requiring that

$$\Delta \mathbf{u}_{n+1}^{\mathrm{T}} \Delta \mathbf{u}_{n+1} = \Delta l^2 \qquad (3.33)$$

These so-called arc-length and spherical path controls are but some of the possible constraints.

Direct addition of the constraint Eq. (3.32) or (3.33) to the system of Eqs (3.31) is now possible and the previously described iterative methods could again be used. However, the 'tangent' equation system would always lose its symmetry so an alternative procedure is generally used.

We note that for a given iteration i we can write quite generally the solution as

$$\begin{aligned}
\boldsymbol{\Psi}_{n+1}^i &= \lambda_{n+1}^i \mathbf{f}_0 - \mathbf{P}(\mathbf{u}_{n+1}^i) \\
\boldsymbol{\Psi}_{n+1}^{i+1} &\approx \boldsymbol{\Psi}_{n+1}^i + d\lambda_{n+1}^i \mathbf{f}_0 - \mathbf{K}_{\mathrm{T}}^i \, d\mathbf{u}_{n+1}^i
\end{aligned} \qquad (3.34)$$

The solution increment for $\mathbf{u}$ may now be given as

$$\begin{aligned}
d\mathbf{u}_{n+1}^i &= \left(\mathbf{K}_{\mathrm{T}}^i \right)^{-1} \left[\boldsymbol{\Psi}_{n+1}^i + d\lambda_{n+1}^i \mathbf{f}_0 \right] \\
d\mathbf{u}_{n+1}^i &= d\breve{\mathbf{u}}_{n+1}^i + d\lambda_{n+1}^i \, d\hat{\mathbf{u}}_{n+1}^i
\end{aligned} \qquad (3.35)$$

where

$$d\breve{\mathbf{u}}_{n+1}^i = \left(\mathbf{K}_T^i\right)^{-1} \mathbf{\Psi}_{n+1}^i$$
$$d\hat{\mathbf{u}}_{n+1}^i = \left(\mathbf{K}_T^i\right)^{-1} \mathbf{f}_0$$

(3.36)

Now an additional equation is cast using the constraint. Thus, for instance, with Eq. (3.33) we have

$$\left(\Delta\mathbf{u}_{n+1}^{i-1} + d\mathbf{u}_{n+1}^i\right)^{\mathrm{T}} \left(\Delta\mathbf{u}_{n+1}^{i-1} + d\mathbf{u}_{n+1}^i\right) = \Delta l^2$$

(3.37)

where $\Delta\mathbf{u}_{n+1}^{i-1}$ is defined by Eq. (3.13). On substitution of Eq. (3.35) into Eq. (3.37) a quadratic equation is available for the solution of the remaining unknown $d\lambda_{n+1}^i$ (which may well turn out to be negative). Additional details may be found in references 13 and 26.

A procedure suggested by Bergan[22,25] is somewhat different from those just described. Here a fixed load increment $\Delta\lambda_{n+1}$ is first assumed and any of the previously introduced iterative procedures are used for calculating the increment $d\mathbf{u}_{n+1}^i$. Now a new increment $\Delta\lambda_{n+1}^\star$ is calculated so that it minimizes a norm of the residual

$$\left[\left(\Delta\lambda_{n+1}^\star\mathbf{f}_0 - \mathbf{P}_{n+1}^{i+1}\right)^{\mathrm{T}} \left(\Delta\lambda_{n+1}^\star\mathbf{f}_0 - \mathbf{P}_{n+1}^{i+1}\right)\right] = \Delta l^2$$

(3.38)

The result is thus computed from

$$\frac{\mathrm{d}\Delta l^2}{\mathrm{d}\Delta\lambda_{n+1}^\star} = 0$$

and yields the solution

$$\Delta\lambda_{n+1}^\star = \frac{\mathbf{f}_0^{\mathrm{T}}\mathbf{P}_{n+1}^{i+1}}{\mathbf{f}_0^{\mathrm{T}}\mathbf{f}_0}$$

(3.39)

This quantity may again well be negative, requiring a load decrease, and it indeed results in a rapid residual reduction in all cases, but precise control of displacement magnitudes becomes more difficult. The interpretation of the Bergan method in a one-dimensional example, shown in Fig. 3.7, is illuminating. Here it gives the exact answers – with a displacement control, the magnitude of which is determined by the initial $\Delta\lambda_{n+1}$ assumed to be the slope $\mathbf{K}_T$ used in the first iteration.

3.2.7 Convergence criteria

In all the iterative processes described the numerical solution is only approximately achieved and some tolerance limits have to be set to terminate the iteration. Since finite precision arithmetic is used in all computer calculations, one can never achieve a better solution than the round-off limit of the calculations.

Frequently, the criteria used involve a norm of the displacement parameter changes $\|d\mathbf{u}_{n+1}^i\|$ or, more logically, that of the residuals $\|\mathbf{\Psi}_{n+1}^i\|$. In the latter case the limit can often be expressed as some tolerance of the norm of forces $\|\mathbf{f}_{n+1}\|$. Thus, we may require that

$$\|\mathbf{\Psi}_{n+1}^i\| \leq \varepsilon\|\mathbf{f}_{n+1}\|$$

(3.40)

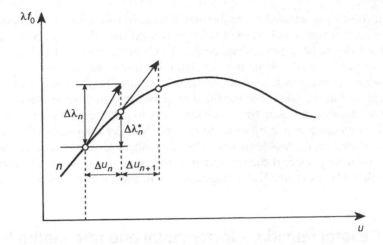

Fig. 3.7   One-dimensional interpretation of the Bergan procedure.

where ε is chosen as a small number, and

$$\|\mathbf{\Psi}\| = \left(\mathbf{\Psi}^{\mathrm{T}}\mathbf{\Psi}\right)^{1/2} \tag{3.41}$$

Other alternatives exist for choosing the comparison norm, and another option is to use the residual of the first iteration as a basis. Thus,

$$\|\mathbf{\Psi}_{n+1}^{i}\| \le \varepsilon \|\mathbf{\Psi}_{n+1}^{1}\| \tag{3.42}$$

The error due to the incomplete solution of the discrete non-linear equations is of course additive to the error of the discretization that we frequently measure in the energy norm.[27] It is possible therefore to use the same norm for bounding of the iteration process. We could, as a third option, require that the error in the energy norm satisfy

$$dE^{i} = \left(d\mathbf{u}_{n+1}^{i,\mathrm{T}}\mathbf{\Psi}_{n+1}^{i}\right)^{1/2} \le \varepsilon \left(d\mathbf{u}_{n+1}^{1,\mathrm{T}}\mathbf{\Psi}_{n+1}^{1}\right)^{1/2}$$
$$\le \varepsilon dE^{1} \tag{3.43}$$

In each of the above forms, problem types exist where the right-hand-side norm is zero. Thus a fourth form, which is quite general, is to compute the norm of the element residuals. If the problem residual is obtained as a sum over elements as

$$\mathbf{\Psi}_{n+1} = \sum_{e} \psi_{n+1}^{e} \tag{3.44}$$

where e denotes an individual element and ψ^{e} the residual from each element, we can express the convergence criterion as

$$\|\mathbf{\Psi}_{n+1}^{i}\| \le \varepsilon \|\psi_{n+1}^{e}\| \tag{3.45}$$

where

$$\|\psi_{n+1}^{e}\| = \sum_{e} \|\left(\psi_{n+1}^{e}\right)^{i}\| \tag{3.46}$$

Once a criterion is selected the problem still remains to choose an appropriate value for ε. In cases where a full Newton scheme is used (and thus asymptotic quadratic convergence should occur) the tolerance may be chosen at half the machine precision. Thus if the precision of calculations is about 16 digits one may choose $\varepsilon = 10^{-8}$ since quadratic convergence assures that the next residual (in the absence of round-off) would achieve full precision. For modified or quasi-Newton schemes such asymptotic rates are not assured, necessitating more iterations to achieve high precision. In these cases it is common practice by some to use much larger tolerance values (say 0.01 to 0.001). However, for problems where large numbers of steps are taken, instability in the solution may occur if the convergence tolerance is too large. We recommend therefore that whenever practical a tolerance of half machine precision be used.

3.3 General remarks – incremental and rate methods

The various iterative methods described provide an essential tool kit for the solution of non-linear problems in which finite element discretization has been used. The precise choice of the optimal methodology is problem dependent and although many comparative solution cost studies have been published[12,17,28] the differences are often marginal. There is little doubt, however, that exact Newton processes (with line search) should be used when convergence is difficult to achieve. Also the advantage of symmetric update matrices in the quasi-Newton procedures frequently make these a very economical candidate. When non-symmetric tangent moduli exist it may be better to consider one of the non-symmetric updates, for example a Broyden method.[13,29]

We have not discussed in the preceding *direct iterative* methods such as the various conjugate direction methods[30–34] or *dynamic relaxation* methods in which an explicit dynamic transient analysis (see Chapter 2) is carried out to achieve a steady-state solution.[35,36] These forms are often characterized by:

1. a diagonal or very sparse form of the matrix used in computing trial increments $d\mathbf{u}$ (and hence very low cost of an iteration) and
2. a significant number of total iterations and hence evaluations of the residual $\mathbf{\Psi}$.

These opposing trends imply that such methods offer the potential to solve large problems efficiently. However, to date such general solution procedures are effective only in certain problems.[37]

One final remark concerns the size of increments $\Delta\mathbf{f}$ or $\Delta\lambda$ to be adopted. First, it is clear that small increments reduce the total number of iterations required per computational step, and in many applications automatic guidance on the size of the increment to preserve a (nearly) constant number of iterations is needed. Here such processes as the use of the 'current stiffness parameter' introduced by Bergan[22] can be effective.

Second, if the behaviour is *path dependent* (e.g. as in plasticity-type constitutive laws) the use of small increments is desirable to preserve accuracy in solution changes. In this context, we have already emphasized the need for calculating such changes by always using the accumulated $\Delta\mathbf{u}_{n+1}^i$ change and not in adding changes arising from each iterative $d\mathbf{u}_{n+1}^i$ step in an increment.

Third, if only a single Newton iteration is used in each increment of $\Delta\lambda$ then the procedure is equivalent to the solution of a standard rate problem incrementally by direct forward integration. Here we note that if Eq. (3.2) is rewritten as

$$\mathbf{P(u)} = \lambda\mathbf{f}_0 \tag{3.47}$$

we can, on differentiation with respect to λ, obtain

$$\frac{\mathrm{d}\mathbf{P}}{\mathrm{d}\mathbf{u}}\frac{\mathrm{d}\mathbf{u}}{\mathrm{d}\lambda} = \mathbf{f}_0 \tag{3.48}$$

and write this as

$$\frac{\mathrm{d}\mathbf{u}}{\mathrm{d}\lambda} = \mathbf{K}_T^{-1}\mathbf{f}_0 \tag{3.49}$$

Incrementally, this may be written in an explicit form by using a Euler method as

$$\Delta\mathbf{u}_{n+1} = \Delta\lambda\mathbf{K}_{Tn}^{-1}\mathbf{f}_0 \tag{3.50}$$

This direct integration is illustrated in Fig. 3.8 and can frequently be divergent as well as being only conditionally stable as a result of the Euler explicit method used. Obviously, other methods can be used to improve accuracy and stability. These include Euler implicit schemes and Runge–Kutta procedures.

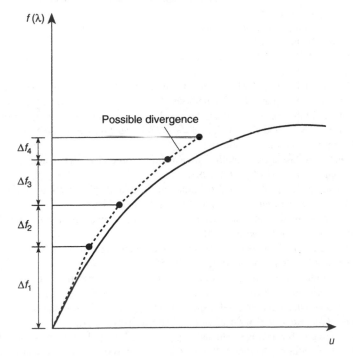

Fig. 3.8 Direct integration procedure.

References

1. G. Strang. *Linear Algebra and its Application*. Academic Press, New York, 1976.
2. J. Demmel. *Applied Numerical Linear Algebra*. Society for Industrial and Applied Mathematics, Philadelphia, PA, 1997.
3. R.M. Ferencz and T.J.R. Hughes. Iterative finite element solutions in nonlinear solid mechanics. In P.G. Ciarlet and J.L. Lions, editors, *Handbook of Numerical Analysis*, volume III, pages 3–178. Elsevier Science Publisher BV, 1999.
4. A. Ralston. *A First Course in Numerical Analysis*. McGraw-Hill, New York, 1965.
5. L. Collatz. *The Numerical Treatment of Differential Equations*. Springer, Berlin, 1966.
6. G. Dahlquist and Å. Björck. *Numerical Methods*. Prentice-Hall, Englewood Cliffs, NJ, 1974. Reprinted by Dover, New York, 2003.
7. H.R. Schwarz. *Numerical Analysis*. John Wiley & Sons, Chichester, 1989.
8. N. Bićanić and K.W. Johnson. Who was 'Raphson'? *International Journal for Numerical Methods in Engineering*, 14:148–152, 1979.
9. J.C. Simo and L. Vu-Quoc. A three-dimensional finite strain rod model. Part II: Geometric and computational aspects. *Computer Methods in Applied Mechanics and Engineering*, 58:79–116, 1986.
10. W.C. Davidon. Variable metric method for minimization. Technical Report ANL-5990, Argonne National Laboratory, 1959.
11. J.E. Dennis and J. More. Quasi-Newton methods – motivation and theory. *SIAM Rev.*, 19:46–89, 1977.
12. H. Matthies and G. Strang. The solution of nonlinear finite element equations. *International Journal for Numerical Methods in Engineering*, 14:1613–1626, 1979.
13. M.A. Crisfield. *Non-linear Finite Element Analysis of Solids and Structures*, volume 1. John Wiley & Sons, Chichester, 1991.
14. M.A. Crisfield. *Non-linear Finite Element Analysis of Solids and Structures*, volume 2. John Wiley & Sons, Chichester, 1997.
15. K.-J. Bathe and A.P. Cimento. Some practical procedures for the solution of nonlinear finite element equations. *cmame*, 22:59–85, 1980.
16. M. Geradin, S. Idelsohn and M. Hogge. Computational strategies for the solution of large nonlinear problems via quasi-Newton methods. *Computers and Structures*, 13:73–81, 1981.
17. M.A. Crisfield. Finite element analysis for combined material and geometric nonlinearity. In W. Wunderlich, E. Stein and K.-J. Bathe, editors, *Nonlinear Finite Element Analysis in Structural Mechanics*. Springer-Verlag, Berlin, 1981.
18. M.A. Crisfield. A fast incremental/iterative solution procedure that handles 'snap through'. *Computers and Structures*, 13:55–62, 1981.
19. T.H.H. Pian and P. Tong. Variational formulation of finite displacement analysis. In *Symp. on High Speed Electronic Computation of Structures*, Liège, 1970.
20. O.C. Zienkiewicz. Incremental displacement in non-linear analysis. *International Journal for Numerical Methods in Engineering*, 3:587–592, 1971.
21. E. Riks. An incremental approach to the solution of snapping and buckling problems. *International Journal of Solids and Structures*, 15:529–551, 1979.
22. P.G. Bergan. Solution algorithms for nonlinear structural problems. In *Int. Conf. on Engineering Applications of the Finite Element Method*, pages 13.1–13.39. Computas, 1979.
23. J.L. Batoz and G. Dhatt. Incremental displacement algorithms for nonlinear problems. *International Journal for Numerical Methods in Engineering*, 14:1261–1266, 1979.
24. E. Ramm. Strategies for tracing nonlinear response near limit points. In W. Wunderlich, E. Stein and K.-J. Bathe, editors, *Nonlinear Finite Element Analysis in Structural Mechanics*, pages 63–89. Springer-Verlag, Berlin, 1981.

25. P. Bergan. Solution by iteration in displacement and load spaces. In W. Wunderlich, E. Stein and K.-J. Bathe, editors, *Nonlinear Finite Element Analysis in Structural Mechanics*. Springer-Verlag, Berlin, 1981.

26. M.A. Crisfield. Incremental/iterative solution procedures for nonlinear structural analysis. In C. Taylor, E. Hinton, D.R.J. Owen and E. Oñate, editors, *Numerical Methods for Nonlinear Problems*. Pineridge Press, Swansea, 1980.

27. O.C. Zienkiewicz, R.L. Taylor and J.Z. Zhu. *The Finite Element Method: Its Basis and Fundamentals*. Butterworth-Heinemann, Oxford, 6th edition, 2005.

28. A. Pica and E. Hinton. The quasi-Newton BFGS method in the large deflection analysis of plates. In C. Taylor, E. Hinton, D.R.J. Owen and E. Oñate, editors, *Numerical Methods for Nonlinear Problems*. Pineridge Press, Swansea, 1980.

29. C.G. Broyden. Quasi-Newton methods and their application to function minimization. *Math. Comp.*, 21:368–381, 1967.

30. M. Hestenes and E. Stiefel. Method of conjugate gradients for solving linear systems. *J. Res. Natl. Bur. Stand.*, 49:409–436, 1954.

31. R. Fletcher and C.M. Reeves. Function minimization by conjugate gradients. *The Computer Journal*, 7:149–154, 1964.

32. E. Polak. *Computational Methods in Optimization. A Unified Approach*. Academic Press, London, 1971.

33. B.M. Irons and A.F. Elsawaf. The conjugate Newton algorithm for solving finite element equations. In K.-J. Bathe, J.T. Oden and W. Wunderlich, editors, *Proc. U.S.-German Symp. on Formulations and Algorithms in Finite Element Analysis*, pages 656–672, Cambridge, Mass., 1977. MIT Press.

34. M. Papadrakakis and P. Ghionis. Conjugate gradient algorithms in nonlinear structural analysis problems. *Computer Methods in Applied Mechanics and Engineering*, 59:11–27, 1986.

35. J.R.H. Otter, E. Cassel and R.E. Hobbs. Dynamic relaxation. *Proc. Inst. Civ. Eng.*, 35:633–656, 1966.

36. O.C. Zienkiewicz and R. Löhner. Accelerated relaxation or direct solution? Future prospects for FEM. *International Journal for Numerical Methods in Engineering*, 21:1–11, 1986.

37. M. Adams. Parallel multigrid algorithms for unstructured 3D large deformation elasticity and plasticity finite element problems. Technical Report UCB//CSD-99–1036, University of California, Berkeley, 1999.

4

Inelastic and non-linear materials

4.1 Introduction

In Chapter 2 we presented a framework for solving general problems in solid mechanics. In this chapter we consider several classical models for describing the behaviour of engineering materials. Each model we describe is given in a *strain-driven* form in which a strain or strain increment obtained from each finite element solution step is used to compute the stress needed to evaluate the internal force, $\int \mathbf{B}^{\mathrm{T}}\sigma \, d\Omega$, as well as a tangent modulus matrix, or its approximation, for use in constructing the tangent stiffness matrix. Quite generally in the study of small deformation and inelastic materials (and indeed in some forms applied to large deformation) the strain (or strain rate) or the stress is assumed to split into an additive sum of parts. We can write this as

$$\varepsilon = \varepsilon^{\mathrm{e}} + \varepsilon^{\mathrm{i}} \tag{4.1}$$

or

$$\sigma = \sigma^{\mathrm{e}} + \sigma^{\mathrm{i}} \tag{4.2}$$

in which we shall generally assume that the elastic part is given by the linear model

$$\varepsilon^{\mathrm{e}} = \mathbf{D}^{-1}\sigma \tag{4.3}$$

in which $\mathbf{D}$ is the matrix of elastic moduli.

In the following sections we shall consider the problems of viscoelasticity, plasticity, and general creep in quite general form. By using these general types it is possible to present numerical solutions which accurately predict many physical phenomena. We begin with viscoelasticity, where we illustrate the manner in which we shall address the solution of problems given in a rate or differential form. This rate form of course assumes time dependence and all viscoelastic phenomena are indeed transient, with time playing an important part. We shall follow this section with a description of plasticity models in which time does not explicitly arise and the problems are time independent. However, we shall introduce for convenience a rate description of the behaviour. This is adopted to allow use of the same kind of algorithms for all forms discussed in this chapter.

4.2 Viscoelasticity – history dependence of deformation

Viscoelastic phenomena are characterized by the fact that the rate at which inelastic strains develop depends not only on the current state of stress and strain but, in general, on the *full history* of their development. Thus, to determine the increment of inelastic strain over a given time interval (or time step) it is necessary to know the state of stress and strain at all *preceding times*. In the computation process these can in fact be obtained and *in principle* the problem presents little theoretical difficulty. Practical limitations appear immediately, however, that each computation point must retain this history information – thus leading to very large storage demands. In the context of linear viscoelasticity, means of overcoming this limitation were introduced by Zienkiewicz *et al.*[1] and White.[2] Extensions to include thermal effects were also included in some of this early work.[3] Further considerations which extend this approach are also discussed in earlier editions of this book.[4,5]

4.2.1 Linear models for viscoelasticity

The representation of a constitutive equation for linear viscoelasticity may be given in the form of either a differential equation or an integral equation.[6,7] In a differential model the constitutive equation may be written as a linear elastic part with an added series of partial strains **q**. Accordingly, we write

$$\sigma(t) = \mathbf{D}_0 \varepsilon(t) + \sum_{m=1}^{M} \mathbf{D}_m \mathbf{q}^{(m)}(t) \qquad (4.4)$$

where for a linear model the partial strains are solutions of the first-order differential equations

$$\dot{\mathbf{q}}^{(m)} + \mathbf{T}_m \mathbf{q}^{(m)} = \dot{\varepsilon} \qquad (4.5)$$

with $\mathbf{T}_m$ a constant matrix of reciprocal *relaxation* times and $\mathbf{D}_0$, $\mathbf{D}_m$ constant moduli matrices. The presence of a split of stress as given by Eq. (4.2) is immediately evident in the above. Each of the forms in Eq. (4.5) represents an elastic response in series with a viscous response and is known as a *Maxwell model*. In terms of a spring–dashpot model, a representation for the Maxwell material is shown in Fig. 4.1(a) for a single stress component. Thus, the sum given by Eq. (4.4) describes a *generalized*

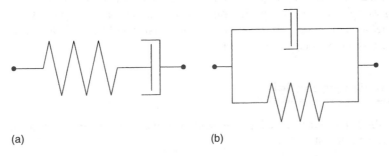

(a)   (b)

Fig. 4.1   Spring–dashpot models for linear viscoelasticity: (a) Maxwell element; (b) Kelvin element.

Maxwell solid in which several elements are assembled in a parallel form and the $\mathbf{D}_0$ term becomes a spring alone.

In an integral form the stress–strain behaviour may be written in a convolution form as

$$\sigma = \mathbf{D}(t)\varepsilon(0) + \int_0^t \mathbf{D}(t - t')\frac{\partial \varepsilon}{\partial t'}\, dt' \tag{4.6}$$

where components of $\mathbf{D}(t)$ are *relaxation moduli* functions.

Inverse relations may be given where the differential model is expressed as

$$\varepsilon(t) = \mathbf{J}_0\sigma(t) + \sum_{m=1}^M \mathbf{J}_m \mathbf{r}^{(m)}(t) \tag{4.7}$$

where for a linear model the partial stresses $\mathbf{r}$ are solutions of

$$\dot{\mathbf{r}}^{(m)} + \mathbf{V}_m \mathbf{r}^{(m)} = \sigma \tag{4.8}$$

in which $\mathbf{V}_m$ are constant reciprocal *retardation* time parameters and $\mathbf{J}_0$, $\mathbf{J}_m$ constant compliance ones (i.e. reciprocal moduli). Each partial stress corresponds to a solution in which a linear elastic and a viscous response are combined in parallel to describe a *Kelvin model* as shown in Fig. 4.1(b). The total model thus is a *generalized Kelvin solid*.

In an integral form the strain–stress constitutive relation may be written as

$$\varepsilon = \mathbf{J}(t)\sigma(0) + \int_0^t \mathbf{J}(t - t')\frac{\partial \sigma}{\partial t'}\, dt' \tag{4.9}$$

where $\mathbf{J}(t)$ are known as *creep compliance* functions.

The parameters in the two forms of the model are related. For example, the creep compliances and relaxation moduli are related through

$$\mathbf{J}(t)\mathbf{D}(0) + \int_0^t \mathbf{J}(t - t')\frac{\partial \mathbf{D}}{\partial t'}\, dt' = \mathbf{D}(t)\mathbf{J}(0) + \int_0^t \mathbf{D}(t - t')\frac{\partial \mathbf{J}}{\partial t'}\, dt' = \mathbf{I} \tag{4.10}$$

as may easily be shown by applying, for example, Laplace transform theory to Eqs (4.6) and (4.9).

The above forms hold for isotropic and anisotropic linear viscoelastic materials. Solutions may be obtained by using standard numerical techniques to solve the constant coefficient differential or integral equations. Here we will proceed to describe a solution for the isotropic case where specific numerical schemes are presented. Generalization of the methods to the anisotropic case may be constructed by using a similar approach and is left as an exercise to the reader.

4.2.2 Isotropic models

To describe in more detail the ideas presented above we consider here isotropic models where we split the stress as

$$\sigma = \mathbf{s} + \mathbf{m}p \qquad \text{with} \qquad p = \tfrac{1}{3}\mathbf{m}^{\mathsf{T}}\sigma \tag{4.11}$$

where $\mathbf{s}$ is the stress deviator, p is the mean (pressure) stress and, for a three-dimensional state of stress, $\mathbf{m}$ is given in Eq. (2.45). Similarly, a split of strain is expressed as

$$\varepsilon = \mathbf{e} + \tfrac{1}{3}\mathbf{m}\theta \quad \text{with} \quad \theta = \mathbf{m}^{T}\varepsilon \tag{4.12}$$

where $\mathbf{e}$ is the strain deviator and θ is the volume change.

In the presentation given here, for simplicity we restrict the viscoelastic response to deviatoric parts and assume pressure–volume response is given by the linear elastic model

$$p = K\theta \tag{4.13}$$

where K is an elastic bulk modulus. A generalization to include viscoelastic behaviour in this component also may be easily performed by using the method described below for deviatoric components.

Differential equation model

The deviatoric part may be stated as differential equation models or in the form of integral equations as described above. In the differential equation model the constitutive equation may be written as

$$\mathbf{s} = 2G\left(\mu_0 \mathbf{e} + \sum_{m=1}^{M} \mu_m \mathbf{q}^{(m)}\right) \tag{4.14}$$

in which μ_m are dimensionless parameters satisfying

$$\sum_{m=0}^{M} \mu_m = 1 \quad \text{with} \quad \mu_m > 0 \tag{4.15}$$

and dimensionless partial deviatoric strains $\mathbf{q}^{(m)}$ are obtained by solving

$$\dot{\mathbf{q}}^{(m)} + \frac{1}{\lambda_m}\mathbf{q}^{(m)} = \dot{\mathbf{e}} \tag{4.16}$$

in which λ_m are *relaxation times*. This form of the representation is again a *generalized Maxwell model* (a set of Maxwell models in parallel).

Each differential equation set may be solved numerically using a one-step time integration method [e.g., GN11 in Sec. 2.7, Eq. (2.66)].[8,9] To solve numerically we first define a set of discrete points, t_k, at which we wish to obtain the solution. For a time t_{n+1} we assume the solution at all previous points up to t_n are known. Using a simple single-step method the solution for each partial stress is given by:

$$\left(1 + \frac{\theta \Delta t}{\lambda_m}\right)\mathbf{q}_{n+1}^{(m)} = \left(1 - \frac{(1-\theta)\Delta t}{\lambda_m}\right)\mathbf{q}_n^{(m)} + \mathbf{e}_{n+1} - \mathbf{e}_n \tag{4.17}$$

in which $\Delta t = t_{n+1} - t_n$.

We note that this form of the solution is given directly in a *strain-driven form*. Accordingly, given the strain from any finite element solution step we can immediately compute the stresses by using Eqs (4.13), (4.14) and (4.17) in Eqs (4.11) and (4.12).

Inserting the above into a Newton-type solution strategy requires the computation of the tangent moduli. The tangent moduli for the viscoelastic model are deduced from

$$
\mathbf{D_T}|_{n+1} = \frac{\partial \boldsymbol{\sigma}_{n+1}}{\partial \boldsymbol{\varepsilon}_{n+1}} = \frac{\partial \mathbf{s}_{n+1}}{\partial \boldsymbol{\varepsilon}_{n+1}} + \mathbf{m} \frac{\partial p_{n+1}}{\partial \boldsymbol{\varepsilon}_{n+1}}
\tag{4.18}
$$

The tangent part for the volumetric term is elastic and given by

$$
\mathbf{m} \frac{\partial p_{n+1}}{\partial \boldsymbol{\varepsilon}_{n+1}} = \mathbf{m} \frac{\partial p_{n+1}}{\partial \theta_{n+1}} \frac{\partial \theta_{n+1}}{\partial \boldsymbol{\varepsilon}_{n+1}} = K \mathbf{m} \mathbf{m}^{\mathrm{T}}
\tag{4.19}
$$

Similarly, the tangent part for the deviatoric term is deduced from Eq. (4.17) as

$$
\frac{\partial \mathbf{s}_{n+1}}{\partial \boldsymbol{\varepsilon}_{n+1}} = \frac{\partial \mathbf{s}_{n+1}}{\partial \mathbf{e}_{n+1}} \frac{\partial \mathbf{e}_{n+1}}{\partial \boldsymbol{\varepsilon}_{n+1}} = 2G \left[\mu_0 + \sum_{m=1}^{M} \frac{\mu_m}{\left(1 + \dfrac{\theta \Delta t}{\lambda_m}\right)} \right] \mathbf{I}_d
\tag{4.20}
$$

where $\mathbf{I}_d$ is defined in Eq. (2.45). Using the above, tangent moduli are expressed as

$$
\mathbf{D_T}|_{n+1} = K \mathbf{m} \mathbf{m}^{\mathrm{T}} + 2G \left[\mu_0 + \sum_{m=1}^{M} \frac{\mu_m}{\left(1 + \dfrac{\theta \Delta t}{\lambda_m}\right)} \right] \mathbf{I}_d
\tag{4.21}
$$

and we note that the only difference from a linear elastic material is the replacement of the elastic shear modulus by the viscoelastic term

$$
G \rightarrow G \left[\mu_0 + \sum_{m=1}^{M} \frac{\mu_m}{\left(1 + \dfrac{\theta \Delta t}{\lambda_m}\right)} \right]
$$

This relation is independent of stress and strain and hence when it is used with a Newton scheme it converges in one iteration (i.e. the residual of a second iteration is numerically zero).

The set of first-order differential equations (4.16) may be integrated exactly for specified strains, $\mathbf{e}$. The integral for each term is given by

$$
\mathbf{q}^{(m)}(t) = \int_{-\infty}^{t} \exp\left[-(t - t')/\lambda_m\right] \frac{\partial \mathbf{e}}{\partial t'} \, dt'
\tag{4.22}
$$

An advantage to the differential equation form, however, is that it may be extended to include *aging* or other *non-linear effects* by making the parameters time or solution dependent. The exact solution to the differential equations for such a situation will then involve integrating factors, leading to more involved expressions. In the following parts of this section we consider the integral equation form and its numerical solution for *linear* viscoelastic behaviour. Models and their solutions for more general cases are left as an exercise for the reader.

Integral equation model

The integral equation form for the deviatoric stresses is expressed in terms of a relaxation modulus function which is defined by an idealized experiment in which, at time zero ($t = 0$), a specimen is subjected to suddenly applied and constant strain, $\mathbf{e}_0$, and the stress response, $\mathbf{s}(t)$, is measured. For a linear material a unique relation is obtained which is independent of the magnitude of the applied strain. This relation may be written as

$$\mathbf{s}(t) = 2G(t)\mathbf{e}_0 \tag{4.23}$$

where $G(t)$ is defined as the *shear relaxation modulus function.* A typical relaxation function is shown in Fig. 4.2. The function is shown on a logarithmic time scale since typical materials have time effects which cover wide ranges in time.

Using linearity and superposition for an arbitrary state of strain yields the integral equation specified as

$$\mathbf{s}(t) = \int_{-\infty}^{t} 2G(t - t')\frac{\partial \mathbf{e}}{\partial t'} \, \mathrm{d}t' \tag{4.24}$$

We note that the above form is a generalization to the Maxwell material. However, the integral equation form may be specialized to the generalized Maxwell model by assuming the shear relaxation modulus function in a Prony series form

$$G(t) = G\left[\mu_0 + \sum_{m=1}^{M} \mu_m \exp(-t/\lambda_m)\right] \tag{4.25}$$

where the μ_m satisfy Eq. (4.15).

Solution to integral equation with Prony series

The solution to the viscoelastic model is performed for a set of discrete points t_k. Thus, again assuming that all solutions are available up to time t_n, we desire to compute the next step for time t_{n+1}. Solution of the general form would require summation over

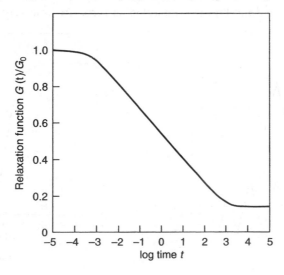

Fig. 4.2 Typical viscoelastic relaxation function.

all previous time steps for each new time; however, by using the generalized Maxwell model we may reduce the solution to a recursion formula in which each new solution is computed by a simple update of the previous solution.

We will consider a special case of the generalized Maxwell material in which the number of terms M is equal to 1 [which defines a *standard linear solid*, Fig. 4.3(a)]. The addition of more terms is easily performed from the one-term solution. Accordingly, we take

$$G(t) = G \left[\mu_0 + \mu_1 \exp(-t/\lambda_1) \right] \tag{4.26}$$

where $\mu_0 + \mu_1 = 1$. For the standard solid only a limited range of time can be considered, as can be observed from Fig. 4.3(b) for the model given by

$$G(t) = G[0.15 + 0.85 \exp(-t)]$$

To consider a wider range it is necessary to use terms in which the λ_m cover the total time by using at least one term for each decade of time (a decade being one unit on the $\log_{10}$ time scale).

Substitution of Eq. (4.26) into Eq. (4.24) yields

$$\mathbf{s}(t) = 2G \int_{-\infty}^{t} \left[\mu_0 + \mu_1 \exp(-(t - t')/\lambda_1) \right] \frac{\partial \mathbf{e}}{\partial t'} \, dt' \tag{4.27}$$

which may be split and expressed as

$$\begin{aligned}
\mathbf{s}(t) &= 2G\mu_0\varepsilon(t) + 2G\mu_1 \int_{-\infty}^{t} \exp(-(t - t')/\lambda_1) \frac{\partial \mathbf{e}}{\partial t'} \, dt' \\
&= 2G[\mu_0\varepsilon(t) + \mu_1\mathbf{q}^{(1)}(t)]
\end{aligned} \tag{4.28}$$

where we note that $\mathbf{q}^{(1)}$ is identical to the form given in Eq. (4.22). Thus use of a Prony series for $G(t)$ is identical to solving the differential equation model exactly.

In applications involving a linear viscoelastic model, it is usually assumed that the material is undisturbed until a time identified as zero. At time zero a strain may be

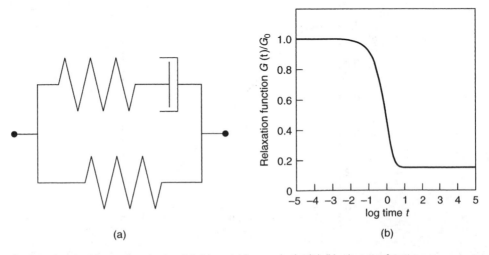

(a)   (b)

Fig. 4.3 Standard linear viscoelastic solid: (a) model for standard solid; (b) relaxation function.

suddenly applied and then varied over subsequent time. To evaluate a solution at time t_{n+1} the integral representation for the model may be simplified by dividing the integral into

$$\int_{-\infty}^{t_{n+1}} (\cdot)\, dt' = \int_{-\infty}^{0^-} (\cdot)\, dt' + \int_{0^-}^{0^+} (\cdot)\, dt' + \int_{0^+}^{t_n} (\cdot)\, dt' + \int_{t_n}^{t_{n+1}} (\cdot)\, dt' \qquad (4.29)$$

In each analysis considered here the material is assumed to be unstrained before the time denoted as zero. Thus, the first term on the right-hand side is zero, the second term includes a jump term associated with $\mathbf{e}_0$ at time zero, and the last two terms cover the subsequent history of strain. The result of this separation when applied to Eq. (4.27) gives the recursion[3]

$$\mathbf{q}_{n+1}^{(1)} = \exp(-\Delta t/\lambda_1)\mathbf{q}_n^{(1)} + \Delta\mathbf{q}^{(1)} \qquad (4.30)$$

where

$$\Delta\mathbf{q}^{(1)} = \int_{t_n}^{t_{n+1}} \exp[-(t_{n+1} - t')/\lambda_1]\frac{\partial\mathbf{e}}{\partial t'}\, dt' \qquad (4.31)$$

and $\mathbf{q}_0^{(1)} = \mathbf{e}_0$.

To obtain a numerical solution, we approximate the strain rate in each time increment by a constant to obtain

$$\Delta\mathbf{q}_{n+1}^{(1)} = \frac{1}{\Delta t}\int_{t_n}^{t_{n+1}} \exp[-(t_{n+1} - t')/\lambda_1]\left[\mathbf{e}_{n+1} - \mathbf{e}_n\right]\, dt' \qquad (4.32)$$

The integral may now be evaluated directly over each time step as[3]

$$\Delta\mathbf{q}_{n+1}^{(1)} = \frac{\lambda_1}{\Delta t}\left[1 - \exp(-\Delta t/\lambda_1)\right](\mathbf{e}_{n+1} - \mathbf{e}_n) = \Delta q_{n+1}^{(1)}(\mathbf{e}_{n+1} - \mathbf{e}_n) \qquad (4.33)$$

This approximation is singular for zero time steps; however, the limit value at $\Delta t = 0$ is one. Thus, for small time steps a series expansion may be used to yield accurate values, giving

$$\Delta q_{n+1}^{(1)} = 1 - \frac{1}{2}\left(\frac{\Delta t}{\lambda_1}\right) + \frac{1}{3!}\left(\frac{\Delta t}{\lambda_1}\right)^2 - \frac{1}{4!}\left(\frac{\Delta t}{\lambda_1}\right)^3 + \cdots \qquad (4.34)$$

Using a few terms for very small time increment ratios yields numerically correct answers (to computer precision). Once the time increment ratio is larger than a certain small value the representation given in Eq. (4.33) is used directly.

The above form gives a recursion which is stable for small and large time steps and produces very smooth transitions under variable time steps.

A numerical approximation to Eq. (4.32) in which the integrand of Eq. (4.31) is evaluated at $t_{n+1/2}$ has also been used with success.[10] In the above recursion we note that a zero and infinite value of a time step produces a correct instantaneous and zero response, respectively, and thus is asymptotically accurate at both limits. The use of finite difference approximations on the differential equation form directly does not produce this property unless $\theta = 1$ and for this value is much less accurate than the solution given by Eq. (4.33).

Using the recursion formula, the constitutive equation now has the simple form

$$s_{n+1} = 2G[\mu_0 e_{n+1} + \mu_1 q_{n+1}^{(1)}] \tag{4.35}$$

The process may also be extended to include effects of temperature on relaxation times for use with thermorheologically simple materials.[3]

The implementation of the above viscoelastic model into a Newton-type solution process again requires the computation of a tangent tensor. Accordingly, for the deviatoric part we need to compute

$$\frac{\partial s_{n+1}}{\partial \varepsilon_{n+1}} = \frac{\partial s_{n+1}}{\partial e_{n+1}} I_d \tag{4.36}$$

The partial derivative with respect to the deviatoric stress follows from Eq. (4.35) as

$$\frac{\partial s}{\partial e} = 2G \left[\mu_0 I + \mu_1 \frac{\partial q^{(1)}}{\partial e} \right] \tag{4.37}$$

Using Eq. (4.33) the derivative of the last term becomes

$$\frac{\partial q_{n+1}^{(1)}}{\partial e_{n+1}} = \Delta q_{n+1}^{(1)}(\Delta t) I \tag{4.38}$$

Thus, the tangent tensor is given by

$$\frac{\partial s_{n+1}}{\partial \varepsilon_{n+1}} = 2G[\mu_0 + \mu_1 \Delta q_{n+1}^{(1)}(\Delta t)] I_d \tag{4.39}$$

Again, the only modification from a linear elastic material is the substitution of the elastic shear modulus by

$$G \rightarrow G[\mu_0 + \mu_1 \Delta q_{n+1}^{(1)}(\Delta t)] \tag{4.40}$$

We note that for zero Δt the full elastic modulus is recovered, whereas for very large increments the equilibrium modulus $\mu_0 G$ is used. Since the material is linear, use of this tangent modulus term again leads to convergence in one iteration (the second iteration produces a *numerically zero* residual).

The inclusion of more terms in the series reduces to evaluation of additional $q_{n+1}^{(m)}$ integral recursions. Computer storage is needed to retain the $q_n^{(m)}$ for each solution (quadrature) point in the problem and each term in the series.

Example 4.1: A thick-walled cylinder subjected to internal pressure

To illustrate the importance of proper element selection when performing analyses in which material behaviour approaches a near incompressible situation we consider the case of internal pressure on a thick-walled cylinder. The material is considered to be isotropic and modelled by viscoelastic response in deviatoric stress–strain only. Material properties are: modulus of elasticity, $E = 1000$; Poisson's ratio, $\nu = 0.3$; $\mu_1 = 0.99$; and $\lambda_1 = 1$. Thus, the viscoelastic relaxation function is given by

$$G(t) = \frac{1000}{2.6}[0.01 + 0.99 \exp(-t)]$$

The ratio of the bulk modulus to shear modulus for instantaneous loading is given by $K/G(0) = 2.167$ and for long time loading by $K/G(\infty) = 216.7$ which indicates a near incompressible behaviour for sustained loading cases (the effective Poisson ratio for infinite time is 0.498). The response for a suddenly applied internal pressure, $p = 10$, is computed to time 20 by using both displacement and the mixed element described in Chapter 2. Quadrilateral elements with four nodes (Q4) and nine nodes (Q9) are considered, and meshes with equivalent nodal forces are shown in Fig. 4.4. The exact solution to this problem is one dimensional and, since all radial boundary conditions are traction ones, the stress distribution should be time independent. During the early part of the solution, when the response is still in the compressible range, the solutions from the two formulations agree well with this exact solution. However, during the latter part of the solution the answers from a displacement element diverge because of near incompressibility effects, whereas those from a mixed element do not. The distribution of quadrature point radial stresses at time $t = 20$ is shown in Fig. 4.5 where

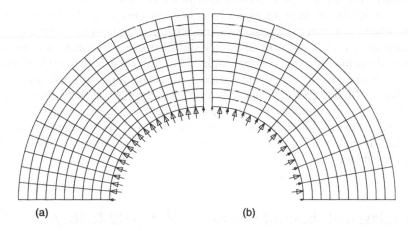

Fig. 4.4 Mesh and loads for internal pressure on a thick-walled cylinder: (a) four-node quadrilaterals; (b) nine-node quadrilaterals.

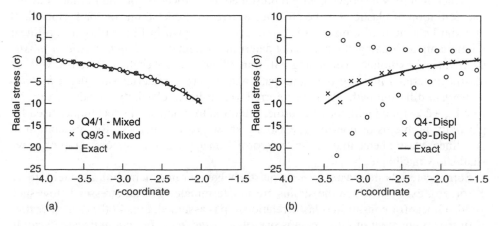

Fig. 4.5 Radial stress for internal pressure on a thick-walled cylinder: (a) mixed model; (b) displacement model.

the highly oscillatory response of the displacement form is clearly evident. We note that extrapolation to 'reduced quadrature' points would avoid these oscillations; however, use of fully reduced integration would lead to singularity in the stiffness matrix and selective reduced integration is difficult to use with general non-linear material behaviour. Thus, for general applications the use of mixed elements is preferred.

4.2.3 Solution by analogies

The labour of step-by-step solutions for linear viscoelastic media can, on occasion, be substantially reduced. In the case of a homogeneous structure with linear isotropic viscoelasticity and constant Poisson ratio operator, the McHenry–Alfrey analogies allow single-step elastic solutions to be used to obtain stresses and displacements at a given time by the use of *equivalent loads*, *displacements* and *temperatures*.[11,12]

Some extensions of these analogies have been proposed by Hilton and Russell.[13] Further, when subjected to steady loads and when strains tend to a constant value at an infinite time, it is possible to determine the final stress distribution even in cases where the above analogies are not applicable. Thus, for instance, where the viscoelastic properties are temperature dependent and the structure is subject to a system of loads and temperatures which remain constant with time, long-term 'equivalent' elastic constants can be found and the problem solved as a single, non-homogeneous elastic one.[14]

The viscoelastic problem is a particular case of a creep phenomenon to which we shall return in Sec. 4.9.3 using some other classical non-linear models to represent material behaviour.

4.3 Classical time-independent plasticity theory

Classical 'plastic' behaviour of solids is characterized by a non-unique stress–strain relationship which is independent of the *rate* of loading but does depend on loading sequence that may be conveniently represented as a process evolving in time. Indeed, one definition of plasticity is the presence of irrecoverable strains on load removal. If uniaxial behaviour of a material is considered, as shown in Fig. 4.6(a), a non-linear relationship on loading alone does not determine whether *non-linear elastic or plastic behaviour* is exhibited. Unloading will immediately discover the difference, with an elastic material following the same path and a plastic material showing a *history-dependent* different path. We have referred to non-linear elasticity already in Sec. 2.4 [see Eq. (2.28)] and will not give further attention to it here as the techniques used for plasticity problems or non-linear elasticity show great similarity. Representation of non-linear elastic behaviour for finite deformation applications is more complex as we shall show in Chapter 5.

Some materials show a nearly *ideal* plastic behaviour in which a limiting yield stress, Y (or σ_y), exists at which the strains are indeterminate. For all stresses below such yield, a linear (or non-linear) elastic relationship is assumed, Fig. 4.6(b) illustrates this. A further refinement of this model is one of a *hardening/softening plastic material* in which the yield stress depends on some parameter κ (such as the accumulated plastic

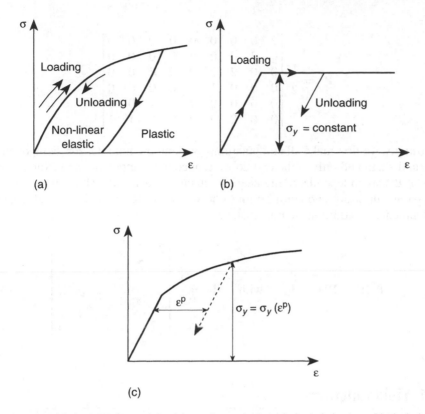

Fig. 4.6 Uniaxial behaviour of materials: (a) non-linear elastic and plastic behaviour; (b) ideal plasticity; (c) strain hardening plasticity.

strain ε^p) [Fig. 4.6(c)]. It is with such kinds of plasticity that this section is concerned and for which much theory has been developed.[15,16]

In a multiaxial rather than a uniaxial state of stress the concept of yield needs to be generalized. It is important to note that in the following development of results in a matrix form all nine tensor components are used instead of the six 'engineering' component form used previously. To distinguish between the two we introduce an underbar on the symbol for all nine-component forms. Thus, we shall use:

$$
\begin{aligned}
\sigma &= \begin{bmatrix} \sigma_x & \sigma_y & \sigma_z & \sigma_{xy} & \sigma_{yz} & \sigma_{zx} \end{bmatrix}^{\mathrm{T}} \\
\underline{\sigma} &= \begin{bmatrix} \sigma_x & \sigma_y & \sigma_z & \sigma_{xy} & \sigma_{yx} & \sigma_{yz} & \sigma_{zy} & \sigma_{zx} & \sigma_{xz} \end{bmatrix}^{\mathrm{T}} \\
\varepsilon &= \begin{bmatrix} \varepsilon_x & \varepsilon_y & \varepsilon_z & \gamma_{xy} & \gamma_{yz} & \gamma_{zx} \end{bmatrix}^{\mathrm{T}} \\
\underline{\varepsilon} &= \begin{bmatrix} \varepsilon_x & \varepsilon_y & \varepsilon_z & \varepsilon_{xy} & \varepsilon_{yx} & \varepsilon_{yz} & \varepsilon_{zy} & \varepsilon_{zx} & \varepsilon_{xz} \end{bmatrix}^{\mathrm{T}}
\end{aligned}
\tag{4.41}
$$

in which $\gamma_{ij} = 2\varepsilon_{ij}$. The transformations between the nine- and six-component forms needed later are obtained by using

$$
\underline{\varepsilon} = \mathbf{P}\varepsilon \quad \text{and} \quad \sigma = \mathbf{P}^{\mathrm{T}}\underline{\sigma}
\tag{4.42}
$$

where

$$\mathbf{P}^{\mathrm{T}} = \frac{1}{2} \begin{bmatrix} 2 & 0 & 0 & 0 & 0 & 0 & 0 & 0 & 0 \\ 0 & 2 & 0 & 0 & 0 & 0 & 0 & 0 & 0 \\ 0 & 0 & 2 & 0 & 0 & 0 & 0 & 0 & 0 \\ 0 & 0 & 0 & 1 & 1 & 0 & 0 & 0 & 0 \\ 0 & 0 & 0 & 0 & 0 & 1 & 1 & 0 & 0 \\ 0 & 0 & 0 & 0 & 0 & 0 & 0 & 1 & 1 \end{bmatrix}$$

Accordingly, we first make all computations by using the nine 'tensor' components of stress and strain and only at the end do we reduce the computations to expressions in terms of the six independent 'engineering' quantities using $\mathbf{P}$. This will permit final expressions for strain and equilibrium to be written in terms of $\mathbf{B}$ as in all previous developments. In addition we note that:

$$\mathbf{P}^{\mathrm{T}}\mathbf{IP} = \mathbf{P}^{\mathrm{T}}\mathbf{P} = \mathbf{I}_0 \quad \text{with} \quad \mathbf{I}_0 = \frac{1}{2} \begin{bmatrix} 2 & & & & & \\ & 2 & & & & \\ & & 2 & & & \\ & & & 1 & & \\ & & & & 1 & \\ & & & & & 1 \end{bmatrix} \tag{4.43}$$

4.3.1 Yield functions

It is quite generally postulated, as an experimental fact, that yielding can occur only if the stress satisfies the general yield criterion

$$F(\underline{\sigma}, \underline{\kappa}, \kappa) = 0 \tag{4.44}$$

where $\underline{\sigma}$ denotes a matrix form with all nine components of stress, $\underline{\kappa}$ represents *kinematic hardening* parameters and κ an *isotropic hardening* parameter.[15] We shall discuss these particular sets of parameters later but, of course, many other types of parameters also can be used to define hardening.

This yield condition can be visualized as a surface in an n-dimensional space of stress with the position and size of the surface dependent on the instantaneous value of the parameters $\underline{\kappa}$ and κ (Fig. 4.7).

4.3.2 Flow rule (normality principle)

Von Mises first suggested that basic behaviour defining the plastic strain increments is related to the yield surface.[17] Heuristic arguments for the validity of the relationship proposed have been given by various workers in the field[18–25] and at the present time the following hypothesis appears to be generally accepted for many materials; if $\underline{\varepsilon}^{\mathrm{p}}$

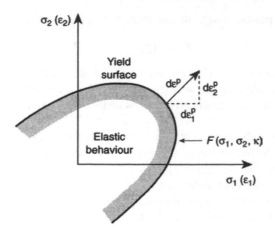

Fig. 4.7 Yield surface and normality criterion in two-dimensional stress space.

denotes the components of the plastic strain tensor the rate of plastic strain is assumed to be given by*

$$\dot{\underline{\varepsilon}}^{p} = \dot{\lambda} F_{,\underline{\sigma}} \tag{4.45}$$

where the notation

$$F_{,\underline{\sigma}} \equiv \frac{\partial F}{\partial \underline{\sigma}} \tag{4.46}$$

is introduced. In the above, $\dot{\lambda}$ is a proportionality constant, as yet undetermined, often referred to as the 'plastic consistency' parameter. During sustained plastic deformation we must have

$$\dot{F} = 0 \quad \text{and} \quad \dot{\lambda} > 0 \tag{4.47}$$

whereas during elastic loading/unloading $\dot{\lambda} = 0$ and $\dot{F} \neq 0$ leading to a general constraint condition in Kuhn–Tucker form[16]

$$\dot{F}\dot{\lambda} = 0 \tag{4.48}$$

The above rule is known as the *normality* principle because relation (4.45) can be interpreted as requiring the plastic strain rate components to be normal to the yield surface in the space of nine stress and strain dimensions.

   Restrictions of the above rule can be removed by specifying separately a *plastic flow rule potential*

$$Q = Q(\underline{\sigma}, \kappa) \tag{4.49}$$

---

* Some authors prefer to write Eq. (4.45) in an incremental form

$$d\underline{\varepsilon}^{p} = d\lambda \, F_{,\underline{\sigma}}$$

where $d\underline{\varepsilon}^{p} \equiv \dot{\underline{\varepsilon}}^{p} dt$, and t is some pseudo-time variable. Here we prefer the rate form to permit use of common solution algorithms in which $d\underline{\varepsilon}$ will denote an increment in a Newton-type solution. (Also note the difference in notation between a small increment 'd' and a differential 'd'.)

which defines the plastic strain rate similarly to Eq. (4.45); that is, giving this as

$$\dot{\underline{\varepsilon}}^{P} = \dot{\lambda} Q_{,\underline{\sigma}}, \qquad \dot{\lambda} \geq 0 \tag{4.50}$$

The particular case of $Q = F$ is known as *associative plasticity*. When this relation is not satisfied the plasticity is *non-associative*. In what follows this more general form will be considered initially (reductions to the associative case follow by simple substitution of $Q = F$).

The satisfaction of the normality rule for the associative case is essential for proving so-called *upper and lower bound* theorems of plasticity as well as uniqueness. In the non-associative case the upper and lower bounds do not exist and indeed it is not certain that the solutions are always unique. This does not prevent the validity of non-associated rules as it is well known that in frictional materials, for instance, uniqueness is seldom achieved but the existence of friction cannot be denied.

4.3.3 Hardening/softening rules

Isotropic hardening

The parameters $\underline{\kappa}$ and κ must also be determined from rate equations and define hardening (or softening) of the plastic behaviour of the material. The evolution of κ, governing the *size* of the yield surface, is commonly related to the rate of plastic work or directly to the consistency parameter. If related to the rate of plastic work κ has dimensions of stress and a relation of the type

$$\dot{\kappa} = \underline{\sigma}^{T} \dot{\underline{\varepsilon}}^{P} = Y(\kappa) \dot{\varepsilon}_{u}^{P} \tag{4.51}$$

is used to match behaviour to a uniaxial tension or compression result. The slope

$$A = \frac{\partial Y}{\partial \kappa} \tag{4.52}$$

provides a modulus defining instantaneous *isotropic hardening*.

In the second approach κ is dimensionless (e.g. an accumulated plastic strain[16]) and is related directly to the consistency parameter using

$$\dot{\kappa} = \left[(\dot{\underline{\varepsilon}}^{P})^{T} \dot{\underline{\varepsilon}}^{P} \right]^{1/2} = \dot{\lambda} [Q_{,\underline{\sigma}}^{T} Q_{,\underline{\sigma}}]^{1/2} \tag{4.53}$$

A constitutive equation is then introduced to match uniaxial results. For example, a simple linear form is given by

$$\sigma_{y}(\kappa) = \sigma_{y0} + H_{i0} \kappa$$

where H_{i0} is a constant isotropic hardening modulus.

Kinematic hardening

A classical procedure to represent kinematic hardening was introduced by Prager[26] and modified by Ziegler.[27] Here the stress in each yield surface is replaced by a linear relation in terms of a 'back stress' $\underline{\kappa}$ as

$$\underline{\varsigma} = \underline{\sigma} - \underline{\kappa} \tag{4.54}$$

with the yield function now given as

$$F(\underline{\sigma} - \underline{\kappa}, \kappa) = F(\underline{\varsigma}, \kappa) = 0 \tag{4.55}$$

during plastic behaviour. We note that with this approach derivatives of the yield surface differ only by a sign and are given by

$$F_{,\underline{\varsigma}} = F_{,\underline{\sigma}} = -F_{,\underline{\kappa}} \tag{4.56}$$

Accordingly, the yield surface will now *translate*, and if isotropic hardening is present will also expand or contract, during plastic loading.

A rate equation may be specified most directly by introducing a conjugate work variable $\underline{\beta}$ from which the hardening parameter $\underline{\kappa}$ is deduced by using a hardening potential $\mathcal{H}$. This may be stated as

$$\underline{\kappa} = -\mathcal{H}_{,\underline{\beta}} \tag{4.57}$$

which is completely analogous to use of an elastic energy to relate $\underline{\sigma}$ and $\underline{\varepsilon}^{e}$. A rate equation may be expressed now as

$$\underline{\dot{\beta}} = \dot{\lambda} Q_{,\underline{\kappa}} \tag{4.58}$$

It is immediately obvious that here also we have two possibilities. Using Q in the above expression defines a *non-associative* hardening, whereas replacing Q by F would give an *associative* hardening. Thus for a fully associative model we require that F be used to define both the plastic potential and the hardening. In such a case the relations of plasticity also may be deduced by using the *principle of maximum plastic dissipation*.[15,16,28,29] A quadratic form for the hardening potential may be adopted and written as

$$\mathcal{H} = \tfrac{1}{2} \underline{\beta}^{\mathrm{T}} \underline{\mathbf{H}}_{k} \underline{\beta} \tag{4.59}$$

in which $\underline{\mathbf{H}}_{k}$ is assumed to be an invertible set of constant hardening parameters. Now $\underline{\beta}$ may be eliminated to give the simple rate form

$$\underline{\dot{\kappa}} = -\dot{\lambda} \underline{\mathbf{H}}_{k} \frac{\partial Q}{\partial \underline{\kappa}} = -\dot{\lambda} \underline{\mathbf{H}}_{k} Q_{,\underline{\kappa}} \tag{4.60}$$

Use of a linear shift in relation (4.54) simplifies this, noting Eq. (4.56), to

$$\underline{\dot{\kappa}} = \dot{\lambda} \underline{\mathbf{H}}_{k} Q_{,\underline{\varsigma}} \tag{4.61}$$

In our subsequent discussion we shall usually assume a general quadratic model for both elastic and hardening potentials. For a more general treatment the reader is referred to references 16 and 30.

Another approach to kinematic hardening was introduced by Armstrong and Frederick[31] and provides a means of retaining smoother transitions from elastic to inelastic behaviour during cyclic loading. Here the hardening is given as

$$\underline{\dot{\kappa}} = \dot{\lambda} [\underline{\mathbf{H}}_{k} Q_{,\underline{\varsigma}} - H_{NL} \underline{\kappa}] \tag{4.62}$$

Applications of this approach are presented by Chaboche[32,33] and numerical compar-isons to a simpler approach using a generalized plasticity model[34,35] are given by Auricchio and Taylor.[36]

Many other approaches have been proposed to represent classical hardening behaviour and the reader is referred to the literature for additional information and discus-sion.[21–23,37–39] A physical procedure utilizing directly the finite element method is available to obtain both ideal plasticity and hardening. Here several ideal plasticity components, each with different yield stress, are put in series and it will be found that both hardening and softening behaviour can be obtained easily retaining the properties so far described. This approach was named by many authors as an 'overlay' model[40,41] and by others is described as a 'sublayer' model.

There are of course many other possibilities to define change in surfaces during the process of loading and unloading. Here frictional soils present one of the most difficult materials to model and for the non-associative case we find it convenient to use the generalized plasticity method described in Sec. 4.6.

4.3.4 Plastic stress–strain relations

To construct a constitutive model for plasticity, the strains are assumed to be divisible into elastic and plastic parts given as

$$\underline{\varepsilon} = \underline{\varepsilon}^e + \underline{\varepsilon}^p \tag{4.63}$$

For linear elastic behaviour, the elastic strains are related to stresses by a symmetric 9×9 matrix of constants $\mathbf{D}$. Differentiating Eq. (4.63) and incorporating the plastic relation (4.50) we obtain

$$\underline{\dot{\varepsilon}} = \mathbf{D}^{-1}\underline{\dot{\sigma}} + \dot{\lambda}Q_{,\sigma} \tag{4.64}$$

The plastic strain (rate) will occur only if the 'elastic' stress changes

$$\underline{\dot{\sigma}}^e \equiv \mathbf{D}\underline{\dot{\varepsilon}} \tag{4.65}$$

tend to put the stress outside the yield surface, that is, is in the *plastic loading* direction. If, on the other hand, this stress change is such that *unloading* occurs then of course no plastic straining will be present, as illustrated for the one-dimensional case in Fig. 4.6. The test of the above relation is therefore crucial in differentiating between loading and unloading operations and underlines the importance of the straining path in computing stress changes.

When plastic loading is occurring the stresses are on the yield surface given by Eq. (4.44). Differentiating this we can therefore write

$$\dot{F} = \frac{\partial F}{\partial \sigma_x}\dot{\sigma}_x + \frac{\partial F}{\partial \sigma_y}\dot{\sigma}_y + \cdots + \frac{\partial F}{\partial \kappa_x}\dot{\kappa}_x + \frac{\partial F}{\partial \kappa_y}\dot{\kappa}_y + \cdots + \frac{\partial F}{\partial \kappa}\dot{\kappa} = 0$$

or

$$\dot{F} = F_{,\sigma}^T\underline{\dot{\sigma}} + F_{,\kappa}^T\underline{\dot{\kappa}} - H_i\dot{\lambda} = 0 \tag{4.66}$$

in which we make the substitution

$$H_i\dot{\lambda} = -\frac{\partial F}{\partial \kappa}\dot{\kappa} = -F_{,\kappa}\dot{\kappa} \tag{4.67}$$

where H_i denotes an isotropic hardening modulus.

For the case where kinematic hardening is introduced, using Eq. (4.54) we can substitute Eq. (4.61) and modify Eq. (4.64) to

$$\mathbf{D}\dot{\varepsilon} = \dot{\varsigma} + (\mathbf{D} + \mathbf{H}_k)\dot{\lambda}Q_{,\varsigma} \qquad (4.68)$$

Similarly, introducing Eq. (4.56) into Eq. (4.66) we obtain

$$\dot{F} = F_{,\varsigma}^{\mathrm{T}}\dot{\varsigma} - H_i\dot{\lambda} = 0 \qquad (4.69)$$

Equations (4.68) and (4.69) now can be written in matrix form as

$$\left\{ \begin{matrix} \mathbf{D}\dot{\varepsilon} \\ 0 \end{matrix} \right\} = \begin{bmatrix} \mathbf{I} & (\mathbf{D} + \mathbf{H}_k)Q_{,\varsigma} \\ F_{,\varsigma}^{\mathrm{T}} & -H_i \end{bmatrix} \left\{ \begin{matrix} \dot{\varsigma} \\ \dot{\lambda} \end{matrix} \right\} \qquad (4.70)$$

The indeterminate constant $\dot{\lambda}$ can now be eliminated (taking care not to multiply or divide by H_i or $\mathbf{H}_k$ which are zero in ideal plasticity). To accomplish the elimination we solve the first set of Eq. (4.70) for $\dot{\varsigma}$, giving

$$\dot{\varsigma} = \mathbf{D}\dot{\varepsilon} - (\mathbf{D} + \mathbf{H}_k)Q_{,\varsigma}\dot{\lambda}$$

and substitute into the second, yielding the expression

$$F_{,\varsigma}^{\mathrm{T}}\mathbf{D}\dot{\varepsilon} - [H_i + F_{,\sigma}^{\mathrm{T}}(\mathbf{D} + \mathbf{H}_k)Q_{,\varsigma}]\dot{\lambda} = 0$$

Equation (4.64) now results in an explicit expansion that determines the *stress changes* in terms of imposed *strain changes*. Using Eq. (4.43) this may now be reduced to a form in which only six independent components are present and expressed as*

$$\dot{\sigma} = \mathbf{D}_{ep}^{\star}\dot{\varepsilon} \qquad (4.71)$$

and

$$\begin{aligned} \mathbf{D}_{ep}^{\star} &= \mathbf{P}^{\mathrm{T}}\mathbf{D}\mathbf{P} - \frac{1}{H^{\star}}\mathbf{P}^{\mathrm{T}}\mathbf{D}Q_{,\varsigma}F_{,\varsigma}^{\mathrm{T}}\mathbf{D}\mathbf{P} \\ &= \mathbf{D} - \frac{1}{H^{\star}}\mathbf{P}^{\mathrm{T}}\mathbf{D}Q_{,\varsigma}F_{,\varsigma}^{\mathrm{T}}\mathbf{D}\mathbf{P} \end{aligned} \qquad (4.72)$$

where

$$H^{\star} = H_i + F_{,\varsigma}^{\mathrm{T}}\left(\mathbf{D} + \mathbf{H}_k\right)Q_{,\varsigma}$$

The elasto-plastic matrix $\mathbf{D}_{ep}^{\star}$ takes the place of the elasticity matrix $\mathbf{D}_{\mathrm{T}}$ in a *continuum* rate formulation. We note that in the absence of kinematic hardening it is possible to make reductions to the six-component form for all the computations at the very beginning. However, the manner in which the back stress enters the computation is not the same as that for the plastic strain and would be necessary to scale the two differently to make the general reduction. Thus, for the developments reported here we prefer to

---

* We shall show this step in more detail below for the J_2 plasticity model. In general, however, the final result involves only the usual form of the **D** matrix and six independent components from the derivative of the yield function.

carry out all calculations using the full nine-component form (or, in the case of plane stress, to follow a four-component form) and make final reductions using Eq. (4.72).

For a generalization of the above concepts to a yield surface possessing 'corners' where $Q_{,\sigma}$ is indeterminate, the reader is referred to the work of Koiter[19] or the multiple surface treatments in Simo and Hughes.[16]

An alternative procedure exists here simply by smoothing the corners. We shall refer to it later in the context of the Mohr–Coulomb surface often used in geomechanics and the procedure can be applied to any form of yield surface.

The continuum elasto-plastic matrix is symmetric only when plasticity is associative *and* when kinematic hardening is symmetric. In general, non-associative materials present stability difficulties, and special care is needed to use them effectively. Similar difficulties occur if the hardening moduli are negative which, in fact, leads to a *softening* behaviour. This is addressed further in Sec. 4.11 and later for large strain in Sec. 6.7.2.

The elasto-plastic matrix given above is defined even for ideal plasticity when H_i and $\underline{\mathbf{H}}_k$ are zero. Direct use of the continuum tangent in an incremental finite element context where the rates are approximated by

$$\dot{\varepsilon}_{n+1}\Delta t \approx \Delta\varepsilon_{n+1} \quad \text{and} \quad \dot{\sigma}_{n+1}\Delta t \approx \Delta\sigma_{n+1}$$

was first made by Yamada *et al.*[42] and Zienkiewicz *et al.*[43] However, this approach does not give quadratic convergence when used in the Newton scheme. For the associative case we can introduce a *discrete time integration algorithm* in order to develop an exact (numerically consistent) tangent which does produce quadratic convergence when used in the Newton iterative algorithm.

4.4 Computation of stress increments

We have emphasized that with the use of iterative procedures within a particular increment of loading, it is important always to compute the stresses as

$$\sigma_{n+1}^k = \sigma_n + \Delta\sigma_n^k \tag{4.73}$$

corresponding to the total change in displacement parameters $\Delta\mathbf{u}_n^k$ and hence the total strain change

$$\Delta\varepsilon_n^k = \mathbf{B}\Delta\mathbf{u}_n^k \quad \Delta\mathbf{u}_n^k = \sum_{i=0}^{k} d\mathbf{u}_n^i \tag{4.74}$$

which has accumulated in all previous iterations within the step. This point is of considerable importance as constitutive models with path dependence (viz. plasticity-type models) have different responses for loading and unloading. If a decision on loading/unloading is based on the increment $d\mathbf{u}_n^k$ erroneous results will be obtained. Such decisions must *always* be performed with respect to the total increment $\Delta\mathbf{u}_n^k$.

In terms of the elasto-plastic modulus matrix given by Eq. (4.72) this means that the stresses have to be integrated as

$$\sigma_{n+1}^k = \sigma_n + \int_0^{\Delta\varepsilon_n^k} \mathbf{D}_{ep}^\star \, d\varepsilon \tag{4.75}$$

incorporating into $\mathbf{D}_{ep}^\star$ the dependence on variables in a manner corresponding to a linear increase of $\Delta\varepsilon_n^k$ (or $\Delta\mathbf{u}_n^k$). Here, of course, all other rate equations have to be suitably integrated, though this generally presents little additional difficulty.

Various procedures for integration of Eq. (4.75) have been adopted and can be classified into explicit and implicit categories.

4.4.1 Explicit methods

In explicit procedures either a direct integration process is used or some form of the Runge–Kutta process is adopted.[44] In the former the known increment $\Delta\varepsilon_n^k$ is subdivided into m intervals and the integral of Eq. (4.75) is replaced by direct summation, writing

$$\Delta\sigma_n^k = \frac{1}{m} \sum_{j=0}^{m-1} \mathbf{D}_{(n+j/m)}^\star \Delta\varepsilon_n^k \tag{4.76}$$

where $\mathbf{D}_{(n+j/m)}^\star$ denotes the tangent matrix computed for stresses and hardening parameters updated from the previous increment in the sum.

This procedure, originally introduced in reference 45 and described in detail in references 46 and 47, is known as *subincrementation*. Its accuracy increases with the number of subincrements, m, used. In general it is difficult *a priori* to decide on this number, and accuracy of prediction is not easy to determine.

Such integration will generally result in the stress change departing from the yield surface by some margin. In problems such as those of ideal plasticity where the yield surface forms a meaningful limit a proportional scaling of stresses (or return map) has been practised frequently to obtain stresses which are on the yield surface at all times.[47,48] In this process the effects of integrating the evolution equation for hardening must also be treated.

A more precise explicit procedure is provided by use of a Runge–Kutta method. Here, first an increment of $\Delta\varepsilon/2$ is applied in a single-step explicit manner to obtain

$$\Delta\sigma_{n+1/2} = \tfrac{1}{2}\mathbf{D}_n^\star \Delta\varepsilon_n \tag{4.77}$$

using the initial elasto-plastic matrix. This increment of stress (and corresponding $\kappa_{n+1/2}$) is evaluated to compute $\mathbf{D}_{n+1/2}^\star$ and finally we evaluate

$$\Delta\sigma_n = \mathbf{D}_{n+1/2}^\star \Delta\varepsilon_n \tag{4.78}$$

This process has a second-order accuracy and, in addition, can give an estimate of errors incurred as

$$\Delta\sigma_n - 2\Delta\sigma_{n+1/2} \tag{4.79}$$

If such stress errors exceed a certain norm the size of the increment can be reduced. This approach is particularly useful for integration of non-associative models or models without yield functions where 'tangent' matrices are simply evaluated (see Sec. 4.6).

4.4.2 Implicit methods: return map algorithm

The integration of Eq. (4.75) can, of course, be written in an implicit form. For instance, we could write in place of Eq. (4.75), during each iteration k, that

$$\Delta\sigma_{n+1}^k = [(1-\theta)\mathbf{D}_n^* + \theta\mathbf{D}_{n+1}^{*,k}]\Delta\varepsilon_{n+1}^k \tag{4.80}$$

where here $\mathbf{D}_n^*$ denotes the value of the tangential matrix at the beginning of the time step and $\mathbf{D}_{n+1}^{*,k}$ the current estimate to the tangential matrix at the end of the step.

This non-linear equation set could be solved by any of the procedures previously described; however, derivatives of the tangent matrix are quite complex and in any case a serious error is committed in the approximate form of Eq. (4.80). Further, there is no guarantee that the stresses do not depart from the yield surface.

Return map algorithm

In 1964 a very simple algorithm was introduced simultaneously by Maenchen and Sacks[49] and by Wilkins.[50] This algorithm uses a two-step process to compute the new stress and was originally implemented in an explicit time integration form, thus requiring no explicit construction of an elasto-plastic tangent matrix; however, later its versatility and robustness was demonstrated for implicit solutions.[51,52] The steps of the algorithm are:

1. Perform a predictor step in which the entire increment of strain (for the present discussion we omit the iteration counter k for simplicity)

$$\underline{\varepsilon}_{n+1} = \underline{\varepsilon}_n + \Delta\underline{\varepsilon}_n$$

   is used to compute *trial* stresses (denoted by superscript TR) assuming elastic behaviour. Accordingly,

$$\underline{\sigma}_{n+1}^{\mathrm{TR}} = \mathbf{D}\left(\underline{\varepsilon}_{n+1} - \underline{\varepsilon}_n^{\mathrm{p}}\right) \tag{4.81}$$

   where only an elastic modulus $\mathbf{D}$ is required.
2. Evaluate the yield function in terms of the trial stress and the values of the plastic parameters at the previous time:

$$F(\underline{\sigma}^{\mathrm{TR}}, \underline{\kappa}_n, \kappa_n) = \begin{cases} \leq 0, & \text{elastic} \\ > 0, & \text{plastic} \end{cases} \tag{4.82}$$

   (a) For an elastic value of F set the current stress to the trial value, accordingly

$$\underline{\sigma}_{n+1} = \underline{\sigma}_{n+1}^{\mathrm{TR}}, \quad \underline{\kappa}_{n+1} = \underline{\kappa}_n \quad \text{and} \quad \kappa_{n+1} = \kappa_n$$

   (b) For a plastic state solve a discretized set of plasticity rate equations (namely, using any appropriate time integration method as described, for example, in reference 8) such that the final value of F_{n+1} is zero.

   A plastic correction can be most easily developed by returning to the original Eq. (4.64) and writing the relation for stress increment as

$$\Delta\underline{\sigma}_n = \mathbf{D}\left(\Delta\underline{\varepsilon}_n - \Delta\underline{\varepsilon}_n^{\mathrm{p}}\right) \tag{4.83}$$

Now integrating the plastic strain relation (4.50) using a form similar to that in Eq. (4.80) yields

$$\Delta\underline{\varepsilon}_n^p = \Delta\lambda[(1-\theta)Q_{,\underline{\sigma}}|_n + \theta Q_{,\underline{\sigma}}|_{n+1}] \tag{4.84}$$

where $\Delta\lambda$ represents an approximation to the change in consistency parameter over the time increment. Kinematic hardening is included by integrating Eq. (4.60) as

$$\Delta\underline{\kappa}_n = -\Delta\lambda\underline{\mathbf{H}}_k[(1-\theta)Q_{,\underline{\kappa}}|_n + \theta Q_{,\underline{\kappa}}|_{n+1}] \tag{4.85}$$

Finally, during the plastic solution we enforce

$$F_{n+1} = 0 \tag{4.86}$$

thus ensuring that final values at t_{n+1} satisfy the yield condition exactly.

The above solution process is particularly simple for $\theta = 1$ (backward difference or Euler implicit) and now, eliminating $\Delta\underline{\varepsilon}_n^p$, we can write the above non-linear system in residual form

$$\underline{\mathbf{R}}_\sigma^i = \Delta\underline{\varepsilon}_n - \underline{\mathbf{D}}^{-1}\Delta\underline{\sigma}_n^i - \Delta\lambda Q_{,\underline{\sigma}}|_{n+1}^i$$

$$\underline{\mathbf{R}}_\kappa^i = -\underline{\mathbf{H}}_k^{-1}\Delta\underline{\kappa}_n^i - \Delta\lambda Q_{,\underline{\kappa}}|_{n+1}^i$$

$$r^i = -F_{n+1}^i$$

and seek solutions which satisfy $\underline{\mathbf{R}}_\sigma^i = 0$, $\underline{\mathbf{R}}_\kappa^i = 0$ and $r^i = 0$. Any of the general iterative schemes described in Chapter 3 can now be used. In particular, the full Newton process is convenient. Noting that $\Delta\underline{\varepsilon}_n$ is treated here as a specified constant (actually, the $\Delta\underline{\varepsilon}_n^k$ from the current finite element solution), we can write, on linearization,

$$\begin{bmatrix} \underline{\mathbf{D}}^{-1} + \Delta\lambda Q_{,\sigma\sigma} & \Delta\lambda Q_{,\sigma\kappa} & Q_{,\underline{\sigma}} \\ \Delta\lambda Q_{,\kappa\sigma} & \underline{\mathbf{H}}_k^{-1} + \Delta\lambda Q_{,\kappa\kappa} & Q_{,\underline{\kappa}} \\ F_{,\underline{\sigma}}^T & F_{,\underline{\kappa}}^T & -H_i \end{bmatrix}_{n+1}^i \begin{Bmatrix} d\underline{\sigma}^i \\ d\underline{\kappa}^i \\ d\lambda^i \end{Bmatrix} = \begin{Bmatrix} \underline{\mathbf{R}}_\sigma^i \\ \underline{\mathbf{R}}_\kappa^i \\ r^i \end{Bmatrix} \tag{4.87}$$

where H_i is the same hardening parameter as that obtained in Eq. (4.67). Some complexity is introduced by the presence of the second derivatives of Q in Eq. (4.87) and the term may be omitted for simplicity (although at the expense of asymptotic quadratic convergence in the Newton iteration). Analytical forms of such second derivatives are available for frequently used potential surfaces.[16,30,51-53] Appendix A also presents results for second derivatives of stress invariants.

It is important to note that the requirement that $F_{n+1} = -r^i$ [Eq. (4.87)] ensures that the r^i residual measures precisely the departure from the yield surface. This measure is not available for any of the tangential forms if $\underline{\mathbf{D}}_{ep}^*$ is adopted.

For the solution it is only necessary to compute $d\lambda^i$ and update as

$$\Delta\lambda^i = \sum_{j=0}^{i} d\lambda^j \tag{4.88}$$

This solution process can be done in precisely the same way as was done in establishing Eq. (4.72). Thus, a solution may be constructed by defining the following:

$$\underline{\mathbf{R}} = \begin{bmatrix} \mathbf{R}_\sigma \\ \mathbf{R}_\kappa \end{bmatrix}, \quad \underline{\zeta} = \begin{bmatrix} \sigma \\ \kappa \end{bmatrix}$$

$$\underline{\nabla}F = \begin{bmatrix} F_{,\sigma} \\ F_{,\kappa} \end{bmatrix}, \quad \underline{\nabla}Q = \begin{bmatrix} Q_{,\sigma} \\ Q_{,\kappa} \end{bmatrix} \tag{4.89}$$

$$\underline{\mathbf{A}} = \begin{bmatrix} \mathbf{D}^{-1} & 0 \\ 0 & \mathbf{H}_k^{-1} \end{bmatrix} + \Delta\lambda^i \begin{bmatrix} Q_{,\sigma\sigma} & Q_{,\sigma\kappa} \\ Q_{,\kappa\sigma} & Q_{,\kappa\kappa} \end{bmatrix}$$

and expressing Eq. (4.87) as

$$d\underline{\zeta} = \underline{\mathbf{A}}^{-1}\underline{\mathbf{R}}^i - \frac{1}{A^\star}\underline{\mathbf{A}}^{-1}\underline{\nabla}Q^i \left[(\underline{\nabla}F^i)^\mathsf{T}\underline{\mathbf{A}}^{-1}\underline{\mathbf{R}}^i - r^i \right] \tag{4.90}$$

where

$$A^\star = H_i + (\underline{\nabla}F^i)^\mathsf{T}\underline{\mathbf{A}}^{-1}\underline{\nabla}Q^i \tag{4.91}$$

Immediately, we observe that at convergence $\underline{\mathbf{R}}^i = \mathbf{0}$ and $r^i = 0$, thus, here we obtain a zero stress increment. At this point we have computed a stress state σ_{n+1} which satisfies the yield condition exactly. However, this stress, when substituted back into the finite element residual [e.g. Eq. (2.20a) or (2.51)], may not satisfy the equilibrium condition and it is now necessary to compute a new iteration k and obtain a new strain increment $d\underline{\varepsilon}_n^k$ from which the process is repeated. We note that inserting this new increment into Eq. (4.87) will again result in a non-zero value for $\underline{\mathbf{R}}_\sigma$, but that $\underline{\mathbf{R}}_\kappa$ and r remain zero until subsequent iterations. Thus, Eq. (4.90) provides directly the required tangent matrix $\tilde{\mathbf{D}}_{ep}^\star$ from

$$\begin{Bmatrix} d\sigma \\ d\underline{\kappa} \end{Bmatrix} = \begin{bmatrix} \underline{\mathbf{A}}^{-1} - \frac{1}{A^\star}\underline{\mathbf{A}}^{-1}\underline{\nabla}Q(\underline{\nabla}F)^\mathsf{T}\underline{\mathbf{A}}^{-1} \end{bmatrix} \begin{Bmatrix} d\varepsilon \\ 0 \end{Bmatrix} = \begin{bmatrix} \tilde{\mathbf{D}}_{ep}^\star & \cdot \\ \cdot & \cdot \end{bmatrix} \begin{Bmatrix} d\varepsilon \\ 0 \end{Bmatrix} \tag{4.92}$$

Thus, we find the tangent matrix $\tilde{\mathbf{D}}_{ep}^\star$ is obtained from the upper diagonal block of Eq. (4.92). We note that this development also follows exactly the procedure for computing $\mathbf{D}_{ep}^\star$ in Eq. (4.72). At this stage the terms may once again be reduced to their six-component form using $\mathbf{P}$ as indicated in Eq. (4.42).

Some remarks on the above algorithm are in order:

1. For non-associative plasticity (namely, $Q \neq F$) the return direction is *not* normal to the yield surface. In this case no solution may exist for some strain increments (in general, arbitrary selection of F and Q forms in non-associative plasticity does not assure stability) and the iteration process will not converge.
2. For associative plasticity the normality principle is valid, requiring a convex yield surface. In this case the above iteration process always converges for a hardening material.
3. Convergence of the finite element equations may not always occur if more than one quadrature point changes from elastic to plastic or from plastic to elastic in subsequent iterations.

Based on these comments it is evident that no universal method exists that can be used with the many alternatives which can occur in practice. In the next several sections we illustrate some formulations which employ the alternatives we have discussed above.

4.5 Isotropic plasticity models

We consider here some simple cases for isotropic plasticity-type models in which both a yield function and a flow rule are used. For an isotropic material linear elastic response may be expressed by moduli defined with two parameters. Here we shall assume these to be the bulk and shear moduli, as used previously in the viscoelastic section (Sec. 4.2). Accordingly, the stress at any discrete time t_{n+1} is computed from elastic strains in matrix form as

$$\underline{\sigma}_{n+1} = p_{n+1}\underline{m} + \underline{s}_{n+1} = K\underline{m}\underline{m}\varepsilon^e_{n+1} + 2G(\mathbf{I} - \tfrac{1}{3}\underline{m}\underline{m}^T)\underline{\varepsilon}^e_{n+1}$$
$$= \mathbf{D}\left(\underline{\varepsilon}_{n+1} - \underline{\varepsilon}^p_{n+1}\right) \tag{4.93}$$

where the elastic modulus matrix for an isotropic material is given in the simple form

$$\mathbf{D} = K\underline{m}\underline{m}^T + 2G(\mathbf{I} - \tfrac{1}{3}\underline{m}\underline{m}^T) \tag{4.94}$$

and $\mathbf{I}$ is the 9×9 identity matrix and $\underline{m}$ is the nine-component matrix

$$\underline{m} = \begin{bmatrix} 1 & 1 & 1 & 0 & 0 & 0 & 0 & 0 & 0 \end{bmatrix}^T$$

Using Eqs (4.42) and (4.43) immediately reduces the above to

$$\mathbf{D} = K\underline{m}\underline{m}^T + 2G(\mathbf{I}_0 - \tfrac{1}{3}\underline{m}\underline{m}^T) \tag{4.95}$$

The above relation yields the stress at the current time provided we know the current total strain and the current plastic strain values. The total strain is available from the finite element equations using the current value of nodal displacements, and the plastic strain is assumed to be computed with use of one of the algorithms given above. In the discussion to follow we consider relations for various classical yield surfaces.

4.5.1 Isotropic yield surfaces

The general procedures outlined in the previous section allow determination of the tangent matrices for almost any yield surface applicable in practice. For an isotropic material all functions can be represented in terms of the three stress invariants:*

$$I_1 = \sigma_{ii} = \underline{m}^T\underline{\sigma}$$
$$2J_2 = s_{ij}s_{ji} = \underline{s}^T\underline{s} = |\underline{s}|^2 \tag{4.96}$$
$$3J_3 = s_{ij}s_{jk}s_{ki} = \det \underline{s}$$

where we can observe that definition of all the invariants is most easily performed in indicial notation.

One useful form of these invariants for use in yield functions is given by[45]

$$3\sigma_m = I_1$$
$$\bar{\sigma} = \sqrt{J_2}$$
$$3\theta = \sin^{-1}\left(-\frac{3\sqrt{3}J_3^{1/3}}{2\bar{\sigma}}\right) \quad \text{with} \quad -\frac{\pi}{6} \le \theta \le \frac{\pi}{6} \tag{4.97}$$

---

* Appendix B presents a summary of invariants and their derivatives.

Using these definitions the surface for several classical yield conditions can be given as:

1. Tresca:

$$F = 2\bar{\sigma}\cos\theta - Y(\kappa) = 0 \qquad (4.98)$$

2. Huber–von Mises:

$$F = \sqrt{2}\bar{\sigma} - \sqrt{\frac{2}{3}}Y(\kappa) = |\underline{s}| - \sqrt{\frac{2}{3}}Y(\kappa) = 0 \qquad (4.99)$$

Both conditions 1 and 2 are well verified in metal plasticity. For soils, concrete and other 'frictional' materials the Mohr–Coulomb or Drucker–Prager surfaces is frequently used.[54]

3. Mohr–Coulomb:

$$F = \sigma_m \sin\phi + \bar{\sigma}\left(\cos\theta - \frac{1}{\sqrt{3}}\sin\phi\sin\theta\right) - c\cos\phi = 0 \qquad (4.100)$$

where $c(\kappa)$ and $\phi(\kappa)$ are the cohesion and the angle of friction, respectively, which can depend on an isotropic strain hardening parameter κ.

4. Drucker–Prager:

$$F = 3\alpha'(\kappa)\sigma_m + \bar{\sigma} - K(\kappa) = 0 \qquad (4.101)$$

where

$$\alpha' = \frac{2\sin\phi}{\sqrt{3}(3 - \sin\phi)} \qquad K = \frac{6\cos\phi}{\sqrt{3}(3 - \sin\phi)}$$

and again c and ϕ can depend on a strain hardening parameter.

These forms lead to a convenient definition of the gradients $F_{,\underline{\sigma}}$ or $Q_{,\underline{\sigma}}$, irrespective of whether the surface is used as a yield condition or a flow potential. Thus we can always write

$$F_{,\underline{\sigma}} = F_{,\sigma_m}\frac{\partial\sigma_m}{\partial\underline{\sigma}} + F_{,\bar{\sigma}}\frac{\partial\bar{\sigma}}{\partial\underline{\sigma}} + F_{,\theta}\frac{\partial\theta}{\partial\underline{\sigma}} \qquad (4.102)$$

and upon noting that

$$\frac{\partial\bar{\sigma}}{\partial\underline{\sigma}} = \frac{\partial\bar{\sigma}}{\partial J_2}\frac{\partial J_2}{\partial\underline{\sigma}} = \frac{1}{2\sqrt{J_2}}\frac{\partial J_2}{\partial\underline{\sigma}}$$

$$\frac{\partial\theta}{\partial\underline{\sigma}} = \frac{\partial\theta}{\partial J_2}\frac{\partial J_2}{\partial\underline{\sigma}} + \frac{\partial\theta}{\partial J_3}\frac{\partial J_3}{\partial\underline{\sigma}} = \tan 3\theta\left[\frac{1}{9}J_3\frac{\partial J_3}{\partial\underline{\sigma}} - \frac{1}{6}J_2\frac{\partial J_2}{\partial\underline{\sigma}}\right] \qquad (4.103)$$

Alternatively, we can always write:

$$F_{,\underline{\sigma}} = F_{,\sigma_m}\frac{\partial\sigma_m}{\partial\underline{\sigma}} + F_{,J_2}\frac{\partial J_2}{\partial\underline{\sigma}} + F_{,J_3}\frac{\partial J_3}{\partial\underline{\sigma}} \qquad (4.104)$$

which can be put into a matrix form as shown in Appendix B.

The values of the three derivatives with respect to the invariants are shown in Table 4.1 for the various yield surfaces mentioned. The form of the various yield surfaces given above is shown with respect to the principal stress space in Fig. 4.8, though many more elaborate ones have been developed, particularly for soil (geomechanics) problems.[55–57]

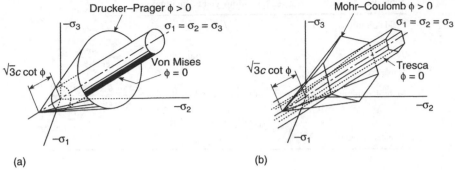

Fig. 4.8 Isotropic yield surfaces in principal stress space: (a) Drucker–Prager and von Mises; (b) Mohr–Coulomb and Trexa.

4.5.2 J_2 model with isotropic and kinematic hardening (Prandtl–Reuss equations)

As noted in Table 4.1 a particularly simple form results if we assume the yield function involves only the second invariant of the deviatoric stresses J_2. Here we present a more detailed discussion of results obtained by using an associated form and the return map algorithm. Since the yield function involves deviatoric quantities only we can initially make all the calculations in terms of these. Accordingly, the elastic deviatoric stress–strain relation is given as

$$\underline{s} = 2G\underline{e}^e = 2G(\underline{e} - \underline{e}^p) \tag{4.105}$$

Continuum rate form
Before constructing the return map solution we first consider the form of the plasticity equations in rate form for this simple model. The plastic deviatoric strain rates are deduced from

$$\dot{\underline{e}}^p = \dot{\lambda}\frac{\partial F}{\partial \underline{s}} = \dot{\lambda}F_{,\underline{s}} \tag{4.106}$$

Including the effects of isotropic and kinematic hardening the Huber–von Mises yield function may be expressed as

$$F = |\underline{s} - \underline{\kappa}| - \sqrt{\tfrac{2}{3}}Y(\kappa) = 0 \tag{4.107}$$

in which $\underline{\kappa}$ are back stresses from kinematic hardening and κ is an isotropic hardening parameter. We assume linear isotropic hardening given by*

$$Y(\kappa) = Y_0 + H_i\kappa \tag{4.108}$$

---

* More general forms of hardening may be approximated by piecewise linear segments, thus making the present formulation quite general.

Table 4.1 Invariant derivatives for various yield conditions

Yield condition	$F_{,\sigma_m}$	$\sqrt{J_2}\,F_{,J_2}$	$J_2\,F_{,J_3}$
Tresca	0	$2\cos\theta(1+\tan\theta\tan 3\theta)$	$\dfrac{\sqrt{3}\sin\theta}{\cos 3\theta}$
Huber–von Mises	0	$\sqrt{3}$	0
Mohr–Coulomb	$\sin\phi$	$\frac{1}{2}\cos\theta\left[1+\tan\theta\sin 3\theta\right.$	$\dfrac{\sqrt{3}\sin\theta+\sin\phi\cos\theta}{2\cos 3\theta}$
		$\left.+\frac{1}{\sqrt{3}}\sin\phi(\tan 3\theta-\tan\theta)\right]$	
Drucker–Prager	$3\alpha'$	1	0

Here a rate of κ is computed from a norm of the plastic strains, by using Eq. (4.53), as

$$\dot{\kappa}=\sqrt{\tfrac{2}{3}}\,\dot{\lambda} \tag{4.109}$$

in which the factor $\sqrt{2/3}$ is introduced to match uniaxial behaviour given by Eq. (4.108). On differentiation of F it will be found that

$$\frac{\partial F}{\partial \underline{s}}=-\frac{\partial F}{\partial \underline{\kappa}}=\underline{n} \quad \text{where} \quad \underline{n}=\frac{\underline{s}-\underline{\kappa}}{|\underline{s}-\underline{\kappa}|} \tag{4.110}$$

Using the above, the plastic strains are given by

$$\dot{\underline{e}}^{\mathrm{P}}=\dot{\lambda}\underline{n} \tag{4.111}$$

and, when substituted into a rate-of-stress relation, yield

$$\dot{\underline{s}}=2G\left[\dot{\underline{e}}-\dot{\lambda}\underline{n}\right] \tag{4.112}$$

A rate form for the kinematic hardening is taken as

$$\dot{\underline{\kappa}}=\tfrac{2}{3}H_k\dot{\lambda}\underline{n} \tag{4.113}$$

The rate of the yield function becomes

$$\dot{F}=\underline{n}^{\mathrm{T}}(\dot{\underline{s}}-\dot{\underline{\kappa}})-\tfrac{2}{3}H_i\dot{\lambda} \tag{4.114}$$

and when combined with the other rate equations gives the expression for the plastic consistency parameter as (noting that with the nine-component form $\underline{n}^{\mathrm{T}}\underline{n}=1$)

$$\dot{\lambda}=\frac{G}{G^{\star}}\underline{n}^{\mathrm{T}}\dot{\underline{e}} \tag{4.115}$$

where

$$G^{\star}=G+\tfrac{1}{3}(H_i+H_k) \tag{4.116}$$

Substitution of Eq. (4.115) into Eq. (4.112), and using Eq. (4.42) to reduce to the six-component form, gives the rate form for stress–strain deviators as

$$\dot{\underline{s}}=2G\left[\mathbf{I}_0-\frac{G}{G^{\star}}\underline{n}\underline{n}^{\mathrm{T}}\right]\dot{\underline{e}} \tag{4.117}$$

We note that for perfect plasticity $H_i = H_k = 0$ leading to $G/G^* = 1$ and, thus, the elastic–plastic tangent for this special case is also here obtained.

Use of Eq. (4.117) in the rate form of Eq. (4.93) gives the final continuum elastic–plastic tangent

$$\mathbf{D}_{ep}^\star = K\mathbf{mm}^T + 2G\left[\mathbf{I}_0 - \frac{1}{3}\mathbf{mm}^T - \frac{2G}{G^\star}\mathbf{nn}^T\right] \qquad (4.118)$$

This then establishes the well-known Prandtl–Reuss stress–strain relations generalized for linear isotropic and kinematic hardening.

Incremental return map form

The return map form for the equations is established by using a backward (Euler implicit) difference form as described previously (see Sec. 4.4.2). Omitting the subscript on the $n + 1$ quantities the plastic strain equation becomes, using Eqs (4.106) and (4.110),

$$\underline{e}^P = \underline{e}_n^P + \Delta\lambda\underline{n} \qquad (4.119)$$

and the accumulated (effective) plastic strain

$$\kappa = \kappa_n + \sqrt{\tfrac{2}{3}}\Delta\lambda \qquad (4.120)$$

Thus, now the discrete constitutive equation is

$$\underline{s} = 2G\,(\underline{e} - \underline{e}^P) \qquad (4.121)$$

the kinematic hardening is

$$\underline{\kappa} = \underline{\kappa}_n + \tfrac{2}{3}H_k\Delta\lambda\underline{n} \qquad (4.122)$$

and the yield function is

$$F = |\underline{s} - \underline{\kappa}| - \sqrt{\tfrac{2}{3}}Y_n - \tfrac{2}{3}H_i\Delta\lambda \qquad (4.123)$$

where $Y_n = Y_0 + \sqrt{2/3}\kappa_n$.

The trial stress, which establishes whether plastic behaviour occurs, is given by

$$\underline{s}^{TR} = 2G\left(\underline{e} - \underline{e}_n^P\right) \qquad (4.124)$$

which for situations where plasticity occurs permits the final stress to be given as

$$\underline{s} = \underline{s}^{TR} - 2G\Delta\lambda\underline{n} \qquad (4.125)$$

Using the definition of $\underline{n}$, we may now combine the stress and kinematic hardening relations as

$$|\underline{s} - \underline{\kappa}|\underline{n} = |\underline{s}^{TR} - \underline{\kappa}_n|\underline{n}^{TR} - \left(2G + \tfrac{2}{3}H_k\right)\Delta\lambda\underline{n} \qquad (4.126)$$

and noting from this that we must have

$$\underline{n}^{TR} = \underline{n} \qquad (4.127)$$

we may solve the yield function directly for the consistency parameter as[16,36]

$$\Delta\lambda = \frac{|\underline{s}^{TR} - \underline{\kappa}_n| - \sqrt{2/3}Y_n}{2G^\star} \qquad (4.128)$$

where $G^\star$ is given by Eq. (4.116).

We can also easily establish the relations for the consistent tangent matrix for this J_2 model. From Eqs (4.121) and (4.119) we obtain the incremental expression

$$d\underline{s} = 2G \left[d\underline{e} - \underline{n}d\lambda - \Delta\lambda d\underline{n} \right] \tag{4.129}$$

The increment of relation (4.127) gives[16]

$$d\underline{n} = d\underline{n}^{TR} = \frac{2G}{|\underline{s} - \underline{\kappa}|} \left[\mathbf{I} - \underline{n}\underline{n}^T \right] d\underline{e} \tag{4.130}$$

and from Eq. (4.128) we have

$$d\lambda = \frac{G}{G^\star} \underline{n}^T d\underline{e} \tag{4.131}$$

Substitution into Eq. (4.129) gives the consistent tangent matrix

$$d\underline{s} = 2G \left[\left(1 - \frac{2G\Delta\lambda}{|\underline{s}^{TR} - \underline{\kappa}_n|} \right) \mathbf{I} - \left(\frac{G}{G^\star} - \frac{2G\Delta\lambda}{|\underline{s}^{TR} - \underline{\kappa}_n|} \right) \underline{n}\underline{n}^T \right] d\underline{e} \tag{4.132}$$

This may now be expressed in terms of the total strains, combined with the elastic volumetric term and reduced to six-components to give

$$\mathbf{D}^\star_{ep} = K\mathbf{m}\mathbf{m}^T + 2G \left[\left(1 - \frac{2G\Delta\lambda}{|\underline{s}^{TR} - \underline{\kappa}_n|} \right) (\mathbf{I}_0 - \tfrac{1}{3}\mathbf{m}\mathbf{m}^T) - \left(\frac{G}{G^\star} - \frac{2G\Delta\lambda}{|\underline{s}^{TR} - \underline{\kappa}_n|} \right) \underline{n}\underline{n}^T \right]$$

$$\tag{4.133}$$

We here note also that when $\Delta\lambda = 0$ the tangent for the return map becomes the continuum tangent, thus establishing consistency of form.

4.5.3 J_2 plane stress

The discussion in the previous part of this section may be applied to solve problems in plane strain, axisymmetry, and general three-dimensional behaviour. In plane strain and axisymmetric problems it is only necessary to note that some strain components are zero. For problems in plane stress, however, it is necessary to modify the algorithm to achieve an efficient solution process. In a plane stress process only the four stresses σ_x, σ_y, τ_{xy} and τ_{yx} need be considered. When considering deviatoric components, however, there are five components, s_x, s_y, s_z, s_{xy} and s_{yx}. The deviators may be expressed in terms of the independent stresses as

$$\underline{s} = \begin{Bmatrix} s_x \\ s_y \\ s_z \\ s_{xy} \\ s_{yx} \end{Bmatrix} = \frac{1}{3} \begin{bmatrix} 2 & -1 & 0 & 0 \\ -1 & 2 & 0 & 0 \\ -1 & -1 & 0 & 0 \\ 0 & 0 & 3 & 0 \\ 0 & 0 & 0 & 3 \end{bmatrix} \begin{Bmatrix} \sigma_x \\ \sigma_y \\ \tau_{xy} \\ \tau_{yx} \end{Bmatrix} = \mathbf{P}_s \underline{\sigma} \tag{4.134}$$

The Huber–von Mises yield function may be written as

$$F = \left[(\underline{\sigma} - \underline{\kappa})^T \mathbf{P}_s^T \mathbf{P}_s (\underline{\sigma} - \underline{\kappa}) \right]^{1/2} - \sqrt{\tfrac{2}{3}} Y(\kappa) \leq 0 \tag{4.135}$$

Expanding Eq. (4.135) gives the plane stress yield function

$$F = \left[\varsigma_x^2 - \varsigma_x\varsigma_y + \varsigma_y^2 + 1.5(\varsigma_{xy}^2 + \varsigma_{yx}^2)\right]^{1/2} - Y(\kappa) \leq 0 \qquad (4.136)$$

where

$$\varsigma_x = \sigma_x - \kappa_x, \qquad \varsigma_y = \sigma_y - \kappa_y, \qquad \varsigma_{xy} = \tau_{xy} - \kappa_{xy}, \qquad \varsigma_{yx} = \tau_{yx} - \kappa_{yx} \quad (4.137)$$

define stresses which are shifted by the kinematic hardening back stress. Plastic strain rates may now be computed by direct differentiation of the yield function, giving

$$\underline{\dot{\varepsilon}}^{\mathrm{P}} = \dot{\lambda}F_{,\underline{\sigma}} = \left\{\begin{array}{c} \dot{\varepsilon}_x \\ \dot{\varepsilon}_y \\ \dot{\varepsilon}_{xy} \\ \dot{\varepsilon}_{yx} \end{array}\right\} = \frac{\dot{\lambda}}{2|\varsigma|} \begin{bmatrix} 2 & -1 & 0 & 0 \\ -1 & 2 & 0 & 0 \\ 0 & 0 & 3 & 0 \\ 0 & 0 & 0 & 3 \end{bmatrix} \left\{\begin{array}{c} \sigma_x \\ \sigma_y \\ \tau_{xy} \\ \tau_{yx} \end{array}\right\} = \frac{\dot{\lambda}}{|\varsigma|}\mathbf{A}_s\underline{\sigma} \qquad (4.138)$$

where $\mathbf{A}_s = \mathbf{P}_s^{\mathrm{T}}\mathbf{P}_s$. Similarly, the rate of the back stress for the kinematic hardening case is given by

$$\frac{1}{H_k}\underline{\dot{\kappa}} = \frac{\dot{\lambda}}{|\varsigma|}\mathbf{A}_s\underline{\sigma} \qquad (4.139)$$

The elastic components are computed by using the plane stress relation. Accordingly, for a plastic step the constitution is given by

$$\underline{\dot{\sigma}} = \mathbf{D}\left(\underline{\dot{\varepsilon}} - \underline{\dot{\varepsilon}}^{\mathrm{P}}\right) \qquad (4.140)$$

where for isotropic behaviour

$$\mathbf{D} = \frac{E}{1 - \nu^2} \begin{bmatrix} 1 & \nu & 0 & 0 \\ \nu & 1 & 0 & 0 \\ 0 & 0 & 1-\nu & 0 \\ 0 & 0 & 0 & 1-\nu \end{bmatrix} \qquad (4.141)$$

with E the modulus of elasticity and ν the Poisson's ratio.

We note that for a J_2 model the volumetric plastic strain must always be zero; consequently, we can complete the determination of plastic strains at any instant by using

$$\varepsilon_z^{\mathrm{P}} = -\varepsilon_x^{\mathrm{P}} - \varepsilon_y^{\mathrm{P}} \qquad (4.142)$$

This may be combined with the elastic strain given by

$$\varepsilon_x^{\mathrm{e}} = -\frac{\nu}{E}\left(\sigma_x + \sigma_y\right) \qquad (4.143)$$

to compute the total strain ε_z and, thus, the thickness change. The solution process now follows the procedures given for the general return mapping case. A procedure which utilizes a spectral transformation on the elastic and plastic parts is given in references 16 and 52. The process given there is more elegant but lacks the clarity of working directly with the stress and plastic strain increments.

4.6 Generalized plasticity

Plastic behaviour characterized by irreversibility of stress paths and the development of permanent strain changes after a stress cycle can be described in a variety of ways. One form of such description has been given in Sec. 4.3. Another general method is presented here.

4.6.1 Non-associative case – frictional materials

This approach assumes *a priori* the existence of a rate process which may be written directly as

$$\dot{\underline{\sigma}} = \mathbf{D}^{\star}\dot{\underline{\varepsilon}} \qquad (4.144)$$

in which the matrix $\mathbf{D}^{\star}$ depends not only on the stress $\underline{\sigma}$ and the state of parameters $\underline{\kappa}$, but also on the direction of the applied stress (or strain) rate $\dot{\underline{\sigma}}$ (or $\dot{\underline{\varepsilon}}$).[58] A slightly less ambitious description arises if we accept the dependence of $\mathbf{D}^{\star}$ only on two directions – those of loading and unloading. If in the general stress space we specify a 'loading' direction by a unit vector $\underline{\mathbf{n}}$ given at every point (and also depending on the state parameters $\underline{\kappa}$), as shown in Fig. 4.9, we can describe plastic loading and unloading by the sign of the projection $\underline{\mathbf{n}}^{\mathrm{T}}\dot{\underline{\sigma}}$. Thus

$$\underline{\mathbf{n}}^{\mathrm{T}}\dot{\underline{\sigma}} \quad \begin{cases} > 0 & \text{for loading} \\ < 0 & \text{for unloading} \end{cases} \qquad (4.145)$$

while $\underline{\mathbf{n}}^{\mathrm{T}}\dot{\underline{\sigma}} = 0$ is a neutral direction in which only elastic straining occurs. One can now write quite generally that

$$\dot{\underline{\sigma}} = \begin{cases} \underline{\mathbf{D}}_{\mathrm{L}}^{\star}\dot{\underline{\varepsilon}} & \text{for loading} \\ \underline{\mathbf{D}}_{\mathrm{U}}^{\star}\dot{\underline{\varepsilon}} & \text{for unloading} \end{cases} \qquad (4.146)$$

where the matrices $\underline{\mathbf{D}}_{\mathrm{L}}^{\star}$ and $\underline{\mathbf{D}}_{\mathrm{U}}^{\star}$ depend only on the state described by $\underline{\sigma}$ and $\underline{\kappa}$.

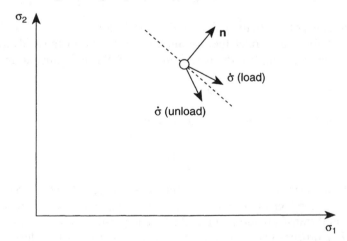

Fig. 4.9 Loading and unloading directions in stress space.

The specification of $\underline{\mathbf{D}}_L^*$ and $\underline{\mathbf{D}}_U^*$ must be such that in the neutral direction of the stress increment $\underline{\dot{\sigma}}$ the strain rates corresponding to this are equal. Thus we require

$$\underline{\dot{\varepsilon}} = (\mathbf{D}_L^*)^{-1}\underline{\dot{\sigma}} = \mathbf{D}_U^{*-1}\underline{\dot{\sigma}} \quad \text{when } \mathbf{n}^T\underline{\dot{\sigma}} = 0 \tag{4.147}$$

A general way to achieve this end is to write

$$(\mathbf{D}_L^*)^{-1} \equiv \underline{\mathbf{D}}^{-1} + \frac{1}{H_L}\mathbf{n}_{gL}\underline{\mathbf{n}}^T \quad \text{and} \quad (\mathbf{D}_U^*)^{-1} \equiv \underline{\mathbf{D}}^{-1} + \frac{1}{H_U}\mathbf{n}_{gU}\underline{\mathbf{n}}^T \tag{4.148}$$

where $\underline{\mathbf{D}}$ is the elastic matrix, $\mathbf{n}_{gL}$ and $\mathbf{n}_{gU}$ are arbitrary unit stress vectors for loading and unloading directions, and H_L and H_U are appropriate plastic moduli which in general depend on $\underline{\sigma}$ and $\underline{\kappa}$.

The value of the tangent matrices $\underline{\mathbf{D}}_L^*$ and $\underline{\mathbf{D}}_U^*$ can be obtained by direct inversion if $H_{L/U} \neq 0$, but more generally can be deduced following procedures given in Sec. 4.3.4 or can be written directly using the *Sherman–Morrison–Woodbury* formula[59] as:

$$\underline{\mathbf{D}}_L^* = \underline{\mathbf{D}} - \frac{1}{H_L^*}\mathbf{D}\mathbf{n}_{gL}\underline{\mathbf{n}}^T\mathbf{D} \quad H_L^* = H_L + \underline{\mathbf{n}}^T\mathbf{D}\mathbf{n}_{gL} \tag{4.149}$$

This form resembles Eq. (4.72) and indeed its derivation is almost identical. We note further that $(\mathbf{D}_L^*)^{-1}$ is now well behaved for H_L zero and a form identical to that of perfect plasticity is represented. Of course, a similar process is used to obtain $\underline{\mathbf{D}}_U^*$.

This simple and general description of *generalized plasticity* was introduced by Mróz and Zienkiewicz.[60,61] It allows:

1. the full model to be specified by a direct prescription of $\mathbf{n}$, $\mathbf{n}_g$ and H for loading and unloading at any point of the stress space;
2. existence of plasticity in both loading and unloading directions;
3. relative simplicity for description of experimental results when these are complex and when the existence of a yield surface of the kind encountered in ideal plasticity is uncertain.

For the above reasons the generalized plasticity forms have proved useful in describing the complex behaviour of soils.[62,62–65] Here other descriptions using various interpolations of $\mathbf{n}$ and moduli form a unique yield surface, known as *bounding surface plasticity* models, are indeed particular forms of the above generalization and have proved to be useful.[66]

Classical plasticity is indeed a special case of the generalized models. Here the yield surface may be used to define a unit normal vector as

$$\underline{\mathbf{n}} = \frac{1}{[F_{,\underline{\sigma}}^T F_{,\underline{\sigma}}]^{1/2}} F_{,\underline{\sigma}} \tag{4.150}$$

and the plastic potential may be used to define

$$\mathbf{n}_g = \frac{1}{[Q_{,\underline{\sigma}}^T Q_{,\underline{\sigma}}]^{1/2}} Q_{,\underline{\sigma}} \tag{4.151}$$

where once again some care must be exercised in defining the matrix notation. Substitution of such values for the unit vectors into Eq. (4.149) will of course retrieve

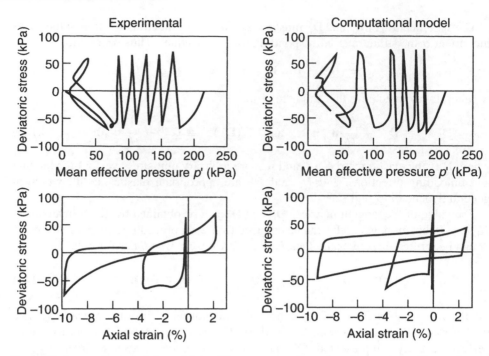

Fig. 4.10 A generalized plasticity model describing a very complex path, and comparison with experimental data. Undrained two-way cyclic loading of Nigata sand.[67] (Note that in an undrained soil test the fluid restrains all volumetric strains, and pore pressures develop; see reference 68.)

the original form of Eq. (4.72). However, interpretation of generalized plasticity in classical terms is more difficult.

The success of generalized plasticity in practical applications has allowed many complex phenomena of soil dynamics to be solved.[69,70] We shall refer to such applications later but in Fig. 4.10 we show how complex cyclic response with plastic loading and unloading can be followed.

While we have specified initially the loading and unloading directions in terms of the total stress rate $\dot{\boldsymbol{\sigma}}$ this definition ceases to apply when strain softening occurs and the plastic modulus H becomes negative. It is therefore more convenient to check the loading or unloading direction by the elastic stress increment $\dot{\boldsymbol{\sigma}}^e$ of Eq. (4.65) and to specify

$$\underline{\mathbf{n}}^T \dot{\underline{\boldsymbol{\sigma}}}^e \begin{cases} > 0 & \text{for loading} \\ < 0 & \text{for unloading} \end{cases} \tag{4.152}$$

This, of course, becomes identical to the previous definition of loading and unloading in the case of hardening.

4.6.2 Associative case – J_2 generalized plasticity

Another modification to the classical rate-independent approach is one in which the transition from an elastic to a fully plastic solution is accomplished with a smooth transition. This approach is useful in improving the match with experimental data for cyclic loading. A particularly simple form applicable to the J_2 model was introduced by Lubliner.[34,35] In this approach, the yield function is modified to a rate form directly and is expressed as

$$h(F)\dot{F} - \dot{\lambda} = 0 \qquad (4.153)$$

where $h(F)$ is given by the function

$$h(F) = \frac{F}{(\beta - F)\delta + H\beta} \qquad (4.154)$$

in which $H = H_i + H_k$, and δ, β are two positive parameters with dimension of stress. In particular, β is a distance between a *limit plastic state* and the current radius of the yield surface, and δ is a parameter controlling the approach to the limit state with increasing accumulated plastic strain.

On discretization and combination with the return map algorithm a rate-independent process is evident and again only minor modifications to the algorithm presented previously are necessary. A full description of the steps involved is given by Auricchio and Taylor.[36] Their paper also includes a development for the non-linear kinematic hardening model given in Eq. (4.62). In the case where the yield function is associative (i.e. $F = Q$) the use of the non-linear kinematic hardening model leads to an unsymmetric tangent stiffness when used with the return map algorithm. On the other hand, the generalized plasticity model is fully symmetric for this case.

In the next section we present further discussion on the use of generalized plasticity to model the behaviour of frictional materials. In general, these involve use of non-associative models where the return map algorithm cannot be used effectively.

4.7 Some examples of plastic computation

The finite element discretization technique in plasticity problems follows precisely the same procedures as those of corresponding elasticity problems. Any of the elements already discussed can be used for problems in plane stress; however, for plane strain, axisymmetry, and three-dimensional problems it is usually necessary to use elements which perform well in *constrained* situations such as encountered for near incompressibility. For this latter class of problems use of mixed elements is generally recommended, although elements and constitutive forms that permit use of reduced integration may also be used.

The use of mixed elements is especially important in metal plasticity as the Huber–von Mises flow rule does not permit any volume changes. As the extent of plasticity spreads at the collapse load the deformation becomes nearly incompressible, and with conventional (fully integrated) displacement elements the system *locks* and a true collapse load cannot be obtained.[71,72]

Finally, we should remark that the possibility of solving plastic problems is not limited to a displacement and mixed formulation alone. Equilibrium fields form a suitable vehicle,[73–75] but owing to their convenient and easy interpretation displacement and mixed forms are most commonly used.

4.7.1 Perforated plate – plane stress solutions

Figure 4.11 shows the configuration and the division into simple triangular and quadrilateral elements. In this example plane stress conditions are assumed and solution is obtained for both ideal plasticity and strain hardening. This problem was studied experimentally by Theocaris and Marketos[76] and was first analysed using finite element methods by Marcal and King[77] and Zienkiewicz et al.[43] (See reference 5 for discussion on these early solutions.) The von Mises criterion is used and, in the case of strain hardening, a constant slope of the uniaxial hardening curve, H, is taken. Data for the problem, from reference 76, are $E = 7000\,\text{kg/mm}^2$, $H = 225\,\text{kg/mm}^2$ and $\sigma_y = 24.3\,\text{kg/mm}^2$. Poisson's ratio is not given but is here taken as in reference 43 as $\nu = 0.3$. To match a configuration considered in the experimental study a strip with 200 mm width and 360 mm length containing a central hole of 200 mm diameter. Using symmetry only one quadrant is discretized as shown in Fig. 4.11. Displacement boundary restraints are imposed for normal components on symmetry boundaries and the top boundary. Sliding is permitted, to impose the necessary zero tangential traction boundary condition. Loading is applied by a uniform non-zero normal displacement

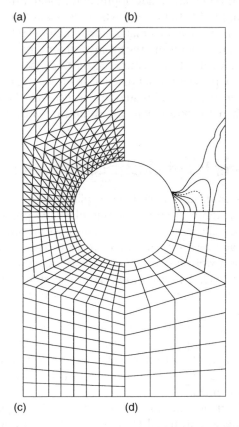

Fig. 4.11 Perforated plane stress tension strip: mesh used and development of plastic zones at loads of 0.55, 0.66, 0.75, 0.84, 0.92, 0.98, 1.02 times σ_y. (a) T3 triangles; (b) plastic zone spread; (c) Q4 quadrilaterals; (d) Q9 quadrilaterals.

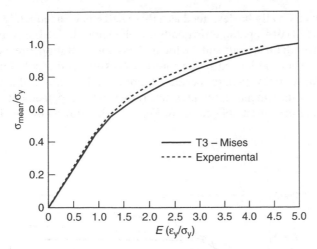

Fig. 4.12 Perforated plane stress tension strip: load deformation for strain hardening case ($H = 225\,\text{kg/mm}^2$).

with equal increments. Displacement elements of type T3, Q4, and Q9 are used with the same nodal layout. Results for the three elements are nearly the same, with the extent of plastic zones indicated for various loads in Fig. 4.11 obtained using the Q4 element. The load–deformation characteristics of the problem are shown in Fig. 4.12 and compared to experimental results. The strain ε_y is the peak value occurring at the hole boundary. This plane stress problem is relatively insensitive to element type and load increment size. Indeed, doubling the number of elements resulted in small changes of all essential quantities.

4.7.2 Perforated plate – plane strain solutions

The problem described above is now analysed assuming a plane strain situation. Data are the same as for the plane stress case except the lateral boundaries are also restrained to create a zero normal displacement boundary condition. This increases the confinement on the mesh and shows more clearly the locking condition cited previously. In Fig. 4.13 we plot the resultant axial load for each load step in the solution. Figure 4.13(a) shows results for the displacement model using T3, Q4, and Q9 elements and it is evident that the T3 and Q4 elements result in an erroneous increasing resultant load after the fully plastic state has developed. The Q9 element shows a clear limit state and indicates that higher order elements are less prone to locking (even though we have shown that for the fully incompressible state the Q9 displacement element will lock!). Figure 4.13(b) presents the same results for the Q4/1 and Q9/3 mixed elements and both give a clear limit load after the fully plastic state is reached.

4.7.3 Steel pressure vessel

This final example, for which test results obtained by Dinno and Gill[78] are available, illustrates a practical application, and the objectives are twofold. First, we show that this problem which can really be described as a thin shell can be adequately represented by a limit number (53) of isoparametric quadratic elements. Indeed, this model simulates well both the overall behaviour and the local stress concentration effects [Fig. 4.14(a)]. Second, this problem is loaded by an internal pressure and a solution is performed up to the 'collapse' point (where, because there is no hardening, the strains increase without limit) by incrementing the pressure rather than displacement. A comparison of calculated and measured deflections in Fig. 4.14(b) shows how well the objectives are achieved.

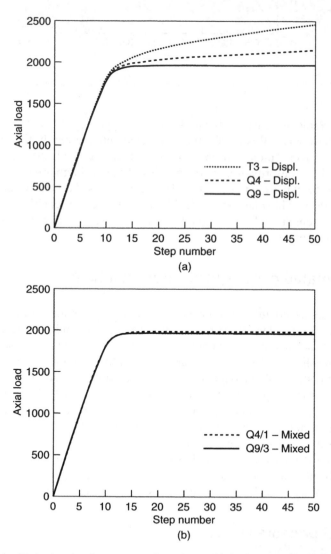

Fig. 4.13 Limit load behaviour for plane strain perforated strip: (a) displacement (displ.) formulation results; (b) mixed formulation results.

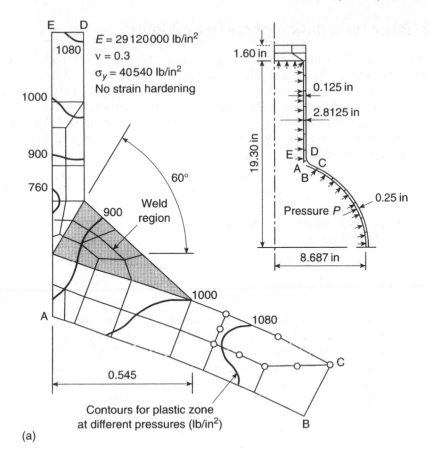

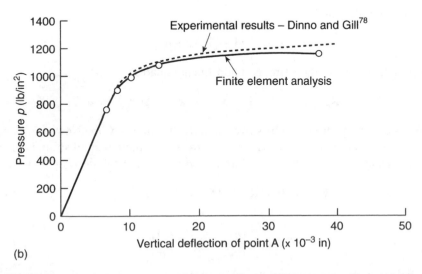

Fig. 4.14 Steel pressure vessel: (a) element subdivision and spread of plastic zones; (b) vertical deflection at point A with increasing pressure.

4.8 Basic formulation of creep problems

The phenomenon of 'creep' is manifested by a time-dependent deformation under a constant stress. Indeed the viscoelastic behaviour described in Sec. 4.2 is a particular model for linear creep. Here we shall deal with some non-linear models. Thus, in addition to an instantaneous strain, the material develops creep strains, ε^c, which generally increase with duration of loading. The constitutive law of creep will usually be of a form in which the *rate of creep strain* is defined as some function of stresses and the total creep strains (ε^c), that is,

$$\dot{\varepsilon}^c \equiv \frac{\partial \varepsilon^c}{\partial t} = \beta(\sigma, \varepsilon^c) \tag{4.155}$$

If we consider the instantaneous strains are elastic (ε^e), the total strain can be written again in an additive form as

$$\varepsilon = \varepsilon^e + \varepsilon^c \tag{4.156}$$

with

$$\varepsilon^e = \mathbf{D}^{-1}\sigma \tag{4.157}$$

where we neglect any initial (thermal) strains or initial (residual) stresses. A special case of this form was considered for linear viscoelasticity in Sec. 4.2. Here we consider a more general non-linear approach commonly used in modelling behaviour of metals at elevated temperatures and in modelling creep in cementitious materials.

We can again use any of the time integration schemes considered above and approximate the constitutive equations in a form similar to that used in plasticity as

$$\begin{aligned}\sigma_{n+1} &= \mathbf{D}\left(\varepsilon_{n+1} - \varepsilon_{n+1}^c\right) \\ \varepsilon_{n+1}^c &= \varepsilon_n^c + \Delta t \beta_{n+\theta}\end{aligned} \tag{4.158}$$

where $\beta_{n+\theta}$ is calculated as

$$\beta_{n+\theta} = (1-\theta)\beta_n + \theta\beta_{n+1}$$

On eliminating $\Delta\varepsilon^c$ we have simply a non-linear equation

$$\mathbf{R}_{n+1} \equiv \varepsilon_{n+1} - \mathbf{D}^{-1}\sigma_{n+1} - \varepsilon_n^c - \Delta t \beta_{n+\theta} = \mathbf{0} \tag{4.159}$$

The system of equations can be solved iteratively using, say, the Newton procedure. Starting from some initial guess, say $\sigma_{n+1} = \sigma_n$ and an increment of strain is given by the finite element process, the general iterative/incremental solution can be written as

$$\mathbf{R}^{i+1} = \mathbf{0} = \mathbf{R}^i - (\mathbf{D}^{-1} + \Delta t \mathbf{C}_{n+1})d\sigma_{n+1}^i \tag{4.160}$$

where

$$\mathbf{C}_{n+1} = \left.\frac{\partial\beta}{\partial\sigma}\right|_{n+\theta} = \theta\left.\frac{\partial\beta}{\partial\sigma}\right|_{n+1} \tag{4.161}$$

Solving this set of equations until the residual $\mathbf{R}$ is zero we obtain a set of stresses σ_{n+1} and tangent matrix

$$\mathbf{D}_{n+1}^{\star} \equiv \left[\mathbf{D}^{-1} + \Delta t \mathbf{C}_{n+1}\right]^{-1} \tag{4.162}$$

which may once again be used to perform any needed iterations on the finite element equilibrium equations. The iterative computation that follows is very similar to that used in plasticity, but here Δt is an actual time and the solution becomes *rate dependent*.

While in plasticity we have generally used implicit (backward difference) procedures; here many simple alternatives are possible. In particular, two schemes with a single iterative step are popular.

4.8.1 Fully explicit solutions

'Initial strain' procedure: $\theta = 0$

Here, from Eqs (4.161) and (4.162) we see that

$$C_{n+1} = 0 \quad \text{and} \quad D_{n+1}^\star = D \tag{4.163}$$

Thus, from Eq. (4.159) we obtain

$$\sigma_{n+1} = D[\varepsilon_{n+1} - \varepsilon_n^c - \Delta t \beta_n] \tag{4.164}$$

which may be used in Eq. (2.11c$_2$) of Chapter 2 to satisfy a discretized equilibrium equation. We note that this form will lead to a standard elastic stiffness matrix. This, of course, is equivalent to evaluating the increment of creep strain from the initial stress values at each time t_n and is exceedingly simple to calculate. While the process has been popular since the earliest days of finite elements[79-81] it is obviously less accurate for a finite step than other alternatives. Of course accuracy will improve if small time steps are used in such calculations. Further, if the time step is too large, unstable results will be obtained. Thus it is necessary for

$$\Delta t \leq \Delta t_{\text{crit}} \tag{4.165}$$

where Δt_{crit} is determined in a suitable manner (see, for example, Chapter 17 in reference 8).

A 'rule of thumb' that proves quite effective in practice is that the increment of creep strain should not exceed one half the total elastic strain[82]

$$\Delta t \left[\beta_n^T \beta_n\right]^{1/2} \leq \tfrac{1}{2} \left[(\varepsilon^e)^T \varepsilon^e\right]^{1/2} \tag{4.166}$$

Fully explicit process with modified stiffness: $1/2 \leq \theta \leq 1$

Here the main difference from the first explicit process is that the matrix C is not equal to zero but within a single step is taken as a constant, that is,

$$C_{n+1} \approx \theta \left. \frac{\partial \beta}{\partial \sigma} \right|_n \tag{4.167}$$

This is equivalent to a modified Newton scheme in which the tangent is held constant at its initial value in the step. Now

$$D_{n+1}^\star = \left[D^{-1} + \Delta t C_{n+1}\right]^{-1}$$

This process is more expensive than the simple explicit one previously mentioned, as the finite element tangent matrix has to be formed and solved for every time step. Further, such matrices can be non-symmetric, adding to computational expense.

Neither of the simplified iteration procedures described above give any attention to errors introduced in the estimates of the creep strain. However, for accuracy the iterative process with $\theta \geq 1/2$ is recommended. Such full iterative procedures were introduced by Cyr and Teter,[83] and later by Zienkiewicz and co-workers.[84,85]

We shall note that the process has much similarity with iterative solutions of plastic problems of Sec. 4.3 in the case of viscoplasticity, which we shall discuss in the next section.

4.9 Viscoplasticity – a generalization

4.9.1 General remarks

The purely plastic behaviour of solids postulated in Sec. 4.3 is probably a fiction as the maximum stress that can be carried is invariably associated with the rate at which this is applied. A purely elasto-plastic behaviour in a uniaxial loading is described in a model of Fig. 4.15(a) in which the plastic strain rate is zero for stresses below yield, that is,

$$\dot{\varepsilon}^{\mathrm{p}} = 0 \quad \text{if } |\sigma - \sigma_y| < 0 \text{ and } |\sigma| > 0$$

and $\dot{\varepsilon}^{\mathrm{p}}$ is indeterminate when $\sigma - \sigma_y = 0$.

An elasto-viscoplastic material, on the other hand, can be modelled as shown in Fig. 4.15(b), where a dashpot is placed in parallel with the plastic element. Now stresses can exceed σ_y for strain rates other than zero.

The viscoplastic (or creep) strain rate is now given by a general expression

$$\dot{\varepsilon}^{\mathrm{vp}} = \gamma \left\langle \phi(\sigma - \sigma_y) \right\rangle \tag{4.168}$$

where the arbitrary function ϕ is such that

$$\begin{aligned} \left\langle \phi(\sigma - \sigma_y) \right\rangle &= 0 & \text{if } |\sigma - \sigma_y| \leq 0 \\ \left\langle \phi(\sigma - \sigma_y) \right\rangle &= \phi(\sigma - \sigma_y) & \text{if } |\sigma - \sigma_y| > 0 \end{aligned} \tag{4.169}$$

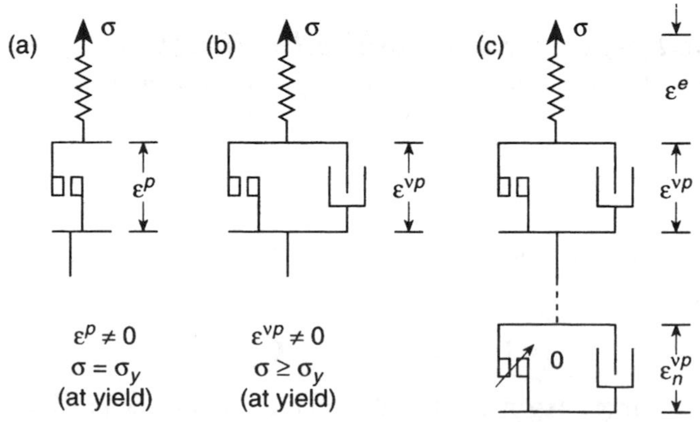

Fig. 4.15 (a) Elasto-plastic; (b) elasto-viscoplastic; (c) series of elasto-viscoplastic models.

The model suggested is, in fact, of a creep-type category described in the previous sections and often is more realistic than that of classical plasticity.

A viscoplastic model for a general stress state is given here and follows precisely the arguments of the plasticity section. In a three-dimensional context ϕ becomes a function of the yield condition $F(\sigma, \kappa, \kappa)$ defined in Eq. (4.44). If this is less than zero, no 'plastic' flow will occur. To include the viscoplastic behaviour we modify Eq. (4.44) as

$$\gamma \langle \phi(F) \rangle - \dot{\lambda} = 0 \tag{4.170}$$

and use Eq. (4.45) to define the plastic strain. Equation (4.170) implies

$$\langle \phi(F) \rangle = \begin{cases} 0 & \text{if } F \le 0 \\ \phi(F) & \text{if } F > 0 \end{cases} \tag{4.171}$$

and γ is some 'viscosity' parameter. Once again *associated* or *non-associated* flows can be invoked, depending on whether $F = Q$ or not. Further, any of the yield surfaces described in Sec. 4.3.1 and hardening forms described in Sec. 4.3.3 can be used to define the appropriate flow in detail. For simplicity, $\phi(F) = F^m$ where m is a positive power often used to define the viscoplastic rate effects in Eq. (4.170).[86,87]

The concept of viscoplasticity in one of its earliest versions was introduced by Bingham in 1922[88] and a survey of such modelling is given in references 86, 89 and 90. The computational procedure for a viscoplastic model can follow any of the general methods described in Sec. 4.4. Early applications commonly used the straightforward Euler (explicit) method.[91–95] The stability requirements for this approach have been considered for several types of yield conditions by Cormeau.[96] A *tangential* process can again be used, but unless the viscoplastic flow is associated ($F = Q$), non-symmetric systems of equations have to be solved at each step. Use of an explicit method will yield solution for the associative and non-associative cases and the system matrix remains symmetric. This process is thus similar to that of a modified Newton method (initial stress method) and is quite efficient. Indeed within the stability limit it has been shown that use of an overrelaxation method leads to rapid convergence.

4.9.2 Iterative solution

The complete iterative solution scheme for viscoplasticity is identical to that used in plasticity except for the use of Eq. (4.170) instead of Eq. (4.44). To underline this similarity we consider the constitutive model without hardening and use the return map implicit algorithm. The linearized relations are identical except for the treatment of relation (4.175). The form becomes

$$\begin{bmatrix} \mathbf{D}^{-1} + \Delta\lambda Q_{,\sigma\sigma} & Q_{,\sigma} \\ \phi' F_{,\sigma}^{\mathrm{T}} & -H_i - \dfrac{1}{\gamma\Delta t} \end{bmatrix}_{n+1}^{i} \begin{Bmatrix} d\sigma^i \\ d\lambda^i \end{Bmatrix} = \begin{Bmatrix} \mathbf{R}_\sigma^i \\ r^i \end{Bmatrix} \tag{4.172}$$

where the discrete residual for Eq. (4.171) is given by

$$r_n = -\phi(F)_n + \frac{1}{\gamma\Delta t}\Delta\lambda_n \tag{4.173}$$

and

$$\phi' = \frac{d\phi}{dF}$$

Now the equations are almost identical to those of plasticity [see Eq. (4.87)], with differences appearing only in the ϕ' and $1/(\gamma\Delta t)$ terms.

Again, a consistent tangent can be obtained by elimination of the $d\lambda^i$ and a general iterative scheme is once more available.

Indeed, as expected, $\gamma\Delta t = \infty$ will now correspond to the exact plasticity solution. This will always be reached by any solution tending to steady state. However, for transient situations this is not the case and use of finite values for $\gamma\Delta t$ will invariably lead to some rate effects being present in the solution.

The viscoplastic laws can easily be generalized to include a series of components, as shown in Fig. 4.15(c). Now we write

$$\dot{\varepsilon}^v = \dot{\varepsilon}_1^v + \dot{\varepsilon}_2^v + \cdots \tag{4.174}$$

and again the standard formulation suffices. If, as shown in the last element of Fig. 4.15(c), the plastic yield is set to zero, a 'pure' creep situation arises in which flow occurs at all stress levels. If a finite value is in a term a corresponding rate equation for the associated λ_j must be used. This is similar to the Koiter treatment for multi-surface plasticity.[19,97]

The use of the Duvaut and Lions[90] approach modifies the return map algorithm for a rate-independent plasticity solution. Once this solution is available a reduction in the value of $\Delta\lambda$ is computed to account for rate effects. The interested reader should consult references 16 and 30 for additional information on this approach.

4.9.3 Creep of metals

If an associated form of viscoplasticity using the von Mises yield criterion of Eq. (4.99) is considered the viscoplastic strain rate can be written as

$$\dot{\varepsilon}^{vp} = \lambda\frac{\partial|s|}{\partial\sigma} = \lambda\mathbf{n} \tag{4.175}$$

with the rate expressed again as

$$\dot{\lambda} = \gamma\langle\bar{\sigma} - \sigma_y\rangle \tag{4.176}$$

If σ_y, the yield stress, is set to zero we can write the above as

$$\dot{\varepsilon}^{vp} = \gamma\bar{\sigma}^m\mathbf{n} \tag{4.177}$$

and we obtain the well-known Norton–Soderberg creep law. In this, generally the parameter γ is a function of time, temperature, and the total creep strain (e.g. the analogue to the plastic strain ε^p). For a survey of such laws the reader can consult specialized references 98 and 99.

An example initially solved using a large number of triangular elements[85] is presented in Fig. 4.16, where a much smaller number of isoparametric quadrilaterals are used in a general viscoplastic program.[95]

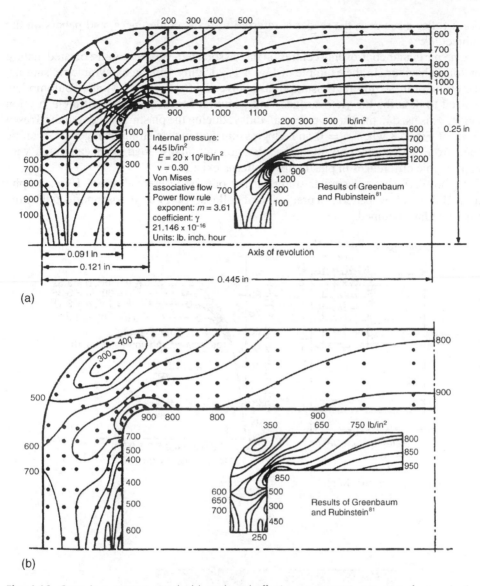

Fig. 4.16 Creep in a pressure vessel: (a) mesh end effective stress contours at start of pressurization; (b) effective stress contours 3 h after pressurization.

4.9.4 Soil mechanics applications

As we have already mentioned, the viscoplastic model provides a simple and effective tool for the solution of plasticity problems in which transient effects are absent. This includes many classical problems which have been solved in references 95 and 68, and the reader is directed there for details. In this section some problems of soil mechanics are discussed in which the facility of the process for solving non-associated behaviour is demonstrated.[100] The whole subject of the behaviour of soils and similar porous media is one in which much yet needs to be done to formulate good constitutive models.

For a fuller discussion the reader is referred to texts, conferences, and papers on the subject.[101,102]

One particular controversy centres on the 'associated' versus 'non-associated' nature of soil behaviour. In the example of Fig. 4.17, dealing with an axisymmetric sample, the effect of these different assumptions is investigated.[86] Here a Mohr–Coulomb law is used to describe the yield surface, and a similar form, but with a different friction angle, $\bar{\phi}$, is used in the plastic potential, thus reducing the plastic potential to the Tresca form of Fig. 4.8 when $\bar{\phi} = 0$ and suppressing volumetric strain changes. As can be seen from the results, only moderate changes in collapse load occur, although very appreciable differences in plastic flow patterns exist.

Figure 4.18 shows a similar study carried out for an embankment. Here, despite quite different flow patterns, a prediction of collapse load was almost unaffected by the flow rate law assumed.

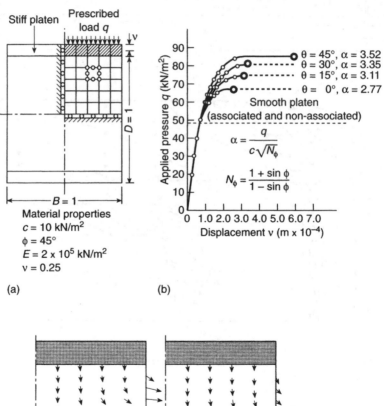

(a)

(b)

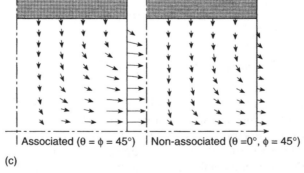

(c)

Fig. 4.17 Uniaxial, axisymmetric compression between rough plates: (a) mesh and problem; (b) pressure displacement result; (c) plastic flow velocity patterns.

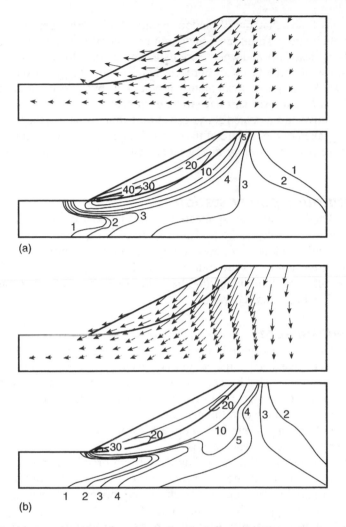

(a)

(b)

Fig. 4.18 Embankment under action of gravity, relative plastic flow velocities at collapse, and effective shear strain rate contours at collapse: (a) associative behaviour; (b) non-associative (zero volume change) behaviour.

The non-associative plasticity, in essence caused by frictional behaviour, may lead to non-uniqueness of solution. The equivalent viscoplastic form is, however, always unique and hence viscoplasticity is on occasion used as a *regularizing* procedure.

4.10 Some special problems of brittle materials

4.10.1 The no-tension material

A hypothetical material capable of sustaining only compressive stresses and straining without resistance in tension is in many respects similar to an ideal plastic material. While in practice such an ideal material does not exist, it gives a reasonable approximation of the behaviour of randomly jointed rock and other granular materials. While an explicit stress–strain relation cannot be generally written, it suffices to carry out the analysis elastically and wherever tensile stresses develop to reduce these to zero. The initial stress (modified Newton) process here is natural and indeed was developed in this context.[103] The steps of calculation are obvious but it is important to remember that the *principal tensile stresses* have to be eliminated.

The 'constitutive' law as stated above can at best approximate to the true situation, no account being taken of the closure of fissures on reapplication of compressive stresses. However, these results certainly give a clear insight into the behaviour of real rock structures.

An underground power station

Figure 4.19(a) and (b) shows an application of this model to a practical problem.[103] In Fig. 4.19(a) an elastic solution is shown for stresses in the vicinity of an underground power station with 'rock bolt' prestressing applied in the vicinity of the opening. The zones in which tension exists are indicated. In Fig. 4.19(b) a *no-tension* solution is given for the same problem, indicating the rather small general redistribution and the zones where 'cracking' has occurred.

Reinforced concrete

A variant on this type of material may be one in which a finite tensile strength exists but when this is once exceeded the strength drops to zero (on fissuring).

Such an analysis was used by Valliappan and Nath[104] in the study of the behaviour of reinforced concrete beams. Good correlation with experimental results for under-reinforced beams (in which development of compressive yield is not important) have been obtained. The beam is one for which test results were obtained by Krahl *et al.*[105] Figure 4.20 shows some relevant results.

Much development work on the behaviour of reinforced concrete has taken place, with various plasticity forms being introduced to allow for compressive failure and procedures that take into account the crack-closing history. References 106 and 107 list some of the basic approaches on this subject.

The subject of analysis of reinforced concrete has proved to be of great importance in recent years and publications in this field are proliferating. References 108–111 guide the reader to current practice in this field.

4.10.2 'Laminar' material and joint elements

Another idealized material model is one that is assumed to be built up of a large number of elastic and inelastic laminae. When under compression, these can transmit shear stress parallel to their direction – providing this does not exceed the frictional resistance. No tensile stresses can, however, be transmitted in the normal direction to the laminae.

This idealized material has obvious uses in the study of rock masses with parallel joints but has much wider applicability. Figure 4.21 shows a two-dimensional situation involving such a material. With a local coordinate axis x' oriented in the direction of

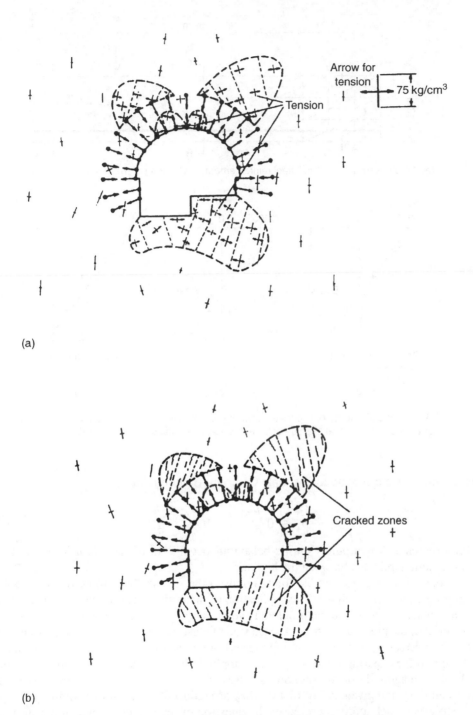

Fig. 4.19 Underground power station: gravity and prestressing loads. (a) Elastic stresses; (b) 'no-tension' stresses.

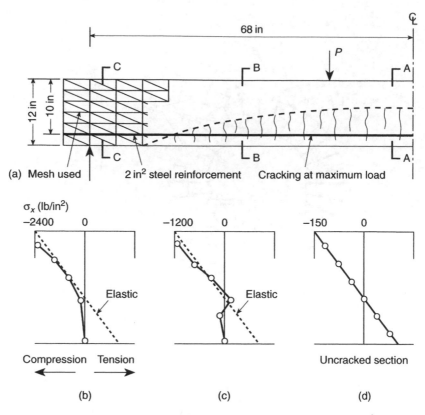

Fig. 4.20 Cracking of a reinforced concrete beam (maximum tensile strength 200 lb/in²). Distribution of stresses at various sections.[104] (a) Mesh used; (b) section AA; (c) section BB; (d) section CC.

the laminae we can write for a simple Coulomb friction joint

$$
\begin{aligned}
|\tau_{x'y'}| < \mu\sigma_{y'} & \qquad \text{if } \sigma_{y'} \leq 0 \\
\sigma_{y'} = 0 & \qquad \text{if } \varepsilon_{y'} > 0
\end{aligned}
\qquad (4.178)
$$

for stresses at which purely elastic behaviour occurs. In the above, μ is the friction coefficient applicable between the laminae.

   If elastic stresses exceed the limits imposed the stresses have to be reduced to the limiting values given above. The application of the initial stress process in this context is again self-evident, and the problem is very similar to that implied in the no-tension material of the previous section. At each step of elastic calculation, first the existence of tensile stresses $\sigma_{y'}$ is checked and, if these develop, a corrective initial stress reducing these and the shearing stresses to zero is applied. If $\sigma_{y'}$ stresses are compressive, the absolute magnitude of the shearing stresses $\tau_{x'y'}$ is checked again; if these stresses exceed the value given by Eq. (4.178) they are reduced to their proper limit.

   However, such a procedure poses the question of the manner in which the stresses are reduced, as two components have to be considered. It is, therefore, preferable to use the statements of relations (4.178) as definitions of plastic yield surfaces (F). The assumption of additional plastic potentials (Q) will now define the flow, and we

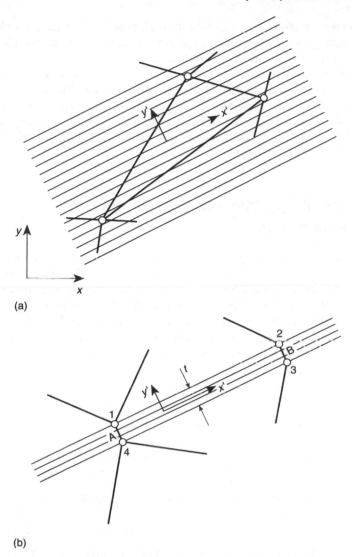

(a)

(b)

Fig. 4.21 'Laminar' material: (a) general laminarity; (b) laminar in narrow joint.

note that associated behaviour, with Eq. (4.178) used as the potential, will imply a simultaneous separation and sliding of the laminae (as the corresponding strain rates $\dot{\gamma}_{x'y'}$ and $\dot{\varepsilon}_{y'}$ are finite). Non-associated plasticity (or viscoplasticity) techniques have therefore to be used. Once again, if stress reversal is possible it is necessary to note the opening of the laminae, that is, the yield surface is made strain dependent.

In some instances the laminar behaviour is confined to a narrow joint between relatively homogeneous elastic masses. This may well be of a nature of a geological fault or a major crushed rock zone. In such cases it is convenient to use narrow, generally rectangular elements whose geometry may be specified by mean coordinates of two ends A and B [Fig. 4.21(b)] and the thickness. The element still has, however, separate points of continuity (1–4) with the adjacent rock mass.[112,113] Such joint elements can be simple rectangles, as shown here, but equally can take more complex shapes if represented by using isoparametric coordinates.

Laminations may not be confined to one direction only – and indeed the interlaminar material itself may possess a plastic limit. The use of such multilaminate models in the context of rock mechanics has proved very effective;[114] with a random distribution of laminations we return, of course, to a typical soil-like material, and the possibilities of extending such models to obtain new and interesting constitutive relations have been highlighted by Pande and Sharma.[115]

4.11 Non-uniqueness and localization in elasto-plastic deformations

In the preceding sections the general processes of dealing with complex, non-linear constitutive relations have been examined and some particular applications were discussed. Clearly, the subject is large and of great practical importance; however, presentation in a single chapter is not practical or possible. For different materials alternate forms of constitutive relations can be proposed and experimentally verified. Once such constitutive relations are available the processes of this chapter serve as a guide for constructing effective numerical solution strategies. Indeed, it is possible to build standard computing systems applicable to a wide variety of material properties in which new specifications of behaviour may be inserted.

What must be restated is that, in non-linear problems:

1. non-uniqueness of solution may arise;
2. convergence can never be, *a priori*, guaranteed;
3. the cost of solution is invariably much greater than it is in linear solutions.

Here, of course, the item of most serious concern is the first one, that is, that of non-uniqueness, which could lead to a physically irrelevant solution even if numerical convergence occurred and possibly large computational expense was incurred. Such non-uniqueness may be due to several reasons in elasto-plastic computations:

1. the existence of corners in the yield (or potential) surfaces at which the gradients are not uniquely defined;
2. the use of a *non-associated* formulation[20,116,117] (to which we have already referred in Sec. 4.9.4);
3. the development of strain softening and localization.[118,119]

The first problem is the least serious and can readily be avoided by modifying the yield (or potential) surface forms to avoid corners. A simple modification of the Mohr–Coulomb (or Tresca) surface expressions [Eq. (4.100)] is easily achieved by writing[55]

$$F = \sigma_m \sin \phi - c \cos \phi + \frac{\bar{\sigma}}{g(\theta)} \qquad (4.179)$$

where

$$g(\theta) = \frac{2K}{(1 + K) - (1 - K) \sin 3\theta}$$

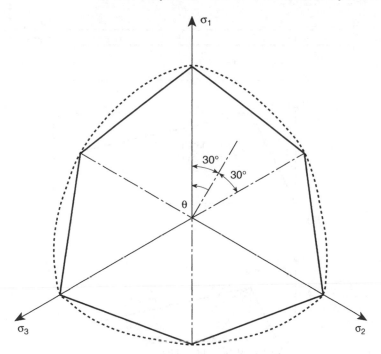

Fig. 4.22 Π plane section of Mohr–Coulomb yield surface in principal stress space, with $\phi = 25°$ (solid line); smooth approximation of Eq. (4.179) (dotted line).

and

$$K = \frac{3 - \sin \phi}{3 + \sin \phi}$$

Figure 4.22 shows how the angular section of the Mohr–Coulomb surface in the Π plane (constant σ_m) now becomes rounded. Similar procedures have been suggested by others.[120] An alternative to smoothing is to introduce a multisurface model and use a solution process which gives unique results for a corner.[16,30]

Much more serious are the second and third possible causes of non-uniqueness mentioned above. Here, theoretical non-uniqueness can be avoided by considering the plastic deformation to be a limit state of viscoplastic behaviour in a manner we have already referred to in Sec. 4.9. Such a process, mathematically known as *regularization*, has allowed us to obtain many realistic solutions for both non-associative and strain softening behaviour in problems which are subjected to steady-state or quasi-static loading, as already shown. For fully transient cases, however, the process is quite delicate and much care is needed to obtain a valid regularization.

However, on occasion (though not invariably), both forms of behaviour can lead to *localization* phenomena where strain (and displacement) discontinuities develop.[117–129] The non-uniqueness can be particularly evident in strain softening plasticity. We illustrate this in an example of Fig. 4.23 where a bar of length L, divided into elements of length h, is subject to a uniformly increasing extension u. The material is initially elastic with a modulus E and after exceeding a stress of σ_y, the yield stress softens (plastically) with a negative modulus H.

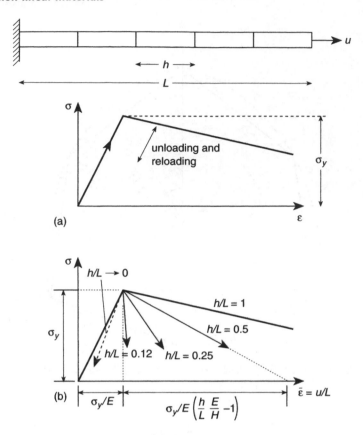

Fig. 4.23 Non-uniqueness: mesh size dependence in extension of a homogeneous bar with a strain softening material. (Peak value of yield stress, σ_y, perturbed in a single element.) (a) Stress σ versus strain ε for material; (b) stress $\bar{\sigma}$ versus average strain $\bar{\varepsilon}$ ($= u/L$) assuming yielding in a single element of length h.

The strain–stress relation is thus [Fig. 4.23(a)]

$$\sigma = E\varepsilon \quad \text{if} \quad \varepsilon < \sigma_y/E = \varepsilon_y \tag{4.180}$$

and for increasing ε only,

$$\sigma = \sigma_y - H\left(\varepsilon - \varepsilon_y\right) \quad \text{if} \quad \varepsilon > \varepsilon_y \tag{4.181}$$

For unloading from any plastic point the material behaves elastically as shown.

One possible solution is, of course, that in which all elements yield identically. Plotting the applied stress versus the elongation strain $\bar{\varepsilon} = u/L$ the material behaviour curve is simply obtained identically as shown in Fig. 4.23(b) ($h/L = 1$). However, it is equally possible that after reaching the maximum stress σ_y only one element (probably one with infinitesimally smaller yield stress owing to computer round-off) continues into the plastic range while all the others unload elastically. The total elongation strain is now given by

$$\bar{\varepsilon} = \frac{u}{L} = \frac{\sigma}{E} - \frac{h(\sigma - \sigma_y)}{LH} \tag{4.182}$$

and as h tends to 0 then $\bar{\varepsilon}$ tends to σ/E. Clearly, a multitude of solutions is possible for any arbitrary element subdivision and in this trivial example a unique finite element solution is impossible (with localization to a single element always occurring). Further, the above simple 'thought experiment' points to another unacceptable paradox implying the inadmissibility of the softening model specified with constant softening modulus. The difficulties are as follows.

1. The behaviour seems to depend on the size (h) of the subdivision chosen (also called a *mesh sensitive* result). Clearly this is unacceptable physically.
2. If the element size falls below a value given by $h = HL/E$ only a catastrophic, brittle, behaviour is possible without involving an unacceptable energy gain.

Similar difficulties can arise with non-associated plasticity which exhibits occasionally an effectively strain softening behaviour in some circumstances (see reference 130).

The computational difficulties can be overcome to some extent by introducing viscoplasticity as a start to any computation. Such *regularization* was introduced as early as 1974[95] and was considered seriously by De Borst and co-authors.[131] However, most of the difficulties remain as steady state is approached.

The problem remains a serious line of research but two possible alternative treatments have emerged. The first of these is physically difficult to accept but is very effective in practice. This is the concept of properties which are labelled as *non-local*. In such an approach the softening modulus is made dependent on the element size. Many authors have contributed here, with the earliest being Bazant and co-workers.[125,126] Other relevant references are 131 and 132.

The second approach, that of a *concentrated discontinuity*, is more elegant but, we believe, computationally more difficult. It was first suggested by Simo, Oliver and Armero in 1993[133] and extended in later publications.[134-137]

Both approaches allow strain and indeed displacement discontinuities to develop following the brittle failure behaviour on which we have already remarked. In the numerical application this limit is *approached* as element size decreases or alternatively when stress singularities, such as corners, trigger this type of behaviour.

In the second approach, continuous plastic behaviour is not permitted and all action is concentrated on discontinuity lines which have to be suitably placed.

A particular form of the non-local approach is illustrated in Fig. 4.24. Here we examine in detail a unit width of an element in which the displacement discontinuity is approximated. In the examples which we shall consider later this discontinuity is a slip one with the 'failure' being modelled as shown. However, an identical approach has been used to model strain softening behaviour of concrete in cracking.[125,126]

The most basic form of non-local behaviour assumes that the work (or energy) expended in achieving the discontinuity must be the same whatever the dimension h of the element. This work is equal to

$$\frac{1}{2}\sigma_y \varepsilon_y h \approx \frac{1}{2}\sigma_y^2 \frac{h}{H} = \frac{1}{2}\sigma_y \Delta U \qquad (4.183)$$

If this work is to be identical in all highly strained elements we will require that

$$\frac{H}{h} = \text{constant} \qquad (4.184)$$

Such a requirement is easy to apply in an adaptive refinement process.

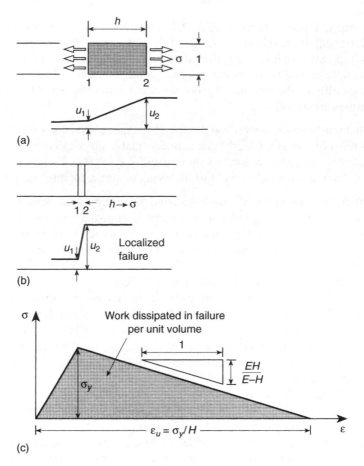

Fig. 4.24 Illustration of a non-local approach (work dissipation in failure is assumed to be constant for all elements): (a) an element in which localization is considered; (b) localization; (c) stress–strain curve showing work dissipated in failure.

At this stage we can comment on the concentrated discontinuity approach of Bazant and co-workers.[125,126] In this we shall simply assume that the displacement increment of Eq. (4.183), that is, ΔU, is permitted to occur only on a discontinuity line and that its magnitude is strictly related to the energy density previously specified in Eq. (4.183). After considering the effects of large deformation we shall show in Sec. 6.7.2 how a very effective treatment and capture of discontinuity can be made adaptively.

4.12 Non-linear quasi-harmonic field problems

Non-linearity may arise in many problems beyond those of solid mechanics, but the techniques described in this chapter are still universally applicable. Here we shall look again at one class of problems which is governed by the quasi-harmonic field equations of Chapter 2.

In some formulations it is assumed that

$$\mathbf{q} = -k(\phi)\nabla\phi \qquad (4.185)$$

which gives, then (with use of definitions from Sec. 2.6),

$$\mathbf{P}_q = \mathbf{H}(\phi)\phi \qquad (4.186)$$

where now $\mathbf{H}$ has the familiar form

$$\mathbf{H} = \int_\Omega (\nabla\mathbf{N})^{\mathrm{T}} k(\phi)\nabla\mathbf{N}\,\mathrm{d}\Omega \qquad (4.187)$$

In this form the general non-linear problem may be solved by direct iteration methods; however, as these often fail to converge it is frequently necessary to use a scheme for which a tangential matrix to Ψ is required, as presented in Sec. 3.2.4 [see Eq. (3.25)]. The tangent for the form given by Eq. (4.185) is generally unsymmetric; however, special forms can be devised which lead to symmetry.[138] In many physical problems, however, the values of k in Eq. (4.185) depend on the absolute value of the gradient of $\nabla\phi$, that is,

$$V = \sqrt{(\nabla\phi)^{\mathrm{T}}\nabla\phi}$$
$$\bar{k}' = \frac{\mathrm{d}k}{\mathrm{d}V} \qquad (4.188)$$

In such cases, we can write

$$\mathbf{H}_{\mathrm{T}} = \frac{\partial\mathbf{H}(\phi)\phi}{\partial\phi} = \mathbf{H} + \mathbf{A} \qquad (4.189)$$

where

$$\mathbf{A} = \int_\Omega (\nabla\mathbf{N})^{\mathrm{T}} \left[(\nabla\phi)^{\mathrm{T}}\bar{k}'\nabla\phi\right] \nabla\mathbf{N},\mathrm{d}\Omega \qquad (4.190)$$

and symmetry is preserved.

Situations of this kind arise in seepage flow where the permeability is dependent on the absolute value of the flow velocity,[140,141] in magnetic fields,[139,142–144] where magnetic response is a function of the absolute field strength, in slightly compressible fluid flow, and indeed in many other physical situations.[145] Figure 4.25 from reference 139 illustrates a typical non-linear magnetic field solution.

While many more interesting problems could be quoted we conclude with one in which the only non-linearity is that due to the heat generation term Q [see Chapter 2, Eq. (2.69)]. This particular problem of spontaneous ignition, in which Q depends exponentially on the temperature, serves to illustrate the point about the possibility of multiple solutions and indeed the non-existence of any solution in certain non-linear cases.[146]

Taking $k = 1$ and $Q = \bar{\delta}\exp\phi$, we examine an elliptic domain in Fig. 4.26. For various values of δ, a Newton iteration is used to obtain a solution, and we find that no convergence (and indeed *no solution*) exists when $\bar{\delta} > \bar{\delta}_{\mathrm{crit}}$ exists; above the critical

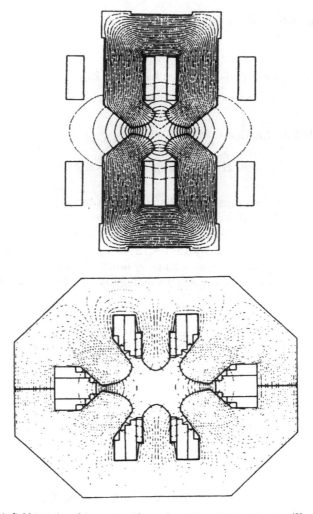

Fig. 4.25 Magnetic field in a six-pole magnet with non-linearity owing to saturation.[139]

value of $\bar{\delta}$ the temperature rises indefinitely and *spontaneous ignition* of the material occurs. For values below this, *two* solutions are possible and the starting point of the iteration determines which one is in fact obtained.

This last point illustrates that an insight into the problem is, in non-linear solutions, even more important than elsewhere.

4.13 Concluding remarks

In this chapter we have considered a number of classical constitutive equations together with numerical algorithms which permit their inclusion in the formulations discussed in Chapter 2. These permit the solution to a wide range of practical problems in solid mechanics and geomechanics. The possibilities for models of real materials is endless

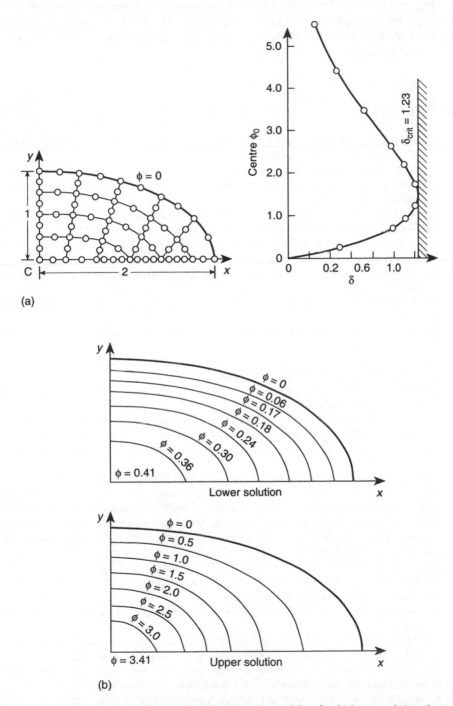

Fig. 4.26 A non-linear heat-generation problem illustrating the possibility of multiple or no solutions depending on the heat generation parameter $\bar{\delta}$; spontaneous combustion.[146] (a) Solution mesh and variation of temperature at point C; (b) two possible temperature distributions for $\bar{\delta} = 0.75$.

and, thus, we have not been able to include many of the extensions available in the literature. For example, the effect of temperature changes will normally affect the material behaviour through both thermal expansions as well as through the change in the material parameters.

References

1. O.C. Zienkiewicz, M. Watson and I.P. King. A numerical method of visco-elastic stress analysis. *Int. J. Mech. Sci*, 10:807–827, 1968.
2. J.L. White. Finite elements in linear viscoelastic analysis. In *Proc. 2nd Conf. Matrix Methods in Structural Mechanics*, volume AFFDL-TR-68-150, pages 489–516. Wright Patterson Air Force Base, Ohio, October 1968.
3. R.L. Taylor, K.S. Pister and G.L. Goudreau. Thermomechanical analysis of viscoelastic solids. *International Journal for Numerical Methods in Engineering*, 2:45–79, 1970.
4. O.C. Zienkiewicz. *The Finite Element Method in Engineering Science*. McGraw-Hill, London, 2nd edition, 1971.
5. O.C. Zienkiewicz and R.L. Taylor. *The Finite Element Method*, volume 2. McGraw-Hill, London, 4th edition, 1991.
6. B. Gross. *Mathematical Structure of the Theories of Viscoelasticity*. Herrmann & Cie, Paris, 1953.
7. R.M. Christensen. *Theory of Viscoelasticity: An Introduction*. Academic Press, New York, 1971 (reprinted 1991).
8. O.C. Zienkiewicz, R.L. Taylor and J.Z. Zhu. *The Finite Element Method: Its Basis and Fundamentals*. Butterworth-Heinemann, Oxford, 6th edition, 2005.
9. T. Belytschko and T.J.R. Hughes, editors. *Computational Methods for Transient Analysis*. North-Holland, Amsterdam, 1983.
10. L.R. Herrmann and F.E. Peterson. A numerical procedure for viscoelastic stress analysis. In *Proceedings 7th ICRPG Mechanical Behavior Working Group*, Orlando, FL, 1968.
11. D. McHenry. A new aspect of creep in concrete and its application to design. *Proc. ASTM*, 43:1064, 1943.
12. T. Alfrey. *Mechanical Behavior of High Polymers*. Interscience, New York, 1948.
13. H.H. Hilton and H.G. Russell. An extension of Alfrey's analogy to thermal stress problems in temperature dependent linear visco-elastic media. *J. Mech. Phys. Solids*, 9:152–164, 1961.
14. O.C. Zienkiewicz, M. Watson and Y.K. Cheung. Stress analysis by the finite element method – thermal effects. In *Proc. Conf. on Prestressed Concrete Pressure Vessels*, London, 1967. Institute of Civil Engineers.
15. J. Lubliner. *Plasticity Theory*. Macmillan, New York, 1990.
16. J.C. Simo and T.J.R. Hughes. *Computational Inelasticity*, volume 7 of *Interdisciplinary Applied Mathematics*. Springer-Verlag, Berlin, 1998.
17. R. von Mises. Mechanik der Plastischen Formänderung der Kristallen. *Z. angew. Math. Mech.*, 8:161–185, 1928.
18. D.C. Drucker. A more fundamental approach to plastic stress–strain solutions. In *Proceedings of the 1st U.S. National Congress on Applied Mechanics*, pages 487–491, 1951.
19. W.T. Koiter. Stress–strain relations, uniqueness and variational theorems for elastic–plastic materials with a singular yield surface. *Q. J. Appl. Math.*, 11:350–354, 1953.
20. R. Hill. *The Mathematical Theory of Plasticity*. Clarenden Press, Oxford, 1950.
21. W. Johnson and P.W. Mellor. *Plasticity for Mechanical Engineers*. Van Nostrand, New York, 1962.
22. W. Prager. *An Introduction to Plasticity*. Addison-Wesley, Reading, Mass., 1959.

23. W.F. Chen. *Plasticity in Reinforced Concrete*. McGraw-Hill, New York, 1982.

24. D.C. Drucker. Conventional and unconventional plastic response and representations. *Appl. Mech. Rev.*, 41:151–167, 1988.

25. G.C. Nayak and O.C. Zienkiewicz. Elasto-plastic stress analysis. Generalization for various constitutive relations including strain softening. *International Journal for Numerical Methods in Engineering*, 5:113–135, 1972.

26. W. Prager. A new method of analyzing stress and strains in work-hardening plastic solids. *J. Appl. Mech.*, 23:493–496, 1956.

27. H. Ziegler. A modification of Prager's hardening rule. *Q. Appl. Math.*, 17:55–65, 1959.

28. J. Mandel. Contribution théorique a l'étude de l'écrouissage et des lois de l'écoulement plastique. In *Proc. 11th Int. Cong. of Appl. Mech.*, pages 502–509, 1964.

29. J. Lubliner. A maximum-dissipation principle in generalized plasticity. *Acta Mechanica*, 52:225–237, 1984.

30. J.C. Simo. Topics on the numerical analysis and simulation of plasticity. In P.G. Ciarlet and J.L. Lions, editors, *Handbook of Numerical Analysis*, volume III, pages 183–499. Elsevier Science Publisher BV, 1999.

31. P.J. Armstrong and C.O. Frederick. A mathematical representation of the multi-axial Bauschinger effect. Technical Report RD/B/N731, C.E.G.B., Berkeley Nuclear Laboratories, R&D Department, 1965.

32. J.L. Chaboche. Constitutive equations for cyclic plasticity and cyclic visco-plasticity. *Int. J. Plasticity*, 5:247–302, 1989.

33. J.L. Chaboche. On some modifications of kinematic hardening to improve the description of ratcheting effects. *Int. J. Plasticity*, 7:661–678, 1991.

34. J.L. Lubliner. A simple model of generalized plasticity. *International Journal of Solids and Structures*, 28:769, 1991.

35. J.L. Lubliner. A new model of generalized plasticity. *International Journal of Solids and Structures*, 30:3171–3184, 1993.

36. F. Auricchio and R.L. Taylor. Two material models for cyclic plasticity: nonlinear kinematic hardening and generalized plasticity. *Int. J. Plasticity*, 11:65–98, 1995.

37. J.F. Besseling. A theory of elastic, plastic and creep deformations of an initially isotropic material. *J. Appl. Mech.*, 25:529–536, 1958.

38. Z. Mróz. An attempt to describe the behaviour of metals under cyclic loads using a more general work hardening model. *Acta Mech.*, 7:199, 1969.

39. O.C. Zienkiewicz, C.T. Chang, N. Bićanić and E. Hinton. Earthquake response of earth and concrete in the partial damage range. In *Proc. 13th Int. Cong. on Large Dams*, pages 1033–1047, New Delhi, 1979.

40. O.C. Zienkiewicz, G.C. Nayak and D.R.J. Owen. Composite and 'Overlay' models in numerical analysis of elasto-plastic continua. In A. Sawczuk, editor, *Foundations of Plasticity*, pages 107–122. Noordhoff, Dordrecht, 1972.

41. D.R.J. Owen, A. Prakash and O.C. Zienkiewicz. Finite element analysis of non-linear composite materials by use of overlay systems. *Computers and Structures*, 4:1251–1267, 1974.

42. Y. Yamada, N. Yishimura and T. Sakurai. Plastic stress–strain matrix and its application for the solution of elasto-plastic problems by the finite element method. *Int. J. Mech. Sci.*, 10:343–354, 1968.

43. O.C. Zienkiewicz, S. Valliappan and I.P. King. Elasto-plastic solutions of engineering problems. Initial stress, finite element approach. *International Journal for Numerical Methods in Engineering*, 1:75–100, 1969.

44. F.B. Hildebrand. *Introduction to Numerical Analysis*. Dover Publishers, 2nd edition, 1987.

45. G.C. Nayak and O.C. Zienkiewicz. Note on the 'alpha'-constant stiffness method for the analysis of nonlinear problems. *International Journal for Numerical Methods in Engineering*, 4:579–582, 1972.

46. G.C. Nayak. *Plasticity and large deformation problems by the finite element method.* PhD thesis, Department of Civil Engineering, University of Wales, Swansea, 1971.

47. E. Hinton and D.R.J. Owen. *Finite Element Programming.* Academic Press, New York, 1977.

48. R.D. Krieg and D.N. Krieg. Accuracy of numerical solution methods for the elastic, perfectly plastic model. *J. Pressure Vessel Tech., Trans. ASME*, 99:510–515, 1977.

49. G. Maenchen and S. Sacks. The tensor code. In B. Alder, editor, *Methods in Computational Physics*, volume 3, pages 181–210. Academic Press, New York, 1964.

50. M.L. Wilkins. Calculation of elastic–plastic flow. In B. Alder, editor, *Methods in Computational Physics*, volume 3, pages 211–263. Academic Press, New York, 1964.

51. J.C. Simo and R.L. Taylor. Consistent tangent operators for rate-independent elastoplasticity. *Computer Methods in Applied Mechanics and Engineering*, 48:101–118, 1985.

52. J.C. Simo and R.L. Taylor. A return mapping algorithm for plane stress elastoplasticity. *International Journal for Numerical Methods in Engineering*, 22:649–670, 1986.

53. I.H. Shames and F.A. Cozzarelli. *Elastic and Inelastic Stress Analysis.* Taylor & Francis, Washington, DC, 1997. (Revised printing.)

54. D.C. Drucker and W. Prager. Soil mechanics and plastic analysis or limit design. *Q. J. Appl. Math.*, 10:157–165, 1952.

55. O.C. Zienkiewicz and G.N. Pande. Some useful forms of isotropic yield surfaces for soil and rock mechanics. In G. Gudehus, editor, *Finite Elements in Geomechanics*, Chapter 5, pages 171–190. John Wiley & Sons, Chichester, 1977.

56. O.C. Zienkiewicz, V.A. Norris and D.J. Naylor. Plasticity and viscoplasticity in soil mechanics with special reference to cyclic loading problems. In *Proc. Int. Conf. on Finite Elements in Nonlinear Solid and Structural Mechanics*, volume 2, Trondheim, Tapir Press, 1977.

57. O.C. Zienkiewicz, V.A. Norris, L.A. Winnicki, D.J. Naylor and R.W. Lewis. A unified approach to the soil mechanics problems of offshore foundations. In O.C. Zienkiewicz, R.W. Lewis and K.G. Stagg, editors, *Numerical Methods in Offshore Engineering*, Chapter 12. John Wiley & Sons, Chichester, 1978.

58. F. Darve. An incrementally nonlinear constitutive law of second order and its application to localization. In C.S. Desai and R.H. Gallagher, editors, *Mechanics of Engineering Materials*, Chapter 9, pages 179–196. John Wiley & Sons, Chichester, 1984.

59. G.H. Golub and C.F. Van Loan. *Matrix Computations.* The Johns Hopkins University Press, Baltimore MD, 3rd edition, 1996.

60. Z. Mróz and O.C. Zienkiewicz. Uniform formulation of constitutive equations for clays and sands. In C.S. Desai and R.H. Gallagher, editors, *Mechanics of Engineering Materials*, Chapter 22, pages 415–450. John Wiley & Sons, Chichester, 1984.

61. O.C. Zienkiewicz and Z. Mróz. Generalized plasticity formulation and applications to geomechanics. In C.S. Desai and R.H. Gallagher, editors, *Mechanics of Engineering Materials*, Chapter 33, pages 655–680. John Wiley & Sons, Chichester, 1984.

62. O.C. Zienkiewicz, K.H. Leung and M. Pastor. Simple model for transient soil loading in earthquake analysis: Part I – basic model and its application. *International Journal for Numerical Analysis Methods in Geomechanics*, 9:453–476, 1985.

63. O.C. Zienkiewicz, S. Qu, R.L. Taylor and S. Nakazawa. The patch test for mixed formulations. *International Journal for Numerical Methods in Engineering*, 23:1873–1883, 1986.

64. M. Pastor and O.C. Zienkiewicz. A generalized plasticity, hierarchical model for sand under monotonic and cyclic loading. In G.N. Pande and W.F. Van Impe, editors, *Proc. Int. Symp. on Numerical Models in Geomechanics*, pages 131–150, Jackson, England. Ghent, 1986.

65. M. Pastor and O.C. Zienkiewicz. Generalized plasticity and modelling of soil behaviour. *International Journal for Numerical Analysis Methods in Geomechanics*, 14:151–190, 1990.

66. Y. Defalias and E.P. Popov. A model of nonlinear hardening material for complex loading. *Acta Mechanica*, 21:173–192, 1975.

67. T. Tatsueka and K. Ishihera. Yielding of sand in triaxial compression. *Soil Found.*, 14:63–76, 1974.
68. O.C. Zienkiewicz, A.H.C. Chan, M. Pastor, B.A. Schrefler and T. Shiomi. *Computational Geomechanics: With Special Reference to Earthquake Engineering*. John Wiley & Sons, Chichester, 1999.
69. O.C. Zienkiewicz, A.H.C. Chan, M. Pastor, D.K. Paul and T. Shiomi. Static and dynamic behaviour of soils: a rational approach to quantitative solutions, I. *Proceedings Royal Society of London*, 429:285–309, 1990.
70. O.C. Zienkiewicz, Y.M. Xie, B.A. Schrefler, A. Ledesma and N. Bićanić. Static and dynamic behaviour of soils: a rational approach to quantitative solutions, II. *Proceedings Royal Society of London*, 429:311–321, 1990.
71. J.C. Nagtegaal, D.M. Parks and J.R. Rice. On numerical accurate finite element solutions in the fully plastic range. *Computer Methods in Applied Mechanics and Engineering*, 4:153–177, 1974.
72. R.M. McMeeking and J.R. Rice. Finite element formulations for problems of large elastic-plastic deformation. *Int. J. Solids Struct.*, 11:601–616, 1975.
73. E.F. Rybicki and L.A. Schmit. An incremental complementary energy method of non-linear stress analysis. *Journal of AIAA*, 8:1105–1112, 1970.
74. R.H. Gallagher and A.K. Dhalla. Direct flexibility finite element elasto-plastic analysis. In *Proceedings 1st International Conference on Structural Mechanics in Reactor Technology*, volume 6, Berlin, 1971.
75. J.A. Stricklin, W.E. Haisler and W. von Reisman. Evaluation of solution procedures for material and/or geometrically non-linear structural analysis. *Journal of AIAA*, 11:292–299, 1973.
76. P.S. Theocaris and E. Marketos. Elastic–plastic analysis of perforated thin strips of a strain-hardening material. *J. Mech. Phys. Solids*, 12:377–390, 1964.
77. P.V. Marçal and I.P. King. Elastic–plastic analysis of two-dimensional stress systems by the finite element method. *Int. J. Mech. Sci.*, 9:143–155, 1967.
78. K.S. Dinno and S.S. Gill. An experimental investigation into the plastic behaviour of flush nozzles in spherical pressure vessels. *Int. J. Mech. Sci.*, 7:817, 1965.
79. A. Mendelson, M.H. Hirschberg and S.S. Manson. A general approach to the practical solution of creep problems. *Trans. ASME, J. Basic Eng.*, 81:85–98, 1959.
80. O.C. Zienkiewicz and Y.K. Cheung. *The Finite Element Method in Structural Mechanics*. McGraw-Hill, London, 1967.
81. G.A. Greenbaum and M.F. Rubinstein. Creep analysis of axisymmetric bodies using finite elements. *Nucl. Eng. Des.*, 7:379–397, 1968.
82. P.V. Marçal. Selection of creep strains for explicit calculations. Private communication, 1972.
83. N.A. Cyr and R.D. Teter. Finite element elastic plastic creep analysis of two-dimensional continuum with temperature dependent material properties. *Computers and Structures*, 3:849–863, 1973.
84. O.C. Zienkiewicz. Visco-plasticity, plasticity, creep and visco-plastic flow (problems of small, large and continuing deformation). In *Computational Mechanics*, TICOM Lecture Notes on Mathematics 461. Springer-Verlag, Berlin, 1975.
85. M.B. Kanchi, D.R.J. Owen and O.C. Zienkiewicz. The visco-plastic approach to problems of plasticity and creep involving geometrically non-linear effects. *International Journal for Numerical Methods in Engineering*, 12:169–181, 1978.
86. P. Perzyna. Fundamental problems in viscoplasticity. *Advances in Applied Mechanics*, 9:243–377, 1966.
87. T.J.R. Hughes and R.L. Taylor. Unconditionally stable algorithms for quasi-static elasto/visco-plastic finite element analysis. *Computers and Structures*, 8:169–173, 1978.
88. E.C. Bingham. *Fluidity and Plasticity*, Chapter VIII, pages 215–218. McGraw-Hill, New York, 1922.

89. P. Perzyna. Thermodynamic theory of viscoplasticity. In *Advances in Applied Mechanics*, volume 11. Academic Press, New York, 1971.

90. G. Duvaut and J.-L. Lions. *Inequalities in Mechanics and Physics*. Springer-Verlag, Berlin, 1976.

91. O.C. Zienkiewicz and I.C. Cormeau. Visco-plasticity solution by finite element process. *Arch. Mech.*, 24:873–888, 1972.

92. J. Zarka. Généralisation de la théorie du potential multiple en viscoplasticité. *J. Mech. Phys. Solids*, 20:179–195, 1972.

93. Q.A. Nguyen and J. Zarka. Quelques méthodes de resolution numérique en elastoplasticité classique et en elasto-viscoplasticité. *Sciences et Technique de l'Armement*, 47:407–436, 1973.

94. O.C. Zienkiewicz and I.C. Cormeau. Visco-plasticity and plasticity. An alternative for finite element solution of material non-linearities. In *Proc. Colloque Methodes Calcul. Sci. Tech.*, pages 171–199, Paris, IRIA, 1973.

95. O.C. Zienkiewicz and I.C. Cormeau. Visco-plasticity, plasticity and creep in elastic solids – a unified numerical solution approach. *International Journal for Numerical Methods in Engineering*, 8:821–845, 1974.

96. I.C. Cormeau. Numerical stability in quasi-static elasto-visco-plasticity. *International Journal for Numerical Methods in Engineering*, 9:109–128, 1975.

97. W.T. Koiter. General theorems for elastic–plastic solids. In I.N. Sneddon and R. Hill, editors, *Progress in Solid Mechanics*, volume 6, Chapter IV, pages 167–221. North Holland, Amsterdam, 1960.

98. F.A. Leckie and J.B. Martin. Deformation bounds for bodies in a state of creep. *Trans. ASME, J. Appl. Mech.*, 34:411–417, 1967.

99. I. Finnie and W.R. Heller. *Creep of Engineering Materials*. McGraw-Hill, New York, 1959.

100. O.C. Zienkiewicz, C. Humpheson and R.W. Lewis. Associated and non-associated visco-plasticity and plasticity in soil mechanics. *Geotechnique*, 25:671–689, 1975.

101. G.N. Pande and O.C. Zienkiewicz, editors. *Soil Mechanics – Transient and Cyclic Loads*. John Wiley & Sons, Chichester, 1982.

102. C.S. Desai and R.H. Gallagher, editors. *Mechanics of Engineering Materials*. John Wiley & Sons, Chichester, 1984.

103. O.C. Zienkiewicz, S. Valliappan and I.P. King. Stress analysis of rock as a 'no tension' material. *Geotechnique*, 18:56–66, 1968.

104. S. Valliappan and P. Nath. Tensile crack propagation in reinforced concrete beams by finite element techniques. In *Int. Conf. on Shear, Torsion and Bond in Reinforced Concrete*, Coimbatore, India, January 1969.

105. N.W. Krahl, W. Khachaturian and C.P. Seiss. Stability of tensile cracks in concrete beams. *Proc. Am. Soc. Civ. Eng.*, 93(ST1):235–254, 1967.

106. D.V. Phillips and O.C. Zienkiewicz. Finite element non-linear analysis of concrete structures. *Proc. Inst. Civ. Eng., Pt. 2*, 61:59–88, 1976.

107. O.C. Zienkiewicz, D.V. Phillips and D.R.J. Owens. Finite element analysis of some concrete non-linearities. Theories and examples. In *IABSE Symp. on Concrete Structures to Triaxial Stress*, Bergamo, 17–19 May 1974.

108. J.G. Rots and J. Blaauwendraad. Crack models for concrete: discrete or smeared? Fixed, multidirectional or rotating? *Heron*, 34:1–59, 1989.

109. J. Isenberg, editor. *Finite Element Analysis of Concrete Structures*. American Society of Civil Engineers, New York, 1993. ISBN 0–87262–983–X.

110. G. Hoffstetter and H. Mang. *Computational Mechanics of Reinforced Concrete Structures*. Vieweg & Sohn, Braunschweig, 1995. ISBN 3–528–06390–4.

111. R. De Borst, N. Bićanić, H. Mang and G. Meschke, editors. *Computational Modelling of Concrete Structures*, volumes 1 and 2. Balkema, Rotterdam, 1998. ISBN 90–5410–9467.

112. R.E. Goodman, R.L. Taylor and T. Brekke. A model for the mechanics of jointed rock. *J. Soil Mech and Foundation Div., ASCE.*, 94(SM3):637–659, 1968.

113. O.C. Zienkiewicz and B. Best. Some non-linear problems in soil and rock mechanics – finite element solutions. In *Conf. on Rock Mechanics*, University of Queensland, Townsville, 1969.

114. O.C. Zienkiewicz and G.N. Pande. Time dependent multilaminate models for rock – a numerical study of deformation and failure of rock masses. *International Journal for Numerical Analysis Methods in Geomechanics*, 1:219–247, 1977.

115. G.N. Pande and K.G. Sharma. Multi-laminate model of clays – a numerical evaluation of the influence of rotation of the principal stress axes. *International Journal for Numerical Analysis Methods in Geomechanics*, 7:397–418, 1983.

116. J. Mandel. Conditions de stabilité et postulat de Drucker. In *Proc. IUTAM Symp. on Rheology and Soil Mechanics*, pages 58–68. Springer-Verlag, Berlin, 1964.

117. J.W. Rudnicki and J.R. Rice. Conditions for the localization of deformations in pressure sensitive dilatant materials. *J. Mech. Phys. Solids*, 23:371–394, 1975.

118. J.R. Rice. The localization of plastic deformation. In W.T. Koiter, editor, *Theoretical and Applied Mechanics*, pages 207–220. North Holland, Amsterdam, 1977.

119. S.T. Pietruszczak and Z. Mròz. Finite element analysis of strain softening materials. *International Journal for Numerical Methods in Engineering*, 10:327–334, 1981.

120. S.W. Sloan and J.R. Booker. Removal of singularities in Tresca and Mohr–Coulomb yield functions. *Comm. App. Num. Meth.*, 2:173–179, 1986.

121. N. Bićanić, E. Pramono, S. Sture and K.J. Willam. On numerical prediction of concrete fracture localizations. In *Proc. NUMETA Conf.*, pages 385–392. Balkema, Rotterdam, 1985.

122. J.C. Simo. Strain softening and dissipation. A unification of approaches. In *Proc. NUMETA Conf.*, pages 440–461, 1985.

123. M. Ortiz, Y. Leroy and A. Needleman. A finite element method for localized failure analysis. *Computer Methods in Applied Mechanics and Engineering*, 61:189–214, 1987.

124. A. Needleman. Material rate dependence and mesh sensitivity in localization problems. *Computer Methods in Applied Mechanics and Engineering*, 67:69–85, 1988.

125. Z.P. Bažant and F.B. Lin. Non-local yield limit degradation. *International Journal for Numerical Methods in Engineering*, 26:1805–1823, 1988.

126. Z.P. Bažant and G. Pijaudier-Cabot. Non linear continuous damage, localization instability and convergence. *J. Appl. Mech.*, 55:287–293, 1988.

127. J. Mazars and Z.P. Basant, editors. *Cracking and Damage*. Elsevier Press, Dordrecht, 1989.

128. Y. Leroy and M. Ortiz. Finite element analysis of strain localization in frictional materials. *International Journal for Numerical Analysis Methods in Geomechanics*, 13:53–74, 1989.

129. T. Belytschko, J. Fish and A. Bayless. The spectral overlay on finite elements for problems with high gradients. *Computer Methods in Applied Mechanics and Engineering*, 81:71–89, 1990.

130. G.N. Pande and S. Pietruszczak. Symmetric tangent stiffness formulation for non-associated plasticity. *Computers and Geotechniques*, 2:89–99, 1986.

131. R. De Borst, L.J. Sluys, H.B. Hühlhaus and J. Pamin. Fundamental issues in finite element analysis of localization of deformation. *Engineering Computations*, 10:99–121, 1993.

132. T. Belytschko and M. Tabbara. *h*-adaptive finite element methods for dynamic problems, with emphasis on localization. *International Journal for Numerical Methods in Engineering*, 36:4245–4265, 1994.

133. J. Simo, J. Oliver and F. Armero. An analysis of strong discontinuities induced by strain-softening in rate-independent inelastic solids. *Comp. Mech.*, 12:277–296, 1993.

134. J. Simo and J. Oliver. A new approach to the analysis and simulations of strong discontinuities. In Z.P. Bažant *et al.*, editors, *Fracture and Damage in Quasi-brittle Structures*, pages 25–39. E & FN Spon, London, 1994.

135. J. Oliver. Modelling strong discontinuities in solid mechanics via strain softening constitutive equations. Part 1 Fundamentals. *International Journal for Numerical Methods in Engineering*, 39:3575–3600, 1996.

136. J. Oliver. Modelling strong discontinuities in solid mechanics via strain softening constitutive equations. Part 2 Numerical simulation. *International Journal for Numerical Methods in Engineering*, 39:3601–3623, 1996.

137. J. Oliver, M. Cervera and O. Manzoli. Strong discontinuities and continuum plasticity models: the strong discontinuity approach. *Int. J. Plasticity*, 15:319–351, 1999.

138. M. Muscat. *The Flow of Homogeneous Fluids through Porous Media*. Edwards, London, 1964.

139. A.M. Winslow. Numerical solution of the quasi-linear Poisson equation in a non-uniform triangle 'mesh'. *J. Comp. Phys.*, 1:149–172, 1966.

140. R.E. Volker. Non-linear flow in porous media by finite elements. *Proc. Am. Soc. Civ. Eng.*, 95(HY6), 1969.

141. H. Ahmed and D.K. Suneda. Non-linear flow in porous media by finite elements. *Proc. Am. Soc. Civ. Eng.*, 95(HY6):1847–1859, 1969.

142. M.V.K. Chari and P. Silvester. Finite element analysis of magnetically saturated D.C. motors. In *IEEE Winter Meeting on Power*, New York, 1971.

143. J.F. Lyness, D.R.J. Owen and O.C. Zienkiewicz. The finite element analysis of engineering systems governed by a non-linear quasi-harmonic equation. *Computers and Structures*, 5:65–79, 1975.

144. O.C. Zienkiewicz, J.F. Lyness and D.R.J. Owen. Three-dimensional magnetic field determination using a scalar potential. A finite element solution. *IEEE Transactions on Magnetics*, MAG-13(5):1649–1656, 1977.

145. D. Gelder. Solution of the compressible flow equations. *International Journal for Numerical Methods in Engineering*, 3:35–43, 1971.

146. C.A. Anderson and O.C. Zienkiewicz. Spontaneous ignition: finite element solutions for steady and transient conditions. *Trans ASME, J. Heat Transfer*, pages 398–404, 1974.

5

Geometrically non-linear problems – finite deformation

5.1 Introduction

In all our previous discussion we have assumed that deformations remained small so that linear relations could be used to represent the strain in a body. We now admit the possibility that deformations can become large during a loading process. In such cases it is necessary to distinguish between the *reference* configuration where initial shape of the body or bodies to be analysed is known and the *current* or *deformed* configuration after loading is applied. Figure 5.1 shows the two configurations and the coordinate frames which will be used to describe each one. We note that the deformed configuration of the body is unknown at the start of an analysis and, therefore, must be determined as part of the solution process – a process that is inherently non-linear. The relationships describing the finite deformation behaviour of solids involve equations related to both the reference and the deformed configurations. We shall generally find that such relations are most easily expressed using indicial notation (e.g. see Chapter 1, Sec. 1.2.1 or Appendix B of reference 1); however, after these indicial forms are developed we shall again return to a matrix form to construct the finite element approximations.

The chapter starts by describing the basic kinematic relations used in finite deformation solid mechanics. This is followed by a summary of different stress and traction measures related to the reference and deformed configurations, a statement of boundary and initial conditions, and a brief overview of material constitution for finite elastic solids. A variational Galerkin statement for the finite elastic material is then given in the reference configuration. Using the variational form the problem is then cast into a matrix form and a standard finite element solution process is indicated. The procedure up to this point is based on equations related to the reference configuration. A transformation to a form related to the current configuration is performed and it is shown that a much simpler statement of the finite element formulation process results – one which again permits separation into a form for treating nearly incompressible situations.

A mixed variational form is introduced and the solution process for problems which can have nearly incompressible behaviour is presented. This follows closely the developments for the small strain form given in Chapter 2. An alternative to the mixed form may also be given in the form of an enhanced strain model.[2-5] Here a fully

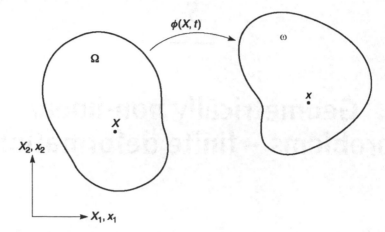

Fig. 5.1 Reference and deformed (current) configuration for finite deformation problems.

mixed construction is shown and leads to a form which performs well in two- and three-dimensional problems.

In finite deformation problems, loads can be given relative to the deformed configuration. An example is a pressure loading which always remains normal to a deformed surface. Here we discuss this case and show that by using finite element-type constructions a very simple result follows. Since the loading is no longer derivable from a potential function (i.e. conservative) the tangent matrix for the formulation is unsymmetric, leading in general to a requirement of an unsymmetric solver in a Newton solution scheme.

The next three chapters concentrate on finite deformation forms for continuum problems where finite elements are used to discretize the problem in all directions modelled. In later chapters we shall consider forms for problems which have one (or more) small dimension(s) and thus can benefit from use of rod, plate and shell formulations of the type we shall discuss in Chapters 10 to 17 of this volume for small deformation situations.

5.2 Governing equations

5.2.1 Kinematics and deformation

The basic equations for finite deformation solid mechanics may be found in standard references on the subject.[6–10] We begin by presenting a summary of the basic equations in three dimensions. Later we will specialize the equations to two dimensions to study problems of plane strain, plane stress and axisymmetry.

A body has material points whose positions are given by the vector $\mathbf{X}$ in a fixed reference configuration, Ω, in a three-dimensional space. In Cartesian coordinates the position vector is described in terms of its components as:

$$\mathbf{X} = X_I \mathbf{E}_I; \qquad I = 1, 2, 3 \tag{5.1}$$

where $\mathbf{E}_I$ are unit orthogonal base vectors and summation convention is used for repeated indices of like kind (e.g. I). After the body is loaded each material point is described by its position vector, $\mathbf{x}$, in the *current* deformed configuration, ω. The position vector in the current configuration is given in terms of its Cartesian components as

$$\mathbf{x} = x_i \mathbf{e}_i; \qquad i = 1, 2, 3 \tag{5.2}$$

where $\mathbf{e}_i$ are unit base vectors for the current time, t, and again summation convention is used.[*] In our discussion, common origins and directions of the reference and current coordinates are used for simplicity. Furthermore, in a Cartesian system base vectors do not change with position and all derivations may be made using components of tensors written in indicial form. Final equations are written in matrix form using standard transformations described in Chapter 2 (see also Appendix B of reference 1).

The position vector at the current time is related to the position vector in the reference configuration through the mapping

$$x_i = \phi_i(X_I, t) \tag{5.3}$$

Determination of ϕ_i is required as part of any solution and is analogous to the displacement vector, which we introduce next. When common origins and directions for the coordinate frames are used, a displacement vector may be introduced as the change between the two frames. Accordingly,

$$x_i = \delta_{iI}(X_I + U_I) \tag{5.4}$$

where summation convention is implied over indices of the same kind and δ_{iI} is a shifter between the two coordinate frames, and is defined by a Kronecker delta quantity such that

$$\delta_{iI} = \begin{cases} 1 & \text{if } i = I \\ 0 & \text{if } i \neq I \end{cases} \tag{5.5}$$

The shifter satisfies the relations

$$\delta_{iI}\delta_{iJ} = \delta_{IJ} \qquad \text{and} \qquad \delta_{iI}\delta_{jI} = \delta_{ij} \tag{5.6}$$

where δ_{IJ} and δ_{ij} are Kronecker delta quantities in the reference and current configuration, respectively. Using the shifter, a displacement component may be written with respect to either the reference configuration or the current configuration and related through

$$u_i = \delta_{iI}U_I \qquad \text{and} \qquad U_I = \delta_{iI}u_i \tag{5.7}$$

and we observe that numerically $u_1 = U_1$, etc. Thus, either may be used equally to express finite element parameters.

A fundamental measure of deformation is described by the deformation gradient relative to X_I given by

$$F_{iI} = \frac{\partial x_i}{\partial X_I} = \frac{\partial \phi_i}{\partial X_I} \tag{5.8a}$$

[*] As much as possible we adopt the notation that upper-case letters refer to quantities defined in the *reference* configuration and lower-case letters to quantities defined in the *current* deformed configuration. Exceptions occur when quantities are related to both the reference and the current configurations.

subject to the constraint

$$J = \det F_{iI} > 0 \tag{5.8b}$$

to ensure that material volume elements remain positive. The deformation gradient is a direct measure which maps a differential line element in the reference configuration into one in the current configuration as (Fig. 5.1)

$$dx_i = \frac{\partial \phi_i}{\partial X_I} dX_I = F_{iI} dX_I \tag{5.9}$$

Thus, it may be used to determine the change in length and direction of a differential line element. The determinant of the deformation gradient also maps a differential volume element in the reference configuration into one in the current configuration, that is

$$d\omega = J d\Omega \tag{5.10}$$

where $d\Omega$ is a differential volume element in the reference configuration and $d\omega$ its corresponding form in the current configuration.

The deformation gradient may be expressed in terms of the displacement as

$$F_{iI} = \delta_{iI} + \frac{\partial u_i}{\partial X_I} = \delta_{iI} + u_{i,I} \tag{5.11}$$

and is a *two-point tensor* since it is referred to both the reference and the current configurations. Expanding the terms in Eq. (5.11), the deformation gradient components are given by

$$F_{iI} = \begin{bmatrix} F_{11} & F_{12} & F_{13} \\ F_{21} & F_{22} & F_{23} \\ F_{31} & F_{32} & F_{33} \end{bmatrix} = \begin{bmatrix} (1 + u_{1,1}) & u_{1,2} & u_{1,3} \\ u_{2,1} & (1 + u_{2,2}) & u_{2,3} \\ u_{3,1} & u_{3,2} & (1 + u_{3,3}) \end{bmatrix} \tag{5.12}$$

Using F_{iI} directly complicates the development of constitutive equations and it is common to introduce deformation measures which are completely related to either the reference or the current configurations. For the reference configuration, the right Cauchy–Green deformation tensor, C_{IJ}, is introduced as

$$C_{IJ} = F_{iI} F_{iJ} \tag{5.13}$$

Alternatively the Green strain tensor, E_{IJ}, given as

$$E_{IJ} = \tfrac{1}{2}(C_{IJ} - \delta_{IJ}) \tag{5.14}$$

may be used. The Green strain may be expressed in terms of the displacements as

$$\begin{aligned} E_{IJ} &= \frac{1}{2}\left[\delta_{iI}\frac{\partial u_i}{\partial X_J} + \delta_{iJ}\frac{\partial u_i}{\partial X_I} + \frac{\partial u_i}{\partial X_I}\frac{\partial u_i}{\partial X_J}\right] \\ &= \frac{1}{2}\left[\delta_{iI} u_{i,J} + \delta_{iJ} u_{i,I} + u_{i,I} u_{i,J}\right] \\ &= \frac{1}{2}\left[U_{I,J} + U_{J,I} + U_{K,I} U_{K,J}\right] \end{aligned} \tag{5.15}$$

In the current configuration a common deformation measure is the left Cauchy–Green deformation tensor, b_{ij}, expressed as

$$b_{ij} = F_{iI} F_{jI} \tag{5.16}$$

The Almansi strain tensor, e_{ij}, is related to the inverse of b_{ij} as

$$e_{ij} = \tfrac{1}{2}(\delta_{ij} - b_{ij}^{-1}) \tag{5.17}$$

or inverting by

$$b_{ij} = (\delta_{ij} - 2e_{ij})^{-1} \tag{5.18}$$

Generally, the Almansi strain tensor will not appear naturally in our constitutive equations and, thus, we will often use expressions in terms of b_{ij} directly.

5.2.2 Stress and traction for reference and deformed states

Stress measures

Stress measures the amount of force per unit of area. In finite deformation problems care must be taken to describe the configuration to which a stress is measured. The Cauchy (true) stress, σ_{ij}, and the Kirchhoff stress, τ_{ij}, are symmetric measures of stress defined with respect to the current configuration. They are related through the determinant of the deformation gradient as

$$\tau_{ij} = J \, \sigma_{ij} \tag{5.19}$$

and often are the stresses used to define general constitutive equations for materials. Notationally, the first stress subscript defines the direction of a normal to the area on which the force acts and the second the direction of the force component. The second Piola–Kirchhoff stress, S_{IJ}, is a symmetric stress measure with respect to the reference configuration and is related to the Kirchhoff stress through the deformation gradient as

$$\tau_{ij} = F_{iI} S_{IJ} F_{jJ} \tag{5.20}$$

Finally, one can introduce the (unsymmetric) first Piola–Kirchhoff stress, P_{iI}, which is related to S_{IJ} through

$$P_{iI} = F_{iJ} S_{JI} \tag{5.21}$$

and to the Kirchhoff stress by

$$\tau_{ij} = P_{iI} F_{jI} \tag{5.22}$$

Traction measures

For the current configuration *traction* is given by

$$t_i = \sigma_{ji} n_j \tag{5.23}$$

where n_j are direction cosines of a unit outward pointing normal to a deformed surface. This form of the traction may be related to a reference surface quantity through *force* relations defined as

$$t_i \, d\gamma = \delta_{iI} T_I \, d\Gamma \tag{5.24}$$

where $d\gamma$ and $d\Gamma$ are surface area elements in the current and reference configurations, respectively, and T_I is traction on the reference configuration. Note that the direction of the traction component is preserved during the transformation and, thus, remains directly related to current configuration forces.

5.2.3 Equilibrium equations

Using quantities related to the current (deformed) configuration, the equilibrium equations for a solid subjected to finite deformation are nearly identical to those for small deformation. The local equilibrium equation (*balance of linear momentum*) is obtained as a force balance on a small differential volume of deformed solid and is given by[8-10]

$$\frac{\partial \sigma_{ij}}{\partial x_i} + \rho b_j^{(m)} = \rho \dot{v}_j \tag{5.25}$$

where ρ is mass density in the current configuration, $b_j^{(m)}$ is body force *per unit mass*, and v_j is the material velocity

$$v_j = \frac{\partial \phi_j}{\partial t} = \dot{x}_j = \dot{u}_j \tag{5.26}$$

The mass density in the current configuration may be related to the reference configuration (initial) mass density, ρ_0, using the *balance of mass* principle[8-10] and yields

$$\rho_0 = J\rho \tag{5.27}$$

Thus differences in the equilibrium equation from those of the small deformation case appear only in the body force and inertial force definitions.

Similarly, moment equilibrium on a small differential volume element of the deformed solid gives the *balance of angular momentum* requirement for the Cauchy stress as

$$\sigma_{ij} = \sigma_{ji} \tag{5.28}$$

which is identical to the result from the small deformation problem given in Eq. (2.9a).

The equilibrium requirements may also be written for the reference configuration using relations between stress measures and the chain rule of differentiation.[8] We will show the form for the balance of linear momentum when discussing the variational form for the problem. Here, however, we comment on the symmetry requirements for stress resulting from angular momentum balance. Using symmetry of the Cauchy stress tensor and Eqs (5.19) and (5.22) leads to the requirement on the first Piola–Kirchhoff stress

$$P_{iI} F_{jI} = F_{iI} P_{jI} \tag{5.29}$$

and subsequently, using Eq. (5.21), to the symmetry of the second Piola–Kirchhoff stress tensor

$$S_{IJ} = S_{JI} \tag{5.30}$$

5.2.4 Boundary conditions

As described in Chapter 2 the basic boundary conditions for a continuum body consist of two types: displacement boundary conditions and traction boundary conditions. Boundary conditions generally are defined on each part of the boundary by specifying components with respect to a local coordinate system defined by the orthogonal basis, e'_i, $i = 1, 2, 3$. Often one of the directions, say e'_3, coincides with the normal to the surface and the other two are in tangential directions along the surface. At each point on the boundary one (and only one) boundary condition must be specified for all three directions of the basis. These conditions can be all for displacements (fixed surface), all for tractions (stress or free surface), or a combination of displacements and tractions (mixed surface).

Displacement boundary conditions may be expressed for a component by requiring

$$x'_i = \bar{x}'_i \tag{5.31a}$$

at each point on the displacement boundary, γ_u. A quantity with a superposed bar, such as $\bar{x}'_i$ again denotes a specified quantity. The boundary condition may also be expressed in terms of components of the displacement vector, u_i. Accordingly, on γ_u

$$u'_i = \bar{u}'_i + \bar{g}'_i \tag{5.31b}$$

where $\bar{g}'_i$ is the initial gap $\delta_{iI}(\bar{X}'_I - X'_I)$. The second type of boundary condition is a traction boundary condition. Using the orthogonal basis described above, the traction boundary conditions may be given for each component by requiring

$$t'_i = \bar{t}'_i \tag{5.31c}$$

at each point on the boundary, γ_t. The boundary condition may be non-linear for loading such as pressure loads, as described later in Sec. 5.7.

Many other types of boundary conditions can exist and in Chapter 7 we discuss one, namely that of *contact*-type conditions.

5.2.5 Initial conditions

Initial conditions describe the state of a body at the start of an analysis. The conditions describe the initial kinematic and stress or strain states with respect to the reference configuration used to define the body. In addition, for constitutive equations with internal variables the initial values of terms which evolve in time must be given (e.g. initial plastic strain).

The initial conditions for the kinematic state consist of specifying the position and velocity at some initial time, commonly taken as zero. Accordingly,

$$x_i(X_I, 0) = \bar{\phi}_i(X_I, 0) \quad \text{or} \quad u_i(X_I, 0) = \bar{u}_i^0(X_I) \tag{5.32a}$$

and

$$v_i(X_I, 0) = \dot{\phi}_i(X_I, 0) = \bar{v}_i^0(X_I) \tag{5.32b}$$

are specified at each point in the body.

The initial conditions for stresses are specified as

$$\sigma_{ij}(X_I, 0) = \bar{\sigma}^0_{ij}(X_I) \tag{5.33}$$

at each point in the body. Finally, as noted above the internal variables in the stress–strain relations that evolve in time must have their initial conditions set. For a finite elastic model, generally there are no internal variables to be set unless initial stress effects are included.

5.2.6 Constitutive equations – hyperelastic material

We shall deal in more detail with constitutive equations for finite deformation materials in Chapter 6. Here we introduce the simplest form for an elastic material which can be used in a finite deformation formulation.

A *hyperelastic material* is one where the stress is determined solely from the current state of deformation as described in Chapter 1, Eq. (1.11). We recall that the form deduces the constitutive behaviour from a stored energy function, W, from which the second Piola–Kirchhoff stress is computed using[8,10]

$$S_{IJ} = 2\frac{\partial W}{\partial C_{IJ}} = \frac{\partial W}{\partial E_{IJ}} \tag{5.34}$$

The simplest representation of the stored energy function is the Saint-Venant–Kirchhoff model given by

$$W(E_{IJ}) = \tfrac{1}{2}C_{IJKL}E_{IJ}E_{KL} \tag{5.35}$$

where C_{IJKL} are *constant elastic moduli* defined in a manner similar to the small deformation ones. Equation (5.34) then gives

$$S_{IJ} = C_{IJKL}E_{KL} \tag{5.36}$$

for the stress–strain relation. While this relation is simple it is not adequate to define the behaviour of elastic finite deformation states. It is useful, however, for the case where strains are small but displacements (rotations) are large and we address this use later in Chapter 17. Other models for representing elastic behaviour at large strain are considered in Chapter 6.

An alternative to the above exists in which we consider the stored energy expressed in terms of the deformation gradient F_{iI} and deduce the first Piola–Kirchhoff stress from

$$P_{iI} = \frac{\partial \hat{W}(F_{jJ})}{\partial F_{iI}} \tag{5.37}$$

Now there are *nine* independent values of stress and deformation that need to be used. The construction of a constitutive relation is also more difficult to construct although one could consider Eq. (5.35) expressed as

$$W(E_{IJ}) = \hat{W}(F_{iI}) = \frac{1}{8}C_{IJKL}(F_{jI}F_{jJ} - 1)(F_{kK}F_{kL} - 1) \tag{5.38}$$

and perform the differentiation to obtain the stress. We will not follow this approach in this volume and leave to the reader the details of computing the derivatives expressed in Eq. (5.38) as well as the subsequent steps in a variational description.

5.3 Variational description for finite deformation

In order to construct finite element approximations for the solution of finite deformation problems it is necessary to write the formulation in a Galerkin (weak) or variational form as illustrated many times previously. Here again we can write these integral forms in either the reference configuration or in the current configuration. The simplest approach is to start from a reference configuration since here integrals are all expressed over *domains which do not change during the deformation process and thus are not affected by variation or linearization steps.* Later the results can be transformed and written in terms of the deformed configuration. Using the reference configuration form variations and linearizations can be carried out in an identical manner as was done in the small deformation case. Thus, all the steps outlined in Chapter 2 can be extended immediately to the finite deformation problem. We shall discover that the final equations obtained by this approach are very different from those of the small deformation problem. However, after all derivation steps are completed a transformation to expressions integrated over the current configuration will yield a form which is nearly identical to the small deformation problem and thus greatly simplifies the development of the final force and stiffness terms as well as programming steps.

To develop a finite element solution to the finite deformation problem we consider first the case of elasticity as a variational problem. Other material behaviour may be considered later by substitution of appropriate constitutive expressions for stress and tangent moduli – identical to the process used in Chapter 4 for the small deformation problem.

5.3.1 Reference configuration formulation

A variational theorem for finite elasticity may be written in the reference configuration as[10,11]

$$\Pi(\mathbf{U}) = \int_\Omega W(C_{IJ}) \, d\Omega - \Pi_{\text{ext}} \tag{5.39a}$$

in which $W(C_{IJ})$ is a stored energy function for a *hyperelastic* material from which the second Piola–Kirchhoff stress is computed using (5.34).[*] When we consider material behaviour in this chapter we restrict attention to the Saint-Venant–Kirchhoff model given by Eq. (5.36); however, the results presented here are general and more complicated behaviour may be used as described, for example, in the next chapter.

The potential for the external work is here assumed to be given by

$$\Pi_{\text{ext}} = \int_\Omega U_I \rho_0 b_I^{(m)} \, d\Omega + \int_{\Gamma_t} U_I \bar{T}_I \, d\Gamma \tag{5.39b}$$

---

[*] The functional in Eq. (5.39a) may also be expressed in terms of the deformation gradient F_{iI} and subsequent steps performed in terms of first Piola–Kirchhoff stress.

where $\bar{T}_I$ denotes specified tractions in the reference configuration and Γ_t is the traction boundary surface in the reference configuration. Taking the variation of Eqs (5.39a) and (5.39b) we obtain

$$\delta\Pi = \int_\Omega \tfrac{1}{2}\delta C_{IJ} S_{IJ} \, d\Omega - \delta\Pi_{\text{ext}} = 0 \tag{5.40a}$$

and

$$\delta\Pi_{\text{ext}} = \int_\Omega \delta U_I \rho_0 b_I^{(m)} \, d\Omega + \int_{\Gamma_t} \delta U_I \bar{T}_I \, d\Gamma \tag{5.40b}$$

where δU_I is a *variation* of the reference configuration displacement (i.e. a virtual displacement) which is arbitrary except at the kinematic boundary condition locations, Γ_u, where, for convenience, it vanishes. Since a virtual displacement is an arbitrary function, satisfaction of the variational equation implies satisfaction of the balance of linear momentum at each point in the body as well as the traction boundary conditions. We note that by using Eq. (5.34) and constructing the variation of C_{IJ}, the first term in the integrand of Eq. (5.40a) can be expressed in alternate forms as

$$\tfrac{1}{2}\delta C_{IJ} S_{IJ} = \delta E_{IJ} S_{IJ} = \delta F_{iI} F_{iJ} S_{IJ} \tag{5.41}$$

where symmetry of S_{IJ} has been used. The variation of the deformation gradient may be expressed directly in terms of the current configuration displacement as

$$\delta F_{iI} = \frac{\partial \delta u_i}{\partial X_I} = \delta u_{i,I} \tag{5.42}$$

Using the above results, after integration by parts using Green's theorem, the variational equation may be written as

$$\delta\Pi = -\int_\Omega \delta u_i \left[(F_{iJ} S_{IJ})_{,I} + \delta_{iI} \rho_0 b_I^{(m)} \right] d\Omega + \int_{\Gamma_t} \delta u_i \left[F_{iJ} S_{IJ} N_I - \delta_{iI} \bar{T}_I \right] d\Gamma = 0 \tag{5.43}$$

giving the Euler equations of (static) equilibrium in the reference configuration as

$$(F_{iJ} S_{IJ})_{,I} + \delta_{iI} \rho_0 b_I^{(m)} = P_{iI,I} + \rho_0 b_i^{(m)} = 0 \tag{5.44}$$

and the reference configuration traction boundary condition

$$S_{IJ} F_{iJ} N_I - \delta_{iI} \bar{T}_I = P_{iI} N_I - \delta_{iI} \bar{T}_I = 0 \tag{5.45}$$

The variational equation (5.40a) is identical to a Galerkin method and, thus, can be used directly to formulate problems with constitutive models different from the hyperelastic behaviour above. In addition, direct use of the variational term (5.40b) permits non-conservative loading forms, such as follower forces or pressures, to be introduced. We shall address such extensions in Sec. 5.7.

Matrix form

At this point we can again introduce matrix notation to represent the stress, strain, and variation of strain. For three-dimensional problems we define the matrix for the second Piola–Kirchhoff stress as

$$\mathbf{S} = \begin{bmatrix} S_{11}, & S_{22}, & S_{33}, & S_{12}, & S_{23}, & S_{31} \end{bmatrix}^T \tag{5.46a}$$

and the Green strain as

$$\mathbf{E} = \begin{bmatrix} E_{11}, & E_{22}, & E_{33}, & 2E_{12}, & 2E_{23}, & 2E_{31} \end{bmatrix}^T \tag{5.46b}$$

where, similar to the small strain problem, the shearing components are doubled to permit the reduction to six components. The variation of the Green strain is similarly given by

$$\delta\mathbf{E} = \begin{bmatrix} \delta E_{11}, & \delta E_{22}, & \delta E_{33}, & 2\delta E_{12}, & 2\delta E_{23}, & 2\delta E_{31} \end{bmatrix}^T \tag{5.46c}$$

which permits Eq. (5.41) to be written as the matrix relation

$$\delta E_{IJ} S_{IJ} = \delta\mathbf{E}^T \mathbf{S} \tag{5.47}$$

The variation of the Green strain is deduced from Eqs (5.13), (5.14) and (5.42) and written as

$$\delta E_{IJ} = \frac{1}{2}\left(\frac{\partial \delta u_i}{\partial X_I} F_{iJ} + \frac{\partial \delta u_i}{\partial X_J} F_{iI} \right) = \frac{1}{2}\left(\delta u_{i,I} F_{iJ} + \delta u_{i,J} F_{iI} \right) \tag{5.48}$$

Substitution of Eq. (5.48) into Eq. (5.46c) we obtain

$$\delta\mathbf{E} = \begin{Bmatrix} F_{i1}\delta u_{i,1} \\ F_{i2}\delta u_{i,2} \\ F_{i3}\delta u_{i,3} \\ F_{i1}\delta u_{i,2} + F_{i2}\delta u_{i,1} \\ F_{i2}\delta u_{i,3} + F_{i3}\delta u_{i,2} \\ F_{i3}\delta u_{i,1} + F_{i1}\delta u_{i,3} \end{Bmatrix} \tag{5.49}$$

as the matrix form of the variation of the Green strain.

Finite element approximation

Using the isoparametric form developed in Chapter 2 and Appendix A (see also Chapters 4 and 5 of reference 1) we represent the reference configuration coordinates as

$$X_I = \sum_a N_a(\boldsymbol{\xi}) \tilde{X}_I^a \tag{5.50a}$$

where $\boldsymbol{\xi}$ are the three-dimensional natural coordinates ξ_1, ξ_2, ξ_3, N_a are standard shape functions (see Appendix A or Chapters 4 and 5 of reference 1), and symbols a, b, c, etc. are introduced to identify uniquely the finite element nodal values from other indices. Similarly, we can approximate the displacement field in each element by

$$u_i = \sum_a N_a(\boldsymbol{\xi}) \tilde{u}_i^a \tag{5.50b}$$

The reference system derivatives are constructed in an identical manner to that described in Chapter 2. Thus,

$$u_{i,I} = N_{a,I} \tilde{u}_i^a \tag{5.51}$$

where explicit writing of the sum is omitted and summation convention for a is again invoked. The derivatives of the shape functions can be established by using standard routines to which the $\tilde{X}_I^a$ coordinates of nodes attached to each element are supplied.

The deformation gradient and Green strain may now be computed with use of Eqs (5.11) and (5.15), respectively. Finally, the variation of the Green strain is given in matrix form as

$$
\delta \mathbf{E} =
\begin{bmatrix}
F_{11}N_{a,1} & F_{21}N_{a,1} & F_{31}N_{a,1} \\
F_{12}N_{a,2} & F_{22}N_{a,2} & F_{32}N_{a,2} \\
F_{13}N_{a,3} & F_{23}N_{a,3} & F_{33}N_{a,3} \\
F_{11}N_{a,2}+F_{12}N_{a,1} & F_{21}N_{a,2}+F_{22}N_{a,1} & F_{31}N_{a,2}+F_{32}N_{a,1} \\
F_{12}N_{a,3}+F_{13}N_{a,2} & F_{22}N_{a,3}+F_{23}N_{a,2} & F_{32}N_{a,3}+F_{33}N_{a,2} \\
F_{13}N_{a,1}+F_{11}N_{a,3} & F_{23}N_{a,1}+F_{21}N_{a,3} & F_{33}N_{a,1}+F_{31}N_{a,3}
\end{bmatrix}
\begin{Bmatrix} \delta\tilde{u}_1^a \\ \delta\tilde{u}_2^a \\ \delta\tilde{u}_3^a \end{Bmatrix}
$$

$$
\tag{5.52}
$$

$$
= \hat{\mathbf{B}}_a \, \delta\tilde{\mathbf{u}}^a \tag{5.53}
$$

where $\hat{\mathbf{B}}_a$ replaces the form previously defined $\mathbf{B}_a$ for the small deformation problem. Expressing the deformation gradient in terms of displacements it is also possible to split this matrix into two parts as

$$
\hat{\mathbf{B}}_a = \mathbf{B}_a + \mathbf{B}_a^{NL} \tag{5.54}
$$

in which $\mathbf{B}_a$ is identical to the small deformation strain–displacement matrix and the remaining non-linear part is given by

$$
\mathbf{B}_a^{NL} =
\begin{bmatrix}
u_{1,1}N_{a,1} & u_{2,1}N_{a,1} & u_{3,1}N_{a,1} \\
u_{1,2}N_{a,2} & u_{2,2}N_{a,2} & u_{3,2}N_{a,2} \\
u_{1,3}N_{a,3} & u_{2,3}N_{a,3} & u_{3,3}N_{a,3} \\
u_{1,1}N_{a,2}+u_{1,2}N_{a,1} & u_{2,1}N_{a,2}+u_{2,2}N_{a,1} & u_{3,1}N_{a,2}+u_{3,2}N_{a,1} \\
u_{1,2}N_{a,3}+u_{1,3}N_{a,2} & u_{2,2}N_{a,3}+u_{2,3}N_{a,2} & u_{3,2}N_{a,3}+u_{3,3}N_{a,2} \\
u_{1,3}N_{a,1}+u_{1,1}N_{a,3} & u_{2,3}N_{a,1}+u_{2,1}N_{a,3} & u_{3,3}N_{a,1}+u_{3,1}N_{a,3}
\end{bmatrix}
\tag{5.55}
$$

It is immediately evident that $\mathbf{B}_a^{NL}$ is zero in the reference configuration and therefore that $\hat{\mathbf{B}}_a \equiv \mathbf{B}_a$. We note, however, that in general no advantage results from this split over the single term expression given in Eq. (5.52).

The variational equation may now be written for the finite element problem by substituting Eqs (5.46a) and (5.52) into Eq. (5.40a) to obtain

$$
\delta\Pi = (\delta\tilde{\mathbf{u}}_a)^T \left(\int_\Omega \hat{\mathbf{B}}_a^T \mathbf{S} \, d\Omega - \mathbf{f}_a \right) = 0 \tag{5.56}
$$

where the external forces are determined from $\delta\Pi_{\text{ext}}$ as

$$
\mathbf{f}_a = \int_\Omega N_a \rho_0 \mathbf{b}^{(m)} \, d\Omega + \int_{\Gamma_t} N_a \bar{\mathbf{T}} \, d\Gamma \tag{5.57}
$$

with $\mathbf{b}^{(m)}$ and $\bar{\mathbf{T}}$ the matrix form of the body and traction force vectors, respectively.

Transient problems

Using the d'Alembert principle we can introduce inertial forces through the body force as

$$
\mathbf{b}^{(m)} \rightarrow \mathbf{b}^{(m)} - \dot{\mathbf{v}} = \mathbf{b}^{(m)} - \ddot{\mathbf{x}} \tag{5.58}
$$

where $\mathbf{v}$ is the material velocity vector defined in Eq. (5.26). Inserting Eq. (5.58) into Eq. (5.57) gives

$$\mathbf{f}_a \rightarrow \mathbf{f}_a - \int_\Omega N_a \rho_0 N_b \, d\Omega \dot{\mathbf{v}}_b \qquad (5.59)$$

This adds an inertial term $\mathbf{M}_{ab}\dot{\mathbf{v}}_b$ to the variational equation where the mass matrix is given in the reference configuration by

$$\mathbf{M}_{ab} = \int_\Omega N_a \rho_0 N_b \, d\Omega \mathbf{I} \qquad (5.60)$$

For the transient problem we can introduce a Newton-type solution and replace Eq. (2.20a) by

$$\Psi_1 = \mathbf{f} - \int_\Omega \hat{\mathbf{B}}^T \mathbf{S} \, d\Omega - \mathbf{M}\dot{\mathbf{v}} = 0 \qquad (5.61)$$

Applying the linearization process defined in Eq. (3.8) to Eq. (5.61) (without the inertia force)[*] we obtain

$$\mathbf{K}_T \, d\mathbf{u}^{(k)} = \Psi_1{}^{(k)} \qquad (5.62a)$$

with updates

$$\mathbf{u}^{(k+1)} = \mathbf{u}^{(k)} + d\mathbf{u}^{(k)} \qquad (5.62b)$$

Iteration continues until convergence is achieved with the process being identical to that introduced in Chapter 2.

The tangent term is given by (omitting the iteration superscript)

$$\mathbf{K}_T = \int_\Omega \hat{\mathbf{B}}^T \hat{\mathbf{D}}_T \hat{\mathbf{B}} \, d\Omega + \int_\Omega \frac{\partial \hat{\mathbf{B}}^T}{\partial \tilde{\mathbf{u}}} \mathbf{S} \, d\Omega - \frac{\partial \mathbf{f}}{\partial \tilde{\mathbf{u}}} = \mathbf{K}_M + \mathbf{K}_G + \mathbf{K}_L \qquad (5.63)$$

where the first term is the material tangent, $\mathbf{K}_M$, in which $\hat{\mathbf{D}}_T$ is the matrix of tangent moduli. For the hyperelastic material we have

$$2\frac{\partial S_{IJ}}{\partial C_{KL}} - 4\frac{\partial^2 W}{\partial C_{IJ}\partial C_{KL}} - \frac{\partial^2 W}{\partial E_{IJ}\partial E_{KL}} = C_{IJKL} \qquad (5.64)$$

which may be transformed to a matrix $\hat{\mathbf{D}}_T$ as described in Chapter 2.

The second term, $\mathbf{K}_G$, defines a tangent term arising from the non-linear form of the strain–displacement equations and is often called the *geometric stiffness*. The derivation of this term is most easily constructed from the indicial form written as

$$\left(\int_\Omega \frac{\partial \delta E_{IJ}}{\partial \tilde{u}_j^b} S_{IJ} \, d\Omega \right) d\tilde{u}_j^b = \delta \tilde{u}_i^a \left(\int_\Omega N_{a,I} \delta_{ij} N_{b,J} S_{IJ} \, d\Omega \right) d\tilde{u}_j^b = \delta \tilde{u}_i^a \, (K_{ij}^{ab})_G \, d\tilde{u}_j^b$$
$$(5.65)$$

Thus, the geometric part of the tangent matrix is given by

$$\mathbf{K}_G^{ab} = G_{ab}\mathbf{I} \qquad (5.66a)$$

---

[*] Extension to transient applications follows directly from the presentation given in Chapter 2.

where

$$G_{ab} = \int_\Omega N_{a,I} S_{IJ} N_{b,J} \, d\Omega \tag{5.66b}$$

The last term in Eq. (5.63) is the tangent relating to loading which changes with deformation (e.g. follower forces, etc.). We assume for the present that the derivative of the force term $\mathbf{f}$ is zero so that $\mathbf{K}_L$ vanishes. In Sec. 5.7 we will consider a follower pressure loading which does give a non-zero $\mathbf{K}_L$ tangent term.

5.3.2 Current configuration formulation

The form of the equations related to the reference configuration presented in the previous section follows from straightforward application of the variational procedures and finite element approximation methods introduced previously in Chapter 2. However, the form of the resulting equations leads to much more complicated strain–displacement matrices, $\hat{\mathbf{B}}$, than previously encountered. To implement such a form it is thus necessary to reprogram completely all the element routines. We will now show that if the equations given above are transformed to the current configuration a much simpler process results.

The transformations to the current configuration are made in two steps. In the first step we replace reference configuration terms by quantities related to the current configuration (e.g. we use Cauchy or Kirchhoff stress). In the second step we convert integrals over the undeformed body to ones in the current configuration.[*]

To transform from quantities in the reference configuration to ones in the current configuration we use the chain rule for differentiation to write

$$\frac{\partial(\cdot)}{\partial X_I} = \frac{\partial(\cdot)}{\partial x_i} \frac{\partial x_i}{\partial X_I} = \frac{\partial(\cdot)}{\partial x_i} F_{iI} \tag{5.67}$$

Using this relationship Eq. (5.48) may be transformed to

$$\delta E_{IJ} = \tfrac{1}{2} \left(\delta u_{i,j} + \delta u_{j,i} \right) F_{iI} F_{jJ} = \delta \varepsilon_{ij} F_{iI} F_{jJ} \tag{5.68}$$

where we have noted that the variation term is identical to the variation of the small deformation strain–displacement relations by again using the notation[†]

$$\delta \varepsilon_{ij} = \tfrac{1}{2} \left(\delta u_{i,j} + \delta u_{j,i} \right) \tag{5.69}$$

Equation (5.41) may now be written as

$$\delta E_{iI} S_{IJ} = \delta \varepsilon_{ij} F_{iI} F_{iJ} S_{IJ} = \delta \varepsilon_{ij} \tau_{ji} = \delta \varepsilon_{ij} \sigma_{ij} J \tag{5.70}$$

and Eq. (5.40a) as

$$\delta \Pi = \int_\Omega \delta \varepsilon_{ij} \sigma_{ij} J \, d\Omega - \delta \Pi_{\text{ext}} = 0 \tag{5.71}$$

---

[*] This latter step need not be done to obtain advantage of the current configuration form of the integrand.
[†] We note that in finite deformation there is no meaning to ε_{ij} itself; only its variation, increment, or rate can appear in expressions.

The second step is now performed easily by noting the transformation of the volume element given in Eq. (5.10) to obtain finally

$$\delta\Pi = \int_\omega \delta\varepsilon_{ij}\sigma_{ij}\mathrm{d}\omega - \delta\Pi_{\text{ext}} = 0 \qquad (5.72a)$$

where ω is the domain in the current configuration.

The external potential Π_{ext} given in Eq. (5.40b) may also be transformed to the current configuration using Eqs (5.24) and (5.27) to obtain

$$\delta\Pi_{\text{ext}} = \int_\omega \delta u_i \rho b_i^{(m)}\,\mathrm{d}\omega + \int_{\gamma_t} \delta u_i \bar{t}_i\,\mathrm{d}\gamma \qquad (5.72b)$$

The computation of the tangent matrix can similarly be transformed to the current configuration. The first term given in Eq. (5.63) is deduced from

$$\int_\Omega \delta E_{IJ} C_{IJKL} \mathrm{d}E_{KL}\,\mathrm{d}\Omega = \int_\Omega \delta\varepsilon_{ij} F_{iI} F_{jJ} C_{IJKL} F_{kK} F_{lL} \mathrm{d}\varepsilon_{kl}\,\mathrm{d}\Omega$$

$$= \int_\omega \delta\varepsilon_{ij} c_{ijkl} \mathrm{d}\varepsilon_{kl}\,\mathrm{d}\omega \qquad (5.73)$$

where

$$J c_{ijkl} = F_{iI} F_{jJ} F_{kK} F_{lL} C_{IJKL} \qquad (5.74)$$

defines the moduli in the current configuration in terms of quantities in the reference state.

Finally, the geometric stiffness term in Eq. (5.63) may be written in the current configuration by transforming Eq. (5.66b) to obtain

$$G_{ab} = \int_\Omega N_{a,I} S_{IJ} N_{b,J}\,\mathrm{d}\Omega = \int_\omega N_{a,i}\sigma_{ij} N_{b,j}\,\mathrm{d}\omega \qquad (5.75)$$

Thus, we obtain a form for the finite deformation problem which is identical to that of the small deformation problem except that a geometric stiffness term is added and integrals and derivatives are to be computed in the deformed configuration. Of course, another difference is the form of the constitutive equations which need to be given in an admissible finite deformation form.

Finite element formulation

The form of the variational problem in the current configuration is easily implemented as a finite element solution process. To obtain the shape functions and their derivatives it is necessary first to obtain the deformed Cartesian coordinates x_i by using Eq. (5.4). After this step standard shape function routines can be used to compute the derivatives of shape functions, $\partial N_a/\partial x_i$. The terms in the variational equation can then be expressed in a form which is identical to that of the small deformation problem. Accordingly, the stress term is written as

$$\int_\omega \delta\varepsilon_{ij}\sigma_{ij}\,\mathrm{d}\omega = \delta\tilde{\mathbf{u}}^T \int_\omega \mathbf{B}^T \boldsymbol{\sigma}\,\mathrm{d}\omega \qquad (5.76)$$

where $\mathbf{B}$ is identical to the form of the small deformation strain–displacement matrix, and Cauchy stress is transformed to matrix form as

$$\boldsymbol{\sigma} = \begin{bmatrix} \sigma_{11}, & \sigma_{22}, & \sigma_{33}, & \sigma_{12}, & \sigma_{23}, & \sigma_{31} \end{bmatrix}^T \tag{5.77}$$

and involves only six independent components.

The residual for the static problem of a Newton solution process is now given by

$$\boldsymbol{\Psi}_1 = \mathbf{f} - \int_\omega \mathbf{B}^T \boldsymbol{\sigma} \, d\omega = \mathbf{0} \tag{5.78}$$

The linearization step of the Newton solution process is performed by computing the tangent stiffness in matrix form. Transforming Eq. (5.73) to matrix form using the relations defined in Chapter 2, the material tangent is given by

$$\mathbf{K}_M = \int_\omega \mathbf{B}^T \mathbf{D}_T \mathbf{B} \, d\omega \tag{5.79}$$

where now the material moduli $\mathbf{D}_T$ are deduced by transforming the c_{ijkl} moduli in the current configuration to matrix form. The form for G_{ab} in Eq. (5.75) may be substituted into Eq. (5.66a) to obtain the geometric tangent stiffness matrix. Thus, the total tangent matrix for the steady-state problem in the current configuration is given by

$$\mathbf{K}_T^{ab} = \int_\omega \mathbf{B}_a^T \mathbf{D}_T \mathbf{B}_b \, d\omega + G_{ab} \mathbf{I} \tag{5.80}$$

and a Newton iterate consists in solving

$$\mathbf{K}_T d\tilde{\mathbf{u}} = \mathbf{f} - \int_\omega \mathbf{B}^T \boldsymbol{\sigma} \, d\omega \tag{5.81}$$

where the external force is obtained from Eq. (5.72b) as

$$\mathbf{f}_a = \int_\omega N_a \rho \mathbf{b}^{(m)} \, d\omega + \int_{\gamma_t} N_a \bar{\mathbf{t}} \, d\gamma \tag{5.82}$$

We can also transform the inertial force to a current configuration form by substituting Eqs (5.10) and (5.27) into Eq. (5.60) to obtain

$$\mathbf{M}_{ab} = \int_\Omega N_a \rho_0 N_b \, d\Omega \mathbf{I} = \int_\omega N_a \rho N_b \, d\omega \mathbf{I} \tag{5.83}$$

and thus, for the transient problem, the residual becomes

$$\boldsymbol{\Psi}_1 = \mathbf{f} - \int_\omega \mathbf{B}^T \boldsymbol{\sigma} \, d\omega - \mathbf{M} \dot{\mathbf{v}} = \mathbf{0} \tag{5.84}$$

Linearization of this term is identical to the small deformation problem and is not given here.

The development of *displacement-based finite element models* for three-dimensional problems may be performed easily merely by adding a few modifications to a standard linear form. These modifications include the following steps.

1. Use current configuration coordinates x_i to compute shape functions and their derivatives. These are computed at nodes by adding current values of displacements $\tilde{u}_i^a$ to reference configuration nodal coordinates $\tilde{X}_I^a$.
2. Add a geometric stiffness matrix to the usual stiffness matrix as indicated in Eq. (5.80).
3. Use the appropriate material constitution for a finite deformation model.
4. Solve the problem by means of an appropriate strategy for non-linear problems.

It should be noted that the presence of the geometric stiffness and non-linear material behaviour may result in a tangent matrix which is no longer always positive definite (indeed, we shall discuss stability problems in Chapter 17 and this is a class of problems for which the tangent matrix can become singular as a result of the geometric stiffness term alone). Furthermore, use of displacement-based elements in finite deformation can lead to locking if the material has internal constraints, such as in nearly incompressible behaviour. It is then necessary again to resort to a mixed formulation to avoid such locking. The advantage of a properly constructed mixed form is that it may be used with equal accuracy for both the nearly incompressible problem as well as any compressible problem. In Sec. 5.5 we consider a mixed form which is a generalization to finite deformation of the one presented in Sec. 2.6 for small deformation problems.

5.4 Two-dimensional forms

The three-dimensional form may be reduced to a two-dimensional form if loading, geometry, and material behaviour do not vary with a third coordinate. As in the linear theory presented in Chapter 1 we have three cases: plane strain, plane stress, and axisymmetric behaviour.

A variational form is an invariant statement for a class of problems. Accordingly, it admits introduction of the basic quantities in the different coordinate frames and dimensions to define the above class of problems for finite deformation applications.

5.4.1 Plane strain

The reduction to the two-dimensional form for plane strain is made by reducing the deformation gradient to the form

$$F_{iI} = \begin{bmatrix} F_{11} & F_{12} & 0 \\ F_{21} & F_{22} & 0 \\ 0 & 0 & 1 \end{bmatrix} = \begin{bmatrix} (1 + u_{1,1}) & u_{1,2} & 0 \\ u_{2,1} & (1 + u_{2,2}) & 0 \\ 0 & 0 & 1 \end{bmatrix} \tag{5.85}$$

in which the displacements u_1 and u_2 are functions of X_1, X_2 only. A current configuration formulation for the finite element problem then follows from Sec. 5.3.2 by restricting the range of indices to two instead of three. Accordingly, the strain–displacement matrix is again identical to that of the small deformation problem given by Eq. (2.9b). Material constitution is specified by the three-dimensional model in which strains in the third direction (normal to the plane of deformation which is here taken as the 12-plane) are set to zero. This gives the same $\mathbf{D}_T$ merely restricted to the terms in the two-dimensional problem (the upper 4×4 part for the ordering given above). Introducing a finite element approximation for u_1 and u_2 in terms of shape functions N_b gives the strain–displacement matrix

$$\mathbf{B}_b = \begin{bmatrix} N_{b,x_1} & 0 & 0 & N_{b,x_2} \\ 0 & N_{b,x_2} & 0 & N_{b,x_1} \end{bmatrix}^T \tag{5.86}$$

for use in the current configuration form. The differential volume for the plane strain problem is given by

$$d\omega = J \, dX_1 \, dX_2 \quad \text{with} \quad J = F_{11} F_{22} - F_{12} F_{21}$$

5.4.2 Plane stress

To consider plane stress we need to account for the change in thickness and this may be included by taking the deformation gradient in the form

$$F_{iI} = \begin{bmatrix} F_{11} & F_{12} & 0 \\ F_{21} & F_{22} & 0 \\ 0 & 0 & F_{33} \end{bmatrix} = \begin{bmatrix} (1+u_{1,1}) & u_{1,2} & 0 \\ u_{2,1} & (1+u_{2,2}) & 0 \\ 0 & 0 & F_{33} \end{bmatrix} \tag{5.87}$$

Here the value for F_{33} is obtained from material constitution in which the stresses in the direction normal to the plane of deformation are zero. Namely,

$$S_{3I} = \sigma_{3i} = 0 \quad \text{for} \quad I, i = 1, 2, 3$$

The resulting problem then gives the same $\mathbf{B}_b$ array as for plane strain but modified $\mathbf{D}_T$ tangent matrix obtained when satisfying the zero stress condition. We will discuss this aspect when we consider constitutive behaviour in Chapter 6, Sec. 6.2.4. The differential volume for the plane stress problem is given by

$$d\omega = H_3 \, J \, dX_1 \, dX_2 \quad \text{with} \quad J = (F_{11} F_{22} - F_{12} F_{21}) F_{33}$$

Here H_3 is the thickness of the slice in the reference configuration.

5.4.3 Axisymmetric with torsion

For the full axisymmetric problem where coordinates are given by $R = X_1$, $Z = X_2$, and Θ in the reference configuration and $r = x_1$, $z = x_2$, and θ in the current configuration, the deformation gradient is given as[12,13]

$$F_{iI} = \begin{bmatrix} r_{,R} & r_{,Z} & 0 \\ z_{,R} & z_{,Z} & 0 \\ r\,\theta_{,R} & r\,\theta_{,Z} & r/R \end{bmatrix} = \begin{bmatrix} (1+u_{r,R}) & u_{r,Z} & 0 \\ u_{z,R} & (1+u_{z,Z}) & 0 \\ r\,\theta_{,R} & r\,\theta_{,Z} & (1+u_r/R) \end{bmatrix}$$

$$= \begin{bmatrix} (1+u_{1,1}) & u_{1,2} & 0 \\ u_{2,1} & (1+u_{2,2}) & 0 \\ x_1\,\phi_{,1} & x_1\,\phi_{,2} & (1+u_1/X_1) \end{bmatrix} \tag{5.88}$$

where the displacements are given by

$$r = x_1 = X_1 + u_1(X_1, X_2)$$
$$z = x_2 = X_2 + u_2(X_1, X_2) \tag{5.89}$$
$$\theta = \Theta + \phi(X_1, X_2)$$

Introduction of a finite element approximation for u_1, u_2, and θ and transformation to the current configuration gives the strain–displacement matrix

$$\mathbf{B}_b = \begin{bmatrix} N_{b,x_1} & 0 & N_b/x_1 & N_{b,x_2} & 0 & 0 \\ 0 & N_{b,x_2} & 0 & N_{b,x_1} & 0 & 0 \\ 0 & 0 & 0 & 0 & (x_1\, N_{b,x_2}) & (x_1\, N_{b,x_1}) \end{bmatrix}^T \tag{5.90}$$

Here the material constitution is identical to the three-dimensional problem. The differential volume for the axisymmetric problem is given by

$$d\omega = 2\pi X_1 J\, dX_1\, dX_2 \quad \text{with} \quad J = (F_{11}\, F_{22} - F_{12}\, F_{21})\, F_{33}$$

5.5 A three-field, mixed finite deformation formulation

A three-field, mixed variational form for the finite deformation hyperelastic problem is given by

$$\Pi(\mathbf{u}, p, \theta) = \int_\Omega \left[W(\bar{C}_{IJ}) + p\,(J - \theta) \right] d\Omega - \Pi_{\text{ext}} \tag{5.91}$$

where p is a pressure in the current (deformed) configuration, J is the determinant of the deformation gradient F_{iI}, θ is the volume in the current configuration for a unit volume in the reference state, W is the stored energy function expressed in terms of a (modified) right Cauchy–Green deformation tensor $\bar{C}_{IJ}$, and Π_{ext} is the functional for the body loading and boundary terms given in Eq. (5.39b).

The (modified) right Green deformation tensor is expressed as

$$\bar{C}_{IJ} = \bar{F}_{iI} \bar{F}_{iJ} \tag{5.92a}$$

where

$$\bar{F}_{iI} = F_{ij}^v F_{jI}^d = (\theta^{1/3} \delta_{ij})(J^{-1/3} F_{jI}) = \left(\frac{\theta}{J} \right)^{1/3} F_{iI} \tag{5.92b}$$

in which F_{ij}^v is a volumetric and F_{jI}^d a deviatoric part. We note that $\det F_{jI}^d = 1$ as required for a deviatoric (constant volume) state.

This form of the variational problem has been used for problems formulated in principal stretches.[14] Here we use the form without referring to the specific structure of the stored energy function. In particular we wish to admit constitutive forms in which the volumetric and deviatoric parts are not split as in reference 14.

The variation of Eq. (5.91) is given by

$$\delta\Pi = \int_\Omega \left[\frac{1}{2} \delta\bar{C}_{IJ} \bar{S}_{IJ} + \delta p\,(J - \theta) + (\delta J - \delta\theta)\, p \right] d\Omega - \delta\Pi_{\text{ext}} \tag{5.93}$$

where a second Piola–Kirchhoff stress based on the modified deformation tensor is defined as

$$\bar{S}_{IJ} = 2\frac{\partial W}{\partial \bar{C}_{IJ}} = \frac{\partial W}{\partial \bar{E}_{IJ}} \quad \text{with} \quad \bar{E}_{IJ} = \tfrac{1}{2}\left(\bar{C}_{IJ} - \delta_{IJ}\right) \tag{5.94}$$

Using Eq. (5.92a) the variation of the modified deformation tensor is given by

$$\delta\bar{C}_{IJ} = \delta\bar{F}_{iI}\bar{F}_{iJ} + \delta\bar{F}_{iJ}\bar{F}_{iI} \tag{5.95}$$

in which

$$\delta\bar{F}_{iI} = \left(\frac{\theta}{J}\right)^{1/3}\left[\delta F_{iI} + \tfrac{1}{3}F_{iI}\left(\frac{\delta\theta}{\theta} - \frac{\delta J}{J}\right)\right]$$

Thus, upon noting that[9,10]

$$\delta J = J F_{jJ}^{-1}\delta F_{jJ}$$

the first term of the integrand in Eq. (5.93) formally may be expanded as

$$\begin{aligned}\delta\bar{C}_{IJ}\bar{S}_{IJ} &= \delta\bar{F}_{iI}\bar{F}_{iJ}\bar{S}_{IJ}\\ &= \frac{1}{3}\left(\frac{\delta\theta}{\theta} - \frac{\delta J}{J}\right)\bar{F}_{iI}\bar{F}_{iJ}\bar{S}_{IJ} + \left(\frac{\theta}{J}\right)^{1/3}\delta F_{iI}\bar{F}_{iJ}\bar{S}_{IJ}\end{aligned} \tag{5.96}$$

This expression again may be simplified by defining current configuration Kirchhoff and Cauchy stresses based on the modified deformation gradient as

$$\bar{\tau}_{ij} = \bar{F}_{iI}\bar{S}_{IJ}\bar{F}_{jJ} = \theta\bar{\sigma}_{ij} \tag{5.97}$$

The definitions introduced for stress are consistent with using standard constitutive models in which the modified deformation tensor is used to compute stresses and material moduli. That is, we need not distinguish whether a standard displacement method or the three-field mixed model as given here is used.

Also, we note from Eq. (5.67) that

$$\delta F_{jJ}F_{jJ}^{-1} = \delta u_{j,k}F_{kJ}F_{jJ}^{-1} = \delta u_{j,k}\delta_{kj} = \delta u_{j,j} \tag{5.98}$$

is the divergence of the variation in displacement. Thus, Eq. (5.96) simplifies to

$$\delta\bar{C}_{IJ}\bar{S}_{IJ} = \frac{1}{3}\left(\frac{\delta\theta}{\theta} - \delta u_{j,j}\right)\bar{\tau}_{ii} + \delta u_{i,j}\bar{\tau}_{ij} = \frac{1}{3}\frac{\delta\theta}{\theta}\bar{\tau}_{kk} + \frac{\partial\delta u_i}{\partial x_j}\left(\bar{\tau}_{ij} - \frac{1}{3}\delta_{ij}\bar{\tau}_{kk}\right) \tag{5.99}$$

Substituting relations deduced above into Eq. (5.93) and noting symmetry of the Kirchhoff stress, a formulation in terms of quantities related to the deformed position may be written as

$$\begin{aligned}\delta\Pi &= \int_{\Omega}\delta\varepsilon_{ij}\left[\bar{\sigma}_{ij} + \delta_{ij}\left(\frac{J}{\theta}p - \bar{p}\right)\right]\theta\,d\Omega + \int_{\Omega}\delta\theta\left(\bar{p} - p\right)d\Omega\\ &\quad + \int_{\Omega}\delta p\left(J - \theta\right)d\Omega - \delta\Pi_{\text{ext}} = 0\end{aligned} \tag{5.100}$$

where $\delta\varepsilon_{ij}$ is given by Eq. (5.69) and $\bar{p} = \bar{\sigma}_{ii}/3$ defines a mean stress based on the Cauchy stress deduced according to Eq. (5.97). This variational equation may be transformed to integrals over the current configuration by replacing $d\Omega$ by $d\omega/J$; however, this step is not an essential transformation.

Finite element equations: matrix notation

The mixed method finite element approximation of the three-field variational form is expressed using deformation measures and stresses related to the current configuration. The development is very similar to that presented in Chapter 2 for the small deformation case.

The reference coordinate and displacement fields are approximated by isoparametric interpolations as indicated in Eqs (5.50a) and (5.50b), respectively. These are used to compute the deformation gradient by means of Eqs (5.11) and (5.51). The pressure and volume are interpolated in a manner which is identical to the small deformation case as

$$p = \mathbf{N}_p \tilde{\mathbf{p}} \quad \text{and} \quad \theta = \mathbf{N}_\theta \tilde{\theta}$$

and for quadrilateral and brick elements are taken to be discontinuous between elements.

Using the above approximation, Eq. (5.100) may be expressed in matrix form as

$$
\delta \Pi = \delta \tilde{\mathbf{u}}^T \int_\Omega \mathbf{B}^T \check{\sigma}\theta \, d\Omega + \delta \tilde{\mathbf{p}}^T \int_\Omega \mathbf{N}_p^T (J - \theta) \, d\Omega
$$
$$
+ \delta \tilde{\theta}^T \int_\Omega \mathbf{N}_\theta^T (\bar{p} - p) \, d\Omega - \delta \Pi_{\text{ext}}
$$
(5.101)

In this form of the finite deformation problem **B** again is identical to the small deformation strain–displacement matrix with a modified stress defined as

$$
\check{\sigma} = \bar{\sigma} + (\check{p} - \bar{p}) \, \mathbf{m} \quad \text{where} \quad \check{p} = \frac{J}{\theta} p
$$
(5.102)

We note that inertia effects may again be included as described for the displacement model and the final result yields the discrete form of Eq. (5.101) given by

$$
\mathbf{P} + \mathbf{M}\dot{\mathbf{v}} = \mathbf{f}
$$
$$
\mathbf{P}_p - \mathbf{K}_{\theta p}\tilde{\mathbf{p}} = \mathbf{0}
$$
$$
-\mathbf{K}_{p\theta}\tilde{\theta} + \mathbf{E}_p = \mathbf{0}
$$
(5.103a)

where the arrays are given as

$$
\mathbf{P} = \int_\Omega \mathbf{B}^T \check{\sigma}\theta \, d\Omega \qquad \mathbf{P}_p = \frac{1}{3} \int_\Omega \mathbf{N}_\theta^T \check{\sigma}\theta \, d\Omega
$$
$$
\mathbf{K}_{\theta p} = \int_\Omega \mathbf{N}_\theta^T \mathbf{N}_p \, d\Omega = \mathbf{K}_{p\theta} \qquad \mathbf{E}_p = \int_\Omega \mathbf{N}_p^T J \, d\Omega
$$
(5.103b)

and force **f** and mass **M** are identical to the terms appearing in the displacement model presented previously.

We can observe that the mixed model reduces to the displacement form if $\theta = J$ and $p = \bar{p}$ at every point in the element. This would occur if our approximations for θ and p contained all the terms appearing in results computed from the displacement-based deformations and, thus, again establishes the principle of limitation.[15] Moreover, if this occurred, any locking tendency in the displacement form would again occur in the mixed approach also.

To obtain a formulation free of locking it is again necessary to select approximations for pressure and volume which satisfy the mixed patch test count conditions as described in Chapters 10 and 11 of reference 1. Here, to approximate p and θ in each element we assume that $\mathbf{N}_\theta = \mathbf{N}_p$ and for four-node quadrilateral and eight-node brick elements of linear order use constant (unit) interpolation. In nine-node quadrilateral and 27-node brick elements of quadratic order we assume linear interpolation. Linear interpolation in ξ_1, ξ_2, ξ_3 or X_1, X_2, X_3 can be used; however, x_1, x_2, x_3 should not be used since the interpolation becomes non-linear (since x_i depend on u_i) and the solution complexity is greatly increased from that indicated above.

The second and third expressions in Eqs (5.103a) are linear in $\tilde{p}$ and $\tilde{\theta}$, respectively, and also are completely formed within a single element. Moreover, the coefficient matrix $\mathbf{K}_{p\theta} = \mathbf{K}_{\theta p}$ is symmetric positive definite when $\mathbf{N}_\theta = \mathbf{N}_p$. Thus, a partial solution can be achieved in each element as

$$\tilde{\mathbf{p}} = \mathbf{K}_{\theta p}^{-1} \mathbf{P}_p$$
$$\tilde{\theta} = \mathbf{K}_{p\theta}^{-1} \mathbf{E}_p$$

$$(5.104)$$

An explicit method in time may be employed to solve the momentum equation: as was indeed used to solve examples shown at the end of Chapter 2. However, here we only consider further an implicit scheme which is applicable to either transient or static problems (see Chapters 2 and 3). A Newton scheme may be employed to solve Eq. (5.101). To construct the tangent matrix $\mathbf{K}_T$ it is necessary to linearize Eq. (5.93). In indicial form, the Newton linearization may be assembled as

$$d\left(\delta\Pi\right) = \int_\Omega \left[\delta\bar{C}_{IJ}\bar{C}_{IJKL}d\bar{C}_{KL} + \tfrac{1}{2}d\left(\delta\bar{C}_{IJ}\right)\bar{S}_{IJ}\right]d\Omega + \int_\Omega pd\left(\delta J\right)d\Omega$$
$$+ \int_\Omega \delta p\left(dJ - d\theta\right)d\Omega + \int_\Omega dp\left(\delta J - \delta\theta\right)d\Omega + d\left(\delta\Pi_{\text{ext}}\right)$$

$$(5.105)$$

where $d\bar{C}_{KL}$, dp, etc., denote incremental quantities and material tangent moduli are denoted by

$$2\frac{\partial\bar{S}_{IJ}}{\partial\bar{C}_{KL}} = \bar{C}_{IJKL}$$

$$(5.106)$$

The above integrals may also be expressed in quantities terms or current configuration terms in an identical manner as for the displacement model presented in Sec. 5.3.2. In this case the reference configuration moduli are transformed to the current configuration using

$$\bar{c}_{ijkl} = \frac{1}{\theta}\bar{F}_{iI}\bar{F}_{jJ}\bar{F}_{kK}\bar{F}_{lL}\bar{C}_{IJKL}$$

$$(5.107)$$

Using standard transformations from indicial to matrix form the moduli for the current configuration may be written in matrix form as $\bar{\mathbf{D}}_T$.

We can now write Eq. (5.105) in matrix form and obtain the set of equations which determine the parameters $d\tilde{\mathbf{u}}$, $d\tilde{\theta}$ and $d\tilde{\mathbf{p}}$ as

$$\begin{bmatrix} \mathbf{K}_{uu} & \mathbf{K}_{u\theta} & \mathbf{K}_{up} \\ \mathbf{K}_{\theta u} & \mathbf{K}_{\theta\theta} & -\mathbf{K}_{\theta p} \\ \mathbf{K}_{pu} & -\mathbf{K}_{p\theta} & 0 \end{bmatrix} \begin{Bmatrix} d\tilde{\mathbf{u}} \\ d\tilde{\theta} \\ d\tilde{\mathbf{p}} \end{Bmatrix} = \begin{Bmatrix} \mathbf{f} - \mathbf{P} \\ 0 \\ 0 \end{Bmatrix}$$

$$(5.108)$$

where

$$\mathbf{K}_{uu} = \int_\Omega \mathbf{B}^T \bar{\mathbf{D}}_{11} \mathbf{B}\theta \, d\Omega + \mathbf{K}_G, \qquad \mathbf{K}_{u\theta} = \int_\Omega \mathbf{B}^T \bar{\mathbf{D}}_{12} \left[\frac{1}{\theta} \mathbf{N}_\theta\right] \theta \, d\Omega = \mathbf{K}_{\theta u}^T$$

$$\mathbf{K}_{up} = \int_\Omega \mathbf{B}^T \mathbf{m} \mathbf{N}_p J \, d\Omega = \mathbf{K}_{pu}^T, \qquad \mathbf{K}_{\theta\theta} = \int_\Omega \left[\frac{1}{\theta}\mathbf{N}_\theta^T\right] \bar{\mathbf{D}}_{22} \left[\frac{1}{\theta}\mathbf{N}_\theta\right]\theta \, d\Omega$$

in which

$$\bar{\mathbf{D}}_{11} = \mathbf{I}_d \bar{\mathbf{D}}_T \mathbf{I}_d - \tfrac{2}{3}\left(\mathbf{m}\bar{\sigma}_d^T + \bar{\sigma}_d \mathbf{m}^T\right) + 2\left(\bar{p} - \tilde{p}\right)\mathbf{I}_0 - \left(\frac{2}{3}\bar{p} - \tilde{p}\right)\mathbf{m}\mathbf{m}^T$$

$$\bar{\mathbf{D}}_{12} = \tfrac{1}{3}\mathbf{I}_d\bar{\mathbf{D}}_T\mathbf{m} + \frac{2}{3}\bar{\sigma}_d = \bar{\mathbf{D}}_{21}^T$$

$$\bar{\mathbf{D}}_{22} = \tfrac{1}{9}\mathbf{m}^T\bar{\mathbf{D}}_T\mathbf{m} - \frac{1}{3}\bar{p}$$

with $\mathbf{I}_0$ defined by Eq. (2.61). Note also that the right-hand side is zero in the second and third rows of Eq. (5.108) since the solution for pressure and volume parameters was determined exactly using Eq. (5.104).

The geometric tangent term is given by

$$\mathbf{K}_G^{ab} = \bar{G}_{ab}\mathbf{I} \quad \text{where} \quad \bar{G}_{ab} = \int_\Omega N_{a,i}\bar{\sigma}_{ij}N_{b,j}\,d\Omega \qquad (5.109)$$

A solution to Eq. (5.108) may be formed by solving the third and second rows as

$$d\tilde{\theta} = \mathbf{K}_{\theta p}^{-1}\mathbf{K}_{pu}d\tilde{u}$$
$$d\tilde{p} = \mathbf{K}_{\theta p}^{-1}\mathbf{K}_{\theta u}d\tilde{u} + \mathbf{K}_{\theta p}^{-1}\mathbf{K}_{\theta\theta}d\tilde{\theta} \qquad (5.110)$$
$$= \left(\mathbf{K}_{\theta p}^{-1}\mathbf{K}_{\theta u} + \mathbf{K}_{\theta p}^{-1}\mathbf{K}_{\theta\theta}\mathbf{K}_{p\theta}^{-1}\mathbf{K}_{pu}\right)d\tilde{u}$$

and substituting the result into the first row to obtain

$$\mathbf{K}_T d\tilde{u} = [\mathbf{K}_{uu} + \mathbf{K}_{u\theta}\mathbf{K}_{p\theta}^{-1}\mathbf{K}_{pu} + \mathbf{K}_{up}\mathbf{K}_{\theta p}^{-1}\mathbf{K}_{\theta u} + \mathbf{K}_{up}\mathbf{K}_{\theta p}^{-1}\mathbf{K}_{\theta\theta}\mathbf{K}_{p\theta}^{-1}\mathbf{K}_{pu}]d\tilde{u} = \mathbf{f} - \mathbf{P} \qquad (5.111)$$

This result is obtained by inverting only the symmetric positive definite matrix $\mathbf{K}_{p\theta}$, which we also note is independent of any specific constitutive model. Alternatively, if we define a mixed volumetric strain–displacement matrix as

$$\mathbf{B}_v = \frac{1}{\theta}\mathbf{N}_\theta\mathbf{W} \quad \text{where} \quad \mathbf{W} = \mathbf{K}_{p\theta}^{-1}\mathbf{K}_{pu}$$

the tangent matrix may be computed directly from

$$\mathbf{K}_T = \int_\Omega \left[\mathbf{B}^T\bar{\mathbf{D}}_{11}\mathbf{B} + \mathbf{B}^T\bar{\mathbf{D}}_{12}\mathbf{B}_v + \mathbf{B}_v^T\bar{\mathbf{D}}_{21}\mathbf{B} + \mathbf{B}_v^T\bar{\mathbf{D}}_{22}\mathbf{B}_v\right]\theta\,d\Omega + \mathbf{K}_G$$

$$= \int_\Omega \begin{bmatrix}\mathbf{B}^T, & \mathbf{B}_v^T\end{bmatrix}\begin{bmatrix}\bar{\mathbf{D}}_{11} & \bar{\mathbf{D}}_{12}\\ \bar{\mathbf{D}}_{21} & \bar{\mathbf{D}}_{22}\end{bmatrix}\begin{bmatrix}\mathbf{B}\\ \mathbf{B}_v\end{bmatrix}d\Omega + \mathbf{K}_G \qquad (5.112)$$

In this form the finite deformation formulation is similar to that developed in Sec. 2.5 for the small strain case.

5.6 A mixed–enhanced finite deformation formulation

An alternative finite element method to that just discussed is the fully mixed method in which strain approximations are *enhanced*. The key idea of the mixed–enhanced formulation is the parameterization of the deformation gradient in terms of a mixed and an enhanced deformation gradient from which a consistent formulation is derived. This methodology allows for a formulation which has standard-order quadrature and variational recoverable stresses, hence circumventing difficulties which arise in earlier *enhanced strain* methods.[2–4,16–18] Recently, an improved form for the enhanced strain hexahedral element has been proposed by Areias *et al.*[5] While the element has improved properties over that given by Simo *et al.*[3] we prefer the approach presented here as a procedure to develop general element types.

There is no need to separate any deformation gradient terms into deviatoric and mean parts as was necessary for the three-field approach discussed in the previous section. The mixed–enhanced formulation discussed here uses a three-field variational form for finite deformation hyperelasticity expressed as

$$\Pi = \int_\Omega [W(\hat{F}_{iI}) + \hat{P}_{iI}(F_{iI} - \hat{F}_{iI})] \, d\Omega - \Pi_{\text{ext}} \tag{5.113}$$

where F_{iI} is the deformation gradient computed from displacements, $\tilde{F}_{iI}$ is an independent deformation gradient, $\hat{P}_{iI}$ is an independent Piola–Kirchhoff stress, W is an objective stored energy function in terms of $\hat{F}_{iI}$, and Π_{ext} is the loading term given by Eq. (5.39b).

The stationary point of Π is obtained by setting to zero the first variation of Eq. (5.113) with respect to the three independent fields. Accordingly,

$$\delta\Pi = \int_\Omega \left[\delta F_{iI} \hat{P}_{iI} + \delta\hat{F}_{iI} \left(\frac{\partial W}{\partial \hat{F}_{iI}} - \hat{P}_{iI} \right) + \delta\hat{P}_{iI} \left(F_{iI} - \hat{F}_{iI} \right) \right] d\Omega - \delta\Pi_{\text{ext}} = 0 \tag{5.114}$$

where $\hat{F}_{iI}$ and $\hat{P}_{iI}$ are *mixed* variables to be approximated directly. The reader will note that we now use the deformation gradient directly instead of the usual C_{IJ}, E_{IJ}, or b_{ij} symmetric forms. We will often use constitutive models which are expressed in these symmetric quantities; however, we note that they are also implicitly functions of the deformation gradient through the definitions given in Sec. 5.2. Once again, at this point we may substitute a first Piola–Kirchhoff stress from any constitutive model in place of the derivative of the stored energy function $\partial W/\partial\hat{F}_{iI}$ in Eq. (5.114). Thus, the present form can be used in a general context.

Finite element approximations to the mixed deformation gradient and the first Piola–Kirchhoff stress are constructed directly in terms of local coordinates of the parent element using standard transformation concepts. Accordingly, we take

$$\hat{F}_{iI} = \bar{F}_{iA} \bar{J}_{\alpha A} \bar{J}_{\beta I} \mathcal{F}_{\alpha\beta}(\boldsymbol{\xi}) \tag{5.115a}$$

and

$$\hat{P}_{iI} = \bar{F}_{iA}^{-1} \bar{J}_{\alpha A}^{-1} \bar{J}_{\beta I}^{-1} \mathcal{P}_{\alpha\beta}(\boldsymbol{\xi}) \tag{5.115b}$$

where ξ denotes the natural coordinates ξ_1, ξ_2, ξ_3, Greek subscripts are associated with the natural coordinates, and $\mathcal{P}_{\alpha\beta}$ and $\mathcal{F}_{\alpha\beta}$ are the first Piola–Kirchhoff stress and deformation gradient approximations in the isoparametric (parent) coordinate space, respectively.[*] The arrays $\bar{J}_{\alpha A}$ and $\bar{F}_{iI}$ used above are average quantities over the element volume, Ω_e. The average quantity $\bar{J}_{\alpha A}$ is defined as

$$\bar{J}_{\alpha A} = \frac{1}{\Omega_e} \int_{\Omega_e} J_{\alpha A}\, d\Omega \qquad (5.116a)$$

where $J_{\alpha A}$ is the standard Jacobian matrix as defined in Eq. (2.6b) (but now written for the reference coordinates), and $\bar{F}_{iI}$ is defined as

$$\bar{F}_{iI} = \frac{1}{\Omega_e} \int_{\Omega_e} F_{iI}\, d\Omega \qquad (5.116b)$$

The above form of approximation will ensure direct inclusion of constant states as well as minimize the order of quadrature needed to evaluate the finite element arrays and eliminate some sensitivity associated with initially distorted elements.

The forms given in Eqs (5.115a) and (5.115b) are constructed so that the energy term of the physical and isoparametric pairs are equal. Accordingly, we observe that

$$\mathcal{P}_{\alpha\beta}\mathcal{F}_{\alpha\beta} = \hat{P}_{iI}\hat{F}_{iI} \qquad (5.117)$$

This greatly simplifies the integrations needed to construct the terms in Eq. (5.114).

To construct the approximations we note that the tensor transformations for the mixed deformation gradient may be written in matrix form as

$$\hat{\mathbf{F}} = \mathbf{A}\mathcal{F} \qquad (5.118a)$$

and

$$\hat{\mathbf{P}} = \mathbf{A}^{-1}\mathcal{P} \qquad (5.118b)$$

where $\mathbf{A}$ is a transformation to matrix form of the fourth-rank tensor given as

$$A_{iI\alpha\beta} = \bar{F}_{iA}\bar{J}_{\alpha A}\bar{J}_{\beta I} \qquad (5.119)$$

The ordering for the matrix–tensor transformation for all the variables is described in Table 5.1.

The approximation for the mixed deformation gradient may now be written as

$$\hat{\mathbf{F}} = \tilde{\gamma}^0 + \frac{1}{j}\mathbf{A}\left[\hat{\mathbf{E}}_1(\xi)\tilde{\gamma} + \hat{\mathbf{E}}_2(\xi)\tilde{\alpha}\right] \qquad (5.120a)$$

and for the mixed stress as

$$\hat{\mathbf{P}} = \tilde{\beta}^0 + \mathbf{A}^{-1}\left[\mathbf{E}_1(\xi)\tilde{\beta}\right] \qquad (5.120b)$$

where $j = \det J_{\alpha A}$ and $\hat{\mathbf{E}}_1$, $\hat{\mathbf{E}}_2$ define the functions to be selected in terms of natural coordinates. The functions suggested in reference 19 are given in Table 5.2. The terms β^0 and γ^0 ensure that constant stress and strain are available in the element.

---

[*] Note the resulting transformed arrays are objective under a superposed rigid body motion.[10]

Table 5.1   Matrix–tensor transformation for the nine-component form

Row or column	1	2	3	4	5	6	7	8	9
i or α	1	2	3	1	2	3	2	3	1
I or β	1	2	3	2	3	1	1	2	3

Table 5.2   Three-dimensional interpolations

α	β	$\hat{\mathbf{E}}_1\gamma$	$\hat{\mathbf{E}}_2\alpha$
1	1	$\xi_2\gamma_1 + \xi_3\gamma_2 + \xi_2\xi_3\gamma_3$	$\xi_1\alpha_1 + \xi_1\xi_2\alpha_2 + \xi_1\xi_3\alpha_3$
2	2	$\xi_1\gamma_4 + \xi_3\gamma_5 + \xi_1\xi_3\gamma_6$	$\xi_2\alpha_4 + \xi_2\xi_3\alpha_5 + \xi_1\xi_2\alpha_6$
3	3	$\xi_1\gamma_7 + \xi_2\gamma_8 + \xi_1\xi_2\gamma_9$	$\xi_3\alpha_7 + \xi_2\xi_3\alpha_8 + \xi_1\xi_3\alpha_9$
1	2	$\xi_3\gamma_{10}$	0
2	3	$\xi_1\gamma_{12}$	0
3	1	$\xi_2\gamma_{14}$	0
2	1	$\xi_3\gamma_{11}$	0
3	2	$\xi_1\gamma_{13}$	0
1	3	$\xi_2\gamma_{15}$	0

The above construction is similar to that used to construct the Pian–Sumihara plane elastic element[20] (see also Sec. 10.4.4 of reference 1).

The enhanced parameters α are added to the normal strains in Table 5.2 such that the resulting strain components are complete polynomials in natural coordinates. This is done to provide the necessary equations to enforce a near incompressibility constraint without loss of rank in the resulting finite element arrays. In addition, the enhanced parameters improve coarse mesh accuracy in bending dominated regimes.

Finite element equations: matrix notation

By isolating the equations associated with the variation of the first Piola–Kirchhoff stress tensor $\delta\hat{\mathbf{P}}$ in Eq. (5.114) some of the element parameters of the mixed–enhanced deformation gradient $\hat{\mathbf{F}}$ may be obtained as

$$\gamma^0 = \frac{1}{\Omega_e} \int_{\Omega_e} \mathbf{F}\,d\Omega = \bar{\mathbf{F}} \tag{5.121a}$$

and

$$\gamma = \left(\int_\square \hat{\mathbf{E}}_1^T \hat{\mathbf{E}}_1\,d\square \right)^{-1} \int_{\Omega_e} \hat{\mathbf{E}}_1^T \mathbf{A}^{-1} \left(\mathbf{F} - \bar{\mathbf{F}} \right) d\Omega \tag{5.121b}$$

where the box denotes integration over the element region defined by the isoparametric coordinates ξ_I. We note that the construction for $\hat{\mathbf{E}}_1$ and $\hat{\mathbf{E}}_2$ is such that integrals have the property

$$\int_\square \hat{\mathbf{E}}_1\,d\square = \int_\square \hat{\mathbf{E}}_2\,d\square = \int_\square \hat{\mathbf{E}}_1^T \hat{\mathbf{E}}_2\,d\square \equiv 0$$

This greatly simplifies the construction of the partial solution given above.

Use of the above definitions for $\hat{\mathbf{F}}$ and $\hat{\mathbf{P}}$ also makes the second term in the integrand of Eq. (5.113) zero, hence the modified functional $\hat{\Pi}$ is expressed as

$$\hat{\Pi} = \int_\Omega W(\tilde{F}_{iI})\,d\Omega - \Pi_{\text{ext}} \tag{5.122}$$

The stationary condition of $\hat{\Pi}$ yields a reduced set of non-linear equations, in terms of the nodal displacements, $\tilde{\mathbf{u}}$, and the enhanced parameters, $\tilde{\alpha}$, expressed as

$$\delta\hat{\Pi} = \int_{\Omega} \frac{\partial W}{\partial \tilde{F}_{iI}} \delta\tilde{F}_{iI}\, d\Omega - \delta\Pi_{\text{ext}} = \left\{ \delta\tilde{\mathbf{u}}^T \quad \delta\tilde{\alpha}^T \right\} \left\{ \begin{array}{c} \mathbf{P}_{\text{int}}(\tilde{\mathbf{u}}, \tilde{\alpha}) - \mathbf{f} \\ \mathbf{P}_{\text{enh}}(\tilde{\mathbf{u}}, \tilde{\alpha}) \end{array} \right\} = 0 \quad (5.123)$$

where $\mathbf{P}_{\text{int}}$ is the internal force vector, $\mathbf{P}_{\text{enh}}$ is the enhanced force vector, and $\mathbf{f}$ is the usual force vector computed from Π_{ext}. Noting that the variations $\delta\tilde{\mathbf{u}}$ and $\delta\tilde{\alpha}$ in Eq. (5.123) are arbitrary the finite element residual vectors are given by

$$\begin{aligned} \boldsymbol{\Psi}_{\mathbf{u}} &= \mathbf{f} - \mathbf{P}_{\text{int}}(\tilde{\mathbf{u}}, \tilde{\alpha}) = 0 \\ \boldsymbol{\Psi}_{\alpha} &= -\mathbf{P}_{\text{enh}}(\tilde{\mathbf{u}}, \tilde{\alpha}) = 0 \end{aligned} \quad (5.124)$$

A solution to these equations may now be constructed in the standard manner discussed in Chapter 2. Using a Newton scheme to linearize Eq. (5.123) we obtain

$$\begin{aligned} d(\delta\hat{\Pi}) &= \int_{\Omega} \delta\hat{F}_{iI} \frac{\partial^2 W}{\partial \hat{F}_{iI} \partial \hat{F}_{jJ}} d\hat{F}_{jJ} + \frac{\partial W}{\partial \hat{F}_{iI}} d(\delta\hat{F}_{iI})\, d\Omega \\ &\equiv \left\{ \delta\tilde{\mathbf{u}}^T \quad \delta\tilde{\alpha}^T \right\} \begin{bmatrix} \hat{\mathbf{K}}_{uu} & \hat{\mathbf{K}}_{u\alpha} \\ \hat{\mathbf{K}}_{\alpha u} & \hat{\mathbf{K}}_{\alpha\alpha} \end{bmatrix} \left\{ \begin{array}{c} d\tilde{\mathbf{u}} \\ d\tilde{\alpha} \end{array} \right\} \end{aligned} \quad (5.125)$$

where $\hat{\mathbf{K}}_{uu}$, etc., are obtained by evaluating all the terms in the integrals in a standard manner, and the process is by now so familiar to the reader we leave it as an exercise.

Using Eqs (5.124)–(5.125) we obtain the system of equations

$$\begin{bmatrix} \hat{\mathbf{K}}_{uu} & \hat{\mathbf{K}}_{u\alpha} \\ \hat{\mathbf{K}}_{\alpha u} & \hat{\mathbf{K}}_{\alpha\alpha} \end{bmatrix} \left\{ \begin{array}{c} d\tilde{\mathbf{u}} \\ d\tilde{\alpha} \end{array} \right\} = \left\{ \begin{array}{c} \boldsymbol{\Psi}_u \\ \boldsymbol{\Psi}_\alpha \end{array} \right\} \quad (5.126)$$

where we note that the parameters $\tilde{\alpha}$ are associated with individual elements. We can use static condensation to perform a partial solution at the element level.[21] Here the situation is slightly different since the equations are non-linear. Thus, it is necessary to use the static condensation process in an iterative manner. Accordingly, given a solution $\tilde{\mathbf{u}}$ for some iterate in a Newton process we can isolate the part for each $\tilde{\alpha}$ and consider a local solution for the equation set

$$d\tilde{\alpha}^{(k)} = \hat{\mathbf{K}}_{\alpha\alpha}^{-1} \boldsymbol{\Psi}_{\alpha}^{(k)} \quad (5.127a)$$

Iteration continues until $\boldsymbol{\Psi}_\alpha$ is zero with updates

$$\tilde{\alpha}^{(k+1)} = \tilde{\alpha}^{(k)} + d\tilde{\alpha}^{(k)} \quad (5.127b)$$

performed on each element separately.

Utilizing the final solution from Eq. (5.127a) an equivalent displacement model involving only the nodal displacement parameters is obtained as

$$\hat{\mathbf{K}}_T^{(k)} d\tilde{\mathbf{u}} = \boldsymbol{\Psi}_u^{(k)} \quad (5.128)$$

where

$$\hat{\mathbf{K}}_T^{(k)} = [\hat{\mathbf{K}}_{uu} - \hat{\mathbf{K}}_{u\alpha} (\hat{\mathbf{K}}_{\alpha\alpha})^{-1} \hat{\mathbf{K}}_{\alpha u}]$$

The system of Eq. (5.128) is solved and the nodal displacements are updated in the usual manner for any displacement problem (see Chapter 2). Additional details and many example solutions using the above formulation, and its specialization to small deformations, may be found in references 19 and 22.

5.7 Forces dependent on deformation – pressure loads

In the derivations presented in the previous sections it was assumed that the forces $\mathbf{f}$ were not themselves dependent on the deformation. In some instances this is not true. For instance, pressure loads on a deforming structure are in this category. Aerodynamic forces are an example of such pressure loads and can induce flutter.

If forces vary with displacement then in relation (5.63) the variation of the forces with respect to the displacements has to be considered. This leads to the introduction of the *load correction matrix* $\mathbf{K}_L$ as originally suggested by Oden[23] and Hibbitt et al.[24]

Here we consider the case where pressure acts on the current configuration and remains normal throughout the deformation history. If the pressure is given by $\bar{p}$ then the surface traction term in $\delta\Pi_{\text{ext}}$ is given by

$$\int_{\gamma_t} \delta u_i \, \bar{t}_i \, d\gamma = \int_{\gamma_t} \delta u_i \, \bar{p} \, n_i \, d\gamma \tag{5.129}$$

where n_i are the direction cosines of an outward pointing normal to the deformed surface. The computation of the nodal forces and tangent matrix terms is most conveniently computed by transforming the above expression to the surface approximated by finite elements.[25-27] In this case we have the approximation to Eq. (5.129) for a three-dimensional problem given in matrix notation by

$$\int_{\gamma_t} \delta u_i \, \bar{p} \, n_i \, d\gamma = \delta\tilde{u}_a \int_{-1}^{1} \int_{-1}^{1} N_a \bar{p}(\xi_1, \xi_2) \left[\left(N_{c,\xi_1} \mathbf{x}_c \right) \times \left(N_{\delta,\xi_2} \mathbf{x}_\delta \right) \right] d\xi_1 \, d\xi_2 \tag{5.130}$$

where ξ_1, ξ_2 are natural coordinates of a two-dimensional finite element surface interpolation, $\bar{p}(\xi_1, \xi_2)$ is a specified nodal pressure at each point on the surface, $\mathbf{x}_c$ are nodal coordinates of the deformed surface, and we have used the relation transforming surface area given in Eq. (2.18). A cross-product may be written in the alternate matrix forms

$$\mathbf{x}_c \times \mathbf{x}_\delta = \hat{\mathbf{x}}_c \mathbf{x}_\delta = -\hat{\mathbf{x}}_\delta \mathbf{x}_c = \hat{\mathbf{x}}_\delta^T \mathbf{x}_c \tag{5.131}$$

where here $\hat{\mathbf{x}}$ denotes a *skew symmetric* matrix given as

$$\hat{\mathbf{x}} = \begin{bmatrix} 0 & -x_3 & x_2 \\ x_3 & 0 & -x_1 \\ -x_2 & x_1 & 0 \end{bmatrix} \tag{5.132}$$

Using the above relations the nodal forces for the 'follower' surface loading are given by

$$\mathbf{f}_a = \int_{-1}^{1} \int_{-1}^{1} N_a \bar{p}(\xi_1, \xi_2) N_{c,\xi_1} N_{\delta,\xi_2} \hat{\mathbf{x}}_c \mathbf{x}_\delta \, d\xi_1 \, d\xi_2 \tag{5.133}$$

Since the nodal forces involve the nodal coordinates in the current configuration explicitly, it is necessary to compute a tangent matrix $\mathbf{K}_L$ for use in a Newton solution scheme. Linearizing Eq. (5.133) we obtain the tangent as

$$\mathbf{K}_L^{ab} = -\frac{\partial \mathbf{f}_a}{\partial \mathbf{u}_b} = \int_{-1}^{1}\int_{-1}^{1} N_a \bar{p}(\xi_1, \xi_2)\left[N_{b,\xi_1}N_{c,\xi_2} - N_{c,\xi_1}N_{b,\xi_2}\right]\hat{\mathbf{x}}_c \, d\xi_1 \, d\xi_2 \quad (5.134)$$

In general the tangent expression is unsymmetric; however, if the pressure loading is applied over a closed surface and is constant the final assembled terms are symmetric.[27]

For cases where the pressure varies over the surface the pressure may be computed by using an interpolation

$$\bar{p}(\xi_1, \xi_2) = N_a(\xi_1, \xi_2)\tilde{p}_a \quad (5.135)$$

in which $\tilde{p}_a$ are values of the known pressure at the nodes. Of course, these could also arise from solution of a problem which generates pressures on the contiguous surfaces and thus leads to the need to solve a coupled problem.[1]

The form for two-dimensional plane problems simplifies considerably since in this case Eq. (5.130) becomes

$$\int_{\gamma_t} \delta u_i \bar{t}_i \, d\gamma = \delta \tilde{u}_a \int_{-1}^{1} N_a \bar{p}(\xi_1)\left(N_{c,\xi_1}\mathbf{x}_c\right) \times \mathbf{e}_3 \, d\xi_1 \quad (5.136)$$

where ξ_1 is a one-dimensional natural coordinate for the surface side, $\mathbf{e}_3$ is the unit vector normal to the plane of deformation (which is constant), and $\bar{p}(\xi_1)$ is now the force per unit length of surface side. For this case the nodal forces for the follower pressure load are given explicitly by

$$\mathbf{f}_a = \int_{-1}^{1} N_a \bar{p}(\xi_1)\begin{Bmatrix} -x_{2,\xi_1} \\ x_{1,\xi_1} \end{Bmatrix} d\xi_1 \quad (5.137)$$

where x_{i,ξ_1} are derivatives computed from the one-dimensional finite element inter-polation used to approximate the element side. The case for axisymmetry involves additional terms and the reader is referred to reference 24 for details.

5.8 Concluding remarks

This chapter presents a unified approach for all finite deformation problems. The various procedures for solution of the resulting non-linear algebraic system have fol-lowed those presented in Chapters 2 and 3. Although not discussed extensively in the chapter, the extension to consider transient (dynamic) situations is easily accomplished. The long-term integration of dynamic problems occasionally presents difficulties using the time integration procedures designed for linear problems (e.g. those discussed in Sec. 2.4 and in reference 1). Here schemes which conserve momentum and energy for hyperelastic materials can be considered as alternatives, and the reader is referred to literature on the subject for additional details.[28–34]

We have also presented some mixed forms for developing elements which per-form well at finite strains and with materials which can exhibit nearly incompressible behaviour. These elements are developed in a form which allow the introduction of finite elastic and inelastic material models without difficulty. Indeed, we have shown that there is no need to decouple the constitutive behaviour between volumetric and deviatoric response as often assumed in many presentations. We usually find that transformation to a current configuration form in which either the Kirchhoff stress or the Cauchy stress is used directly will lead to a form which admits a simple extension of existing small deformation finite element procedures for developing the necessary residual (force) and stiffness matrices. An exception here is the presentation of the mixed–enhanced form in which all basic development is shown using the deformation gradient and first Piola–Kirchhoff stress. Here we could express final answers in a current configuration form also, but we leave these steps for the reader to perform.

References

1. O.C. Zienkiewicz, R.L. Taylor and J.Z. Zhu. *The Finite Element Method: Its Basis and Fundamentals*. Butterworth-Heinemann, Oxford, 6th edition, 2005.
2. J.C. Simo and F. Armero. Geometrically non-linear enhanced strain mixed methods and the method of incompatible modes. *International Journal for Numerical Methods in Engineering*, 33:1413–1449, 1992.
3. J.C. Simo, F. Armero and R.L. Taylor. Improved versions of assumed enhanced strain tri-linear elements for 3d finite deformation problems. *Computer Methods in Applied Mechanics and Engineering*, 110:359–386, 1993.
4. S. Glaser and F. Armero. On the formulation of enhanced strain finite elements in finite deformation. *Engineering Computations*, 14:759–791, 1996.
5. P.M.A. Areias, J.M.A. César de Sá, C.A. Conceição Antoónio and A.A. Fernandes. Analysis of 3D problems using a new enhanced strain hexahedral element. *International Journal for Numerical Methods in Engineering*, 58:1637–1682, 2003.
6. L.E. Malvern. *Introduction to the Mechanics of a Continuous Medium*. Prentice-Hall, Englewood Cliffs, NJ, 1969.
7. P. Chadwick. *Continuum Mechanics*. John Wiley & Sons, New York, 1976.
8. M.E. Gurtin. *An Introduction to Continuum Mechanics*. Academic Press, New York, 1981.
9. I.H. Shames and F.A. Cozzarelli. *Elastic and Inelastic Stress Analysis*. Taylor & Francis, Washington, DC, 1997. (Revised printing.)
10. J. Bonet and R.D. Wood. *Nonlinear Continuum Mechanics for Finite Element Analysis*. Cambridge University Press, Cambridge, 1997. ISBN 0–521–57272–X.
11. J.C. Simo and T.J.R. Hughes. *Computational Inelasticity*, volume 7 of *Interdisciplinary Applied Mathematics*. Springer-Verlag, Berlin, 1998.
12. C.C. Celigoj. An assumed enhanced displacement gradient ring-element for finite deformation axisymmetric and torsional problems. *International Journal for Numerical Methods in Engineering*, 43:1369–1382, 1998.
13. C.C. Celigoj. An improved 'assumed enhanced displacement gradient' ring-element for finite deformation axisymmetric and torsional problems. *International Journal for Numerical Methods in Engineering*, 50:899–918, 2001.
14. J.C. Simo and R.L. Taylor. Quasi-incompressible finite elasticity in principal stretches: continuum basis and numerical algorithms. *Computer Methods in Applied Mechanics and Engineering*, 85:273–310, 1991.
15. B. Fraeijs de Veubeke. Displacement and equilibrium models in finite element method. In O.C. Zienkiewicz and G.S. Holister, editors, *Stress Analysis*, Chapter 9, pages 145–197. John Wiley & Sons, Chichester, 1965.

16. U. Andelfinger, E. Ramm and D. Roehl. 2d- and 3d-enhanced assumed strain elements and their application in plasticity. In D. Owen, E. Oñate and E. Hinton, editors, *Proceedings of the 4th International Conference on Computational Plasticity*, pages 1997–2007. Pineridge Press, Swansea, 1992.

17. P. Wriggers and G. Zavarise. Application of augmented Lagrangian techniques for non-linear constitutive laws in contact interfaces. *Communications in Numerical Methods in Engineering*, 9:813–824, 1993.

18. M. Bischoff, E. Ramm and D. Braess. A class of equivalent enhanced assumed strain and hybrid stress finite elements. *Computational Mechanics*, 22:443–449, 1999.

19. E.P. Kasper and R.L. Taylor. A mixed-enhanced strain method: Part 1 – Geometrically linear problems. *Computers and Structures*, 75(3):237–250, 2000.

20. T.H.H. Pian and K. Sumihara. Rational approach for assumed stress finite elements. *International Journal for Numerical Methods in Engineering*, 20:1685–1695, 1985.

21. E.L. Wilson. The static condensation algorithm. *International Journal for Numerical Methods in Engineering*, 8:199–203, 1974.

22. E.P. Kasper and R.L. Taylor. A mixed–enhanced strain method: Part 2 – Geometrically nonlinear problems. *Computers and Structures*, 75(3):251–260, 2000.

23. J.T. Oden. Discussion on 'Finite element analysis of non-linear structures' by R.H. Mallett and P.V. Marçal. *Proc. Am. Soc. Civ. Eng.*, 95(ST6):1376–1381, 1969.

24. H.D. Hibbitt, P.V. Marçal and J.R. Rice. A finite element formulation for problems of large strain and large displacement. *International Journal of Solids and Structures*, 6:1069–1086, 1970.

25. J.C. Simo, R.L. Taylor and P. Wriggers. A note on finite element implementation of pressure boundary loading. *Communications in Applied Numerical Methods*, 7:513–525, 1991.

26. K. Schweizerhof. *Nichtlineare Berechnung von Tragwerken unter verformungsabhangiger belastung mit finiten Elementen*. Doctoral dissertation, U. Stuttgart, Stuttgart, Germany, 1982.

27. K. Schweizerhof and E. Ramm. Displacement dependent pressure loads in non-linear finite element analysis. *Computers and Structures*, 18(6):1099–1114, 1984.

28. T.A. Laursen and V. Chawla. Design of energy conserving algorithms for frictionless dynamic contact problems. *International Journal for Numerical Methods in Engineering*, 40:863–886, 1997.

29. J.C. Simo and N. Tarnow. The discrete energy-momentum method. Conserving algorithm for nonlinear elastodynamics. *Zeitschrift für Mathematik und Physik*, 43:757–793, 1992.

30. J.C. Simo and N. Tarnow. Exact energy-momentum conserving algorithms and symplectic schemes for nonlinear dynamics. *Computer Methods in Applied Mechanics and Engineering*, 100:63–116, 1992.

31. N. Tarnow. *Energy and momentum conserving algorithms for Hamiltonian systems in the nonlinear dynamics of solids*. PhD thesis, Department of Mechanical Engineering, Stanford University, Stanford, California, 1993.

32. O. González. *Design and analysis of conserving integrators for nonlinear Hamiltonian systems with symmetry*. PhD thesis, Department of Mechanical Engineering, Stanford University, Stanford, California, 1996.

33. M.A. Crisfield and J. Shi. An energy conserving co-rotational procedure for non-linear dynamics with finite elements. *Nonlinear Dynamics*, 9:37–52, 1996.

34. U. Galvanetto and M.A. Crisfield. An energy-conserving co-rotational procedure for thedynamics of planar beam structures. *International Journal for Numerical Methods in Engineering*, 39:2265–2282, 1996.

6

Material constitution for finite deformation

6.1 Introduction

In order to complete the finite element development for the finite deformation problem it is necessary to describe how the material behaves when subjected to deformation or deformation histories. In the previous chapter we considered elastic behaviour without introducing details on how to model specific material behaviour. Clearly, restriction to elastic behaviour is inadequate to model the behaviour of many engineering materials as we have already shown in previous applications. The modelling of engineering materials at finite strain is a subject of much research and any complete summary on the state of the art is clearly outside the scope of what can be presented here. In this chapter we present only some classical methods which may be used to model elastic, viscoelastic and elastic–plastic-type behaviours. The reader is directed to literature for details on other constitutive models (e.g. see references 1 and 2).

We first consider some methods which may be used to describe the behaviour of isotropic elastic materials which undergo finite deformation. In this section we restrict attention to those materials in which a stored energy function is used; such behaviour is often called *hyperelastic*. Later we will extend this to permit the use of viscoelastic and elastic–plastic models and show that much of the material presented in Chapter 4 is here again useful. Finally, to permit the modelling of materials which are not isotropic or cannot be expressed as an extension to elastic behaviour (e.g. generalized plasticity models of Chapter 4) we introduce a rate form – here again many options are possible. This latter form is heuristic and such an approach should be used with caution and only when experimental data are available to verify the behaviour obtained.

6.2 Isotropic elasticity

6.2.1 Isotropic elasticity – formulation in invariants

We consider a finite deformation form for *hyperelasticity* in which a stored energy density function, W, is used to compute stresses. For a stored energy density expressed in terms of right Cauchy–Green deformation tensor, C_{IJ}, the second Piola–Kirchhoff stress is computed by using Eq. (5.34). Through standard transformation we can also obtain the Kirchhoff stress as[1,3,4]

$$\tau_{ij} = 2b_{ik}\frac{\partial W}{\partial b_{kj}} \tag{6.1}$$

and thus, by using Eq. (5.19), also obtain directly the Cauchy stress.

For an isotropic material the stored energy density depends only on three invariants of the deformation. Here we consider the three invariants (noting they also are equal to those for b_{ij}) expressed as[1,4]

$$
\begin{aligned}
I &= C_{KK} = b_{kk} \\
II &= \tfrac{1}{2}(I^2 - C_{KL}C_{LK}) = \tfrac{1}{2}(I^2 - b_{kl}b_{lk}) \\
III &= \det C_{KL} = \det b_{kl} = J^2 \quad \text{where} \quad J = \det F_{KL}
\end{aligned}
\tag{6.2}
$$

and write the strain energy density as

$$W(C_{KL}) = W(b_{kl}) \equiv W(I, II, J) \tag{6.3}$$

where we select J instead of III as the measure of the volume change. In this form the second Piola–Kirchhoff stress is computed as

$$S_{IJ} = 2\left[\frac{\partial W}{\partial I}\frac{\partial I}{\partial C_{IJ}} + \frac{\partial W}{\partial II}\frac{\partial II}{\partial C_{IJ}} + \frac{\partial W}{\partial J}\frac{\partial J}{\partial C_{IJ}}\right] \tag{6.4}$$

The derivatives of the invariants may be evaluated as (see Appendix B)

$$\frac{\partial I}{\partial C_{IJ}} = \delta_{IJ}, \quad \frac{\partial II}{\partial C_{IJ}} = I\delta_{IJ} - C_{IJ}, \quad \frac{\partial J}{\partial C_{IJ}} = \frac{1}{2}JC_{IJ}^{-1} \tag{6.5}$$

Thus, the stress is given by

$$S_{IJ} = 2\left[\delta_{IJ} \quad (I\delta_{IJ} - C_{IJ}) \quad \tfrac{1}{2}JC_{IJ}^{-1}\right]\left\{\begin{array}{c}\dfrac{\partial W}{\partial I} \\[2mm] \dfrac{\partial W}{\partial II} \\[2mm] \dfrac{\partial W}{\partial J}\end{array}\right\} \tag{6.6}$$

Using Eq. (5.20) the second Piola–Kirchhoff stress may be transformed to the Cauchy stress and gives

$$\sigma_{ij} = \frac{2}{J}\left[b_{ij} \quad (Ib_{ij} - b_{im}b_{mj}) \quad \tfrac{1}{2}J\delta_{ij}\right]\left\{\begin{array}{c}\dfrac{\partial W}{\partial I} \\[2mm] \dfrac{\partial W}{\partial II} \\[2mm] \dfrac{\partial W}{\partial J}\end{array}\right\} \tag{6.7}$$

A Newton-type solution process requires computation of the elastic moduli for the finite elasticity model. The elastic moduli with respect to the reference configuration are deduced from[1,4]

$$C_{IJKL} = 4\frac{\partial^2 W}{\partial C_{IJ}\partial C_{KL}} = 2\frac{\partial S_{IJ}}{\partial C_{KL}} \tag{6.8}$$

thus from Eq. (6.6) the general form for the elastic moduli of an isotropic material is given by

$$C_{IJKL} = 4\left[\delta_{IJ}, \; (I\delta_{IJ} - C_{IJ}), \; \tfrac{1}{2}JC_{IJ}^{-1}\right]$$

$$\times \begin{bmatrix} \dfrac{\partial^2 W}{\partial I^2} & \dfrac{\partial^2 W}{\partial I\partial II} & \dfrac{\partial^2 W}{\partial I\partial J} \\[2mm] \dfrac{\partial^2 W}{\partial II\partial I} & \dfrac{\partial^2 W}{\partial II^2} & \dfrac{\partial^2 W}{\partial II\partial J} \\[2mm] \dfrac{\partial^2 W}{\partial J\partial I} & \dfrac{\partial^2 W}{\partial J\partial II} & \dfrac{\partial^2 W}{\partial J^2} \end{bmatrix} \begin{Bmatrix} \delta_{KL} \\[2mm] (I\delta_{KL} - C_{KL}) \\[2mm] \tfrac{1}{2}JC_{KL}^{-1} \end{Bmatrix}$$

$$+ \left[\delta_{IJ}\delta_{KL} - \tfrac{1}{2}(\delta_{IK}\delta_{JL} + \delta_{IL}\delta_{JK}), \;\; J(C_{IJ}^{-1}C_{KL}^{-1} - 2C_{IJKL}^{-1})\right] \begin{Bmatrix} 4\dfrac{\partial W}{\partial II} \\[3mm] \dfrac{\partial W}{\partial J} \end{Bmatrix} \tag{6.9}$$

where

$$C_{IJKL}^{-1} = \tfrac{1}{2}\left[C_{IK}^{-1}C_{JL}^{-1} + C_{IL}^{-1}C_{JK}^{-1}\right] \tag{6.10}$$

The spatial elasticities related to the Cauchy stress are obtained by the *push forward* transformation

$$Jc_{ijkl} = F_{iI}F_{jJ}F_{kK}F_{lL}C_{IJKL} \tag{6.11}$$

which, applied to Eq. (6.9), gives

$$c_{ijkl} = \frac{4}{J}[b_{ij}, \; (Ib_{ij} - b_{im}b_{mj}), \; \tfrac{1}{2}J\delta_{ij}]$$

$$\times \begin{bmatrix} \dfrac{\partial^2 W}{\partial I^2} & \dfrac{\partial^2 W}{\partial I\partial II} & \dfrac{\partial^2 W}{\partial I\partial J} \\[2mm] \dfrac{\partial^2 W}{\partial II\partial I} & \dfrac{\partial^2 W}{\partial II^2} & \dfrac{\partial^2 W}{\partial II\partial J} \\[2mm] \dfrac{\partial^2 W}{\partial J\partial I} & \dfrac{\partial^2 W}{\partial J\partial II} & \dfrac{\partial^2 W}{\partial J^2} \end{bmatrix} \begin{Bmatrix} b_{kl} \\[2mm] (Ib_{kl} - b_{kn}b_{nl}) \\[2mm] \tfrac{1}{2}J\delta_{kl} \end{Bmatrix}$$

$$+ \frac{1}{J}\left[b_{ij}b_{kl} - \tfrac{1}{2}(b_{ik}b_{jl} + b_{il}b_{jk}), \;\; J[\delta_{ij}\delta_{kl} - 2\mathcal{I}_{ijkl}]\right] \begin{Bmatrix} 4\dfrac{\partial W}{\partial II} \\[3mm] \dfrac{\partial W}{\partial J} \end{Bmatrix} \tag{6.12}$$

where

$$\mathcal{I}_{ijkl} = \frac{1}{2}\left[\delta_{ik}\delta_{jl} + \delta_{il}\delta_{jk}\right] \tag{6.13}$$

The above expressions describe completely the necessary equations to construct a finite element model for any isotropic hyperelastic material written in terms of invariants. All that remains is to select a specific form for the stored energy function W. Here, many options exist and we include below only a few very simple models. For others the reader is referred to literature on the subject.

Example 6.1   Volumetric behaviour

If we let the stored energy function be given by

$$W = k_U U(J)$$

where k_U is a scalar material parameter and J measures the volumetric deformation. Using Eq. (6.7) we obtain the stress in the current configuration given by

$$\sigma_{ij}^{(U)} = k_U \frac{\partial U}{\partial J} \delta_{ij} \tag{6.14}$$

which is a pure hydrostatic stress (i.e. a *pressure*). Using Eq. (6.12) the elastic tangent moduli for this model are given by

$$c_{ijkl}^{(U)} = k_U \left[\left(J \frac{\partial^2 U}{\partial J^2} + \frac{\partial U}{\partial J} \right) \delta_{ij} \delta_{kl} - 2 \mathcal{I}_{ijkl} \frac{\partial U}{\partial J} \right] \tag{6.15}$$

In the models given below we will assume that the volumetric behaviour is proportional to $U(J)$ in which one of the following models is used:

$$U(J) = \begin{cases} \frac{1}{4}(J^2 - 1) - \frac{1}{2}\ln J \\ \frac{1}{2}(J - 1)^2 \\ \frac{1}{2}(\ln J)^2 \end{cases} \tag{6.16}$$

The derivatives of these give

$$\frac{\partial U}{\partial J} = \begin{cases} \frac{1}{2}(J - \frac{1}{J}) \\ J - 1 \\ \frac{1}{J}\ln J \end{cases} \quad \text{and} \quad \frac{\partial^2 U}{\partial J^2} = \begin{cases} \frac{1}{2}(1 + \frac{1}{J^2}) \\ 1 \\ \frac{1}{J^2}(1 - \ln J) \end{cases} \tag{6.17}$$

We note that only the first of these models gives a pressure which approaches an infinite value when $J \to 0$ and when $J \to \infty$. However, when J is near unity it may be approximated by

$$J \approx 1 + \frac{\partial u_i}{\partial x_i}$$

and all models give

$$\frac{\partial U}{\partial J} \approx \frac{\partial u_i}{\partial x_i} \quad \text{and} \quad \frac{\partial^2 U}{\partial J^2} \approx 1$$

Thus, any of the models may be used when the deformations remain moderate.

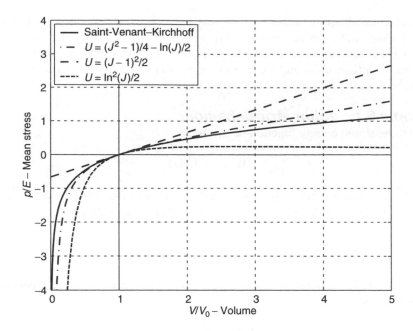

Fig. 6.1 Volumetric deformation.

In Fig. 6.1 we show the behaviour of the above forms together with the results from the Saint-Venant–Kirchhoff model presented in the previous chapter. The material property for k_U is chosen to match the small strain bulk modulus for small strain, with $k_U = K = E/(1 - 2\nu)/3$. Results are normalized by E and ν is equal to 0.25. It is clear that the model using $U = \ln^2(J)/2$ is not useful in problems which have large dilation since the mean stress is getting smaller. For this stress state the other models are acceptable. However, we shall see next that when uniaxial stress states are considered the Saint-Venant–Kirchhoff model also becomes unacceptable for large strain cases.

Example 6.2 Compressible neo-Hookean material

As an example, we consider the case of a neo-Hookean material[5] that includes a compressibility effect. The stored energy density is expressed as

$$W(I, J) = W^{(1)}(I, J) + \lambda U(J)$$
$$W^{(1)}(I, J) = \tfrac{1}{2}\mu(I - 3 - 2\ln J) \tag{6.18}$$

where the material constants $k_U = \lambda$ and μ are selected to give the same response in small deformations as a linear elastic material using classical Lamé parameters[1] and $U(J)$ is one of the volumetric stored energy functions given in (6.16).

The non-zero derivatives of $W^{(1)}$ are given by:

$$\left\{ \begin{array}{c} \dfrac{\partial W^{(1)}}{\partial I} \\[2mm] \dfrac{\partial W^{(1)}}{\partial J} \end{array} \right\} = \dfrac{1}{2}\mu \left\{ \begin{array}{c} 1 \\[2mm] -\dfrac{1}{J} \end{array} \right\} \qquad \text{and} \qquad \dfrac{\partial^2 W}{\partial J^2} = \dfrac{\mu}{J^2}$$

Substitution of these into Eqs (6.7) and (6.12) gives

$$\sigma_{ij}^{(1)} = \frac{\mu}{J}(b_{ij} - \delta_{ij}) \quad \text{and} \quad c_{ijkl}^{(1)} = \frac{2\mu}{J}\mathcal{I}_{ijkl} \tag{6.19}$$

The final stress is obtained from the sum of Eqs (6.14) and (6.19) and, similarly, the tangent moduli from the sum of Eqs (6.15) and (6.19).

We note that when $J \approx 1$ the small deformation result

$$c_{ijkl} \approx \lambda \delta_{ij}\delta_{kl} + 2\mu\mathcal{I}_{ijkl} \tag{6.20}$$

is obtained and thus matches the usual linear elastic relations in terms of the Lamé parameters. This permits the finite deformation formulation to be used directly for analyses in which the small strain assumptions hold as well as for situations in which deformations are finite.

Example 6.3   A modified compressible neo-Hookean material

As an alternative example, we consider the case of a modified neo-Hookean material which in the small strain limit is identical to the isotropic linear elastic model given in terms of bulk K ($= k_U$) and shear G moduli.

The stored energy density for this case is expressed as

$$\begin{aligned} W(I, J) &= W^{(2)}(I, J) + K\,U(J) \\ W^{(2)}(I, J) &= \tfrac{1}{2}\,G\,(J^{-2/3}I - 3) \end{aligned} \tag{6.21}$$

For this model, the non-zero derivatives of $W^{(2)}$ are given by:

$$\left\{ \begin{array}{c} \dfrac{\partial W^{(2)}}{\partial I} \\[2mm] \dfrac{\partial W^{(2)}}{\partial J} \end{array} \right\} = \frac{1}{2}\,G \left\{ \begin{array}{c} J^{-2/3} \\[2mm] -\tfrac{2}{3}\,J^{-5/3}I \end{array} \right\} \quad \text{and} \quad \left[\begin{array}{cc} \dfrac{\partial^2 W^{(2)}}{\partial I^2} & \dfrac{\partial^2 W^{(2)}}{\partial I\,\partial J} \\[3mm] \dfrac{\partial^2 W^{(2)}}{\partial J\,\partial I} & \dfrac{\partial^2 W^{(2)}}{\partial J^2} \end{array} \right]$$

$$= \frac{1}{2}\,G\,J^{-2/3} \left[\begin{array}{cc} 0 & -\tfrac{2}{3}\,J \\[2mm] -\tfrac{2}{3}\,J & \tfrac{10}{9}\,J^2 I \end{array} \right]$$

Substitution of these results into Eq. (6.7) gives

$$\sigma_{ij}^{(2)} = \frac{G}{J}(\tilde{b}_{ij} - \tfrac{1}{3}\delta_{ij}\tilde{I}) \tag{6.22}$$

where $\tilde{b}_{ij} = J^{-2/3}b_{ij}$ and $\tilde{I} = J^{-2/3}I = \tilde{b}_{kk}$. We note that the term multiplying G/J is a *deviatoric quantity*, that is

$$\tilde{b}_{ii} - \tfrac{1}{3}\delta_{ii}\tilde{b}_{kk} = 0,$$

we thus define

$$\tilde{b}_{ij}^d = \tilde{b}_{ij} - \tfrac{1}{3}\delta_{ij}\tilde{b}_{kk}$$

and simplify Eq. (6.22) to

$$\sigma_{ij}^{(2)} = \frac{G}{J}\tilde{b}_{ij}^d \tag{6.23}$$

Using the definitions for the derivatives of the stored energy function and introducing the above deformation measures the material moduli for the current spatial configuration are given as

$$c_{ijkl}^{(2)} = \frac{2G}{3J} \left[\tilde{b}_{mm} (\mathcal{I}_{ijkl} - \tfrac{1}{3} \delta_{ij} \delta_{kl}) - \delta_{ij} \tilde{b}_{kl}^d - \tilde{b}_{ij}^d \delta_{kl} \right] \tag{6.24}$$

Again, the results for the total stress and material moduli are obtained by combining the above with a model for volumetric behaviour.

We note that when $J \approx 1$ the small deformation result becomes

$$c_{ijkl} \approx K \, \delta_{ij} \delta_{kl} + 2 \, G \, (\mathcal{I}_{ijkl} - \tfrac{1}{3} \delta_{ij} \delta_{kl}) \tag{6.25}$$

and thus matches the usual linear elastic relations in terms of the bulk and shear moduli.

In Fig. 6.2 we show the uniaxial response of the two forms of the neo-Hookean material together with those for the Saint-Venant–Kirchhoff model. The properties of the models are picked to match the small strain case for a modulus of elasticity E. For the neo-Hookean models we note that uniaxial behaviour involves both volumetric and distortional deformations, thus, it is necessary to use a model for $U(J)$ in addition to each of the forms for $W^{(i)}$. The parameters for λ, K and $\mu = G$ are selected to match the small strain values in terms of E and ν with $\nu = 0.25$.

While the above models are classical and easily treated, they are not accurate for hyperelastic materials which exhibit increased stiffness with stretch. In order to treat such materials it is necessary to consider alternative models.

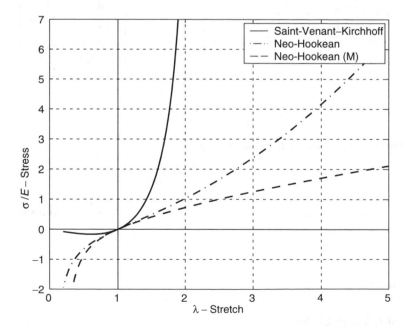

Fig. 6.2 Uniaxial stretch. Saint-Venant-Kirchhoff and neo-Hookean material. (M) denotes a modified compressible material.

6.2.2 Isotropic elasticity – formulation in modified invariants

In the previous example we introduced invariants based on the modified deformation gradient given by

$$\tilde{F}_{iI} = J^{-1/3} F_{iI} \tag{6.26}$$

which yields the modified right and left Cauchy deformation tensors

$$\tilde{C}_{IJ} = \tilde{F}_{iI}\tilde{F}_{iJ} \quad \text{and} \quad \tilde{b}_{ij} = \tilde{F}_{iI}\tilde{F}_{jI} \tag{6.27}$$

respectively. The modified invariants given by

$$\tilde{I} = J^{-2/3} I \quad \text{and} \quad \widetilde{II} = J^{-4/3} II \tag{6.28}$$

together with J may be used to construct stored energy functions as

$$W = W(\tilde{I}, \widetilde{II}, J) \tag{6.29}$$

In addition to the modified neo-Hookean model given above, several forms in terms of $W(\tilde{I})$ have been introduced to represent the behaviour of hyperelastic materials subjected to large stretch. A general form for such models depending only on $\tilde{I}$ is given by

$$\left\{ \begin{array}{c} \dfrac{\partial W}{\partial I} \\[2mm] \dfrac{\partial W}{\partial J} \end{array} \right\} = \frac{\partial W}{\partial \tilde{I}} \left\{ \begin{array}{c} J^{-2/3} \\[2mm] -\frac{2}{3} J^{-1}\tilde{I} \end{array} \right\} \tag{6.30}$$

which may be used in Eq. (6.7) to compute Cauchy stresses. Similarly, we recover

$$\begin{bmatrix} \dfrac{\partial^2 W}{\partial I^2} & \dfrac{\partial^2 W}{\partial I \partial J} \\[3mm] \dfrac{\partial^2 W}{\partial J \partial I} & \dfrac{\partial^2 W}{\partial J^2} \end{bmatrix} = \frac{\partial W}{\partial \tilde{I}} \begin{bmatrix} 0 & -\frac{2}{3} J^{-5/3} \\[2mm] -\frac{2}{3} J^{-5/3} & \frac{10}{9} J^{-2}\tilde{I} \end{bmatrix} + \frac{\partial^2 W}{\partial \tilde{I}^2} \begin{bmatrix} J^{-4/9} & -\frac{2}{3} J^{-5/3}\tilde{I} \\[2mm] -\frac{2}{3} J^{-5/3}\tilde{I} & \frac{4}{9} J^{-2}\tilde{I}^2 \end{bmatrix} \tag{6.31}$$

which may be used in Eq. (6.12) to compute the tangent matrix.

In the next two examples we consider two forms based on the first invariant $\tilde{I}$ and the volumetric effects from J.

Example 6.4 Yeoh model

The first extended form was proposed by Yeoh[6] and uses the stored energy function

$$W^{(3)} = \frac{1}{2} \mu \left[(\tilde{I} - 3) + k_1(\tilde{I} - 3)^2 + k_2(\tilde{I} - 3)^3 \right] \tag{6.32}$$

for the deviatoric part and $k_U\, U(J)$ given in Example 6.1 for the volumetric part.

The derivative with respect to the modified invariant for the Yeoh model is given by

$$\frac{\partial W^{(3)}}{\partial \tilde{I}} = \frac{1}{2} \mu \left[1 + 2 k_1 (\tilde{I} - 3) + 3 k_2 (\tilde{I} - 3)^2 \right] \tag{6.33}$$

Similarly the second derivative is given by

$$\frac{\partial^2 W^{(3)}}{\partial \tilde{I}^2} = \frac{1}{2} \mu \left[k_1 + 6 k_2 (\tilde{I} - 3) \right] \tag{6.34}$$

These may be used in Eqs (6.30) and (6.7) to compute the Cauchy stress and in Eqs (6.31) and (6.12) to compute the tangent tensor. Indeed, the modifications to the expressions given for the modified neo-Hookean model in Example 6.3 are quite trivial.

Example 6.5   Arruda–Boyce model

The second extended form is due to Arruda and Boyce[7,8] and the stored energy expression is given by

$$W^{(4)} = \frac{1}{2} \bar{\mu} \left[(\tilde{I} - 3) + \frac{1}{10n} (\tilde{I}^2 - 9) + \frac{11}{525\, n^2} (\tilde{I}^3 - 27) \right] \tag{6.35}$$

where

$$\bar{\mu} = \frac{\mu}{1 + \frac{3}{5n} + \frac{99}{175n^2}}$$

for the deviatoric part and also uses $k_U\, U(J)$ given in Example 6.1 for the volumetric part. Typically $\mu = G$, $k_U = K$, the linear elastic shear and bulk moduli, and n is the number of segments in the chain of the material molecular network structure.

The derivatives of the stored energy function are

$$\frac{\partial W^{(4)}}{\partial \tilde{I}} = \frac{1}{2} \mu \left[1 + \frac{1}{5n} \tilde{I} + \frac{11}{175n^2} \tilde{I}^2 \right] \tag{6.36}$$

and

$$\frac{\partial^2 W^{(4)}}{\partial \tilde{I}^2} = \frac{1}{2} \mu \left[\frac{1}{5n} + \frac{22}{175n^2} \tilde{I} \right] \tag{6.37}$$

which may be used in Eqs (6.30) and (6.7) to compute the Cauchy stress and in Eqs (6.31) and (6.12) to compute the tangent tensor.

In Fig. 6.3 we show the results for a uniaxial stretch with the properties E, $\nu = 0.49$ and $n = 2$. The stress is normalized by E. Note the properties are now set to give a nearly incompressible state which is where the model is most applicable.

The above models may be used in either the displacement model or with the three-field mixed form described in the previous chapter for situations where the ratio λ/μ or K/G is large (i.e. nearly incompressible behaviour). Indeed this was an early use of the model. In addition, using Eqs (6.7) and (6.12) it is a simple task to develop any material model for which the stored energy function is expressed in terms of invariants.

6.2.3 Isotropic elasticity – formulation in principal stretches

Other forms of elastic constitutive equations may be introduced by using appropriate expansions of the stored energy density function. As an alternative, an elastic formulation expressed in terms of principal stretches (which are the square root of the eigenvalues of C_{IJ} or b_{ij}) may be introduced. This approach has been presented by Ogden[9] and by Simo and Taylor.[10]

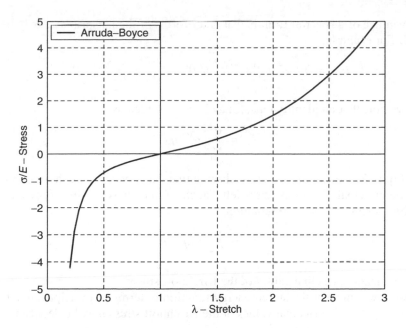

Fig. 6.3 Uniaxial stretch. Arruda–Boyce model.

We first consider a change of coordinates given by (see Appendix B, Volume 1)

$$x_i = \Lambda_{m'i} x_{m'} \tag{6.38}$$

where $\Lambda_{m'i}$ are direction cosines between two Cartesian systems. The transformation equations for a second-rank tensor, say b_{ij}, may then be written in the form

$$b_{ij} = \Lambda_{m'i} b_{m'n'} \Lambda_{n'j} \tag{6.39}$$

To compute specific relations for the transformation array we consider the solution of the eigenproblem

$$b_{ij} q_j^{(n)} = q_i^{(n)} b_n; \quad n = 1, 2, 3 \quad \text{with} \quad q_k^{(m)} q_k^{(n)} = \delta_{mn} \tag{6.40}$$

where b_n are the *principal values* of b_{ij}, and $q_i^{(n)}$ are direction cosines for the principal directions. The principal values of b_{ij} are equal to the square of the *principal stretches*, λ_n, that is,

$$b_n = \lambda_n^2 \tag{6.41}$$

If we assign the direction cosines in the transformation equation (6.39) as

$$\Lambda_{n'j} \equiv q_j^{(n)} \tag{6.42}$$

the spectral representation of the deformation tensor results and may be expressed as

$$b_{ij} = \sum_m \lambda_m^2 q_i^{(m)} q_j^{(m)} \tag{6.43}$$

An advantage of a spectral form is that other forms of the tensor may easily be represented. For example,

$$b_{ik}b_{kj} = \sum_m \lambda_m^4 q_i^{(m)} q_j^{(m)} \quad \text{and} \quad b_{ik}^{-1} = \sum_m \lambda_m^{-2} q_i^{(m)} q_j^{(m)} \tag{6.44}$$

Also, we note that an identity tensor may be represented as

$$\delta_{ij} = \sum_m q_i^{(m)} q_j^{(m)} \tag{6.45}$$

From Eq. (6.7) we can immediately observe that Cauchy and Kirchhoff stresses have the same principal directions as the left Cauchy–Green tensor. Thus, for example, the Kirchhoff stress has the representation

$$\tau_{ij} = \sum_m \tau_m q_i^{(m)} q_j^{(m)} \tag{6.46}$$

where τ_m denote principal values of the Kirchhoff stress.

If we now represent the stored energy function in terms of principal stretch values as $\hat{w}(\lambda_1, \lambda_2, \lambda_3)$ the principal values of the Kirchhoff stress may be deduced from[2,9]

$$\tau_m = \lambda_m \frac{\partial \hat{w}}{\partial \lambda_m} \quad \text{(no sum)} \tag{6.47}$$

The reader is referred to the literature for a more general discussion on formulations in principal stretches for use in general elasticity problems.[2,9,10] Here we wish to consider one form which is useful to develop solution algorithms for finite elastic–plastic behaviour of isotropic materials in which elastic strains are quite small. Such a form is useful, for example, in modelling metal plasticity.

Example 6.6   Logarithmic principal stretch form

A particularly simple result is obtained by writing the stored energy function in terms of logarithmic principal stretches. Accordingly, we take

$$\hat{w}(\lambda_1, \lambda_2, \lambda_3) = w(\varepsilon_1, \varepsilon_2, \varepsilon_3) \quad \text{where} \quad \varepsilon_m = \log(\lambda_m) \tag{6.48}$$

From Eq. (6.47) it follows that

$$\tau_m = \frac{\partial w}{\partial \varepsilon_m} \tag{6.49}$$

which is now identical to the form from linear elasticity, but expressed in principal directions. It also follows that the elastic moduli may be written as[2,9] (summation convention is not used to write this expression)

$$
\begin{aligned}
J c_{ijkl} = &\sum_{m=1}^{3} \sum_{n=1}^{3} [c_{mn} - 2\tau_m \delta_{mn}] q_i^{(m)} q_j^{(m)} q_k^{(n)} q_l^{(n)} \\
&+ \frac{1}{2} \sum_{m=1}^{3} \sum_{\substack{n=1 \\ n \neq m}}^{3} g_{mn} [q_i^{(m)} q_j^{(n)} q_k^{(m)} q_l^{(n)} + q_i^{(m)} q_j^{(n)} q_k^{(n)} q_l^{(m)}]
\end{aligned}
\tag{6.50}
$$

where

$$c_{mn} = \frac{\partial^2 w}{\partial \varepsilon_m \partial \varepsilon_n} \quad \text{and} \quad g_{mn} = \begin{cases} \dfrac{\tau_m \lambda_n^2 - \tau_n \lambda_m^2}{\lambda_m^2 - \lambda_n^2}; & \lambda_m \neq \lambda_n \\ \dfrac{\partial(\tau_m - \tau_n)}{\partial \varepsilon_m}; & \lambda_m = \lambda_n \end{cases} \quad (6.51)$$

In practice the equal root form is used whenever differences are less than a small tolerance (say 10^{-8}).

Use of a quadratic form for w given by

$$w = \frac{1}{2}(K - \frac{2}{3}G)[\varepsilon_1 + \varepsilon_2 + \varepsilon_3]^2 + G[\varepsilon_1^2 + \varepsilon_2^2 + \varepsilon_3^2] \quad (6.52)$$

yields principal Kirchhoff stresses given by

$$\begin{Bmatrix} \tau_1 \\ \tau_2 \\ \tau_3 \end{Bmatrix} = \begin{bmatrix} K + \frac{4}{3}G & K - \frac{2}{3}G & K - \frac{2}{3}G \\ K - \frac{2}{3}G & K + \frac{4}{3}G & K - \frac{2}{3}G \\ K - \frac{2}{3}G & K - \frac{2}{3}G & K + \frac{4}{3}G \end{bmatrix} \begin{Bmatrix} \varepsilon_1 \\ \varepsilon_2 \\ \varepsilon_3 \end{Bmatrix} \quad (6.53)$$

in which the 3×3 elasticity matrix is given by a *constant* coefficient matrix which is identical to the usual linear elastic expression in terms of bulk and shear moduli. We also note that when roots are equal

$$\frac{\partial(\tau_m - \tau_n)}{\partial \varepsilon_m} = (K + \frac{4}{3}G) - (K - \frac{2}{3}G) = 2G \quad (6.54)$$

which defines the usual shear modulus form in isotropic linear elasticity.

6.2.4 Plane stress applications

The above constitutive models may be used directly for the two-dimensional plane strain and axisymmetric problems; however, for plane stress it is necessary to modify the constitutive terms to enforce the $\sigma_{33} = 0$ condition and thus account for the F_{33} term of the deformation gradient described in Sec. 5.4.2. A local iteration form based on the second Piola–Kirchhoff stress (S) and the Green–Lagrange strain (E) was proposed by Klinkel and Govindjee.[11] The local iteration is based on a linearized form for the S_{33} stress and a Newton method. Accordingly, for each computation point in the element (i.e. the quadrature point used to compute the arrays) we obtain the deformation gradient given by

$$F_{iI} = \begin{bmatrix} F_{11} & F_{12} & 0 \\ F_{21} & F_{22} & 0 \\ 0 & 0 & F_{33} \end{bmatrix} = \begin{bmatrix} F_m & 0 \\ 0 & F_{33} \end{bmatrix} \quad (6.55)$$

in which the components F_{11}, F_{12}, F_{21} and F_{22} (F_m) are computed from the current displacements as described in Eq. (5.87) and F_{33} is computed from the material constitution as described next. A linearization of the second Piola–Kirchhoff stress S_{IJ} in terms of the Green–Lagrange strain E_{IJ} may be expressed as

$$S_{IJ}^{(k+1)} = S_{IJ}^{(k)} + dS_{IJ}^{(k)}$$

where

$$dS_{IJ}^{(k)} = C_{IJKL}^{(k)} dE_{KL}^{(k)}$$

where C_{IJKL} are the tangent moduli in the reference configuration for the given constitutive model.

We can partition the above relations into two parts and write the result in matrix notation. The in-plane components are given by

$$\mathbf{S}_m = \begin{bmatrix} S_{11} & S_{22} & S_{12} \end{bmatrix}^T$$

and the stress normal to the plane of deformation S_{33}.* Accordingly, the linearization may be written as

$$\begin{Bmatrix} \mathbf{S}_m^{(k+1)} \\ S_{33}^{(k+1)} \end{Bmatrix} = \begin{Bmatrix} \mathbf{S}_m^{(k)} \\ S_{33}^{(k)} \end{Bmatrix} + \begin{Bmatrix} d\mathbf{S}_m^{(k)} \\ dS_{33}^{(k)} \end{Bmatrix} \tag{6.56a}$$

where

$$\begin{Bmatrix} d\mathbf{S}_m^{(k)} \\ dS_{33}^{(k)} \end{Bmatrix} = \begin{bmatrix} \mathbf{C}_{mm}^{(k)} & \mathbf{C}_{m3}^{(k)} \\ \mathbf{C}_{3m}^{(k)} & C_{3333}^{(k)} \end{bmatrix} \begin{Bmatrix} d\mathbf{E}_m^{(k)} \\ dE_{33}^{(k)} \end{Bmatrix} \tag{6.56b}$$

where $\mathbf{E}_m$ contains the in-plane Green–Lagrange strains in the same order as for stresses. The tangent moduli are given by $\mathbf{C}_{mm}$ for the in-plane components, C_{3333} associated with S_{33} and E_{33} and $\mathbf{C}_{m3}$, $\mathbf{C}_{3m}$ are the moduli which couple the two.

To obtain the value of E_{33} for the current solution step Klinkel and Govindjee propose a local Newton iteration given by

$$S_{33}^{(k+1)} = S_{33}^{(k)} + C_{3333}^{(k)} dE_{33}^{(k)} = 0 \tag{6.57a}$$

with the update

$$E_{33}^{(k+1)} = E_{33}^{(k)} + dE_{33}^{(k)} \tag{6.57b}$$

where $dE_{33}^{(k)} = F_{33} \, dF_{33}$. Iteration continues until convergence to a specified tolerance is achieved. This form only involves the solution of the scalar equation as

$$dE_{33}^{(k)} = -\frac{S_{33}^{(k)}}{C_{3333}^{(k)}} \tag{6.58}$$

After convergence is achieved the tangent moduli $\hat{\mathbf{D}}_T$ for the next equilibrium iteration [viz. Eqs (5.62a) and (5.63)] are given by

$$\hat{\mathbf{D}}_T = \mathbf{C}_{mm} - \mathbf{C}_{m3} C_{3333}^{-1} \mathbf{C}_{3m} \tag{6.59}$$

The above form is efficient for problems formulated in terms of the reference configuration as described in Sec. 5.3.1; however, for spatial forms where the Cauchy stress is computed directly the above requires the constitutive model to be transformed to the reference configuration in order to perform the local iteration.

---

* We restrict attention here to cases where the 3 direction defines a plane of symmetry such that given $E_{13} = E_{23} = 0$ the stress components S_{13}, S_{23} are also zero.

An alternative is to transform Eq. (6.57a) to the current configuration by noting

$$\sigma_{33}^{(k)} = \frac{1}{J^{(k)}} F_{33}^{(k)} S_{33}^{(k)} F_{33}^{(k)} \quad \text{and} \quad c_{3333}^{(k)} = \frac{1}{J^{(k)}} F_{33}^{(k)} F_{33}^{(k)} C_{3333}^{(k)} F_{33}^{(k)} F_{33}^{(k)} \tag{6.60}$$

This gives the local iteration Newton method written as

$$\sigma_{33}^{(k)} + c_{3333}^{(k)} d\varepsilon_{33}^{(k)} = 0 \tag{6.61}$$

where the incremental normal strain for plane stress is given by

$$d\varepsilon_{33}^{(k)} = \frac{d F_{33}^{(k)}}{F_{33}^{(k)}} \tag{6.62}$$

The iteration thus is achieved by solving the scalar equation to give

$$d F_{33}^{(k)} = -\frac{\sigma_{33}^{(k)}}{c_{3333}^{(k)}} F_{33}^{(k)} \tag{6.63}$$

with the update

$$F_{33}^{(k+1)} = F_{33}^{(k)} + d F_{33}^{(k)} \tag{6.64}$$

This local iteration is continued until convergence is achieved, that is

$$|d F_{33}^{(k)}| \leq tol \cdot |F_{33}^{(k)}|$$

where *tol* is a specified tolerance.

In the current configuration form given in Sec. 5.3.2 the current volume element depends on the determinant of the deformation gradient, which for the plane stress problem is given by

$$J = (F_{11} F_{22} - F_{12} F_{21}) F_{33} \tag{6.65}$$

which must be used to compute the element stiffness and residual arrays. We note that F_{33} also is used to compute the thickness of the deformed slab as

$$h_3 = F_{33} H_3 \tag{6.66}$$

where H_3 is the thickness in the reference configuration.

After convergence the tangent moduli of the current configuration are used in the next iteration. We can transform Eq. (6.56b) to obtain

$$\begin{Bmatrix} d\sigma_m \\ d\sigma_{33} \end{Bmatrix} = \begin{bmatrix} \mathbf{D}_{mm} & \mathbf{D}_{m3} \\ \mathbf{D}_{3m} & \mathbf{D}_{33} \end{bmatrix} \begin{Bmatrix} d\varepsilon_m \\ d\varepsilon_{33} \end{Bmatrix} \tag{6.67}$$

where $D_{33} = c_{3333}$ and the remaining terms are the standard transformation of moduli to matrix form. The reduced plane stress moduli are now given by

$$\mathbf{D}_T = \mathbf{D}_{mm} - \mathbf{D}_{m3} D_{33}^{-1} \mathbf{D}_{3m} \tag{6.68}$$

The above form is superior to other approaches based on a linearization of the entire stress with the zero stress condition enforced at the element level (e.g. see references 12–14). The approach given above may be applied to any constitutive model. The only requirements are an available tangent array and a computation of the σ_{33} stress component (see reference 11 for addition details and applications).

6.3 Isotropic viscoelasticity

The theory of linear viscoelasticity presented in Sec. 4.2 can be easily extended to a finite deformation form.[15,16] The isotropic form given by Eqs (4.13) to (4.16) is based on a split into deviatoric and mean stress–strain response where the stress is given by

$$\sigma_{ij} = \sigma_{ij}^v + \sigma_{ij}^d$$
$$\sigma_{ij} = p\,\delta ij + s_{ij} \tag{6.69}$$

with

$$p = \tfrac{1}{3}\sigma_{kk} \quad \text{and} \quad s_{ij} = \sigma_{ij} - \tfrac{1}{3}\sigma_{kk}\,\delta_{ij}$$

To develop a finite deformation form we use a split of the second Piola–Kirchhoff stress into two parts as

$$S_{IJ} = S_{IJ}^v + S_{IJ}^d \tag{6.70}$$

where

$$S_{IJ}^v = \tfrac{1}{3}(S_{KL}C_{KL})\,C_{IJ}^{-1}$$
$$S_{IJ}^d = S_{IJ} - \tfrac{1}{3}(S_{KL}C_{KL})\,C_{IJ}^{-1} \tag{6.71}$$

Applying the transformation given in Eqs (5.19) and (5.20) to (6.71) gives immediately Eq. (6.69) and thus represents a split of stress into mean and deviatoric parts in the current configuration.

In the finite deformation generalization we shall assume that an elastic model is available in the form

$$W = k_U\,U(J) + W^d \tag{6.72}$$

where U yields an elastic stress with the form S_{IJ}^v and W^d an elastic deviatoric stress S_{IJ}^d given by

$$S_{IJ}^v = 2\,k_U\,\frac{\partial U}{\partial C_{IJ}}$$

$$S_{IJ}^d = 2\,\frac{\partial W^d}{\partial C_{IJ}} \tag{6.73}$$

A viscoelastic model is then given by

$$S_{IJ}^d = \mu_0\,S_{IJ}^d + \sum_{m=1}^{M} \mu_m\,Q_{IJ}^{(m)} \tag{6.74}$$

where

$$\sum_{m=0}^{M} \mu_m = 1 \quad \text{with} \quad \mu_m > 0 \tag{6.75}$$

and the partial stress Q_{IJ} satisfies the differential equation

$$\dot{Q}_{IJ}^{(m)} + \frac{1}{\lambda_m}\,Q_{IJ}^{(m)} = \dot{S}_{IJ}^d \tag{6.76}$$

Here λ_m is a relaxation time similar to that described in Sec. 4.2.

A numerical solution to the differential equation for $Q_{IJ}^{(m)}$ is given by Eqs (4.22) and (4.29) to (4.34). In addition the tangent moduli are given by those of the elastic model used to define W multiplied by appropriate constants, again as described for the linear model.

6.4 Plasticity models

For isotropic materials, the modelling of elastic–plastic behaviour in which the total deformations are large may be performed by an extension of a hyperelastic formulation. In this case the deformation gradient is decomposed in a *product form* (instead of the additive form assumed in Chapter 4) written as[17–19]

$$F_{iI} = F_{ij}^e F_{jI}^p \tag{6.77}$$

where F_{ij}^e is the elastic part and F_{jI}^p the plastic part. The deformation picture is often shown as three parts, a reference state, a deformed state, and an *intermediate* state. The intermediate state is assumed to be the state of a point in a stress-free condition.* From this decomposition deformation tensors may be defined as

$$b_{ij}^e = F_{iK}^e F_{jK}^e \quad \text{and} \quad C_{IJ}^p = F_{KI}^p F_{KJ}^p \tag{6.78}$$

which when combined with Eq. (6.77) give the alternate representation

$$b_{ij}^e = F_{iI} \left(C_{IJ}^p \right)^{-1} F_{jJ} \tag{6.79}$$

An incremental setting may now be established that obtains a solution for a time t_{n+1} given the state at time t_n. The steps to establish the algorithm are too lengthy to include here and the interested reader is referred to literature for details.[2,16,20,21]

The components $(b_{ij}^e)_n$ denote values of the converged elastic deformation tensor at time t_n. We assume at the start of a new load step a *trial value* of the elastic tensor is determined from

$$(b_{ij}^e)_{n+1}^{tr} = f_{ik}(b_{kl}^e)_n f_{jl} \tag{6.80}$$

where an incremental deformation gradient is computed as

$$f_{ij} = (F_{iK})_{n+1}(F_{jK}^{-1})_n \tag{6.81}$$

A spectral representation of the trial tensor is then determined by using Eq. (6.43) giving

$$(b_{ij}^e)_{n+1} = \sum_m (\lambda_m^e)_{n+1}^2 q_i^{(m),tr} q_j^{(m),tr} \tag{6.82}$$

Owing to isotropy $q_i^{(m),tr}$ can be shown to equal the final directions $q_i^{(m)}$.[2]
Trial logarithmic strains are computed as

$$(\varepsilon_m^{tr})_{n+1} = \log(\lambda_m^e)_{n+1} \tag{6.83}$$

---

* The intermediate state is not a configuration, as it is generally discontinuous across interfaces between elastic and inelastic response.

and used with the stored energy function $W(b_{ij}^e)$ to compute trial values of the principal Kirchhoff stress $(\tau_m^{tr})_{n+1}$. This may be used in conjunction with the return map algorithm (see Sec. 3.4.2) and a yield function written in principal stresses τ_m to compute a final stress state and any internal hardening variables. This part of the algorithm is identical to the small strain form and needs no additional description except to emphasize that only the normal stress is included in the calculation of yield and flow directions. We note in particular that any of the yield functions for isotropic materials which we discussed in Chapter 4 may be used. The use of the return map algorithm also yields the consistent elastic–plastic tangent in principal space which can be transformed by means of Eq. (6.50) for subsequent use in the finite element matrix form.

The last step in the algorithm is to compute the final elastic deformation tensor. This is accomplished from the spectral form and final elastic logarithmic strains resulting from the return map solution as

$$(b_{ij}^e)_{n+1} = \sum_{m=1}^{3} \exp[2(\varepsilon_m^e)_{n+1}]q_i^{(m)}q_j^{(m)} \tag{6.84}$$

The advantages of the above algorithm are numerous. The form again permits a consistent linearization of the algorithm resulting in optimal performance when used with a Newton solution scheme. Most important, all the steps previously developed for the small deformation case are used. For example, although not discussed here, extension to viscoplastic and generalized plastic forms for isotropic materials is again given by results contained in Secs 3.6.2 and 3.9. The primary difficulty is an inability to easily treat materials which are anisotropic. Here recourse to a rate form of the constitutive equation is possible, as discussed next.

6.5 Incremental formulations

In the previous sections we have assumed that the deformation gradient can be computed from Eq. (5.8a). In some formulations it is convenient to use an *updated form* (sometimes referred to as an *updated Lagrangian method*) as shown in Fig. 6.4. Here a time parameter t may be introduced to distinguish between the various configurations. Thus, when t is zero we describe the body in its initial configuration Ω by the coordinates X_I. We assume then that the solution is carried out at a set of discrete times t_i with the last known solution defined at t_n, configuration $\omega^{(n)}$ shown in Fig. 6.4. It is then desired to compute the solution at t_{n+1}, which will be the new current configuration, $\omega^{(n+1)}$. In this format the known 'reference configuration' may be defined as the body at t_n, with coordinates $x_i^{(n)}$. The deformation gradient at t_{n+1} may now be defined as

$$F_{iI}^{(n+1)} = \frac{\partial \phi_i^{(n+1)}}{\partial x_j^{(n)}} \frac{\partial x_j^{(n)}}{\partial X_I} = f_{ij}^{(n)} F_{jI}^{(n)} \tag{6.85}$$

where $f_{ij}^{(n)}$ is an *incremental deformation gradient* associated with the reference configuration at $x_j^{(n)}$ for the incremental time $\Delta t = t_{n+1} - t_n$.

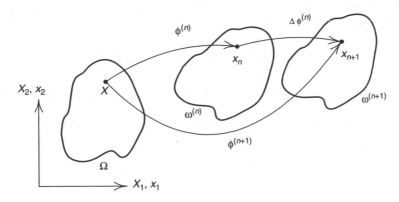

Fig. 6.4 Incremental deformation motions and configurations.

To compute volumetric changes we use the determinant of F_{ij} which is given by

$$\det(F_{iI}^{(n+1)}) = \det(f_{ij}^{(n)})\,\det(F_{jI}^{(n)})$$
$$J^{(n+1)} = j^{(n)}\,J^{(n)} \tag{6.86}$$

where $j^{(n)}$ is the determinant of the incremental deformation gradient.

The forms given in Eqs (6.85) and (6.86) also permit an incremental update to advance the solution from t_n to t_{n+1}. For example, the right Cauchy–Green deformation tensor given in Eq. (5.13) may be expressed in terms of the incremental quantities as

$$C_{IJ}^{(n+1)} = F_{iI}^{(n+1)}\,F_{iJ}^{(n+1)} = f_{ij}^{(n)}\,f_{ik}^{(n)}\,F_{jI}^{(n)}\,F_{kJ}^{(n)}$$
$$= c_{jk}^{(n)}\,F_{jI}^{(n)}\,F_{kJ}^{(n)} \tag{6.87}$$

where $c_{jk}^{(n)} = f_{ij}^{(n)}\,f_{ik}^{(n)}$ is the incremental right Cauchy–Green deformation tensor.

Example 6.7   Incremental Saint-Venant–Kirchhoff model

For the Saint-Venant–Kirchhoff model the stress may now be determined from (5.36) by substituting Eq. (6.87) into Eq. (5.14). The final result may be written in the form

$$S_{IJ}^{(n+1)} = S_{IJ}^{(n)} + C_{IJKL}\Delta E_{KL}^{(n)} \tag{6.88}$$

where

$$\Delta E_{KL}^{(n)} = E_{KL}^{(n+1)} - E_{KL}^{(n)}$$
$$= \tfrac{1}{2}(C_{KL}^{(n+1)} - C_{KL}^{(n)}) = \tfrac{1}{2}(F_{mK}^{(n+1)}\,F_{mL}^{(n+1)} - F_{mK}^{(n)}\,F_{mL}^{(n)}) \tag{6.89}$$
$$= \tfrac{1}{2}F_{kK}^{(n+1)}\,F_{lL}^{(n+1)}(\delta_{kl} - f_{mk}^{-1(n)}\,f_{ml}^{-1(n)}) = F_{kK}^{(n+1)}\,F_{lL}^{(n+1)}\,\Delta e_{kl}^{(n)}$$

where $\Delta e_{kl}^{(n)}$ is an incremental Almansi strain measure. Using this Eq. (6.88) may be transformed to the current configuration giving the Kirchhoff stress as

$$\tau_{ij}^{(n+1)} = f_{ik}^{(n)}\,\tau_{kl}^{(n)}\,f_{jl}^{(n)} + J^{(n+1)}\,c_{ijkl}^{(n+1)}\,\Delta e_{kl}^{(n)} \tag{6.90a}$$

and the Cauchy stress as

$$\sigma_{ij}^{(n+1)} = \frac{1}{j^{(n)}} f_{ik}^{(n)} \sigma_{kl}^{(n)} f_{jl}^{(n)} + c_{ijkl}^{(n+1)} \Delta e_{kl}^{(n)} \tag{6.90b}$$

Now all the steps to compute the finite element arrays follow identically those given in Chapter 5.

The left Cauchy–Green deformation tensor may also be written in terms of the incremental deformation gradient as

$$\begin{aligned} b_{ij}^{(n+1)} &= F_{iI}^{(n+1)} F_{jI}^{(n+1)} = f_{ik}^{(n)} F_{kI}^{(n)} f_{jl}^{(n)} F_{lI}^{(n)} \\ &= f_{ik}^{(n)} f_{jl}^{(n)} b_{kl}^{(n)} \end{aligned} \tag{6.91}$$

and, thus, may be updated directly and used directly in any of the elastic models included in Sec. 6.2 in their current configuration form.

6.6 Rate constitutive models

The construction of a rate form for elastic constitutive equations deduced from a stored energy function is easily performed in the reference configuration by taking a time derivative of Eq. (5.34), which gives

$$\dot{S}_{IJ} = C_{IJKL} \dot{E}_{KL} \tag{6.92}$$

where, as before, C_{IJKL} are moduli given by Eq. (6.8). The above result follows naturally from the notion of a derivative since

$$\dot{S}_{IJ} = \lim_{\eta \to 0} \frac{S_{IJ}(t+\eta) - S_{IJ}(t)}{\eta} \tag{6.93}$$

Such a definition is clearly not appropriate for the Cauchy or Kirchhoff stress since they are related to different configurations at time $t+\eta$ and t and thus would not satisfy the requirements of objectivity.[4,22] A definition of an *objective time derivative* may be computed for the Kirchhoff stress by using Eq. (5.20) and is sometimes referred to as the Truesdell rate[23] or equivalently a Lie derivative form.[24] Accordingly, to deduce an objective rate of the Kirchhoff stress we differentiate Eq. (5.20) with respect to time, obtaining

$$\dot{\tau}_{ij} = F_{iI} \dot{S}_{IJ} F_{jJ} + \dot{F}_{iI} S_{IJ} F_{jJ} + F_{iI} S_{IJ} \dot{F}_{jJ} \tag{6.94}$$

Introducing the rate of deformation tensor l_{ij} defined as

$$\dot{F}_{iI} = \frac{\partial \dot{x}_i}{\partial X_I} = \frac{\partial \dot{x}_i}{\partial x_j} \frac{\partial x_j}{\partial X_I} = \frac{\partial v_i}{\partial x_j} \frac{\partial x_j}{\partial X_I} = l_{ij} F_{jI} \tag{6.95}$$

in which $\dot{x}_j = v_j$ is material velocity, the stress rate may be written as

$$\dot{\tau}_{ij} = F_{iI} \dot{S}_{IJ} F_{jJ} + l_{ik} \tau_{kj} + \tau_{ik} l_{jk} \tag{6.96}$$

The objective stress, denoted as $\overset{\circ}{\tau}_{ij}$, is then given by

$$\overset{\circ}{\tau}_{ij} = \dot{\tau}_{ij} - l_{ik}\tau_{kj} - \tau_{ik}l_{jk} = F_{iI}\dot{S}_{IJ}F_{jJ} \tag{6.97}$$

The rate of the second Piola–Kirchhoff stress may be transformed by noting

$$\dot{E}_{KL} = \tfrac{1}{2}\left(F_{kL}\dot{F}_{kK} + F_{kK}\dot{F}_{kL}\right) = \tfrac{1}{2}(F_{lK}F_{kL}l_{kl} + F_{kK}F_{lL}l_{kl}) = F_{kK}F_{lL}\dot{\varepsilon}_{kl} \tag{6.98}$$

where

$$\dot{\varepsilon}_{kl} = \frac{1}{2}\left(l_{kl} + l_{lk}\right) = \frac{1}{2}\left(\frac{\partial v_k}{\partial x_l} + \frac{\partial v_l}{\partial x_k}\right) \tag{6.99}$$

The form $\dot{\varepsilon}_{kl}$ is identical to the rate of small strain. Furthermore we have upon grouping terms the rate of stress expression

$$\overset{\circ}{\tau}_{ij} = Jc_{ijkl}\dot{\varepsilon}_{kl} \tag{6.100}$$

in which c_{ijkl} is computed now by means of Eq. (5.74). Incremental forms may be deduced by integrating the rate equation. These involve objective approximations for the Lie derivative.[2,16] For example an approximation to the 'strain rate' may be computed from[16]

$$(\dot{\varepsilon}_{ij})_{n+1/2} \approx \frac{1}{\Delta t}(f_{ik}^{-1})_{n+1/2}\Delta E_{kl}(f_{jl}^{-1})_{n+1/2} \tag{6.101}$$

$$\Delta E_{kl} = \tfrac{1}{2}\left[(f_{km})_{n+1}(f_{lm})_{n+1} - \delta_{kl}\right]$$

where

$$(f_{ij})_{n+\alpha} = \delta_{ij} + \alpha\frac{\partial\Delta(u_i)_{n+1}}{\partial(x_j)_n} \tag{6.102}$$

with $\Delta(u_i)_{n+1} = (u_i)_{n+1} - (u_i)_n$. Similarly, an approximation to the Lie derivative of Kirchhoff stress may be taken as

$$(\overset{\circ}{\tau}_{ij})_{(n+1/2)} \approx \frac{1}{\Delta t}(f_{ik})_{n+1/2}\left[(f_{km}^{-1})_{n+1}(\tau_{mn})_{(n+1)}(f_{ln}^{-1})_{n+1} - (\tau_{kl})_{(n)}\right](f_{jl})_{n+1/2} \tag{6.103}$$

Often simpler approximations are used to approximate the integral of the velocity gradient. Here

$$\int_t^{t+\Delta t} l_{ij}dt \approx \Delta t\,\frac{\partial v_i^{(n+1/2)}}{\partial x_j^{(n+1/2)}} = \frac{\partial\Delta u_i^{(n+1/2)}}{\partial x_j^{(n+1/2)}} = \Delta l_{ij}^{(n+1/2)} \tag{6.104}$$

where $\Delta u_i^{(n+1)} = x_i^{(n+1)} - x_i^{(n)}$ and

$$x_i^{(n+\alpha)} = x_i^{(n)} + \alpha\,(x_i^{(n+1)} - x_i^{(n)})$$
$$\Delta u_i^{(n+\alpha)} = \alpha\,\Delta u_i^{(n+1)} \tag{6.105}$$

with $0 \leq \alpha \leq 1$ and also we note $\Delta u_i^{(n)} = 0$.

An explicit update for the Lie derivative may then be approximated as

$$\Delta t \, (\overset{\circ}{\tau}_{ij})^{(n+1)} \approx \tau_{ij}^{(n+1)} - \tau_{ij}^{(n)} - \Delta l_{ik}^{(n+1/2)} \tau_{kj}^{(n)} - \tau_{ik}^{(n)} \Delta l_{jk}^{(n+1/2)}$$
$$= J^{(n+1/2)} C_{ijkl}^{(n+1/2)} \, \Delta \varepsilon_{kl}^{(n+1/2)} \tag{6.106}$$

in which $\Delta \varepsilon_{kl}$ is the symmetric part of Δl_{kl}. The Kirchhoff stress at t_{n+1} may now be determined by solving Eq. (6.106). Other approximations may be used; however, the above are quite convenient. In the approximation a modulus array c_{ijkl} must also be obtained. Here there is no simple form which is always consistent with the tangent needed for a full Newton solution scheme and, often, a constant array is used based on results from linear elasticity. Such models based on the rate form are referred to as *hypoelastic* and in cyclic loading can create or lose energy.

Extension of the above to include general material constitution may be performed by replacing the strain rate by an additive form given, for example, as

$$\dot{\varepsilon} = \dot{\varepsilon}^e + \dot{\varepsilon}^p \tag{6.107}$$

for an elastic–plastic material. In this form we can again use all the constitutive equations discussed in Chapter 4 (including those which are not isotropic) to construct a finite element model for the large strain problem. Here, since approximations not consistent with a Newton scheme are generally used for the moduli, convergence generally does not achieve an asymptotic quadratic rate. Use of quasi-Newton schemes and line search, as described in Chapter 3, can improve the convergence properties and leads to excellent performance in many situations.

Many other stress rates may be substituted for the Lie derivative. For example, the Jaumann–Zaremba stress rate form given as

$$\overset{\triangledown}{\tau}_{ij} = \dot{\tau}_{ij} - \dot{\omega}_{ik} \tau_{kj} - \tau_{ik} \dot{\omega}_{jk} = \dot{\tau}_{ij} - \dot{\omega}_{ik} \tau_{kj} + \tau_{ik} \dot{\omega}_{kj} = J c_{ijkl} \dot{\varepsilon}_{kl} \tag{6.108}$$

may be used. This form is deduced by noting that the rate of deformation tensor may be split into a symmetric and skew-symmetric form as

$$l_{ij} = \dot{\varepsilon}_{ij} + \dot{\omega}_{ij} \tag{6.109}$$

where $\dot{\omega}_{ij}$ is the rate of spin or vorticity. It is then assumed that the symmetric part of l_{ij} is small compared to the rate of spin. This form was often used in early developments of finite element solutions to large strain problems and enjoys considerable popularity even today.

6.7 Numerical examples

6.7.1 Necking of circular bar

In this example we consider the three-dimensional behaviour of a cylindrical bar subjected to tension. In the presence of plastic deformation an unstable plastic *necking* will occur at some location along a bar of mild steel, or similar elastic–plastic behaving

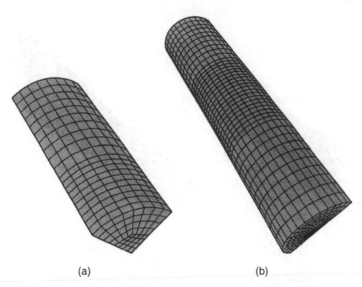

(a)   (b)

Fig. 6.5 Necking of a cylindrical bar: eight-node elements. (a) Finite element model; (b) half-bar by symmetry.

material. This is easily observed from the tension test of a cylindrical specimen which tapers by a small amount to a central location to ensure that the location of necking will occur in a specified location. A finite element model is constructed having the same taper, and here only one-eighth of the bar need be modelled as shown in Fig. 6.5(a). In Fig. 6.5(b) we show the half-bar model which is projected by symmetry and reflection and on which the behaviour will be illustrated.

This problem has been studied by several authors and here the properties are taken as described by Simo and co-workers.[2,16,25] The one-eighth quadrant model consists of 960 eight-node hexahedra of the mixed type discussed in Sec. 10.5. The radius at the loading end is taken as $R = 6.413$ and a uniform taper to a central radius of $R_c = 0.982 \times R$ is used. The total length of the bar is $L = 53.334$ (giving a half length of 25.667). The mesh along the length is uniform between the centre (0) and a distance of 10, and again from 10 to the end. A blending function mesh generation is used (see Sec. 9.12, Volume 1) to ensure that exterior nodes lie exactly on the circular radius. This ensures that, as much as possible for the discretization employed, the response will be axisymmetric.

The finite deformation plasticity model based on the logarithmic stretch elastic behaviour from Sec. 6.2.3 and the finite plasticity as described in Sec. 6.4 is used for the analysis. The material properties used are as follows: elastic properties are $K = 164.21$ and $G = 80.1938$; a J_2 plasticity model in terms of principal Kirchhoff stresses τ_i with an initial yield in tension of $\tau_y = 0.45$ is used. Only isotropic hardening is included and a saturation-type model defined by

$$\kappa = H_i \, \varepsilon^p + \left[\tau_y^\infty - \tau_y\right]\left(1 - \exp\left[-\bar{\delta}\varepsilon^p\right]\right)$$

with the parameters

$$H_i = 0.12924, \qquad \tau_y^\infty = 0.715, \qquad \text{and} \qquad \bar{\delta} = 16.93$$

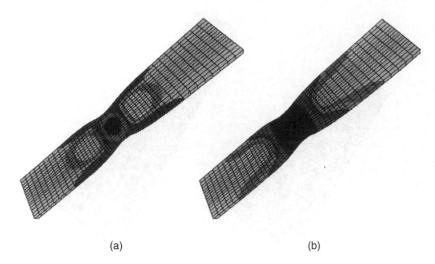

(a)   (b)

Fig. 6.6   Deformed configuration and contours for necking of bar: (a) first invariant (J_1); (b) second invariant (J_2).

is employed. An alternative to this is a piecewise linear behaviour as suggested by some authors; however, the above model is very easy to implement and gives a smooth behaviour with increase in the accumulated plastic strain ε^p as the hardening parameter κ.

In Fig. 6.6 we show the deformed configuration of the bar at an elongation of 22.5 per cent (elongation = 6 units). Figure 6.6(a) has the contours of the first invariant of Cauchy stress superposed and Fig. 6.6(b) those for the second invariant of the deviator stresses. It is apparent that considerable variation in pressure (first invariant divided by 3) occurs in the necked region, whereas the values of the second deviator invariant vary more smoothly in this region. A plot of the radius at the centre of the bar is shown in Fig. 6.7 for different elongation values.

This example is quite sensitive to solve as the response involves an unstable behaviour of the necking process. Use of a full Newton scheme was generally ineffective in this regime and here a modified Newton scheme together with a BFGS (Broyden–Fletcher–Goldfarb–Shanno) secant update was employed (see Sec. 3.2.4). When near convergence was achieved the algorithm was then switched to a full Newton process and during the last iterations quadratic convergence was obtained when used with an algorithmic consistent tangent matrix as described in Secs 6.2.3 and 6.4.

6.7.2 Adaptive refinement and localization (slip-line) capture

The simple discussion of localization phenomena given in Sec. 4.11 is sufficient we believe to convince the reader that with softening plastic behaviour localization and indeed rapid failure will occur inevitably. Similar behaviour will often be observed with ideal plasticity especially if large deformations are present. Here, however, 'brittle type' of failure will be replaced by collapse in which displacement can continue to increase without any increase of load. It is well known that during such continuing displacement

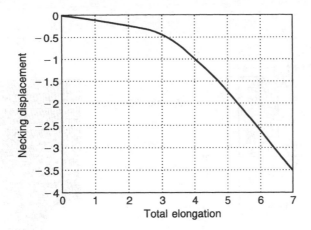

Fig. 6.7 Neck radius versus elongation displacement for a half-bar.

1. the elastic strains will remain unchanged;
2. all displacement is confined to *plastic mechanisms*. Such mechanisms will (fre-quently) involve discontinuous displacements, such as sliding, and will therefore involve localization.

To control and minimize the errors of the analysis it will be necessary to estimate errors and adaptively remesh in each step of an elasto-plastic computation. This, of course, implies a difficult and costly process. Nevertheless, many attempts to use adaptive refinement were made and references 26–34 provide a list of some successful attempts.

Adaptive refinement based on energy norm error estimates

A comprehensive survey of error estimation and *h* refinement in adaptivity is given in reference 35 (see Chapters 13 and 14). Most of the procedures there described could be applied with success to elasto-plastic analysis. One in particular, the *recovery* procedures for stress and strain, can be used very efficiently. Indeed, in references 36 and 37 the Superconvergent Patch Recovery (SPR) and Recovery by Equilibrium Patch (REP) methods are used successfully to estimate the errors. (The SPR method was introduced by Zienkiewicz and Zhu and described fully in references 38, 36 and 39. The REP method was presented by Boroomand and Zienkiewicz in references 40 and 41.) In Figs 6.8(a) and 6.8(b) we show an analysis of a tension strip using these procedures.

It will be observed that as the load increases the refined mesh tends to capture a solution in which displacements are localized. In this solution triangular elements using quadratic displacements with three added bubble modes together with linear discontinuous pressures (T6/3B/3D) were used as these have an excellent performance in incompressibility and may be incorporated easily into an adaptive process. In this problem we have attempted here to keep the error to 5% in the relative energy norm.

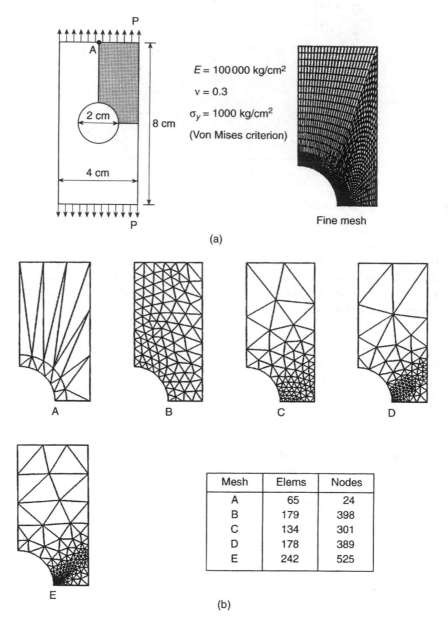

Fig. 6.8(a) Adaptive refinement applied to the problem of a perforated strip: (a) the geometry of the strip and a very fine mesh are used to obtain an 'exact' solution; (b) various stages of refinement aiming to achieve a 5% energy norm, relative, error at each load increment (quadratic elements T6/3B/3D were used); (c) local displacement results.

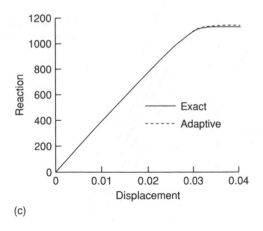

(c)

Fig. 6.8(b) Continued

Alternate refinement using error indicators: discontinuity capture

In the illustrative example of the previous section we have shown how a refinement on the test of a specified energy norm error can indicate and capture discontinuity and slip lines. Nevertheless the process is not economical and may require the use of a very large number of elements. More direct processes have been developed for adaptive refinement in high-speed fluid dynamics where shocks presenting very similar discontinuity properties form (e.g. see reference 42). A brief summary of the procedure is given below.

The processes developed are based on the recognition that in certain directions the unknown function that we are attempting to model exhibits higher gradient or curvatures. High degree of refinement can be achieved economically in high gradient areas with elongated elements. In such areas the smaller side of elements (h_{min}) is placed across the discontinuity, and the larger side (h_{max}) in the direction parallel to the discontinuity. We show such a directionality in Fig. 6.9. For determination of gradients and curvatures, we shall require a scalar function to be considered. The scalar variable which we frequently use in plasticity problems could be the absolute displacement value

$$U = (\mathbf{u}^T\mathbf{u})^{1/2} \tag{6.110}$$

The original refinement indicator of the type we are describing attempts to achieve an equal interpolation error ensuring that in the major and minor direction the equality

$$h_{min}^2 \left.\frac{\partial^2 U}{\partial x_1^2}\right|_{max} = h_{max}^2 \left.\frac{\partial^2 U}{\partial x_2^2}\right|_{min} = \text{constant} \tag{6.111}$$

By fixing the value of the constant in the above equation and evaluating the approximate curvatures and the ratio of stretching h_{max}/h_{min} of the function U immediately we have sufficient data to design a new mesh from the existing one. The procedures for such

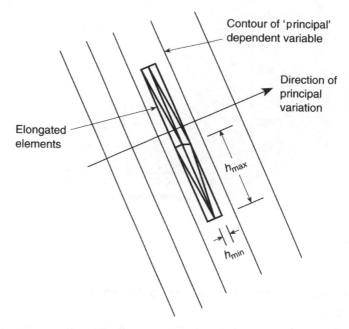

Fig. 6.9   Elongation of elements used to model the nearly one-dimensional behaviour and the discontinuity.

mesh generation are given in references 43 and 44, although other methods can be adopted.

As an alternative to the above-mentioned procedure we can aim at limiting the first derivative of U by making

$$h\frac{\partial U}{\partial x_1} = \text{constant} \tag{6.112}$$

For this procedure it is not easy to evaluate the stretching ratio. However, the first-derivative condition is useful for guiding the refinement.

A plastic localization calculation based on Eq. (6.111) is shown in Fig. 6.10. This is an early example taken from reference 29. Here, purely plastic flow is shown, ignoring the elastic effects and the refinement is based on the second derivatives. Such a flow formulation (rigid plastic flow) is also frequently used in metal forming calculations.

In the next example we shall use adaptivity based on the first derivatives. Figure 6.11 illustrates a load on a rigid footing over a vertical cut. Here a T6/3C element (triangular with quadratic displacement and continuous linear pressure) is used. Both coarse and adaptively refined meshes give nearly exact answers for the case of ideal plasticity.

For strain softening with a plastic modulus $H = -5000$ answers appear to be mesh dependent. Here, we show how answers become almost mesh independent if H is varied with element size in the manner discussed in Sec. 4.11 [see Eq. (4.184)]. Figures 6.12 and 6.13 show, respectively, the behaviour of a rigid footing placed on an embankment and on a flat foundation with eccentric loading. All cases illustrate the excellent discontinuity capturing properties of the adaptive refinement.

6.8 Concluding remarks

This chapter presents a summary of some models for use in finite deformation analyses. The scope of the presentation is limited but provides a description of requirements for use with many finite element formulations. We have not presented any discussion on coupled phenomena which can occur in problems, e.g. such effects of temperature, diffusion, creep, etc. For this, the reader is referred to the extensive literature on the subject for additional details.

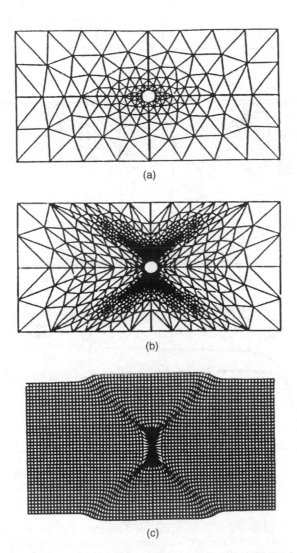

(a)

(b)

(c)

Fig. 6.10 Adaptive analysis of plastic flow deformation in a perforated plate: (a) initial mesh, 273 degrees of freedom; (b) final adapted mesh; (c) displacement of an initially uniform grid embedded in the material.

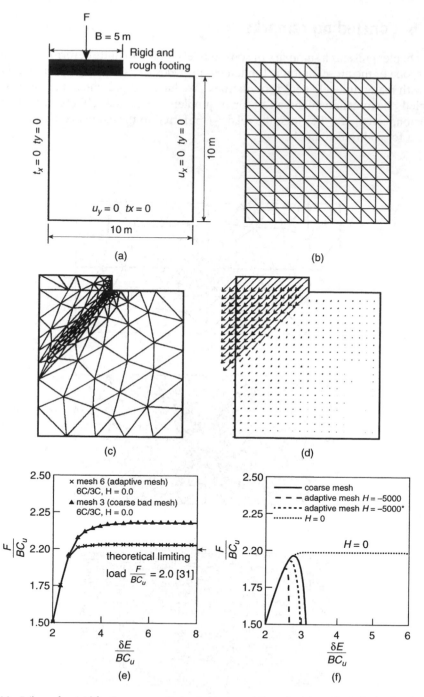

Fig. 6.11 Failure of a rigid footing on a vertical cut. Ideal von Mises plasticity and quadratic triangles with linear variation for pressure (T6/3C) elements are assumed. (a) Geometrical data; (b) coarse mesh; (c) final adapted mesh; (d) displacements after failure; (e) displacement–load diagrams for adaptive mesh and ideal plasticity ($H = 0$); (f) softening behaviour. Coarse mesh and adapted mesh results are with a constant H of -5000 and a variable H starting from -5000 at coarse mesh size.

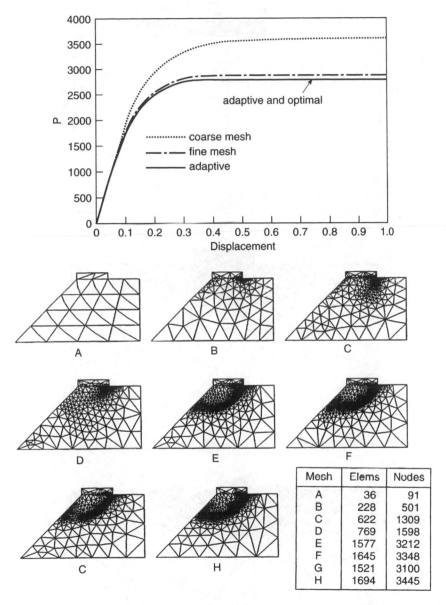

Fig. 6.12 A $p - \delta$ diagram of elasto-plastic slope aiming at 2.5% error in ultimate load (15% incremental energy error) with use of quadratic triangular elements (T6/3B). Mesh A: $u = 0.0$ (coarse mesh). Mesh B: $u = 0.025$. Mesh C: $u = 0.15$. Mesh D: $u = 0.3$. Mesh E: $u = 0.45$. Mesh F: $u = 0.6$. Mesh G: $u = 0.75$. Mesh H: $u = 0.9$. The last mesh (mesh H, named as the 'optimal mesh') is used for the solution of the problem from the first load step, without further refinement.

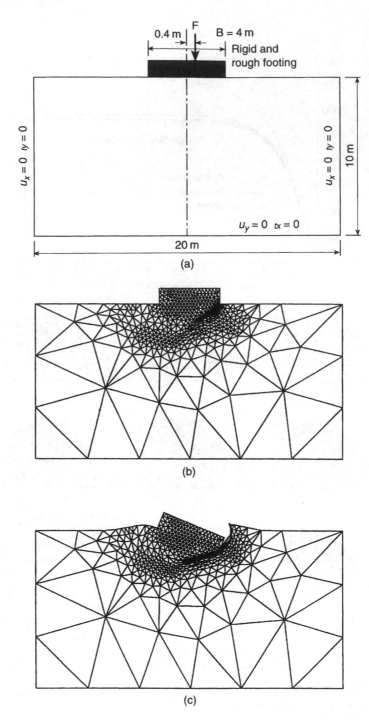

Fig. 6.13 Foundation (eccentric loading); ideal von Mises plasticity. (a) Geometry and boundary conditions; (b) adaptive mesh; (c) deformed mesh using T6/1D elements ($H = 0$, $\nu = 0.49$).

References

1. I.H. Shames and F.A. Cozzarelli. *Elastic and Inelastic Stress Analysis*. Taylor & Francis, Washington, DC, 1997. (Revised printing.)
2. J.C. Simo. Topics on the numerical analysis and simulation of plasticity. In P.G. Ciarlet and J.L. Lions, editors, *Handbook of Numerical Analysis*, volume III, pages 183–499. Elsevier Science Publisher BV, 1999.
3. M.E. Gurtin. *An Introduction to Continuum Mechanics*. Academic Press, New York, 1981.
4. J. Bonet and R.D. Wood. *Nonlinear Continuum Mechanics for Finite Element Analysis*. Cambridge University Press, Cambridge, 1997. ISBN 0–521–57272–X.
5. M. Mooney. A theory of large elastic deformation. *J. Appl. Physics*, 1:582–592, 1940.
6. O.H. Yeoh. Characterization of elastic properties of carbon-black filled rubber vulcanizates. *Rubber Chemistry and Technology*, 63:792–805, 1990.
7. E.M. Arruda and M.C. Boyce. A three-dimensional constitutive model for the large stretch behavior of rubber elastic materials. *Journal of the Mechanics and Physics of Solids*, 41:389–412, 1993.
8. G.A. Holzapfel. *Nonlinear Solid Mechanics. A Continuum Approach for Engineering*. John Wiley & Sons, Chichester, 2000.
9. R.W. Ogden. *Non-linear Elastic Deformations*. Ellis Horwood, Limited (reprinted by Dover, 1997), Chichester, England, 1984.
10. J.C. Simo and R.L. Taylor. Quasi-incompressible finite elasticity in principal stretches: continuum basis and numerical algorithms. *Computer Methods in Applied Mechanics and Engineering*, 85:273–310, 1991.
11. S. Klinkel and S. Govindjee. Using finite strain 3D-material models in beam and shell elements. *Engineering Computations*, 19:902–921, 2002.
12. T.J.R. Hughes and E. Carnoy. Nonlinear finite element formulation accounting for large membrane stress. *Computer Methods in Applied Mechanics and Engineering*, 39:69–82, 1983.
13. E. Dvorkin, D. Pantuso and E. Repetto. A formulation of the MITC4 shell element for finite strain elasto-plastic analysis. *Engineering Computations*, 1:17–40, 1995.
14. R. De Borst. The zero normal stress condition in plane stress and shell elastoplasticity. *Communications in Applied Numerical Methods*, 7:29–33, 1991.
15. J.C. Simo. On a fully three-dimensional finite-strain viscoelastic damage model: formulation and computational aspects. *Computer Methods in Applied Mechanics and Engineering*, 60:153–173, 1987.
16. J.C. Simo and T.J.R. Hughes. *Computational Inelasticity*, volume 7 of *Interdisciplinary Applied Mathematics*. Springer-Verlag, Berlin, 1998.
17. J. Mandel. Contribution théorique a l'étude de l'écrouissage et des lois de l'écoulement plastique. In *Proc. 11th Int. Cong. of Appl. Mech.*, pages 502–509, 1964.
18. E.H. Lee and D.T. Liu. Finite strain elastic–plastic theory particularly for plane wave analysis. *J. Appl. Phys.*, 38, 1967.
19. E.H. Lee. Elastic–plastic deformations at finite strains. *J. Appl. Mech.*, 36:1–6, 1969.
20. J.C. Simo. Algorithms for multiplicative plasticity that preserve the form of the return mappings of the infinitesimal theory. *Computer Methods in Applied Mechanics and Engineering*, 99:61–112, 1992.
21. F. Auricchio and R.L. Taylor. A return-map algorithm for general associative isotropic elastoplastic materials in large deformation regimes. *Int. J. Plasticity*, 15:1359–1378, 1999.
22. L.E. Malvern. *Introduction to the Mechanics of a Continuous Medium*. Prentice-Hall, Englewood Cliffs, NJ, 1969.
23. C. Truesdell and W. Noll. The non-linear field theories of mechanics. In S. Flügge, editor, *Handbuch der Physik III/3*. Springer-Verlag, Berlin, 1965.
24. J.E. Marsden and T.J.R. Hughes. *Mathematical Foundations of Elasticity*. Dover, New York, 1994.

25. J.C. Simo and F. Armero. Geometrically non-linear enhanced strain mixed methods and the method of incompatible modes. *International Journal for Numerical Methods in Engineering*, 33:1413–1449, 1992.

26. M. Ortiz, Y. Leroy and A. Needleman. A finite element method for localized failure analysis. *Computer Methods in Applied Mechanics and Engineering*, 61:189–214, 1987.

27. R. De Borst, L.J. Sluys, H.B. Hühlhaus and J. Pamin. Fundamental issues in finite element analysis of localization of deformation. *Engineering Computations*, 10:99–121, 1993.

28. O.C. Zienkiewicz, H.C. Huang and M. Pastor. Localization problems in plasticity using the finite elements with adaptive remeshing. *International Journal for Numerical Analysis Methods in Geomechanics*, 19:127–148, 1995.

29. O.C. Zienkiewicz, M. Pastor and M. Huang. Softening, localization and adaptive remeshing. Capture of discontinuous solutions. *Comp. Mech.*, 17:98–106, 1995.

30. J. Yu, D. Peric and D.R.J. Owen. Adaptive finite element analysis of strain localization problem for elasto plastic Cosserat continuum. In D.R.J. Owen *et al.*, editors, *Computational Plasticity III: Models, Software and Applications*, pages 551–566. Pineridge Press, Swansea, 1992.

31. P. Steinmann and K. Willam. Adaptive techniques for localization analysis. In *Proc. ASME – Winter Annual Meeting 92*, ASME, New York, 1992.

32. M. Pastor, J. Peraire and O.C. Zienkiewicz. Adaptive remeshing for shear band localization problems. *Archive of Applied Mechanics*, 61:30–91, 1991.

33. M. Pastor, C. Rubio, P. Mira, J. Peraire and O.C. Zienkiewicz. Numerical analysis of localization. In G.N. Pande and S. Pietruszczak, editors, *Numerical Models in Geomechanics*, pages 339–348. A.A. Balkema, 1991.

34. R. Larsson, K. Runesson and N.S. Ottosen. Discontinuous displacement approximations for capturing localization. *International Journal for Numerical Methods in Engineering*, 36:2087–2105, 1993.

35. O.C. Zienkiewicz, R.L. Taylor and J.Z. Zhu. *The Finite Element Method: Its Basis and Fundamentals*. Butterworth-Heinemann, Oxford, 6th edition, 2005.

36. O.C. Zienkiewicz and J.Z. Zhu. The superconvergent patch recovery (SPR) and *a posteriori* error estimates. Part 1: The recovery technique. *International Journal for Numerical Methods in Engineering*, 33:1331–1364, 1992.

37. O.C. Zienkiewicz, B. Boroomand and J.Z. Zhu. Recovery procedures in error estimation and adaptivity: adaptivity in linear problems. In P. Ladevèze and J.T. Oden, editors, *Advances in Adaptive Computational Mechanics in Mechanics*, pages 3–23. Elsevier Science Ltd, 1998.

38. O.C. Zienkiewicz and J.Z. Zhu. The superconvergent patch recovery (SPR) and adaptive finite element refinement. *Computer Methods in Applied Mechanics and Engineering*, 101:207–224, 1992.

39. O.C. Zienkiewicz and J.Z. Zhu. The superconvergent patch recovery (SPR) and *a posteriori* error estimates. Part 2: Error estimates and adaptivity. *International Journal for Numerical Methods in Engineering*, 33:1365–1382, 1992.

40. B. Boroomand and O.C. Zienkiewicz. Recovery by equilibrium patches (REP). *International Journal for Numerical Methods in Engineering*, 40:137–154, 1997.

41. B. Boroomand and O.C. Zienkiewicz. An improved REP recovery and the effectivity robustness test. *International Journal for Numerical Methods in Engineering*, 40:3247–3277, 1997.

42. O.C. Zienkiewicz, R.L. Taylor and P. Nithiarasu. *The Finite Element Method for Fluid Dynamics*. Butterworth-Heinemann, Oxford, 6th edition, 2005.

43. J. Peraire, M. Vahdati, K. Morgan and O.C. Zienkiewicz. Adaptive remeshing for compressible flow computations. *Journal of Computational Physics*, 72:449–466, 1987.

44. O.C. Zienkiewicz and J. Wu. Automatic directional refinement in adaptive analysis of compressible flows. *International Journal for Numerical Methods in Engineering*, 37:2189–2219, 1994.

7

Treatment of constraints – contact and tied interfaces

7.1 Introduction

In many problems situations arise where the position of parts of the boundary of one body coincide with those of another part of the boundary of the same or another body. Such problems are commonly called *contact problems*. Finite element methods have been used for many years to solve contact problems.[1–32] The patch test has also been extended to test consistency of contact developments.[33] Contact problems are inherently non-linear since, prior to contact, boundary conditions are given by traction conditions (often the traction being simply zero) whereas during 'contact' kinematic constraints must be imposed to prevent penetration of one boundary through the other, called the *impenetrability condition*. In addition, the constraints must enforce traction continuity between the bodies during persistent contacts.

In this chapter we consider modelling of the interaction between one or more bodies that come into contact with each other. Such *contact problems* are among the most difficult to model by finite elements and we summarize here only some of the approaches which have proved successful in practice. In general, the finite element discretization process itself leads to surfaces which are not smooth and, thus, when large sliding occurs, the transition from one element to the next leads to discontinuities in the response – and in transient applications can induce non-physical inertial discontinuities also. For quasi-static response such discontinuity leads to difficulties in defining a unique solution and here methods of multi-surface plasticity prove useful.[34]

We include in the chapter some illustrations of performance for many of the formulations and problem classes discussed; however, the range is so broad that it is not possible to cover a comprehensive set. Here again the reader is referred to cited literature for additional insight and results.

The solution of a contact problem involves: (a) using a search algorithm to identify points on a boundary segment that interact with those on another boundary segment and (b) the insertion of appropriate conditions to prevent the penetration and correctly transmit the traction between the bodies. Figure 7.1 shows a typical situation in which one body is being pressed into a second body. In Fig. 7.1(a) the two objects are not in contact and the boundary conditions are specified by zero traction conditions for both bodies. In Fig. 7.1(b) the two objects are in contact along a part of the boundary

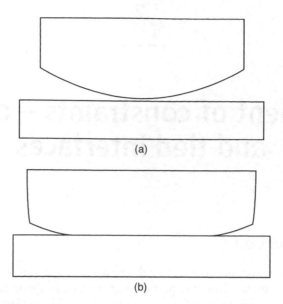

(a)

(b)

Fig. 7.1 Contact between two bodies: (a) no contact condition; (b) contact state.

segment and here conditions must be inserted to ensure that penetration does not occur and traction is consistent. Along this boundary different types of contact interaction can be modelled, a simple case being a *frictionless* condition in which the only non-zero traction component is normal to the contact surface. A more complex condition occurs in which traction tangential to the surface can be generated by *frictional* conditions. The simplest frictional condition is a Coulomb model where, in a *slip* condition, the tangential traction is directly proportional to the normal traction. If the magnitude of the tangential traction is less than the limit condition the points on the surface are assumed to *stick*. Overall the frictional problem leads to a *stick–slip*-type response. We shall consider this condition in more detail later; however, first we consider the process of imposing a contact condition for the frictionless problem. Even in this form there are several aspects to consider for the finite element problem.

In modelling contact problems by finite element methods immediate difficulties arise. First, it is not possible to model contact at every point along a boundary. This is primarily because of the fact that the finite element representation of the boundary is not *smooth*. For example, in the two-dimensional case in which boundaries of individual elements are straight line segments as shown in Fig. 7.2 nodes A and B are in contact with the lower body but the segment between the nodes is not in contact. Also finite element modelling results in non-unique representation of a normal between the two bodies and, again because of finite element discretization, the normals are not continuous between elements. This is illustrated also in Fig. 7.2 where it is evident that the normal to the segment between nodes A and B is not the same as the negative normal of the facets around node C (which indeed are not unique at node C). Before proceeding to methods appropriate for large deformation problems we consider the case where the nodes on one surface interact directly with those on the other surface and describe conditions that may be introduced to prevent penetration of the bodies.

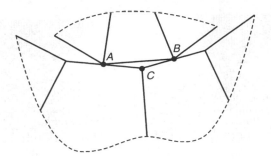

Fig. 7.2 Contact by finite elements.

7.2 Node–node contact: Hertzian contact

7.2.1 Geometric modelling

For applications in which displacements on the contact boundary are *small* it is some-times possible to model the contact by means of nodes (this form is applicable to Hertzian contact problems[35]). For this to be possible, the finite element mesh must be constructed such that boundary nodes on one body, here referred to as *slave* nodes, align with the location of the boundary nodes on the other body, referred to as *master* nodes, to within conditions acceptable for small deformation analysis. Such conditions may also be extended for cases where the boundary of one body is treated as flat and rigid (unilateral contact). A problem in which such conditions may be used is the inter-action between two half discs (or hemispheres) which are pressed together along the line of action between their centres. A simple finite element model for such a problem is shown in Fig. 7.3(a) where it is observed that the horizontal alignment of potential contact nodes on the boundary of each disc are identical. The solution after pressing the bodies together is indicated in Fig. 7.3(b) and contours for the vertical normal stress are shown in Fig. 7.3(c). It is evident that the contours do not match perfectly along the vertical axis owing to lack of alignment of the nodes in the deformed position. How-ever, the mismatch is not severe, and useful engineering results are possible. Later we will consider methods which give a more accurate representation; however, before doing so we consider the methods available to prevent penetration.

The determination of which nodes are in contact for the problem shown in Fig. 7.3 can be monitored simply by comparing the vertical position of each node pair and, thus, a finite element model may be treated as a simple two-node element. Denoting the upper disc as slave body 's' and the lower one as master body 'm' we can monitor the vertical *gap* given by

$$g = \tilde{x}_2^s - \tilde{x}_2^m = (\tilde{X}_2^s + \tilde{u}_2^s) - (\tilde{X}_2^m + \tilde{u}_2^m) = (\tilde{X}_2^s - \tilde{X}_2^m) + (\tilde{u}_2^s - \tilde{u}_2^m) \qquad (7.1)$$

Thus, the solution of each contact constraint is treated as

$$g = \begin{cases} > 0; & \text{No contact} \\ 0; & \text{Contact} \\ < 0; & \text{Penetration} \end{cases} \qquad (7.2)$$

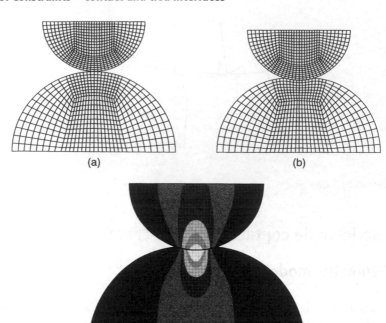

Fig. 7.3 Contact between semicircular discs: node–node solution. (a) Undeformed mesh; (b) deformed mesh; (c) vertical stress contours.

We note that penetration can exist for any solution iteration in which the constraint condition is not imposed. Thus, the next step is to insert a constraint condition for any nodal pair (element) in which the gap g is negative or zero (here some tolerance may be necessary to define 'zero'). There are many approaches which can be used to insert a constraint. Here we discuss use of a Lagrange multiplier form, penalty approaches, and an augmented Lagrangian approach.[7,36]

7.2.2 Contact models

Lagrange multiplier form

A Lagrange multiplier approach is given simply by multiplying the gap condition given in Eq. (7.1) by the multiplier. Accordingly, we can write for each nodal pair for which a contact constraint is assigned a variational term

$$\Pi_c = \int_{\Gamma_c} \mathbf{t}_\Gamma^T (\mathbf{x}^s - \mathbf{x}^m)\, d\Gamma \approx \lambda_n g \tag{7.3}$$

where $\mathbf{t}_\Gamma$ is the surface traction, $\mathbf{x}^s$ is the position on the surface of the slave body, $\mathbf{x}^m$ is the position on the surface of the master body, λ_n is a Lagrange multiplier *force* and g is the *gap* given by Eq. (7.1). We then add the first variation of Π_c to the variational equations being used to solve the problem. The first variation to Eq. (7.3) is given as

$$\delta \Pi_c = \delta \lambda_n g + (\delta \tilde{u}_2^s - \delta \tilde{u}_2^m) \lambda_n = \begin{bmatrix} \delta \tilde{u}_2^s & \delta \tilde{u}_2^m & \delta \lambda_n \end{bmatrix} \begin{Bmatrix} \lambda_n \\ -\lambda_n \\ g \end{Bmatrix} \tag{7.4}$$

and thus we identify λ_n as a 'force' applied to each node to prevent penetration. Linearization of Eq. (7.4) produces a tangent matrix term for use in a Newton solution process. The final tangent and residual for the nodal contact element may be written as

$$\begin{bmatrix} 0 & 0 & 1 \\ 0 & 0 & -1 \\ 1 & -1 & 0 \end{bmatrix} \begin{Bmatrix} d\tilde{u}_2^s \\ d\tilde{u}_2^m \\ d\lambda_n \end{Bmatrix} = \begin{Bmatrix} -\lambda_n \\ \lambda_n \\ -g \end{Bmatrix} \tag{7.5}$$

and is accumulated into the global equations in a manner identical to any finite element assembly process. It is evident that the equations in this form introduce a new unknown for each contact pair. Also, as for any Lagrange multiplier approach, the equations have a zero diagonal for each multiplier term, thus, special care is needed in the solution process to avoid division by the zero diagonal.

Of course in a contact state, one could select one of the parameters, say $\tilde{x}_2^s$, as a primary variable and directly satisfy the gap constraint by making $\tilde{x}_2^m = \tilde{x}_2^s$. This approach is called *constraint elimination* and may be used to reduce the number of overall unknowns. In the simple frictionless node-to-node contact case it is simple to implement as no transformations are needed to write the constraint equation. In a general case, however, the approach can become quite cumbersome and it is often simpler to use the Lagrange multiplier form directly or to consider other related approaches.

If the global tangent matrix has its non-zero sparse structure defined for the case when all the specified contact elements are active (e.g. the tangent matrix defined by Eq. (7.5) can be inserted without adding new non-zero terms) then a full contact analysis may be performed using Eq. (7.5) when $g \leq 0$ and using the alternate tangent matrix and residual

$$\begin{bmatrix} 0 & 0 & 0 \\ 0 & 0 & 0 \\ 0 & 0 & 1 \end{bmatrix} \begin{Bmatrix} d\tilde{u}_2^s \\ d\tilde{u}_2^m \\ d\lambda_n \end{Bmatrix} = \begin{Bmatrix} 0 \\ 0 \\ 0 \end{Bmatrix} \tag{7.6}$$

for nodal pairs when $g > 0$. However, if a large number of possible contact pairs are inactive (i.e. $g > 0$) it is more efficient to recompute the sparse structure of the global tangent matrix to just accommodate the active contact pairs (i.e. those for which $g \leq 0$). This step can be performed by determining all the active pairs prior to computing the tangent arrays.

Perturbed Lagrangian

The problem related to the zero diagonal may be resolved by considering a *perturbed Lagrangian* form where

$$\Pi_c = \lambda_n g - \frac{1}{2\kappa} \lambda_n^2 \tag{7.7}$$

in which κ is a parameter to be selected. As $\kappa \to \infty$ the perturbed Lagrangian method converges to the same functional as the standard Lagrange multiplier method. The first variation of Π_c becomes

$$\delta\Pi_c = \delta\lambda_n \left(g - \frac{1}{\kappa}\lambda_n \right) + [\delta\tilde{u}_2^s - \delta\tilde{u}_2^m]\lambda_n \tag{7.8}$$

and again we identify λ_n as a 'force' applied to each node to prevent penetration. Linearization of Eq. (7.4) produces a tangent matrix term for use in a Newton solution process. The final tangent and residual for the nodal contact element may be written as

$$\begin{bmatrix} 0 & 0 & 1 \\ 0 & 0 & -1 \\ 1 & -1 & -1/\kappa \end{bmatrix} \begin{Bmatrix} d\tilde{u}_2^s \\ d\tilde{u}_2^m \\ d\lambda_n \end{Bmatrix} = \begin{Bmatrix} -\lambda_n \\ \lambda_n \\ -g + \lambda_n/\kappa \end{Bmatrix} \tag{7.9}$$

which is added into the equations in a manner identical to the Lagrange multiplier form. It is also possible to eliminate λ_n directly from Eq. (7.8) giving

$$\lambda_n = \kappa\,g = \kappa\,(\tilde{x}_2^s - \tilde{x}_2^m) \tag{7.10}$$

Substitution into Eq. (7.9) and eliminating $d\lambda_n$ gives the reduced form

$$\begin{bmatrix} \kappa & -\kappa \\ -\kappa & \kappa \end{bmatrix} \begin{Bmatrix} d\tilde{u}_2^s \\ d\tilde{u}_2^m \end{Bmatrix} = \begin{Bmatrix} -\lambda_n \\ \lambda_n \end{Bmatrix} \tag{7.11}$$

In a perturbed Lagrangian approach the final gap will not be zero but becomes a small number depending on the value of the parameter κ selected. Thus, the advantage of the perturbed Lagrangian method is somewhat offset by a need to identify a value of the parameter that gives an acceptable answer. Indeed, in a complex problem this is not a trivial task, especially for problems involving contact between structural elements (e.g. rods, plates, or shells) and solid elements. This can be avoided in part by modifying Eq. (7.9) to read

$$\begin{bmatrix} 0 & 0 & 1 \\ 0 & 0 & -1 \\ 1 & -1 & -1/\kappa \end{bmatrix} \begin{Bmatrix} d\tilde{u}_2^s \\ d\tilde{u}_2^m \\ d\lambda_n \end{Bmatrix} = \begin{Bmatrix} -\lambda_n \\ \lambda_n \\ -g \end{Bmatrix} \tag{7.12}$$

Here this form is called a *perturbed tangent* method and is a combination of the perturbed Lagrangian tangent matrix with the Lagrange multiplier residual. As such it is not a consistent linearization of any functional and there is some loss in convergence rate in solving the overall non-linear problem. Moreover, it is not possible to directly solve for λ_n in each element and an iterative update must be used with

$$d\lambda_n = \kappa\left(g + d\tilde{u}_2^s - d\tilde{u}_2^m \right)$$
$$\lambda_n \leftarrow \lambda_n + d\lambda_n \tag{7.13}$$

In the above the incremental displacements are those from the last global solution; however, the same form may be used in Eq. (7.12) to give the reduced problem

$$\begin{bmatrix} \kappa & -\kappa \\ -\kappa & \kappa \end{bmatrix} \begin{Bmatrix} d\tilde{u}_2^s \\ d\tilde{u}_2^m \end{Bmatrix} = \begin{Bmatrix} -\lambda_n - \kappa g \\ \lambda_n + \kappa g \end{Bmatrix} \tag{7.14}$$

The method does, however, converge to a solution in which g approaches zero when κ is large enough. Thus the difficulty of selecting an appropriate value for κ is still not fully resolved.

Penalty function form

An alternative approach which avoids the difficulties of dealing with a zero diagonal from a Lagrange multiplier method is the classical *penalty method*. In this method the contact term is given by

$$\Pi = \tfrac{1}{2} \kappa g^2 \qquad\qquad (7.15)$$

where κ is a penalty parameter. The matrix equation for a nodal pair is now given by

$$\begin{bmatrix} \kappa & -\kappa \\ -\kappa & \kappa \end{bmatrix} \begin{Bmatrix} d\tilde{u}_2^s \\ d\tilde{u}_2^m \end{Bmatrix} = \begin{Bmatrix} -\kappa g \\ \kappa g \end{Bmatrix} \qquad\qquad (7.16)$$

For the scalar problem considered here the penalty and the perturbed Lagrangian methods lead to identical reduced problems. However, when multi-point constraints are considered and independent approximations are taken for λ_n and $\mathbf{u}$ the two methods are different unless the limitation principle is satisfied (i.e. the λ_n includes all the terms in the expression for g). Thus, in practice, the use of the perturbed Lagrangian method is preferred. This is especially crucial for more complex methods in treating contact problems, such as mortar methods or other surface-to-surface treatments.[31,32,37]

Augmented Lagrangian form

A compromise between the perturbed Lagrangian or penalty methods and the Lagrange multiplier method may be achieved by using an iterative update for the multiplier combined with a penalty-like form. We write the augmented form as[38]

$$\begin{bmatrix} \kappa & -\kappa \\ -\kappa & \kappa \end{bmatrix} \begin{Bmatrix} d\tilde{u}_2^s \\ d\tilde{u}_2^m \end{Bmatrix} = \begin{Bmatrix} -\lambda_n^k - \kappa g \\ \lambda_n^k + \kappa g \end{Bmatrix} \qquad\qquad (7.17)$$

where an update to the Lagrange multiplier is computed by using[7]

$$\lambda_n^{k+1} = \lambda_n^k + \kappa g \qquad\qquad (7.18)$$

Such an update may be computed after each Newton iteration or in an added iteration loop after convergence of the Newton iteration. In either case a loss of quadratic convergence in solving the global non-linear problem results for the simple augmented strategy shown. Improvements to super-linear convergence are possible as shown by Zavarise and Wriggers,[39] and a more complex approach which restores the quadratic convergence rate may be introduced at the expense of retaining an added variable.[20] In general, however, use of a fairly large value of the penalty parameter in the simple scheme shown above is sufficient to achieve good solutions with few added iterations.

In summary, we find the Lagrange multiplier form to be the only one that does not require the identification of an appropriate value for the κ parameter. Furthermore, in a Newton solution algorithm the Lagrange multiplier form leads to optimal satisfaction of the impenetrability condition in a minimum number of iterations.

7.3 Tied interfaces

Before describing generalizations to the above treatment of nodal contact problems we consider a technique to connect regions in which the finite element mesh is different

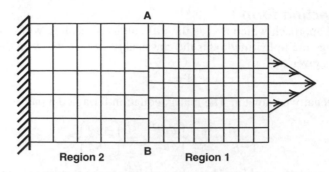

Fig. 7.4 Tied interface for a two region problem.

in each region. A simple case of this type of situation is shown in Fig. 7.4 for a beam loaded by an end traction. The region near the load is described by a finer mesh than that used for the more remote portions. The model now requires the introduction of an *interface* to 'tie' the two parts together. Thus along the boundary 'AB' it is necessary to have

$$\mathbf{x}^1\Big|_{AB} = \mathbf{x}^2\Big|_{AB} \quad \text{and} \quad \mathbf{t}^1\Big|_{AB} + \mathbf{t}^2\Big|_{AB} = \mathbf{0} \tag{7.19}$$

where $\mathbf{x}^i$ is the deformed position and $\mathbf{t}^i$ the traction for the interface between the regions. To impose these conditions we may introduce the Lagrange multiplier functional

$$\Pi_I = \int_{\Gamma_I} \boldsymbol{\lambda}^T \left(\mathbf{x}^1 - \mathbf{x}^2 \right) d\Gamma^1 \tag{7.20}$$

in which we identify the multiplier as

$$\boldsymbol{\lambda} = \mathbf{t}^1 = -\mathbf{t}^2$$

We use a standard finite element approximation to define the positions $\mathbf{x}^i$. For the approximation of $\boldsymbol{\lambda}$ we can consider several alternative approximations. Here we consider the approximation

$$\boldsymbol{\lambda} \approx \sum_a N_a(\boldsymbol{\xi})\tilde{\boldsymbol{\lambda}}_a \tag{7.21}$$

where N_a are also standard shape functions. An approximation to the integral (7.20) may be given using quadrature points located at the nodes of region 1. Accordingly, we have

$$\Pi_I \approx \tilde{\boldsymbol{\lambda}}_a^T \left(\tilde{\mathbf{x}}_a^1 - \sum_b N_b(\boldsymbol{\xi}_a)\tilde{\mathbf{x}}_b^2 \right) A_a \tag{7.22}$$

where $\boldsymbol{\xi}_a$ is the location on the surface of region 2 for the node a of region 1, A_a is the surface Jacobian for node a, and we assume unit quadrature weights.

The location for each $\boldsymbol{\xi}_a$ can be obtained using a *closest point projection* where

$$c = \tfrac{1}{2} \left(\tilde{\mathbf{x}}_a^1 - \sum_b N_b(\boldsymbol{\xi})\tilde{\mathbf{x}}_b^2 \right)^T \left(\tilde{\mathbf{x}}_a^1 - \sum_c N_c(\boldsymbol{\xi})\tilde{\mathbf{x}}_c^2 \right) = \min \tag{7.23}$$

For an interface of a three-dimensional problem this gives the two equations

$$\frac{\partial c}{\partial \xi} = \left(\tilde{\mathbf{x}}_a^1 - \sum_b N_b(\xi)\tilde{\mathbf{x}}_b^2 \right)^T \sum_c \frac{\partial N_c}{\partial \xi}\tilde{\mathbf{x}}_c^2 = \mathbf{g}^T \mathbf{T}_\xi^2 = 0$$

$$\frac{\partial c}{\partial \eta} = \left(\tilde{\mathbf{x}}_a^1 - \sum_b N_b(\xi)\tilde{\mathbf{x}}_b^2 \right)^T \sum_c \frac{\partial N_c}{\partial \eta}\tilde{\mathbf{x}}_c^2 = \mathbf{g}^T \mathbf{T}_\eta^2 = 0$$

(7.24)

where ξ and η are the parent coordinates of the surface facet and $\mathbf{g} = \tilde{\mathbf{x}}_a^1 - \sum_b N_b\tilde{\mathbf{x}}_b^2$. For a two-dimensional problem we omit the second equation and consider only the parent coordinate ξ. We note that $\mathbf{T}_\xi^2$ and $\mathbf{T}_\eta^2$ are tangent vectors to the facet describing the surface of region 2.

The equations (7.24) are in general non-linear and may be solved using a Newton method given as

$$\mathbf{A}\, d\boldsymbol{\xi} = -\mathbf{R}$$

where $d\boldsymbol{\xi} = (d\xi,\, d\eta)^T$, $\mathbf{R} = (c_{,\xi},\, c_{,\eta})^T$ and

$$\mathbf{A} = \begin{bmatrix} (\mathbf{T}_\xi^T \mathbf{T}_\xi + k_{\xi\xi}) & (\mathbf{T}_\xi^T \mathbf{T}_\eta + k_{\xi\eta}) \\ (\mathbf{T}_\eta^T \mathbf{T}_\xi + k_{\eta\xi}) & (\mathbf{T}_\eta^T \mathbf{T}_\eta + k_{\eta\eta}) \end{bmatrix}$$

with

$$k_{\xi\xi} = \mathbf{g}^T \sum_c N_{c,\xi\xi}\tilde{\mathbf{x}}_c^2, \quad k_{\xi\eta} = \mathbf{g}^T \sum_c N_{c,\xi\eta}\tilde{\mathbf{x}}_c^2 = k_{\eta\xi} \quad \text{and} \quad k_{\eta\eta} = \mathbf{g}^T \sum_c N_{c,\eta\eta}\tilde{\mathbf{x}}_c^2$$

Once ξ_a is known the functional given in Eq. (7.20) may be satisfied using the Lagrange multiplier form or any of the other methods described above for the node–node contact problem.

Example 7.1 Two-dimensional tied interface using linear elements

Consider the example of a two-dimensional problem in which the edges of elements are linear segments. The interpolation for the positions is given by

$$\mathbf{x}^2 = N_1(\xi)\tilde{\mathbf{x}}_1^2 + N_2(\xi)\tilde{\mathbf{x}}_2^2 = \tfrac{1}{2}(1 - \xi)\,\tilde{\mathbf{x}}_1^2 + \tfrac{1}{2}(1 + \xi)\,\tilde{\mathbf{x}}_2^2$$

The tangent vector for the edge is given by

$$\mathbf{T}_\xi^2 = \tfrac{1}{2}\left(\tilde{\mathbf{x}}_2^2 - \tilde{\mathbf{x}}_1^2 \right)$$

and is constant over the whole element edge. The closest point projection gives

$$\frac{\partial c}{\partial \xi} = \left(\tilde{\mathbf{x}}_a^1 - \tfrac{1}{2}(1 - \xi)\tilde{\mathbf{x}}_1^2 - \tfrac{1}{2}(1 + \xi)\tilde{\mathbf{x}}_2^2 \right)^T \mathbf{T}_\xi^2 = 0$$

is a linear relation in ξ and gives the solution

$$\xi_a = \frac{\left(\tilde{\mathbf{x}}_a^1 - \tfrac{1}{2}(\tilde{\mathbf{x}}_1^2 + \tilde{\mathbf{x}}_2^2) \right)^T \mathbf{T}_\xi}{\mathbf{T}_\xi^T \mathbf{T}_\xi}$$

The interface functional for node a is given by

$$\Pi_I = \tilde{\boldsymbol{\lambda}}_a^T \left(\tilde{\mathbf{x}}_a^1 - N_1(\xi_a)\tilde{\mathbf{x}}_1^2 - N_2(\xi_a)\tilde{\mathbf{x}}_2^2 \right) A_a$$

where A_a is half the area of the one or two elements adjacent to node a. The variation of the functional gives

$$\delta\Pi_I = \begin{bmatrix} \delta\tilde{\mathbf{x}}_1^{2,T} & \delta\tilde{\mathbf{x}}_1^{2,T} & \delta\tilde{\mathbf{x}}_a^{1,T} & \delta\tilde{\boldsymbol{\lambda}}_a^T \end{bmatrix} \begin{Bmatrix} -N_1(\xi_a)\tilde{\boldsymbol{\lambda}}_a \\ -N_2(\xi_a)\tilde{\boldsymbol{\lambda}}_a \\ \tilde{\boldsymbol{\lambda}}_a \\ \mathbf{g} \end{Bmatrix} A_a$$

and is used to defined the element residual vector. Similarly,

$$d(\delta\Pi_I) = \begin{bmatrix} \delta\tilde{\mathbf{x}}_1^{2,T} & \delta\tilde{\mathbf{x}}_1^{2,T} & \delta\tilde{\mathbf{x}}_a^{1,T} & \delta\tilde{\boldsymbol{\lambda}}_a^T \end{bmatrix} \begin{bmatrix} \mathbf{0} & \mathbf{0} & \mathbf{0} & -N_1(\xi_a)\mathbf{I} \\ \mathbf{0} & \mathbf{0} & \mathbf{0} & -N_2(\xi_a)\mathbf{I} \\ \mathbf{0} & \mathbf{0} & \mathbf{0} & \mathbf{I} \\ -N_1(\xi_a)\mathbf{I} & -N_2(\xi_a)\mathbf{I} & \mathbf{I} & \mathbf{0} \end{bmatrix}$$

$$\times \begin{bmatrix} d\tilde{\mathbf{x}}_1^2 \\ d\tilde{\mathbf{x}}_1^2 \\ d\tilde{\mathbf{x}}_a^1 \\ d\tilde{\boldsymbol{\lambda}}_a \end{bmatrix} A_a$$

defines the tangent array. We note the structure of the arrays is identical to that obtained for the node–node contact treatment. Accordingly, any of the other solution methods may also be employed to formulate the interface arrays.

7.4 Node–surface contact

7.4.1 Geometric modelling

A simple form for contact between bodies in which nodes on the surface of one body do not interact directly with nodes on a second body may be defined by a *node–surface* treatment similarly to that used in the previous section for a tied interface. A two-dimensional treatment for this case is shown in Fig. 7.5 where a node, called the *slave node*, with deformed position $\tilde{\mathbf{x}}_s$ can contact a segment, called the *master surface*, defined by an interpolation

$$\mathbf{x}^m = \sum_a N_a(\boldsymbol{\xi})\tilde{\mathbf{x}}_a^m \tag{7.25}$$

where $\boldsymbol{\xi}$ is equal to ξ, η in three dimensions and to ξ in two dimensions. This interpolation may be treated either as the usual interpolation along the boundary facets of elements describing the target body as shown in Fig. 7.5(a) or by an interpolation which

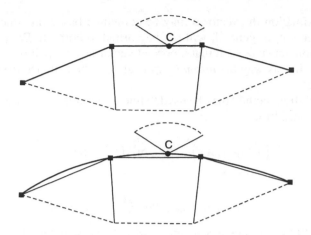

Fig. 7.5 Node-to-surface contact: (a) contact using element interpolations; (b) contact using 'smoothed' interpolations.

smooths the slope discontinuity between adjacent element surface facets as shown in Fig. 7.5(b).

A contact between the two bodies occurs when g_n, the *gap* shown in Fig. 7.6, becomes zero. The determination of a contact state requires a search to find which target facet is a potential contact point on the master surface and computation of the associated gap g_n and contact position ξ for each one.[6,40,41] If the gap is positive no contact condition exists and, thus, no modification to the governing equations is required. If the gap is negative a 'penetration' of the two bodies has occurred and it is necessary to modify the equilibrium equations to make the gap zero and to define the contact tractions (or nodal forces) that occur.

The determination of the gap g_n and the point ξ on the target (master) facet may be obtained by solving the constraint equation

$$c(g_n, \xi) = \tilde{\mathbf{x}}_s - \mathbf{x}^m(\xi) - g_n \, \mathbf{n} = \tilde{\mathbf{x}}_s - \sum_a N_a(\xi)\tilde{\mathbf{x}}_a^m - g_n \, \mathbf{n} = \mathbf{0} \qquad (7.26)$$

where $\mathbf{n}$ is a unit 'normal' vector (i.e. $\mathbf{n}^T\mathbf{n} = 1$) that is defined relative to the master contact surface. This is an alternative to the closest point projection presented in the previous section and permits a more general treatment of contact. Thus, in a finite

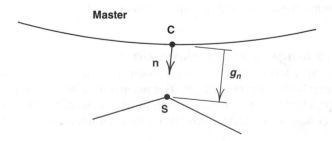

Fig. 7.6 Node-to-surface contact: gap and normal definition.

element approximation the vector $\mathbf{n}$ may be computed based on either the master or the slave surface but, in general, will not be normal to both. In Eq. (7.26), however, we define the contact state such that a normal vector always points outward from the master surface. In this way the *normal gap* is always the distance from the master to the slave surface.

A Newton solution method may be used to find a solution to Eq. (7.26). Linearizing, we solve for iterates from

$$\left(\mathbf{x}^m_{,\xi} + g_n \, \mathbf{n}_{,\xi}\right)^k d\xi^k + \mathbf{n}^k \, dg^k_n = \mathbf{c}(g^k, \xi^k) \tag{7.27}$$

with updates

$$g^{k+1}_n = g^k_n + dg^k_n \quad \text{and} \quad \xi^{k+1} = \xi^k + d\xi^k$$

until the constraint equation $\mathbf{c}$ is satisfied to within a specified tolerance.

Alternatively, the problem may be split into two parts by premultiplying Eq. (7.26) by the transpose of $\mathbf{n}_m$ and $\mathbf{t}_i$ where $\mathbf{n}_m$ is a unit normal to the master surface and $\mathbf{t}_i$ are unit tangent vectors orthogonal to $\mathbf{n}$ (when defined on the slave surface $\mathbf{n} \neq \mathbf{n}_m$). It is not necessary that these vectors be unit vectors but they must satisfy the orthogonality relations

$$\mathbf{t}^T_i \mathbf{n} = \mathbf{n}^T_m \mathbf{x}^m_{,\xi}\Big|_{\xi=\xi_c} = \mathbf{n}^T_m \mathbf{x}^m_{,\eta}\Big|_{\xi=\xi_c} = 0$$

where

$$\mathbf{x}^m_{,\xi} = \sum_a N_{a,\xi} \, \tilde{\mathbf{x}}^m_a \quad \text{and} \quad \mathbf{x}^m_{,\eta} = \sum_a N_{a,\eta} \, \tilde{\mathbf{x}}^m_a \tag{7.28}$$

Also note that $\mathbf{t}_1$ does not need to be orthogonal to $\mathbf{t}_2$. In this approach we obtain

$$g_n = \mathbf{n}^T_m \left(\tilde{\mathbf{x}}_s - \sum_a N_a(\xi_c)\tilde{\mathbf{x}}^m_a\right) \Big/ \left(\mathbf{n}^T_m \mathbf{n}\right) \tag{7.29a}$$

and ξ_c is found by solving

$$\mathbf{t}^T_i \tilde{\mathbf{c}} = \mathbf{t}^T_i \left(\tilde{\mathbf{x}}_s - \sum_a N_a(\xi)\tilde{\mathbf{x}}^m_a\right) \tag{7.29b}$$

for $i = 1, 2$ in three dimensions and $i = 1$ in two dimensions. In a general setting the equation for ξ is non-linear and a Newton method may be used to find a solution; however, once the contact point ξ_c is determined the expression (7.29a) for the gap g_n is linear.

Normal and tangent vector definitions

As noted above $\mathbf{n}$ is defined here to be a unit normal vector that always points outward from the *master* body contact surface. The normal to a master surface for a three-dimensional problem may be computed from the cross product of two vectors that are tangent to the contact surface. Accordingly, for a three-dimensional finite element facet we can use

$$\mathbf{N}_m = \mathbf{x}^m_{,\xi} \times \mathbf{x}^m_{,\eta} \tag{7.30a}$$

and in two dimensions

$$\mathbf{N}_m = \mathbf{x}^m_{,\xi} \times \mathbf{e}_z = \mathbf{T}_m \times \mathbf{e}_z = -\mathbf{e}_z \times \mathbf{T}_m = \mathbf{E}_z \, \mathbf{T}_m; \quad \mathbf{E}_z = \begin{bmatrix} 0 & 1 \\ -1 & 0 \end{bmatrix} \qquad (7.30\mathrm{b})$$

where $\mathbf{e}_z$ is a unit vector normal to the plane of deformation and '$\times$' denotes the vector cross product. The unit normal is then defined by

$$\mathbf{n}_m = \frac{\mathbf{N}_m}{\|\mathbf{N}_m\|} \quad \text{where} \quad \|\mathbf{N}_m\| = (\mathbf{N}_m^T \mathbf{N}_m)^{1/2} \qquad (7.31)$$

A 'normal' to the slave surface (with outward direction relative to the master body) may be determined by summing the normals to the elements surrounding a slave node s as

$$\mathbf{N}_s = -\sum_{e=1}^{e_s} \mathbf{x}^{s,e}_{,\xi} \times \mathbf{x}^{s,e}_{,\eta} \Big|_{\xi=\xi_s} \qquad (7.32\mathrm{a})$$

for three dimensions or

$$\mathbf{N}_s = -\sum_{e=1}^{2} \mathbf{x}^{s,e}_{,\xi} \Big|_{\xi=\xi_s} \times \mathbf{e}_z = \mathbf{T}_s \times \mathbf{e}_z = \mathbf{E}_z \, \mathbf{T}_s \qquad (7.32\mathrm{b})$$

for two dimensions. The unit normal is then obtained using Eq. (7.31) with $\mathbf{N}_s$ now replacing $\mathbf{N}_m$.

The variation of a unit normal vector is computed from Eq. (7.31) and given by

$$\delta\mathbf{n} = \frac{1}{\|\mathbf{N}\|} \left(\mathbf{I} - \mathbf{n}\mathbf{n}^T \right) \delta\mathbf{N} \qquad (7.33)$$

where $\mathbf{I}$ is an identity matrix and $\mathbf{N}$ is computed from Eq. (7.30a) or (7.32a) for three-dimensional problems and Eq. (7.30b) or (7.32b) for two-dimensional ones. The variation, $\delta\mathbf{N}$, is computed directly from the definition of $\mathbf{N}$. We note from Eq. (7.33) that the unit normal is orthogonal to its variation so that

$$\mathbf{n}^T \delta\mathbf{n} = 0 \qquad (7.34)$$

For two-dimensional problems the variation of $\mathbf{n}$ may be simplified by noting the spectral decomposition of the identity matrix may be written as

$$\mathbf{I} = \mathbf{n}\mathbf{n}^T + \mathbf{t}\mathbf{t}^T \qquad (7.35)$$

where $\mathbf{t}$ is the unit tangent orthogonal to $\mathbf{n}$ and $\mathbf{e}_z$. Thus, the variation of the unit normal vector in two dimensions may be written as

$$\delta\mathbf{n} = \frac{1}{\|\mathbf{N}\|} \mathbf{t}\mathbf{t}^T \delta\mathbf{N} = -\frac{1}{\|\mathbf{N}\|} \mathbf{t}\mathbf{n}^T \delta\mathbf{T} \qquad (7.36)$$

where from either Eq. (7.30b) or (7.32b) we note that

$$\delta\mathbf{N} = \mathbf{E}_z \, \delta\mathbf{T} \qquad (7.37)$$

In computing $\delta\xi$ and tangent arrays for a finite element representation of contact it is necessary to obtain variations and increments of tangent vectors. These are computed in an identical manner to the normal vector. Thus, for a unit tangent we can write

$$\delta\mathbf{t}_i = \delta\left(\frac{\mathbf{T}_i}{\|\mathbf{T}_i\|}\right) = \frac{1}{\|\mathbf{T}_i\|}(\mathbf{I} - \mathbf{t}_i\mathbf{t}_i^T)\,\delta\mathbf{T}_i \tag{7.38a}$$

In two dimensions the use of the spectral representation of the identity simplifies the representation to

$$\delta\mathbf{t} = \frac{1}{\|\mathbf{T}\|}\mathbf{n}\,\mathbf{n}^T\,\delta\mathbf{T} \tag{7.38b}$$

Example 7.2   Normal vector to 2D linear master facets

Consider a two-dimensional problem modelled by four-node quadrilateral or three-node triangular elements in which the edges are linear segments and the interpolations for N_a are given by

$$N_1 = \tfrac{1}{2}(1 - \xi) \quad \text{and} \quad N_2 = \tfrac{1}{2}(1 + \xi)$$

The tangent vector to a master facet shown in Fig. 7.7(a) may be defined as

$$\mathbf{T}_m = \mathbf{x}_{,\xi}^m = \tfrac{1}{2}(\tilde{\mathbf{x}}_2^m - \tilde{\mathbf{x}}_1^m)$$

and, thus, the interpolation for $\mathbf{x}^m$ may be written as

$$\mathbf{x}^m(\xi) = \tfrac{1}{2}(\tilde{\mathbf{x}}_1^m + \tilde{\mathbf{x}}_2^m) + \mathbf{T}_m\,\xi$$

The normal vector

$$\mathbf{N}_m = \mathbf{T}_m \times \mathbf{e}_z = \mathbf{E}_z\mathbf{T}_m$$

gives components $N_1^m = T_2^m$ and $N_2^m = -T_1^m$. We note that in this case the vectors are constant over the entire facet. Unit normal and tangent vectors are given by

$$\mathbf{n}_m = \frac{\mathbf{N}_m}{\|\mathbf{N}_m\|} \quad \text{and} \quad \mathbf{t}_m = \frac{\mathbf{T}_m}{\|\mathbf{T}_m\|}$$

where we note that $\|\mathbf{T}_m\| = \|\mathbf{N}_m\|$. Thus, the relation (7.26) for $\mathbf{c}$ is linear in g_n and ξ and the solution is given by

$$g_n = \mathbf{n}_m^T\mathbf{g}_0 \quad \text{where} \quad \mathbf{g}_0 = \tilde{\mathbf{x}}_s - \tfrac{1}{2}(\tilde{\mathbf{x}}_1^m + \tilde{\mathbf{x}}_2^m)$$

and

$$\xi_c = \frac{\mathbf{t}_m^T\mathbf{g}_0}{\mathbf{t}_m^T\mathbf{T}_m} = \frac{1}{\|\mathbf{T}_m\|}\mathbf{t}_m^T\mathbf{g}_0 = \frac{1}{\|\mathbf{N}_m\|}\mathbf{t}_m^T\mathbf{g}_0$$

Example 7.3   Normal vector to 2D linear slave facets

If we consider the two-dimensional problem with edges of slave elements as shown in Fig. 7.7(b) a normal vector may be written using Eq. (7.32b) as

$$\mathbf{N}_s = [\tfrac{1}{2}(\tilde{\mathbf{x}}_1^s - \tilde{\mathbf{x}}_s) + \tfrac{1}{2}(\tilde{\mathbf{x}}_s - \tilde{\mathbf{x}}_2^s)] \times \mathbf{e}_z = \tfrac{1}{2}(\tilde{\mathbf{x}}_1^s - \tilde{\mathbf{x}}_1^2) \times \mathbf{e}_z = \mathbf{T}_s \times \mathbf{e}_z = \mathbf{E}_z\mathbf{T}_s$$

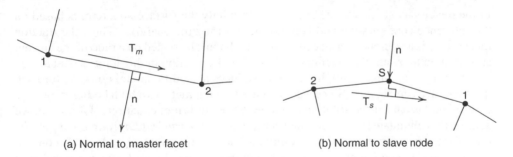

Fig. 7.7 Node-to-surface contact: normal vector description.

which gives components $N_1^s = T_2^s$ and $N_2^s = -T_1^s$. Again, we obtain the vectors

$$\mathbf{n}_s = \frac{\mathbf{N}_s}{\|\mathbf{N}_s\|} \quad \text{and} \quad \mathbf{t}_s = \frac{\mathbf{T}_s}{\|\mathbf{T}_s\|}$$

which are independent of ξ.

The solution for the gap g_n and contact point ξ_c are now given by

$$g_n = \frac{\mathbf{n}_m^T(\tilde{\mathbf{x}}_s - N_1(\xi_c)\tilde{\mathbf{x}}_1^m - N_2(\xi_c)\tilde{\mathbf{x}}_2^m))}{\mathbf{n}_m^T\mathbf{n}_s} = \frac{\mathbf{n}_m^T\mathbf{g}_0}{\mathbf{n}_m^T\mathbf{n}_s}$$

and

$$\xi_c = \frac{\mathbf{t}_s^T\mathbf{g}_0}{\mathbf{t}_s^T\mathbf{T}_m}$$

where $\mathbf{n}_m$, $\mathbf{T}_m$ and $\mathbf{g}_0$ are identical to the vectors defined in Example 7.2.

7.4.2 Contact modelling – frictionless case

For a *frictionless* contact only normal tractions are involved on the surfaces between the two bodies; thus sliding can occur without generation of tangential forces and the contact traction is given by

$$\mathbf{t}_\Gamma = \lambda_n \mathbf{n} \tag{7.39}$$

where λ_n is the magnitude of a normal traction applied to the contact target and $\mathbf{n}$ is a unit normal directed outward relative to the master surface. This case can be included by appending the variation of a Lagrange multiplier term to the Galerkin (weak) form describing equilibrium of the problem for each contact slave node. Accordingly, again using nodal quadrature on the slave surface we obtain

$$\Pi_c = \int_{\Gamma_c} \mathbf{g}^T\mathbf{t}\,d\Gamma \approx (\lambda_n\mathbf{n}^T)\,(g_n\mathbf{n})\,A_c = \lambda_n\,g_n\,A_c \tag{7.40}$$

where A_c is a surface area associated with the slave node and for the solution point of Eq. (7.26), denoted by ξ_c, the gap is given by

$$g_n\,\mathbf{n} = \tilde{\mathbf{x}}_s - \mathbf{x}^m(\xi_c) \tag{7.41}$$

In the development summarized here, for simplicity the surface area term is based on the reference configuration and kept constant during the analysis. Thus, the traction measure λ_n is a reference surface measure which must be scaled by a ratio of the current surface area to obtain the magnitude of the Cauchy traction in the deformed state.

Use of the Lagrange multiplier form introduces an additional unknown λ_n for each master–slave contact pair. Since a contact traction interacts with both bodies it must be determined as part of the solution of the global equilibrium equations. Of course, we again could eliminate the contact tractions by using a *perturbed Lagrangian* or *penalty* form for the constraint in a manner similar to that used for treating node–node contact. However, even then the problem is more complex as we do not know *a priori* which master facet interacts with a specified contact slave node. Thus, for each contact state a search to establish the active set of pairs is necessary in order to compute the non-zero structure of the problem tangent matrix. This implies that the non-zero structure of the tangent matrix will change during the solution of any contact problem and continual updates are required to describe the sparse structure (or profile) of the global matrix.

Contact residual

The variation of the potential given in Eq. (7.40) may be expressed as

$$\delta \Pi_c = (\delta g_n \, \lambda_n + \delta \lambda_n \, g_n) \, A_c \tag{7.42}$$

and used to compute the contact residual array. The result appears to be nearly identical to that obtained for the node-to-node contact. However, the relationship between the δg_n and $\delta \xi$ terms with the $\delta \tilde{\mathbf{x}}$ (or $\delta \tilde{\mathbf{u}}$) is more complex and must be determined from a variation of the gap relation (7.26). Formally, this is given by

$$\mathbf{n} \, \delta g_n + \mathbf{T}^m_\xi \, \delta \xi + \mathbf{T}^m_\eta \, \delta \eta = \delta \tilde{\mathbf{x}}_s - \sum_a N_a(\xi) \delta \tilde{\mathbf{x}}^m - g_n \, \delta \mathbf{n} \tag{7.43}$$

where

$$\mathbf{T}^m_\xi = \mathbf{x}^m_{,\xi} = \sum_a N_{a,\xi} \, \tilde{\mathbf{x}}^m_a \quad \text{and} \quad \mathbf{T}^m_\eta = \mathbf{x}^m_{,\eta} = \sum_a N_{a,\eta} \, \tilde{\mathbf{x}}^m_a$$

with $\mathbf{T}^m_\xi$ and $\mathbf{T}^m_\eta$ being tangent vectors to the master surface at point ξ.

The expression for δg_n may now be obtained in the same manner used to compute g, thus,

$$(\mathbf{n}^T_m \mathbf{n}) \, \delta g_n = \mathbf{n}^T_m \left(\delta \tilde{\mathbf{x}}_s - \sum_a N_a(\xi) \delta \tilde{\mathbf{x}}^m_a - g_n \, \delta \mathbf{n} \right) \tag{7.44}$$

This may be written in matrix notation as

$$\delta g_n = \delta \tilde{\mathbf{x}}^T \mathbf{U}_n \tag{7.45}$$

where $\mathbf{U}_n$ is a vector that defines the distribution of contact forces and $\tilde{\mathbf{x}}$ is a vector of coordinates for the set of nodes involved in the contact constraint equation (7.26). Thus the variation of Π_c may be written in matrix form as

$$\delta \Pi_c = \begin{bmatrix} \delta \tilde{\mathbf{x}}^T, & \delta \lambda_n \end{bmatrix} \begin{Bmatrix} \lambda_n \, \mathbf{U}_n \, A_c \\ g_n \, A_c \end{Bmatrix} \tag{7.46}$$

An alternate form to compute the variation of the gap may be given by multiplying Eq. (7.43) by the transpose of **n** to give

$$\delta g_n = \mathbf{n}^T \left(\delta \tilde{\mathbf{x}}_s - \sum_a N_a(\xi) \delta \tilde{\mathbf{x}}^m - \mathbf{T}_\xi^m \, \delta\xi - \mathbf{T}_\eta^m \, \delta\eta \right) \qquad (7.47)$$

where we have used Eq. (7.34). In constructing models in which the normal is based on the slave surface geometry we will make use of this alternate form.

The computation of $\delta\xi$ proceeds by premultiplying Eq. (7.43) by the unit tangent vectors $\mathbf{t}_i$ to obtain

$$\mathbf{t}_i^T \mathbf{T}_\xi^m \, \delta\xi + \mathbf{t}_i^T \mathbf{T}_\eta^m \, \delta\eta = \mathbf{t}_i^T \left(\delta \tilde{\mathbf{x}}_s - \sum_a N_a(\xi) \delta \tilde{\mathbf{x}}^m \right) - g_n \mathbf{t}_i^T \delta\mathbf{n}; \quad i = 1, 2 \qquad (7.48)$$

We note that when $\mathbf{n} = \mathbf{n}(\xi)$, its variation will contribute to the coefficients of $\delta\xi$ and $\delta\eta$.

To make the steps clearer we consider next the formulation in two dimensions, first for the case where the normal **n** is computed from the master surface geometry and, second, the case where it is computed from the slave surface geometry.

Example 7.4   Contact forces for normal to 2D master surface

As an example we consider the case where the edge of a two-dimensional element is a linear segment as shown in Fig. 7.7(a). In this case the tangent vector may be taken as the $\mathbf{T}_m$ defined in Example 7.2 and gives the normal vector

$$\mathbf{N}_m = \mathbf{E}_z \mathbf{x}^m_{,\xi} = \tfrac{1}{2} \mathbf{E}_z (\tilde{\mathbf{x}}^m_2 - \tilde{\mathbf{x}}^m_1) = \mathbf{E}_z \mathbf{T}_m$$

With this definition the unit normal vector $\mathbf{n}_m$ given by Eq. (7.31) is orthogonal to $\mathbf{T}_m$ and, thus, δg_n may be determined from

$$\delta g_n = \mathbf{n}_m^T \left(\delta \tilde{\mathbf{x}}_s - \sum_{a=1}^{2} N_a \, \delta \tilde{\mathbf{x}}^m_a \right) = \mathbf{U}_n^T \delta\mathbf{x}$$

where

$$\mathbf{U}_n = \left\{ \begin{array}{c} -N_1(\xi)\mathbf{n}_m \\ -N_2(\xi)\mathbf{n}_m \\ \mathbf{n}_m \end{array} \right\} \qquad \text{and} \qquad \delta\mathbf{x} = \left\{ \begin{array}{c} \delta\tilde{\mathbf{x}}^m_1 \\ \delta\tilde{\mathbf{x}}^m_2 \\ \delta\tilde{\mathbf{x}}_s \end{array} \right\}$$

Indeed $\delta\xi$ need not be computed at this point. We will find, however, it is required in order to compute the tangent.

The variation of Π_c is now given by

$$\delta\Pi_c = \begin{bmatrix} \delta\tilde{\mathbf{x}}^T & \delta\lambda_n \end{bmatrix} \left\{ \begin{array}{c} \lambda_n \mathbf{U}_n A_c \\ g_n A_c \end{array} \right\}$$

Example 7.5   Contact forces for normal to 2D slave surface

As a second example we consider the case where the edges of a two-dimensional element are linear segments as shown in Fig. 7.7(b). In this case the tangent vector may be taken as $\mathbf{T}_s$ as defined in Example 7.3 and gives the normal vector

$$\mathbf{N}_s = (\tilde{\mathbf{x}}^s_1 - \tilde{\mathbf{x}}^s_2) \times \mathbf{e}_z = \mathbf{E}_z (\tilde{\mathbf{x}}^s_1 - \tilde{\mathbf{x}}^s_2) = \mathbf{E}_z \mathbf{T}_s$$

Since $\mathbf{T}_s$ is independent of ξ, its variation is given by

$$\delta \mathbf{T}_s = \delta \tilde{\mathbf{x}}_1^s - \delta \tilde{\mathbf{x}}_2^s$$

Using this and previously defined quantities, we can define the variation of the unit normal as

$$\delta \mathbf{n}_s = -\frac{1}{\|\mathbf{N}_s\|} \mathbf{t}_s \, \mathbf{n}_s^T \delta \mathbf{T}_s = -\mathbf{t}_s \, \mathbf{W}_n^T \delta \tilde{\mathbf{x}}$$

where

$$\mathbf{W}_n = \frac{1}{\|\mathbf{N}_s\|} \left\{ \begin{array}{c} \mathbf{0} \\ \mathbf{0} \\ \mathbf{0} \\ \mathbf{n}_s \\ -\mathbf{n}_s \end{array} \right\} \quad \text{and} \quad \delta \tilde{\mathbf{x}} = \left\{ \begin{array}{c} \delta \tilde{\mathbf{x}}_1^m \\ \delta \tilde{\mathbf{x}}_2^m \\ \delta \tilde{\mathbf{x}}_s \\ \delta \tilde{\mathbf{x}}_1^s \\ \delta \tilde{\mathbf{x}}_2^s \end{array} \right\}$$

From Eq. (7.44) we obtain the variation of the gap as

$$\delta g_n = \delta \tilde{\mathbf{x}} \left(\mathbf{V}_n + g_n \, k_n \, \mathbf{W}_n \right) = \delta \tilde{\mathbf{x}}^T \, \mathbf{U}_n$$

where

$$\mathbf{V}_n = \frac{1}{\mathbf{n}_s^T \mathbf{n}_m} \left\{ \begin{array}{c} -N_1(\xi)\mathbf{n}_m \\ -N_2(\xi)\mathbf{n}_m \\ \mathbf{n}_m \\ \mathbf{0} \\ \mathbf{0} \end{array} \right\} \quad \text{and} \quad k_n = \frac{\mathbf{t}_s^T \mathbf{n}_m}{\mathbf{n}_s^T \mathbf{n}_m}$$

The variation of Π_c may be expressed as

$$\delta \Pi_c = \begin{bmatrix} \delta \tilde{\mathbf{x}}^T & \delta \lambda_n \end{bmatrix} \left\{ \begin{array}{c} \lambda_n \, \mathbf{U}_n \, A_c \\ g_n \, A_c \end{array} \right\}$$

Example 7.6   Contact forces for normal to 2D slave surface – alternative form

In the previous two examples we obtained a form in which the distribution of the contact forces depends on the normal to the master surface $\mathbf{N}_m$. For the two-dimensional problem in which element edges are straight line segments the normal to the master surface can have a sudden discontinuity between adjacent elements. This will lead to a discontinuity in the normal traction λ_n when sliding from one facet to the next occurs. The effects of the discontinuity may be mitigated by modifying the definition of the variation of the gap in Example 7.5. Multiplying Eq. (7.43) by $\mathbf{n}_s^T$ we obtain

$$\delta g_n = \mathbf{n}_s^T \left(\delta \tilde{\mathbf{x}}_s - N_1(\xi)\delta \tilde{\mathbf{x}}_1^m - N_2(\xi)\delta \tilde{\mathbf{x}}_2^m \right) - \mathbf{n}_s^T \mathbf{T}_m \, \delta \xi$$

Computing $\delta \xi$ from the expression given in Example 7.3 yields

$$\delta \xi = \frac{1}{\mathbf{t}_s^T \mathbf{T}_m} \left[\delta \mathbf{t}_s^T \left(\tilde{\mathbf{x}}_s - N_1(\xi)\tilde{\mathbf{x}}_1^m - N_2(\xi)\tilde{\mathbf{x}}_2^m \right) + \left(\delta \tilde{\mathbf{x}}_s^T - N_1(\xi)\delta \tilde{\mathbf{x}}_1^{mT} - N_2(\xi)\delta \tilde{\mathbf{x}}_2^{mT} \right) \mathbf{t}_s \right]$$

$$= \frac{1}{\mathbf{t}_s^T \mathbf{T}_m} \delta \tilde{\mathbf{x}}^T \left(\mathbf{V}_t + g_n \, \mathbf{W}_n \right)$$

where $\mathbf{W}_n$ is defined in Example 7.5 and

$$\mathbf{V}_t = \left\{ \begin{array}{c} -N_1(\xi)\mathbf{t}_s \\ -N_2(\xi)\mathbf{t}_s \\ \mathbf{t}_s \\ 0 \\ 0 \end{array} \right\}$$

The expression for the variation of g_n may now be written in the matrix form

$$\delta g_n = \delta \tilde{\mathbf{x}}^T [\mathbf{V}_s + k_t (\mathbf{V}_t + g_n \mathbf{W}_n)] = \delta \tilde{\mathbf{x}}^T \mathbf{U}_n$$

where

$$\mathbf{V}_s = \left\{ \begin{array}{c} -N_1(\xi)\mathbf{n}_s \\ -N_2(\xi)\mathbf{n}_s \\ \mathbf{n}_s \\ 0 \\ 0 \end{array} \right\} \quad \text{and} \quad k_t = \frac{\mathbf{n}_s^T \mathbf{T}_m}{\mathbf{t}_s^T \mathbf{T}_m}$$

This form may be used to give the contact residual as above and when linearized leads to a symmetric tangent matrix for use in a Newton solution algorithm. If we restrict the variation to one where we set $\delta \xi$ to zero, which is also an admissible variation, the distribution of contact forces is given by the form

$$\delta g_n = \delta \tilde{\mathbf{x}}^T \bar{\mathbf{U}}_n \quad \text{where} \quad \bar{\mathbf{U}}_n = \mathbf{V}_s$$

which yields the variation of Π_c as

$$\delta \Pi_c = \begin{bmatrix} \delta \tilde{\mathbf{x}}^T & \delta \lambda_n \end{bmatrix} \left\{ \begin{array}{c} \lambda_n \bar{\mathbf{U}}_n A_c \\ g_n A_c \end{array} \right\}$$

This form has a distribution dependent on the normal to the slave surface and with no forces acting on the slave nodes adjacent to s that are used to compute the normal. This is physically more realistic; however, we shall find that the tangent matrix will not be symmetric. This is a distinct disadvantage unless friction is also included where, normally, the tangent matrix also will be unsymmetric during slip.

Contact tangent

To compute the contact tangent array we linearize the variation of the potential $\delta \Pi_c$ and obtain

$$d(\delta \Pi_c) = \Big(\delta g_n \, d\lambda_n + \delta \lambda_n \, dg_n + \lambda_n \, d(\delta g_n) \Big) A_c \qquad (7.49)$$

Except for the last term the structure is identical to that obtained for the node-to-node contact problem.

The increment to the normal gap, dg_n, is obtained from Eq. (7.44) by replacing the variation δ by the increment d. To obtain a symmetric tangent the computation of $d(\delta g_n)$ proceeds from Eq. (7.43) as

$$(\mathbf{n}_m^T \mathbf{n}) \, d(\delta g_n) = - (\delta \mathbf{n}^T \mathbf{n}_m) \, dg_n - \delta g_n \, (\mathbf{n}_m^T d\mathbf{n}) - g_n \, (\mathbf{n}^T d(\delta \mathbf{n}))$$

$$- \sum_a (\delta \xi^T N_{a,\xi}) (\mathbf{n}_m^T \, d\tilde{\mathbf{x}}_a^m) - \sum_a (\delta \tilde{\mathbf{x}}_a^{mT} \mathbf{n}_m) (N_{a,\xi}^T d\xi) \qquad (7.50)$$

$$- \delta \xi^T (\mathbf{n}_m^T \mathbf{x}_{,\xi\xi}^m) \, d\xi$$

where

$$\delta \boldsymbol{\xi}^T N_{a,\xi} = [\delta\xi, \delta\eta] \left\{ \begin{array}{c} N_{a,\xi} \\ N_{a,\eta} \end{array} \right\}$$

and

$$\delta \boldsymbol{\xi}^T (\mathbf{n}_m^T \mathbf{x}_{,\xi\xi}) \, \mathrm{d}\boldsymbol{\xi} = [\delta\xi, \delta\eta] \left[\begin{array}{cc} (\mathbf{n}_m^T \mathbf{x}_{,\xi\xi}^m) & (\mathbf{n}_m^T \mathbf{x}_{,\xi\eta}^m) \\ (\mathbf{n}_m^T \mathbf{x}_{,\eta\xi}^m) & (\mathbf{n}_m^T \mathbf{x}_{,\eta\eta}^m) \end{array} \right] \left\{ \begin{array}{c} \mathrm{d}\xi \\ \mathrm{d}\eta \end{array} \right\}$$

In matrix form

$$\mathrm{d}(\delta g_n) = \delta \tilde{\mathbf{x}}^T \, \mathbf{K}_G \, \mathrm{d}\tilde{\mathbf{x}} \tag{7.51}$$

The final form for the tangent may be written as

$$\mathrm{d}(\delta \Pi_c) = [\delta \tilde{\mathbf{x}}^T, \delta \lambda_n] \left[\begin{array}{cc} \lambda_n \, \mathbf{K}_G & \bar{\mathbf{U}}_n \\ \mathbf{U}_n^T & 0 \end{array} \right] \left\{ \begin{array}{c} \mathrm{d}\tilde{\mathbf{x}} \\ \mathrm{d}\lambda_n \end{array} \right\} \tag{7.52}$$

where

$$\delta g_n = \delta \tilde{\mathbf{x}}^T \, \bar{\mathbf{U}}_n \quad \text{and} \quad \mathrm{d} g_n = \mathbf{U}_n^T \, \mathrm{d}\tilde{\mathbf{x}}$$

For cases where $\mathbf{U}_n \equiv \bar{\mathbf{U}}_n$ the tangent will be symmetric. To illustrate the process we again consider the two dimensional problems.

Example 7.7   Tangent for 2D linear master surface

For this case we note that

$$\mathbf{x}_{,\xi\xi}^m = \mathbf{0} \quad \text{and} \quad \delta\mathbf{n}^T \mathbf{n}_m = \mathbf{n}_m \mathrm{d}\mathbf{n} = 0$$

so that Eq. (7.50) simplifies to

$$\mathrm{d}(\delta g_n) = - \delta\xi \, \mathbf{n}_m^T \sum_a (N_{a,\xi} \, \mathrm{d}\tilde{\mathbf{x}}_a^m) - \sum_a (\delta\tilde{\mathbf{x}}_a^{mT} \, N_{a,\xi}) \, \mathbf{n}_m \, \mathrm{d}\xi - g_n \, \mathbf{n}_m^T \, \mathrm{d}(\delta\mathbf{n}_m)$$

$$= -\delta\xi \, \mathbf{n}_m^T \, \mathrm{d}\mathbf{T}_m - \delta\mathbf{T}_m^T \mathbf{n}_m \, \mathrm{d}\xi - g_n \, \mathbf{n}_m^T \, \mathrm{d}(\delta\mathbf{n}_m)$$

where since the $N_{a,\xi}$ are constants

$$\delta\mathbf{T}_m = \sum_a N_{a,\xi} \, \delta\tilde{\mathbf{x}}_a^m \quad \text{and} \quad \mathrm{d}\mathbf{T}_m = \sum_a N_{a,\xi} \, \mathrm{d}\tilde{\mathbf{x}}_a^m$$

Using Eqs (7.33) and (7.34) it is easy to show that

$$\mathbf{n}_m^T \mathrm{d}(\delta\mathbf{n}_m) = - \frac{1}{\|\mathbf{N}_m\|^2} \, \delta\mathbf{N}_m^T \, (\mathbf{t}_m \, \mathbf{t}_m^T) \, \mathrm{d}\mathbf{N}_m = - \frac{1}{\|\mathbf{N}_m\|^2} \, \delta\mathbf{T}_m^T \, (\mathbf{n}_m \, \mathbf{n}_m^T) \, \mathrm{d}\mathbf{T}_m$$

In matrix form $\delta\xi$ is given by

$$\delta\xi = \frac{1}{\|\mathbf{N}_m\|} \left(\mathbf{V}_t + g_n \, \mathbf{D}_n \right) \delta\tilde{\mathbf{x}} = \frac{1}{\|\mathbf{N}_m\|} \, \mathbf{V}_h \, \delta\tilde{\mathbf{x}}$$

where

$$\mathbf{V}_t = \left\{ \begin{array}{c} -N_1(\xi)\mathbf{t}_m \\ -N_2(\xi)\mathbf{t}_m \\ \mathbf{t}_m \end{array} \right\}; \quad \mathbf{D}_n = \frac{1}{2\,\|\mathbf{N}_m\|} \left\{ \begin{array}{c} -\mathbf{n}_m \\ \mathbf{n}_m \\ \mathbf{0} \end{array} \right\}$$

and $\tilde{\mathbf{x}}$ is as defined in Example 7.4. Thus, the geometric stiffness term may be written as

$$\mathbf{K}_G = -\mathbf{V}_h \, \mathbf{D}_n^T - \mathbf{D}_n \, \mathbf{V}_h^T + g_n \, \mathbf{D}_n \, \mathbf{D}_n^T$$

which is symmetric. The final tangent is given by Eq. (7.52) with $\bar{\mathbf{U}}_n$ equal to the $\mathbf{U}_n$ given in Example 7.4.

Example 7.8   Tangent for normal to 2D slave surface

For this case Eq. (7.50) simplifies to

$$(\mathbf{n}_m^T \mathbf{n}_s) \, d(\delta g_n) = -\delta \xi \, \mathbf{n}_m^T d\mathbf{T}_m - \delta \mathbf{T}_m^T \mathbf{n}_m \, d\xi - g_n \, \mathbf{n}_m^T d(\delta \mathbf{n}_s)$$
$$- \delta g_n \mathbf{n}_m^T d\mathbf{n}_s - \delta \mathbf{n}_s^T \mathbf{n}_m \, dg_n$$

Using the definitions for $\mathbf{V}_n$ and $\mathbf{W}_n$ from Example 7.5 we obtain

$$\frac{1}{\mathbf{n}_m^T \mathbf{n}_s} \, \mathbf{n}_m^T d\mathbf{n}_s = k_n \, \mathbf{W}_n^T \, d\tilde{\mathbf{x}}$$

$$\frac{1}{\mathbf{n}_m^T \mathbf{n}_s} \, \mathbf{n}_m^T d\mathbf{T}_m = \mathbf{D}_m^T \, d\tilde{\mathbf{x}}$$

and

$$d\xi = \left(\mathbf{V}_t^T + g_n \, k_n \, \mathbf{W}_n\right) d\tilde{\mathbf{x}} = \mathbf{V}_h^T \, d\tilde{\mathbf{x}}$$

with similar expressions for the terms with variations. In the above

$$\mathbf{V}_t = \frac{1}{\mathbf{n}_m^T \mathbf{n}_s} \begin{Bmatrix} -N_1(\xi)\mathbf{t}_s \\ -N_2(\xi)\mathbf{t}_s \\ \mathbf{t}_s \\ 0 \\ 0 \end{Bmatrix} \quad \text{and} \quad \mathbf{D}_m = \frac{1}{2\,\mathbf{n}_m^T \mathbf{n}_s} \begin{Bmatrix} -\mathbf{n}_m \\ \mathbf{n}_m \\ 0 \\ 0 \\ 0 \end{Bmatrix}$$

and

$$\frac{1}{\mathbf{n}_m^T \mathbf{n}_s} \, \mathbf{n}_m^T d(\delta \mathbf{n}_s) = -\frac{1}{\|\mathbf{N}_s\|^2} \, \delta \mathbf{T}_s^T \left(\mathbf{n}_s \, \mathbf{n}_s^T - k_n \, (\mathbf{t}_s \, \mathbf{n}_s^T + \mathbf{n}_s \, \mathbf{t}_s^T)\right) d\mathbf{T}_s$$

In matrix notation we may write the geometric term as

$$\mathbf{K}_G = -\mathbf{D}_m \, \mathbf{V}_h^T - \mathbf{V}_h \, \mathbf{D}_m^T + k_n \, (\mathbf{W}_n \, \mathbf{U}_n^T - \mathbf{U}_n \, \mathbf{W}_n^T)$$
$$+ g_n \left(\mathbf{W}_n \, \mathbf{W}_n^T - k_n \, (\mathbf{W}_t \, \mathbf{W}_n + \mathbf{W}_n \, \mathbf{W}_t)\right)$$

where

$$\mathbf{W}_t = \frac{1}{\|\mathbf{N}_s\|} \begin{Bmatrix} 0 \\ 0 \\ 0 \\ -\mathbf{t}_s \\ \mathbf{t}_s \end{Bmatrix}$$

Again the geometric term is symmetric and with $\bar{\mathbf{U}}_n = \mathbf{U}_n$ the tangent given by Eq. (7.52) is also symmetric.

Example 7.9   Tangent for normal to 2D slave surface – alternative form

The tangent follows from Example 7.6 where

$$\delta g_n = \mathbf{n}_s^T \left(\delta \tilde{\mathbf{x}}_s - N_1(\xi)\delta\tilde{\mathbf{x}}_1^m - N_2(\xi)\delta\tilde{\mathbf{x}}_2^m \right) = \delta\tilde{\mathbf{x}}^T \bar{\mathbf{U}}_n$$

and gives

$$d(\delta g_n) = d\mathbf{n}_s^T \left(\delta \tilde{\mathbf{x}}_s - N_1(\xi)\delta\tilde{\mathbf{x}}_1^m - N_2(\xi)\delta\tilde{\mathbf{x}}_2^m \right) - \tfrac{1}{2} \mathbf{n}_s^T \left(\delta\tilde{\mathbf{x}}_2^m - \delta\tilde{\mathbf{x}}_1^m \right) d\xi$$

For this case we have

$$d\mathbf{n}_s = - \frac{1}{\|\mathbf{N}_s\|} \mathbf{t}_s \, \mathbf{n}_s^T \, d\mathbf{T}_s = - \mathbf{t}_s \, \mathbf{W}_n^T \, d\tilde{\mathbf{x}}$$

with $\mathbf{W}_n$ given in Example 7.5 and $d\xi = \mathbf{V}_h^T d\tilde{\mathbf{x}}$ as described in Example 7.8. Thus the geometric term is given by

$$\mathbf{K}_G = -\mathbf{V}_t \, \mathbf{W}_n^T - \mathbf{D}_s \, \mathbf{V}_h^T$$

where

$$\mathbf{V}_t = \left\{ \begin{array}{c} -N_1(\xi)\mathbf{t}_s \\ -N_2(\xi)\mathbf{t}_s \\ \mathbf{t}_s \\ 0 \\ 0 \end{array} \right\} \qquad \text{and} \qquad \mathbf{D}_s = \frac{1}{2} \left\{ \begin{array}{c} -\mathbf{n}_s \\ \mathbf{n}_s \\ 0 \\ 0 \\ 0 \end{array} \right\}$$

with $\mathbf{V}_t$ defined in Example 7.8. The tangent matrix is given by Eq. (7.52) and is now clearly unsymmetric due to the form of $\mathbf{U}_n$ and $\bar{\mathbf{U}}_n$ as well as the form of $\mathbf{K}_G$.

All of the above forms suffer some solution irregularity during sliding from one facet to another. In the form where normals are defined relative to the master surface an expedient solution is to use concepts from multi-surface plasticity to define a 'continuous' approximation for the normal. This leads to additional considerations which are not given here and are left for the reader to develop (see reference 21, 40 or 41). It is also possible to 'smooth' the master surface using continuous interpolation across facets.[41–43]

Extension to three-dimensional problems is straightforward and involves addition terms related to second derivatives of shape functions unless three-node triangular facets are used. Extension to include frictional effects is described next for a simple Coulomb model. For other models the reader is referred to the literature for additional details.[34,40,41,44–57]

7.4.3 Contact modelling – frictional case

For *frictional* contact both normal and tangential tractions are involved on the surfaces between the two bodies. Thus, the contact traction for a three-dimensional problem is expressed as

$$\lambda = \lambda_n \, \mathbf{n} + \sum_{i=1}^{2} \lambda_{ti} \, \mathbf{t}_i = \lambda_n \, \mathbf{n} + \lambda_t \tag{7.53a}$$

where t_i form a pair of tangent vectors on the contact surface. For a two-dimensional case this simplifies to

$$\lambda = \lambda_n \mathbf{n} + \lambda_t \mathbf{t} \tag{7.53b}$$

In a finite element setting the tangent vectors are obtained during the computation of the normal as described above. For example, in a three-dimensional problem in which the normal is defined for a master facet the tangent vectors may be taken as

$$\mathbf{T}_1^m = \mathbf{x}_{,\xi}^m \quad \text{and} \quad \mathbf{T}_2^m = \mathbf{x}_{,\eta}^m \tag{7.54}$$

with unit vectors defined in the usual way.

In general the tangential traction components are dependent upon the magnitude of the normal traction as well as upon the amount of sliding, lubrication, temperature and other effects that occur. For a detailed description of various models the reader is referred to references 58, 40 and 41. Here we restrict our attention to a simple Coulomb model in which the tangential forces satisfy a *stick–slip* behaviour.

For a *stick* condition the solution is obtained using the contact functional

$$\Pi_c^{st} = \boldsymbol{\lambda}^T \mathbf{g}^{st} \, A_c = \left(\lambda_n \mathbf{n}^T \mathbf{g}^{st} + \sum_{i=1}^{2} \lambda_{ti} \, \mathbf{t}_i^T \mathbf{g}^{st} \right) A_c \tag{7.55}$$

In the stick case the total gap, $\mathbf{g}^{st}$, is determined from the constraint equation

$$\mathbf{c} = \tilde{\mathbf{x}}_s - \sum_a N_a(\boldsymbol{\xi}_0) \, \tilde{\mathbf{x}}_a^m - \mathbf{g}^{st} = \mathbf{0} \tag{7.56}$$

in which $\boldsymbol{\xi}_0$ defines a *fixed* point on the surface at which either initial contact is made or a previously sliding state stops. In a discrete setting in which a solution is sought at time t_{n+1} the value of $\boldsymbol{\xi}_0 = \boldsymbol{\xi}_n$, the value obtained from the solution at time t_n. For a stick state a variation of the gap is given by

$$\delta \mathbf{g}^{st} = \delta \tilde{\mathbf{x}}_s - \sum_a N_a(\boldsymbol{\xi}_0) \, \delta \tilde{\mathbf{x}}_a^m \tag{7.57}$$

The residual equations for a three-dimensional problem are obtained from

$$\delta \Pi_c^{st} = \left(\delta \lambda_n \, (\mathbf{n}^T \mathbf{g}^{st}) + \lambda_n \, \delta(\mathbf{n}^T \mathbf{g}^{st}) + \sum_{i=1}^{2} \delta \lambda_{ti} \, (\mathbf{t}_i^T \mathbf{g}^{st}) + \sum_{i=1}^{2} \lambda_{ti} \, \delta(\mathbf{t}_i^T \mathbf{g}^{st}) \right) A_c$$

$$= \begin{bmatrix} \delta \tilde{\mathbf{x}}^T & \delta \lambda_n & \delta \lambda_{t1} & \delta \lambda_{t2} \end{bmatrix} \left\{ \begin{matrix} \lambda_n \bar{\mathbf{U}}_n^{st} + \lambda_{t1} \bar{\mathbf{U}}_{t1}^{st} + \lambda_{t2} \bar{\mathbf{U}}_{t2}^{st} \\ g_n \\ g_{t1} \\ g_{t2} \end{matrix} \right\} A_c$$

$$\tag{7.58a}$$

in which λ_n and λ_{ti} are Lagrange multipliers. For the two-dimensional case this simplifies to

$$\delta \Pi_c^{st} = \begin{bmatrix} \delta \tilde{\mathbf{x}}^T & \delta \lambda_n & \delta \lambda_t \end{bmatrix} \left\{ \begin{matrix} \lambda_n \bar{\mathbf{U}}_n^{st} + \lambda_t \bar{\mathbf{U}}_t^{st} \\ g_n \\ g_t \end{matrix} \right\} A_c \tag{7.58b}$$

Alternatively, a perturbed Lagrangian form may be written and the contact forces may then be computed directly in terms of the gap relations. Since the multiplier forces are scalars this often leads to a form which is identical to a penalty method.

In a simple Coulomb friction model a stick state persists whenever

$$\|\boldsymbol{\lambda}_t\| = (\boldsymbol{\lambda}_t^T \boldsymbol{\lambda}_t)^{1/2} < \mu \, |\lambda_n| \tag{7.59}$$

The 'friction' parameter μ is a positive material constant that depends on the properties of the contacting surfaces. When $\|\boldsymbol{\lambda}_t\|$ reaches its limit value 'sliding' occurs and the tangential tractions are given by

$$\boldsymbol{\lambda}_t = -\mu \, \lambda_n \, \frac{\dot{\mathbf{g}}_t^{sl}}{\|\dot{\mathbf{g}}_t^{sl}\|} \approx -\mu \, \lambda_n \, \frac{\Delta \mathbf{g}_t^{sl}}{\|\Delta \mathbf{g}_t^{sl}\|} \tag{7.60}$$

in which for a discrete time increment $\Delta t = t_{n+1} - t_n$, the gap increment is defined by $\Delta \mathbf{g}_t^{sl} = \mathbf{g}_t^{sl}(t_{n+1}) - \mathbf{g}_t^{sl}(t_n)$. As shown in Fig. 7.8 an increment of tangential slip may be defined by[41]

$$d\mathbf{g}_t^{sl} = \mathbf{x}_{,\xi}^m \, d\xi + \mathbf{x}_{,\eta}^m \, d\eta = \mathbf{T}_1^m \, d\xi + \mathbf{T}_2^m \, d\eta \tag{7.61}$$

The incremental slip is then given by

$$\Delta \mathbf{g}_t^{sl} = \int_{t_n}^{t_{n+1}} d\mathbf{g}_t^{sl} \tag{7.62}$$

Thus, the movement of the contact point on the surface is described by the parameters ξ and η. The location for every contact point is described by Eq. (7.26). Accordingly, during persistent contact we will always have

$$\mathbf{t}_i^T \left[\mathbf{x}^s - \mathbf{x}^m(\xi) \right] = 0 \tag{7.63}$$

from which differentials, variations and increments may be computed. For a finite element surface approximation we can compute $\delta\xi$ from Eq. (7.43) and the other quantities using the same expression with variations replaced by the appropriate terms.

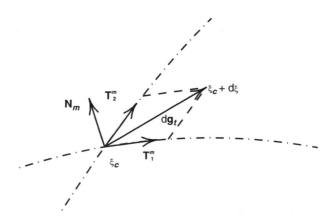

Fig. 7.8   Increment of tangential slip.

In a discrete setting the first iteration in each time (load) increment is assumed to be a stick state. If at the end of the iteration $\|\lambda_t\| > \mu|\lambda_n|$ the state is changed to slip for the next iteration. In many instances some of the conditions assumed as slip will give solutions in which it is necessary to change the state back to stick. Indeed, due to the discrete nature of the solution process and that of the finite element approximation for the contacting surfaces it may occur that there is no fully consistent solution. In such cases it may be necessary to 'accept' a solution after a set number of iterations (provided the 'error' is sufficiently small).

A contact functional Π_c^{sl} does not exist for a sliding state; however, a Galerkin equation for the contact equations may be written directly as

$$\delta\Pi_c^{sl} = \left(\delta\lambda_n\, g_n^{sl} + \delta g_n^{sl}\, \lambda_n + \sum_{i=1}^{2}\sum_{j=1}^{2} \delta g_{ti}^{sl}\, A_{ij}\, \lambda_{tj}\right) A_c$$

$$= \begin{bmatrix} \delta\tilde{\mathbf{x}}^T & \delta\lambda_n \end{bmatrix} \left\{ \begin{matrix} \lambda_n \bar{\mathbf{U}}_n^{sl} + \lambda_{t1}\bar{\mathbf{U}}_{t1}^{sl} + \lambda_{t2}\bar{\mathbf{U}}_{t2}^{sl} \\ g_n \end{matrix} \right\} A_c \tag{7.64}$$

where $A_{ij} = \mathbf{t}_i^T\mathbf{t}_j$ and

$$\delta g_{t1}^{sl} A_{1i} + \delta g_{t2}^{sl} A_{2i} = \delta\tilde{\mathbf{x}}^T \mathbf{U}_{ti}^{sl}; \quad i = 1, 2$$

The normal component terms depend on the current value of ξ and are computed in an identical manner to that described above for the frictionless case. When sliding occurs the tangential tractions λ_{ti} are not independent Lagrange multipliers but are computed from the Coulomb model in terms of the normal component and the geometric properties defining Δg_t^{sl}.

The tangent matrix for a stick–slip behaviour is computed in two parts. For the stick state we compute the tangent terms from Eq. (7.58a) and for the slip state from Eq. (7.64). The tangent for a stick state has the form

$$d(\delta\Pi) = \begin{bmatrix} \delta\tilde{\mathbf{x}}^T & \delta\lambda_n & \delta\lambda_{t1} & \delta\lambda_{t2} \end{bmatrix} \begin{bmatrix} (\lambda \cdot \mathbf{K}_G^{st}) & \bar{\mathbf{U}}_n^{st} & \bar{\mathbf{U}}_{t1}^{st} & \bar{\mathbf{U}}_{t2}^{st} \\ \mathbf{U}_n^{T,st} & 0 & 0 & 0 \\ \mathbf{U}_{t1}^{T,st} & 0 & 0 & 0 \\ \mathbf{U}_{t2}^{T,st} & 0 & 0 & 0 \end{bmatrix} \left\{ \begin{matrix} d\tilde{\mathbf{x}} \\ d\lambda_n \\ d\lambda_{t1} \\ d\lambda_{t2} \end{matrix} \right\} A_c \tag{7.65}$$

where

$$\lambda \cdot \mathbf{K}_G^{st} = \lambda_n \mathbf{K}_{Gn}^{st} + \lambda_{t1} \mathbf{K}_{G1}^{st} + \lambda_{t2} \mathbf{K}_{G2}^{st}$$

with

$$d(\delta(\mathbf{n}^T\mathbf{g})) = \delta\tilde{\mathbf{x}}^T \mathbf{K}_{Gn}^{st}\, d\tilde{\mathbf{x}}^T \quad \text{and} \quad d(\delta(\mathbf{t}_i^T\mathbf{g})) = \delta\tilde{\mathbf{x}}^T \mathbf{K}_{Gi}^{st}\, d\tilde{\mathbf{x}}^T$$

For the slip case the final tangent has the form

$$d(\delta\Pi) = \begin{bmatrix} \delta\tilde{\mathbf{x}}^T & \delta\lambda_n \end{bmatrix} \begin{bmatrix} (\lambda \cdot \mathbf{K}_G^{sl}) & (\bar{\mathbf{U}}_n^{sl} + \alpha\mu\bar{\mathbf{U}}_{t1}^{sl} + \beta\mu\bar{\mathbf{U}}_{t2}^{sl}) \\ \mathbf{U}_n^{T,sl} & 0 \end{bmatrix} \left\{ \begin{matrix} d\tilde{\mathbf{x}} \\ d\lambda_n \end{matrix} \right\} A_c \tag{7.66}$$

where α and β are proportions of the λ_{t1} and λ_{t2} components satisfying Coulomb relation for slip. In a two-dimensional problem $\beta = 0$ and $\alpha = \pm 1$ with the sign depending on the direction of sliding. To illustrate the process of computing the tangent for each state we again consider some two-dimensional examples.

Example 7.10   Residual and tangent for stick – normal to master surface

The gap for the stick case is given by

$$\mathbf{g}^{st} = \tilde{\mathbf{x}}_s - N_1(\xi_0)\tilde{\mathbf{x}}_1^m - N_2(\xi_0)\tilde{\mathbf{x}}_2^m$$

and its variation by

$$\delta\mathbf{g}^{st} = \delta\tilde{\mathbf{x}}_s - N_1(\xi_0)\,\delta\tilde{\mathbf{x}}_1^m - N_2(\xi_0)\,\delta\tilde{\mathbf{x}}_2^m$$

with a similar expression for $d\mathbf{g}^{st}$. From Eqs (7.58a) and (7.58b) we obtain for the normal residual

$$\delta(\mathbf{n}_m^T\mathbf{g}) = \delta\mathbf{g}^T\mathbf{n}_m - \frac{1}{\|\mathbf{N}_m\|}\delta\mathbf{T}_m^T\mathbf{n}_m\,\mathbf{t}_m^T\mathbf{g} = \delta\tilde{\mathbf{x}}^T(\mathbf{V}_{n0} - g_t\,\mathbf{D}_n) = \delta\tilde{\mathbf{x}}^T\mathbf{U}_{n0}$$

where $\mathbf{D}_n$ is defined in Example 7.7. Similarly, for the tangential residual we obtain

$$\delta(\mathbf{t}_m^T\mathbf{g}) = \delta\mathbf{g}^T\mathbf{t}_m + \frac{1}{\|\mathbf{N}_m\|}\delta\mathbf{T}_m^T\mathbf{n}_m\,\mathbf{n}_m^T\mathbf{g} = \delta\tilde{\mathbf{x}}^T(\mathbf{V}_{t0} + g_n\,\mathbf{D}_n) = \delta\tilde{\mathbf{x}}^T\mathbf{U}_{t0}$$

where in the above

$$\mathbf{V}_{n0} = \left\{ \begin{array}{c} -N_1(\xi_0)\,\mathbf{n}_m \\ -N_2(\xi_0)\,\mathbf{n}_m \\ \mathbf{n}_m \end{array} \right\} \quad \text{and} \quad \mathbf{V}_{t0} = \left\{ \begin{array}{c} -N_1(\xi_0)\,\mathbf{t}_m \\ -N_2(\xi_0)\,\mathbf{t}_m \\ \mathbf{t}_m \end{array} \right\}$$

The geometric term for the normal tangent is given by

$$d(\delta\mathbf{n}_m^T\mathbf{g}) = \delta\mathbf{n}_m^T d\mathbf{g} + \delta\mathbf{g}^T d\mathbf{n}_m + \mathbf{g}^T d(\delta\mathbf{n}_m)$$

where

$$\delta\mathbf{n}_m^T d\mathbf{g} = -\frac{1}{\|\mathbf{N}_m\|}\delta\mathbf{T}_m^T\mathbf{n}_m\,\mathbf{t}_m^T d\mathbf{g} = -\delta\tilde{\mathbf{x}}^T\mathbf{D}_n\mathbf{V}_{t0}^T d\tilde{\mathbf{x}}$$

and

$$d(\delta\mathbf{n}_m) = \frac{1}{\|\mathbf{N}_m\|^2}\left(\mathbf{t}_m\,\delta\mathbf{T}_m^T(\mathbf{n}_m\,\mathbf{t}_m^T + \mathbf{n}_m\,\mathbf{t}_m^T)\,d\mathbf{T}_m - \mathbf{n}_m\,\delta\mathbf{T}_m^T\mathbf{n}_m\,\mathbf{n}_m^T d\mathbf{T}_m \right)$$

$$= \mathbf{t}_m\,\delta\tilde{\mathbf{x}}^T(\mathbf{D}_n\mathbf{D}_t^T + \mathbf{D}_t\mathbf{D}_n^T)\,d\tilde{\mathbf{x}} - \mathbf{n}_m\,\delta\tilde{\mathbf{x}}^T\mathbf{D}_n\mathbf{D}_n^T d\tilde{\mathbf{x}}$$

where

$$\mathbf{D}_t = \frac{1}{2\,\|\mathbf{N}_m\|} \left\{ \begin{matrix} -\mathbf{t}_m \\ \mathbf{t}_m \\ \mathbf{0} \end{matrix} \right\}$$

A similar computation for the tangential geometric stiffness gives

$$d(\delta\mathbf{t}_m^T\mathbf{g}) = \delta\mathbf{t}_m^T d\mathbf{g} + \delta\mathbf{g}^T d\mathbf{t}_m + \mathbf{g}^T d(\delta\mathbf{t}_m)$$

with

$$\delta\mathbf{t}_m^T d\mathbf{g} = \delta\tilde{\mathbf{x}}^T \mathbf{D}_n \mathbf{V}_{n0}^T d\tilde{\mathbf{x}}$$

and

$$d(\delta\mathbf{t}_m) = -\mathbf{n}_m\,\delta\tilde{\mathbf{x}}^T(\mathbf{D}_n\mathbf{D}_t^T + \mathbf{D}_t\mathbf{D}_n^T)\,d\tilde{\mathbf{x}} - \mathbf{t}_m\,\delta\tilde{\mathbf{x}}^T\mathbf{D}_n\mathbf{D}_n^T\,d\tilde{\mathbf{x}}$$

Thus the final form of the geometric stiffness is given by

$$\boldsymbol{\lambda}\cdot\mathbf{K}_G^{st} = -\lambda_n(\mathbf{D}_n\mathbf{V}_{t0}^T + \mathbf{V}_{t0}\mathbf{D}_n) + \lambda_t(\mathbf{D}_n\mathbf{V}_{n0}^T + \mathbf{V}_{n0}\mathbf{D}_n)$$
$$+ (\lambda_n g_t - \lambda_t g_n)(\mathbf{D}_n\mathbf{D}_t^T + \mathbf{D}_t\mathbf{D}_n^T) - (\lambda_n g_n + \lambda_t g_t)\mathbf{D}_n\mathbf{D}_n^T$$

The residual for stick may be written in matrix form as

$$\delta\Pi_c = \begin{bmatrix} \delta\tilde{\mathbf{x}}^T & \delta\lambda_n & \delta\lambda_t \end{bmatrix} \left\{ \begin{matrix} \lambda_n\,\mathbf{U}_{n0} + \lambda_t\,\mathbf{U}_{t0} \\ g_n \\ g_t \end{matrix} \right\}$$

and the tangent as

$$d(\delta\Pi_c) = \begin{bmatrix} \delta\tilde{\mathbf{x}}^T & \delta\lambda_n & \delta\lambda_t \end{bmatrix} \begin{bmatrix} \boldsymbol{\lambda}\cdot\mathbf{K}_G^{st} & \mathbf{U}_{n0} & \mathbf{U}_{t0} \\ \mathbf{U}_{n0}^T & 0 & 0 \\ \mathbf{U}_{t0}^T & 0 & 0 \end{bmatrix} \left\{ \begin{matrix} d\tilde{\mathbf{x}} \\ d\lambda_n \\ d\lambda_t \end{matrix} \right\}$$

Example 7.11   Residual and tangent for slip – normal to master surface

The residual and tangent for the normal gap and normal traction component is identical to that for a frictionless behaviour. Thus, the residual is the same as that presented in Example 7.4 and the tangent as given in Example 7.7. For the tangential behaviour, the variation of the tangential slip is given by

$$\delta g_t^{sl} = \delta\xi\,\|\mathbf{T}_m\| = \delta\mathbf{t}_m^T(\tilde{\mathbf{x}}_s - \sum_a N_a\tilde{\mathbf{x}}_a^m) = (\delta\tilde{\mathbf{x}}_s - \sum_a N_a\delta\tilde{\mathbf{x}}_a^m)^T\mathbf{t}_m$$

$$= \delta\tilde{\mathbf{x}}^T\mathbf{V}_h = \delta\tilde{\mathbf{x}}^T(\mathbf{V}_t + g_n\,\mathbf{D}_n)$$

where, noting that $\|\mathbf{N}_m\| = \|\mathbf{T}_m\|$, the vectors for $\mathbf{V}_h$, $\mathbf{V}_t$ and $\mathbf{D}_n$ are defined in Example 7.7. Expanding this relation gives

$$d(\delta g_t^{sl}) = \delta \mathbf{t}_m^T (d\tilde{\mathbf{x}}_s - \sum_a N_a d\tilde{\mathbf{x}}_a^m - T_m \, d\xi) + (\delta \tilde{\mathbf{x}}_s - \sum_a N_a \delta \tilde{\mathbf{x}}_a^m)^T dt_m$$
$$+ (\tilde{\mathbf{x}}_s - \sum_a N_a \tilde{\mathbf{x}}_a^m)^T d(\delta \mathbf{t}_m)$$

Inserting the definitions for terms and writing in matrix notation gives

$$d(\delta g_t^{sl}) = \delta \tilde{\mathbf{x}}^T \mathbf{K}_{Gt}^{sl} d\tilde{\mathbf{x}}$$

with

$$\mathbf{K}_{Gt}^{sl} = \mathbf{D}_n \mathbf{V}_n^T + \mathbf{V}_n \mathbf{D}_n^T - \mathbf{D}_t \mathbf{V}_h^T - g_n (\mathbf{D}_n \mathbf{D}_t^T + \mathbf{D}_t \mathbf{D}_n^T)$$

in which

$$\mathbf{D}_t = \frac{1}{\|\mathbf{N}_m\|} \left\{ \begin{array}{c} -\mathbf{t}_m \\ \mathbf{t}_m \\ \mathbf{0} \end{array} \right\}$$

Thus, the final form of the residual for slip is given by

$$\delta g_t^{sl} = \begin{bmatrix} \delta \tilde{\mathbf{x}}^T & \delta \lambda_n \end{bmatrix} \left\{ \begin{array}{c} \lambda_n (\mathbf{U}_n + \alpha \, \mu \mathbf{U}_t) \\ g_n \end{array} \right\}$$

and that for the tangent is

$$d(\delta \Pi_c) = \begin{bmatrix} \delta \tilde{\mathbf{x}}^T & \delta \lambda_n \end{bmatrix} \begin{bmatrix} \boldsymbol{\lambda} \cdot \mathbf{K}_G^{sl} & (\mathbf{U}_n + \alpha \, \mu \mathbf{U}_t) \\ \mathbf{U}_n^T & 0 \end{bmatrix} \left\{ \begin{array}{c} d\tilde{\mathbf{x}} \\ d\lambda_n \end{array} \right\}$$

where $\alpha = \pm 1$ depending on the direction of sliding and

$$\boldsymbol{\lambda} \cdot \mathbf{K}_G^{sl} = \lambda_n (\mathbf{K}_{Gn}^{sl} + \alpha \, \mu \, \mathbf{K}_{Gt}^{sl})$$

7.5 Surface–surface contact

The above treatment for contact may be generalized to a surface-to-surface treatment in which behaviour over a facet on the slave body interacts with one or more facets on the master body. An early attempt at defining appropriate segments for two-dimensional applications was presented by Simo et al.[9] and for three-dimensional ones by Papadopoulos and Taylor.[59] Both of these approaches had practical difficulties in large deformation and sliding situations. More recently developments have utilized so-called *mortar methods*. Mortar methods have their roots in domain decomposition in which subdomains are joined using appropriate tied interfaces as introduced in Sec. 7.3. Mortaring relates to how the Lagrange multiplier is approximated such that accuracy and stability are maintained. A brief description for tied interfaces is given in Chapter 12 of reference 60 and a detailed mathematical presentation is given in reference 61. The basic treatment for contact problems relies on a proper definition of the gap relation and appropriate quadrature over the contact surface facets. At the present time the work of Puso and Laursen[31,32,62] presents the most comprehensive treatment for this approach. The details of the construction are quite involved and the reader is referred to the cited papers for the additional details and results.

7.6 Numerical examples

7.6.1 Contact between two discs

As a first example here we consider the contact problem previously solved using a node–node approach. In that case we observed a small but significant discontinuity between the contours of vertical stress between the bodies, indicating that traction is not correctly transmitted across the section. Here we use the node–surface method given above in which the contact area of each body is taken as the boundary of elements. The solution is achieved by using a penalty method and a *two-pass* solution procedure where on the first pass one body is the slave and the other the master and on the second pass the designation is reversed. This approach has been shown to be necessary in order to satisfy the mixed patch test for contact.[33] The results using this approach are shown in Fig. 7.9. For the solution, the two-dimensional plane strain finite deformation displacement element described in Sec. 5.3.2 is used with material behaviour given by the neo-Hookean hyperelastic model described in Sec. 6.2.1. Properties are: $E = 100\,000$ for the upper body and $E = 1000$ for the lower body. A Poisson ratio of $\nu = 0.25$ is used to compute Lamé parameters λ_n and μ. As can readily be seen in the figure the results obtained are significantly better than those from the node-to-node analysis.

7.6.2 Contact between a disk and a block

As a second example we consider the interaction between a semicircular disk and a rectangular block. The disk has a radius of 10 units with the material modelled by a neo-Hookean material with initial modulus $E = 100$ and Poisson ratio $\nu = 0.25$.

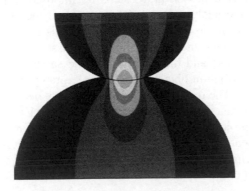

Fig. 7.9 Contact between semicircular disks: vertical contours for node-to-surface solution.

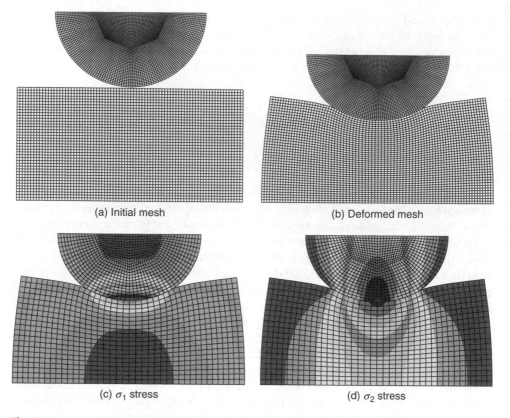

(a) Initial mesh   (b) Deformed mesh

(c) σ_1 stress   (d) σ_2 stress

Fig. 7.10 Contact between a disk and a block – frictionless solution.

The block has a width of 30 units and a height of 15 units. The block is also modelled by a neo-Hookean material with initial modulus $E = 10$ and Poisson ratio $\nu = 0.45$. An initial gap of 0.2 units exists between the disk and the block. The disk is punched into the block by an overall vertical motion of 4 units, thus leading to a large deformation of the block. A solution assuming frictionless contact was achieved using the Lagrange multiplier approach in 40 steps by imposed displacements along the top of the disk. In the analysis the disk is treated as the slave body. The initial and final configuration of the mesh used in the analysis is shown in Fig. 7.10 along with contours for the major and minor principal stresses (a coarser mesh is used to allow display of the mesh and the contours on the same figure). In Fig. 7.11 we show the pressure distribution obtained from the Lagrange multipliers acting on the slave nodes in contact. The result has a small oscillatory behaviour despite the rather fine discretization on the contact surface. This is characteristic of the action between slave nodes and the projection points on the master surface and emphasizes the need for improved treatment by surface-to-surface methods. Nevertheless, it is possible to obtain necessary engineering results for design from such treatment and today the node-to-surface form is the one available in most research and commercial computer programs.

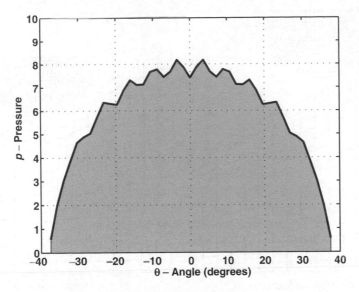

Fig. 7.11 Contact between a disk and a block – contact pressure.

7.6.3 Frictional sliding of a flexible disk on a sloping block

As an example which includes friction we consider the forced motion of a flexible half disk along a sloping disk. The disk is fixed along its top edge which is subjected to the imposed displacement

$$u_1 = t; \ 0 \le t \quad \text{and} \quad u_2 = \begin{cases} t; & 0 \le t \le 4 \\ 4; & 4 < t \end{cases}$$

The disk has a radius of 3 units and is composed of a neo-Hookean material with material parameters $E = 8$ and $\nu = 0.35$. The sloping block has a width of 10 units with a height of 10 units at the left side and 5 units at the right side and is also modelled by a neo-Hookean material with parameters $E = 20$ and $\nu = 0.25$. The centre of the disk is initially located at $x_1 = 4$ and $x_2 = 11.6$. The solution is obtained using Lagrange multiplier form with a Coulomb friction model of $\mu = 0.0$ (frictionless) and $\mu = 0.2$. A set of views of the deformed positions is shown in Fig. 7.12 and the resultant force history in Fig. 7.13. The peak normal force occurs at the time where the vertical motion becomes constant. Note that friction slightly increases the value of the peak load.

7.6.4 Upsetting of a cylindrical billet

To illustrate performance in highly strained regimes, we consider large compression of a three-dimensional cylindrical billet. The initial configuration is a cylinder with radius

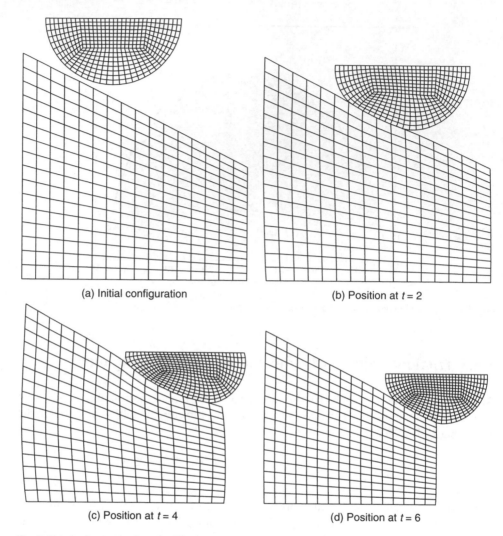

(a) Initial configuration   (b) Position at $t = 2$

(c) Position at $t = 4$ (d) Position at $t = 6$

Fig. 7.12 Configurations for a frictional sliding.

$r = 10$ and height $h = 15$. The mesh consists of 459 eight-node hexahedral elements based on the mixed–enhanced formulation presented in Sec. 5.6. The billet is loaded via displacement control on the upper surface, while the lower edge is fully restrained. A Newton solution process is used in which the upper displacement is increased by increments of displacement equal to 0.25 units.

To prevent penetration with the rigid base during large deformations a simple node-on-node penalty formulation with a penalty parameter $k = 10^6$ is defined for nodes on the lower part of the cylindrical boundary. A neo-Hookean material model with $\lambda_n = 10^4$ and $\mu = 10$ is used for the simulation. Figure 7.14 depicts the initial mesh and progression of deformation.

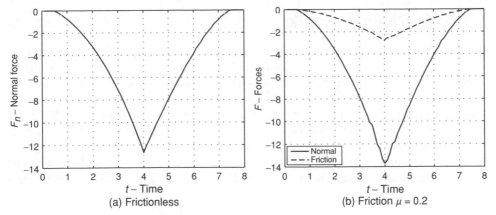

Fig. 7.13 Resultant force history for sliding disk.

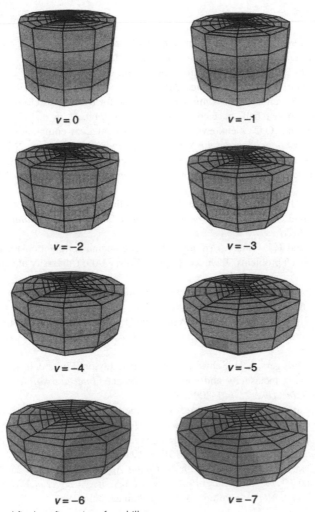

Fig. 7.14 Initial and final configurations for a billet.

7.7 Concluding remarks

In the preceding three chapters we have presented methods to implement basic strategies for solving general finite deformation problems in solid mechanics. A variational structure easily implemented for both two- and three-dimensional problems which can include nearly incompressible behaviour has been given. In addition we have shown how various constitutive models can be included to represent elastic, viscoelastic and elastic–plastic behaviour. Finally in this chapter we have included an introduction to constraining interactions resulting from intermittent contact between contiguous bodies. The formulation of methods to model contact is an active area of research and many new procedures are being introduced. The reader is encouraged to consult recent research on the topic.

References

1. S.K. Chan and I.S. Tuba. A finite element method for contact problems of solid bodies – Part I. Theory and validation. *Int. J. Mech. Sci.*, 13:615–625, 1971.
2. S.K. Chan and I.S. Tuba. A finite element method for contact problems of solid bodies – Part II. Application to turbine blade fastenings. *Int. J. Mech. Sci.*, 13:627–639, 1971.
3. J.J. Kalker and Y. van Randen. A minimum principle for frictionless elastic contact with application to non-Hertzian half-space contact problems. *J. Engr. Math.*, 6:193–206, 1972.
4. A. Francavilla and O.C. Zienkiewicz. A note on numerical computation of elastic contact problems. *International Journal for Numerical Methods in Engineering*, 9:913–924, 1975.
5. T.J.R. Hughes, R.L. Taylor, J.L. Sackman, A. Curnier and W. Kanoknukulchai. A finite element method for a class of contact–impact problems. *Computer Methods in Applied Mechanics and Engineering*, 8:249–276, 1976.
6. J.O. Hallquist, G.L. Goudreau and D.J. Benson. Sliding interfaces with contact–impact in large scale Lagrangian computations. *Computer Methods in Applied Mechanics and Engineering*, 51:107–137, 1985.
7. J.A. Landers and R.L. Taylor. An augmented Lagrangian formulation for the finite element solution of contact problems. Technical Report SESM 85/09, University of California, Berkeley, 1985.
8. P. Wriggers and J.C. Simo. A note on tangent stiffness for fully nonlinear contact problems. *Comm. Appl. Num. Meth.*, 1:199–203, 1985.
9. J.C. Simo, P. Wriggers and R.L. Taylor. A perturbed Lagrangian formulation for the finite element solution of contact problems. *Computer Methods in Applied Mechanics and Engineering*, 50: 163–180, 1985.
10. J.C. Simo, P. Wriggers, K.H. Schweizerhof and R.L Taylor. Finite deformation post-buckling analysis involving inelasticity and contact constraints. *International Journal for Numerical Methods in Engineering*, 23:779–800, 1986.
11. K.-J. Bathe and A.B. Chaudhary. A solution method for planar and axisymmetric contact problems. *International Journal for Numerical Methods in Engineering*, 21:65–88, 1985.
12. K.-J. Bathe and P.A. Bouzinov. On the constraint function method for contact problems. *Computers and Structures*, 64(5/6):1069–1085, 1997.
13. J.J. Kalker. Contact mechanical algorithms. *Comm. Appl. Num. Meth.*, 4:25–32, 1988.
14. N. Kikuchi and J.T. Oden. *Contact Problems in Elasticity: A Study of Variational Inequalities and Finite Element Methods*, volume 8. SIAM, Philadelphia, 1988.

15. H. Parisch. A consistent tangent stiffness matrix for three dimensional non-linear contact analysis. *International Journal for Numerical Methods in Engineering*, 28:1803–1812, 1989.

16. D.J. Benson and J.O. Hallquist. A single surface contact algorithm for the post-buckling analysis of shell structures. *Comp. Meth. Appl. Mech. Engr.*, 78:141–163, 1990.

17. T. Belytschko and M.O. Neal. Contact–impact by the pinball algorithm with penalty and Lagrangian methods. *International Journal for Numerical Methods in Engineering*, 31:547–572, 1991.

18. R.L. Taylor and P. Papadopoulos. On a finite element method for dynamic contact–impact problems. *International Journal for Numerical Methods in Engineering*, 36:2123–2139, 1992.

19. Z. Zhong and J. Mackerle. Static contact problems – a review. *Engineering Computations*, 9:3–37, 1992.

20. J.-H. Heegaard and A. Curnier. An augmented Lagrangian method for discrete large-slip contact problems. *International Journal for Numerical Methods in Engineering*, 36:569–593, 1993.

21. T.A. Laursen and S. Govindjee. A note on the treatment of frictionless contact between non-smooth surfaces in fully non-linear problems. *Communications in Numerical Methods in Engineering*, 10:869–878, 1994.

22. T.A. Laursen and V. Chawla. Design of energy conserving algorithms for frictionless dynamic contact problems. *International Journal for Numerical Methods in Engineering*, 40:863–886, 1997.

23. P. Papadopoulos and R.L. Taylor. A mixed formulation for the finite element solution of contact problems. *Computer Methods in Applied Mechanics and Engineering*, 94:373–389, 1992.

24. P. Papadopoulos, R.E. Jones and J.M. Solberg. A novel finite element formulation for frictionless contact problems. *International Journal for Numerical Methods in Engineering*, 38:2603–2617, 1995.

25. P. Papadopoulos and J.M. Solberg. A Lagrange multiplier method for the finite element solution of frictionless contact problems. *Mathematical and Computer Modelling*, 28:373–384, 1998.

26. J.M. Solberg and P. Papadopoulos. A finite element method for contact/impact. *Finite Elements in Analysis and Design*, 30:297–311, 1998.

27. E. Bittencourt and G.J. Creus. Finite element analysis of three-dimensional contact and impact in large deformation problems. *Computers and Structures*, 69:219–234, 1998.

28. M. Cuomo and G. Ventura. Complementary energy approach to contact problems based on consistent augmented Lagrangian formulation. *Mathematical and Computer Modelling*, 28:185–204, 1998.

29. C. Kane, E.A. Repetto, M. Ortiz and J.E. Marsden. Finite element analysis of non smooth contact. *Computer Methods in Applied Mechanics and Engineering*, 180:1–26, 1999.

30. I. Paczelt, B.A. Szabo and T. Szabo. Solution of contact problem using the *hp*-version of the finite element method. *Computers and Mathematics with Applications*, 38:49–69, 1999.

31. M.A. Puso and T.A. Laursen. A mortar segment-to-segment contact method for large deformation solid mechanics. *Computer Methods in Applied Mechanics and Engineering*, 193:601–629, 2004.

32. M.A. Puso and T.A. Laursen. A mortar segment-to-segment frictional contact method for large displacements. *Computer Methods in Applied Mechanics and Engineering*, 193:4891–4913, 2004.

33. R.L. Taylor and P. Papadopoulos. A patch test for contact problems in two dimensions. In P. Wriggers and W. Wagner, editors, *Nonlinear Computational Mechanics*, pages 690–702. Springer, Berlin, 1991.

34. T.A. Laursen and V.G. Oancea. Automation and assessment of augmented lagrangian algorithms for frictional contact problems. *J. Appl. Mech*, 61:956–963, 1994.

35. S.P. Timoshenko and J.N. Goodier. *Theory of Elasticity*. McGraw-Hill, New York, 3rd edition, 1969.

36. D.G. Luenberger. *Linear and Nonlinear Programming*. Addison-Wesley, Reading, Mass., 1984.

37. B.I. Wohlmuth. A mortar finite element method using dual spaces for the Lagrange multiplier. *Society for Industrial and Applied Mathematics*, 38:989–1012, 2000.
38. K.J. Arrow, L. Hurwicz and H. Uzawa. *Studies in Non-Linear Programming*. Stanford University Press, Stanford, CA, 1958.
39. G. Zavarise and P. Wriggers. A superlinear convergent augmented Lagrangian procedure for contact problems. *Engineering Computations*, 16:88–119, 1999.
40. T.A. Laursen. *Computational Contact and Impact Mechanics*. Springer, Berlin, 2002.
41. P. Wriggers. *Computational Contact Mechanics*. John Wiley & Sons, Chichester, 2002.
42. R.L. Taylor and P. Wriggers. Smooth surface discretization for large deformation frictionless contact. Technical Report UCB/SEMM-99/04, University of California, Berkeley, February 1999.
43. M.A. Puso and T.A. Laursen. A 3d contact smoothing algorithm using Gregory patches. *International Journal for Numerical Methods in Engineering*, 54:1161–1194, 2002.
44. A.B. Chaudhary and K.-J. Bathe. A solution method for static and dynamic analysis of three-dimensional contact problems with friction. *Computers and Structures*, 24:855–873, 1986.
45. J.-W. Ju and R.L. Taylor. A perturbed Lagrangian formulation for the finite element solution of nonlinear frictional contact problems. *Journal de Mécanique Théorique et Appliquée*, 7(Supplement, 1):1–14, 1988.
46. A. Currier and P. Alart. A generalized Newton method for contact problems with friction. *Journal de Mécanique Théorique et Appliquée*, 7:67–82, 1988.
47. P. Wriggers, T. Vu Van and E. Stein. Finite element formulation of large deformation impact–contact problems with friction. *Computers and Structures*, 37:319–331, 1990.
48. P. Alart and A. Currier. A mixed formulation for frictional contact problems prone to Newton like solution methods. *Computer Methods in Applied Mechanics and Engineering*, 92:353–375, 1991.
49. J.C. Simo and T.A. Laursen. An augmented Lagrangian treatment of contact problems involving friction. *Computers and Structures*, 42:97–116, 1992.
50. T.A. Laursen and J.C. Simo. A continuum-based finite element formulation for the implicit solution of multibody, large-deformation, frictional, contact problems. *International Journal for Numerical Methods in Engineering*, 36:3451–3486, 1993.
51. T.A. Laursen and J.C. Simo. Algorithmic symmetrization of Coulomb frictional problems using augmented Lagrangians. *Computer Methods in Applied Mechanics and Engineering*, 108:133–146, 1993.
52. A. Heege and P. Alart. A frictional contact element for strongly curved contact problems. *International Journal for Numerical Methods in Engineering*, 39:165–184, 1996.
53. C. Agelet de Saracibar. A new frictional time integration algorithm for large slip multi-body frictional contact problems. *Computer Methods in Applied Mechanics and Engineering*, 142:303–334, 1997.
54. W. Ling and H.K. Stolarski. On elasto-plastic finite element analysis of some frictional contact problems with large sliding. *Engineering Computations*, 14:558–580, 1997.
55. F. Jourdan, P. Alart and M. Jean. A gauss-seidel like algorithm to solve frictional contact problems. *Computer Methods in Applied Mechanics and Engineering*, 155:31–47, 1998.
56. C. Agelet de Saracibar. Numerical analysis of coupled thermomechanical frictional contact. Computational model and applications. *Archives of Computational Methods in Engineering*, 5(3):243–301, 1998.
57. G. Pietrzak and A. Currier. Large deformation frictional contact mechanics: continuum formulation and augmented Lagrangian treatment. *Computer Methods in Applied Mechanics and Engineering*, 177:351–381, 1999.
58. J.T. Oden and J.A.C. Martins. Models and computational methods for dynamic friction phenomena. *Computer Methods in Applied Mechanics and Engineering*, 52:527–634, 1985.
59. P. Papadopoulos and R.L. Taylor. A mixed formulation fo the finite element solution of contact problems. *Computer Methods in Applied Mechanics and Engineering*, 94:373–389, 1992.

60. O.C. Zienkiewicz, R.L. Taylor and J.Z. Zhu. *The Finite Element Method: Its Basis and Funda-mentals*. Butterworth-Heinemann, Oxford, 6th edition, 2005.
61. B.I. Wohlmuth. *Discretization Methods and Iterative Solvers Based on Domain Decomposition*. Springer-Verlag, Heidelberg, 2001.
62. M.A. Puso and T.A. Laursen. Mesh tying on curved interfaces in 3d. *Engineering Computations*, 20:305–319, 2003.

8

Pseudo-rigid and rigid–flexible bodies

8.1 Introduction

Many situations are encountered where treatment of the entire system as deformable bodies is neither necessary nor practical. For example, the frontal impact of a vehicle against a barrier requires a detailed modelling of the front part of the vehicle but the primary function of the engine and the rear part is to provide inertia, deformation being negligible for purposes of modelling the frontal impact. A second example, from geotechnical engineering, is the modelling of rock mass landslides or interaction between rocks on a conveyor belt where deformation of individual blocks is secondary. In this chapter we consider briefly the study of the first class of problems and in the next chapter the second type in much more detail.

The above problem classes divide themselves into two further subclasses: one where it is necessary to include some simple mechanisms of deformation in each body (e.g. an individual rock piece) and the second in which the individual bodies have no deformation at all. The first class is called *pseudo-rigid* body deformation[1] and the second *rigid body* behaviour.[2] Here we wish to illustrate how such behaviour can be described and combined in a finite element system. For the modelling of pseudo-rigid body analyses we follow closely the work of Cohen and Muncaster[1] and the numerical implementation proposed by Solberg and Papadopoulos.[3] The literature on rigid body analysis is extensive, and here we refer the reader to papers for additional details on methods and formulations beyond those covered here.[4–21]

8.2 Pseudo-rigid motions

In this section we consider the analysis of systems which are composed of many small bodies, each of which is assumed to undergo large displacements and a uniform deformation.[*] The individual bodies which we consider are of the types shown in Fig. 8.1. In particular, a faceted shape can be constructed directly from a finite element discretization in which the elements are designated as all belonging to a single solid object or the individual bodies can be described by simple geometric forms such as discs or ellipsoids.

[*] Higher-order approximations can be included using polynomial approximation for the deformation of each body.

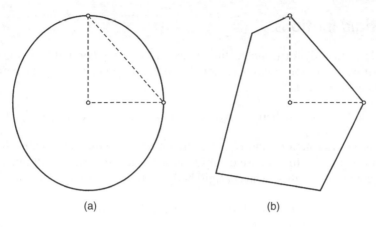

(a)   (b)

Fig. 8.1 Shapes for pseudo-rigid and rigid body analysis: (a) ellipsoid; (b) faceted body.

A *homogeneous* motion of a body may be written as

$$\phi_i(X_I, t) = r_i(t) + F_{iI}(t)[X_I - R_I] \tag{8.1}$$

in which X_I is position, t is time, R_I is some reference point in the undeformed body, r_i is the position of the same point in the deformed body, and F_{iI} is a constant deformation gradient. We note immediately that at time zero the deformation gradient is the identity tensor (matrix) and Eq. (8.1) becomes

$$\phi_i(X_I, 0) = r_i(0) + \delta_{iI}[X_I - R_I] = r_i(0) + \delta_{iI}X_I - \delta_{iI}R_I \equiv \delta_{iI}X_I \tag{8.2}$$

where $r_i(0) = \delta_{iI}R_I$ by definition. The behaviour of solids which obey the above description is sometimes referred to as analysis of *pseudo-rigid bodies*.[1] A treatment by finite elements has been considered by Solberg and Papadopoulos,[3] and an alternative expression for motions restricted to incrementally linear behaviour has been developed by Shi, and the method is commonly called *discontinuous deformation analysis* (DDA).[22] The DDA form, while widely used in the geotechnical community, is usually combined with a simple linear elastic constitutive model and linear strain-displacement forms which can lead to large errors when finite rotations are encountered.

Once the deformation gradient is computed, the procedures for analysis follow the methods described in Chapter 5. It is, of course, necessary to include the inertial term for each body in the analysis. No difficulties are encountered once a shape of each body is described and a constitutive model is introduced. For elastic behaviour it is not necessary to use a complicated model, and here use of the St Venant–Kirchhoff relation is adequate – indeed, if large deformations occur within an individual body the approximation of homogeneous deformation generally is not adequate to describe the solution. The primary difficulty for this class of problems is modelling the large number of interactions between bodies by contact phenomena and here the reader is referred to Chapters 7 and 9 and references on the subject for additional information on contact and other details.[22,23]

8.3 Rigid motions

The pseudo-rigid body form can be directly extended to rigid bodies by using the polar decomposition on the deformation tensor. The polar decomposition of the deformation gradient may be given as[24–26]

$$F_{iI} = \Lambda_{iJ} U_{JI} \quad \text{where} \quad \Lambda_{iI} \Lambda_{iJ} = \delta_{IJ} \quad \text{and} \quad \Lambda_{iI} \Lambda_{jI} = \delta_{ij} \tag{8.3}$$

Here Λ_{iI} is a rigid rotation[*] and U_{IJ} is a stretch tensor (which has eigenvalues λ_m as defined in Chapter 5). In the case of rigid motions the stretches are all unity and U_{IJ} simply becomes an identity. Thus, a rigid body motion may be specified as

$$\phi_i(X_I, t) = r_i(t) + \Lambda_{iI}(t)[X_I - R_I] \tag{8.4}$$

or, in matrix form, as

$$\phi(\mathbf{X}, t) = \mathbf{r}(t) + \mathbf{\Lambda}(t)[\mathbf{X} - \mathbf{R}] \tag{8.5}$$

Alternatively, we can express the rigid motion using Eq. (8.1) and impose constraints to make the stretches unity. For example, in two dimensions we can represent the motion in terms of the displacements of the vertices of a triangle and apply constraints that the lengths of the triangle sides are unchanged during deformation. The constraints may be added as Lagrange multipliers or other constraint methods and the analysis may proceed directly from a standard finite element representation of the triangle. Such an approach has been used in reference 27 with a penalty method used to impose the constraints. Here we do not pursue this approach further and instead consider direct use of rigid body motions to construct the formulation.

For subsequent use we note the form of the variation of a rigid motion and its incremental part. These may be expressed as

$$\delta\phi = \delta\mathbf{r} + \widehat{\delta\boldsymbol{\theta}}\mathbf{\Lambda}[\mathbf{X} - \mathbf{R}]$$
$$d\phi = d\mathbf{r} + \widehat{d\boldsymbol{\theta}}\mathbf{\Lambda}[\mathbf{X} - \mathbf{R}]$$

Using Eq. (8.5) these may be simplified to

$$\delta\phi = \delta\mathbf{r} - \hat{\mathbf{y}}\delta\boldsymbol{\theta} \quad \text{where} \quad \mathbf{y} = \mathbf{x} - \mathbf{r}$$
$$d\phi = d\mathbf{r} - \hat{\mathbf{y}}d\boldsymbol{\theta} \tag{8.6}$$

where $d\phi$ and $\delta\boldsymbol{\theta}$ are incremental and variational rotation vectors, respectively.

In a similar manner we obtain the velocity for the rigid motion as

$$\dot{\phi} = \dot{\mathbf{r}} - \hat{\mathbf{y}}\omega \tag{8.7}$$

in which $\dot{\mathbf{r}}$ is translational velocity and ω angular velocity, both at the centre of mass. The angular velocity is obtained by solving

$$\dot{\mathbf{\Lambda}} = \hat{\omega}\mathbf{\Lambda} \tag{8.8}$$

---

[*] Often literature denotes this rotation as R_{iI}; however, here we use R_I as a position of a point in the body and to avoid confusion use Λ_{iI} to denote rotation.

or

$$\dot{\Lambda} = \Lambda \hat{\Omega} \tag{8.9}$$

where Ω is the reference configuration angular velocity.[8] This is clearer by writing the equations in indicial form given by

$$\dot{\Lambda}_{iI} = \omega_{ij} \Lambda_{jI} = \Lambda_{iJ} \Omega_{JI} \tag{8.10}$$

where the velocity matrices are defined in terms of vector components and give the skew symmetric form

$$\omega_{ij} = \begin{bmatrix} 0 & -\omega_3 & \omega_2 \\ \omega_3 & 0 & -\omega_1 \\ -\omega_2 & \omega_1 & 0 \end{bmatrix} \tag{8.11}$$

and similarly for Ω_{IJ}. The above form allows for the use of either the material angular velocity or the spatial one. Transformation between the two is easily performed since the rigid rotation must satisfy the orthogonality conditions

$$\Lambda^T \Lambda = \Lambda \Lambda^T = I \tag{8.12}$$

at all times. Using Eqs (8.8) and (8.9) we obtain

$$\hat{\omega} = \Lambda \hat{\Omega} \Lambda^T \tag{8.13}$$

or by transforming in the opposite way

$$\hat{\Omega} = \Lambda^T \hat{\omega} \Lambda \tag{8.14}$$

8.3.1 Equations of motion for a rigid body

If we consider a single rigid body subjected to concentrated loads $\mathbf{f}_a$ applied at points whose current position is $\mathbf{x}_a$ and locate the reference position for $\mathbf{R}$ at the centre of mass, the equations of equilibrium are given by conservation of linear momentum

$$\dot{\mathbf{p}} = \sum_a \mathbf{f}_a = \mathbf{f}; \quad \mathbf{p} = m\dot{\mathbf{r}} \tag{8.15}$$

where $\mathbf{p}$ defines a *linear momentum*, $\mathbf{f}$ is a resultant force and total mass of the body is computed from

$$m = \int_\Omega \rho_0 \, dV \tag{8.16}$$

and conservation of angular momentum

$$\dot{\boldsymbol{\pi}} = \sum_a (\mathbf{x}_a - \mathbf{r}) \times \mathbf{f}_a = \mathbf{m}; \quad \boldsymbol{\pi} = \mathbb{I}\boldsymbol{\omega} \tag{8.17}$$

where $\boldsymbol{\pi}$ is the *angular momentum* of the rigid body, $\mathbf{m}$ is a resultant couple and $\mathbb{I}$ is the spatial inertia tensor.

The spatial inertia tensor (matrix) $\mathbb{I}$ is computed from

$$\mathbb{I} = \Lambda \mathbb{J} \Lambda^{\mathrm{T}} \tag{8.18}$$

where $\mathbb{J}$ is the inertia tensor (matrix) computed from an integral on the reference configuration and is given by

$$\mathbb{J} = \int_{\Omega} \rho_0 \left[(\mathbf{Y}^{\mathrm{T}} \mathbf{Y}) \mathbf{I} - \mathbf{Y} \mathbf{Y}^{\mathrm{T}} \right] dV \quad \text{where} \quad \mathbf{Y} = \mathbf{X} - \mathbf{R} \tag{8.19}$$

Thus, description of an individual rigid body requires locating the centre of mass $\mathbf{R}$ and computing the total mass m and inertia matrix $\mathbb{J}$. It is then necessary to integrate the equilibrium equations to define the position $\mathbf{r}$ and the orientation of the body Λ.

8.3.2 Construction from a finite element model

If we model a body by finite elements, as described throughout this volume, we can define individual bodies or parts of bodies as being rigid. For each such body (or part of a body) it is then necessary to define the total mass, inertia matrix, and location of the centre of mass.

This may be accomplished by computing the integrals given by Eqs (8.16) and (8.19) together with the relation to determine the centre of mass given by

$$m\mathbf{R} = \int_{\Omega} \rho_0 \mathbf{X} \, dV \tag{8.20}$$

In these expressions it is necessary only to define each point in the volume of an element by its reference position interpolation $\mathbf{X}$. For solid (e.g. brick or tetrahedral) elements such interpolation is given by Eq. (5.50a) which in matrix form becomes (omitting the summation symbol)

$$\mathbf{X} = N_\alpha \tilde{\mathbf{X}}_\alpha \tag{8.21}$$

This interpolation may be used to determine the volume element necessary to carry out all the integrals numerically by quadrature.[28]

The total mass may now be computed as

$$m = \sum_e \left(\int_{\Omega_e} \rho_0 \, dV \right) \tag{8.22}$$

where Ω_e is the reference volume of each element e. Use of Eq. (8.21) in Eq. (8.20) to determine the centre of mass now gives

$$\mathbf{R} = \frac{1}{m} \sum_e \left(\int_{\Omega_e} \rho_0 N_\alpha \, dV \right) \mathbf{X}_\alpha \tag{8.23}$$

and finally the reference inertia tensor (matrix) as

$$\mathbb{J} = \sum_e M_{\alpha\beta} \left[(\mathbf{Y}_\alpha^{\mathrm{T}} \mathbf{Y}_\beta) \mathbf{I} - \mathbf{Y}_\alpha \mathbf{Y}_\beta^{\mathrm{T}} \right]; \quad \mathbf{Y}_\alpha = \mathbf{X}_\alpha - \mathbf{R} \tag{8.24}$$

where

$$M_{\alpha\beta}^e = \int_{\Omega_e} \rho_0 N_\alpha N_\beta \, dV \qquad (8.25)$$

The above definition of Y_α tacitly assumes that $\sum_\alpha N_\alpha = 1$. If other interpolations are used to define the shape functions (e.g. hierarchical shape functions) it is necessary to modify the above procedure to determine the mass and inertia matrix.

8.3.3 Transient solutions

The integration of the translational rigid term $\mathbf{r}$ may be performed using any of the methods described in reference 28 or indeed by other methods described in the literature. The integration of the rotational part can also be performed by many schemes; however, it is important that updates of the rotation produce discrete time values for rigid rotations which retain an orthonormal character, that is, the Λ_n must satisfy the orthogonality condition given by Eq. (8.12). One procedure to obtain this is to assume that the angular velocity within a time increment is constant, being measured as

$$w(t) \approx w_{n+\alpha} = \frac{1}{\Delta t}\theta \qquad (8.26)$$

in which Δt is the time increment between t_n and t_{n+1}, θ is the increment of rotation during the time step, and $0 \le \alpha \le 1$. The approximation

$$w_{n+\alpha} = (1 - \alpha)w_n + \alpha w_{n+1} \qquad (8.27)$$

is used to define intermediate values in terms of those at t_n and t_{n+1}. Equation (8.8) now becomes a constant coefficient ordinary differential equation which may be integrated exactly, yielding the solution

$$\Lambda(t) = \exp[\hat{\theta}(t - t_n)/\Delta t]\Lambda_n \qquad t_n \le t \le t_{n+1} \qquad (8.28)$$

In particular at $t_{n+\alpha}$ we obtain

$$\Lambda_{n+\alpha} = \exp[\alpha\hat{\theta}]\Lambda_n$$

This may also be performed using the material angular velocity Ω.[8] Many algorithms exist to construct the exponential of a matrix, and the *closed-form expression* given by the classical formula of Euler and Rodrigues (e.g. see Wittaker[29]) is quite popular. This is given by

$$\exp[\hat{\theta}] = \mathbf{I} + \frac{\sin|\theta|}{|\theta|}\hat{\theta} + \frac{1}{2}\frac{\sin^2 |\theta|/2}{[|\theta|^2/2]}\hat{\theta}^2 \qquad \text{where} \qquad |\theta| = \left[\theta^{\mathrm{T}}\theta\right]^{1/2} \qquad (8.29)$$

This update may also be given in terms of quaternions and has been used for integration of both rigid body motions as well as for the integration of the rotations appearing in three-dimensional beam formulations (see Chapter 17).[8,30,31] Another alternative to the direct use of the exponential update is to use the Cayley transform to perform updates for Λ which remain orthonormal.

Once the form for the update of the rigid rotation is defined any of the integration procedures defined in reference 28 may be used to advance the incremental rotation by noting that θ or Θ (the material counterpart) are in fact the change from time t_n to t_{n+1}. The reader also is referred to reference 8 for additional algorithms directly based on the GN11 and GN22 methods.[28] Here forms for conservation of linear and angular momentum are of particular importance.

8.4 Connecting a rigid body to a flexible body

In some analyses the rigid body is directly attached to flexible body parts of the problem [Fig. 8.2(a)]. Consider a rigid body that occupies the part of the domain denoted as Ω_r and is 'bonded' to a flexible body with domain Ω_f. In such a case the formulation to 'bond' the surface may be performed in a concise manner using Lagrange multiplier constraints. We shall find that these multiplier constraints can be easily eliminated from the analysis by a local solution process, as opposed to the need to carry them to the global solution arrays as was the case in their use in contact problems (see Sec. 5.3).

8.4.1 Lagrange multiplier constraints

A simple two-dimensional rigid–flexible body problem is shown in Fig. 8.2(a) in which the interface will involve only three–nodal points. In Fig. 8.2(b) we show an exploded view between the rigid body and one of the elements which lies along the rigid–flexible interface. Here we need to enforce that the position of the two interface nodes for the element will have the same deformed position as the corresponding point on the rigid body. Such a constraint can easily be written using Eq. (8.4) as

$$\mathbf{C}_\alpha = \mathbf{r}(t) + \Lambda(t)\,[\mathbf{X}_\alpha - \mathbf{R}] - \mathbf{x}_\alpha(t) = \mathbf{0} \tag{8.30}$$

in which the subscript α denotes a node number. We can now modify a functional to include the constraint using a classical Lagrange multiplier approach in which we add the term

$$\Pi_{rf} = \lambda_\alpha \mathbf{C}_\alpha = \lambda_\alpha\,[\mathbf{x}_\alpha(t) - \mathbf{r}(t) - \Lambda(t)\,[\mathbf{X}_\alpha - \mathbf{R}]] \tag{8.31}$$

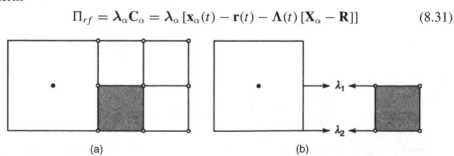

(a)   (b)

Fig. 8.2   Lagrange multiplier constraint between flexible and rigid bodies: (a) rigid–flexible body; (b) Lagrange multipliers.

Taking the variation we obtain

$$\delta\Pi_{rf} = \delta\lambda_\alpha \left[\mathbf{x}_\alpha - \mathbf{r} - \mathbf{\Lambda}\left[\mathbf{X}_\alpha - \mathbf{R}\right]\right] + \lambda_\alpha \left[\delta\mathbf{x}_\alpha - \delta\mathbf{r} - \delta\boldsymbol{\theta}\mathbf{\Lambda}\left[\mathbf{X}_\alpha - \mathbf{R}\right]\right] \qquad (8.32)$$

From this we immediately obtain the constraint equation and a modification to the equilibrium equations for each flexible node and the rigid body. Accordingly, the modified variational principle may now be written for a typical node α on the interface of the rigid body as

$$\delta\Pi + \delta\Pi_{rf} = \begin{bmatrix} \delta\mathbf{x}_\mu & \delta\mathbf{x}_\alpha & \delta\mathbf{r} & \delta\boldsymbol{\theta} & \delta\lambda_\mu \end{bmatrix}$$
$$\times \left\{ \begin{array}{l} \mathbf{M}_{\mu\nu}\dot{\mathbf{v}}_\nu + \mathbf{M}_{\mu\beta}\dot{\mathbf{v}}_\beta + \mathbf{P}_\mu - \mathbf{f}_\mu \\ \mathbf{M}_{\alpha\nu}\dot{\mathbf{v}}_\nu + \mathbf{M}_{\alpha\beta}\dot{\mathbf{v}}_\beta + \mathbf{P}_\alpha - \mathbf{f}_\alpha + \lambda_\alpha \\ \dot{\mathbf{p}} - \mathbf{f} - \lambda_\alpha \\ \dot{\boldsymbol{\pi}} - \mathbf{m} - \hat{\mathbf{y}}_\alpha^\mathrm{T}\lambda_\alpha \\ \mathbf{x}_\alpha - \mathbf{r} - \mathbf{\Lambda}\left[\mathbf{X}_\alpha - \mathbf{R}\right] \end{array} \right\} = 0 \qquad (8.33)$$

where $\mathbf{y}_\alpha = \mathbf{x}_\alpha - \mathbf{r}$ are the nodal values of $\mathbf{y}$, β are any other rigid body nodes connected to node α and μ, ν are flexible nodes connected to node α.

Since the parameters $\mathbf{x}_\alpha$ enter the equations in a linear manner we can use the constraint equation to eliminate their appearance in the equations. Accordingly, from the variation of the constraint equation we may write

$$\delta\mathbf{x}_\alpha = [\mathbf{I}, \; \hat{\mathbf{y}}_\alpha]\left\{ \begin{array}{l} \delta\mathbf{r} \\ \delta\boldsymbol{\theta} \end{array} \right\} \qquad (8.34)$$

which permits the remaining equations in Eq. (8.33) to be rewritten as

$$\delta\Pi + \delta\Pi_{rf} = \begin{bmatrix} \delta\mathbf{x}_\mu & \delta\mathbf{r} & \delta\boldsymbol{\theta} \end{bmatrix}\left\{ \begin{array}{l} \mathbf{M}_{\mu\nu}\dot{\mathbf{v}}_\nu + \mathbf{M}_{\mu\beta}\dot{\mathbf{v}}_\beta + \mathbf{P}_\mu - \mathbf{f}_\mu \\ \dot{\mathbf{p}} - \mathbf{f} + \mathbf{M}_{\alpha\nu}\dot{\mathbf{v}}_\nu + \mathbf{M}_{\alpha\beta}\dot{\mathbf{v}}_\beta + \mathbf{P}_\alpha - \mathbf{f}_\alpha \\ \dot{\boldsymbol{\pi}} - \mathbf{m} - \hat{\mathbf{y}}_\alpha^\mathrm{T}(\mathbf{M}_{\alpha\nu}\dot{\mathbf{v}}_\nu + \mathbf{M}_{\alpha\beta}\dot{\mathbf{v}}_\beta + \mathbf{P}_\alpha - \mathbf{f}_\alpha) \end{array} \right\} = 0 \qquad (8.35)$$

For use in a Newton solution scheme it is necessary to linearize Eq. (8.35). This is easily achieved

$$d(\delta\Pi) + d(\delta\Pi_{rf}) = \begin{bmatrix} \delta\mathbf{x}_\mu & \delta\mathbf{r} & \delta\boldsymbol{\theta} \end{bmatrix}$$
$$\times \begin{bmatrix} (\mathbf{K}_{\mu\nu})_\mathrm{T} & (\mathbf{K}_{\mu\beta})_\mathrm{T} & \mathbf{0} & \mathbf{0} \\ (\mathbf{K}_{\alpha\nu})_\mathrm{T} & (\mathbf{K}_{\alpha\beta})_\mathrm{T} & \mathbf{K}_\mathrm{T}^p & \mathbf{0} \\ -\hat{\mathbf{y}}_\alpha^\mathrm{T}(\mathbf{K}_{\alpha\nu})_\mathrm{T} & -\hat{\mathbf{y}}_\alpha^\mathrm{T}(\mathbf{K}_{\alpha\beta})_\mathrm{T} & \mathbf{0} & \mathbf{K}_\mathrm{T}^\theta \end{bmatrix}\left\{ \begin{array}{l} d\mathbf{x}_\nu \\ d\mathbf{x}_\beta \\ d\mathbf{r} \\ d\boldsymbol{\theta} \end{array} \right\} \qquad (8.36)$$

Once again this form may be reduced using the equivalent of Eq. (8.34) for an incremental $d\mathbf{x}_\beta$ to obtain

$$d(\delta\Pi) + d(\delta\Pi_{rf}) = \begin{bmatrix} \delta\mathbf{x}_\mu & \delta\mathbf{r} & \delta\boldsymbol{\theta} \end{bmatrix}$$
$$\times \begin{bmatrix} (\mathbf{K}_{\mu\nu})_T & (\mathbf{K}_{\mu\beta})_T & -(\mathbf{K}_{\mu\beta})_T\hat{\mathbf{y}}_\beta \\ (\mathbf{K}_{\alpha\nu})_T & \mathbf{K}_T^p + (\mathbf{K}_{\alpha\beta})_T & -(\mathbf{K}_{\alpha\beta})_T\hat{\mathbf{y}}_\beta \\ -\hat{\mathbf{y}}_\alpha^T(\mathbf{K}_{\alpha\nu})_T & -\hat{\mathbf{y}}_\alpha^T(\mathbf{K}_{\alpha\beta})_T & [\mathbf{K}_T^\theta + \hat{\mathbf{y}}_\alpha^T(\mathbf{K}_{\alpha\beta})_T\hat{\mathbf{y}}_\beta] \end{bmatrix}$$
$$\times \begin{Bmatrix} d\mathbf{x}_\nu \\ d\mathbf{r} \\ d\boldsymbol{\theta} \end{Bmatrix} \tag{8.37}$$

Combining all the steps we obtain the set of equations for each rigid body as

$$\begin{bmatrix} (\mathbf{K}_{\mu\nu})_T & (\mathbf{K}_{\mu\beta})_T & -(\mathbf{K}_{\mu\beta})_T\hat{\mathbf{y}}_\beta \\ (\mathbf{K}_{\alpha\nu})_T & [\mathbf{K}_T^p + (\mathbf{K}_{\alpha\beta})_T] & -(\mathbf{K}_{\alpha\beta})_T\hat{\mathbf{y}}_\beta \\ -\hat{\mathbf{y}}_\alpha^T(\mathbf{K}_{\alpha\nu})_T & -\hat{\mathbf{y}}_\alpha^T(\mathbf{K}_{\alpha\beta})_T & [\mathbf{K}_T^\theta + \hat{\mathbf{y}}_\alpha^T(\mathbf{K}_{\alpha\beta})_T\hat{\mathbf{y}}_\beta] \end{bmatrix} \begin{Bmatrix} d\mathbf{x}_\nu \\ d\mathbf{r} \\ d\boldsymbol{\theta} \end{Bmatrix}$$
$$= \begin{Bmatrix} \boldsymbol{\Psi}_\mu \\ \mathbf{f} - \dot{\mathbf{p}} + \boldsymbol{\Psi}_\alpha \\ \mathbf{m} - \dot{\boldsymbol{\pi}} + \hat{\mathbf{y}}_\alpha^T\boldsymbol{\Psi}_\alpha \end{Bmatrix} \tag{8.38}$$

in which $\boldsymbol{\Psi}_\alpha$ and $\boldsymbol{\Psi}_\mu$ are the residuals from the finite element calculation at node α and μ, respectively. We recall from Chapter 5 that each is given by a form

$$\psi_\alpha = \mathbf{f}_\alpha - \mathbf{P}_\alpha(\boldsymbol{\sigma}) - \mathbf{M}_{\alpha\nu}\dot{\mathbf{v}}_\nu - \mathbf{M}_{\alpha\beta}\dot{\mathbf{v}}_\beta \tag{8.39}$$

which is now not zero since total balance of momentum includes the addition of the λ_α.

The above steps to compute the residual and the tangent can be performed in each element separately by noting that

$$\lambda_\alpha = \sum_e \lambda_\alpha^e \tag{8.40}$$

where λ_α^e denotes the contribution from element e. Thus, the steps to constrain a flexible body to a rigid body are once again a standard finite element assembly process and may easily be incorporated into a solution system.

The above discussion has considered the connection between a rigid body and a body which is modelled using solid finite elements (e.g. quadrilateral and hexahedral elements in two and three dimensions, respectively). It is also possible directly to connect beam elements which have nodal parameters of translation and rotation. This is easily performed if the rotation parameters of the beam are also defined in terms of the rigid rotation $\boldsymbol{\Lambda}$. In this case one merely transforms the rotation to be defined relative to the reference description of the rigid body rotation and assembles the result directly into the rotation terms of the rigid body. If one uses a rotation for both the beam and the rigid body which is defined in terms of the global Cartesian reference configuration no transformation is required. Shells can be similarly treated; however, it is best then to define the shell directly in terms of three rotation parameters instead of only two at points where connection is to be performed.[32,33]

8.5 Multibody coupling by joints

Often it is desirable to have two (or more) rigid bodies connected in some specified manner. For example, in Fig. 8.3 we show a disc connected to an arm. Both are treated as rigid bodies but it is desired to have the disc connected to the arm in such a way that it can rotate freely about the axis normal to the page. This type of motion is characteristic of many rotating machine connections and it as well as many other types of connections are encountered in the study of rigid body motions.[4,34] This type of interconnection is commonly referred to as a *joint*. In quite general terms joints may be constructed by a combination of two types of simple constraints: *translational constraints* and *rotational constraints*.

8.5.1 Translation constraints

The simplest type of joint is a spherical connection in which one body may freely rotate around the other but relative translation is prevented. Such a situation is shown in Fig. 8.3 where it is evident the spinning disc must stay attached to the rigid arm at its axle. Thus it may not translate relative to the arm in any direction (additional constraints are necessary to ensure it rotates only about the one axis – these are discussed in Sec. 8.5.2). If a full translation constraint is imposed a simple relation may be introduced as

$$\mathbf{C}_j = \mathbf{x}^{(a)} - \mathbf{x}^{(b)} = \mathbf{0} \tag{8.41}$$

where a and b denote two rigid bodies. Thus, addition of the Lagrange multiplier constraint

$$\Pi_j = \boldsymbol{\lambda}_j^{\mathrm{T}}[\mathbf{x}^{(a)} - \mathbf{x}^{(b)}] \tag{8.42}$$

imposes the spherical joint condition. It is necessary only to define the location for the spherical joint in the reference configuration. Denoting this as $\mathbf{X}_j$ (which is common

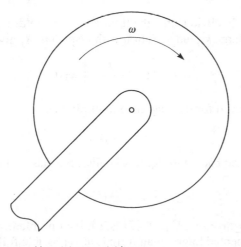

Fig. 8.3 Spinning disc constrained by a joint to a rigid arm.

to the two bodies) and introducing the rigid motion yields a constraint in terms of the rigid body positions as

$$\Pi_j = \lambda_j^T[\mathbf{r}^{(a)} + \Lambda^{(a)}(\mathbf{X}_j - \mathbf{R}^{(a)}) - \mathbf{r}^{(b)} - \Lambda^{(b)}(\mathbf{X}_j - \mathbf{R}^{(b)})] \tag{8.43}$$

The variation and subsequent linearization of this relation yields the contribution to the residual and tangent matrix for each body, respectively. This is easily performed using relations given above and is left as an exercise for the reader.

If the translation constraint is restricted to be in one direction with respect to, say, body a it is necessary to track this direction and write the constraint accordingly. To accomplish this the specific direction of the body a in the reference configuration is required. This may be computed by defining two points in space $\mathbf{X}_1$ and $\mathbf{X}_2$ from which a unit vector $\mathbf{V}$ is defined by

$$\mathbf{V} = \frac{\mathbf{X}_2 - \mathbf{X}_1}{|\mathbf{X}_2 - \mathbf{X}_1|} \tag{8.44}$$

The direction of this vector in the current configuration, $\mathbf{v}$, may be obtained using the rigid rotation for body a

$$\mathbf{v} = \Lambda^{(a)}\mathbf{V} \tag{8.45}$$

A constraint can now be introduced into the variational problem as

$$\Pi_j = \lambda_j\{\mathbf{V}^T(\Lambda^{(a)})^T[\mathbf{r}^{(a)} + \Lambda^{(a)}(\mathbf{X}_j - \mathbf{R}^{(a)}) - \mathbf{r}^{(b)} - \Lambda^{(b)}(\mathbf{X}_j - \mathbf{R}^{(b)})]\} \tag{8.46}$$

where, owing to the fact there is only a single constraint direction, the Lagrange multiplier is a scalar λ_j and, again, $\mathbf{X}_j$ denotes the reference position where the constraint is imposed.

The above constraints may also be imposed by using a penalty function. The most direct form is to *perturb* each Lagrange multiplier form by a penalty term. Accordingly, for each constraint we write the variational problem as

$$\Pi_j = \lambda_j C_j - \frac{1}{2k_j}\lambda_j^2 \tag{8.47}$$

where it is immediately obvious that the limit $k_j \rightarrow \infty$ yields exact satisfaction of the constraint. Use of a large k_j and variation with respect to λ_j give

$$\delta\lambda_j\left[C_j - \frac{1}{k_j}\lambda_j\right] = 0 \tag{8.48}$$

and may easily be solved for the Lagrange multiplier as

$$\lambda_j = k_j C_j \tag{8.49}$$

which when substituted back into Eq. (8.47) gives the classical form

$$\Pi_j = \frac{k_j}{2}[C_j]^2 \tag{8.50}$$

The reader will recognize that Eq. (8.47) is a mixed problem, whereas Eq. (8.50) is irreducible. An augmented Lagrangian form is also possible following the procedures described in reference 28 and used in Chapter 7 for contact problems.

8.5.2 Rotation constraints

A second kind of constraint that needs to be considered relates to rotations. We have already observed in Fig. 8.3 that the disc is free to rotate around only one axis. Accordingly, constraints must be imposed which limit this type of motion. This may be accomplished by constructing an orthogonal set of unit vectors $\mathbf{V}_I$ in the reference configuration and tracking the orientation of the deformed set of axes for each body as

$$\mathbf{v}_i^{(c)} = \delta_{iI}\mathbf{\Lambda}^{(c)}\mathbf{V}_I \quad \text{for} \quad c = a, b \quad \mathbf{v}_I^{\mathrm{T}}\mathbf{V}_J = \delta_{IJ} \tag{8.51}$$

A rotational constraint which imposes that axis i of body a remains perpendicular to axis j of body b may then be written as

$$(\mathbf{v}_i^{(a)})^{\mathrm{T}}\mathbf{v}_j^{(b)} = \mathbf{V}_I^{\mathrm{T}}(\mathbf{\Lambda}^{(a)})^{\mathrm{T}}\mathbf{\Lambda}^{(b)}\mathbf{V}_J = 0 \tag{8.52}$$

Example 8.1: Revolute joint

As an example, consider the situation shown for the disc in Fig. 8.3 and define the axis of rotation in the reference configuration by the Cartesian unit vectors $\mathbf{E}_I$ (i.e. $\mathbf{V}_I = \mathbf{E}_I$). If we let the disc be body a and the arm body b the set of constraints can be written as (where $\mathbf{v}_3$ is axis of rotation)

$$\mathbf{C}_j = \left\{ \begin{array}{c} \mathbf{x}^{(a)} - \mathbf{x}^{(c)} \\ (\mathbf{v}_1^{(a)})^{\mathrm{T}}\mathbf{v}_3^{(b)} \\ (\mathbf{v}_2^{(a)})^{\mathrm{T}}\mathbf{v}_3^{(b)} \end{array} \right\} = \mathbf{0} \tag{8.53}$$

and included in a formulation using a Lagrange multiplier form

$$\Pi_j = \boldsymbol{\lambda}_j^{\mathrm{T}}\mathbf{C}_j \tag{8.54}$$

The modifications to the finite element equations are obtained by appending the variation and linearization of Eq. (8.54) to the usual equilibrium equations. Here five Lagrange multipliers are involved to impose the three translational constraints (spherical joint) and the angle constraints for the rotating disc. The set of constraints is known as a *revolute* joint.[2]

8.5.3 Library of joints

Translational and rotational constraints may be combined in many forms to develop different types of constraints between rigid bodies. For the development it is necessary to have only the three types of constraints described above. Namely, the spherical joint, a single translational constraint, and a single rotational constraint. Once these are available it is possible to combine them to form classical constraint joints and here the reader is referred to the literature for the many kinds commonly encountered.[2,4,7,35]

The only situation that requires special mention is the case when a series of rigid bodies is connected together to form a *closed loop*. In this case the method given above can lead to situations in which some of the joints are redundant. Using Lagrange multipliers this implies the resulting tangent matrix will be singular and, thus, one cannot obtain solutions. Here a penalty method provides a viable method to circumvent this problem. The penalty method introduces *elastic deformation* in the joints and in this way removes the singular problem. If necessary an augmented Lagrangian method can be used to keep the deformation in the joint within required small tolerances. An alternative to this is to extract the closed loop rigid equations from the problem and use singular valued decomposition[36] to identify the redundant equations. These may then be removed by constructing a pseudo-inverse for the tangent matrix of the closed loop. This method has been used successfully by Chen to solve single loop problems.[35]

8.6 Numerical examples

8.6.1 Rotating disc

As a first example we consider a problem for the rotating disc on a rigid arm which is attached to a deformable base as shown in Fig. 8.4. The finite element model is constructed from four-node displacement elements in which a St Venant–Kirchhoff material model is used for the elastic part. The elastic properties in the model are $E = 10\,000$ and $\nu = 0.25$, with a uniform mass density $\rho_0 = 5$ throughout. The disc and arm are made rigid by using the procedures described in this chapter. The disc is attached to the arm by means of a revolute joint with the constraints imposed using the Lagrange multiplier method. The rigid arm is constrained to the elastic support by using the local Lagrange multiplier method described in Sec. 8.4. The problem is excited by a constant vertical load applied at the revolute joint and a torque applied to spin the disc. Each load is applied for the first 10 units of time.

The mesh and configuration are shown in Fig. 8.4(a). Deformed positions of the model are shown at 2.5 unit intervals of time in Fig. 8.4(b)–(h). A marker element shows the position of the rotating disc. The displacements at the revolute joint and the radial exterior point at the marker element location are shown in Fig. 8.5.

8.6.2 Beam with attached mass

As a second example we consider an elastic cantilever beam with an attached end mass of rectangular shape. The beam is excited by a horizontal load applied at the top as a triangular pulse for two units of time. The rigid mass is attached to the top of the beam by using the Lagrange multiplier method described in Sec. 8.4 and here it is necessary to constrain both the translation and the rotation parameters of the beam. The beam is three dimensional and has an elastic modulus of $E = 100\,000$ and a moment of inertia in both directions of $I_{11} = I_{22} = 12$. The beam mass density is low, with a value of $\rho_0 = 0.02$. The tip mass is a cube with side lengths 4 units and mass density $\rho_0 = 1$. The shape of the beam at several instants of time is shown in Fig. 8.6 and it is clear that large translation and rotation is occurring and also that the rigid block is correctly following a constrained rigid body motion.

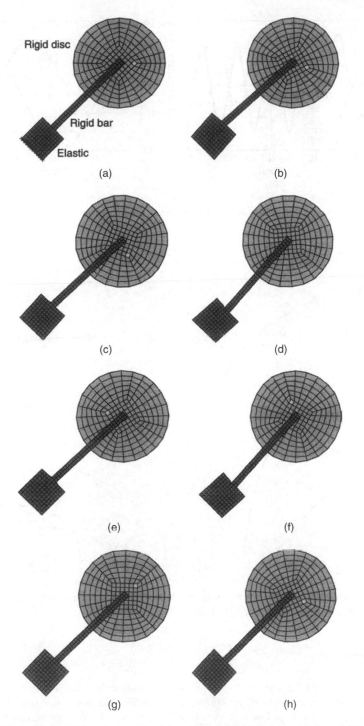

Fig. 8.4 Rigid–flexible model for spinning disc: (a) problem definition, solutions at time; (b) $t = 2.5$ units; (c) $t = 5.0$ units; (d) $t = 7.5$ units; (e) $t = 10.0$ units; (f) $t = 12.5$ units; (g) $t = 15.0$ units; (h) $t = 17.5$ units.

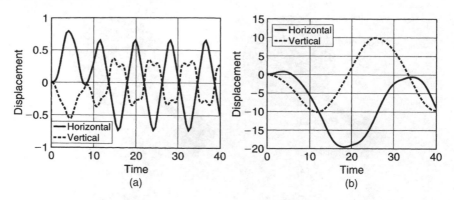

Fig. 8.5 Displacements for rigid–flexible model for spinning disc. Displacement at: (a) revolute; (b) disc rim.

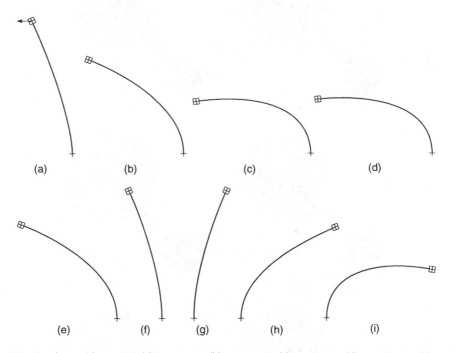

Fig. 8.6 Cantilever with tip mass: (a) $t = 2$ units; (b) $t = 4$ units; (c) $t = 6$ units; (d) $t = 10$ units; (e) $t = 12$ units; (f) $t = 14$ units; (g) $t = 16$ units; (h) $t = 18$ units; (i) $t = 20$ units.

References

1. H. Cohen and R.G. Muncaster. *The Theory of Pseudo-rigid Bodies*. Springer, New York, 1988.
2. A.A. Shabana. *Dynamics of Multibody Systems*. John Wiley & Sons, New York, 1989.
3. J.M. Solberg and P. Papadopoulos. A simple finite element-based framework for the analysis of elastic pseudo-rigid bodies. *International Journal for Numerical Methods in Engineering*, 45:1297–1314, 1999.
4. D.J. Benson and J.O. Hallquist. A simple rigid body algorithm for structural dynamics programs. *International Journal for Numerical Methods in Engineering*, 22:723–749, 1986.

5. R.A. Wehage and E.J. Haug. Generalized coordinate partitioning for dimension reduction in analysis of constrained dynamic systems. *Journal of Mechanical Design*, 104:247–255, 1982.

6. A. Cardona and M. Geradin. Beam finite element nonlinear theory with finite rotations. *International Journal for Numerical Methods in Engineering*, 26:2403–2438, 1988.

7. A. Cardona, M. Geradin and D.B. Doan. Rigid and flexible joint modelling in multibody dynamics using finite elements. *Computer Methods in Applied Mechanics and Engineering*, 89:395–418, 91.

8. J.C. Simo and K. Wong. Unconditionally stable algorithms for rigid body dynamics that exactly conserve energy and momentum. *International Journal for Numerical Methods in Engineering*, 31:19–52, 1991. [Addendum: 33:1321–1323, (1992).]

9. H.T. Clark and D.S. Kang. Application of penalty constraints for multibody dynamics of large space structures. *Advances in the Astronautical Sciences*, 79:511–530, 1992.

10. G.M. Hulbert. Explicit momentum conserving algorithms for rigid body dynamics. *Computers and Structures*, 44:1291–1303, 1992.

11. M. Geradin, D.B. Doan and I. Klapka. MECANO: a finite element software for flexible multibody analysis. *Vehicle System Dynamics*, 22:87–90, 1993. Supplement issue.

12. S.N. Atluri and A. Cazzani. Rotations in computational solid mechanics. *Archives of Computational Methods in Engineering*, 2:49–138, 1995.

13. O.A. Bauchau, G. Damilano and N.J. Theron. Numerical integration of nonlinear elastic multibody systems. *International Journal for Numerical Methods in Engineering*, 38:2727–2751, 1995.

14. J.A.C. Ambrósio. Dynamics of structures undergoing gross motion and nonlinear deformations: a multibody approach. *Computers and Structures*, 59:1001–1012, 1996.

15. R.L. Huston. Multibody dynamics since 1990. *Applied Mechanics Reviews*, 49:S35–S40, 1996.

16. O.A. Bauchau and N.J. Theron. Energy decaying scheme for non-linear beam models. *Computer Methods in Applied Mechanics and Engineering*, 134:37–56, 1996.

17. O.A. Bauchau and N.J. Theron. Energy decaying scheme for non-linear elastic multi-body systems. *Computers and Structures*, 59:317–331, 1996.

18. C. Bottasso and M. Borri. Energy preserving/decaying schemes for nonlinear beam dynamics using the helicoidal approximation. *Computer Methods in Applied Mechanics and Engineering*, 143:393–415, 1997.

19. O.A. Bauchau. Computational schemes for flexible, nonlinear multi-body systems. *Multibody System Dynamics*, 2:169–222, 1998.

20. O.A. Bauchau and C.L. Bottasso. On the design of energy preserving and decaying schemes for flexible nonlinear multi-body systems. *Computer Methods in Applied Mechanics and Engineering*, 169:61–79, 1999.

21. O.A. Bauchau and T. Joo. Computational schemes for non-linear elasto-dynamics. *International Journal for Numerical Methods in Engineering*, 45:693–719, 1999.

22. G.-H. Shi. *Block System Modelling by Discontinuous Deformation Analysis*. Computational Mechanics Publications, Southampton, 1993.

23. E.G. Petocz. *Formulation and analysis of stable time-stepping algorithms for contact problems*. PhD thesis, Department of Mechanical Engineering, Stanford University, Stanford, California, 1998.

24. M.E. Gurtin. *An Introduction to Continuum Mechanics*. Academic Press, New York, 1981.

25. L.E. Malvern. *Introduction to the Mechanics of a Continuous Medium*. Prentice-Hall, Englewood Cliffs, NJ, 1969.

26. J. Bonet and R.D. Wood. *Nonlinear Continuum Mechanics for Finite Element Analysis*. Cambridge University Press, Cambridge, 1997. ISBN 0–521–57272–X.

27. J.C. García Orden and J.M. Goicolea. Dynamic analysis of rigid and deformable multibody systems with penalty methods and energy-momentum schemes. *Computer Methods in Applied Mechanics and Engineering*, 188:789–804, 2000.

28. O.C. Zienkiewicz, R.L. Taylor and J.Z. Zhu. *The Finite Element Method: Its Basis and Fundamentals*. Butterworth-Heinemann, Oxford, 6th edition, 2005.

29. E.T. Wittaker. *A Treatise on Analytical Dynamics*. Dover Publications, New York, 1944.

30. J.H. Argyris and D.W. Scharpf. Finite elements in time and space. *Nuclear Engineering and Design*, 10:456–469, 1969.

31. A. Ibrahimbegovic and M. Al Mikdad. Finite rotations in dynamics of beams and implicit time-stepping schemes. *International Journal for Numerical Methods in Engineering*, 41:781–814, 1998.

32. J.C. Simo. On a stress resultant geometrically exact shell model. Part VII: Shell intersections with 5/6 DOF finite element formulations. *Computer Methods in Applied Mechanics and Engineering*, 108:319–339, 1993.

33. P. Betsch, F. Gruttmann and E. Stein. A 4–node finite shell element for the implementation of general hyperelastic 3d-elasticity at finite strains. *Computer Methods in Applied Mechanics and Engineering*, 130:57–79, 1996.

34. H. Goldstein. *Classical Mechanics*. Addison-Wesley, Reading, 2nd edition, 1980.

35. A.J. Chen. *Energy-momentum conserving methods for three dimensional dynamic nonlinear multibody systems*. PhD thesis, Department of Mechanical Engineering, Stanford University, Stanford, California, 1998. (Also SUDMC Report 98–01.)

36. G.H. Golub and C.F. Van Loan. *Matrix Computations*. The Johns Hopkins University Press, Baltimore MD, 3rd edition, 1996.

9

Discrete element methods*

9.1 Introduction

In the previous chapters we have considered the solution of solid mechanics problems from the view of a continuum. A computational framework which allows for some level of displacement field discontinuity to be represented *a priori* might be better suited to model particular phenomena, for example the behaviour of jointed rock or the granular material flow in silos. The treatment of these classes of problems is more naturally related to *discrete element methods*, in which distinctly separate material regions interacting with other discrete elements in some way are considered. A number of more complex models for both solid materials as well as for contact interactions have been formulated in the context of a discrete element methodology, with successful applications in many fields of science and engineering.

Moreover, most media are discontinuous at some level of observation (*nano, micro, meso, macro*), where the continuum assumptions cease to apply. This happens when the scale of the problem becomes similar to the characteristic length scale of the associated material structure and surface interaction laws between bodies or particles are invoked instead of a homogenized continuum constitutive law. The computational modelling of inherently discontinuous media requires the discrete nature of discontinuities to be taken into account. Discontinuities can be either pre-existing (e.g. joints, bedding planes, interfaces, planes of weakness, construction joints) or they evolve (e.g. in the case of cohesive frictional materials, where the growth and coalescence of micro cracks eventually appear in a form of a macro-crack). Many structures, structural systems or structural components comprise discrete discontinuities, appearing either in a highly regular or structured manner or they are of a heterogeneous nature. An obvious example of structured discontinua is brick masonry or jointed rock structures where the displacement discontinuities commonly occur at block interfaces without necessarily rendering structures unsafe. Such problems are best considered with discrete element methods.

The term *discrete element methods* will here be understood to comprise different techniques suitable for a simulation of dynamic behaviour of systems of multiple rigid, simply deformable (pseudo-rigid) or fully deformable separated bodies of simplified or arbitrary shapes, subject to continuous changes in the contact status and varying contact forces, which in turn influence the subsequent movement of the bodies. Such problems

---

* This chapter was contributed by Professor Nenad Bićanić, University of Glasgow, UK.

are non-smooth in space (separate bodies) and in time (jumps in velocities upon collisions) and the unilateral constraints (non-penetrability) need to be considered. A system of bodies changes its position continuously under the action of external forces and interaction forces between bodies, which may eventually lead to a steady-state configuration, once static equilibrium is achieved. For rigid bodies, the contact interaction law is the only constitutive law considered, while the continuum constitutive law (e.g. elasticity, plasticity, damage, fracturing) needs to be included for deformable bodies.

Computational modelling of multi-body contacts (both the contact detection and contact resolution) represents the dominant feature in discrete element methods, as the number of bodies considered may be very large. If the number of potential contact surfaces is relatively small (e.g. non-linear finite element analysis of contact problems) it is convenient to define groups of nodes, segments or surfaces which belong to a possible contact set *a priori*. These geometric attributes can then be continuously checked against one another and the kinematic resolution can be treated in a very rigorous manner. Bodies which are possibly in contact may be internally discretized by finite elements (Fig. 9.1), and their material behaviour can essentially be of any complexity (viz. Chapter 5).

The category of discrete element methods specifically refers to simulations involving a large number of bodies where the contact locations and conditions cannot be defined in advance and need to be continuously updated as the solution progresses. Discrete element methods are most frequently applied to macroscopically discrete system of bodies (jointed rock, granular flow) but have also been successfully utilized in a microscopic setting, where very simple interaction laws between individual particles provide the material behaviour observed at a homogenized, macroscopic level.

The discrete element method is most commonly defined[1] as a computational modelling framework which

1. allows finite displacements and rotations of discrete bodies, including complete detachment;
2. recognizes new contacts automatically, as the calculation progresses.

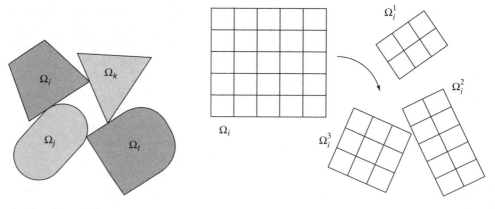

Fig. 9.1 System of rigid or deformable bodies, discretization of bodies into finite elements, continuously changing configurations and possible fragmentation.

There exist many methods (e.g. DEM *discrete element method*, RBSM *rigid block spring method*, DDA *discontinuous deformation analysis*, DEM/FEM *combined discrete/finite elements*, NSCD *non-smooth contact dynamics*), which belong to a broad family of discrete element methods.[2-4] Although these methods appear under different names and each of them is developing in its own right, there are many unifying aspects and a more general framework is emerging, which allows for an equivalence between these apparently different methodologies to be recognized. Possible classification may be based on the manner these methods address: (a) detection of contacts, (b) treatment of contacts (rigid, deformable), (c) deformability (constitutive law) of bodies in contact (rigid, deformable, elastic, elasto-plastic, etc.), (d) large displacements and large rotations, (e) number (small or large) and/or distribution (loose or dense packing) of interacting bodies considered, (f) consideration of the model boundaries, (g) possible subsequent fracturing or fragmentation and (h) time-stepping integration schemes (explicit, implicit). Discrete element methods are also used for problems where the discrete nature of the emerging discontinuities needs to be taken into account. Application ranges from modelling problems of a discontinuous behaviour *a priori* (granular and particulate materials, silo flow, sediment transport, jointed rocks, stone or brick masonry) to problems where the modelling of transition from a continuum to a discontinuum is more important. Increased complexity of different discontinuous models is achieved by incorporating the deformability of solid material and/or by more complex contact interaction laws, as well as by the introduction of some failure or fracturing criteria controlling the solid material behaviour and the emergence of new discontinuities.

9.2 Early DEM formulations

The initial formulation of the *discrete element method*, originally termed *distinct element method* or DEM,[5] was based on the assumption of rigid circular bodies in two dimensions with deformable contacts. The overall solution scheme for the DEM is straightforward, typically formulated in an explicit time-stepping format. Movement of bodies is driven by external forces (Fig. 9.2) and varying contact forces (normal and tangential contact forces, proportional to current overlap and viscous contact forces proportional to the relative velocities in the normal and tangential direction). The method considers each body in turn and at any given time determines all forces (external or contact) acting at it. Out of balance forces (or moments) induce accelerations (translational or rotational), which then determine the movement of that body during the next time step.

The simplest computational sequence for the DEM (often formulated in an explicit 'leap-frog' format, see Table 9.1) typically proceeds by solving the equations of motion of a given discrete element and updating contact force histories as a consequence of contacts between different discrete elements and/or resulting from contacts with model boundaries.

The Rigid Bodies Spring Model, RBSM,[7] was proposed early as a generalized limit plastic analysis framework. Solid structures are assumed to be assemblies of rigid blocks, interconnected by discrete deformable interfaces with distributed (elastic) normal and tangential springs.

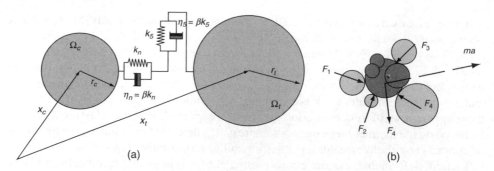

Fig. 9.2 (a) Discrete element bodies (particles) in contact, giving rise to axial and tangential contact forces. Force magnitudes related to the relative normal and tangential velocity and to relative normal and tangential velocity at the contact point; (b) arbitrary particle shapes as assemblies of clustered particles of simple shapes.

The stiffness matrix is obtained by considering rigid bodies to be connected by distributed normal and tangential springs with stiffness values k_n and k_s per unit length, respectively. The rigid displacement field within an arbitrary two-dimensional block is expressed in terms of the centroid displacements and rotation $(u, v, \theta)^T$.

Centroid degree of freedoms $(u_i, v_i, \theta_i)^T$ and $(u_j, v_j, \theta_j)^T$ of the two neighbouring blocks (Fig. 9.3) with centroids located at (x_i^0, y_i^0) and (x_j^0, y_j^0), respectively, define independently the displacements at a common interface point P

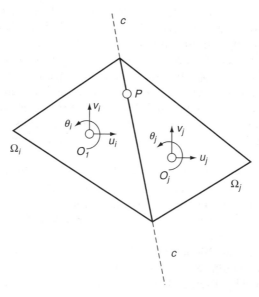

Fig. 9.3 Two rigid blocks with an elastic interface contact in RBSM.

Table 9.1 Simple discrete element method algorithm (after Cundall and Strack[6])

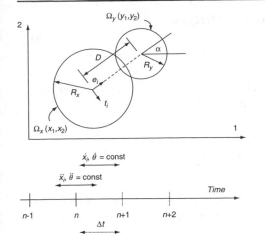

$$e_i = \frac{y_i - x_i}{D} = (\cos\alpha, \sin\alpha)$$
$$t_i = (\sin\alpha, -\cos\alpha)$$

Body centre
$(x_1, x_2), (y_1, y_2)$

Translational velocity
$\dot{x}_i, \dot{y}_i$

Rate of rotation
θ_x, θ_y

A. FORCE-DISPLACEMENT LAW
(1) Relative velocities
(2) Relative displacements

(3) Contact force increments

(4) Total forces
(5) Check for slip

(6) Compute moments

$$\dot{X}_i = (\dot{x}_i - \dot{y}_i) - (\dot\theta_x R_x + \dot\theta_y R_y)t_i$$
$$\dot{n} = \dot{X}_i e_i, \quad \dot{s} = \dot{X}_i t_i$$
$$\Delta n = \dot{n}\Delta t, \quad \Delta s = \dot{s}\Delta t$$
$$\Delta F_n = k_n(\Delta n + \beta\dot{n}),$$
$$\Delta F_s = k_s(\Delta s + \beta\dot{s})$$
$$F_n = F_n^{n-1} + \Delta F_n, \quad F_s = F_s^{n-1} + \Delta F_s$$
$$F_s = \min(F_s, C + F_n \tan\phi)$$
If not converged go to (1), else
$$M_x = \sum F_x R_x, \quad M_y = \sum F_y R_y$$

B. EQUATIONS OF MOTION
(1) Assume force and moment
constant over Δt
(2) Acceleration
(3) Velocity

(4) Assume velocities constant
over Δt
(5) Displacements
(6) Rotation

$$\Delta t = (t^{n+1/2} - t^{n-1/2})$$

$$m\ddot{x}_i = \sum F_i, \quad I\ddot\theta = \sum M_i$$
$$\dot{x}_i^{n+1/2} = \dot{x}_i^{n-1/2} + \ddot{x}_i\Delta t,$$
$$\dot\theta^{n+1/2} = \dot\theta^{n-1/2} + \ddot\theta\Delta t$$
$$\Delta t = (t^{n+1} - t^n)$$

$$x_i^{n+1} = x_i^n + \dot{x}_i^{n+1/2}\Delta t$$
$$\theta^{n+1} = \theta^n + \dot\theta^{n+1/2}\Delta t$$
If not converged go to A (6), else
Next time step: Go to A

C. TIME INCREMENT

$$\mathbf{U}_P = \mathbf{Q}_P\mathbf{u}$$
$$\mathbf{U}_P = \begin{bmatrix} U_P^i, & V_P^i, & U_P^j, & V_P^j \end{bmatrix}^T$$
$$\mathbf{u} = \begin{bmatrix} u_i & v_i & \theta_i & u_j & v_j & \theta_j \end{bmatrix}^T \tag{9.1}$$
$$\mathbf{Q}_P = \begin{bmatrix} 1 & 0 & -(y_P - y_i^0) & 0 & 0 & 0 \\ 0 & 1 & (x_P - x_i^0) & 0 & 0 & 0 \\ 0 & 0 & 0 & 1 & 0 & -(y_P - y_i^0) \\ 0 & 0 & 0 & 0 & 1 & (x_P - x_i^0) \end{bmatrix}$$

and the relative displacements at the location P are expressed as

$$\delta_P = \begin{bmatrix} \delta_n^P & \delta_s^P \end{bmatrix}^T = \mathbf{M}\tilde{\mathbf{U}}_P = \mathbf{M}\tilde{\mathbf{R}}\mathbf{Q}_P\,\mathbf{u} = \mathbf{B}\,\mathbf{u}$$

$$\text{with} \quad \mathbf{M} = \begin{bmatrix} -1 & 0 & 1 & 0 \\ 0 & -1 & 0 & 1 \end{bmatrix} \quad \text{and} \quad \tilde{\mathbf{U}}_P = \tilde{\mathbf{R}}\mathbf{U}_P = \tilde{\mathbf{R}}\mathbf{Q}_P\mathbf{u} \tag{9.2}$$

after $\mathbf{U}_P$ is projected $\tilde{\mathbf{U}}_P$ aligned to the local coordinate system along the interface. The constitutive relation in plane stress can be written as

$$\boldsymbol{\sigma} = \mathbf{D}\boldsymbol{\delta} \qquad \boldsymbol{\sigma} = \begin{bmatrix} \sigma_n, & \tau_s \end{bmatrix}^T$$

$$\text{with} \quad \mathbf{D} = \begin{bmatrix} k_n & 0 \\ 0 & k_s \end{bmatrix} \qquad k_n = \frac{(1-\nu)E}{h(1-2\nu)(1+\nu)} \qquad k_s = \frac{E}{h(1+\nu)} \tag{9.3}$$

where h is the sum of shortest distances between the two block centroids to the contact line. This projected distance h is also used to evaluate approximate normal and shear strain components through

$$\boldsymbol{\varepsilon}_P = \begin{bmatrix} \varepsilon_n \\ \gamma_s \end{bmatrix}_P = \frac{1}{h}\begin{bmatrix} \delta_n \\ \delta_s \end{bmatrix} = \frac{1}{h}\boldsymbol{\delta}_P \tag{9.4}$$

Applying the virtual work principle along the interface leads to

$$\mathbf{f} = \left[\int_S \mathbf{B}^T \mathbf{D}\mathbf{B}\,\mathrm{d}s \right]\mathbf{u} = \mathbf{K}\mathbf{u} \tag{9.5}$$

Generalization to the three-dimensional situation is straightforward and the method can clearly be interpreted as a similar method to the finite element method with joint (interface) elements of zero thickness, the only difference coming from the assumption that the overall elastic behaviour is represented only by the distributed stiffness springs along interfaces.

Given some criterion, the contact springs may be deactivated, so that an interface becomes a discontinuity and the progressive failure is modelled by following evolving discontinuities, through cracks and/or slipping at the interfaces between rigid blocks.

9.3 Contact detection

A principal algorithmic issue of the discrete element method represents a detection of bodies in contact followed by the evaluation of the contact forces (both a magnitude and a direction) emanating from the contact. The contact detection problem generally can be stated as one of finding a contact or overlap of a given contactor body with a number of bodies from a target set of N bodies in R^n space but the strategies for contact detection are intimately related to the geometric characterization and topological attributes of interacting bodies. If the interacting bodies are of very simple geometry (e.g. circular in two dimensions or spherical in three dimensions) an algorithmic check for a possible overlap is simple and the definition of the tangential contact plane is unambiguous. Rigid bodies of more complex shapes can be approximated by forming convenient *clusters* of rigidly connected circular or spherical shapes, while the contact detection

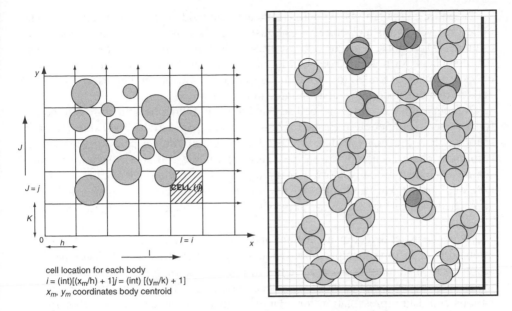

cell location for each body
$i = (int)[(x_m/h) + 1]j = (int) [(y_m/k) + 1]$
x_m, y_m coordinates body centroid

Fig. 9.4 Hashing or binning algorithm for simple particle shapes and clustered particles.

and resolution of the tangential contact plane remain the same as for individual bodies. However, this may be a crude approximation, and when interacting bodies of arbitrary geometry are considered, the algorithmic complexity of the contact detection and the associated definition of the contact plane between the two bodies increase significantly. The efficiency of these algorithms is crucial, as the conceptually simple procedure to test the possibility of contact of a body with *all other bodies at every time step* becomes highly uneconomical, once the number of bodies becomes large. Contact search algorithms are typically based on so-called *body-based search* or a *space-based search*. In the former, only the space in the vicinity of the specified discrete element is searched (and the search repeated only after a number of time steps), whereas the latter implies a subdivision of the total searching space into a number of overlapping *windows*.

For arbitrary geometric shapes, most algorithms typically employ a *two phase strategy*, where the bodies are first approximated by simpler geometric constructs (*bounding boxes* or *bounding spheres*) which encircle the actual body and a list of possible contact pairs is established via an efficient *global neighbour* or *region search* algorithm. This is then followed by a detailed *local contact resolution* phase, where the potential contact pairs are examined by considering the actual body geometries. This phase is strongly linked with the manner that the geometry of actual bodies is characterized.

9.3.1 Global neighbour or region search

An example of the *region search* represents the *boxing algorithm*,[8] also referred to as *hashing* or *binning* algorithm.

The entire computational domain is typically subdivided into regular cells, and a list of bodies overlapping a given cell is established via the contact detection of square (2D) or cube (3D) regions. The contact resolution phase for a given body then comprises a detailed check for possible contact with all bodies which share the same cell and the check is usually extended to a list of bodies associated with neighbouring cells (i.e. 8 cells in two dimensions, 26 cells in three dimensions).

The relationship between the cell size and the maximum size of a body is important for the overall efficiency. If the cell size is large compared to the body size, the initial search is fast, but many bodies are listed as potential contact pairs and the contact resolution phase is thereby extensive. However, if the cell size is small, the initial search is computationally more demanding, but this results in a smaller number of potential contact pairs and consequently a faster contact resolution phase. A balance is reached with cell sizes that are approximately of the size of the largest body in the system.

Efficient contact detection algorithms and powerful data representation concepts are often borrowed from other disciplines, notably computer graphics,[9] with compact data representation techniques to describe the current geometric position of a discrete element – e.g. nodes, sides or faces. The decomposition of the computational space and various cell data representation for a large number of contactor objects (binary tree, quad tree, direct evidence, combination of direct evidence, rooted trees, alternating data trees) are usually adopted.[8,10–15] Algorithmic issues and details of the associated data structures are quite involved and there is a non-linear relationship between the number of cells and the total number of bodies. Linear complexity contact detection algorithms are desirable and have been shown to be essential for simulations involving a very large number of bodies – e.g. the NBS (no binary search) algorithm for bodies of similar sizes has a total contact detection time proportional to the total number of bodies, irrespective of the particle packing density.[11]

For simple shapes very efficient data structures map a minimum set of parameters which uniquely define a domain in R_n into a representative point in an associated R_{2n} space (Fig. 9.5) – for example, a one-dimensional segment $(a-b)$ is mapped into a representative point in two-dimensional space, with coordinates (a, b), or a two-dimensional rectangle of a size $(x_{min} - x_{max})$ and $(y_{min} - y_{max})$ is mapped to a representative point in a four-dimensional space $(x_{min}, y_{min}, x_{max}, x_{max})$. Alternative representation schemes are also possible, e.g. by characterizing a rectangular domain in R^2 by the starting point coordinates (x_{min}, y_{min}) and the two rectangle sizes (h_x, h_y) followed by a mapping into an associated R^4 space $(x_{min}, y_{min}, h_x, h_y)$. As the representation of the physical domain is reduced to a point, region search algorithms are more efficient in the mapped R^{2n} spaces than in the physical R^n space.

9.3.2 Contact resolution

After the list of potential contact pairs is established through the global neighbour search, a detailed contact resolution algorithm is required, which will in turn depend on the way the detailed body geometries are defined. The contact resolution phase searches through potential pairs and if the actual contact is established, the algorithm needs to define the orientation of the contact plane, so that a local (n, t, s) coordinate

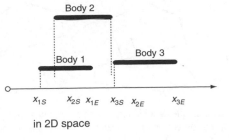

in 2D space

Body 1 → $[x_{1S}, x_{1E}]$ Body 2 → $[x_{2S}, x_{2E}]$ Body 3 → $[x_{3S}, x_{3E}]$

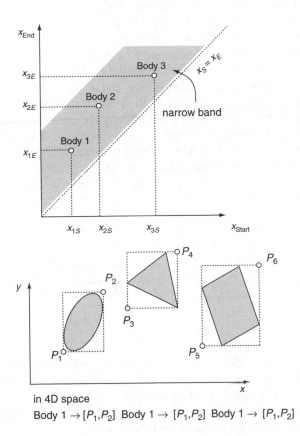

in 4D space
Body 1 → $[P_1, P_2]$ Body 1 → $[P_1, P_2]$ Body 1 → $[P_1, P_2]$

Fig. 9.5 Mapping of a segment from one-dimensional space into a point in an associated two-dimensional space and mapping of a box in two dimensions into an associated four-dimensional space.

system can be determined and the conditions for impenetrability or sliding can be properly applied.

Body geometry characterization can be categorized[16] into three main groups: (a) polygon or polyhedron representation, (b) implicit continuous function representation (elliptical or general superquadrics) and (c) discrete function representation (DFR).

Polygonal representation in two dimensions defines a body in terms of corners and edges, and there are a number of algorithms to determine an intersection of two

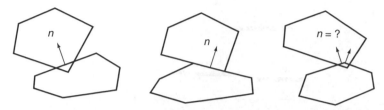

Fig. 9.6 Definition of the contact plane: a unique definition for the corner to edge case (a), the edge to edge case (b) and an ambiguous situation for the corner-to-corner (c) contact problem (after Hogue[16]).

coplanar polygons. Convex polygons simplify the algorithm, as concave corners imply a possibility of multiple contact points. There are no difficulties to define the orientation of the contact plane when considering the corner-to-edge or the edge-to-edge contact, as the contact plane normal is uniquely defined by the edge normal. Difficulties arise when a corner-to-corner contact needs to be resolved (Fig. 9.6), as the orientation of the contact plane is not uniquely defined. This problem can be regularized by the rounding of corners[17] to ensure continuous changes of the contact plane outer normals. The generalization in three dimensions can be realized by the common plane concept,[17] which 'hovers' between the two bodies coming into a corner-to-corner contact, and the actual orientation of this common plane is found by a local optimization problem of maximizing the gap between the plane and a set of closest corners.

Another possible procedure (restricted to a two-dimensional situation) utilizes an optimum triangularization of the space between the polygons,[18] whereby a collapse of a triangle indicates an occurrence of contact.

A continuous implicit function representation of bodies, e.g. elliptical particles in two dimensions,[19,20] ellipsoids in three dimensions,[21,22] or superquadrics (Fig. 9.7) in two and three dimensions,[23-25] provides an opportunity to employ a simple analytical check (i.e. inside–outside) to identify whether a given point lies inside or on the boundary ($\phi(x, y) \leq 0$) or outside ($\phi(x, y) > 0$) of the body where

$$\phi(x, y) = \left(\frac{x}{a}\right)^{\beta_1} + \left(\frac{y}{b}\right)^{\beta_2} - 1 \tag{9.6}$$

Unlike a polygonal representation, it is now significantly more difficult to solve for a complete intersection of overlapping superquadrics, and the solution is normally found by discretizing one of the surfaces into facets and nodes, and the contact for a specific node on one body can be verified through the inside–outside analytical check with respect to the functional representation of the other body.

A discrete functional representation (DFR) describes the body boundary with a parametric function in one parameter. The concept of the DFR[26] replaces the continuous implicit function representation of bodies by the set of pre-evaluated function values on a background grid for the inside–outside check, which is used as an algorithmic look-up table. As such discrete function values at the grid nodes can be arbitrary, a grid (or cage) of cells can also be used to model an arbitrarily shaped body – including bodies with holes. The DFR concept in contact detection is illustrated through the polar DFR descriptor in two dimensions[27] (Fig. 9.8) where, following the global region search for possible neighbours, the local contact is established by transforming the local coordinates of the approaching corner P_i of a body i into the polar coordinates

Fig. 9.7 Superquadrics in three dimensions (reproduced from Hogue[16]).

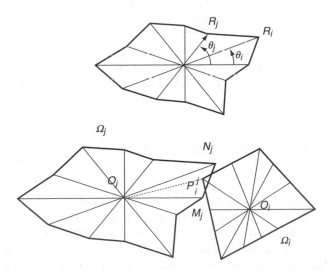

Fig. 9.8 Contact detection and the polar discrete functional representation (DFR) of bodies' geometry.

of the other body P_i^j and checking if an intersection between the segments $(\overline{O_j P_i^j})$ and $(\overline{M_j N_j})$ can be found.

9.4 Contact constraints and boundary conditions

In the case of rigid bodies of simple shapes, an event-by-event simulation strategy can be applied in order to strictly satisfy the condition of impenetrability. In such cases, if the contact time is considered to be infinitely short, time instants of collisions can be calculated exactly and momentum exchange methodologies used to determine post-impact velocities, with possible energy loss accounted for by a coefficient of restitution or by friction losses. In either case there are non-smooth step changes and reversals in velocities. Such methodologies are used for molecular dynamics (MD) simulations with a very large number of particles. Although the event-driven algorithms work well for loose (gas-like) assemblies of particles, for dense configurations these lead to an effective solution locking or inelastic collapse,[28] manifested in critically slow simulations.

9.4.1 Regularization of non-smooth contact

In the case of deformable bodies the contact time is finite and contact forces vary for the duration of the contact. In computational simulations it is necessary to regularize the non-smooth nature for the impenetrability and friction conditions. Constraints on impenetrability during the contact between the two bodies Ω_c and Ω_t require that the *gap* between them must be non-negative. In frictionless and cohesionless contacts, only a compressive interaction force F_n exists and this interaction force vanishes for an inactive contact described by the Signorini condition $g > 0$. The infinitely steep 'non-smooth' graph (viz. Fig. 9.9) is regularized by assuming that the interaction force F_n is a function of the gap violation and is replaced by a penalty formulation, with a linear or non-linear penalty coefficient.

Non-smooth relations also exist if the interaction law considers a tangential friction force F_s, related and opposed to the relative sliding velocity $\dot{s}$. For a Coulomb friction law, there exists a threshold tangential force, proportional to the normal interaction force $F_s = \mu F_n$, before any sliding can occur, corresponding again to an infinitely steep graph.

9.4.2 Contact constraints between bodies

Once the contact between discrete elements is detected as a geometric overlap, the actual contact forces have to be evaluated, which are then used for the subsequent motion of the discrete elements controlled by the governing dynamic equilibrium equations. Contact forces follow from an imposition of contact constraints at contacting points. In variational formulations, a constraint functional Π_c can therefore be added to the functional of the unconstrained system using a penalty function, Lagrange multiplier, augmented Lagrangian or perturbed Lagrangian form, as discussed in Sec. 7.2.

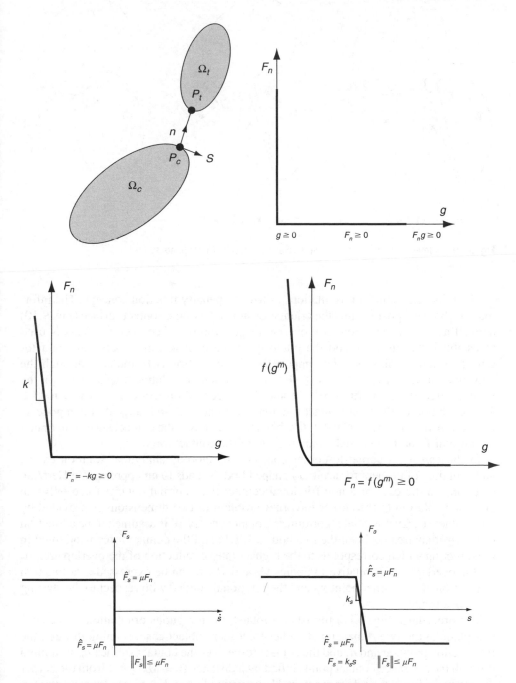

Fig. 9.9 Non-smooth treatment of normal contact, regularized treatment of normal contact (linear and non-linear penalty term) and non-smooth and regularized treatment of frictional contact.

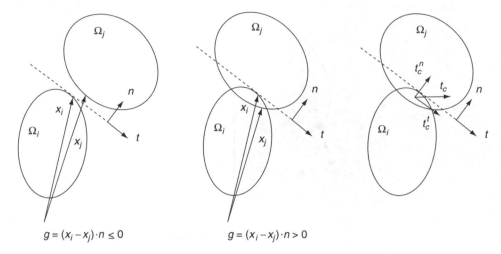

$$g = (x_i - x_j) \cdot n \le 0 \qquad\qquad\qquad g = (x_i - x_j) \cdot n > 0$$

Fig. 9.10 Determination of the contact surface and its local n–t coordinate system.

Most discrete element formulations utilize the penalty function concept. The information about the position and the orientation normal $\mathbf{n}$ of the contact surface (Fig. 9.10) as well as the current geometric overlap or penetration of contactor objects is used to establish the direction and the intensity of the contact forces between contactor objects at any given time. The impenetrability condition is formulated through the gap function $g = [\mathbf{x}^i - \mathbf{x}^j] \cdot \mathbf{n} \le 0$, which defines the relative displacement in the normal direction $u_n = [\mathbf{u}_i - \mathbf{u}_j] \cdot \mathbf{n}$ and the tangential direction $u_t = [\mathbf{u}_i - \mathbf{u}_j] \cdot \mathbf{t}$. The resolution of the total contact traction $\mathbf{t}_c$ into normal and tangential components $\mathbf{t}_c = (\mathbf{t}_c \cdot \mathbf{n})\mathbf{n} + (\mathbf{t}_c \cdot \mathbf{t})\mathbf{t} = t_c^n \, \mathbf{n} + t_c^t \, \mathbf{t}$ is then integrated over the contact surface to obtain the normal F_n and tangential component F_t of the contact force.

In the case of a corner-to-corner contact, no rigorous analytical solution exists, and rounding of corners for arbitrary shaped bodies leads to an approximate Hertzian solution. In the case of a non-frictional contact (i.e. normal contact force only) an elegant resolution to the corner-to-corner problem in two dimensions is provided by the contact energy potential algorithm.[29] Contact energy W is assumed to be a function of the overlap area between the two bodies $W(A)$ and the contact force is oriented in the direction which corresponds to the highest rate of reduction of the overlap area A. As the overlap area is relative to bodies Ω_i and Ω_j, it can be expressed as a function of position of the corner point $\mathbf{x}_P$ and the rotational angle θ with respect to the starting reference frame.

The procedure (Fig. 9.11) furnishes a robust, unambiguous orientation of the contact plane in the two-dimensional corner-to-corner contact case, running through the intersection points g and h, and the contact force over the contact surface b_w is applied through the reference contact point shifted by a distance $d = M_\theta / \|F_n\|$ from the corner P, where the force F_n and the moment M_θ are defined from the contact energy potential as

$$F_n = \frac{\partial W(A)}{\partial A} \frac{\partial A(\mathbf{x}_P, \theta)}{\partial \mathbf{x}_P} \quad \text{and} \quad M_\theta = \frac{\partial W(A)}{\partial A} \frac{\partial A(\mathbf{x}_P, \theta)}{\partial \theta}$$

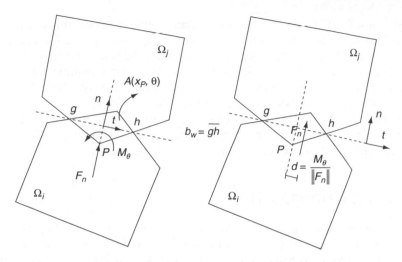

Fig. 9.11 Corner-to-corner contact, based on energy potential (after Feng and Owen[29]).

Different choices for the potential function (Table 9.2) are capable of reproducing various traditional models for contact forces.

9.4.3 Contact constraints on model boundaries

Treatment of model boundaries represents an important aspect in the discrete element methods. Boundaries can be formulated either as real physical boundaries, or through virtual constraints. In the so-called periodic boundary, often adopted in the DEM analysis of granular media, a virtual constraint implies that the particles (or bodies) 'exiting' on one side of the computational domain with a certain velocity are reintroduced on the other side, with the same, now 'incoming', velocity. The use of periodic boundaries excludes capturing of any localization phenomena. In cases of particle assemblies, flexible and hydrostatic boundaries have also been employed. A flexible boundary[30] framework is equivalent to physically 'stringing' together particles on the perimeter of the discrete element assembly, forming a boundary network. Flexible boundaries are mostly used to simulate the controlled stress boundary condition $\sigma_{ij} = \sigma_{ij}^C$, where the traction vector $t_i^m = A_m \sigma_{ij}^C n_j^m$ is distributed over the centroids of a triangular facet A_m, that is formed by three boundary particles. The hydrostatic boundary[31] can be interpreted as a virtual wall of pressurized fluid imagined to surround granular material particles and the desired hydrostatic pressure at the centroid of the intersection area between the particle and virtual wall.

Table 9.2 Contact energy potential functions (after Feng and Owen[29])

	$W(A)$	$\|F_n\|$
Linear form	$k_n A$	$k_n b_w$
Hertz-type form	$\frac{2}{3} k_n A^{3/2}$	$k_n A^{1/2} b_w$
General power form	$\frac{1}{m} k_n A^m$	$k_n A^{m-1} b_w$

The simplest form of a physical boundary representation in two dimensions is given by the definition of line segments (often referred to as 'walls' in the DEM context), and the kinematics of the contact between the particle and the wall is again resolved in the penalty format. Frequently, individual particles are declared as immovable, thereby creating an efficient way of representing and characterizing a rigid boundary, without any changes in the contact detection algorithm. An interesting idea is the finite wall (FW) method,[32,33] with the boundary surface triangulated into a number of rigid planar elements, which are then represented by inscribed circles and subsequently used in the contact detection analysis between the particles and the boundary.

9.5 Block deformability

Consideration of deformability increases the complexity for the analysis of multi-body systems, where the bodies represent fully deformable media or belong to a class of constrained media, corresponding to restricted deformability. The deformability of individual discrete elements was initially dealt with by subdividing the body into triangular constant strain elements,[34] which can be identified as an early precursor of today's combined finite/discrete element modelling.

Further developments of the discrete element methods include simple body deformability, so that a displacement at any point within a simply deformable element can be expressed by $u_i = u_i^0 + \omega_{ij} x_j^r + \varepsilon_{ij} x_j^r$, where u_i^0 are the rigid body displacements of the element centroid, ω_{ij}, ε_{ij} are the rotation and strain tensor respectively and x_j^r represent local coordinates of the point, relative to element centroid. It should be noted that this deformability statement implies a spatially constant deformation gradient that is equivalent to the class of pseudo-rigid bodies[35] discussed in Chapter 8. Displacements of body centroids follow from balance equations for the translation of the centre of mass in direction $\sum_i F_i = m\ddot{u}$, and the rotation about the centre of mass $\sum_i M_i^C = I\ddot{\theta}_i$. The *simply deformable* discrete elements[17] comprise generalized strain modes ε_k, independent of the rigid body modes $m^k \ddot{x} = \sigma_A^k - \sigma_I^k$, where m^k is the generalized mass, σ_A^k is a generalized applied stresses and σ_I^k generalized internal stresses, corresponding to strain modes. The discrete element deformation field (displacements relative to the centroid) can also be expanded in terms of the eigenmodes of the generalized eigenvalue problem, associated with the discrete element stiffness and mass matrix, giving rise to the Modal Expansion Discrete Element Method,[36,37] where the 'deformability' equations are uncoupled (due to the orthogonality of eigenvectors) and appear as modal equations written with respect to a non-inertial frame of reference. In the practical realization and implementation of the discrete element methodology the deformability of a discrete element of an arbitrary shape is either described by an internal division into finite elements (discrete finite elements and/or combined finite/discrete elements) or by the polynomial expansion of a given order for the displacement field (discontinuous deformations analysis).

9.5.1 Combined finite/discrete element method

A combination of discrete and finite elements was employed in the discrete finite element approach by Ghaboussi.[38] The combined finite/discrete element approach of

Munjiza et al.[39,40] introduces bodies deformability through finite element expansion of the displacement field within a discrete element, while the contact between the discrete elements is solved again in an explicit transient dynamic setting. The over-all algorithmic framework for a combined finite element/discrete element framework (Table 9.3) proceeds by solving the balance equations, while updating force histories as a consequence of contacts between different discrete element domains, internally subdivided into finite elements. A combined finite element/discrete element method is easily extended to problems comprising progressive fracturing and fragmentation, as complex constitutive relations can be utilized within discrete elements.

Large displacements and rotations of discrete domains, internally discretized by finite elements, have been formulated in a generalized updated Lagrangian (UL) format by Barbosa and Ghaboussi.[41] The contact forces were obtained using a penalty form (concentrated and distributed contacts) and transformed into equivalent nodal forces on the finite element mesh. The equations of motion for each of the deformable discrete elements (assuming also a presence of mass proportional damping ($\mathbf{C} = \alpha \mathbf{M}^t$)) are then expressed as

$$\mathbf{M}^t \ddot{\mathbf{U}} + \alpha \mathbf{M}^t \dot{\mathbf{U}} = \mathbf{f}_{\text{ext}}^t + \mathbf{f}_{\text{cont}}^t - \mathbf{f}_{\text{int}}^t = \mathbf{f}_{\text{ext}}^t + \mathbf{f}_{\text{cont}}^t - \sum_k \int_{\Omega_k^t} (\mathbf{B}_k^t)^T \sigma_k^t \, d\Omega_k \qquad (9.7)$$

In the computational solution Barbosa and Ghaboussi[41] used a central difference scheme for integrating the incremental updated Lagrangian formulation, which neglects the non-linear part of the stress–strain relationship. Evaluation of the internal force vector

$$\mathbf{f}_{\text{int}}^{t+\Delta t} = \sum_k \int_{\Omega_k^{t+\Delta t}} (\mathbf{B}_k^{t+\Delta t})^T \sigma_k^{t+\Delta t} \, d\Omega_k$$

at the new time station $t + \Delta t$ recognizes the continuous changing of the configuration, as the Cauchy stress at $t + \Delta t$ cannot be evaluated by simply adding a stress increment

Table 9.3 Pseudo code for the combined discrete/finite element method, small displacement analysis, including material non-linearity (after Petrinić[12])

---

(1) Increment from the time station $t = t_n$

    current displacement state $\mathbf{u}_n$

    external load vector, contact forces $\mathbf{F}_n^{\text{ext}}$, $\mathbf{F}_n^c \to \hat{\mathbf{F}}^{\text{ext}}$

    internal force, e.g. $\mathbf{F}_n^{\text{int}} = \int_\Omega \mathbf{B}^T \sigma_n \, d\Omega$

(2) Solve for the displacement increment from

    $\mathbf{M} \ddot{u}_n + \mathbf{F}_n^{\text{int}} = \hat{\mathbf{F}}_n^{\text{ext}}$

    $\dot{u}_{n+1/2} = \mathbf{M}^{-1}(\hat{\mathbf{F}}_n^{\text{ext}} - \mathbf{F}_n^{\text{int}}) \Delta t + \dot{u}_{n-1/2}$

    $u_{n+1} = u_n + \dot{u}_{n+1/2} \Delta t$

    for an explicit time-stepping scheme

(3) Compute the strain increment $\Delta \varepsilon_{n+1} = \mathbf{f}(\Delta \mathbf{u}_{n+1})$

(4) Check the total stress predictor $\sigma_{n+1}^* = \sigma_n + \mathbf{D} \Delta \varepsilon_{n+1}$ against a failure criterion, e.g. hardening plasticity $\phi(\sigma_{n+1}, \kappa) = 0$

(4) Compute inelastic strain increment $\Delta \varepsilon_{n+1}^{\text{inel}}$, e.g. using an associated plastic flow rule

(5) Update stress state $\sigma_{n+1} = \mathbf{D}(\Delta \varepsilon_{n+1} - \Delta \varepsilon_{n+1}^{\text{inel}})$

(6) Establish contact states between discrete element domains at t_{n+1} and the associated contact forces $\mathbf{F}_{n+1}^c$

(7) $n \to n + 1$, go to step (1)

---

due to straining of the material to the Cauchy stress at t and the effects of the rigid body rotation on the components of the Cauchy stress tensor need to be accounted for [viz. incremental form (6.90b) or Jaumann–Zaremba rate form (6.109)].

For inelastic analyses, care needs be given to the objectivity of the adopted constitutive law. Advanced combined discrete/finite element frameworks[42] include a rigorous treatment of changes in configuration, evaluation of the deformation gradient and the objective stress measures.

9.5.2 Discontinuous deformation analysis

The discontinuous deformation analysis (DDA) employs a general polynomial approximation of the displacement field superimposed over the centroid movement for each discrete body.[43,44] The DDA development followed the formulation of the Keyblock Theory[45] and was used for simulating the behaviour of a jointed deformable rock. Blocks of arbitrary shapes with convex or concave boundaries, including holes, are considered. The original framework under the name the 'DDA method' comprises a number of distinct features: (a) the assumption of the order of the displacement approximation field over the whole block domain, (b) the derivation of the incremental equilibrium equations on the basis of the minimization of potential energy, (c) the block interface constitutive law (Mohr–Coulomb) with tension cutoff and (d) use of a special implicit time-stepping algorithm. The original implementation has been expanded and modified by many other researchers. The method is realized in an incremental form and it deals with the large displacements and deformations as an accumulation of small displacements and deformations. The issue of inaccuracies when large rotations are occurring has been recognized and several partial remedies have been proposed.[46,47]

The DDA method represents an alternative way of introducing solid deformability into the discrete element framework, and block sliding and separation is considered along predetermined discontinuity planes at block boundaries. The initial formulation was restricted to simply deformable blocks (constant strain state over the entire block of arbitrary shapes in two dimensions, similar to a constant deformation gradient as in pseudo-rigid bodies, Fig. 9.12). The first-order polynomial displacement field for the block $[u , v]_{1st}^{iT}$ is equivalent to the three displacement components of the block centroid, augmented by the displacement field with a clear physical meaning of three constant strain states, i.e. it represents a constrained medium capable only of sustaining a spatially constant displacement gradient. The six deformation variables are denoted by the block deformation vector $\mathbf{D}_{1st}^i$.

$$\begin{bmatrix} u \\ v \end{bmatrix}_{1st}^i = \begin{bmatrix} 1 & 0 & -(y-y_0) & (x-x_0) & 0 & (y-y_0)/2 \\ 0 & 1 & (x-x_0) & 0 & (y-y_0) & (x-x_0)/2 \end{bmatrix}_i \mathbf{D}_{1st}^i$$

$$\mathbf{D}_{1st}^i = \begin{bmatrix} u_0 & v_0 & \phi_0 & \varepsilon_x & \varepsilon_y & \gamma_{xy} \end{bmatrix}_i^T \tag{9.8}$$

$$\begin{bmatrix} u \\ v \end{bmatrix}_{1st}^i = \mathbf{T}_{1st}^i \mathbf{D}_{1st}^i$$

An increase of block deformability characterization can be achieved by increasing the order of the displacement polynomial used, leading to a correspondingly larger number

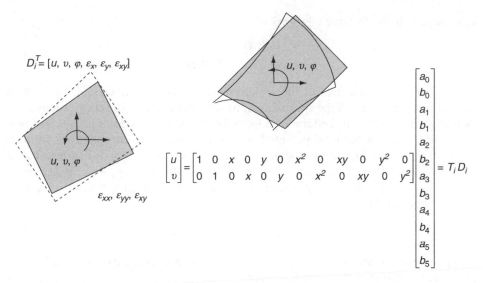

Fig. 9.12 Deformation variables for the first- and second-order polynomial approximation in discontinuous deformation analysis.

of block deformation variables (higher-order DDA, where higher-order strain fields are added for blocks of arbitrary shape). By increasing the order of the displacement field, the block medium becomes less constrained and gradually moves towards full deformability.

A second-order approximation for the block displacement field requires 12 deformation variables, which can again be given a recognizable physical meaning. The formulation implies a linearly varying displacement gradient across the block domain and the deformation parameters comprise the centroid displacements and rotation, strain tensor components at the centroid, as well as the spatial gradients of the strain tensor components, i.e.

$$\mathbf{D}_{2nd}^i = \begin{bmatrix} u_0 & v_0 & \phi_0 & \varepsilon_x^0 & \varepsilon_y^0 & \gamma_{xy}^0 & \varepsilon_{x,x} & \varepsilon_{y,x} & \gamma_{xy,x} & \varepsilon_{x,y} & \varepsilon_{y,y} & \gamma_{xy,y} \end{bmatrix}^{iT}$$

$$(9.9)$$

For a higher-order approximation of order n for the block displacement field,[48] the spatial distribution of the deformation gradient is of the order $(n-1)$ and a clear physical interpretation of the deformation variables is no longer plausible. The generalized block deformation variables are defined as

$$u = d_1 + d_3 x + d_5 y + d_7 x^2 + d_9 xy + d_{11} y^2 + \ldots + d_{r-1} x^n + \ldots + d_{m-1} y^n$$

$$v = d_2 + d_4 x + d_6 y + d_8 x^2 + d_{10} xy + d_{12} y^2 + \ldots + d_r x^n + \ldots + d_m y^n$$

$$\begin{bmatrix} u \\ v \end{bmatrix}_{nth}^i = \mathbf{T}_{nth}^i \mathbf{D}_{nth}^i \qquad \mathbf{D}_{nth}^i = \begin{bmatrix} d_1 & d_2 & d_3 & \ldots & d_{m-1} & d_m \end{bmatrix}^T \qquad (9.10)$$

Once the block displacement field is approximated with a finite number of generalized deformation variables, the associated block strain and block stress field can be expressed in a manner similar to finite elements as

$$\varepsilon_i = \mathbf{B}_i \mathbf{D}_i$$
$$\sigma_i = \mathbf{E}_i \varepsilon_i = \mathbf{E}_i \mathbf{B}_i \mathbf{D}_i \tag{9.11}$$

For a system of N blocks, all block deformation variables (n variables per block, depending on the order of the approximation) are assembled into a set of system deformation variables ($N \times n$) and the incremental equilibrium equations are derived from the minimization of the total potential energy Π comprising contributions from the block strain energy Π_e, energy from the external concentrated and distributed loads Π_P, Π_q, interblock contact energy Π_c, block initial stress energy Π_σ, as well as the energy associated with an imposition of displacement boundary conditions using a penalty approach Π_b

$$\Pi = \Pi_e + \Pi_P + \Pi_q + \Pi_c + \Pi_\sigma + \Pi_b \tag{9.12}$$

Components of the stiffness matrix and the load vector are obtained by the usual process of the minimization of the potential energy

$$\mathbf{K}_{ii} = \int_{\Omega^i} \mathbf{B}_i^T \mathbf{E}_i \mathbf{B}_i \, d\Omega \tag{9.13}$$

The global system stiffness matrix (Fig. 9.13) contains ($n \times n$) submatrices $\mathbf{K}_{ii}$ and non-zero submatrices $\mathbf{K}_{ij}$ are included when the blocks i and j are in *active* contact and $\mathbf{D}$ comprises displacement variables of all blocks considered in the system.

The interblock contact conditions of impenetrability and Mohr–Coulomb friction can be interpreted as block displacement constraints, which are algorithmically reduced to an interaction problem between a vertex of one block with the edge of another block. Denoting the increments of deformation variables of the two blocks in contact by $\mathbf{D}_i$ and $\mathbf{D}_j$ respectively, the penetration of the vertex in the direction normal to the block edge can be expressed as a function of these deformation increments and various algorithmic approaches can be adopted for an implicit imposition of the impenetrability condition (Fig. 9.14).

If the *penalty format* is adopted to impose contact constraints, additional terms appear both in the global DDA stiffness matrix as well as in the RHS load vector (Table 9.4),

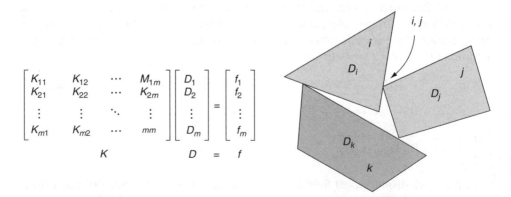

Fig. 9.13 Assembly process in the DDA analysis.

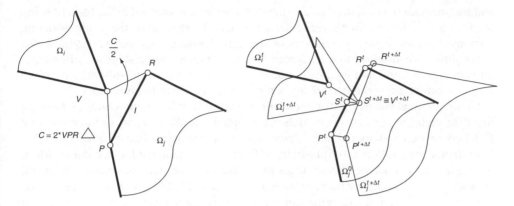

Fig. 9.14 Non-Penetration and frictional contact constraint in DDA, point-to-edge contact.

Table 9.4 Additional terms in the DDA stiffness matrix and the load vectors as a result of contact between bodies i and j

Normal non-penetration constraint	
$\mathbf{K}_{ii} = \mathbf{K}_{ii} + p\mathbf{\Pi}_i\mathbf{H}_i^T$	
$\mathbf{K}_{ij} = \mathbf{H}_i\mathbf{G}_j^T$	$\mathbf{f}_i = \mathbf{f}_i - pC\,\mathbf{H}_i$
$\mathbf{K}_{ji} = \mathbf{G}_j\mathbf{H}_i^T$	$\mathbf{f}_j = \mathbf{f}_j - pC\,\mathbf{G}_j$
$\mathbf{K}_{jj} = \mathbf{K}_{jj} + p\mathbf{G}_j\mathbf{G}_j^T$	

Frictional constraint	
$\mathbf{K}_{ii} = \mathbf{K}_{ii} + p\mathbf{H}_i^{fr}\mathbf{H}_i^{fr,T}$	
$\mathbf{K}_{ij} = \mathbf{H}_i^{fr}\mathbf{G}_i^{fr,T}$	$\mathbf{f}_i = \mathbf{f}_i - pC^{fr}\,\mathbf{H}_i^{fr}$
$\mathbf{K}_{ji} = \mathbf{G}_j^{fr}\mathbf{H}_i^{fr,T}$	$\mathbf{f}_j = \mathbf{f}_j - pC^{fr}\,\mathbf{G}_j^{fr}$
$\mathbf{K}_{jj} = \mathbf{K}_{jj} + p\mathbf{G}_j^{fr}\mathbf{G}_j^{fr,T}$	

and the terms differ depending on the nature of the constraint (normal gap or frictional constraint).

The DDA method typically adopts an implicit algorithm and uses so-called open–close iterations which proceed until the global equilibrium is satisfied (norm of the out-of-balance forces within some tolerance) and a near zero penetration condition satisfied at all active contact positions. For the normal gap condition convergence implies that the identified set of contacts does not change between iterations, whereas for the frictional constraint it implies that the changes in the location of the projected contact point remains within a given tolerance. For complex block shapes, the convergence may be very slow, as both activation and deactivation of contacts during the iteration process are possible.

9.5.3 Block fracturing and fragmentation

Consideration of block deformability allows for a more precise determination of stress states throughout the discrete element. Block interface and through-block fracturing and fragmentation were introduced to the discrete element method in the form of brittle fracturing.[49–52] Later models recognized the need to regularize the strain softening response for quasi-brittle materials, prior to eventual separation through cracking and/or shear slip). Constitutive models adopted for a pre-fragmentation stage are often based on concepts of continuum damage mechanics, regularized strain softening plasticity formulations or are formulated using some higher-order continuum theory.[53]

Computational issues for discrete element methods, when more complex formulations for inelastic constitutive models are adopted, are the same as in the continuum FEM context computational inelasticity, where the consistent linearization procedure has affected the application of plasticity of fracture criteria in the DEM context. Moreover, every time a partial fracture takes place, the discrete element changes its overall geometry while a complete fracture leads to a creation of two or more discrete elements, there is a need for automatic remeshing of the newly obtained domains (Fig. 9.15). An unstructured remeshing technique can be applied, where the new mesh orientation and density is decided upon based on the distribution of the residual material strength, or some other state variable.[39]

An additional algorithmic problem arises upon separation, as the 'book-keeping' of neighbours and updating of the discrete element list are needed whenever a new (partial or complete) failure occurs. In addition, there is also a need to transfer state variables (plastic strains, equivalent plastic strain, dissipated energy, damage variables) from the original deformable discrete element to the newly created deformable discrete elements.

Fragmentation frameworks are also used with DEM implementations which consider clusters of particles bonded together to represent a solid body of a complex shape. The bond stiffness terms are derived on the basis of equivalent continuum strain energy.[54,55] In these lattice-like models of solids, fracturing (Fig. 9.16) is introduced through breakage of interparticle lattice bonds, which may be treated as a simple normal bond (truss elements) and/or parallel bond (beam elements), which can be seen as a microscopic representation of the Cosserat continuum. Bond failure is typically considered on the basis of limited strength, but some softening lattice models for quasi-brittle material have also considered a gradual reduction of strength, i.e. softening, before a complete

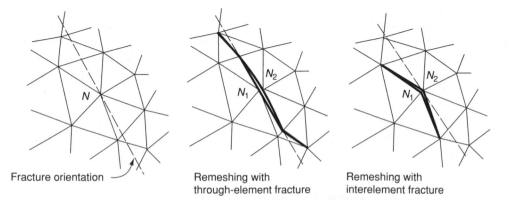

Fracture orientation

Remeshing with
through-element fracture

Remeshing with
interelement fracture

Fig. 9.15 Element and node-based local remeshing algorithm in a two-dimensional context (Munjiza et al.;[39] Feng and Owen[29]), following the weighted local residual strength concept.

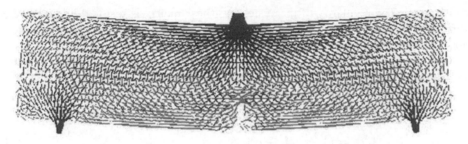

Fig. 9.16 Fracturing of the notched concrete beam modelled by jointed particulate assembly, with normal contact bond of limited strength (Itasca PFC 2D[56]).

breakage of the bond takes place. Despite use of very simple bond failure rules, bonded particulate assemblies have been shown to reproduce macroscopic manifestations of softening, dilation, and progressive fracturing.

Simulations using bonded particle assemblies typically use an explicit time-stepping scheme, i.e. the overall stiffness matrix is never assembled and steady-state solutions are obtained through dynamic relaxation (local or global damping by viscous or kinetic means). A jointed particulate medium is very similar to the more recent developments in lattice models for heterogeneous fracturing media.[57–59] Adaptive continuum/discontinuum strategies are also envisaged, where the discrete elements are adaptively introduced into a continuum model, if and when conditions for fracturing are met.

Discrete element methods are also coupled with fluid flow methods and a number of combined finite/discrete element simulations of various multi-field problems have been reported.

9.6 Time integration for discrete element methods

Governing balance equations are integrated in time using a time-stepping scheme, such as the GN22 method discussed in Chapter 2 and in reference 60. Both the usual DEM and the DDA time-stepping schemes can be interpreted as members of the GN22 family of algorithms and only the issues pertaining to their use in a discrete element context will be discussed here.

The traditional DEM framework implies a conditionally stable explicit time-stepping scheme. Local numerical dissipation is sometimes introduced to avoid any artificial increase in the contact energy.[61] Such modified temporal operators do not affect the size of the critical time step, and the choice of the actual computational time step is controlled by accuracy requirements resulting from the numerical energy dissipation added to the system. An energy balance check is desired, for the purpose of monitoring possible creation of spurious energy, as well as monitoring energy dissipation during fracturing, where the inclusion of softening imposes severe limits on the admissible time step.

When inertial effects are omitted in the DDA context, the resulting system of equilibrium equations may be singular when blocks are separated or have insufficient constraints. The regularity of the system stiffness matrix is restored by adding very soft spring stiffness to the block centroid deformation variables. Such modification is not necessary when inertial effects are included.

The DDA framework typically utilizes a particular type of the generalized collocation time integration scheme[62]

$$
\begin{aligned}
&\mathbf{M}\,\ddot{\mathbf{x}}_{n+\alpha} + \mathbf{C}\,\dot{\mathbf{x}}_{n+\alpha} + \mathbf{K}\,\mathbf{x}_{n+\alpha} = \mathbf{f}_{n+\alpha} \\
&\ddot{\mathbf{x}}_{n+\alpha} = (1-\alpha)\,\ddot{\mathbf{x}}_n + \alpha\,\ddot{\mathbf{x}}_{n+1} \\
&\mathbf{f}_{n+\alpha} = (1-\alpha)\,\mathbf{f}_n + \alpha\,\mathbf{f}_{n+1} \\
&\mathbf{x}_{n+\alpha} = \mathbf{x}_n + \alpha\Delta t\,\dot{\mathbf{x}}_n + \tfrac{1}{2}\,(\alpha\Delta t)^2[(1-\beta_2)\,\ddot{\mathbf{x}}_n + \beta_2\,\ddot{\mathbf{x}}_{n+1}] \\
&\dot{\mathbf{x}}_{n+\alpha} = \dot{\mathbf{x}}_n + \alpha\,\Delta t[(1-\beta_1)\,\ddot{\mathbf{x}}_n + \beta_1\,\ddot{\mathbf{x}}_{n+1}]
\end{aligned}
\tag{9.14}
$$

which, for the case $\alpha = 1$, leads to a recursive algorithm

$$
\begin{aligned}
&\left[\frac{2}{\beta_2\Delta t^2}\mathbf{M} + \frac{2\beta_1}{\beta_2\Delta t}\mathbf{C} + \mathbf{K}\right]\mathbf{x}_{n+1} = \mathbf{f}_{n+1} + \left[\frac{2}{\beta_2\Delta t^2}\mathbf{M} + \frac{2\beta_1}{\beta_2\Delta t}\mathbf{C}\right]\mathbf{x}_n \\
&+\left[\frac{2}{\beta_2\Delta t}\mathbf{M} + \left(\frac{2\beta_1}{\beta_2} - 1\right)\mathbf{C}\right]\dot{\mathbf{x}}_n + \left[\left(\frac{1}{\beta_2} - 1\right)\mathbf{M} + \Delta t\left(\frac{\beta_1}{\beta_2} - 1\right)\mathbf{C}\right]\ddot{\mathbf{x}}_n
\end{aligned}
\tag{9.15}
$$

A specific choice of the time integration parameters $\beta_1 = \beta_2 = 1$ represents an implicit, unconditionally stable scheme. The DDA time-stepping algorithm is also referred to as *Right Riemann* in accelerations,[63] as both the new displacement and velocity depend only on the acceleration at the end of the time increment as

$$
\begin{aligned}
&\mathbf{x}_{n+1} = \mathbf{x}_n + \Delta t\,\dot{\mathbf{x}}_n + \tfrac{1}{2}\,\Delta t^2\,\ddot{\mathbf{x}}_{n+1} \\
&\dot{\mathbf{x}}_{n+1} = \dot{\mathbf{x}}_n + \Delta t\,\ddot{\mathbf{x}}_{n+1}
\end{aligned}
\tag{9.16}
$$

which may also be expressed as

$$
\begin{aligned}
&\mathbf{x}_{n+1} = \mathbf{x}_n + \tfrac{1}{2}\,\Delta t\,[\dot{\mathbf{x}}_n + \dot{\mathbf{x}}_{n+1}] \\
&\ddot{\mathbf{x}}_{n+1} = \frac{1}{\Delta t}\,[\dot{\mathbf{x}}_{n+1} - \dot{\mathbf{x}}_n]
\end{aligned}
\tag{9.17}
$$

For this choice of time integration parameters the coefficient matrix associated with the acceleration vector $\ddot{\mathbf{x}}_n$ vanishes, hence

$$
\left[\frac{2}{\Delta t^2}\mathbf{M} + \frac{2}{\Delta t}\mathbf{C} + \mathbf{K}\right]\mathbf{x}_{n+1} = \mathbf{f}_{n+1} + \left[\frac{2}{\Delta t^2}\mathbf{M} + \frac{2}{\Delta t}\mathbf{C}\right]\mathbf{x}_n + \left[\frac{2}{\Delta t}\mathbf{M} + \mathbf{C}\right]\dot{\mathbf{x}}_n
$$
$$
\hat{\mathbf{K}}\,\mathbf{x}_{n+1} = \hat{\mathbf{f}}_{n+1}
\tag{9.18}
$$

The *effective stiffness matrix* $\hat{\mathbf{K}}$ includes the inertia and damping terms and the *effective load vector* $\hat{\mathbf{f}}_{n+1}$ accounts for the velocity at the start of the increment. Thus, the next solution $\mathbf{x}_{n+1}$ is obtained from

$$
\mathbf{x}_{n+1} = \hat{\mathbf{K}}^{-1}\hat{\mathbf{f}}_{n+1}
\tag{9.19}
$$

Using Eq. (9.17) the next rate of deformation vector (velocity) then equals

$$
\dot{\mathbf{x}}_{n+1} = \frac{2}{\Delta t}\,[\mathbf{x}_{n+1} - \mathbf{x}_n] - \dot{\mathbf{x}}_n
\tag{9.20}
$$

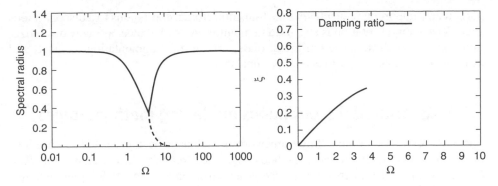

Fig. 9.17   Spectral radius and algorithmic damping for the DDA time integration scheme (generalized Newmark scheme GN22 $\beta_1 = \beta_2 = 1$) (reproduced from Doolin and Sitar[63]).

which is then used for the start of the next time increment. We note that in this form it is never necessary to compute the acceleration during the solution process. If necessary, the acceleration may be computed to interpret the inertial effects in the solution.

The effective stiffness matrix is regular due to the presence of inertia terms and the block separation can be accounted for without a need to add artificial springs to block centroid variables, even when blocks are separated. The time-stepping procedure can also be used to obtain a steady-state solution through a dynamic relaxation process. The steady-state solution can be obtained by the use of the so-called kinetic damping, i.e. by simply setting all block velocities to zero at the beginning of every increment.

From the analysis of spectral radii for the above time integrator (Fig. 9.17), it is clear that the scheme is associated with a very substantial numerical damping,[63,64] which is otherwise absent in the central difference scheme.

Moreover, predictor–corrector or even predictor–multiple corrector schemes are adopted in simulations of granular materials,[65] in order to capture the high frequency events, like the collapse of arching or avalanching.

The discrete element method simulations are computationally intensive and there is a need for the development of parallel processing procedures. The explicit time integration procedure is a naturally concurrent process and can be implemented on parallel hardware in a highly efficient form.[66,67] Much of the computational effort devoted to contact detection and the search algorithms, which is formulated principally for sequential implementation, has to be restructured to suit parallel computation. Efficient solution procedures are employed to minimize communication requirements and maintain a load balanced processor configuration.

Parallel implementations of DEM are typically made on both workstation clusters as well as multi-processor workstations. In this way the effort of porting software from sequential to parallel hardware is usually minimized. The program development often utilizes the sequential programming language with the organization of data structure suited to multi-processor workstation environment, where every processor independently performs the same basic DEM algorithm on a subdomain and only communicates with other subdomains via interface data. In view of the large mass of time dependent data produced by a typical FEM/DEM simulation, it is essential that some means be available of continually visualizing the solution process. Visualization is particularly important in displaying the transition from a continuous to a discontinuous state. Visual representation is also used to monitor energy balance, as many discretization concepts both in space (stiffness, mass, fracturing, fragmentation, contact) and time can contribute to spurious energy imbalances.

9.7 Associated discontinuous modelling methodologies

There are several other modelling frameworks which have been proposed as variations of the existing discrete element methods, e.g. the *modified distinct element method* (MDEM).[68] It is also interesting to observe a convergence of methodologies designed to deal with a transition from a continuum to a discontinuum. For example, in a continuum setting, the pre-existing macro discontinuities are traditionally accounted for through the use of interface (or joint) elements,[34] which may be used to model crack formation as well as shearing at joints or predetermined planes of weakness.[69] Joint planes are assumed to be of a discrete nature, their location and orientation are either given, or are fixed once they are formed. The term discrete cracking is adopted, as opposed to the term smeared cracking, where the localized failure at an integration point level is considered in a locally averaged sense. Natural extensions are interface constitutive models which account for a combination of cracking and Coulomb friction, which have been frequently formulated as two parameter failure surfaces in the context of computational plasticity.

The *non-smooth contact dynamics method* (NSCDM)[70–72] is closely related to both the combined FEM/DEM and the DDA, but it comprises significant differences, as the unilateral Signorini condition and the dry friction Mohr–Coulomb model are employed without resorting to contact regularization. For multi-body contact, with jumps in the velocity field, it is not possible to define the acceleration as a usual derivative of a smooth function and the non-smooth time discretized form of dynamic equations is obtained by integrating the balance equations, so that the velocities become the primary unknowns

$$\mathbf{M}\,(\dot{\mathbf{x}}_{n+1} - \dot{\mathbf{x}}_n) = \int_{t_n}^{t_{n+1}} \mathbf{f}(t, \mathbf{x}, \dot{\mathbf{x}})\,\mathrm{d}t + \int_{t_n}^{t_{n+1}} \mathbf{r}\,\mathrm{d}t$$

$$\mathbf{x}_{n+1} = \mathbf{x}_n + \int_{t_n}^{t_{n+1}} \dot{\mathbf{x}}\,\mathrm{d}t \tag{9.21}$$

The force impulse on the RHS is split into two parts, where the first part

$$\int_{t_n}^{t_{n+1}} \mathbf{f}(t, \mathbf{x}, \dot{\mathbf{x}})\,\mathrm{d}t$$

(which excludes the contact forces) is considered continuous, whereas the second part

$$\int_{t_n}^{t_{n+1}} \mathbf{r}\,\mathrm{d}t$$

(representing contact force contributions to the total impulse) is replaced by the mean value impulse

$$\mathbf{r}_{n+1} = \frac{1}{\Delta t} \int_{t_n}^{t_{n+1}} \mathbf{r}\,\mathrm{d}t$$

over the finite time interval. In physical terms this implies that the actual contact interaction history is only accounted for in an averaged sense over the time interval. This has a consequence that the fine details of the contact history are discarded, which are either impossible to characterize due to insufficient data available, or inconsequential in terms of their effect on the overall behaviour of the multi-body system.

Different time-stepping algorithms may be adopted and typically a low-order implicit time-stepping scheme is used. Resolution for contact kinematics then considers the relationship between the relative velocities of contacting bodies and the mean value impulse at discrete contact locations

$$\dot{\mathbf{x}}_{rel} = \dot{\mathbf{x}}_{rel\text{-free}} + \Delta t \ \mathbf{K}_{eff}^{-1} \mathbf{r}_{n+1} \tag{9.22}$$

where the first part represents the 'free' relative velocity (without the influence of contact forces) and the second part comprises 'corrective' velocities emanating from contacts. The actual algorithm is realized in a similar manner to the closest point projection stress return schemes in computational plasticity – a 'free predictor' for relative velocities is followed by iterations to obtain 'iterative corrector' values for the mean value impulses, such that the inequality constraints (both Signorini and Mohr–Coulomb) are satisfied in a similar manner as the plastic consistency condition is iteratively satisfied in computational plasticity algorithms. The admissible domains in contact problems are generally non-convex and it is argued that it is necessary to treat the contact forces in a fully implicit manner, whereas other forces can be considered either explicitly or implicitly,[73] leading to either implicit/implicit or explicit/implicit algorithms.

9.8 Unifying aspects of discrete element methods

There are many similarities in the apparently different discrete element methodologies, which partly stem from the fact that a given methodology is often presented as a package of specific choices, e.g. comprising a choice for the manner the bodies' deformability is treated, a way the governing equations are arrived at, a specific time integration scheme or a special way of dealing with unilateral constraints. For example, the DDA framework is often perceived as a methodology which specifically deals with simply deformable (constant strain) domains of arbitrary shapes, uses a very special case of implicit time-stepping scheme and treats contact constraints through a series of open–close iterations. Conversely, the DEM framework is often presented as a methodology for rigid particles of simple shapes, where the balance equations are integrated by an explicit scheme, with a penalty format for contacts. These perceived restrictions seem arbitrary, as there is nothing wrong, for example, in integrating the DEM equations with an implicit scheme or solving the DDA equations by an explicit method. It is also interesting to observe that the choice of a contact detection scheme was never perceived fixed and many contact detection algorithms have been applied.

It can be observed that the major developments for all discrete element methods have primarily been associated with a characterization of bodies' enhanced deformability. While the higher-order DDA increases the order of polynomial approximation for the displacement field, other attributes of the method remain the same. In the same spirit the combined finite/discrete element again enhances the description of deformability, while the other algorithmic features remain the same.

It is therefore perhaps most appropriate to leave all other attributes aside and concentrate on the characterization of bodies' deformability and the treatment of geometrically non-linear motions as the basis for any comparison or equivalence and a potential identification of common, unifying concepts of different discrete element methodologies. In that context it is interesting to return to the background of the theory of pseudo-rigid bodies,[35] as discussed in Chapter 8. As in all discrete element methodologies, the theory is concerned with large-scale motions of deformable bodies. It was presented simultaneously as a generalization of the classical rigid body mechanics for bodies with some added deformability, as well as a 'restriction' (or coarse version in the terminology of reference 35) of a fully deformable continuum. Clearly, a hierarchy of theories emerges, depending on the degree of deformability added to the rigid body or on the class of restrictions introduced to the fully deformable continuum body. This is where the relationship between the theory of pseudo-rigid bodies and the discrete element methods provides a powerful link that allows for a rational comparison of different assumptions and an assessment of consequences of approximations adopted.

In that spirit, the equivalence of the equations of motion between the elastic pseudo-rigid bodies and constant strain finite element approximations was confirmed and it was argued[74] that the pseudo-rigid bodies can be viewed as a generalization of the low-order DDA in the regime of finite (as opposed to stepwise linearized as in DDA) kinematics. The crucial equivalence stems from the spatially homogeneous deformation gradient present in both cases, which can again be viewed as an enrichment of deformability given to the rigid body, or as a restriction imposed to the fully deformable continuum.

The availability of a rational theory has an additional benefit by fully explaining manifestations of excessive volume changes for large rotations, discussed earlier in Sec. 9.7.

Similarly, the higher-order DDA and higher-order pseudo-rigid bodies[75] can be unified in the sense that the deformation gradient $\mathbf{F}$ varies linearly over the body, i.e.

$$\mathbf{F} = \mathbf{F}^{(1)}\left(\mathbf{X} - \mathbf{X}_0\right) + \mathbf{F}^{(0)} \tag{9.23}$$

where $\mathbf{F}^{(0)}$ is the deformation gradient at $\mathbf{X} = \mathbf{X}_0$ and $\mathbf{F}^{(1)}$ is the derivative of the deformation gradient with respect to $\mathbf{X}$, assumed constant. Clearly, if $\mathbf{F}^{(1)} = \mathbf{0}$, the simple pseudo-rigid body (or the lower-order DDA) is recovered. The extension to general higher-order approximations is plausible and again can be interpreted as enrichments of deformability or restrictions of the fully deformable continuum. Another rational interpretation of discrete element methods was developed based on Cosserat points theory[76] which moves away from algorithmic differences and concentrates on similarities between the DEM and DDA methods when they are recast using Hamilton's principle of least action.

9.9 Concluding remarks

There is a degree of commonality of novel ideas in terms of describing the block deformability in discrete element methods and novel developments in the continuum-based techniques. The *numerical manifold method* of Shi[77] and Chen *et al.*[78] advocates similar ideas to the ones advocated in the meshless[79] or the partition of unity methods[80] in dealing with emerging discontinuities. Similar to the meshless methods, the manifold method identifies a cover displacement function C_i and the cover weighting function w_i, where the geometry of the actual blocks Ω_i is utilized for numerical integration purposes over a background grid. The treatment of any emerging discontinuities is envisaged by introducing the concept of effective cover regions, where there is a need to introduce n independent covers, if a cover intersects n disconnected domains. These concepts point to a range of possibilities in the simulation of progressive discontinuities in quasi-brittle materials.

Discrete element methods increasingly appear in formulations and applications in multi-field or multi-physics problems, in particular in the area of the coupled fluid flow in discontinuous, jointed media. These discontinuous modelling frameworks are promising especially in the context of fragmentation and in the microscopic simulation of the behaviour of heterogeneous materials, where simple constitutive laws at the macro, meso or nano level directly generate manifestations of complex macroscopic behaviour, such as plasticity or fracture.[81] Increased computing power and efficient contact detection algorithms will not only allow modelling of progressive fracturing of continua and a transition to discontinuum including a fragmented state but will also encourage the development of discrete micro-structural models where an internal length scale may be intrinsically incorporated into the model. Moreover the large-scale simulations with adaptive multi-scale material models are increasingly feasible, where different regions are accounted for at a different scale of observation. Some of these aspects are considered in the next chapter.

References

1. P.A. Cundall. Numerical experiments on localization in frictional materials. *Ingenieur-Archiv*, 59:148–159, 1989.
2. P.A. Cundall and R.D. Hart. Numerical modelling of discontinua. *Engineering Computations*, 9:101–114, 1992.
3. J.R. Williams and G.C.W. Mustoe, editors. *2nd International Conference on Discrete Element Methods (DEM)*, Cambridge, 1993. Massachusetts Institute of Technology, MIT IESL Publications.
4. A. Munjiza. *The Combined Finite-Discrete Element Method*. John Wiley & Sons, Chichester, 2004.
5. P.A. Cundall. A computer model for simulating progressive large scale movements in blocky rock systems. In *Proceedings Symp. Intl. Society of Rock Mechanics*, volume 1, Nancy, France, 1971. Paper II-8.
6. P.A. Cundall and D.L. Strack. A discrete numerical model for granular assemblies. *Geotechnique*, 29:47–65, 1979.
7. T. Kawai. New element models in discrete structural analysis. *Journal of Society of Naval Architecture in Japan*, 141:187–193, 1977.
8. L.M. Taylor and D.S. Preece. Simulation of blasting induced rock motion using spherical element models. In G.G.W. Mustoe, M. Henriksen and H.P. Huttelmaier, editors, *1st U.S. Conference on Discrete Element Methods (DEM)*. Colorado School of Mines Press, Golden, 1989.
9. D. Baraff. Fast contact force computation for non-penetrating rigid bodies. In *SIGGRAPH 94 Computer Graphics Proceedings*, Orlando, FL, 1994.

10. J. Bonet and J. Peraire. An alternating digit tree (ADT) algorithm for 3d geometric search and intersection problems. *International Journal for Numerical Methods in Engineering*, 31:1–17, 1991.

11. A. Munjiza and K.R.F. Andrews. NBS contact detection algorithm for bodies of similar size. *International Journal for Numerical Methods in Engineering*, 43:131–149, 1998.

12. N. Petrinić. *Aspects of Discrete Element Modelling Involving Facet-to-Facet Contact Detection and Interaction*. PhD thesis, University of Wales, Swansea, 1996.

13. J.R. Williams and R. O'Connor. Discrete element simulation and the contact problem. *Archives of Computational Methods in Engineering*, 6:279–304, 1999.

14. E. Perkins and J.R. Williams. A fast contact detection algorithm insensitive to object sizes. *Engineering Computations*, 18:48–61, 2001.

15. Y.T. Feng and D.R.J. Owen. An augmented spatial digital tree algorithm for contact detection in computational mechanics. *International Journal for Numerical Methods in Engineering*, 55: 159–176, 2002.

16. C. Hogue. Shape representation and contact detection for discrete element simulations of arbitrary geometries. *Engineering Computations*, 3:374–390, 1998.

17. P.A. Cundall. Formulation of a three-dimensional distinct element model – Part I: A scheme to detect and represent contacts in a system composed of many polyhedral blocks. *Int. J. Rock Mech., Min. Sci. and Geomech. Abstr.*, 25:107–116, 1988.

18. D. Müller. *Techniques Infirmatiques Efficaces pour la Simulation de Mêlieux Granulaires par des Méthodes d'Eléments Distincs*. These 1545, EFF Lausanne, Lausanne, 1996.

19. J.M. Ting. A robust algorithm for ellipse based discrete element modelling of granular materials. *Computers and Geotechnics*, 13:175–186, 1992.

20. L. Vu-Quoc, X. Zhang and O.R. Walton. A 3-discrete-element method for dry granular flows of ellipsoidal particles. *Computer Methods in Applied Mechanics and Engineering*, 187:483–528, 2000.

21. C. Lin. *Extensions to the DDA for Jointed Rock Masses and other Blocky Systems*. PhD thesis, University of Colorado, Boulder, 1995.

22. X. Lin and T.T. Ng. Contact detection algorithm for three dimensional ellipsoids in discrete element modelling. *International Journal for Numerical and Analytical Methods in Geomechanics*, 19:653–659, 1995.

23. A.P. Pentland and J.R. Williams. Good vibrations: modal dynamics for graphics and animation. *Computer Graphics (ACM)*, 23(3):215–222, 1989.

24. J.R. Williams and A.P. Pentland. Superquadrics and modal dynamics for discrete elements in interactive design. *Engineering Computations*, 9:115–127, 1992.

25. R. Wait. Discrete element models of particle flows. *Mathematical Modelling and Analysis*, 6:156–164, 2001.

26. R. O'Connor, J.J. Gill and J.R. Williams. A linear complexity contact detection algorithm for multi-body simulation. In *2nd International Conference on Discrete Element Methods*, MIT IESL Publication, 1993.

27. C. Hogue and D. Newland. Efficient computer simulation of moving granular particles. *Powder Technology*, 78:51–66, 1994.

28. S. McNamara and R. Young. Inelastic collapse in two dimensions. *Physical Review E*, 50:28–31, 1994.

29. Y.T. Feng and D.R.J. Owen. An energy based corner to corner contact algorithm. In B.K. Cook and P.J. Jensen, editors, *Discrete Element Methods: Numerical Modeling of Discontinua, 3rd International Conference on Discrete Element Methods*, volume ASCE Geotechnical Special Publication No. 117. The Geo-Institute of the American Society of Civil Engineers, 2002.

30. M. Kuhn. A flexible boundary for three dimensional DEM particle assemblies. In *2nd International Conference on Discrete Element Methods*, MIT IESL Publication, 1993.

31. T.T. Ng. Hydrostatic boundaries in discrete element methods. In B.K. Cook and P.J. Jensen, editors, *Discrete Element Methods: Numerical Modeling of Discontinua, 3rd International Conference on Discrete Element Methods*, volume ASCE Geotechnical Special Publication No. 117. The Geo-Institute of the American Society of Civil Engineers, 2002.

32. M. Kremmer and J.F. Favier. A method for representing boundaries in discrete element modelling – Part I: Geometry and contact detection. *International Journal for Numerical Methods in Engineering*, 51:1407–1421, 2001.

33. M. Kremmer and J.F. Favier. A method for representing boundaries in discrete element modelling – Part II: Kinematics. *International Journal for Numerical Methods in Engineering*, 51: 1423–1436, 2001.

34. R.E. Goodman, R.L. Taylor and T. Brekke. A model for the mechanics of jointed rock. *J. Soil Mech and Foundation Div., ASCE.*, 94(SM3):637–659, 1968.

35. H. Cohen and R.G. Muncaster. *The Theory of Pseudo-rigid Bodies*. Springer, New York, 1988.

36. J.R. Williams, G. Hocking and G.G.W. Mustoe. The theoretical basis of the discrete element method. In *NUMETA 85, Numerical Methods of Engineering, Theory and Applications*. A.A. Balkema, Rotterdam, 1985.

37. J.R. Williams and G.G.W. Mustoe. Modal methods for the analysis of discrete systems. *Computers and Geotechnics*, 4:1–19, 1987.

38. J. Ghaboussi. Fully deformable discrete element analysis using a finite element approach. *Intl. J. of Comp. and Geotech.*, 5:175–195, 1988.

39. A. Munjiza, D.R.J. Owen and N. Bićanić. A combined finite/discrete element method in transient dynamics of fracturing solids. *Engineering Computations*, 12:145–174, 1995.

40. A. Munjiza, J.P. Latham and N.W.M. John. 3d dynamics of discrete element systems comprising irregular discrete elements – integration solution for finite rotations in 3D. *International Journal for Numerical Methods in Engineering*, 56:35–55, 2003.

41. R. Barbosa and J. Ghaboussi. Discrete finite element method. *Engineering Computations*, 9:253–266, 1992.

42. D.R.J. Owen, D. Peric, E.A. de Souza Neto, A.J.L. Crook, J. Yu and P.A. Klerck. Computational strategies for discrete systems and multi-fracturing materials. In *ECCM European Conference on Computational Mechanics*, Munchen, 1999.

43. G.H. Shi. *Discontinous Deformation Analysis – A New Numerical Model for Statics and Dynamics of Block Systems*. PhD dissertation, University of California, Berkeley, 1988.

44. G.-H. Shi. *Block System Modelling by Discontinuous Deformation Analysis*. Computational Mechanics Publications, Southampton, 1993.

45. G.H. Shi and R.E. Goodman. A new concept for support of underground and surface excavations in discontinuous rocks based on a keystone principle. In *22nd U.S. Symposium on Rock Mechanics*, MIT, 1981.

46. M.M. McLaughlin and N. Sitar. Rigid body rotations in DDA. In M.R. Salami and D. Banks, editors, *First Intl Forum on DDA and Simulation of Discontinuous Media*. TSI Press, University of California, Berkeley, 1996.

47. T.C. Ke. The issue of rigid body rotation in DDA. In M.R. Salami and D. Banks, editors, *First Intl Forum on DDA and Simulation of Discontinuous Media*. TSI Press, University of California, Berkeley, 1996.

48. M.Y. Ma, M. Zaman and J.H. Zhou. Discontinuous deformation analysis using the third order displacement function. In M.R. Salami and D. Banks, editors, *First Intl Forum on DDA and Simulation of Discontinuous Media*. TSI Press, University of California, Berkeley, 1996.

49. G.G.W. Mustoe, J.R. Williams and G. Hocking. Penetration and fracturing of brittle plates under dynamic impact. In G.N. Pande and J. Middleton, editors, *NUMETA 87, Numerical Methods in Engineering, Theory and Applications*. Martinus Nijhoff Publishers, Dordrecht, 1987.

50. G. Hocking, J.R. Williams and G.G.W. Mustoe. Dynamics analysis for three dimensional contact and fracturing of multiple bodies. In G.N. Pande and J. Middleton, editors, *NUMETA 87, Numerical Methods in Engineering, Theory and Applications*. Martinus Nijhoff Publishers, Dordrecht, 1987.

51. J.R. Williams, G.G.W. Mustoe and G. Hocking. Three dimensional analysis of multiple bodies including automatic fracturing. In D.R.J. Owen, E. Hinton and E. Onate, editors, *COMPLAS, 1st International Conference on Computational Plasticity*. Pineridge Press, Swansea, 1987.

52. G. Hocking. The DEM for analysis of fragmentation of discontinua. In G.G.W. Mustoe, M. Henriksen and H.P. Huttelmaier, editors, *1st U.S. Conference on Discrete Element Methods (DEM)*. Colorado School of Mines Press, Golden, 1989.

53. R. De Borst. Some recent issues in computational failure mechanics. *International Journal for Numerical Methods in Engineering*, 52:63–96, 2001.

54. H. Morikawa and Y. Sawamoto. Local fracture analysis of a reinforced slab by the discrete element method. In *2nd International Conference on Discrete Element Methods*, MIT IESL Publication, 1993.

55. D.V. Griffiths and G.G.W. Mustoe. Modelling of elastic continua using a grillage of structural elements based on discrete element concepts. *International Journal for Numerical Methods in Engineering*, 50:1759–1775, 2001.

56. T. Groger, U. Tuzun and D. Heyes. Shearing of wet particle systems discrete element simulations. In *1st International PFC Symposium*. Itasca, 2002.

57. E.J. Schlangen and J. van Mier. Experimental and numerical analysis of micromechanisms of fracture of cement based composites. *Cement and Concrete Composites*, 6:81–88, 1992.

58. M. Jirasek and Z.P. Bažant. Macroscopic fracture characteristics of random particle systems. *International Journal of Fracture*, 69:201–228, 1995.

59. G.A. D'Adetta, E. Ramm and F. Kun. Fracture simulations of cohesive frictional materials by discrete element models. In N. Bićanić, editor, *ICADD-4, 4th Conference on Analysis of Discontinuous Deformation*, University of Glasgow, 2001.

60. O.C. Zienkiewicz, R.L. Taylor and J.Z. Zhu. *The Finite Element Method: Its Basis and Fundamentals*. Butterworth-Heinemann, Oxford, 6th edition, 2005.

61. A. Munjiza, D.R.J. Owen and A.L. Crook. An $m(m^{-1}k)^m$ proportional damping in explicit integration of dynamic structural systems. *International Journal for Numerical Methods in Engineering*, 41:1277–1296, 1998.

62. T.J.R. Hughes. Analysis of transient algorithms with particular reference to stability behavior. In T. Belytschko and T.J.R. Hughes, editors, *Computational Methods for Transient Analysis*. North-Holland, Amsterdam, 1983.

63. D.M. Doolin and N. Sitar. Time integration in discontinuous deformation analysis. *Journal of Engineering Mechanics, ASCE*, 130:249–258, 2004.

64. C.S. Chang and C.B. Acheampong. Accuracy and stability for static analysis using dynamic formulation in discrete element methods. In *2nd International Conference on Discrete Element Methods*, MIT IESL Publication, 1993.

65. A. Anandarajah. Multiple time-stepping for the discrete element analysis of colloidal particles. *Powder Technology*, 106:132–141, 1999.

66. J. Ghaboussi, M.M. Basole and S. Ranjithan. Three dimensional discrete element analysis on massively parallel computers. In *2nd International Conference on Discrete Element Methods*, MIT IESL Publication, 1993.

67. D.R.J. Owen and Y.T. Feng. Parallelised finite/discrete element simulation of multi fracture solids and discrete systems. *Engineering Computations*, 18:557–576, 2001.

68. K. Meguro and H. Tagel-Din. A new simplified and efficient technique for fracture behaviour analysis of concrete structures. In *FRAMCOS-3 Intl Conference Fracture Mechanics of Concrete Structures*, 1999.

69. J. Rots. Computational modelling of concrete fracture. *Heron*, 34:1–59, 1988.

70. J.J. Moreau. Numerical aspects of the sweeping process. *Computer Methods in Applied Mechanics and Engineering*, 177:329–349, 1999.
71. M. Jean. The non-smooth contact dynamics method. *Computer Methods in Applied Mechanics and Engineering*, 177:235–257, 1999.
72. M. Jean, V. Acary and Y. Monerie. Non-smooth contact dynamics approach of cohesive materials. *Phil. Trans. The Royal Society of London*, 359:1–22, 2001.
73. C. Kane, E.A. Repetto, M. Ortiz and J.E. Marsden. Finite element analysis of non smooth contact. *Computer Methods in Applied Mechanics and Engineering*, 180:1–26, 1999.
74. J.M. Solberg and P. Papadopoulos. A simple finite element-based framework for the analysis of elastic pseudo-rigid bodies. *International Journal for Numerical Methods in Engineering*, 45:1297–1314, 1999.
75. P. Papadopoulos. On a class of higher order pseudo rigid bodies. *Mathematics and Mechanics of Solids*, 6:631–640, 2001.
76. D.M. Doolin. Preliminary results unifying discontinuous deformation analysis (DDA) and the distinct element method (DEM). In Ming Lu, editor, *Development and Application of Discontinuous Modelling for Rock Engineering, ICADD-6*. Swets & Zeitlinger, Lisse, 2004.
77. G.H. Shi. Numerical manifold method. In Y. Ohnishi, editor, *CAD-2 Conference on Analysis of Discontinuous Deformation*, Kyoto University, 1997.
78. G. Chen, Y. Ohnishi and T. Ito. Development of high order manifold method. In Y. Ohnishi, editor, *CAD-2 Conference on Analysis of Discontinuous Deformation*, Kyoto University, 1997.
79. T. Belytschko, Y. Krongauz, D. Organ, M. Fleming and P. Krysl. Meshless methods: an overview and recent developments. *Computer Methods in Applied Mechanics and Engineering*, 139:3–47, 1996
80. J.M. Melenk and I. Babuška. The partition of unity finite element method: basic theory and applications. *Computer Methods in Applied Mechanics and Engineering*, 139:289–314, 1996.
81. P.A. Cundall. A discontinuous future for numerical modelling in soil and rock. In B.K. Cook and P.J. Jensen, editors, *Discrete Element Methods: Numerical Modeling of Discontinua, 3rd International Conference on Discrete Element Methods*, volume ASCE Geotechnical Special Publication No. 117. The Geo-Institute of the American Society of Civil Engineers, 2002.

10

Structural mechanics problems in one dimension – rods

10.1 Introduction

In the previous chapters we considered the domain to be a continuum, a rigid multi-body system or a set of discrete elements. In the study of continuum problems we developed finite element approximations based on approximation of the displacement, stress and strain fields at each point in the domain. While such approximation is general there are instances when it is difficult to obtain viable solutions economically. Many such situations arise when one or two dimensions of the domain are small compared to the others. For example, when two dimensions are small we have a very slender cross-section which is translated along a one-dimensional axis as shown in Fig. 10.1. Such a form is herein called a *rod* and consists of a member which carries axial, shear, moment and torsion force resultants. When one dimension is small compared to the other two we have either a plate theory for initially flat surfaces or a shell theory for general curved surfaces. In this chapter we consider the behaviour of rods. Plate and shell problems will be considered in subsequent chapters.

Bending of rods is generally associated with a *beam* theory such as the classical Euler–Bernoulli theory studied in introductory strength of materials.[1–3] If one attempts to model a rod with a standard three-dimensional finite element model there are two aspects which give difficulty. One is purely numerical and associated with large round-off errors when attempting to solve the simultaneous equations.[4] The other is a new form of *locking* in interactions between bending, shear and axial behaviour when low-order elements are used. Often a much more economical solution is to use a *structural mechanics* approach in which the problem is formulated as a one-dimensional problem along the axis of the rod. Using this approach and appropriate interpolation forms one can avoid numerical difficulties associated with round-off and locking.

In this chapter we consider approximation for two classical rod theories. The first combines the Euler–Bernoulli theory of bending with axial and torsion theories. In this theory the deformation field is restricted to axial, torsion and bending strains. The application of the Euler–Bernoulli theory is usually restricted to situations where dimensions along the axis of the rod are at least ten times those of the transverse dimensions. In the second form we consider the Timoshenko theory of bending together with the axial and torsion theories. The Timoshenko theory adds a transverse shear deformation to the other strains and is applicable when the length to cross-section dimensions are above five (when smaller than this the continuum theory becomes viable).

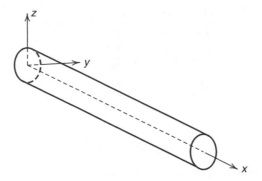

Fig. 10.1 Slender rod.

Care should always be used when measuring the length parameter. Distances between sudden changes in cross-section or loads should be considered. Also, when transient effects are included the frequency content of the solution will establish which natural 'modes' are active and the length between modal zeros also needs to be considered.

Use of the Timoshenko theory can lead to 'locking' effects when the theory is applied to cases where the Euler–Bernoulli theory also could be used. However, it is desirable to have a single formulation which remains valid throughout the range of length to cross-section considerations and for this the Timoshenko theory should be used. In this chapter we show how finite element approximations, which are free from locking effects, can be developed for the Timoshenko theory. These are also useful when we consider plate and shell problems in the later chapters. Indeed one of our goals for presenting the rod theory is to develop an understanding of the delicate nature of 'locking' when structural mechanics formulations, including those of plate and shell problems, are solved by finite element methods.

10.2 Governing equations

We consider a straight rod with the axis of definition in the x direction and the cross-section A in the $y–z$ plane (Fig. 10.1).[*]

In the study of rods we will assume that the primary stress components are the normal stress σ_x and shear stresses τ_{xy}, τ_{xz} acting on each cross-section. The remaining stresses, while they can exist, are of less importance and their effects are either ignored or are included as applied boundary loads to the rod as indicated next.

10.2.1 Equilibrium equations

We first consider the static behaviour of a rod where the local balance of momentum equations at each point of a body are expressed as

---

[*] In the study of structural mechanics formulations we shall revert to the notation where coordinates are denoted by x, y, z. A similar type of notation will also be introduced for displacements, strains and stresses.

$$\frac{\partial \sigma_x}{\partial x} + \frac{\partial \tau_{yx}}{\partial y} + \frac{\partial \tau_{zx}}{\partial z} + b_x = 0$$

$$\frac{\partial \tau_{xy}}{\partial x} + \frac{\partial \sigma_y}{\partial y} + \frac{\partial \tau_{zy}}{\partial z} + b_y = 0 \qquad (10.1a)$$

$$\frac{\partial \tau_{xz}}{\partial x} + \frac{\partial \tau_{yz}}{\partial y} + \frac{\partial \sigma_z}{\partial z} + b_z = 0$$

and

$$\tau_{xy} = \tau_{yx}; \quad \tau_{xz} = \tau_{zx} \quad \text{and} \quad \tau_{yz} = \tau_{zy} \qquad (10.1b)$$

The balance of momentum equations for a rod are obtained by integrations of Eq. (10.1a) over the cross-section A. Accordingly, for the equation in the x direction we write

$$\int_A \left[\frac{\partial \sigma_x}{\partial x} + \frac{\partial \tau_{yx}}{\partial y} + \frac{\partial \tau_{zx}}{\partial z} + b_x \right] dA = 0 \qquad (10.2)$$

We consider the case for which the cross-section is either constant or varies slowly enough along the x direction such that we can use

$$\int_A \frac{\partial f}{\partial x} \, dA \approx \frac{\partial}{\partial x} \int_A f \, dA; \quad \int_A \frac{\partial f}{\partial y} \, dA = \oint_S f \, n_y \, dS \quad \text{and} \quad \int_A \frac{\partial f}{\partial z} \, dA = \oint_S f \, n_z \, dS \qquad (10.3)$$

where S is the perimeter of the cross-section. Thus, after integration over the cross-section Eq. (10.2) is given by

$$\frac{\partial P}{\partial x} + q_x = 0 \qquad (10.4a)$$

where P is an *axial force* resultant and q_x an applied load in the x direction defined by

$$P = \int_A \sigma_x \, dA \quad \text{and} \quad q_x = \int_A b_x \, dA + \oint_S \left(\tau_{yx} n_y + \tau_{zx} n_z \right) dS, \qquad (10.4b)$$

respectively. Similarly, integrating the equations for the y and z directions gives

$$\frac{\partial S_y}{\partial x} + q_y = 0 \quad \text{and} \quad \frac{\partial S_z}{\partial x} + q_z = 0 \qquad (10.5a)$$

where

$$S_y = \int_A \tau_{yx} \, dA; \quad q_y = \int_A b_y \, dA + \oint_S \left(\sigma_y n_y + \tau_{zy} n_z \right) dS$$

$$S_z = \int_A \tau_{zx} \, dA; \quad q_z = \int_A b_z \, dA + \oint_S \left(\tau_{yz} n_y + \sigma_z n_z \right) dS \qquad (10.5b)$$

In the above S_y, S_z are *transverse shear force* resultants and q_y, q_z are applied loads per unit length.

Moment resultants acting on the rod are recovered from the pair of equations

$$\int_A \left\{ \begin{matrix} -y \\ z \end{matrix} \right\} \left[\frac{\partial \sigma_x}{\partial x} + \frac{\partial \tau_{yx}}{\partial y} + \frac{\partial \tau_{zx}}{\partial z} + b_x \right] dA = \left\{ \begin{matrix} 0 \\ 0 \end{matrix} \right\} \qquad (10.6)$$

Evaluating the integrals using Eq. (10.3) yields the relations

$$\frac{\partial M_z}{\partial x} + S_y + m_z = 0 \quad \text{and} \quad \frac{\partial M_y}{\partial x} - S_z + m_y = 0 \tag{10.7a}$$

where

$$M_z = -\int_A y\,\sigma_x\,dA; \quad m_z = -\int_A y\,b_x\,dA - \oint_S y\left(\tau_{yx}n_y + \tau_{zx}n_z\right)\,dS \tag{10.7b}$$

$$M_y = \int_A z\,\sigma_x\,dA; \quad m_y = \int_A z\,b_x\,dA + \oint_S z\left(\tau_{yx}n_y + \tau_{zx}n_z\right)\,dS$$

define the *bending moment* resultants M_y, M_z and loading couple per unit length m_y, m_z.

One final couple, a torque, can also act on a cross-section. The *torsion* resultant $M_z = T$ is computed from shear stresses acting on the cross-section as

$$T = \int_A \left(y\,\tau_{xz} - z\,\tau_{xy}\right)\,dA \tag{10.8}$$

Guided by the solution of the Saint-Venant torsion problem (see, for example, reference 5 and Sec. 7.5.1 of reference 6) we neglect σ_y, σ_z and τ_{yz} in the equilibrium equations, thus the integrated equilibrium equation can be obtained from the local form

$$\int_A \left[y \left(\frac{\partial \tau_{xz}}{\partial x} + b_z \right) - z \left(\frac{\partial \tau_{xy}}{\partial x} + b_y \right) \right] dA = 0 \tag{10.9}$$

This yields the result

$$\frac{\partial T}{\partial x} + m_x = 0 \tag{10.10a}$$

where m_x is a distributed loading torque per unit length given by

$$m_z = \int_A \left[y\,b_z - z\,b_y \right] dA \tag{10.10b}$$

10.2.2 Kinematics

The development of a structural mechanics theory of rods is based on the assumed kinematic field for each cross-section. Here we will consider only the simplest theories where it is assumed that a plane cross-section remains plane throughout the deformation process. Accordingly, we assume a displacement field expressed as

$$u_x(x, y, z) = u(x) + z\,\theta_y(x) - y\,\theta_z(x)$$
$$u_y(x, y, z) = v(x) - z\,\theta_x(x) \tag{10.11}$$
$$u_z(x, y, z) = w(x) + y\,\theta_x(x)$$

where u, v, w are displacements of the axis defining the rod and θ_x, θ_y, θ_z are small rotation angles about the coordinate axes. A warping function can be added to improve the kinematic approximation as shown by Klinkel and Govindjee;[7] however, this complicates the formulation and we choose instead to use the simple form shown above. Based on the displacements given in Eq. (10.11) we obtain the non-zero strains

$$
\varepsilon_x = \frac{\partial u}{\partial x} + z \frac{\partial \theta_y}{\partial x} - y \frac{\partial \theta_z}{\partial x} = \varepsilon(x) + z \chi_y(x) - y \chi_z(x)
$$

$$
2\,\varepsilon_{xy} = \left(\frac{\partial v}{\partial x} - \theta_z \right) - z \frac{\partial \theta_x}{\partial x} = \gamma_y(x) - z \chi_x(x) \tag{10.12}
$$

$$
2\,\varepsilon_{xz} = \left(\frac{\partial w}{\partial x} + \theta_y \right) + y \frac{\partial \theta_x}{\partial x} = \gamma_z(x) + y \chi_x(x)
$$

The strain measures χ_y and χ_z are related to the reciprocal radii of curvature of the initially straight bar in y and z directions, respectively. Since the radii of an initially straight bar are infinite, the χ_y, χ_z terms are in fact changes in curvature of the rod.

In the displacements assumed above we have ignored the actual distortion of the cross-section due to transverse shear and torsion resultants. It will be necessary to account for this omission later through the use of correction factors.

10.2.3 Transient behaviour

Using the assumed form for the displacement field, inertia effects may be included in the theory by replacing the body forces using the d'Alembert principle. Accordingly,

$$
\begin{aligned}
b_x &\to b_x - \rho\,(\ddot{u} + z\,\ddot{\theta}_y - y\,\ddot{\theta}_z) \\
b_y &\to b_y - \rho\,(\ddot{v} - z\,\ddot{\theta}_x) \\
b_z &\to b_z - \rho\,(\ddot{w} + y\,\ddot{\theta}_x)
\end{aligned} \tag{10.13}
$$

may be inserted into the relations defined in Sec. 10.2.1 to yield the momentum balance equations for the transient case as

$$
\begin{aligned}
\frac{\partial P}{\partial x} + q_x &= \rho A \left(\ddot{u} + \bar{z}\,\ddot{\theta}_y - \bar{y}\,\ddot{\theta}_z \right) \\
\frac{\partial S_y}{\partial x} + q_y &= \rho A \left(\ddot{v} - \bar{z}\,\ddot{\theta}_x \right) \\
\frac{\partial S_z}{\partial x} + q_z &= \rho A \left(\ddot{w} + \bar{y}\,\ddot{\theta}_x \right)
\end{aligned}
$$

for the force resultants. Placing the x axis at the centroid of the cross-section gives $\bar{y} = \bar{z} = 0$ and simplifies the above to

$$
\begin{aligned}
\frac{\partial P}{\partial x} + q_x &= \rho A\,\ddot{u} \\
\frac{\partial S_y}{\partial x} + q_y &= \rho A\,\ddot{v} \\
\frac{\partial S_z}{\partial x} + q_z &= \rho A\,\ddot{w}
\end{aligned} \tag{10.14}
$$

The balance of momentum for the couples is given by

$$\frac{\partial M_z}{\partial x} + S_y + m_z = \rho \left(I_z \ddot{\theta}_z - I_{yz} \ddot{\theta}_y \right)$$

$$\frac{\partial M_y}{\partial x} - S_z + m_y = \rho \left(I_y \ddot{\theta}_y - I_{yz} \ddot{\theta}_z \right) \tag{10.15}$$

$$\frac{\partial T}{\partial x} + m_x = \rho J \ddot{\theta}_x$$

where the area inertia array is given by

$$I_y = \int_A z^2 \, dA; \quad I_z = \int_A y^2 \, dA \quad \text{and} \quad I_{yz} = \int_A y \, z \, dA$$

and the polar inertia by

$$J = I_y + I_z$$

The orientation of the axes may always be placed such that $I_{yz} = 0$; in the sequel we assume this form to simplify the presentation of numerical approximations.

10.2.4 Elastic constitutive relations

Based on the assumption that v_y, σ_z, and τ_{yz} are zero and x defines a material symmetry direction, the constitutive equations for an orthotropic linear elastic material are given in matrix form by

$$\begin{Bmatrix} \sigma_x \\ \tau_{xy} \\ \tau_{xz} \end{Bmatrix} = \begin{bmatrix} E & 0 & 0 \\ 0 & G_y & 0 \\ 0 & 0 & G_z \end{bmatrix} \begin{Bmatrix} \varepsilon_x \\ 2\varepsilon_{xy} \\ 2\varepsilon_{xz} \end{Bmatrix} \tag{10.16}$$

where E is the elastic modulus in the x direction and G_y, G_z are the shear moduli for the y and z directions. Here we are interested in procedures for developing a finite element approximation to the theory and, thus, we assume the properties are constant in the rod.

Inserting the constitutive relations in the expressions for the axial force resultant and assuming the x axis at the cross-section centroid we obtain

$$P = \int_A E \left(\varepsilon(x) + z \chi_y(x) - y \chi_z(x) \right) dA = EA \, \varepsilon(x) \tag{10.17}$$

where EA is the *elastic axial stiffness* of the rod cross-section.

Repeating the above for the bending moments we obtain (assuming $I_{yz} = 0$)

$$M_y = EI_y \chi_y(x) \quad \text{and} \quad M_z = EI_z \chi_z(x) \tag{10.18}$$

In the above EI_y and EI_z are the *elastic bending stiffness* of the rod cross-section computed with respect to the centroid of the cross-section.

The relation for the torsion is expressed as

$$T = \kappa_x \, GJ \, \chi_x(x) \tag{10.19}$$

where κ_x is a *correction factor* accounting for the difference between the assumed displacement field and the true warping of the cross-section and

$$GJ = \int_A \left(G_z \, y^2 + G_y \, z^2 \right) \, dA$$

The term $\kappa_x GJ$ is the *elastic torsional stiffness* of the cross-section.

Finally, the constitutive behaviour for the shear resultants is given by

$$\begin{aligned} S_y &= \kappa_y \, G_y A \, \gamma_y \\ S_z &= \kappa_z \, G_z A \, \gamma_z \end{aligned} \tag{10.20}$$

where κ_y and κ_z are correction factors accounting for the cross-section distortion.

For the assumed form of the constitution and location of the x axis the behaviour of the rod separates into three distinct types of problems as shown in Table 10.1.

The equations given above for elastic rods constitute those for the Timoshenko theory of beams[8] combined with the axial and torsion problems. The Timoshenko theory includes transverse shearing strains as well as changes in curvature in the behaviour of the beam.

An alternative theory exists for which the transverse shear deformation becomes negligible and from $\gamma_y = \gamma_z \approx 0$ we may use

$$\theta_y = -\frac{\partial w}{\partial x} \quad \text{and} \quad \theta_z = \frac{\partial v}{\partial x} \tag{10.21}$$

In addition for the slender case the couple loading m_y, m_z and the rotatory inertia terms $\rho I_y \ddot{\theta}_y$, $\rho I_z \ddot{\theta}_z$ are also usually ignored in the equilibrium equations. This is generally called the *Euler–Bernoulli* theory of beams. Noting that the shear forces are now computed from the simple equilibrium relations

$$S_y = -\frac{\partial M_z}{\partial x} \quad \text{and} \quad S_z = \frac{\partial M_y}{\partial x}$$

the relations for the beam may be written as shown in Table 10.2.

In the next section we consider the weak form for each type.

Table 10.1 Equations for a rod with Timoshenko beam theory for bending

Type	Momentum balance	Strain–displacement	Constitution
1. Axial:	$\dfrac{\partial P}{\partial x} + q_x = \rho A \, \ddot{u};$	$\varepsilon(x) = \dfrac{\partial u}{\partial x};$	$P = EA \, \varepsilon(x)$
2. Torsion:	$\dfrac{\partial T}{\partial x} + m_x = \rho J \, \ddot{\theta}_x;$	$\chi_x = \dfrac{\partial \theta_x}{\partial x};$	$T = \kappa_x \, GJ \, \chi_x(x)$
3. Bending:	$\dfrac{\partial S_y}{\partial x} + q_y = \rho A \, \ddot{v};$	$\gamma_y = \dfrac{\partial v}{\partial x} - \theta_z;$	$S_y = \kappa_y \, G_y A \, \gamma_y$
	$\dfrac{\partial M_z}{\partial x} + S_y + m_z = \rho I_z \, \ddot{\theta}_z;$	$\chi_z = \dfrac{\partial \theta_z}{\partial x};$	$M_z = EI_z \, \chi_z(x)$
	$\dfrac{\partial S_z}{\partial x} + q_z = \rho A \, \ddot{w};$	$\gamma_z = \dfrac{\partial w}{\partial x} + \theta_y;$	$S_z = \kappa_z \, G_z A \, \gamma_z$
	$\dfrac{\partial M_y}{\partial x} - S_z + m_y = \rho I_y \, \ddot{\theta}_y;$	$\chi_y = \dfrac{\partial \theta_y}{\partial x};$	$M_y = EI_y \, \chi_y(x)$

Table 10.2 Equations for Euler–Bernoulli beam theory

Momentum balance	Strain–displacement	Constitution
$-\dfrac{\partial^2 M_z}{\partial x^2} + q_y = \rho\, A\, \ddot{v};$	$\chi_z = \dfrac{\partial^2 v}{\partial x^2};$	$M_z = E I_z\, \chi_z(x)$
$\dfrac{\partial^2 M_y}{\partial x} + q_z = \rho\, A\, \ddot{w};$	$\chi_y = -\dfrac{\partial^2 w}{\partial x^2};$	$M_y = E I_y\, \chi_y(x)$

10.3 Weak (Galerkin) forms for rods

For the form given above we can construct weak (Galerkin) forms individually for each of the problem types. Each of the forms given below is initially given in terms of a three-field variational form of Hu–Washizu type. Although each of the forms given could include displacement boundary conditions as natural conditions [i.e. see Eq. (1.38)], we choose to enforce these as essential conditions as described in Chapter 2. The three-field form will be used to construct *mixed* finite element approximations for the rod. From the three-field form a two-field mixed form of Hellenger–Reissner type and an irreducible form of displacement type may also be obtained for use in finite element approximations.

10.3.1 Axial weak form

For the axial behaviour we write a three-field variational form as

$$\delta\Pi_\varepsilon(u, P, \varepsilon) = \int_L \left[\delta\varepsilon\, (\hat{P}(\varepsilon) - P) + \delta u\, (\rho A\, \ddot{u} - q_x) \right] dx$$

$$+ \int_L \left[\delta P \left(\frac{\partial u}{\partial x} - \varepsilon \right) + \frac{\partial \delta u}{\partial x}\, P \right] dx - \left(\delta u\, \bar{P} \right)\Big|_{\partial L_P} = 0$$

(10.22)

where δu, $\delta\varepsilon$ and δP are arbitrary functions over the length of the rod L; ∂L_P denotes the ends of the bar at which the axial force is specified and $P = \hat{P}(\varepsilon)$ is the constitutive equation. After integration by parts we recover all the equations for the axial behaviour except the displacement boundary condition $u = \bar{u}$ on ∂L_u and the initial conditions $u(x, 0) = \bar{u}_0$ and $\dot{u}(x, 0) = \bar{v}_0$.

As noted in Chapter 1 we can recover a Hellinger–Reissner type form provided we can find the inverse constitutive relation such that

$$P = \hat{P}(\varepsilon) \quad \rightarrow \quad \varepsilon = \hat{\varepsilon}(P)$$

(10.23)

With this substitution Eq. (10.22) becomes

$$\delta\Pi_P(u, P) = \int_L \left[\delta P \frac{\partial u}{\partial x} + \frac{\partial \delta u}{\partial x}\, P + \delta u\, (\rho A\, \ddot{u} - q_x) \right] dx$$

$$- \int_L \delta P\, \hat{\varepsilon}(P)\, dx - \left(\delta u\, \bar{P} \right)\Big|_{\partial L_P} = 0$$

(10.24)

An irreducible form for the axial behaviour is obtained from Eq. (10.22) by enforcing the strain-displacement equation at each point of the rod. Accordingly, we substitute

$$\varepsilon = \frac{\partial u}{\partial x} = \breve{\varepsilon} \quad \text{and} \quad \delta\varepsilon = \frac{\partial \delta u}{\partial x} = \delta\breve{\varepsilon}$$

into Eq. (10.22) and obtain

$$\delta\Pi_u(u) = \int_L \left[\frac{\partial \delta u}{\partial x} \hat{P}(\breve{\varepsilon}) + \delta u \left(\rho A\, \ddot{u} - q_x \right) \right] dx - \left(\delta u\, \bar{P} \right)\Big|_{\partial L_P} = 0 \qquad (10.25)$$

Boundary conditions

The boundary conditions for the axial response involve specification of the force condition $P = \bar{P}$ or the displacement condition $u = \bar{u}$ at each end of the bar. In addition it is often required to impose a symmetry condition as $u = 0$ or an asymmetry condition $P = 0$ when solving problems.

10.3.2 Torsion weak form

For the torsion behaviour we write a three-field variational form as

$$\delta\Pi_T(\theta_x, T, \chi_x) = \int_L \left[\delta\chi_x \left(\hat{T}(\chi_x) - T \right) + \delta\theta_x \left(\rho J\, \ddot{\theta}_x - m_x \right) \right] dx$$

$$+ \int_L \left[\delta T \left(\frac{\partial \theta_x}{\partial x} - \chi_x \right) + \frac{\partial \delta\theta_x}{\partial x} T \right] dx - \left(\delta\theta_x\, \bar{T} \right)\Big|_{\partial L_T} = 0$$

$$(10.26)$$

where $\delta\theta_x$, $\delta\chi_x$ and δT are arbitrary functions over the length of the rod L; and ∂L_T denotes the ends of the bar at which the torque is specified. After integration by parts we recover all the equations for the torsion behaviour except the displacement boundary condition $\theta_x = \bar{\theta}_x$ on ∂L_x and the initial conditions $\theta_x(x, 0) = \bar{\theta}_{x0}$ and $\dot{\theta}_x(x, 0) = \bar{\theta}_{x0}$.

Similar to the axial problem, we may recover a Hellinger–Reissner form by finding an inverse constitutive relation of the form

$$\chi_x = \hat{\chi}_x(T) \qquad (10.27)$$

Alternatively, an irreducible form for the torsion behaviour may be obtained by enforcing the strain–displacement equation at each point of the rod. Accordingly, we substitute

$$\chi_x = \frac{\partial \theta_x}{\partial x} = \breve{\chi}_x \quad \text{and} \quad \delta\chi_x = \frac{\partial \delta\theta_x}{\partial x} = \delta\breve{\chi}_x$$

into Eq. (10.26) and obtain

$$\delta\Pi_{\theta_x}(\theta_x) = \int_L \left[\frac{\partial \delta\theta_x}{\partial x} \hat{T}(\breve{\chi}_x) + \delta\theta_x \left(\rho J\, \ddot{\theta}_x - q_x \right) \right] dx - \left(\delta\theta_x\, \bar{T} \right)\Big|_{\partial L_T} = 0 \quad (10.28)$$

Boundary conditions

The boundary conditions for the torsion response are identical to those for the axial problem, namely a force condition $T = \bar{T}$, a displacement condition $\theta_x = \bar{\theta}_x$, a symmetry condition as $\theta_x = 0$ or an asymmetry condition $T = 0$.

10.3.3 Bending weak forms

For the bending problem we have two cases to consider: the Euler–Bernoulli beam theory and the Timoshenko beam theory.

As given above the bending of the rod occurs in two planes but, with the exception of the signs, leads to identical sets of equations. Accordingly, in the following we consider only the bending problem for the x–z plane in terms of the displacements w, θ_y, the forces M_y, S_z and strains χ_y, γ_z (with $\gamma_z = 0$ for the Euler–Bernoulli problem). The behaviour in the x–y plane is obtained by making appropriate changes in the notation and signs, but otherwise is identical.

Euler–Bernoulli beam

We begin by considering the Euler–Bernoulli theory since, like axial force and torsion problems above, the weak form depends on one resultant M_y, one strain measure χ_y, and one displacement component w. The three-field weak form for the Euler–Bernoulli theory is given by

$$\delta\Pi_{\chi_y}(w, M_y, \chi_y) = \int_L \left[\delta\chi_y\, (\hat{M}_y(\chi_y) - M_y) + \delta w\, (\rho A\, \ddot{w} - q_z) \right] dx$$

$$- \int_L \left[\delta M_y \left(\frac{\partial^2 w}{\partial x^2} + \chi_y \right) + \frac{\partial^2 \delta w}{\partial x^2}\, M_y \right] dx \qquad (10.29)$$

$$+ \left(\frac{\partial \delta w}{\partial x}\, \bar{M}_y \right)\Big|_{\partial M_y} - \left(\delta w\, \bar{S}_z \right)\Big|_{\partial S_z} = 0$$

This weak form differs from the previous forms in two ways. First, the strain measure involves second derivatives and hence for the form given in Eq. (10.29) will require C^1 continuous interpolation for the finite element solution. Second, there are two boundary conditions at each end of a beam of length L.

A Hellinger–Reissner form may be obtained by expressing the curvature in terms of the bending moment as

$$\chi_y = \hat{\chi}_y(M_y) \qquad (10.30)$$

and substituting the result into Eq. (10.29) to give

$$\delta\Pi_{M_y}(w, M_y) = \int_L \left[-\delta M_y\, \frac{\partial^2 w}{\partial x^2} - \frac{\partial^2 \delta w}{\partial x^2}\, M_y + \delta w\, (\rho A\, \ddot{w} - q_z) \right] dx$$

$$- \int_L \delta M_y\, \hat{\chi}_y(M_y)\, dx + \left(\frac{\partial \delta w}{\partial x}\, \bar{M}_y \right)\Big|_{\partial M_y} - \left(\delta w\, \bar{S}_z \right)\Big|_{\partial S_z} = 0$$

$$(10.31)$$

An irreducible form for the Euler–Bernoulli beam is obtained by substituting

$$\chi_y = -\frac{\partial^2 w}{\partial x^2} = \check{\chi}_y \quad \text{and} \quad \delta\chi_y = -\frac{\partial^2 \delta w}{\partial x^2} = \delta\check{\chi}_y$$

into Eq. (10.29) to obtain

$$\delta\Pi_w(w) = -\int_L \left[\frac{\partial^2 \delta w}{\partial x^2}\,\hat{M}_y(\check{\chi}_y) + \delta w\,(\rho A\,\ddot{w} - q_z)\right] dx$$

$$+ \left(\frac{\partial \delta w}{\partial x}\,\bar{M}_y\right)\Big|_{\partial M_y} - \left(\delta w\,\bar{S}_z\right)\Big|_{\partial S_z} = 0$$

(10.32)

Note for the linear elastic form described in Eq. (10.18) and Table 10.1

$$M_y = \hat{M}_y(\check{\chi}_y) = EI_y\,\check{\chi}_y = -EI_y\frac{\partial^2 w}{\partial x^2}$$

and, thus, when $\delta w = w$ the first term in Eq. (10.32) is positive as required for a form obtained from the minimum potential energy principle.

The options for boundary conditions for bending are more numerous than for the axial or torsion problem since there are two conditions to be imposed at each end of the bar. The options are

Natural condition   Essential condition

$$S_z = \bar{S}_z; \qquad w = \bar{w}$$

$$M_y = \bar{M}_y; \qquad \frac{\partial w}{\partial x} = -\bar{\theta}_y$$

These conditions may be combined to form typical boundary types described in Table 10.3.

Table 10.3   Typical boundary conditions for Euler–Bernoulli and Timoshenko beam bending

BC type	Symbol used	Euler–Bernoulli condition		Timoshenko condition	
Fixed:		$w = \bar{w};$	$\frac{\partial w}{\partial x} = -\bar{\theta}_y$	$w = \bar{w};$	$\theta_y = \bar{\theta}_y$
Pinned:		$w = \bar{w};$	$M_y = \bar{M}_y$	$w = \bar{w};$	$M_y = \bar{M}_y$
Roller:		$w = \bar{w};$	$M_y = \bar{M}_y$	$w = \bar{w};$	$M_y = \bar{M}_y$
Free:		$S_z = \bar{S}_z;$	$M_y = \bar{M}_y$	$S_z = \bar{S}_z;$	$M_y = \bar{M}_y$
Symmetry:		$S_z = 0;$	$\frac{\partial w}{\partial x} = -\bar{\theta}_y$	$S_z = 0;$	$\theta_y = \bar{\theta}_y$
Asymmetry:		$w = 0;$	$M_y = 0$	$w = 0;$	$M_y = 0$

Timoshenko beam

The three-field weak form for the Timoshenko beam theory involves two displacement components, two forces and two strains as given by

$$
\delta\Pi_{\gamma_z}(w,\theta_y,S_z,M_y,\gamma_z,\chi_y) = \int_L \left[\delta\chi_y\,(\hat{M}_y(\chi_y,\gamma_z) - M_y) + \delta\gamma_z\,(\hat{S}_z(\chi_y,\gamma_z) - S_z)\right]\mathrm{d}x
$$
$$
+ \int_L \left[\delta w\,(\rho A\,\ddot{w} - q_z) + \delta\theta_y\,(\rho I_y\,\ddot{\theta}_y - m_y)\right]\mathrm{d}x
$$
$$
+ \int_L \left[\delta M_y\left(\frac{\partial\theta_y}{\partial x} - \chi_y\right) + \delta S_z\left(\frac{\partial w}{\partial x} + \theta_y - \gamma_z\right)\right]\mathrm{d}x
$$
$$
+ \int_L \left[\frac{\partial\delta\theta_y}{\partial x}\,M_y + \left(\frac{\partial\delta w}{\partial x} + \delta\theta_y\right)S_z\right]\mathrm{d}x
$$
$$
- \left(\delta\theta_y\,\bar{M}_y\right)\Big|_{\partial M_y} - \left(\delta w\,\bar{S}_z\right)\Big|_{\partial S_z} = 0
$$

$$(10.33)$$

For this formulation the reduction to a Hellinger–Reissner form requires determination of a pair of inverse relations such that

$$
\chi_y = \hat{\chi}_y(M_y,S_z) \quad\text{and}\quad \gamma_z = \hat{\gamma}_z(M_y,S_z) \tag{10.34}
$$

This greatly increases the complexity of the problem, and except for the linear elastic problem, has not been used extensively. For this form

$$
\delta\Pi_{\gamma_z}(w,\theta_y,S_z,M_y,\gamma_z,\chi_y) = \int_L \left[\delta w\,(\rho A\,\ddot{w} - q_z) + \delta\theta_y\,(\rho I_y\,\ddot{\theta}_y - m_y)\right]\mathrm{d}x
$$
$$
+ \int_L \left[\delta M_y\,\frac{\partial\theta_y}{\partial x} + \delta S_z\left(\frac{\partial w}{\partial x} + \theta_y\right)\right]\mathrm{d}x
$$
$$
+ \int_L \left[\frac{\partial\delta\theta_y}{\partial x}\,M_y + \left(\frac{\partial\delta w}{\partial x} + \delta\theta_y\right)S_z\right]\mathrm{d}x \tag{10.35}
$$
$$
- \int_L \left[\delta M_y\,\hat{\chi}_y(M_y,S_z) + \delta S_z\,\hat{\gamma}_z(M_y,s_z)\right]\mathrm{d}x
$$
$$
- \left(\delta\theta_y\,\bar{M}_y\right)\Big|_{\partial M_y} - \left(\delta w\,\bar{S}_z\right)\Big|_{\partial S_z} = 0
$$

An irreducible form is obtained by an exact satisfaction of the strain–displacement equations, as described previously for other problem forms. Accordingly, we obtain for the Timoshenko beam the relation

$$
\delta\Pi_{\theta_y}(w,\theta_y) = \int_L \left[\frac{\partial\delta\theta_y}{\partial x}\,\hat{M}_y(\tilde{\chi}_y,\tilde{\gamma}_z) + \left(\frac{\partial\delta w}{\partial x} + \delta\theta_y\right)\hat{S}_z(\tilde{\chi}_y,\tilde{\gamma}_z)\right]\mathrm{d}x
$$
$$
+ \int_L \left[\delta w\,(\rho A\,\ddot{w} - q_z) + \delta\theta_y\,(\rho I_y\,\ddot{\theta}_y - m_y)\right]\mathrm{d}x \tag{10.36}
$$
$$
- \left(\delta\theta_y\,\bar{M}_y\right)\Big|_{\partial M_y} - \left(\delta w\,\bar{S}_z\right)\Big|_{\partial S_z} = 0
$$

where $\tilde{\chi}_y = \partial\theta_y/\partial x$ and $\tilde{\gamma}_z = \partial w/\partial x + \theta_y$.

Boundary conditions

The options for boundary conditions are similar to those for the Euler–Bernoulli theory except θ_y replaces $\partial w/\partial x$. The options are

$$\begin{array}{cc}
\text{Natural condition} & \text{Essential condition} \\
S_z = \bar{S}_z; & w = \bar{w} \\
M_y = \bar{M}_y; & \theta_y = \bar{\theta}_y
\end{array}$$

These conditions may be combined to form the same boundary condition types as in the Euler–Bernoulli theory as described in Table 10.3.

10.4 Finite element solution: Euler–Bernoulli rods

The finite element solution of rod problems involves finding one-dimensional interpolations to approximate each of the functions appearing in the weak forms. We begin by considering the Euler–Bernoulli bending problem since, as we will show later, it has implications on the approximations for the axial behaviour of the rod when either non-linear material behaviour or large displacements are considered.

10.4.1 Beam bending – irreducible form

The Euler–Bernoulli beam is the first case we have considered which requires C^1 continuity in the approximation for the transverse displacement w. In all previous problems considered we were able to use C^0 approximation. If we first consider the static behaviour where $\ddot{w}$ is negligible we can find interpolations which give *exact* solutions at the interelement nodes. Subsequently these may be used in transient analysis; however then the solution will no longer be exact at the interelement nodes.

In order to obtain exact interelement nodal solutions for ordinary differential equations the interpolation functions for the *weight function* δw must satisfy the exact solution of the adjoint differential equation of the weak form (see Tong[9] or Appendix H in reference 6). The adjoint equation is obtained by integrations on the weak form until all derivatives of the solution variable w are removed. For the case where the linear elastic constitutive equation (10.18) is used and EI_y is constant the adjoint equation is given by

$$EI_y \frac{\mathrm{d}^4 \delta w}{\mathrm{d}x^4} = 0 \tag{10.37}$$

and the exact solution is a cubic polynomial in x. In a Galerkin solution we will use the same form of interpolation for both δw and w.

Hermite interpolation for beam displacement

To construct cubic functions for each element we express the interpolation in the natural coordinate ξ and use the coordinate interpolation

$$x = \tfrac{1}{2}(1-\xi)\tilde{x}_1 + \tfrac{1}{2}(1+\xi)\tilde{x}_2 = N_1(\xi)\tilde{x}_1 + N_2(\xi)\tilde{x}_2 \tag{10.38}$$

with $-1 \le \xi \le 1$ and $\tilde{x}_a$ the coordinate for node a.

Since second derivatives of w appear in the weak form (10.32) the interpolation for w must have C^1 continuity, indicating that both the function and its first derivative must be continuous in the entire domain L. The most logical manner to satisfy this requirement is to use *Hermite interpolation* forms.[10] A Hermitian polynomial of order $2n + 1$ is written for a segment in terms of end values of the function and its first n-order derivatives, thus making it trivial to satisfy C^n continuity. The lowest-order Hermite interpolation involves a complete polynomial of cubic order with the desired C^1 continuity needed for the Euler–Bernoulli beam shape functions. The interpolation in terms of ξ is given by

$$w(\xi, t) = \sum_{a=1}^{2} \left[H_a^{(0)}(\xi)\, \tilde{w}_a(t) + H_a^{(1)}(\xi)\, \frac{\partial \tilde{w}_a}{\partial x}(t) \right] = \mathbf{N}^b(\xi)\, \tilde{\mathbf{u}}^b(t) \tag{10.39}$$

where the superscript b denotes bending,

$$H_a^{(0)}(\xi) = \frac{1}{4} \begin{cases} 2 - 3\xi + \xi^3; & a = 1 \\ 2 + 3\xi - \xi^3; & a = 2 \end{cases};$$

$$\tag{10.40a}$$

$$H_a^{(1)}(\xi) = \frac{L_e}{8} \begin{cases} 1 - \xi - \xi^2 + \xi^3; & a = 1 \\ -1 - \xi + \xi^2 + \xi^3; & a = 2 \end{cases}$$

with

$$\mathbf{N}^b(\xi) = \left[H_1^{(0)}, H_1^{(1)}, H_2^{(0)}, H_2^{(1)} \right]; \quad \mathbf{u}^b = \left[\tilde{w}_1, \frac{\partial \tilde{w}_1}{\partial x}, \tilde{w}_2, \frac{\partial \tilde{w}_2}{\partial x} \right]^T$$

and $L_e = \tilde{x}_2 - \tilde{x}_1$ the element length.

The chain rule

$$\frac{\partial w}{\partial \xi} = \frac{\partial w}{\partial x} \frac{\partial x}{\partial \xi} = \frac{1}{2} L_e \frac{\partial w}{\partial x}$$

may be used to compute derivatives of w as

$$\frac{\partial w}{\partial x} = \frac{2}{L_e} \frac{\partial w}{\partial \xi}; \quad \frac{\partial^2 w}{\partial x^2} = \frac{4}{L_e^2} \frac{\partial^2 w}{\partial \xi^2} \quad \text{and} \quad \frac{\partial^3 w}{\partial x^3} = \frac{8}{L_e^3} \frac{\partial^3 w}{\partial \xi^3}$$

and yields

$$\frac{\partial H_a^{(0)}}{\partial x} = \frac{3}{2 L_e} \begin{cases} -1 + \xi^2; & a = 1 \\ 1 - \xi^2; & a = 2 \end{cases} \quad \text{and} \quad \frac{\partial H_a^{(1)}}{\partial x} = \frac{1}{4} \begin{cases} -1 - 2\xi + 3\xi^2; & a = 1 \\ -1 + 2\xi + 3\xi^2; & a = 2 \end{cases}$$

$$\tag{10.40b}$$

for first derivatives,

$$\frac{\partial^2 H_a^{(0)}}{\partial x^2} = \frac{6}{L_e^2} \begin{cases} \xi; & a = 1 \\ -\xi; & a = 2 \end{cases} \quad \text{and} \quad \frac{\partial^2 H_a^{(1)}}{\partial x^2} = \frac{1}{L_e} \begin{cases} -1 + 3\xi; & a = 1 \\ 1 + 3\xi; & a = 2 \end{cases}$$

$$\tag{10.40c}$$

for second derivatives and

$$\frac{\partial^3 H_a^{(0)}}{\partial x^3} = \frac{12}{L_e^3} \left\{ \begin{array}{ll} 1; & a = 1 \\ -1; & a = 2 \end{array} \right. \quad \text{and} \quad \frac{\partial^3 H_a^{(1)}}{\partial x^3} = \frac{6}{L_e^2} \left\{ \begin{array}{ll} 1; & a = 1 \\ 1; & a = 2 \end{array} \right. \tag{10.40d}$$

for third derivatives. The third derivatives are only needed to compute shear force in the element. While the C^1 continuity is easily observed from Figures 10.2 and 10.3, we note that the second and third derivatives are discontinuous and can adversely affect the accuracy of moments and shears in the elements.

Strain-displacement relations

Using the Hermite interpolations in Eq. (10.40c) the finite element interpolation for the bending strain is given by

$$\chi_y = -\frac{\partial^2 w}{\partial x^2} = -\sum_{a=1}^{2} \left[\frac{\partial^2 H_a^{(0)}}{\partial x^2} \quad \frac{\partial^2 H_a^{(1)}}{\partial x^2} \right] \left\{ \begin{array}{c} \tilde{w}_a \\ \partial \tilde{w}_a \\ \overline{\partial x} \end{array} \right\} = \mathbf{B}^b \, \tilde{\mathbf{u}}^b \tag{10.41}$$

in which $\mathbf{B}^b$ is the strain–displacement matrix for bending.

Stiffness and load arrays

If we consider the weak form (10.28) for an element of length L_e we obtain

$$\delta \Pi_w^e = \delta \tilde{\mathbf{u}}^{bT} \int_{L_e} \left[\mathbf{B}^{bT} \, \hat{M}_y (\mathbf{B}^b \tilde{\mathbf{u}}^b) \, dx + \mathbf{N}^{bT} \left(\rho A \, \mathbf{N}^b \ddot{\tilde{\mathbf{u}}}^b - q_z \right) dx \right]$$
$$= \delta \tilde{\mathbf{u}}^{bT} \left[\mathbf{P}^b + \mathbf{M}^b \, \ddot{\tilde{\mathbf{u}}}^b - \mathbf{f}^b \right] \tag{10.42}$$

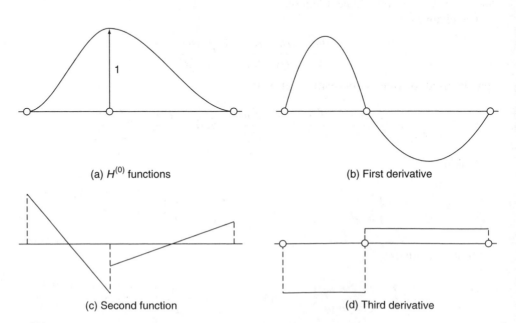

(a) $H^{(0)}$ functions

(b) First derivative

(c) Second function

(d) Third derivative

Fig. 10.2   Hermite interpolation on two elements – $H^{(0)}$ functions.

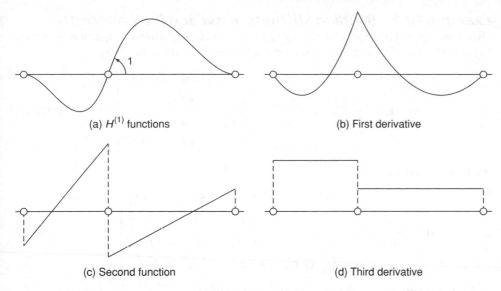

(a) $H^{(1)}$ functions   (b) First derivative

(c) Second function   (d) Third derivative

Fig. 10.3 Hermite interpolation on two elements – $H^{(1)}$ functions.

where $\mathbf{P}^b$ is the stress divergence vector, $\mathbf{M}^b$ is the mass matrix and $\mathbf{f}^b$ is the force vector for the element. For a given constitutive behaviour a Newton method may be used to linearize the expression for moment. For a local constitutive equation that has the linearized form

$$d\sigma_x = E_T\, d\varepsilon_x$$

where E_T is the tangent modulus, the linearized moment is given by

$$dM_y = -EI_y \frac{\partial^2 dw}{\partial x^2} = EI_y \mathbf{B}^b d\tilde{\mathbf{u}}^b$$

Thus applying a Newton method to Eq. (10.42) yields the form

$$d(\delta\Pi)^e_w = \delta\tilde{\mathbf{u}}^{bT} \left[\int_{L_e} \left(\mathbf{B}^{bT}\, EI_y \mathbf{B}^b + c_M\, \mathbf{N}^{bT}\, \rho A\, \mathbf{N}^b \right) dx \right] d\tilde{\mathbf{u}}^b$$

$$= \delta\tilde{\mathbf{u}}^{bT} \left[\mathbf{K}^b_T + c_M\, \mathbf{M}^b \right] d\tilde{\mathbf{u}}^b$$

where

$$d\ddot{\tilde{\mathbf{u}}} = c_M\, d\tilde{\mathbf{u}}$$

is obtained from the time integration scheme used (e.g. GN22 in Chapter 2).

Example 10.1   Bending stiffness, mass and load matrices

Using the approximation (10.42) for a linear elastic constitutive equation given by Eq. (10.18) with EI_y constant, the element stiffness matrix is given by

$$\mathbf{K}^e = \frac{2\,EI_y}{L_e^3} \begin{bmatrix} 6 & -3\,L_e & -6 & -3\,L_e \\ -3\,L_e & 2\,L_e^2 & 3\,L_e & L_e^2 \\ -6 & 3\,L_e & 6 & 3\,L_e \\ -3\,L_e & L_e^2 & 3\,L_e & 2\,L_e^2 \end{bmatrix} \tag{10.43a}$$

the mass matrix by

$$\mathbf{M}^e = \frac{1}{420}\,\rho A L_e \begin{bmatrix} 156 & 22 L_e & 54 & -13\,L_e \\ 22\,L_e & 4\,L_e^2 & 13\,L_e & -3\,L_e^2 \\ 54 & 13\,L_e & 156 & 22\,L_e \\ -13\,L_e & -3\,L_e^2 & 22\,L_e & 4\,L_e^2 \end{bmatrix} \tag{10.43b}$$

and for constant q_z over the element the load vector by

$$\mathbf{f}^e = \frac{1}{12}\,q_z\,L_e^2 \begin{bmatrix} 6 L_e, & -1, & 6 L_e, & 1 \end{bmatrix}^T \tag{10.43c}$$

10.4.2 Axial deformation – irreducible form

For a linearly elastic rod the axial constitutive behaviour is given by

$$P = EA\,\frac{\partial u}{\partial x}$$

For the case where EA is constant the adjoint differential equation obtained for the static problem using Eq. (10.25) is given by

$$EA\,\frac{\partial^2 \delta u}{\partial x^2} = 0 \tag{10.44}$$

which yields a linear solution in the rod. This is adequate provided the rod is linearly elastic. However, we note that both bending and axial behaviour of a rod depend on the axial stress σ_x, consequently it is important that the approximation for axial behaviour has strains which are of the same order as those for bending. In the development of the irreducible solution for the Euler–Bernoulli theory the bending strains χ_y varied linearly along the length of each element. To have compatible behaviour we need to have the axial strain ε_x vary linearly also. Accordingly, we use the hierarchic interpolation

$$\delta u = \tfrac{1}{2}(1 - \xi)\,\delta\tilde{u}_1 + \tfrac{1}{2}(1 + \xi)\,\delta\tilde{u}_2 + (1 - \xi^2)\,\Delta(\delta\tilde{u}_3) = N^a\,\delta\tilde{u}^a \tag{10.45}$$

where $\tilde{u}^a = (\tilde{u}_1, \tilde{u}_2, \tilde{u}_3)^T$. Equation (10.38) is again employed for the x coordinate interpolation. The same form of interpolation is used for u. Using this interpolation the axial strain in an element of length L_e is given by

$$\varepsilon = \frac{1}{L_e}(\tilde{u}_2 - \tilde{u}_1 - 4\xi \Delta\tilde{u}_3) = \mathbf{B}^a \tilde{\mathbf{u}}^a \qquad (10.46)$$

Substituting the above results into the weak form (10.25) yields

$$\delta\Pi_u^e = \delta\tilde{\mathbf{u}}^{aT} \left[\left(\int_{L_e} \mathbf{B}^{aT} \mathbf{D}^a \mathbf{B}^a \, dx \right) \tilde{\mathbf{u}}^a - \int_{L_e} \mathbf{N}^{aT} q_x \, dx \right] \qquad (10.47)$$

in which $\mathbf{D}^a = [EA]$ is the axial stiffness matrix.

Example 10.2   Axial stiffness and load arrays
If EA is constant in L_e the element stiffness is given by

$$\mathbf{K}^a = \frac{EA}{3L_e} \begin{bmatrix} 3 & 0 & 0 \\ 0 & 3 & 0 \\ 0 & 0 & 16 \end{bmatrix}$$

and for constant q_x the element load by

$$\mathbf{f}^a = \tfrac{1}{6} q_x L_e \begin{Bmatrix} 3 \\ 3 \\ 6 \end{Bmatrix}$$

We note for the static linearly elastic problem that $\Delta\tilde{u}_3$ depends only on the distribution of the load on the element and does not affect the values $\tilde{u}_1$ and $\tilde{u}_2$, thus for uniform cross-sections the interelement nodal values will be exact. If EA varies along the element or the material is non-linear this will not be true and $\Delta\tilde{u}_3$ will interact with the end values, which will then no longer be exact.

10.4.3 Torsion deformation – irreducible form

The constitutive equation for the torsional behaviour of a linear elastic rod is given by

$$T = \kappa_x GJ \frac{\partial\theta_x}{\partial x}$$

For a constant $\kappa_x GJ$ along the length of an element the adjoint differential equation obtained for the static behaviour is determined from Eq. (10.28) to be

$$\kappa_x GJ \frac{\partial^2 \delta\theta_x}{\partial x^2} = 0$$

which has the linear solution given by

$$\delta\theta_x = \tfrac{1}{2}(1-\xi)\,\delta\tilde{\theta}_{x1} + \tfrac{1}{2}(1+\xi)\,\delta\tilde{\theta}_{x2}$$

An identical expression is used to represent θ_x in terms of $\tilde{\theta}_{x1}$ and $\tilde{\theta}_{x2}$. Note that a linear form gives a constant strain which is identical to the form obtained for shear strains from the Hermite interpolation in the Euler–Bernoulli theory.

Inserting the above interpolations into the weak form (10.28) for a single element gives

$$\delta\Pi_{\theta_x}^e = \begin{bmatrix} \delta\tilde{\theta}_{x1} & \delta\tilde{\theta}_{x2} \end{bmatrix} \left\{ \frac{1}{2L_e} \begin{bmatrix} -1 \\ 1 \end{bmatrix} \left(\int_{-1}^{1} \kappa_x GJ(\xi)\, d\xi \right) \begin{bmatrix} -1 & 1 \end{bmatrix} \right\} \begin{Bmatrix} \tilde{\theta}_{x1} \\ \tilde{\theta}_{x2} \end{Bmatrix}$$

$$- \begin{bmatrix} \delta\tilde{\theta}_{x1} & \delta\tilde{\theta}_{x2} \end{bmatrix} \begin{bmatrix} \tfrac{1}{4} L_e \int_{-1}^{1} \begin{Bmatrix} (1-\xi) \\ (1+\xi) \end{Bmatrix} m_x(\xi)\, d\xi \end{bmatrix}$$

(10.48)

where integrals are carried out over the coordinate ξ.

Example 10.3   Torsion stiffness and load arrays

For an element of length L_e and constant torsional stiffness the stiffness matrix from Eq. (10.48) is given by

$$\mathbf{K}^t = \frac{\kappa_x GJ}{L_e} \begin{bmatrix} 1 & 0 \\ 0 & 1 \end{bmatrix}$$

and for constant m_x the element load by

$$\mathbf{f}^a = \tfrac{1}{2} m_x L_e \begin{Bmatrix} 1 \\ 1 \end{Bmatrix}$$

10.4.4 Beam bending – mixed form

The three-field weak form given by Eq. (10.29) may be used to obtain exact interelement nodal solutions of the finite element formulation of elastic problems in which the EI_y *varies* along the length of the element. Accordingly, from the weak form we obtain the adjoint differential equation for δM_y given by

$$\frac{\partial^2 \delta M_y}{\partial x^2} = 0$$

(10.49)

for which the exact solution is a *linear function* over the element length that may be given by the interpolation

$$\delta M_y = N_1(\xi)\, \delta\tilde{M}_{y1} + N_2(\xi)\, \delta\tilde{M}_{y2}$$

(10.50)

where $N_a(\xi)$ are defined in Eq. (10.38).

Considering an individual element of length L_e and performing an integration by parts on the derivatives of w and δw gives the weak form for the static problem

$$\delta\Pi_{M_y}^e(w, M_y, \chi_y) = \int_{L_e} \left[\delta\chi_y\, (\hat{M}_y(\chi_y) - M_y) - \delta M_y\, \chi_y(M_y) \right] dx$$

$$- \left(\delta M_y \frac{\partial w}{\partial x} \right)\Big|_{\partial L_e} + \left(\frac{\partial \delta M_y}{\partial x} w \right)\Big|_{\partial L_e}$$

(10.51)

$$- \left(\frac{\partial \delta w}{\partial x} M_y \right)\Big|_{\partial L_e} + \left(\delta w \frac{\partial M_y}{\partial x} \right)\Big|_{\partial L_e} - \int_{L_e} \delta w \left(\frac{\partial^2 M_y}{\partial x^2} + q_z \right) dx$$

If we use a curvature $\hat{\chi}_y(M_y)$ which satisfies the constitutive equation

$$\hat{M}_y(\chi_y) = M_y$$

and a real moment field which satisfies

$$\frac{\partial^2 M_y}{\partial x^2} + q_z = 0$$

the weak form for the static problem simplifies to

$$\delta\Pi^e_{M_y}(w, M_y) = -\left.\left(\delta M_y \frac{\partial w}{\partial x}\right)\right|_{\partial L_e} + \left.\left(\frac{\partial \delta M_y}{\partial x} w\right)\right|_{\partial L_e}$$
$$-\left.\left(\frac{\partial \delta w}{\partial x} M_y\right)\right|_{\partial L_e} + \left.\left(\delta w \frac{\partial M_y}{\partial x}\right)\right|_{\partial L_e} - \int_{L_e} \delta M_y \chi_y(M_y)\,\mathrm{d}x$$

$$(10.52)$$

Using the shape functions given in Eq. (10.38) the real moment field may be expressed as

$$M_y = N_1(\xi)\,\tilde{M}_{y1} + N_2(\xi)\,\tilde{M}_{y2} + \bar{M}_y(q_z) \tag{10.53}$$

where $\bar{M}_y(q_z)$ is a particular solution due to the loading q_z acting on the element. In this form only end values for w_a and $\partial w_a/\partial x$ and their variation appear and no internal interpolation is needed to obtain exact interelement solutions at nodes. Note that in this form the moments are not continuous across element ends and, thus, $\tilde{M}_{y1}$ and $\tilde{M}_{y2}$ may be eliminated at the element level.

For the linear elastic case the determination of $\hat{\chi}_y(M_y)$ is trivial and is given by the simple relation

$$\hat{\chi}_y(M_y) = \frac{1}{EI_y}M_y$$

If the material behaviour is non-linear then the determination is more complex and, indeed, may not be possible to obtain. We consider this further in Sec. 10.4.5. Also, when inertia effects are included an interpolation for w must be used; however, this becomes a case for which the interelement solution is no longer exact. The form given above has been used by several authors, see, for example, references 11–25.

Example 10.4   Element array for variable EI_y

Let us consider the example of a uniformly loaded element with length L_e that has a variable moment of inertia given by

$$\frac{1}{EI_y} = N_1(\xi)\,\frac{1}{EI_{y1}} + N_2(\xi)\,\frac{1}{EI_{y2}}$$

A particular solution for a uniformly loaded element may be expressed as

$$\bar{M}_y = \tfrac{1}{8}q_z L_e^2 (1 - \xi^2)$$

With the above forms

$$\int_{L_e} \delta M_y \chi_y(M_y)\,\mathrm{d}x = \delta\tilde{M}_a \int_{L_e} N_a \frac{1}{EI_y}\left(N_b \tilde{M}_b + \bar{M}_y\right)\mathrm{d}x$$

$$= \begin{bmatrix} \delta\tilde{M}_{y1} & \delta\tilde{M}_{y1} \end{bmatrix} \left(\begin{bmatrix} V_{11} & V_{12} \\ V_{21} & V_{22} \end{bmatrix} \begin{Bmatrix} \tilde{M}_{y1} \\ \tilde{M}_{y1} \end{Bmatrix} + \begin{Bmatrix} f_{M1} \\ f_{M2} \end{Bmatrix} \right)$$

$$= \delta\tilde{M}_y^T \left(\mathbf{V}\,\tilde{\mathbf{M}}_y + \mathbf{f}_M\right)$$

where

$$V = \frac{L_e}{12EI_{y1}} \begin{bmatrix} 3 & 1 \\ 1 & 1 \end{bmatrix} + \frac{L_e}{12EI_{y2}} \begin{bmatrix} 1 & 1 \\ 1 & 3 \end{bmatrix} \quad \text{and} \quad f_M = \frac{q_z L_e^2}{120} \left\{ \begin{array}{l} \dfrac{3}{EI_{y1}} + \dfrac{2}{EI_{y2}} \\ \dfrac{2}{EI_{y1}} + \dfrac{3}{EI_{y2}} \end{array} \right\}$$

The element boundary terms give

$$-\left(\delta M_y \frac{\partial w}{\partial x}\right)\bigg|_{\partial L} + \left(\frac{\partial \delta M_y}{\partial x} w\right)\bigg|_{\partial L} = \frac{1}{L_e} \begin{bmatrix} \delta \tilde{M}_{y1} & \delta \tilde{M}_{y2} \end{bmatrix} \begin{bmatrix} 1 & -L_e & -1 & 0 \\ -1 & 0 & 1 & L_e \end{bmatrix} \left\{ \begin{array}{l} \tilde{w}_1 \\ \tilde{\theta}_{y1} \\ \tilde{w}_2 \\ \tilde{\theta}_{y2} \end{array} \right\}$$

$$= \delta \tilde{M}_y^T G \tilde{u}^b$$

The other element boundary terms give

$$-\left(\frac{\partial \delta w}{\partial x} M_y\right)\bigg|_{\partial L} + \left(\delta w \frac{\partial M_y}{\partial x}\right)\bigg|_{\partial L} = \delta \tilde{u}^{bT} \left[G^T \tilde{M}_y + f_u \right]$$

where f_u is a force vector resulting from $\bar{M}_y$ and for the particular solution given above becomes

$$f_u = \tfrac{1}{2} q_z L_e \begin{bmatrix} 1 & 0 & 1 & 0 \end{bmatrix}^T$$

When substituted back into Eq. (10.52) the above results give an element weak form

$$\delta \Pi^e_{M_y} = \begin{bmatrix} \delta \tilde{u}^{bT} & \delta \tilde{M}_y^T \end{bmatrix} \left(\begin{bmatrix} 0 & G^T \\ G & -V \end{bmatrix} \left\{ \begin{array}{l} \tilde{u}^b \\ \tilde{M}_y \end{array} \right\} - \left\{ \begin{array}{l} f_u \\ f_M \end{array} \right\} \right) \tag{10.54}$$

After $\tilde{M}_y$ is eliminated for each element we obtain the 'stiffness' matrix

$$K^e = G^T V^{-1} G$$

and load vector

$$f^e = f_u + G^T V^{-1} f_M$$

which may be assembled in an identical manner as in the irreducible form.

Example 10.5 Cantilever tapered bar – Euler–Bernoulli theory

To illustrate the performance of the irreducible and mixed forms presented we consider a cantilever beam that is loaded by a uniform load q_z as shown in Fig. 10.4. The bar has a variable bending stiffness given in reciprocal form by

$$\frac{1}{EI_y} = \frac{1}{EI_0} + \left(\frac{1}{EI_1} - \frac{1}{EI_1} \right)(4\eta - \eta^2)$$

where $\eta = x/L$ and the stiffness at the support and free end are EI_0 and EI_1, respectively.

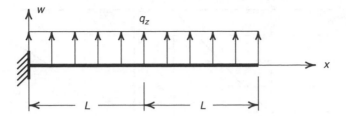

Fig. 10.4 Tapered cantilever beam.

The exact solution for the displacement and slope are given by

$$w(x) = \frac{q_z L^4}{24 E I_0} \left[8\eta^3 - \eta^4 - 24\eta^2 \right] + \frac{1}{240} q_z L^4 \left(\frac{1}{E I_1} - \frac{1}{E I_1} \right)$$
$$\times \left[\eta^6 - 12\eta^5 + 50\eta^4 - 80\eta^3 \right]$$

$$\frac{\partial w}{\partial x} = \frac{q_z L^3}{6 E I_0} \left[6\eta^2 - \eta^3 - 12\eta \right] + \frac{1}{120} q_z L^4 \left(\frac{1}{E I_1} - \frac{1}{E I_1} \right)$$
$$\times \left[3\eta^5 - 30\eta^4 + 100\eta^3 - 120\eta^2 \right]$$

and those for moment and shear by

$$M = \tfrac{1}{2} q_z L^2 \left[4\eta - \eta^2 - 4 \right]$$
$$S = q_z L \left[2 - \eta \right]$$

In Fig. 10.5 we present the solution using two equal length elements and an irreducible (displacement) and mixed solution as described above. The results for the mixed solution are exact whereas those for the irreducible solution have error in all quantities, although those for the displacement and slope are very small. Using more elements will lead to convergent results for the irreducible form.

10.4.5 Inelastic behaviour of rods

The inelastic behaviour of a three-dimensional rod requires simultaneous consideration of the axial, bending and torsion behaviour. For the two-dimensional case the torsion behaviour can be omitted and we can consider only axial and bending interaction.

Let us consider the case where the constitutive behaviour is elastic–plastic as described in Chapter 4. The constitutive equation in one dimension may be given as

$$\sigma_x = E \left(\varepsilon_x - \varepsilon_x^p \right) \tag{10.55a}$$

where E is the elastic modulus and ε_x^p is the plastic strain.

We assume an associative form and a yield function given by

$$F(\sigma, \kappa) = \left[\sigma_x^2 - \sigma_y^2(\kappa) \right]^{1/2} \le 0 \tag{10.55b}$$

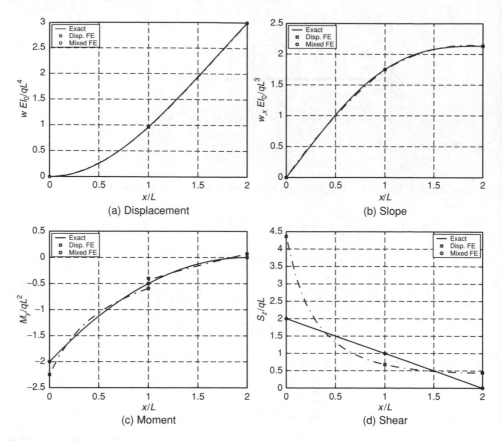

Fig. 10.5 Solution for tapered beam using irreducible (displacement) and mixed solutions – Euler–Bernoulli theory.

where σ_y is the uniaxial yield stress and κ an isotropic hardening parameter. Thus, the plastic strain is given by

$$\dot{\varepsilon}^p = \dot{\lambda}\,\frac{\partial F}{\partial \sigma_x} = \dot{\lambda}\,\frac{\sigma_x}{|\sigma_x|} \tag{10.55c}$$

where $|\sigma_x| = [\sigma_x^2]^{1/2}$. For simplicity we consider the linear hardening form given by

$$\sigma_y = \sigma_0 + H_i\,\kappa \tag{10.55d}$$

where H_i is the hardening modulus and κ is computed from

$$\dot{\kappa} = \left[(\dot{\varepsilon}^p)^2\right]^{1/2} = \dot{\lambda} \tag{10.55e}$$

A numerical solution to the plasticity equations may be constructed using the return map algorithm presented in Sec. 4.4.2. This will give a sequence of stresses at each 'time' step t_n given by

$$\sigma_x^{(n)} = E(\varepsilon_x^{(n)} - \varepsilon_x^{p(n)})$$

which linearized gives the incremental form

$$d\sigma_x^{(n)} = E_T^{(n)}\,d\varepsilon_x^{(n)} \tag{10.56}$$

Irreducible form for inelastic problems

For a two-dimensional problem based on an irreducible form the above result for inelastic stress gives the force resultants

$$
\begin{aligned}
P^{(n)} &= \int_A E(\varepsilon_x^{(n)} - \varepsilon_x^{p(n)}) \, \mathrm{d}A \\
M_y^{(n)} &= -\int_A z \, E(\varepsilon_x^{(n)} - \varepsilon_x^{p(n)}) \, \mathrm{d}A
\end{aligned}
\tag{10.57}
$$

where

$$
\varepsilon_x^{(n)} = \varepsilon_0^{(n)} + z \, \chi_y^{(n)}
\tag{10.58}
$$

For a Newton solution a linearization of Eq. (10.57) will give the tangent matrix for the cross-section as

$$
\mathbf{D}_T = \int_A \begin{bmatrix} E_T & z \, E_T \\ z \, E_T & z^2 \, E_T \end{bmatrix}
\tag{10.59}
$$

where E_T is the material tangent modulus obtained from the local constitutive equation [e.g. from Eq. (10.55a)] at each computation point on the cross-section. Note that the tangent matrix will no longer be diagonal in all situations and, thus, it is necessary to consider the axial and bending problems simultaneously. The arrays for each element of the coupled problem are deduced from

$$
\delta \Pi^e = \delta \mathbf{u}^T \int_{L_e} \left(\mathbf{B}^T \mathbf{P} - \mathbf{N}^T \mathbf{q} \right) \, \mathrm{d}x = -\delta \mathbf{u}^T \boldsymbol{\psi}
\tag{10.60}
$$

where

$$
\mathbf{B} = \begin{bmatrix} \mathbf{B}^a & \mathbf{0} \\ \mathbf{0} & \mathbf{B}^b \end{bmatrix}, \quad
\mathbf{P} = \left\{ \begin{matrix} P \\ M_y \end{matrix} \right\}, \quad
\mathbf{N} = \begin{bmatrix} \mathbf{N}^a & \mathbf{0} \\ \mathbf{0} & \mathbf{N}^b \end{bmatrix}, \quad
\mathbf{q} = \left\{ \begin{matrix} q_x \\ q_z \end{matrix} \right\} \quad \text{and} \quad
\tilde{\mathbf{u}}_a = \left\{ \begin{matrix} \tilde{u}_a \\ \tilde{w}_a \\ \partial \tilde{w}_a / \partial x \end{matrix} \right\}
$$

A linearization for a Newton method gives the element tangent matrix

$$
\mathbf{K}_T^e = \int_{L_e} \mathbf{B}^T \mathbf{D}_T \mathbf{B} \, \mathrm{d}x
$$

The solution process now follows a standard procedure for solving non-linear problems. For each time t_n

1. Initialize an equilibrium iteration parameter $k = 0$ and assume an elastic behaviour at all points to compute a trial displacement $\tilde{\mathbf{u}}_{(n)}^{(0)}$ where the superscript indicates an iteration number k.

2. For each quadrature point used to compute element arrays perform a numerical integration over the cross-section to compute P, M_y and $\mathbf{D}_T$. This requires storage of the plastic solution parameters for each solution point on each cross-section.

3. Use P, M_y and $\mathbf{D}_T$ to compute the element arrays $\mathbf{K}^{(k)}$ and residual $\boldsymbol{\psi}^{(k)}$. Solve the problem

$$
\begin{aligned}
\mathbf{K}^{(k)} \, d\mathbf{u}^{(k)} &= \boldsymbol{\psi}^{(k)} \\
\mathbf{u}^{(k+1)} &= \mathbf{u}^{(k)} + d\mathbf{u}^{(k)}
\end{aligned}
$$

4. Check convergence. If converged set $n = n + 1$ and go to step 1; else set $k = k + 1$ and go to step 2.

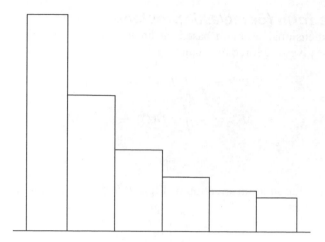

Fig. 10.6 Typical discontinuous strain distribution in rod element.

Mixed form for inelastic problems

The development of the inelastic finite element approximation given above leads to a very simple implementation; however, when used with just a few elements the inaccuracy in force resultants can lead to significant solution error. As shown above for the elastic bending case very accurate force prediction is obtained when we are able to exactly determine the curvatures from the constitutive equation, thus leading to a Hellinger–Reissner-type weak form. In the inelastic case it is necessary to determine the exact axial strain and curvature over the entire element in order to reduce the problem to an equivalent two-field form and this is not possible in general. Thus it is necessary to return to the three-field weak form and introduce an approximation for the axial strain and curvature in addition to force, moment and displacement variables.

The approximation of strains may be performed in several ways; however, dividing the element into segments in which the axial strain and curvature are assumed constant in each segment as shown in Fig. 10.6 is simple and gives good results. The shape functions for the strains may be expressed as

$$\varepsilon = \sum_j N_j(\xi)\,\tilde{\varepsilon}_j; \quad j = 1, N$$

$$\chi_y = \sum_j N_j(\xi)\,\tilde{\chi}_{yj}; \quad j = 1, N \tag{10.61}$$

where

$$N_j(\xi) = 1 \quad \text{for} \quad \xi_j \le \xi \le \xi_{j+1} \tag{10.62}$$

The values of ξ_j are selected with $\xi_1 = -1$ and $\xi_{N+1} = 1$.

The three-field weak form for the axial and bending behaviour may be written in matrix form as

$$\delta\Pi^e = \begin{bmatrix} \delta\tilde{\mathbf{u}}^T & \delta\tilde{\mathbf{s}}^T & \delta\tilde{\mathbf{e}}^T \end{bmatrix} \left(\begin{bmatrix} \mathbf{0} & \mathbf{G}^T & \mathbf{0} \\ \mathbf{G} & \mathbf{0} & \mathbf{B}_e^T \\ \mathbf{0} & \mathbf{B}_e & \mathbf{0} \end{bmatrix} \begin{Bmatrix} \tilde{\mathbf{u}} \\ \tilde{\mathbf{s}} \\ \tilde{\mathbf{e}} \end{Bmatrix} - \begin{Bmatrix} \mathbf{f} \\ \mathbf{0} \\ \mathbf{s}_p - \hat{\mathbf{s}}(\mathbf{e}) \end{Bmatrix} \right) \tag{10.63}$$

$$= -\delta\tilde{\mathbf{U}}^T\,\mathbf{\Psi}$$

where

$$G = \begin{bmatrix} G^a & 0 \\ 0 & G^b \end{bmatrix}; \quad b_j = \frac{1}{2} L_e \int_{\xi_j}^{\xi_{j+1}} \begin{bmatrix} 1 & 0 & 0 \\ 0 & \frac{1}{2}(1-\xi) & \frac{1}{2}(1-\xi) \end{bmatrix} d\xi; \quad B_e = \begin{bmatrix} b_1 \\ \vdots \\ b_N \end{bmatrix}$$

$$\tilde{U} = \begin{Bmatrix} \tilde{u} \\ \tilde{s} \\ \tilde{e} \end{Bmatrix}; \quad \tilde{u}_a = \begin{Bmatrix} \tilde{u}_a \\ \tilde{w}_a \\ \partial \tilde{w}_a / \partial x \end{Bmatrix}; \quad \tilde{s} = \begin{Bmatrix} \tilde{N} \\ \tilde{M}_{y1} \\ \tilde{M}_{y2} \end{Bmatrix}; \quad \tilde{e}_j = \begin{Bmatrix} \tilde{\varepsilon}_j \\ \tilde{\chi}_{yj} \end{Bmatrix}; \quad \tilde{e} = \begin{bmatrix} e_1 \\ \vdots \\ e_N \end{bmatrix}$$

s_p denotes the particular solution for the resultants and $\hat{s}(e)$ denotes the constitutive equation expressions in terms of the strain approximations. The constitutive equations are computed from

$$s = \int_{L_e} \begin{bmatrix} 1 & 0 \\ 0 & \frac{1}{2}(1-\xi) \\ 0 & \frac{1}{2}(1+\xi) \end{bmatrix} \left(\int_A \begin{Bmatrix} 1 \\ z \end{Bmatrix} \sigma_x(\varepsilon_x) \, dA \right) dx \tag{10.64}$$

which may be evaluated using trapezoidal quadrature* over each segment of the element to give

$$s = \frac{1}{2} L_e \sum_{l=1}^N \begin{bmatrix} 1 & 0 \\ 0 & \frac{1}{2}(1-\xi_l) \\ 0 & \frac{1}{2}(1+\xi_l) \end{bmatrix} \left(\int_A \begin{Bmatrix} 1 \\ z \end{Bmatrix} \sigma_x(\varepsilon_x(\xi_l)) \, dA \right) W_l \tag{10.65}$$

where $\xi_l = (x_j + x_{j+1})/2$, $W_l = 2/N$ and

$$\varepsilon_x(\xi_l) = \varepsilon_l + z \chi_{yl}$$

Linearizing Eq. (10.64) using the incremental stress given by Eq. (10.56) gives the element tangent matrix

$$K_{hw}^e = \begin{bmatrix} 0 & G^T & 0 \\ G & 0 & B_e^T \\ 0 & B_e & k \end{bmatrix} \tag{10.66}$$

where k is the constitutive tangent given by

$$k = \begin{bmatrix} k_1 & \cdots & 0 \\ \vdots & \ddots & \vdots \\ 0 & \cdots & k_n \end{bmatrix} \quad \text{and} \quad k_j = \int_A \begin{Bmatrix} 1 \\ z \end{Bmatrix} E_T \begin{bmatrix} 1 & z \end{bmatrix} dA$$

The integral over A is performed using numerical integration.[24] The constitutive parameters may be eliminated to give

$$\hat{K}^e = \begin{bmatrix} 0 & G^T \\ G & -V \end{bmatrix} \quad \text{where } V = B_e^T \begin{bmatrix} k_1 & \cdots & 0 \\ \vdots & \ddots & \vdots \\ 0 & \cdots & k_n \end{bmatrix} B_e^T \tag{10.67}$$

---

* Other forms based on ξ_j placed by Gauss or Gauss–Lobatto quadrature[10] may also be used.

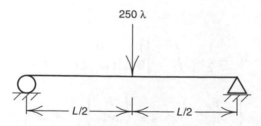

Fig. 10.7 Simply supported beam with central point load.

Finally, the tangent matrix for the element is obtained by eliminating the force parameters to give

$$\mathbf{K}^e = \mathbf{G}^T \mathbf{V}^{-1} \mathbf{G}$$

Similar reductions may be made for any non-zero force components in $\mathbf{\Psi}$.

In the above formulation the only displacement parameters obtained are at the ends of each element. For plotting displacements along the length of the element may be obtained by integrating the curvature values.[11,12,15]

Example 10.6   Simply supported beam with point load

As an example problem we consider a simply supported beam with a central point load as shown in Fig. 10.7. To show the advantages of the mixed formulation presented above we allow the beam to have elasto-plastic behaviour. The entire beam is modelled with two elements (one element for each symmetric half); five curvature stations per element are used along the beam axis. A rectangular cross-section is considered with 10 Gauss–Lobatto points through the depth to permit modelling of the spread of the plastic zone. The properties for the analysis are shown in Table 10.4.

The central load is allowed to vary using a load control strategy (e.g. see Chapter 2 or reference 26). For the comparison we also consider the solution using an irreducible model with cubic Hermite polynomial shape functions. Solutions for two, four, and eight elements for the length are used (one, two and four on each half length). In Fig. 10.8 we show the force–displacement relation, deformed shape and distribution of moment and curvature along the length of the beam at the last computed load state for each analysis. The displacement model permits only linear change, whereas the mixed model presented here allows for arbitrary change at each axial station used (five in the present case). The superiority of the mixed form is evident in the force–displacement, the computed deformed shape and the moment and the curvature distribution.

Table 10.4   Properties for inelastic beam

Length	$L = 180$
Depth	$h = 10$
Width	$b = 10$
Elastic modulus	$E = 29\,000$
Yield stress	$\sigma_y = 50$
Hardening modulus	$H_{iso} = 290\ (1\%)$

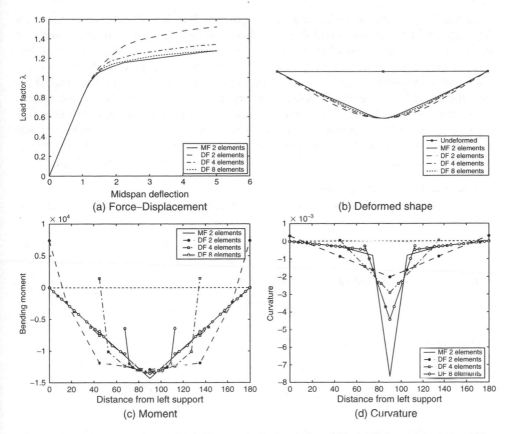

Fig. 10.8 Simply supported, point loaded beam – inelastic solution. DF = displacement formulation; MF = mixed formulation.

10.5 Finite element solution: Timoshenko rods

The development of finite element models for the Timoshenko theory, which includes the primary effects from both axial and shearing strains, is developed from the functional (10.36) for an irreducible theory or from Eq. (10.33) for a mixed theory. The highest derivative in either form is first order, consequently C^0 functions may be used in the approximation. However, the approximation can be quite sensitive if it is desired that the Timoshenko form accurately solve problems in which the transverse shear deformations are small compared to the bending strains. Below we present some successful mixed approximations which are free from locking. We begin by considering the irreducible form for bending.

10.5.1 Beam bending – irreducible form

Since the highest derivatives of w and θ_y in the functional (10.36) are only first order, both may be interpolated by C^0 functions. The use of the Timoshenko theory permits the solution of problems for which the length to cross-section dimensions are somewhat larger than for the Euler–Bernoulli theory. It is desirable, however, to have a finite element implementation of the Timoshenko theory which permits the solution of problems at the 'thin' limit where the Euler–Bernoulli theory is applicable. The choice of appropriate functions is quite delicate if we wish to avoid 'locking' between shear and bending response as the beam approaches the thin limit. For example, as we show next the use of low-order/equal-order C^0 interpolation for the transverse displacement w and rotation θ_y will lead to an element which *locks* as the beam becomes slender (i.e. $L/d > 5$, where L is a characteristic length and d a cross-section dimension). Later we consider an approach which avoids such locking. Moreover these developments are also useful when we consider the study of plate problems in the next chapters.

Equal-order interpolation

Use of an equal-order interpolation for the transverse displacement and the rotation is expressed as

$$\begin{Bmatrix} w \\ \theta_y \end{Bmatrix} = \sum_a N_a(\xi) \begin{Bmatrix} \tilde{w}_a \\ \tilde{\theta}_{ya} \end{Bmatrix} = \mathbf{N}\,\tilde{\mathbf{u}}^b \tag{10.68}$$

where $\tilde{w}_a$ and $\tilde{\theta}_{ya}$ are the values at node a with

$$\tilde{\mathbf{u}}_a^b = \begin{Bmatrix} \tilde{w}_a \\ \tilde{\theta}_{ya} \end{Bmatrix}$$

The curvature and shear strain in each element are given by

$$\begin{Bmatrix} \chi_y \\ \gamma_z \end{Bmatrix} = \sum_a \left\{ \begin{array}{c} \dfrac{dN_a}{dx} \tilde{\theta}_{ya} \\[2mm] \left[\dfrac{dN_a}{dx} \tilde{w}_a + N_a \tilde{\theta}_{ya} \right] \end{array} \right\} = \mathbf{B}^b \,\tilde{\mathbf{u}}^b \tag{10.69}$$

where the strain displacement matrix $\mathbf{B}^b$ is given by

$$\mathbf{B}_a^b = \begin{bmatrix} \dfrac{dN_a}{dx} & 0 \\[2mm] N_a & \dfrac{dN_a}{dx} \end{bmatrix} \tag{10.70}$$

The material elastic property array may be written as

$$\mathbf{D}^b = \begin{bmatrix} EI_y & 0 \\ 0 & \kappa_z G_z A \end{bmatrix} \tag{10.71}$$

Using the above arrays in Eq. (10.36) the element stiffness and load array of the static problem are computed from

$$\delta\Pi_w = \delta\tilde{\mathbf{u}}^{bT} \left[\left(\int_{L_e} \mathbf{B}^{bT} \mathbf{D}^b \,\mathbf{B}^b \, dx \right) \tilde{\mathbf{u}}^b - \int_{L_e} \mathbf{N}^T q_z \, dx \right] \tag{10.72}$$

Example 10.7   Linear interpolation

As an example we consider the case where w and θ_y in each element are given by the linear interpolation

$$\begin{Bmatrix} w \\ \theta_y \end{Bmatrix} = N_1(\xi) \begin{Bmatrix} \tilde{w}_1 \\ \tilde{\theta}_{y1} \end{Bmatrix} + N_2(\xi) \begin{Bmatrix} \tilde{w}_2 \\ \tilde{\theta}_{y2} \end{Bmatrix}$$

where $N_1 = (1 - \xi)/2$ and $N_2 = (1 + \xi)/2$.

Evaluating the integrals in Eq. (10.72) analytically (or by a 2-point Gauss quadrature) gives the element stiffness

$$\mathbf{K}^e = \frac{EI_y}{L_e}\begin{bmatrix} 0 & 0 & 0 & 0 \\ 0 & 1 & 0 & -1 \\ 0 & 0 & 0 & 0 \\ 0 & -1 & 0 & 1 \end{bmatrix} + \frac{\kappa_y G_y A}{6L_e}\begin{bmatrix} 6 & 3L_e & -6 & 3L_e \\ 3L_e & 2L_e^2 & 3L_e & L_e^2 \\ -6 & 3L_e & 6 & 3L_e \\ 3L_e & L_e^2 & 3L_e & 2L_e^2 \end{bmatrix}$$

and for a uniform load q_z the element force vector

$$\mathbf{f}^e = \tfrac{1}{2} q_z L_e \begin{bmatrix} 1 & 0 & 1 & 0 \end{bmatrix}^T$$

Evaluating the element stiffness by a reduced 1-point quadrature gives

$$\mathbf{K}^e = \frac{EI_y}{L_e}\begin{bmatrix} 0 & 0 & 0 & 0 \\ 0 & 1 & 0 & -1 \\ 0 & 0 & 0 & 0 \\ 0 & -1 & 0 & 1 \end{bmatrix} + \frac{\kappa_y G_y A}{4L_e}\begin{bmatrix} 4 & -2L_e & -4 & -2L_e \\ -2L_e & L_e^2 & 2L_e & L_e^2 \\ -4 & 2L_e & 4 & 2L_e \\ -2L_e & L_e^2 & 2L_e & L_e^2 \end{bmatrix}$$

The only difference between exact and reduced integration appears in the K_{22}, K_{24}, K_{42} and K_{44} terms in the shear modulus matrix. These seemingly small differences have a significant effect on the element performance as shown in Fig. 10.9 in which we solve a cantilever beam using 20 elements for different span L to depth h ratios for a rectangular cross-section. The use of exact integration leads to a solution which 'locks' as the beam becomes slender, whereas the reduced integration form shows no locking for the range plotted.

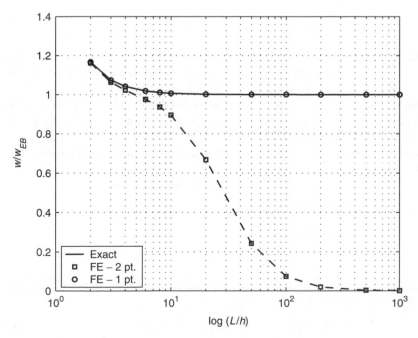

Fig. 10.9 Cantilever beam solved using equal-order interpolation on linear elements. w/w_{EB} is ratio of tip displacement for Timoshenko beam theory to that of Euler–Bernoulli theory.

Exact nodal solution form

As in the Euler–Bernoulli theory we first consider the static behaviour for the linear elastic Timoshenko beam. Our goal is to determine the interpolation functions which give exact interelement nodal solutions when the material properties are constant along the element length. The requirement is for δw and $\delta\theta_y$ to satisfy the homogeneous adjoint equations of the functional given in Eq. (10.36). For EI_y and $\kappa_z G_z A$ constant the equations are given by

$$EI_y \frac{d^2 \delta\theta_y}{dx^2} + \kappa_z G_z A \left(\frac{d\delta w}{dx} + \delta\theta_y \right) = 0$$

$$\kappa_z G_z A \frac{d}{dx} \left(\frac{d\delta w}{dx} + \delta\theta_y \right) = 0$$

(10.73)

Differentiating the first equation and combining with the second gives the simpler requirements

$$\frac{d^3 \delta\theta_y}{dx^3} = 0 \quad \text{and} \quad \frac{d^2 \delta w}{dx^2} + \frac{d\delta\theta_y}{dx} = 0$$

which imply $\delta\theta_y$ is a quadratic polynomial and δw a cubic one.

The expression for $\delta\theta_y$ may be given as a Lagrange interpolation in the natural coordinates ξ as

$$\delta\theta_y = \tfrac{1}{2} (\xi^2 - \xi) \delta\tilde\theta_{y1} + \tfrac{1}{2} (\xi^2 + \xi) \delta\tilde\theta_{y2} + (1 - \xi^2) \delta\tilde\theta_{y3}$$

(10.74)

where $\delta\tilde\theta_{y1}$ and $\delta\tilde\theta_{y2}$ are the interelement nodal parameters and $\delta\tilde\theta_{y3}$ is a mid-length parameter. Using the above interpolation for $\delta\theta_y$ in the differential equation for δw gives an interpolation

$$\delta w = \tfrac{1}{2}(1 - \xi) \delta\tilde w_1 + \tfrac{1}{2}(1 + \xi) \delta\tilde w_2$$
$$- L_e (1 - \xi^2) \left[\left(\tfrac{1}{8} - \tfrac{1}{12}\xi \right) \delta\tilde\theta_{y1} - \left(\tfrac{1}{8} + \tfrac{1}{12}\xi \right) \delta\tilde\theta_{y2} + \tfrac{1}{6}\xi \, \delta\tilde\theta_{y3} \right]$$

(10.75)

The parametric relation for x given in Eq. (10.38) is then used to construct the derivatives. We note that the interpolation for δw is *linked* to the parameters of $\delta\theta_y$. The earliest appearance of linked interpolation appears in a paper by Fraeijs de Veubeke.[27] Alternative expressions have been deduced by Tessler and Dong,[28] Crisfield[29] and Stolarski *et al.*[30] We shall find such *linked interpolation* forms are also very useful in the construction of plate bending elements.

The interpolation for the variables may also be given in hierarchical form by substituting

$$\delta\tilde\theta_{y3} = \delta(\Delta\tilde\theta_{y3}) + \tfrac{1}{2}(\delta\tilde\theta_{y1} + \delta\tilde\theta_{y2})$$

into Eqs (10.74) and (10.75) to obtain

$$\delta\theta_y = \tfrac{1}{2} (1 - \xi) \delta\tilde\theta_{y1} + \tfrac{1}{2} (1 + \xi) \delta\tilde\theta_{y2} + (1 - \xi^2) \delta(\Delta\tilde\theta_{y3})$$
$$= N_1(\xi) \delta\tilde\theta_{y1} + N_2(\xi) \delta\tilde\theta_{y2} + N_3(\xi) \delta(\Delta\tilde\theta_{y2}) = \mathbf{N}_\theta \, \delta\tilde{\boldsymbol\theta}$$

(10.76a)

and

$$\delta w = \tfrac{1}{2}(1 - \xi) \delta\tilde w_1 + \tfrac{1}{2}(1 + \xi) \delta\tilde w_2$$
$$- L_e (1 - \xi^2) \left[\tfrac{1}{8} (\delta\tilde\theta_{y1} - \delta\tilde\theta_{y2}) + \tfrac{1}{6}\xi \delta(\Delta\tilde\theta_{y3}) \right]$$
$$= N_1 \delta\tilde w_1 + N_{w1} \delta\tilde\theta_{y1} + N_2 \delta\tilde w_2 + N_{w2} \delta\tilde\theta_{y2} + N_{w3} \delta(\Delta\tilde\theta_{y3}) = \mathbf{N}_w \tilde{\mathbf{u}}^b$$

(10.76b)

where

$$\mathbf{N}_w = \begin{bmatrix} N_1 & N_{w1} & N_2 & N_{w2} & N_{w3} \end{bmatrix}$$
$$\tilde{\mathbf{u}}^b = \begin{bmatrix} \tilde{w}_1 & \tilde{\theta}_{y1} & \tilde{w}_2 & \tilde{\theta}_{y2} & \Delta\tilde{\theta}_{y3} \end{bmatrix}^T$$

(10.77)

This form has some useful advantages which we will exploit later. The same form of interpolation is used for w and θ_y in a Galerkin solution.

Using the hierarchical interpolations in the strain–displacement relations gives

$$\chi_y = \frac{1}{L_e} \left[\tilde{\theta}_{y2} - \tilde{\theta}_{y1} - 4\xi \Delta\tilde{\theta}_{y3} \right]$$

$$\gamma_z = \frac{1}{L_e}(\tilde{w}_2 - \tilde{w}_1) + \frac{1}{2}(\tilde{\theta}_{y1} + \tilde{\theta}_{y2}) + \frac{2}{3}\Delta\tilde{\theta}_{y3}$$

(10.78)

The finite element strain–displacement relations may be written in the standard matrix form

$$\begin{Bmatrix} \chi_y \\ \gamma_z \end{Bmatrix} = \frac{1}{L_e} \begin{bmatrix} 0, & -1, & 0, & 1, & -4\xi \\ -1, & \frac{1}{2}L_e, & 1, & \frac{1}{2}L_e, & \frac{2}{3}L_e \end{bmatrix} \tilde{\mathbf{u}}^b = \mathbf{B}^b \, \tilde{\mathbf{u}}^b$$

(10.79)

Using the above arrays in Eq. (10.36) the functional of the static problem for an individual element is given by

$$\delta\Pi_w = \delta\tilde{\mathbf{u}}^{bT} \left[\left(\int_{L_e} \mathbf{B}^{bT}\mathbf{D}^b\,\mathbf{B}^b \, dx \right) \tilde{\mathbf{u}}^b - \int_{L_e} \mathbf{N}_w^T q_z \, dx \right]$$

(10.80)

Example 10.8 Stiffness and load arrays for uniform Timoshenko beam

If we evaluate Eq. (10.80) for an element of length L_e with EI_y and $\kappa_z G_z A$ constant the element stiffness matrix is given by

$$\hat{\mathbf{K}}^e = \frac{EI_y}{3L_e} \begin{bmatrix} 0 & 0 & 0 & 0 & 0 \\ 0 & 3 & 0 & -3 & 0 \\ 0 & 0 & 0 & 0 & 0 \\ 0 & -3 & 0 & 3 & 0 \\ 0 & 0 & 0 & 0 & 16 \end{bmatrix}$$

$$+ \frac{\kappa_z G_z A}{36L_e} \begin{bmatrix} 36 & -18L_e & -36 & -18L_e & -24L_e \\ -18L_e & 9L_e^2 & 18L_e & 9L_e^2 & 12L_e^2 \\ -36 & 18L_e & 36 & 18L_e & 24L_e \\ -18L_e & 9L_e^2 & 18L_e & 9L_e^2 & 12L_e^2 \\ -24L_e & 12L_e^2 & 24L_e & 12L_e^2 & 16L_e^2 \end{bmatrix}$$

(10.81)

and for constant q_z the element load vector

$$\hat{\mathbf{f}}^e = \frac{q_z L^2}{12} \begin{bmatrix} 6L_e, & -1, & 6L_e, & 1, & 0 \end{bmatrix}^T$$

(10.82)

which, except for the last 0, is identical to the load vector of the Euler–Bernoulli theory.

Since the $\Delta\theta_{y3}$ is associated with each element individually, it may be eliminated and the reduced stiffness may be written as

$$
\mathbf{K}^e = \frac{12EI_y}{L_e^3}
\begin{bmatrix}
\alpha & -\frac{1}{2}\alpha L_e & -\alpha & -\frac{1}{2}\alpha L_e \\
-\frac{1}{2}\alpha L_e & \frac{1}{3}\alpha L_e^2(\alpha+\beta) & \frac{1}{2}\alpha L_e & \frac{1}{6}\alpha L_e^2(\alpha-2\beta) \\
-\alpha & \frac{1}{2}\alpha L_e & \alpha & \frac{1}{2}\alpha L_e \\
-\frac{1}{2}\alpha L_e & \frac{1}{6}\alpha L_e^2(\alpha-2\beta) & \frac{1}{2}\alpha L_e & \frac{1}{3}\alpha L_e^2(\alpha+\beta)
\end{bmatrix}
\tag{10.83}
$$

where

$$
\alpha = \frac{1}{1+\gamma}; \quad \beta = \frac{\gamma}{4(1+\gamma)} \quad \text{and} \quad \gamma = \frac{12EI_y}{\kappa_z G_z A L_e^2}
$$

As the beam becomes 'thin' $\gamma \to 0$ and this gives $\alpha \to 1$ and $\beta \to 0$ which when substituted into Eq. (10.83) we immediately observe that the stiffness converges to that of the Euler–Bernoulli theory and, thus, the element will not lock.

Timoshenko theory – constant strain form

It is possible to use the displacements given by Eqs (10.76a) and (10.76b) in which the $\Delta\bar{\theta}_3$ terms are omitted. Accordingly, with the interpolation given by

$$
\begin{aligned}
w &= \tfrac{1}{2}(1-\xi)\,\tilde{w}_1 + \tfrac{1}{2}(1+\xi)\,\tilde{w}_2 - \tfrac{1}{8}\,L_e\,(1-\xi^2)\,(\tilde{\theta}_{y1}-\tilde{\theta}_{y2}) \\
\theta_y &= \tfrac{1}{2}(1-\xi)\,\tilde{\theta}_{y1} + \tfrac{1}{2}(1+\xi)\,\tilde{\theta}_{y2}
\end{aligned}
\tag{10.84}
$$

the strain–displacement relations are given by the constant relations

$$
\begin{aligned}
\chi_y &= \frac{1}{L_e}\,(\tilde{\theta}_{y2}-\tilde{\theta}_{y1}) \\
\gamma_z &= \frac{1}{L_e}\,(\tilde{w}_2-\tilde{w}_1) + \frac{1}{2}\,(\tilde{\theta}_{y1}+\tilde{\theta}_{y2})
\end{aligned}
\tag{10.85}
$$

The finite element strain–displacement relations again may be written in the standard matrix form

$$
\begin{Bmatrix} \chi_y \\ \gamma_z \end{Bmatrix} = \frac{1}{L_e}
\begin{bmatrix}
0, & -1, & 0, & 1, \\
-1, & \frac{1}{2}L_e, & 1, & \frac{1}{2}L_e
\end{bmatrix} \tilde{\mathbf{u}}^b = \mathbf{B}^b\,\tilde{\mathbf{u}}^b
\tag{10.86}
$$

where now the $\tilde{\mathbf{u}}^b$ contains only the end values of w and θ_y.

Inserting the above arrays in Eq. (10.80) yields the element stiffness and load relations

$$
\mathbf{K}^e = \mathbf{B}^{bT}\left(\int_{L_e} \mathbf{D}^b\,dx\right)\mathbf{B}^b \quad \text{and} \quad \mathbf{f}^e = \int_{L_e} \mathbf{N}_w^T q_z\,dx
\tag{10.87}
$$

where $\mathbf{N}_w$ is given by Eq. (10.77) with the last entry removed.

Example 10.9   Stiffness/load for uniform constant strain Timoshenko beam

If we evaluate Eq. (10.87) for an element of length L_e with EI_y and $\kappa_z G_z A$ constant the element stiffness matrix is given by

$$
\hat{\mathbf{K}}^e = \frac{EI_y}{3L_e}
\begin{bmatrix}
0 & 0 & 0 & 0 \\
0 & 3 & 0 & -3 \\
0 & 0 & 0 & 0 \\
0 & -3 & 0 & 3
\end{bmatrix}
+ \frac{\kappa_z G_z A}{4L_e}
\begin{bmatrix}
4 & -2L_e & -4 & -2L_e \\
-2L_e & L_e^2 & 2L_e & L_e^2 \\
-4 & 2L_e & 4 & 2L_e \\
-2L_e & L_e^2 & 2L_e & L_e^2
\end{bmatrix}
\tag{10.88}
$$

and for constant q_z the element load vector

$$\mathbf{f}^e = \frac{q_z L^2}{12} \begin{bmatrix} 6L_e, & -1, & 6L_e, & 1 \end{bmatrix}^T \tag{10.89}$$

We note that the above stiffness is identical to that obtained in Example 10.7 by reduced integration; however, in the present case we were able to use exact integration and hence the linked form will be more useful when we consider plate bending problems. In addition, the linked form gives forces on the nodal rotations which improves the solution obtained.

Timoshenko theory – enhanced assumed strain form

The constant strain form of the displacement interpolation has the advantage that the transient behaviour involves the interelement nodes only, whereas use of the added internal form complicates the dynamics. The advantages of the form which gives exact nodal answers has significant accuracy advantages. Using an enhanced assumed strain form[31-33] can restore the advantage of the full interpolation without complicating the dynamics. In the enhanced assumed strain form we use the constant strain form of the displacements w and θ_y given by Eq. (10.84) and *enhance* the strains with the effects which arise from the $\Delta\tilde{\theta}_{y3}$. Thus, we express the strains as

$$\varepsilon^b = \begin{Bmatrix} \chi_y \\ \gamma_z \end{Bmatrix} = \frac{1}{L_e} \begin{Bmatrix} (\theta_{y2} - \theta_{y1}) \\ (w_2 - w_1) + \frac{1}{2} L_e (\theta_{y1} + \theta_{y2}) \end{Bmatrix} + \begin{Bmatrix} 4\xi \\ \frac{2}{3} \end{Bmatrix} \tilde{\alpha} \tag{10.90}$$

$$= \mathbf{B}^b \tilde{\mathbf{u}}^b + \mathbf{B}^b_{en} \tilde{\alpha}$$

where $\mathbf{B}^b$ is the strain–displacement matrix given in Eq. (10.86) and $\tilde{\alpha}$ is the enhanced strain parameter. The elimination of the enhanced parameter at the element level now proceeds as given for the static case and is not affected by inertial effects.

10.5.2 Timoshenko – mixed form

A mixed form approximation for the Timoshenko beam which gives exact interelement nodal behaviour may be constructed in a similar way as we performed for the Euler–Bernoulli theory in Sec. 10.4.1. Accordingly, considering the static problem the approximations are deduced from

$$\frac{d\delta S_z}{dx} = 0 \quad \text{and} \quad \frac{d\delta M_y}{dx} + \delta S_z = 0 \tag{10.91}$$

which are satisfied by

$$\delta M_y = N_1(\xi) \, \delta \tilde{M}_{y1} + N_2(\xi) \, \delta \tilde{M}_{y2} \quad \text{and} \quad \delta S_z = \frac{1}{L_e} (\delta M_{y2} - \delta M_{y1}) \tag{10.92}$$

where $N_a(\xi)$ are defined in Eq. (10.38).

In order to obtain exact interelement nodal displacements for the mixed formulation we let

$$M_y = N_1(\xi)\,\tilde{M}_{y1} + N_2(\xi)\,\tilde{M}_{y2} + \bar{M}_y(\xi)$$

$$S_z = \frac{1}{L_e}(M_{y2} - M_{y1}) + \bar{S}_z(\xi)$$
(10.93)

where $\bar{M}_y$ and $\bar{S}_z$ are particular solutions which satisfy

$$\frac{d\bar{S}_z}{dx} + q_z = 0 \quad \text{and} \quad \frac{d\bar{M}_y}{dx} - \bar{S}_z + m_y = 0$$

In addition we employ curvature approximations which satisfy the constitutive equations. Accordingly, for the linear elastic problem we use

$$\chi_y = \frac{1}{EI_y}M_y \quad \text{and} \quad \gamma_z = \frac{1}{\kappa_z G_z A}S_z$$

With these approximations, after integrating Eq. (10.33) by parts we obtain the weak form for the static problem

$$\delta\Pi_{M_y}^e(w, \theta_y, M_y) = \left(\delta M_y\,\theta_y\right)\Big|_{\partial L_e} + \left(\delta\theta_y\,M_y\right)\Big|_{\partial L_e} + \left(\delta S_z\,w\right)\Big|_{\partial L_e} + \left(\delta w\,S_z\right)\Big|_{\partial L_e}$$

$$- \int_{L_e}\left[\delta M_y\,\frac{1}{EI_y}\,M_y + \delta S_z\,\frac{1}{\kappa_z G_z A}\,S_z\right]dx$$
(10.94)

which is a similar form to that deduced for the Euler–Bernoulli beam in Eq. (10.52).

Example 10.10　Timoshenko theory mixed stiffness matrix

Let us consider the example of a uniformly loaded element with length L_e that has a variable moment of inertia given in Example 10.4 and a shear stiffness given by

$$\frac{1}{\kappa_z G_z A} = N_1(\xi)\,\frac{1}{\kappa_z G_z A_1} + N_2(\xi)\,\frac{1}{\kappa_z G_z A_2}$$

The particular solutions for the uniformly loaded element again may be expressed as

$$\bar{M}_y = \frac{1}{8}q_z\,L_e^2\,(1 - \xi^2) \quad \text{and} \quad \bar{S}_z = -\frac{1}{2}q_z\,L_e\,\xi$$

The arrays for V, G, f_M and f_u for the bending response are identical to those of the weak form (10.54) given in Example 10.4. The remaining term from the shear response is given by

$$\int_{L_e}\delta S_z\gamma_z(S_z)\,dx = \frac{1}{L_e}(\delta M_{y2} - \delta M_{y1})\left[\int_{L_e}\frac{1}{\kappa_z G_z A}\,dx\,(M_{y2} - M_{y1})\right.$$

$$\left. - \int_{L_e}\frac{1}{\kappa_z G_z A}\,\bar{S}_z\,dx\right]$$

$$= \begin{bmatrix}\delta\tilde{M}_{y1} & \delta\tilde{M}_{y1}\end{bmatrix}\left(\begin{bmatrix}W_{11} & W_{12} \\ W_{21} & W_{22}\end{bmatrix}\begin{Bmatrix}\tilde{M}_{y1} \\ \tilde{M}_{y1}\end{Bmatrix} + \begin{Bmatrix}f_{S1} \\ f_{S2}\end{Bmatrix}\right)$$

$$= \delta\tilde{M}_y^T\left(W\,\tilde{M}_y + f_S\right)$$

where

$$\mathbf{W} = \frac{1}{2L_e}\left(\frac{1}{\kappa_z G_z A_1} + \frac{1}{\kappa_z G_z A_2}\right)\begin{bmatrix} 1 & -1 \\ -1 & 1 \end{bmatrix} \quad \text{and}$$

$$\mathbf{f}_S = \tfrac{1}{12} q_z L_e^2 \left(\frac{1}{\kappa_z G_z A_2} - \frac{1}{\kappa_z G_z A_1}\right)\begin{Bmatrix} 1 \\ -1 \end{Bmatrix}$$

Thus, the weak form for an element is given by

$$\delta\Pi^e_{M_y} = \begin{bmatrix} \tilde{\mathbf{u}}^{bT} & \tilde{\mathbf{M}}^T_y \end{bmatrix}\left(\begin{bmatrix} \mathbf{0} & \mathbf{G}^T \\ \mathbf{G} & -(\mathbf{V}+\mathbf{W}) \end{bmatrix}\begin{Bmatrix} \tilde{\mathbf{u}}^b \\ \tilde{\mathbf{M}}_y \end{Bmatrix} - \begin{Bmatrix} \mathbf{f}_u \\ \mathbf{f}_M + \mathbf{f}_S \end{Bmatrix}\right)$$

After $\tilde{\mathbf{M}}_y$ is eliminated for each element we obtain the 'stiffness' matrix

$$\mathbf{K}^e = \mathbf{G}^T(\mathbf{V}+\mathbf{W})^{-1}\mathbf{G}$$

and load vector

$$\mathbf{f}^e = \mathbf{f}_u + \mathbf{G}^T(\mathbf{V}+\mathbf{W})^{-1}(\mathbf{f}_M + \mathbf{f}_S)$$

which may be assembled in an identical manner as in the irreducible form.

We note that the construction of the element array for the Timoshenko mixed form merely adds the effects of the shear to the terms of the Euler–Bernoulli theory. Another difference, however, is the nodal rotation θ_y is no longer directly related to the derivative of the transverse displacement w.

Example 10.11 Cantilever tapered bar – Timoshenko theory

To illustrate the performance of the irreducible and mixed forms presented for the Timoshenko beam theory we consider a cantilever beam that is loaded by a uniform load q_z as shown in Fig. 10.10. The bar has a variable bending and shear stiffness given by

$$EI_y = \frac{2EI_0}{\eta} \quad \text{and} \quad \kappa_z G_z A = \frac{2kGA_0}{\eta}$$

where $\eta = x/L$ and the stiffness values at the support are twice the free end value of EI_0.

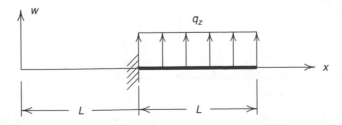

Fig. 10.10 Tapered cantilever beam.

The exact solution for the displacement and slope are given by

$$w(x) = \frac{q_z L^4}{240 E I_0} \left[3\eta^5 - 20\eta^4 + 40\eta^3 - 55\eta + 32 \right]$$
$$+ \frac{q_z L^2}{6 k G A_0} \left[3\eta^2 - \eta^3 - 2 \right]$$
$$\theta_y = \frac{q_z L^3}{48 E I_0} \left[3\eta^4 - 16\eta^3 + 24\eta^2 - 11 \right]$$

and those for moment and shear by

$$M = \tfrac{1}{2} q_z L^2 \left[4\eta - \eta^2 - 4 \right]$$
$$S = q_z L \left[2 - \eta \right]$$

In Fig. 10.11 we present the solution using two equal length elements and an irreducible (displacement) solution with cubic w and quadratic θ_y [FE(3)], an irreducible solution with quadratic w and linear θ_y [FE(2)] and the mixed solution as described above. The results for the mixed solution are exact whereas those for the other solu-

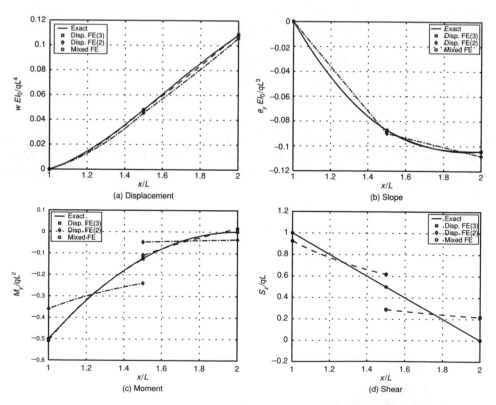

Fig. 10.11 Solution for tapered beam using irreducible (displacement) and mixed solutions – Timoshenko theory.

tions have error in all quantities, although those for the displacement and slope are quite small.

10.5.3 Discrete Kirchhoff constraints

In the previous discussion we have considered formulations for the Euler–Bernoulli and Timoshenko theories of beams. Using appropriate interpolations in the Timoshenko theory we have shown that it is possible to obtain a viable solution for situations in which the Euler–Bernoulli theory is accurate. We consider here an alternative to the previous approach by considering a reduced form of the functional for the Timoshenko theory to directly solve the thin beam case. In this form we use an irreducible form for the bending strains, set the constitutive behaviour for shear to zero, and introduce constraints to enforce zero shear strains. Initially this approach was applied to the study of thin plates based on the Kirchhoff theory and, thus, the approach is termed a *discrete Kirchhoff* method. In this section we illustrate the approach by considering a beam example.

The reduced functional is given by

$$
\delta\Pi(w,\theta_y) = \int_L \frac{\partial \delta\theta_y}{\partial x}\, \hat{M}_y(\chi_y(\theta_y))\, \mathrm{d}x
$$
$$
- \left.\left(\delta\theta_y\, \bar{M}_y\right)\right|_{\partial M_y} - \left.\left(\delta w\, \bar{S}_z\right)\right|_{\partial S_z} = 0 \qquad (10.95)
$$

and is valid if we introduce appropriate constraints to satisfy (discretely)

$$
\frac{\partial w}{\partial x} + \theta_y = 0 \qquad (10.96)
$$

To solve the problem posed by Eqs (10.95) and (10.96) we can

1. approximate w and θ_y by independent interpolations of C_0 continuity as

$$
w = \mathbf{N}_w \tilde{\mathbf{w}} \quad \text{and} \quad \theta_y = \mathbf{N}_\theta \tilde{\theta} \qquad (10.97)
$$

2. impose a discrete approximation to the constraint of Eq. (10.96) and solve the problem resulting from substitution of Eq. (10.96) into Eq. (10.95) by either discrete elimination, use of suitable Lagrangian multipliers, or penalty procedures.

In the application of the so-called *discrete Kirchhoff constraints*, Eq. (10.96) is approximated by point (or subdomain) *collocation* and direct elimination is used to *reduce the number of nodal parameters*. Of course, other means of imposing the constraints could be used with *identical* effect and we shall return to these in the next chapter. However, direct elimination is advantageous in reducing the final total number of variables and can be used effectively.

Example 10.12   One-dimensional beam example
We illustrate the process to impose discrete constraints on a simple, one-dimensional, example of a beam shown in Fig. 10.12. In this, initially the displacements and rotations

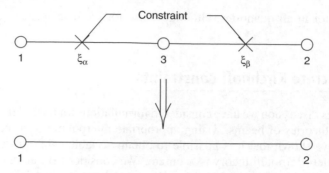

Fig. 10.12 A beam element with independent, Lagrangian, interpolation of w and θ_y with constraint $\partial w/\partial x + \theta_y = 0$ applied at points ×.

are taken as determined by a quadratic interpolation of an identical kind and we write in place of Eq. (10.97),

$$\left\{ \begin{matrix} w \\ \theta_y \end{matrix} \right\} = \sum_{a=1}^{3} N_a \left\{ \begin{matrix} \tilde{w}_a \\ \tilde{\theta}_a \end{matrix} \right\} \tag{10.98}$$

where a are the three element nodes.

The constraint is now applied by point collocation at coordinates x_α and x_β of the beam; that is, we require that at these points

$$\frac{\partial w}{\partial x} + \theta_y = 0 \tag{10.99}$$

This can be written by using the interpolation of Eq. (10.98) as two simultaneous equations

$$\begin{aligned} \sum_{a=1}^{3} N_a'(x_\alpha)\tilde{w}_a + \sum_{a=1}^{3} N_a(x_\alpha)\tilde{\theta}_a = 0 \\ \sum_{a=1}^{3} N_a'(x_\beta)\tilde{w}_a + \sum_{a=1}^{3} N_a(x_\beta)\tilde{\theta}_a = 0 \end{aligned} \tag{10.100}$$

where

$$N_a(x_\alpha) = N_a(x)\Big|_{x=x_\alpha} \qquad \text{and} \qquad N_a'(x_\alpha) = \frac{\mathrm{d}N_i}{\mathrm{d}x}\Big|_{x=x_\alpha}$$

Equations (10.100) can be used to eliminate $\tilde{w}_3$ and $\tilde{\theta}_3$. Writing Eqs (10.100) explicitly we have

$$\sum_{a=1}^{3} \mathbf{A}_a \left\{ \begin{matrix} \tilde{w}_a \\ \tilde{\theta}_a \end{matrix} \right\} = \mathbf{0} \quad \text{where} \quad \mathbf{A}_a = \begin{bmatrix} N_a'(x_\alpha), & N_a(x_\alpha) \\ N_a'(x_\beta), & N_a(x_\beta) \end{bmatrix} \tag{10.101}$$

Substitution of the above into Eq. (10.98) results directly in shape functions from which the centre node has been eliminated, that is,

$$\left\{ \begin{matrix} w \\ \theta_y \end{matrix} \right\} = \sum_{a=1}^{2} \bar{N}_a \left\{ \begin{matrix} \tilde{w}_a \\ \tilde{\theta}_a \end{matrix} \right\} \tag{10.102}$$

with

$$\bar{N}_a = N_a \mathbf{I} - N_3 \mathbf{A}_3^{-1}\mathbf{A}_a; \quad a = 1, 2$$

where $\mathbf{I}$ is a 2×2 identity matrix.

If these functions are used for the beam, we arrive at an element that is convergent. Indeed, in the particular case where x_α and x_β are chosen to coincide with the two Gauss quadrature points the element stiffness coincides with that given by a displacement formulation involving a cubic w interpolation as described above for the irreducible form. In fact, the agreement is *exact* for a uniform beam.

10.6 Forms without rotation parameters

It is possible to formulate the Timoshenko beam theory without direct use of rotation parameters. Such an approach has advantages for problems with large rotations where use of rotation parameters leads to introduction of trigonometric functions (e.g. see Chapter 17). Here we consider the case of a straight beam in two dimensions where each element is defined by coordinates at the two ends. Starting from a four-node rectangular element in which the origin of a local Cartesian coordinate system passes through the centroid of the element we may write interpolations as (Fig. 10.13)

$$
\begin{aligned}
x &= N_i(\xi, \eta)\tilde{x}_i + N_j(\xi, \eta)\tilde{x}_j + N_k(\xi, \eta)\tilde{x}_k + N_l(\xi, \eta)\tilde{x}_l \\
y &= N_i(\xi, \eta)\tilde{y}_i + N_j(\xi, \eta)\tilde{y}_j + N_k(\xi, \eta)\tilde{y}_k + N_l(\xi, \eta)\tilde{y}_l
\end{aligned}
\tag{10.103}
$$

in which N_i, etc. are the usual four-node bilinear shape functions. Noting the rectangular form of the element, these interpolations may be rewritten in terms of alternative parameters [Fig. 10.13(b)] as

$$
\begin{aligned}
x &= N_1(\xi)\tilde{x}_1 + N_2(\xi)\tilde{x}_2 \\
y &= \frac{\eta}{2}[N_1(\xi)\tilde{t}_1 + N_2(\xi)\tilde{t}_2]
\end{aligned}
\tag{10.104}
$$

where shape functions are

$$N_1(\xi) = \tfrac{1}{2}(1 - \xi), \qquad N_2(\xi) = \tfrac{1}{2}(1 + \xi) \tag{10.105}$$

and new nodal parameters are related to the original ones through

$$
\begin{aligned}
\tilde{x}_1 &= \tfrac{1}{2}(\tilde{x}_i + \tilde{x}_l) \quad \text{and} \quad \tilde{t}_1 = \tilde{y}_l - \tilde{y}_k \\
\tilde{x}_2 &= \tfrac{1}{2}(\tilde{x}_j + \tilde{x}_k) \quad \text{and} \quad \tilde{t}_2 = \tilde{y}_k - \tilde{y}_j
\end{aligned}
\tag{10.106}
$$

Since the element is rectangular $\tilde{t}_1 = \tilde{t}_2 = \tilde{t}$ [Fig. 10.13(b)]; however, the above interpolations can be generalized easily to elements which are tapered. We can now use isoparametric concepts to write the displacement field for the element as

$$
\begin{aligned}
u &= N_1(\xi)\left[\tilde{u}_1 + \frac{\eta\tilde{t}}{2\tilde{t}}\Delta\tilde{u}_1\right] + N_2(\xi)\left[\tilde{u}_2 + \frac{\eta\tilde{t}}{2\tilde{t}}\Delta\tilde{u}_2\right] \\
v &= N_1(\xi)\left[\tilde{v}_1 + \frac{\eta\tilde{t}}{2\tilde{t}}\Delta\tilde{v}_1\right] + N_2(\xi)\left[\tilde{v}_2 + \frac{\eta\tilde{t}}{2\tilde{t}}\Delta\tilde{v}_2\right]
\end{aligned}
\tag{10.107}
$$

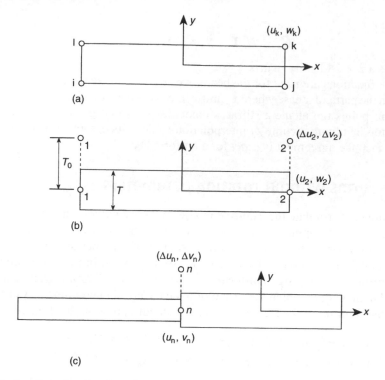

Fig. 10.13 One-dimensional bending of planar beams: (a) geometry, Q4 element; (b) geometry, no rotation parameters; (c) joining elements with different thickness.

in which $\tilde{t}$ is a 'thickness' parameter chosen to permit elements of different cross-section to be joined at a common node n [Fig. 10.13(c)].

It is evident that the above interpolations are identical to those originally written for the quadrilateral element. Only the parameters are different. Based on results for the incompressible problem, we also know the element will not perform well in bending situations because of 'shear locking', especially when the aspect ratio of the element length to depth becomes very large. In order to improve the behaviour we introduce a three-field approximation by using the enhanced strain concept described in reference 6. Accordingly, the mixed strain approximation will be taken as

$$
\left\{ \begin{array}{c} \varepsilon_x \\ \varepsilon_y \\ \gamma_{xy} \end{array} \right\} = \left[\begin{array}{cccc} N_{i,x} & \dfrac{\eta \tilde{t}}{2\tilde{t}} N_{i,x} & 0 & 0 \\ 0 & 0 & 0 & \dfrac{1}{\tilde{t}} N_i \\ 0 & \dfrac{1}{\tilde{t}} N_i & N_{i,x} & \dfrac{\eta \tilde{t}}{2\tilde{t}} N_{i,x} \end{array} \right] \left\{ \begin{array}{c} \tilde{u}_i \\ \Delta \tilde{u}_i \\ \tilde{v}_i \\ \Delta \tilde{v}_i \end{array} \right\} + \left[\begin{array}{cccc} \xi & 0 & 0 & 0 \\ 0 & \eta & 0 & 0 \\ 0 & 0 & \xi & \eta \end{array} \right] \left\{ \begin{array}{c} \beta_1 \\ \beta_2 \\ \beta_3 \\ \beta_4 \end{array} \right\}
$$

(10.108)

where β_i are parameters of the enhanced strains.[6] The remainder of the development is straightforward and is left as an exercise for the reader. We do note that here it is not necessary to use a constitutive equation which has been reduced to give zero stress in the through-thickness (y) direction. By including additional enhanced terms in the thickness direction one may use the three-dimensional constitutive equations directly. Such developments have been pursued for plate and shell applications.[34-37] We note that while the above form can be used for flat surfaces and easily extended for smoothly curved surfaces it has difficulties when 'kinks' or multiple branches are encountered as then there is no unique 'thickness' direction. Thus, considerable additional work remains to be done to make this a generally viable approach.

10.7 Moment resisting frames

The formulation given above may be used to solve moment resisting frame problems such as the plane frame shown in Fig. 10.14(a). Using a global coordinate system $\mathbf{x}' = (x', y', z')$ a transformation between the local frame $\mathbf{x}$ used in the above description of axial, bending and torsion is given by [Fig. 10.14(b)]

$$\mathbf{x} = \mathbf{L}\mathbf{x}' \tag{10.109}$$

where $\mathbf{L}$ is an array of direction cosines. The same form may be used to transform the degrees of freedom for the problem. In a general three-dimensional setting we have

$$\begin{Bmatrix} \tilde{\mathbf{u}} \\ \tilde{\boldsymbol{\theta}} \end{Bmatrix} = \begin{bmatrix} \mathbf{L} & \mathbf{0} \\ \mathbf{0} & \mathbf{L} \end{bmatrix} \begin{Bmatrix} \tilde{\mathbf{u}}' \\ \tilde{\boldsymbol{\theta}}' \end{Bmatrix} \tag{10.110}$$

where

$$\tilde{\mathbf{u}}_a = \begin{Bmatrix} \tilde{u}_a \\ \tilde{v}_a \\ \tilde{w}_a \end{Bmatrix}, \quad \tilde{\boldsymbol{\theta}}_a = \begin{Bmatrix} \tilde{\theta}_{xa} \\ \tilde{\theta}_{ya} \\ \tilde{\theta}_{za} \end{Bmatrix}$$

with similar relations for $\tilde{\mathbf{u}}'$ and $\tilde{\boldsymbol{\theta}}'$.

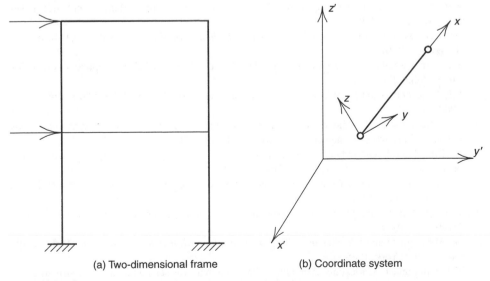

(a) Two-dimensional frame   (b) Coordinate system

Fig. 10.14 Moment resisting frame and coordinate transformation.

Using the above transformation, the residuals for each element are transformed by

$$\begin{Bmatrix} \psi_u' \\ \psi_\theta' \end{Bmatrix} = \begin{bmatrix} \mathbf{L}^T & \mathbf{0} \\ \mathbf{0} & \mathbf{L}^T \end{bmatrix} \begin{Bmatrix} \psi_u \\ \psi_\theta \end{Bmatrix} \tag{10.111}$$

and the tangent matrix by

$$\begin{bmatrix} \mathbf{K}_{uu}' & \mathbf{K}_{u\theta}' \\ \mathbf{K}_{\theta u}' & \mathbf{K}_{\theta\theta}' \end{bmatrix} = \begin{bmatrix} \mathbf{L}^T & \mathbf{0} \\ \mathbf{0} & \mathbf{L}^T \end{bmatrix} \begin{bmatrix} \mathbf{K}_{uu} & \mathbf{K}_{u\theta} \\ \mathbf{K}_{\theta u} & \mathbf{K}_{\theta\theta} \end{bmatrix} \begin{bmatrix} \mathbf{L} & \mathbf{0} \\ \mathbf{0} & \mathbf{L} \end{bmatrix} \tag{10.112}$$

10.8 Concluding remarks

In this chapter we have summarized the basic steps needed to formulate and solve rod problems by the finite element method. We have focused attention on the development of accurate interpolation of the variables which, in certain circumstances, give exact solutions at interelement nodes. In addition, for the Timoshenko beam theory, we have shown how to obtain interpolation functions which do not 'lock' in applications to thin beam problems. This latter aspect will be exploited in the solution of plate and shell problems in the next chapters.

References

1. G. Wempner. *Mechanics of Solids, with Applications to Thin Bodies.* McGraw-Hill, New York, 1973.
2. E.P. Popov and T.A. Balan. *Engineering Mechanics of Solids.* Prentice-Hall, Upper Saddle River, NJ, 2nd edition, 1999.
3. F.P. Beer, Jr., E.R. Johnson and W.E. Clausen. *Mechanics of Materials.* McGraw-Hill, Boston, 3rd edition, 2002.
4. J. Demmel. *Applied Numerical Linear Algebra.* Society for Industrial and Applied Mathematics, Philadelphia, PA, 1997.
5. S.P. Timoshenko and J.N. Goodier. *Theory of Elasticity.* McGraw-Hill, New York, 3rd edition, 1969.
6. O.C. Zienkiewicz, R.L. Taylor and J.Z. Zhu. *The Finite Element Method: Its Basis and Fundamentals.* Butterworth-Heinemann, Oxford, 6th edition, 2005.
7. S. Klinkel and S. Govindjee. Anisotropic bending-torsion coupling for warping in a non-linear beam. *Computational Mechanics,* 31:78–87, 2003.
8. S.P. Timoshenko and S. Woinowski-Krieger. *Theory of Plates and Shells.* McGraw-Hill, New York, 2nd edition, 1959.
9. P. Tong. Exact solution of certain problems by the finite element method. *Journal of AIAA,* 7:179–180, 1969.
10. M. Abramowitz and I.A. Stegun, editors. *Handbook of Mathematical Functions.* Dover Publications, New York, 1965.
11. V. Ciampi and L. Carlesimo. A nonlinear beam element for seismic analysis of structures. In *Proc. European Conference on Earthquake Engineering,* pages 73–80, Lisbon, June 1986.

12. E. Spacone, V. Ciampi and F.C. Filippou. Mixed formulation of nonlinear beam finite element. *Computers and Structures*, 58:71–83, 1995.

13. E. Spacone, F.C. Filippou and F.F. Taucer. Fiber beam–column model for non-linear analysis of R/C frames. 1. Formulation. *Earthquake Engineering and Structural Dynamics*, 25:711–725, 1996.

14. E. Spacone, F.C. Filippou and F.F. Taucer. Fiber beam–column model for non-linear analysis of R/C frames. 2. Applications. *Earthquake Engineering and Structural Dynamics*, 25:727–742, 1996.

15. A. Neuenhofer and F.C. Filippou. Evaluation of nonlinear frame finite element models. *J. Structural Engineering, ASCE*, 123:958–966, 1997.

16. A. Neuenhofer and F.C. Filippou. Geometrically nonlinear flexibility-based frame finite element. *J. Structural Engineering, ASCE*, 124:704–711, 1998.

17. M. Petrangeli and V. Ciampi. Equilibrium based iterative solution for the non-linear beam problem. *International Journal for Numerical Methods in Engineering*, 40:423–437, 1997.

18. M. Petrangeli, P.E. Pinto and V. Ciampi. Fiber element for cyclic bending and shear of RC structures. I. Theory. *J. Engineering Mechanics, ASCE*, 125:994–1001, 1999.

19. M. Schultz and F.C. Filippou. Non-linear spatial Timoshenko beam element with curvature interpolation. *International Journal for Numerical Methods in Engineering*, 50:761–785, 2001.

20. M. Schultz and F.C. Filippou. Generalized warping torsion formulation. *J. Structural Engineering, ASCE*, 124:339–347, 1998.

21. A. Ayoub and F.C. Filippou. Nonlinear finite-element analysis of RC shear panels and walls. *J. Structural Engineering, ASCE*, 124:298–308, 1998.

22. A. Ayoub and F.C. Filippou. Mixed formulation of bond-slip problems under cyclic loads. *J. Structural Engineering, ASCE*, 125(ST6):661–671, 1999.

23. A. Ayoub and F.C. Filippou. Mixed formulation of nonlinear steel–concrete composite beam element. *J. Structural Engineering, ASCE*, 126:371–381, 2000.

24. R.L. Taylor, F.C. Filippou and A. Saritas. Finite element solution of beam problems. *Computational Mechanics*, 31, 2003.

25. K.D. Hjelmstad and E. Taciroglu. Mixed variational methods for finite element analysis of geometrically non-linear, inelastic Bernoulli–Euler beams. *Communications in Numerical Methods of Engineering*, 19:809–832, 2003.

26. M.A. Crisfield. *Non-linear Finite Element Analysis of Solids and Structures*, volume 2. John Wiley & Sons, Chichester, 1997.

27. B. Fraeijs de Veubeke. Displacement and equilibrium models in finite element method. In O.C. Zienkiewicz and G.S. Holister, editors, *Stress Analysis*, Chapter 9, pages 145–197. John Wiley & Sons, Chichester, 1965.

28. A. Tessler and S.B. Dong. On a hierarchy of conforming Timoshenko beam elements. *Computers and Structures*, 14:335–344, 1981.

29. M.A. Crisfield. *Finite Elements and Solution Procedures for Structural Analysis, Vol. 1, Linear Analysis*. Pineridge Press, Swansea, 1986.

30. H.K. Stolarski, N. Carpenter and T. Belytschko. Bending and shear mode decomposition in C^0 structural elements. *Journal of Structural Mechanics, ASCE*, 11(2):153–176, 1983.

31. J.C. Simo and M.S. Rifai. A class of mixed assumed strain methods and the method of incompatible modes. *International Journal for Numerical Methods in Engineering*, 29:1595–1638, 1990.

32. U. Andelfinger, E. Ramm and D. Roehl. 2d- and 3d-enhanced assumed strain elements and their application in plasticity. In D. Owen, E. Oñate and E. Hinton, editors, *Proceedings of the 4th International Conference on Computational Plasticity*, pages 1997–2007. Pineridge Press, Swansea, 1992.

33. M. Bischoff, E. Ramm and D. Braess. A class of equivalent enhanced assumed strain and hybrid stress finite elements. *Computational Mechanics*, 22:443–449, 1999.

34. N. Büchter, E. Ramm and D. Roehl. Three-dimensional extension of non-linear shell formulations based on the enhanced assumed strain concept. *International Journal for Numerical Methods in Engineering*, 37:2551–2568, 1994.

35. M. Braun, M. Bischoff and E. Ramm. Nonlinear shell formulations for complete three-dimensional constitutive laws include composites and laminates. *Computational Mechanics*, 15:1–18, 1994.

36. P. Betsch, F. Gruttmann and E. Stein. A 4-node finite shell element for the implementation of general hyperelastic 3d-elasticity at finite strains. *Computer Methods in Applied Mechanics and Engineering*, 130:57–79, 1996.

37. M. Bischoff and E. Ramm. Shear deformable shell elements for large strains and rotations. *International Journal for Numerical Methods in Engineering*, 40:4427–4449, 1997.

Plate bending approximation: thin (Kirchhoff) plates and C_1 continuity requirements

11.1 Introduction

The subject of bending of plates and indeed its extension to shells was one of the first to which the finite element method was applied in the early 1960s. At that time the various difficulties that were to be encountered were not fully appreciated and for this reason the topic remains one in which research is active to the present day. Although the subject is of direct interest only to applied mechanicians and structural engineers there is much that has more general applicability, and many of the procedures which we shall introduce can be directly translated to other fields of application.

Plates and shells are but a particular form of a three-dimensional solid, the treatment of which presents no theoretical difficulties, at least in the case of elasticity. However, the thickness of such structures (denoted throughout this and later chapters as t) is very small when compared with other dimensions, and complete three-dimensional numerical treatment is not only costly but in addition often leads to serious numerical ill-conditioning problems. To ease the solution, even long before numerical approaches became possible, several classical assumptions regarding the behaviour of such structures were introduced. Clearly, such assumptions result in a series of approximations. Thus numerical treatment will, in general, concern itself with the approximation to an already approximate theory (or mathematical model), the validity of which is restricted. On occasion we shall point out the shortcomings of the original assumptions, and indeed modify these as necessary or convenient. This can be done simply because now we are granted more freedom than that which existed in the 'pre-computer' era.

The *thin plate* theory is based on the assumptions formalized by Kirchhoff in 1850,[1] and indeed his name is often associated with this theory, though an early version was presented by Sophie Germain in 1811.[2–4] A relaxation of the assumptions was made by Reissner in 1945[5] and in a slightly different manner by Mindlin[6] in 1951. These modified theories extend the field of application of the theory to *thick plates* and we shall associate this name with the Reissner–Mindlin postulates.

It turns out that the thick plate theory is simpler to implement in the finite element method, though in the early days of analytical treatment it presented more difficulties. As it is more convenient to introduce first the thick plate theory and by imposition of additional assumptions to limit it to thin plate theory we shall follow this path in the present chapter. However, when discussing numerical solutions we shall reverse the process and follow the historical procedure of dealing with the thin plate situations first in this chapter. The extension to thick plates and to what turns out always to be a *mixed* formulation will be the subject of Chapter 12.

In the thin plate theory it is possible to represent the state of deformation by one quantity w, the lateral displacement of the middle plane of the plate. Thus we find that thin plates share some of the same characteristics as the Euler–Bernoulli beam theory considered in the previous chapter. Clearly, such a formulation is *irreducible*. The achievement of this irreducible form introduces second derivatives of w in the strain definition and continuity conditions between elements have now to be imposed not only on this quantity but also on its derivatives (C_1 continuity). This is to ensure that the plate remains continuous and does not 'kink'.[*] Thus at nodes on element interfaces it will always be necessary to use both the value of w and its slopes (first derivatives of w) to impose continuity, again this is similar to the treatment of Euler–Bernoulli beam theory.

Determination of suitable shape functions is now much more complex than those needed for C_0 continuity or for beams. Indeed, as complete slope continuity is required on the interfaces between various elements, the mathematical and computational difficulties often rise disproportionately fast. It is, however, relatively simple to obtain shape functions which, while preserving continuity of w, may violate its slope continuity between elements, though normally not at a node where such continuity is imposed.[†] If such chosen functions satisfy the 'patch test' then convergence will still be found.[7] The first part of this chapter will be concerned with such 'non-conforming' or 'incompatible' shape functions. In later parts new functions will be introduced by which continuity can be restored. The solution with such 'conforming' shape functions will now give bounds to the energy of the correct solution, but, on many occasions, will yield inferior accuracy to that achieved with non-conforming elements. Thus, for practical usage the methods of the first part of the chapter are often recommended.

The shape functions for rectangular elements are the simplest to form for thin plates and will be introduced first. Shape functions for triangular and quadrilateral elements are more complex and will be introduced later for solutions of plates of *arbitrary* shape or, for that matter, for dealing with shell problems where such elements are essential.

The problem of thin plates is associated with *fourth-order differential equations* leading to a potential energy function which contains *second derivatives* of the unknown function. It is characteristic of a large class of physical problems and, although the chapter concentrates on the structural problem, the reader will find that the procedures developed will also be equally applicable to any problem which is of fourth order.

The difficulty of imposing C_1 continuity on the shape functions has resulted in many alternative approaches to the problems in which this difficulty is side-stepped. Several possibilities exist. Two of the most important are:

---

[*] If 'kinking' occurs the second derivative or curvature becomes infinite and squares of infinite terms occur in the energy expression.

[†] Later we show that even slope discontinuity at the node may be used.

1. independent interpolation of rotations ϕ and displacement w, imposing continuity as a special constraint, often applied at discrete points only;
2. the introduction of Lagrange multiplier variables or indeed other variables to avoid the necessity of C_1 continuity.

Both approaches fall into the class of mixed formulations and we shall discuss these briefly at the end of the chapter. However, a fuller statement of mixed approaches will be made in the next chapter where both thick and thin approximations will be dealt with simultaneously.

11.2 The plate problem: thick and thin formulations

11.2.1 Governing equations

The mechanics of plate action can be illustrated in one dimension by considering a plate of infinite extent in one dimension (here assumed the y) and considering equations similar to those developed for a beam in Sec. 10.2. Here we consider the problem of cylindrical bending of plates.[2] In this problem the plate is to be loaded and supported by conditions independent of y. In this case we may analyse a strip of unit width subjected to some *stress resultants* M_x, P_x, and S_x, which denote x-direction bending moment, axial force and transverse shear force, respectively,[*] as shown in Fig. 11.1. For cross-sections that are originally normal to the middle plane of the plate we can use the approximation that at some distance from points of support or concentrated loads plane sections will remain plane during the deformation process. The postulate that sections normal to the middle plane remain plane during deformation is thus the *first* and most important assumption of the theory of plates (and indeed shells). To this is added the *second* assumption. This simply observes that the direct stresses in the normal direction, z, are small, that is, of the order of applied lateral load intensities, q, and hence direct strains in that direction can be neglected. This 'inconsistency' in approximation is compensated for by assuming a plane stress condition in each lamina.

With these two assumptions it is easy to see that the total state of deformation can be described by displacements u and w of the middle surface ($z = 0$) and a rotation ϕ_x of the normal (Fig. 11.1). Thus the local displacements in the directions of the x and z axes are taken as

$$u_1(x, z) = u(x) + z\phi_x(x) \qquad \text{and} \qquad u_3(x, z) = w(x) \qquad (11.1)$$

Immediately the strains in the x and z directions are available as

$$\varepsilon_x = \frac{\partial u_1}{\partial x} = \frac{\partial u}{\partial x} + z\frac{\partial \phi_x}{\partial x}$$

$$\varepsilon_z = 0 \qquad (11.2)$$

$$\gamma_{xz} = \frac{\partial u_1}{\partial z} + \frac{\partial u_3}{\partial x} = \frac{\partial w}{\partial x} + \phi_x$$

---

[*] Here we change our notation slightly from that used for beams in order to conform to the notation commonly used for plates.

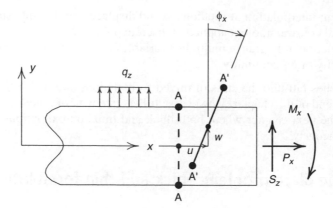

Fig. 11.1 Displacements and force resultants for cylindrical bending of a plate.

The non-zero strains are identical to those obtained for bending of beams. For the cylindrical bending problem a state of linear elastic, plane stress for each lamina yields the stress–strain relations

$$\sigma_x = \frac{E}{1 - \nu^2}\varepsilon_x \quad \text{and} \quad \tau_{xz} = G\gamma_{xz}$$

The stress resultants are obtained as

$$P_x = \int_{-t/2}^{t/2} \sigma_x \, dz = B\frac{\partial u}{\partial x}$$
$$S_x = \int_{-t/2}^{t/2} \tau_{xz} \, dz = \kappa Gt \left(\frac{\partial w}{\partial x} + \phi_x \right) \qquad (11.3a)$$
$$M_x = \int_{-t/2}^{t/2} \sigma_x z \, dz = D\frac{\partial \phi_x}{\partial x}$$

where B is the in-plane plate stiffness and D the bending stiffness for an isotropic elastic material and are computed from

$$B = \frac{Et}{1 - \nu^2} \quad \text{and} \quad D = \frac{Et^3}{12(1 - \nu^2)} \qquad (11.3b)$$

with ν Poisson's ratio, E and G direct and shear elastic moduli, respectively.[*]
   Three equations of equilibrium complete the basic formulation. These equilibrium equations may be computed directly from a differential element of the plate or by integration over the thickness of the local equilibrium equations as was performed for

---

[*] A constant κ has been added here to account for the fact that the shear stresses are not constant across the section. A value of $\kappa = 5/6$ is exact for a rectangular, homogeneous section and corresponds to a parabolic shear stress distribution.

the beam in Sec. 10.2.1. Using the latter approach and assuming zero inertial forces we have for the axial resultant

$$\int_{-t/2}^{t/2} \left[\frac{\partial \sigma_x}{\partial x} + \frac{\partial \tau_{xz}}{\partial z} + b_x \right] dz = \frac{\partial}{\partial x} \int_{-t/2}^{t/2} \sigma_x \, dz + \int_{-t/2}^{t/2} b_x \, dz + \tau_{xz} \Big|_{t/2} - \tau_{xz} \Big|_{-t/2} = 0$$

$$\frac{\partial P_x}{\partial x} + q_x = 0$$

(11.4a)

where q_x is an axial load similar to that obtained for the beam. Similarly, the shear resultant follows from

$$\int_{-t/2}^{t/2} \left[\frac{\partial \tau_{xz}}{\partial x} + \frac{\partial \sigma_z}{\partial z} + b_z \right] dz = \frac{\partial}{\partial x} \int_{-t/2}^{t/2} \tau_{xz} \, dz + \int_{-t/2}^{t/2} b_x \, dz + \sigma_z \Big|_{t/2} - \sigma_z \Big|_{-t/2} = 0$$

$$\frac{\partial S_x}{\partial x} + q_z = 0$$

(11.4b)

where the transverse loading q_z arises from the body force and the resultant of the normal traction on the top and/or bottom surfaces. Finally, the moment equilibrium is deduced from

$$\int_{-t/2}^{t/2} z \left[\frac{\partial \sigma_x}{\partial x} + \frac{\partial \tau_{xz}}{\partial z} + b_x \right] dz = \frac{\partial}{\partial x} \int_{-t/2}^{t/2} z \, \sigma_x \, dz - \int_{-t/2}^{t/2} \tau_{xz} \, dz + \int_{-t/2}^{t/2} z \, b_x \, dz = 0$$

$$\frac{\partial M_x}{\partial x} - S_x + m_x = 0$$

(11.4c)

Generally, m_x loads are not included in plate theory except as an artifice to introduce the d'Alembert inertial forces. As we found was true for beams, in the elastic case of a plate it is easy to see that the in-plane displacements and forces, u and P_x, decouple from the other terms and the problem of lateral deformations can be dealt with separately. We shall thus only consider bending in the present chapter, returning to the combined problem, characteristic of shell behaviour, in later chapters.

Equations (11.1)–(11.4c) are typical for thick plates, and the thin plate theory adds an additional assumption. This simply neglects the shear deformation and puts $G = \infty$ (or $\gamma_{xz} = 0$). Equation (11.3a) thus becomes

$$\frac{\partial w}{\partial x} + \phi_x = 0 \qquad\qquad (11.5)$$

This thin plate assumption is equivalent to stating that the normals to the middle plane remain normal to it during deformation and is the same as the Bernoulli–Euler assumption for thin beams considered in Chapter 10. The thin, constrained theory is very widely used in practice and proves adequate for a large number of structural problems, though, of course, should not be taken literally as the true behaviour near supports or where local load action is important and is three dimensional.

In Fig. 11.2 we illustrate some of the boundary conditions imposed on plates and immediately note that the diagrammatic representations of simple support as a knife

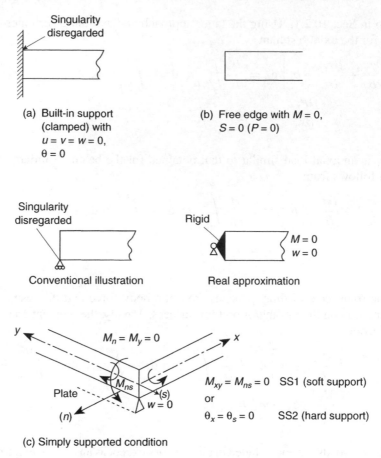

(a) Built-in support
    (clamped) with
    $u = v = w = 0$,
    $\theta = 0$

(b) Free edge with $M = 0$,
    $S = 0$ $(P = 0)$

Conventional illustration

Real approximation

(c) Simply supported condition

Fig. 11.2 Support (end) conditions for a plate. Note: the conventionally illustrated simple support leads to infinite displacement – reality is different.

edge would lead to infinite displacements and stresses. Of course, if a rigid bracket is added in the manner shown this will alter the behaviour to that which we shall generally assume.

The one-dimensional problem of plates and the introduction of thick and thin assumptions translate directly to the general theory of plates. In Fig. 11.3 we illustrate the extensions necessary and write, in place of Eq. (11.1) (assuming u_0 and v_0 to be zero),

$$u_x = z\,\phi_x(x, y) \quad u_y = z\,\phi_y(x, y) \quad u_z = w(x, y) \tag{11.6}$$

where we note that displacement parameters are now functions of x and y. It is sometimes advantageous to replace ϕ_x and ϕ_y by rotations about the x and y coordinates in a manner used for the beam developments. Thus,

$$\begin{Bmatrix} \phi_x \\ \phi_y \end{Bmatrix} = \begin{bmatrix} 0 & 1 \\ -1 & 0 \end{bmatrix} \begin{Bmatrix} \theta_x \\ \theta_y \end{Bmatrix} \quad \text{or} \quad \begin{Bmatrix} \theta_x \\ \theta_y \end{Bmatrix} = \begin{bmatrix} 0 & -1 \\ 1 & 0 \end{bmatrix} \begin{Bmatrix} \phi_x \\ \phi_y \end{Bmatrix} \tag{11.7}$$

$$\phi = \mathbf{T}\,\theta$$

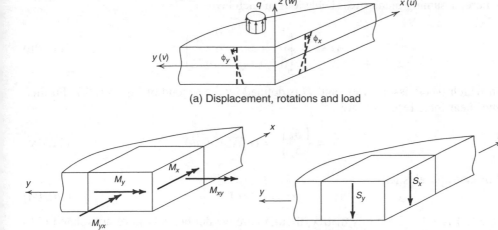

(a) Displacement, rotations and load

(b) Force resultants (on-faces)

Fig. 11.3 Definitions of variables for plate approximations.

may be substituted at any time in the development. We do not make this change here as it complicates the expressions for the governing equations. We normally would make the substitution on nodal parameters after arrays are formulated.

The strains may now be separated into bending (in-plane components) and transverse shear groups and we have, in place of Eq. (11.2),

$$
\boldsymbol{\varepsilon} = \begin{Bmatrix} \varepsilon_x \\ \varepsilon_y \\ \gamma_{xy} \end{Bmatrix} = z \begin{bmatrix} \dfrac{\partial}{\partial x} & 0 \\ 0 & \dfrac{\partial}{\partial y} \\ \dfrac{\partial}{\partial y} & \dfrac{\partial}{\partial x} \end{bmatrix} \begin{Bmatrix} \phi_x \\ \phi_y \end{Bmatrix} \equiv z \, \mathcal{L}\boldsymbol{\phi} \tag{11.8a}
$$

and

$$
\boldsymbol{\gamma} = \begin{Bmatrix} \gamma_{xz} \\ \gamma_{yz} \end{Bmatrix} = \begin{Bmatrix} \dfrac{\partial w}{\partial x} \\ \dfrac{\partial w}{\partial y} \end{Bmatrix} + \begin{Bmatrix} \phi_x \\ \phi_y \end{Bmatrix} = \boldsymbol{\nabla} w + \boldsymbol{\phi} \tag{11.8b}
$$

We note that now in addition to normal bending moments M_x and M_y, now defined by expression (11.3a) for the x and y directions, respectively, a twisting moment arises defined by

$$
M_{xy} = \int_{-t/2}^{t/2} \tau_{xy} \, dz \tag{11.9}
$$

Introducing appropriate constitutive relations, all moment components can be related to displacement derivatives. For isotropic elasticity we can thus write, in place of Eq. (11.3a),

$$
\mathbf{M} = \begin{Bmatrix} M_x \\ M_y \\ M_{xy} \end{Bmatrix} = \mathbf{D}\,\mathcal{L}\boldsymbol{\phi} \tag{11.10a}
$$

where, assuming plane stress behaviour in each layer,

$$\mathbf{D} = D \begin{bmatrix} 1 & \nu & 0 \\ \nu & 1 & 0 \\ 0 & 0 & (1-\nu)/2 \end{bmatrix} \tag{11.10b}$$

in which ν is Poisson's ratio and D is defined by the second of Eqs (11.3b). Further, the shear force resultants are

$$\mathbf{S} = \begin{Bmatrix} S_x \\ S_y \end{Bmatrix} = \alpha\,(\nabla w + \phi) \tag{11.10c}$$

For isotropic elasticity

$$\alpha = \kappa\,G\,t\,\mathbf{I} \tag{11.10d}$$

where $\mathbf{I}$ is a 2×2 identity matrix (though here we deliberately have not related G to E and ν to allow for possibly different shear rigidities).

Of course, the constitutive relations can be simply generalized to anisotropic or inhomogeneous behaviour such as can be manifested if several layers of materials are assembled to form a *composite*. The only apparent difference is the structure of the $\mathbf{D}$ and α matrices, which can always be found by simple integration.

The governing equations of thick and thin plate behaviour are completed by writing the equilibrium relations. Again omitting the 'in-plane' behaviour we have, in place of Eq. (11.4b),

$$\begin{bmatrix} \dfrac{\partial}{\partial x}, & \dfrac{\partial}{\partial y} \end{bmatrix} \begin{Bmatrix} S_x \\ S_y \end{Bmatrix} + q \equiv \nabla^{\mathrm{T}}\mathbf{S} + q = 0 \tag{11.11a}$$

and, in place of Eq. (11.4c),

$$\begin{bmatrix} \dfrac{\partial}{\partial x} & 0 & \dfrac{\partial}{\partial y} \\ 0 & \dfrac{\partial}{\partial y} & \dfrac{\partial}{\partial x} \end{bmatrix} \begin{Bmatrix} M_x \\ M_y \\ M_{xy} \end{Bmatrix} + \begin{Bmatrix} S_x \\ S_y \end{Bmatrix} \equiv \mathcal{L}^{\mathrm{T}}\mathbf{M} + \mathbf{S} = \mathbf{0} \tag{11.11b}$$

Equations (11.10a)–(11.11b) are the basis from which the solution of both thick and thin plates can start. For thick plates any (or all) of the independent variables can be approximated independently, leading to a mixed formulation which we shall discuss in Chapter 12 and also briefly in Sec. 11.16 of this chapter.

For thin plates in which the shear deformations are suppressed Eq. (11.10c) is rewritten as

$$\nabla w + \phi = \mathbf{0} \tag{11.12}$$

and the strain–displacement relations (11.8a) become

$$\varepsilon = -z\,\mathcal{L}\nabla w = -z \begin{Bmatrix} \dfrac{\partial^2 w}{\partial x^2} \\[2mm] \dfrac{\partial^2 w}{\partial y^2} \\[2mm] 2\dfrac{\partial^2 w}{\partial x \partial y} \end{Bmatrix} = -z\,\chi \tag{11.13}$$

where χ is the matrix of changes in curvature of the plate. Using the above form for the thin plate, both irreducible and mixed forms can now be written. In particular, it is an easy matter to eliminate $\mathbf{M}$, $\mathbf{S}$ and ϕ and leave only w as the variable.

Applying the operator ∇^{T} to expression (11.11a), inserting Eqs (11.10a) and (11.11a) and finally replacing ϕ by the use of Eq. (11.12) gives a scalar equation

$$(\mathcal{L}\nabla)^{\mathrm{T}}\mathbf{D}\mathcal{L}\nabla w + q = 0 \qquad (11.14a)$$

where, using Eq. (11.13),

$$(\mathcal{L}\nabla) = \left[\frac{\partial^2}{\partial x^2}, \ \frac{\partial^2}{\partial y^2}, \ 2\frac{\partial^2}{\partial x\partial y}\right]^{\mathrm{T}}$$

In the case of isotropy with constant bending stiffness D this becomes the well-known biharmonic equation of plate flexure

$$D\left(\frac{\partial^4 w}{\partial x^4} + 2\frac{\partial^4 w}{\partial x^2\partial y^2} + \frac{\partial^4 w}{\partial y^4}\right) + q = 0 \qquad (11.14b)$$

11.2.2 The boundary conditions

The boundary conditions which have to be imposed on the problem (see Figs 11.2 and 11.4) include the following classical conditions.

1. *Fixed boundary*, where displacements on restrained parts of the boundary are given specified values.[*] These conditions are expressed as

$$w = \bar{w}; \quad \phi_n = \bar{\phi}_n \quad \text{and} \quad \phi_s = \bar{\phi}_s$$

Here n and s are directions normal and tangential to the boundary curve of the middle surface. A *clamped edge* is a special case with zero values assigned.
2. *Traction boundary*, where stress resultants M_n, M_{ns} and S_n (conjugate to the displacements θ_n, θ_s and w) are given prescribed values.

$$M_n = \bar{M}_n; \quad M_{ns} = \bar{M}_{ns} \quad \text{and} \quad S_n = \bar{S}_n$$

A *free edge* is a special case with zero values assigned.
3. *'Mixed' boundary conditions*, where both traction and displacement components can be specified. Typical here is the *simply supported edge* (see Fig. 11.2). For this, clearly, $M_n = 0$ and $w = 0$, but it is less clear whether M_{ns} or ϕ_s needs to be given. Specification of $M_{ns} = 0$ is *physically* a more acceptable condition. This should be adopted for thick plates. Thus options for simply supported edges are

Type	Conditions		
SS1	$w = 0,$	$M_n = 0,$	$M_{ns} = 0$
SS2	$w = 0,$	$M_n = 0,$	$\phi_s = 0$

[*] Note that in thin plates the specification of w along s automatically specifies ϕ_s by Eq. (11.12), but this is not the case in thick plates where the quantities are independently prescribed.

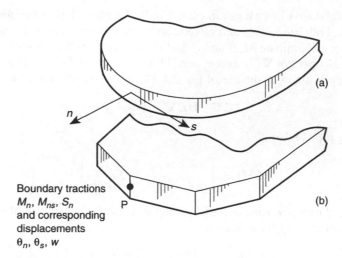

Boundary tractions
M_n, M_{ns}, S_n
and corresponding
displacements
θ_n, θ_s, w

Fig. 11.4 Boundary traction and conjugate displacement. Note: the simply supported condition requiring M_n = 0, ϕ_s = 0 and w = 0 is identical at a corner node to specifying $\phi_n = \phi_s = 0$, that is, a clamped support. This leads to a paradox if a curved boundary (a) is modelled as a polygon (b).

In thin plates ϕ_s is automatically specified from w and we shall find certain difficulties, and indeed anomalies, associated with this assumption.[8,9] For instance, in Fig. 11.4 we see how a specification of $\phi_s = 0$ at corner nodes implicit in thin plates formally leads to the prescription of all boundary parameters, which is identical to boundary conditions of a clamped plate for this point. Thus, for a curved simply supported edge a unique normal to each node must be specified and used to specify one of the mixed conditions given above.

11.2.3 The irreducible, thin plate approximation

The thin plate approximation when cast in terms of a single variable w is clearly irreducible and is in fact typical of a displacement formulation. The equations (11.11a) and (11.11b) can be written together as

$$- (\mathcal{L}\nabla)^{\mathrm{T}}\mathbf{M} + q = 0 \tag{11.15}$$

and the constitutive relation (11.10a) can be recast by using Eq. (11.12) as

$$\mathbf{M} = - \mathbf{D}\mathcal{L}\nabla w \tag{11.16}$$

The derivation of the finite element equations can be obtained either from a weak form of Eq. (11.15) obtained by weighting with an arbitrary function (say $v = \mathbf{N}\tilde{v}$) and integration by parts (done twice) or, more directly, by application of the virtual work equivalence. Using the latter approach we may write the internal virtual work for the plate as

$$\delta\Pi_{\mathrm{int}} = \int_\Omega (\delta\varepsilon)^{\mathrm{T}}\mathbf{D}\varepsilon \, \mathrm{d}\Omega = \int_\Omega \delta w (\mathcal{L}\nabla)^{\mathrm{T}}\mathbf{D}(\mathcal{L}\nabla)w \, \mathrm{d}\Omega \tag{11.17}$$

where Ω denotes the area of the plate reference (middle) surface and $\mathbf{D}$ is the plate stiffness, which for isotropy is given by Eq. (11.10b).

Similarly the external work is given by[2]

$$\delta\Pi_{\text{ext}} = \int_\Omega \delta w \, q \, d\Omega + \int_{\Gamma_n} \delta\phi_n \, \bar{M}_n \, d\Gamma + \int_{\Gamma_t} \delta\phi_s \, \bar{M}_{ns} \, d\Gamma + \int_{\Gamma_s} \delta w \, \bar{S}_n \, d\Gamma \quad (11.18)$$

where $\bar{M}_n$, $\bar{M}_{ns}$, $\bar{S}_n$ are specified values and Γ_n, Γ_t and Γ_s are parts of the boundary where each component is specified. For thin plates with straight edges Eq. (11.12) gives immediately $\phi_s = -\partial w/\partial s$ and thus the last two terms above may be combined as

$$\int_{\Gamma_t} \delta\phi_s \bar{M}_{ns} \, d\Gamma + \int_{\Gamma_s} \delta w \bar{S}_n \, d\Gamma = \int_{\Gamma_s} \delta w \left(\bar{S}_n + \frac{\partial \bar{M}_{ns}}{\partial s} \right) d\Gamma + \sum_i \delta w_i R_i \quad (11.19)$$

where R_i are concentrated forces arising at locations where corners exist (see Fig. 11.2).[2] Substituting into Eqs (11.17) and (11.18) the discretization

$$w = \mathbf{N}\tilde{\mathbf{u}} \quad (11.20)$$

where $\tilde{\mathbf{u}}$ are appropriate parameters, we can obtain for a linear case standard displacement approximation equations

$$\mathbf{K}\tilde{\mathbf{u}} = \mathbf{f} \quad (11.21)$$

with

$$\mathbf{K}\tilde{\mathbf{u}} = \left(\int_\Omega \mathbf{B}^\mathsf{T} \mathbf{D} \mathbf{B} \, d\Omega \right) \tilde{\mathbf{u}} \equiv - \int_\Omega \mathbf{B}^\mathsf{T} \mathbf{M} \, d\Omega \quad (11.22)$$

and

$$\mathbf{f} = \int_\Omega \mathbf{N}^\mathsf{T} q \, d\Omega + \mathbf{f}_b \quad (11.23)$$

where $\mathbf{f}_b$ is the boundary contribution to be discussed later and

$$\mathbf{M} = -\mathbf{D}\mathbf{B}\tilde{\mathbf{u}} \quad (11.24)$$

with

$$\mathbf{B} = (\mathcal{L}\nabla)\mathbf{N} \quad (11.25)$$

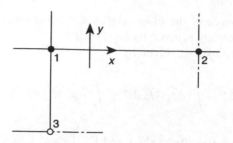

Fig. 11.5 Continuity requirement for normal slopes.

It is of interest, and indeed important to note, that when tractions are prescribed to non-zero values the force term $\mathbf{f}_b$ includes all prescribed values of M_n, M_{ns} and S_n irrespective of whether the thick or thin formulation is used. The reader can verify that this term is

$$\mathbf{f}_b = \int_\Gamma (\mathbf{N}_n^T \bar{M}_n + \mathbf{N}_s^T \bar{M}_{ns} + \mathbf{N}^T \bar{S}_n) \, d\Gamma \tag{11.26a}$$

where $\bar{M}_n$, $\bar{M}_{ns}$ and $\bar{S}_n$ are prescribed values and for thin plates [though, of course, relation (11.26a) is valid for thick plates also]:

$$\mathbf{N}_n = \frac{\partial \mathbf{N}}{\partial n} \quad \text{and} \quad \mathbf{N}_s = \frac{\partial \mathbf{N}}{\partial s} \tag{11.26b}$$

The reader will recognize in the above the well-known ingredients of a displacement formulation (see Chapter 2 and reference 7) and the procedures are almost automatic once $\mathbf{N}$ is chosen.

11.2.4 Continuity requirement for shape functions (C_1 continuity)

In Sec. 11.3–11.13 we will be concerned with the above formulation [starting from Eqs (11.17) and (11.18)], and the presence of the second derivatives indicates quite clearly that we shall need C_1 continuity of the shape functions for the irreducible, thin plate, formulation. This continuity is difficult to achieve and reasons for this are given below.

To ensure the continuity of both w and its normal slope across an interface we must have both w and $\partial w/\partial n$ uniquely defined by values of nodal parameters along such an interface. Consider Fig. 11.5 depicting the side 1–2 of a rectangular element. The normal direction n is in fact that of y and we desire w and $\partial w/\partial y$ to be uniquely determined by values of w, $\partial w/\partial x$, $\partial w/\partial y$ at the nodes lying along this line.

To show the C_1 continuity along side 1–2, we write

$$w = A_1 + A_2 x + A_3 y + \cdots \tag{11.27}$$

and

$$\frac{\partial w}{\partial y} = B_1 + B_2 x + B_3 y + \cdots \tag{11.28}$$

with a number of constants in each expression just sufficient to determine a unique solution for the nodal parameters associated with the line.

Thus, for instance, if only two nodes are present a cubic variation of w should be permissible noting that $\partial w/\partial x$ and w are specified at each node. Similarly, only a linear, or two-term, variation of $\partial w/\partial y$ would be permissible.

Note, however, that a similar exercise could be performed along the side placed in the y direction preserving continuity of $\partial w/\partial x$ along this. Along side 1–2 we thus have $\partial w/\partial y$, depending on nodal parameters of line 1–2 only, and along side 1–3 we have $\partial w/\partial x$, depending on nodal parameters of line 1–3 only. Differentiating the first with respect to x, on line 1–2 we have $\partial^2 w/\partial x \partial y$, depending on nodal parameters of line 1–2 only, and similarly, on line 1–3 we have $\partial^2 w/\partial y \partial x$, depending on nodal parameters of line 1–3 only.

At the common point, 1, an inconsistency arises immediately as we cannot automatically have there the necessary identity for continuous functions

$$\frac{\partial^2 w}{\partial x \partial y} \equiv \frac{\partial^2 w}{\partial y \partial x} \tag{11.29}$$

for arbitrary values of the parameters at nodes 2 and 3. *It is thus impossible to specify simple polynomial expressions for shape functions ensuring full compatibility when only w and its slopes are prescribed at corner nodes.*[10]

Thus if any functions satisfying the compatibility are found with the three nodal variables, they must be such that at corner nodes these functions are not continuously differentiable and the cross-derivative is not unique. Some such functions are discussed in the second part of this chapter.[11–16]

The above proof has been given for a rectangular element. Clearly, the arguments can be extended for any two arbitrary directions of interface at the corner node 1.

A way out of this difficulty appears to be obvious. We could specify the cross-derivative as one of the nodal parameters. For an assembly of rectangular elements, this is convenient and indeed permissible. Simple functions of that type have been suggested by Bogner *et al.*[17] and used with some success. Unfortunately, the extension to nodes at which a number of element interfaces meet with different angles (Fig. 11.6) is not, in general, permissible. Here, the continuity of cross-derivatives in several sets of orthogonal directions implies, in fact, a specification of *all second derivatives at a node*.

This, however, violates physical requirements if the plate stiffness varies abruptly from element to element, for then equality of moments normal to the interfaces cannot

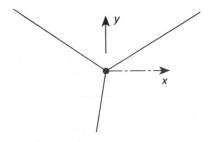

Fig. 11.6 Nodes where elements meet in arbitrary directions.

be maintained. However, this process has been used with some success in homogeneous plate situations[18-25] although Smith and Duncan[18] comment adversely on the effect of imposing such *excessive continuities* on several orders of higher derivatives.

The difficulties of finding compatible displacement functions have led to many attempts at ignoring the complete slope continuity while still continuing with the other necessary criteria. Proceeding perhaps from a naive but intuitive idea that the imposition of slope continuity at nodes only must, in the limit, lead to a complete slope continuity, several successful, 'non-conforming', elements have been developed.[12,26-40]

The convergence of such elements is not obvious but can be proved either by application of the patch test or by comparison with finite difference algorithms. We have discussed the importance of the patch test extensively in reference 7 and additional details are available in Sec. 11.7 and in references 41–43.

In plate problems the importance of the patch test in both design and testing of elements is paramount and this test should never be omitted. In the first part of this chapter, dealing with non-conforming elements, we shall repeatedly make use of it. Indeed, we shall show how some successful elements have developed via this analytical interpretation.[44-49]

Non-conforming shape functions

11.3 Rectangular element with corner nodes (12 degrees of freedom)

Shape functions

Consider a rectangular element of a plate 1234 coinciding with the xy plane as shown in Fig. 11.7. At each node, a, displacements $\tilde{\mathbf{u}}_a$ are introduced. These have three components: the first a displacement in the z direction, w_a, the second a rotation about the x axis, $(\tilde{\theta}_x)_a$, and the third a rotation about the y axis, $(\tilde{\theta}_y)_a$.*

The nodal displacement vectors are defined below as $\tilde{\mathbf{u}}_a$. The element displacement will, as usual, be given by a listing of the nodal displacements, now totalling twelve:

$$\tilde{\mathbf{u}}^e = \begin{Bmatrix} \tilde{\mathbf{u}}_1 \\ \tilde{\mathbf{u}}_2 \\ \tilde{\mathbf{u}}_3 \\ \tilde{\mathbf{u}}_4 \end{Bmatrix} \quad \text{with} \quad \tilde{\mathbf{u}}_a = \begin{Bmatrix} \tilde{w}_a \\ \tilde{\theta}_{xa} \\ \tilde{\theta}_{ya} \end{Bmatrix} \qquad (11.30)$$

A polynomial expression is conveniently used to define the shape functions in terms of the 12 parameters. Certain terms must be omitted from a complete fourth-order

---

* Note that we have changed here the convention from that of Fig. 11.3 in this chapter by using Eq. (11.7). This allows transformations needed for shells to be carried out in an easier manner. However, when manipulating the equations of Chapter 12 we shall return to the original definitions using the ϕ of Fig. 11.3. Similar difficulties are discussed by Hughes.[50]

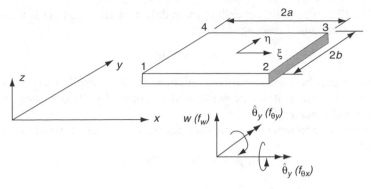

Forces and corresponding
displacements

Fig. 11.7 A rectangular plate element.

polynomial. Writing

$$w = \alpha_1 + \alpha_2 x + \alpha_3 y + \alpha_4 x^2 + \alpha_5 xy + \alpha_6 y^2 + \alpha_7 x^3 + \alpha_8 x^2 y$$
$$+ \alpha_9 xy^2 + \alpha_{10} y^3 + \alpha_{11} x^3 y + \alpha_{12} xy^3 \qquad (11.31)$$
$$\equiv \mathbf{P}\boldsymbol{\alpha}$$

has certain advantages. In particular, along any x constant or y constant line, the displacement w will vary as a cubic. The element boundaries or interfaces are composed of such lines. As a cubic is uniquely defined by four constants, the two end values of slopes and the two displacements at the ends will therefore define the displacements along the boundaries uniquely. As such end values are common to adjacent elements continuity of w will be imposed along any interface.

It will be observed that the gradient of w normal to any of the boundaries also varies along it in a cubic way. (Consider, for instance, values of the normal $\partial w/\partial x$ along a line on which x is constant.) As on such lines only two values of the normal slope are defined, the cubic is not specified uniquely and, in general, a discontinuity of normal slope will occur. The function is thus 'non-conforming'.

The constants α_1 to α_{12} can be evaluated by writing down the 12 simultaneous equations linking the values of w and its slopes at the nodes when the coordinates take their appropriate values. For instance,

$$\tilde{w}_a = \alpha_1 + \alpha_2 x_a + \alpha_3 y_a + \cdots$$
$$\frac{\partial w}{\partial y}\bigg|_a = \tilde{\theta}_{xa} = \alpha_3 + \alpha_5 x_a + \cdots$$
$$-\frac{\partial w}{\partial x}\bigg|_a = \tilde{\theta}_{ya} = -\alpha_2 - \alpha_5 y_a - \cdots$$

Listing all 12 equations, we can write, in matrix form,

$$\tilde{\mathbf{u}}^e = \mathbf{C}\boldsymbol{\alpha} \qquad (11.32)$$

where $\mathbf{C}$ is a 12×12 matrix depending on nodal coordinates, and $\boldsymbol{\alpha}$ is a vector of the 12 unknown constants. Inverting we have

$$\boldsymbol{\alpha} = \mathbf{C}^{-1}\tilde{\mathbf{u}}^e \tag{11.33}$$

This inversion can be carried out by computer or, if an explicit expression for the stiffness, etc., is desired, it can be performed algebraically. This was in fact done by Zienkiewicz and Cheung.[26]

It is now possible to write the expression for the displacement within the element in a standard form as

$$\mathbf{u} \equiv w = \mathbf{N}\tilde{\mathbf{u}}^e = \mathbf{P}\mathbf{C}^{-1}\tilde{\mathbf{u}}^e \tag{11.34}$$

where

$$\mathbf{P} = (1, x, y, x^2, xy, y^2, x^3, x^2y, xy^2, y^3, x^3y, xy^3)$$

The form of the $\mathbf{B}$ is obtained directly from Eqs (11.20) and (11.25). We thus have

$$\mathbf{L}\nabla w = \left\{ \begin{array}{ll} \alpha_4 & +6x\,\alpha_7 & +2y\,\alpha_8 + 6xy\,\alpha_{11} \\ \alpha_6 & +2x\,\alpha_9 & +6y\,\alpha_{10} + 6xy\,\alpha_{12} \\ \alpha_5 & +4x\,\alpha_8 & +4y\,\alpha_9 + 6x^2\alpha_{11} + 6y^2\alpha_{12} \end{array} \right\}$$

We can write the above as

$$\mathbf{L}\nabla w = \mathbf{Q}\boldsymbol{\alpha} = \mathbf{Q}\mathbf{C}^{-1}\tilde{\mathbf{u}}^e = \mathbf{B}\tilde{\mathbf{u}}^e \quad \text{and thus} \quad \mathbf{B} = \mathbf{Q}\mathbf{C}^{-1} \tag{11.35a}$$

in which

$$\mathbf{Q} = \begin{bmatrix} 0 & 0 & 0 & 2 & 0 & 0 & 6x & 2y & 0 & 0 & 6xy & 0 \\ 0 & 0 & 0 & 0 & 0 & 2 & 0 & 0 & 2x & 6y & 0 & 6xy \\ 0 & 0 & 0 & 0 & 2 & 0 & 0 & 4x & 4y & 0 & 6x^2 & 6y^2 \end{bmatrix} \tag{11.35b}$$

It is of interest to remark now that the displacement function chosen does in fact permit a state of constant strain (curvature) to exist and therefore satisfies one of the criteria of convergence stated in reference 7.[*]

An explicit form of the shape function $\mathbf{N}$ was derived by Melosh[36] and can be written simply in terms of normalized coordinates. Thus, we can write for any node

$$\mathbf{N}_a^T = \tfrac{1}{8}(1+\xi_0)(1+\eta_0) \left\{ \begin{array}{l} 2+\xi_0+\eta_0-\xi^2-\eta^2 \\ b\eta_a(1-\eta^2) \\ -a\xi_a(1-\xi^2) \end{array} \right\} \tag{11.36}$$

with normalized coordinates defined as:

$$\xi = \frac{x-x_c}{a} \quad \text{where} \quad \xi_0 = \xi\xi_a$$

$$\eta = \frac{y-y_c}{b} \quad \text{where} \quad \eta_0 = \eta\eta_a$$

This form avoids the explicit inversion of $\mathbf{C}$; however, for simplicity we pursue the direct use of polynomials to deduce the stiffness and load matrices.

---

[*] If α_7 to α_{12} are zero, then the 'strain' defined by second derivatives is constant. By Eq. (11.32), the corresponding $\tilde{\mathbf{u}}^e$ can be found. As there is a unique correspondence between $\tilde{\mathbf{u}}^e$ and $\boldsymbol{\alpha}$ such a state is therefore unique. All this presumes that $\mathbf{C}^{-1}$ does in fact exist. The algebraic inversion shows that the matrix $\mathbf{C}$ is never singular.

Stiffness and load matrices

Standard procedures can now be followed, and it is almost superfluous to recount the details. The stiffness matrix relating the nodal *forces* (given by lateral force and two moments at each node) to the corresponding nodal displacement is

$$\mathbf{K}^e = \int_{\Omega^e} \mathbf{B}^T \mathbf{D} \mathbf{B} \, dx \, dy \tag{11.37a}$$

or, substituting Eq. (11.35a) into this expression,

$$\mathbf{K}^e = \mathbf{C}^{-T} \left(\int_{-b}^{b} \int_{-a}^{a} \mathbf{Q}^T \mathbf{D} \mathbf{Q} \, dx \, dy \right) \mathbf{C}^{-1} \tag{11.37b}$$

The terms not containing x and y have now been moved outside the integration operation. If $\mathbf{D}$ is constant the terms within the integration can be multiplied together and integrated explicitly without difficulty.

The external forces at nodes arising from distributed loading can be computed from expression (11.23). The contribution of these forces to each of the nodes is

$$\mathbf{f}_a = \left\{ \begin{array}{c} f_{w_a} \\ f_{\theta_{x_a}} \\ f_{\theta_{y_a}} \end{array} \right\} = \int_{-b}^{b} \int_{-a}^{a} \mathbf{N}_a^T q \, dx \, dy \tag{11.38a}$$

or, by Eq. (11.34),

$$\mathbf{f} = -\mathbf{C}^{-T} \int_{-b}^{b} \int_{-a}^{a} \mathbf{P}^T q \, dx \, dy \tag{11.38b}$$

The integral is again evaluated simply.

Example 11.1   Uniform load on 12 DOF rectangle

If we consider a rectangular element with the dimensions a and b shown in Fig. 11.7 and subjected to uniform loading q, evaluation of Eq. (11.38a) (or Eq. [11.38b)] yields the element nodal load vector given by

$$\mathbf{f}_1 = \tfrac{1}{12}qab \left\{ \begin{array}{c} 3 \\ b \\ -a \end{array} \right\}, \quad \mathbf{f}_2 = \tfrac{1}{12}qab \left\{ \begin{array}{c} 3 \\ -b \\ -a \end{array} \right\}, \quad \mathbf{f}_3 = \tfrac{1}{12}qab \left\{ \begin{array}{c} 3 \\ b \\ a \end{array} \right\}, \quad \mathbf{f}_4 = \tfrac{1}{12}qab \left\{ \begin{array}{c} 3 \\ -b \\ a \end{array} \right\}$$

We note that the force associated with w_a coincides with physical intuition as one-fourth the total load on the element. The moment type forces associated with the rotations is similar to that found for the Euler–Bernoulli beam in Example 10.1 of Sec. 10.4.1. If a physically 'lumped'-type loading is considered these moment forces are simply ignored.

The vector of nodal plate forces due to initial strains and initial stresses can also be found in a similar way. It is necessary to remark in this connection that initial strains, such as may be due to a temperature rise, is seldom confined in its effects on curvatures. Usually, direct (in-plane) strains in the plate are introduced additionally, and the complete problem can be solved only by consideration of a plane stress (membrane) problem as well as that of bending.

11.4 Quadrilateral and parallelogram elements

The rectangular element developed in the preceding section passes the patch test[41] and is always convergent. However, it cannot be easily generalized into a quadrilateral shape. Transformation of coordinates of the isoparametric type described in Chapter 2 can be performed but unfortunately now it will be found that the constant curvature criterion is violated. As expected, such elements behave badly but convergence may still occur providing the patch test is passed in the curvilinear coordinates.[7] Henshell *et al.*[40] studied the performance of such an element (and also some of a higher order) and concluded that reasonable accuracy is attainable. Their paper gives all the details of transformations required for an isoparametric mapping and the resulting need for numerical integration.

Only for the case of a parallelogram is it possible to achieve states of constant curvature exclusively using functions of ξ and η and for this case the patch test is satisfied. For a parallelogram the local coordinates can be related to the global ones by the explicit expression (Fig. 11.8)

$$
\begin{aligned}
\xi &= \frac{x - y \cot \alpha}{a} \\
\eta &= \frac{y \csc \alpha}{b}
\end{aligned}
\tag{11.39}
$$

and all expressions for the stiffness and loads can therefore also be derived directly. Such an element is suggested in the discussion in reference 26, and the stiffness matrices have been worked out by Dawe.[28] A somewhat different set of shape functions was suggested by Argyris.[29]

11.5 Triangular element with corner nodes (9 degrees of freedom)

At first sight, it would seem that once again a simple polynomial expansion could be used in a manner identical to that of the previous section. As only nine independent movements are imposed, only nine terms of the expansion are permissible. Here an immediate difficulty arises as the full cubic expansion contains 10 terms [Eq. (11.31)

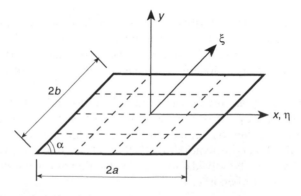

Fig. 11.8 Parallelogram element and skew coordinates.

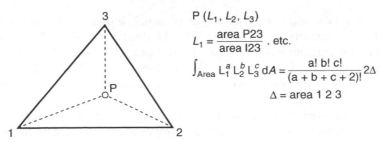

Fig. 11.9 Area coordinates.

with $\alpha_{11} = \alpha_{12} = 0$] and any omission has to be made arbitrarily. To retain a certain symmetry of appearance all 10 terms could be retained and two coefficients made equal (for example, $\alpha_8 = \alpha_9$) to limit the number of unknowns to nine. Several such possibilities have been investigated but a further, much more serious, problem arises. The matrix corresponding to **C** of Eq. (11.32) becomes singular for certain orientations of the triangle sides. This happens, for instance, when two sides of the triangle are parallel to the x and y axes respectively.

An 'obvious' alternative is to add a central node to the formulation and eliminate this by static condensation. This would allow a complete cubic to be used, but again it was found that an element derived on this basis does not converge to correct answers.

Difficulties of asymmetry can be avoided by the use of area coordinates described in reference 7. These are indeed nearly always a natural choice for triangles, see (Fig. 11.9).

Shape functions
As before we shall use polynomial expansion terms, and it is worth remarking that these are given in area coordinates in an unusual form. For instance,

$$\alpha_1 L_1 + \alpha_2 L_2 + \alpha_3 L_3 \tag{11.40}$$

gives the three terms of a *complete* linear polynomial and

$$\alpha_1 L_1^2 + \alpha_2 L_2^2 + \alpha_3 L_3^2 + \alpha_4 L_1 L_2 + \alpha_5 L_2 L_3 + \alpha_6 L_3 L_1 \tag{11.41}$$

gives all six terms of a quadratic (containing within it the linear terms).[*] The 10 terms of a cubic expression are similarly formed by the products of all possible cubic combinations, that is,

$$L_1^3, L_2^3, L_3^3, L_1^2 L_2, L_1^2 L_3, L_2^2 L_3, L_2^2 L_1, L_3^2 L_1, L_3^2 L_2, L_1 L_2 L_3 \tag{11.42}$$

For a 9 degree-of-freedom element any of the above terms can be used in a suitable combination, remembering, however, that only nine independent functions are needed

---

[*] However, it is also possible to write a complete quadratic as

$$\alpha_1 L_1 + \alpha_2 L_2 + \alpha_3 L_3 + \alpha_4 L_1 L_2 + \alpha_5 L_2 L_3 + \alpha_6 L_3 L_1$$

and so on, for higher orders. This has the advantage of explicitly stating all retained terms of polynomials of lower order.

and that constant curvature states have to be obtained. Figure 11.10 shows some functions that are of importance. The first [Fig. 11.10(a)] gives one of three functions representing a simple, unstrained rotation of the plate. Obviously, these must be available to produce the rigid body modes. Further, functions of the type $L_1^2 L_2$, of which there are six in the cubic expression, will be found to take up a form similar (though not identical) to Fig. 11.10(b).

The cubic function $L_1 L_2 L_3$ is shown in Fig. 11.10(c), illustrating that this is a purely internal (bubble) mode with zero values and slopes at all three corner nodes (though slopes are not zero along edges). This function could thus be useful for a nodeless or internal variable but will not, in isolation, be used as it cannot be prescribed in terms of corner variables. It can, however, be added to any other basic shape in any proportion, as indicated in Fig. 11.10(b).

The functions of the second kind are of special interest. They have zero values of w at all corners and indeed always have zero slope in the direction of one side. A linear combination of two of these (for example, $L_2^2 L_1$ and $L_2^2 L_3$) are capable of providing any desired slopes in the x and y directions at one node while maintaining all other nodal slopes at zero.

For an element envisaged with 9 degrees of freedom we must ensure that all six quadratic terms are present. In addition we select three of the cubic terms. The

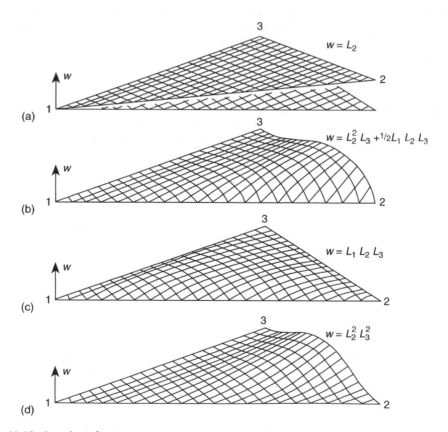

Fig. 11.10 Some basic functions in area coordinate polynomials.

quadratic terms ensure that a constant curvature, necessary for patch test satisfaction, is possible. Thus, the polynomials we consider are

$$\mathbf{P} = \left[L_1, L_2, L_3, L_1 L_2, L_2 L_3, L_3 L_1, L_1^2 L_2, L_2^2 L_3, L_3^2 L_1 \right]$$

and we write the interpolation as

$$w = \mathbf{P}\alpha \tag{11.43}$$

where α are parameters to be expressed in terms of nodal values. The nine nodal values are denoted as

$$\left[\tilde{w}_a, \ \tilde{\theta}_{xa}, \ \tilde{\theta}_{ya} \right] = \left[\tilde{w}_a, \ \left. \frac{\partial \tilde{w}}{\partial y} \right|_a, \ -\left. \frac{\partial \tilde{w}}{\partial x} \right|_a \right]; \quad a = 1, 2, 3$$

Upon noting that

$$\left\{ \begin{array}{c} \frac{\partial}{\partial x} \\ \frac{\partial}{\partial y} \end{array} \right\} = \left[\begin{array}{ccc} \frac{\partial L_1}{\partial x} & \frac{\partial L_2}{\partial x} & \frac{\partial L_3}{\partial x} \\ \frac{\partial L_1}{\partial y} & \frac{\partial L_2}{\partial y} & \frac{\partial L_3}{\partial y} \end{array} \right] \left\{ \begin{array}{c} \frac{\partial}{\partial L_1} \\ \frac{\partial}{\partial L_2} \\ \frac{\partial}{\partial L_3} \end{array} \right\} = \frac{1}{2\Delta} \left[\begin{array}{ccc} b_1 & b_2 & b_3 \\ c_1 & c_2 & c_3 \end{array} \right] \left\{ \begin{array}{c} \frac{\partial}{\partial L_1} \\ \frac{\partial}{\partial L_2} \\ \frac{\partial}{\partial L_3} \end{array} \right\} \tag{11.44}$$

where

$$2\Delta = b_1 c_2 - b_2 c_1; \quad b_a = y_b - y_c; \quad c_a = x_c - x_b$$

with a, b, c a cyclic permutation of indices, we now determine the shape function by a suitable inversion [see Sec. 11.3, Eq. (11.34)], and write for node a

$$\mathbf{N}_a^T = \left\{ \begin{array}{c} 3L_i^2 - 2L_i^3 \\ L_i^2(b_j L_k - b_k L_j) + \frac{1}{2}(b_j - b_k)L_1 L_2 L_3 \\ L_i^2(c_j L_k - c_k L_j) + \frac{1}{2}(c_j - c_k)L_1 L_2 L_3 \end{array} \right\} \tag{11.45}$$

Here the term $L_1 L_2 L_3$ is added to permit constant curvature states.

The computation of stiffness and load matrices can again follow the standard patterns, and integration of expressions (11.22) and (11.23) can be done exactly using the general integrals given in Fig. 11.9. However, numerical quadrature is generally used and proves equally efficient (see Sec. 2.3). The stiffness matrix requires computation of second derivatives of shape functions and these may be conveniently obtained from

$$\left[\begin{array}{cc} \frac{\partial^2 N_a}{\partial x^2} & \frac{\partial^2 N_a}{\partial x \partial y} \\ \frac{\partial^2 N_a}{\partial y \partial x} & \frac{\partial^2 N_a}{\partial y^2} \end{array} \right] = \frac{1}{4\Delta^2} \left[\begin{array}{ccc} b_1 & b_2 & b_3 \\ c_1 & c_2 & c_3 \end{array} \right] \left[\begin{array}{ccc} \frac{\partial^2 N_a}{\partial L_1^2} & \frac{\partial^2 N_a}{\partial L_1 \partial L_2} & \frac{\partial^2 N_a}{\partial L_1 \partial L_3} \\ \frac{\partial^2 N_a}{\partial L_2 \partial L_1} & \frac{\partial^2 N_a}{\partial L_2^2} & \frac{\partial^2 N_a}{\partial L_2 \partial L_3} \\ \frac{\partial^2 N_a}{\partial L_3 \partial L_1} & \frac{\partial^2 N_a}{\partial L_3 \partial L_2} & \frac{\partial^2 N_a}{\partial L_3^2} \end{array} \right] \left[\begin{array}{cc} b_1 & c_1 \\ b_2 & c_2 \\ b_3 & c_3 \end{array} \right]$$

$$\tag{11.46}$$

in which N_a denotes any of the shape functions given in Eq. (11.45).

The element just derived is one first developed in reference 12. Although it satisfies the constant strain criterion (being able to produce constant curvature states) it unfortunately does not pass the patch test for arbitrary mesh configurations. Indeed, this was pointed out in the original reference (which also was the one in which the patch test was mentioned for the first time). However, the patch test is fully satisfied with this element for meshes of triangles created by three sets of equally spaced straight lines. In general, the performance of the element, despite this shortcoming, made the element quite popular in some practical applications.[38]

It is possible to amend the element shape functions so that the resulting element passes the patch test in all configurations. An early approach was presented by Kikuchi and Ando[51] by replacing boundary integral terms in the virtual work statement of Eq. (11.18) by

$$\delta\Pi_{\text{ext}} = \int_\Omega \delta w q \, d\Omega + \sum_e \delta \left[\int_{\Gamma_e} \left(\frac{\partial w}{\partial n} - \frac{\partial \breve{w}}{\partial n} \right) M_n(w) \, d\Gamma \right]$$
$$+ \int_{\Gamma_s} \left[\delta w \bar{S}_n + \frac{\partial \delta w}{\partial s} \bar{M}_{ns} \right] d\Gamma + \int_{\Gamma_n} \frac{\partial \delta \breve{w}}{\partial n} \bar{M}_n \, d\Gamma$$

(11.47)

in which Γ_e is the boundary of each element e, $M_n(w)$ is the normal moment computed from second derivatives of the w interpolation, and s is the tangent direction along the element boundaries. The interpolations given by Eq. (11.45) are C_0 conforming and have slopes which match those of adjacent elements at nodes. To correct the slope incompatibility between nodes, a simple interpolation is introduced along each element boundary segment as

$$\frac{\partial \breve{w}}{\partial n} = (1 - s') \left[\frac{\partial w}{\partial x} \Big|_a n_y + \frac{\partial w}{\partial y} \Big|_a n_x \right] + s' \left[\frac{\partial w}{\partial x} \Big|_b n_y + \frac{\partial w}{\partial y} \Big|_b n_x \right]$$

(11.48)

where s' is 0 at node a and 1 at node b, and n_x and n_y are direction cosines of the ab side with respect to the x and y axes, respectively. The above modification requires boundary integrals in addition to the usual area integrals; however, the final result is one which passes the patch test.

Bergen[44,46,47] and Samuelsson[48] also show a way of producing elements which pass the patch test, but a successful modification useful for general application with elastic and inelastic material behaviour is one derived by Specht.[49] This modification uses three fourth-order terms in place of the three cubic terms of the equation preceding Eq. (11.43). The particular form of these is so designed that the patch test criterion which we shall discuss in detail later in Sec. 11.7 is identically satisfied. We consider now the nine polynomial functions given by

$$\mathbf{P} = \big[L_1, L_2, L_3, L_1 L_2, L_2 L_3, L_3 L_1,$$
$$L_1^2 L_2 + \tfrac{1}{2} L_1 L_2 L_3 \left\{ 3(1 - \mu_3) L_1 - (1 + 3\mu_3) L_2 + (1 + 3\mu_3) L_3 \right\},$$
$$L_2^2 L_3 + \tfrac{1}{2} L_1 L_2 L_3 \left\{ 3(1 - \mu_1) L_2 - (1 + 3\mu_1) L_3 + (1 + 3\mu_1) L_1 \right\},$$
$$L_3^2 L_1 + \tfrac{1}{2} L_1 L_2 L_3 \left\{ 3(1 - \mu_2) L_3 - (1 + 3\mu_2) L_1 + (1 + 3\mu_2) L_2 \right\} \big]$$

(11.49)

where

$$\mu_a = \frac{l_c^2 - l_b^2}{l_a^2}$$

(11.50)

and l_a is the length of the triangle side opposite node a and a, b, c is a cyclic permutation.[*]

The modified interpolation for w is taken as

$$w = \mathbf{P}\alpha \qquad (11.51)$$

and, on identification of nodal values and inversion, the shape functions can be written explicitly in terms of the components of the vector $\mathbf{P}$ defined by Eq. (11.49) as

$$\mathbf{N}_a^T = \left\{ \begin{array}{c} P_a - P_{a+3} + P_{c+3} + 2\,(P_{a+6} - P_{c+6}) \\ -b_b\,(P_{c+6} - P_{c+3}) - b_c P_{a+6} \\ -c_b\,(P_{c+6} - P_{c+3}) - c_c P_{a+6} \end{array} \right\} \qquad (11.52)$$

where a, b, c are the cyclic permutations of 1, 2, 3.

Once again, stiffness and load matrices can be determined either explicitly or using numerical quadrature. The element derived above passes all the patch tests and performs excellently.[41] Indeed, if the quadrature is carried out in a 'reduced' manner using the three quadrature points and weights[7]

L_1	L_2	L_3	W
1/2	1/2	0	1/3
1/2	0	1/2	1/3
0	1/2	1/2	1/3

then the element is one of the best thin plate triangles with 9 degrees of freedom that is currently available, as we shall show in the section dealing with numerical comparisons.

11.6 Triangular element of the simplest form (6 degrees of freedom)

If conformity at nodes (C_1 continuity) is to be abandoned, it is possible to introduce even simpler elements than those already described by reducing the element inter-connections. A very simple element of this type was first proposed by Morley.[30] In this element, illustrated in Fig. 11.11, the interconnections require continuity of the displacement w at the triangle vertices and of normal slopes at the element mid-sides.

With 6 degrees of freedom the expansion can be limited to quadratic terms alone, which one can write as

$$w = \begin{bmatrix} L_1, & L_2, & L_3, & L_1 L_2, & L_2 L_3, & L_3 L_1 \end{bmatrix} \alpha \qquad (11.53a)$$

Identification of nodal variables and inversion leads to the following shape functions: for corner nodes

$$N_a = L_a - L_a(1 - L_a) - \frac{b_a b_c - c_a c_c}{b_b^2 + c_b^2} L_b(1 - L_b) - \frac{b_a b_b - c_a c_b}{b_c^2 + c_c^2} L_c(1 - L_c)$$

$$(11.53b)$$

---

[*] The constants μ_a are geometric parameters occurring in the expression for normal derivatives. Thus on side l_a the normal derivative is given by

$$\frac{\partial}{\partial n} = \frac{l_a}{4\Delta} \left[\frac{\partial}{\partial L_b} + \frac{\partial}{\partial L_c} - 2 \frac{\partial}{\partial L_a} + \mu_a \left(\frac{\partial}{\partial L_c} - \frac{\partial}{\partial L_b} \right) \right]$$

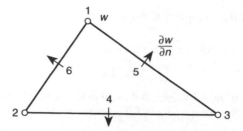

Fig. 11.11 The simplest non-conforming triangle, from Morley,[30] with 6 degrees of freedom.

and for 'normal gradient' nodes

$$N_{a+3} = \frac{2\Delta}{\sqrt{b_a^2 + c_a^2}} L_a (1 - L_a) \qquad (11.53c)$$

where the symbols are identical to those used in Eq. (11.44) and subscripts a, b, c are a cyclic permutation of 1, 2, 3.

Establishment of stiffness and load matrices follows the standard pattern and we find that once again the element passes fully all the patch tests required. This simple element performs reasonably, as we shall show later, though its accuracy is, of course, less than that of the preceding ones.

It is of interest to remark that the moment field described by the element satisfies exactly interelement equilibrium conditions on the normal moment M_n, as the reader can verify. Indeed, originally this element was derived as an equilibrating one using the complementary energy principle,[52] and for this reason it always gives an upper bound on the strain energy of flexure. This is the simplest possible element as it simply represents the minimum requirements of a constant moment field. An explicit form of stiffness routines for this element is given by Wood.[31]

11.7 The patch test – an analytical requirement

The patch test in its different forms is generally applied numerically to test the final form of an element.[7] However, the basic requirements for its satisfaction by shape functions that violate compatibility can be forecast accurately if certain conditions are satisfied in the choice of such functions. These conditions follow from the requirement that for constant strain states the virtual work done by internal forces acting at the discontinuity must be zero. Thus if the tractions acting on an element interface of a plate are (see Fig. 11.4)

$$M_n, \qquad M_{ns} \qquad \text{and} \qquad S_n \qquad (11.54)$$

and if the corresponding mismatch of virtual displacements are

$$\Delta \phi_n \equiv \Delta \left(\frac{\partial w}{\partial n} \right), \qquad \Delta \phi_s \equiv \Delta \left(\frac{\partial w}{\partial s} \right) \qquad \text{and} \qquad \Delta w \qquad (11.55)$$

then ideally we would like the integral given below to be zero, as indicated, at least for the constant stress states:

$$\int_{\Gamma_e} M_n \Delta\phi_n \, d\Gamma + \int_{\Gamma_e} M_{ns} \Delta\phi_s \, d\Gamma + \int_{\Gamma_e} S_n \Delta w \, d\Gamma = 0 \qquad (11.56)$$

The last term will always be zero identically for constant M_x, M_y, M_{xy} fields as then $S_x = S_y = 0$ [in the absence of applied couples, see Eq. (11.11b)] and we can ensure the satisfaction of the remaining conditions if

$$\int_{\Gamma_e} \Delta\phi_n \, d\Gamma = 0 \quad \text{and} \quad \int_{\Gamma_e} \Delta\phi_s \, d\Gamma = 0 \qquad (11.57)$$

is satisfied for each straight side Γ_e of the element.

For elements joining at vertices where $\partial w/\partial n$ is prescribed, these integrals will be identically zero only if anti-symmetric cubic terms arise in the departure from linearity and a quadratic variation of normal gradients is absent, as shown in Fig. 11.12(a). This is the motivation for the rather special form of shape function basis chosen to describe the incompatible triangle in Eq. (11.49), and here the first condition of Eq. (11.57) is automatically satisfied. The satisfaction of the second condition of Eq. (11.57) is always ensured if the function w and its first derivatives are prescribed at the corner nodes.

For the purely quadratic triangle of Sec. 11.6 the situation is even simpler. Here the gradients can only be linear, and if their value is prescribed at the element mid-side as shown in Fig. 11.11(b) the integral is identically zero.

The same arguments apparently fail when the rectangular element with the function basis given in Eq. (11.34) is examined. However, the reader can verify by direct algebra that the integrals of Eqs (11.57) are identically satisfied. Thus, for instance,

$$\int_{-a}^{a} \frac{\partial w}{\partial y} \, dx = 0 \quad \text{when} \quad y = \pm b$$

and $\partial w/\partial y$ is taken as zero at the two nodes (i.e. departure from prescribed linear variations only is considered).

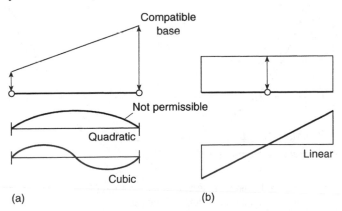

(a)   (b)

Fig. 11.12 Continuity condition for satisfaction of patch test [$\int (\partial w / \partial n) \, ds = 0$]; variation of $\partial w / \partial n$ along side. (a) Definition by corner nodes (linear component compatible); (b) definition by one central node (constant component compatible).

The remarks of this section are verified in numerical tests and lead to an intelligent, *a priori*, determination of conditions which make shape functions convergent for incompatible elements.

11.8 Numerical examples

The various plate bending elements already derived – and those to be derived in subsequent sections – have been used to solve some classical plate bending problems. We first give two specific illustrations and then follow these with a general convergence study of elements discussed.

Example 11.2   Deflections and moments for clamped square plate

The solution of a clamped plate subjected to uniform loading q was a topic of considerable study during the early 1900s.[53,54] An accurate numerical solution in series form was given by Hencky[55,56] and recently evaluated using the form given by Wojtaszak[57] by Taylor and Govindjee[58] to obtain correct solution values to several significant figures.

Figure 11.13 shows the deflections and moments in a square plate clamped along its edges and solved by the use of the rectangular element derived in Sec. 11.3 and a uniform mesh.[26] Table 11.1 gives numerical results for a set of similar examples solved

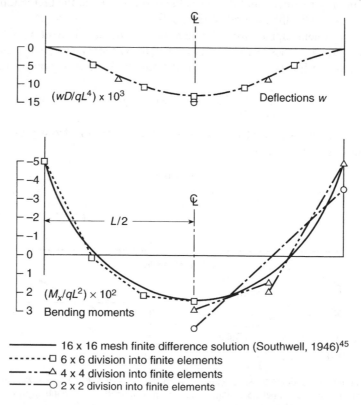

——————— 16 x 16 mesh finite difference solution (Southwell, 1946)[45]
- - - - - - ☐ 6 x 6 division into finite elements
— · · —△ 4 x 4 division into finite elements
— · —○ 2 x 2 division into finite elements

Fig. 11.13 A square plate with clamped edges; uniform load q; square elements.

Table 11.1 Computed central deflection of a square plate for several meshes (rectangular elements)[39]

Mesh	Total number of nodes	Simply supported plate		Clamped plate	
		α*	β†	α*	β†
2×2	9	0.003446	0.013784	0.001480	0.005919
4×4	25	0.003939	0.012327	0.001403	0.006134
8×8	81	0.004033	0.011829	0.001304	0.005803
16×16	169	0.004050	0.011715	0.001283	0.005710
Series (Timoshenko)		0.004062	0.01160	0.001265	0.00560

* $w_{max} = \alpha q L^4 / D$ for uniformly distributed load q.
† $w_{max} = \beta P L^2 / D$ for central concentrated load P.
Note: Subdivision of whole plate given for mesh.

Table 11.2 Corner supported square plate

Method	Mesh	Point 1		Point 2	
		w	M_x	w	M_x
Finite element	2×2	0.0126	0.139	0.0176	0.095
	4×4	0.0165	0.149	0.0232	0.108
	6×6	0.0173	0.150	0.0244	0.109
Marcus[59]		0.0180	0.154	0.0281	0.110
Ballesteros and Lee		0.0170	0.140	0.0265	0.109
Multiplier		qL^4/D	qL^2	qL^4/D	qL^2

Note: Point 1, centre of side; point 2, centre of plate.

with the same element, and Table 11.2 presents another square plate with more complex boundary conditions.[39] Exact results are available here and comparisons are made.[59,60]

Example 11.3 Skewed slab bridge
Figures 11.14 and 11.15 show practical engineering applications to more complex shapes of slab bridges. In both examples the requirements of geometry necessitate the use of a triangular element – with that of reference 12 being used here. Further, in both examples, beams reinforce the slab edges and these are simply incorporated in the analysis on the assumption of concentric behaviour.

Example 11.4 Comparison of convergence behaviour
In Fig. 11.16(a)–(d) we show the results of a convergence study for a square plate with simply supported and clamped edge conditions using various triangular and rectangular elements and two load types. This type of diagram is commonly used for assessing the behaviour of various elements, and we show on it the performance of the elements already described as well as others to which we shall refer to later. Table 11.3 gives the key to the various element 'codes' which include elements yet to be described.[63–66]

Example 11.5 Energy convergence in a skew plate
The comparisons in Example 11.4 single out only one displacement and each plot uses the number of mesh divisions in a quarter of the plate as abscissa. It is therefore difficult to deduce the convergence rate and the performance of elements with multiple nodes. A more convenient plot gives the energy norm $||\mathbf{u}||$, versus the number of degrees of freedom N on a logarithmic scale (see also reference 7). We show such a comparison

Text continued on page 356

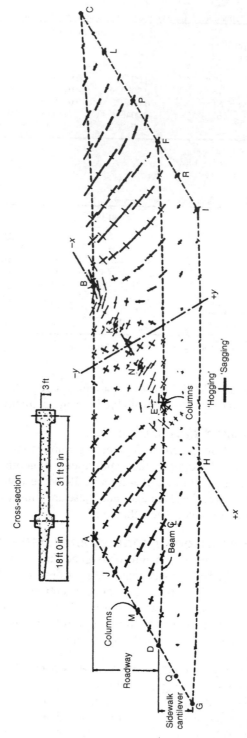

Fig. 11.14 A skew, curved, bridge with beams and non-uniform thickness; plot of principal moments under dead load.

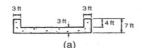

(a)

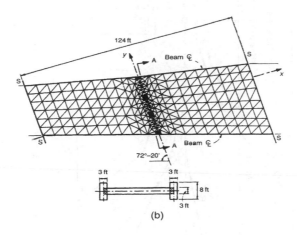

(b)

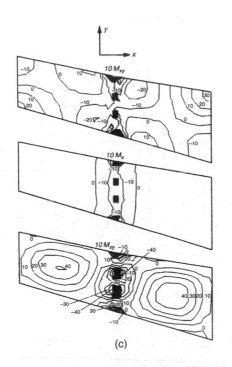

(c)

Fig. 11.15 Castleton railway bridge: general geometry and details of finite element subdivision. (a) Typical actual section; (b) idealization and meshing; (c) moment components (ton ft^{-1}) under uniform load of 150 lb ft^{-2} with computer plot of contours.

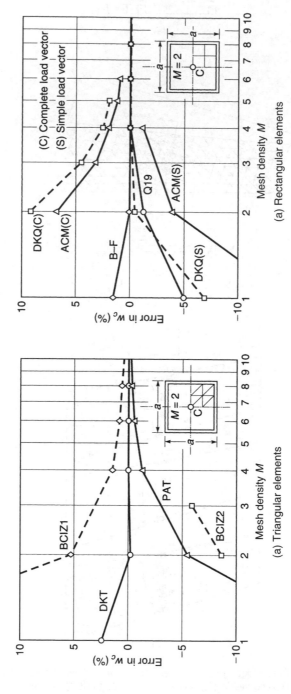

Fig. 11.16 (a) Simply supported uniformly loaded square plate.

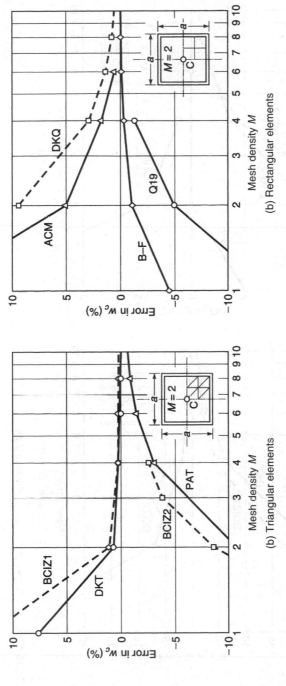

Fig. 11.17 *Cont.* (b) simply supported square plate with concentrated central load.

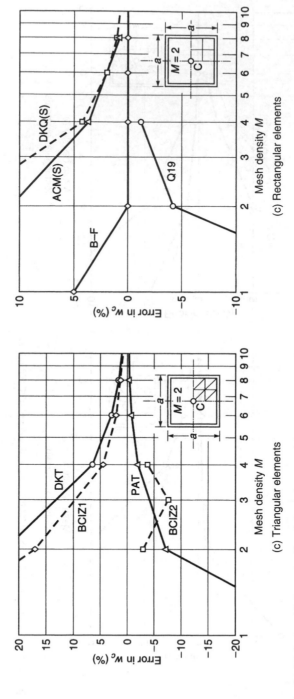

Fig. 11.16 *Cont.* (c) clamped uniformly loaded square plate.

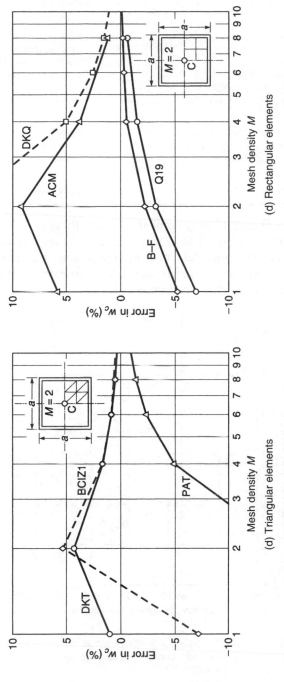

Fig. 11.16 *Cont.*   (d) clamped square plate with concentrated central load. Percentage error in central displacement (see Table 11.3 for key).

Table 11.3 List of elements for comparison of performance in Fig. 11.16: (a) 9 degree-of-freedom triangles; (b) 12 degree-of-freedom rectangles; (c) 16 degree-of-freedom rectangle

Code	Reference	Symbol	Description and comment
(a)			
BCIZ 1	Bazeley et al.[12]	◇	Displacement, non-conforming (fails patch test)
PAT	Specht[49]	△	Displacement, non-conforming
BCIZ 2	Bazeley et al.[12]	□	Displacement, conforming
(HCT)	Clough and Tocher[10]		
DKT	Stricklin et al.[61] and Dhatt[62]	○	Discrete Kirchhoff
(b)			
ACM	Zienkiewicz and Cheung[26]	△	Displacement, non-conforming
Q19	Clough and Felippa[15]	○	Displacement, conforming
DKQ	Batoz and Ben Tohar[61]	□	Displacement, conforming
(c)			
BF	Bogner et al.[17]	◇	Displacement, conforming

for some elements in Fig. 11.17 for a problem of a slightly skewed, simply supported plate.[9] It is of interest to observe that, owing to a singularity in the obtuse angle corners, both high- and low-order elements converge at almost identical rates (though, of course, the former give better overall accuracy).[7] Different rates of convergence would, of course, be obtained if no singularity existed.

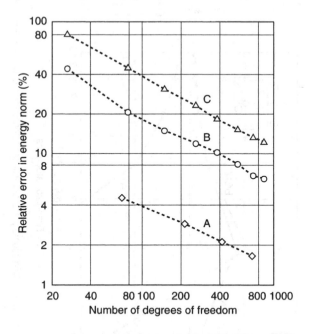

A Fifth-order conforming triangle[19-23]
B Low-order conforming element
      $(p = 2)$[10,11]
C Hybrid[82]

Fig. 11.17 Rate of convergence in energy norm versus degree of freedom for three elements: the problem of a slightly skewed, simply supported plate (80°) with uniform mesh subdivision.[9]

Conforming shape functions with nodal singularities

11.9 General remarks

It has already been demonstrated in Sec. 11.3 that it is impossible to devise a simple polynomial function with only three nodal degrees of freedom that will be able to satisfy slope continuity requirements at all locations along element boundaries. The alternative of imposing curvature parameters at nodes has the disadvantage, however, of imposing excessive conditions of continuity (although we will investigate some of the elements that have been proposed from this class). Furthermore, it is desirable from many points of view to limit the nodal variables to three quantities only. These, with simple physical interpretation, allow the generalization of plate elements to shells to be easily interpreted also.

It is, however, possible to achieve C_1 continuity by provision of additional shape functions for which, in general, *second-order derivatives have non-unique values at nodes*. Providing the patch test conditions are satisfied, convergence is again assured.

Such shape functions will be discussed now in the context of triangular and quadrilateral elements. The simple rectangular shape will be omitted as it is a special case of the quadrilateral.

11.10 Singular shape functions for the simple triangular element

Consider for instance either of the following sets of functions:

$$\varepsilon_{bc} = \frac{L_a L_b^2 L_c^2 (L_c - L_b)}{(L_a + L_b)(L_b + L_c)} \tag{11.58}$$

or

$$\varepsilon_{bc} = \frac{L_a L_b^2 L_c^2 (1 + L_a)}{(L_a + L_b)(L_b + L_c)} \tag{11.59}$$

in which once again a, b, c are a cyclic permutation of 1, 2, 3. Both have the property for $a = 1$ that along two sides (1–2 and 1–3) of a triangle (Fig. 11.18) their values and the values of their normal slope are zero. On the third side (2–3) the function is zero but a normal slope exists. In both, its variation is parabolic. Now, all the functions used to define the non-conforming triangle [see Eq. (11.43)] were cubic and hence permit also a parabolic variation of the normal slope which is not uniquely defined by the two end nodal values (and hence resulted in non-conformity). However, if we specify as an additional variable the *normal slope of w* at a mid-point of each side then, by combining the new functions ε_{bc} with the other functions previously given, a *unique parabolic variation of normal slope* along interelement faces is achieved and a compatible element results.

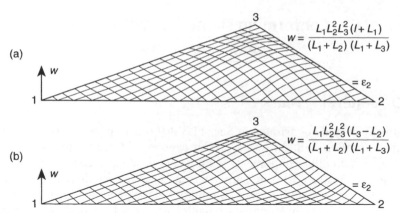

Fig. 11.18 Some singular area coordinate functions.

Apparently, this can be achieved by adding three such additional degrees of freedom to expression (11.43) and proceeding as described above. This will result in an element shown in Fig. 11.19(a), which has six nodes, three corner ones as before and three additional ones at which only normal slope is specified. Such an element requires the definition of a node (or an alternative) to define the normal slope and also involves assembly of nodes with differing numbers of degrees of freedom. It is necessary to define a unique normal slope for the parameter associated with the mid-point of adjacent elements. One simple solution is to use the direction of increasing node number of the adjacent vertices to define a unique normal.

Another alternative, which avoids the above difficulties, is to constrain the mid-side node degree of freedom. For instance, we can assume that the normal slope at the centre-point of a line is given as the average of the two slopes at the ends. This, after suitable transformation, results in a compatible element with exactly the same degrees of freedom as that described in previous sections [see Fig. 11.19(b)].

The algebra involved in the generation of suitable shape functions along the lines described here is quite extensive and will not be given fully.

First, the normal slopes at the mid-sides are calculated from the basic element shape functions [Eq. (11.45)] as

$$\left[\frac{\partial w}{\partial n}\bigg|_4 \quad \frac{\partial w}{\partial n}\bigg|_5 \quad \frac{\partial w}{\partial n}\bigg|_6\right]^T = \mathbf{Z}\tilde{\mathbf{u}}^e \tag{11.60}$$

Similarly, the average values of the nodal slopes in directions normal to the sides are calculated from these functions:

$$\left[\frac{\partial w}{\partial n}\bigg|_4^a \quad \frac{\partial w}{\partial n}\bigg|_5^a \quad \frac{\partial w}{\partial n}\bigg|_6^a\right]^T = \bar{\mathbf{Z}}\tilde{\mathbf{u}}^e \tag{11.61}$$

The contribution of the ε functions to these slopes is added in proportions of $\varepsilon_{bc} - \gamma_a$ and is simply (as these give unit normal slope)

$$\boldsymbol{\gamma} = \begin{bmatrix} \gamma_1 & \gamma_2 & \gamma_3 \end{bmatrix}^T \tag{11.62}$$

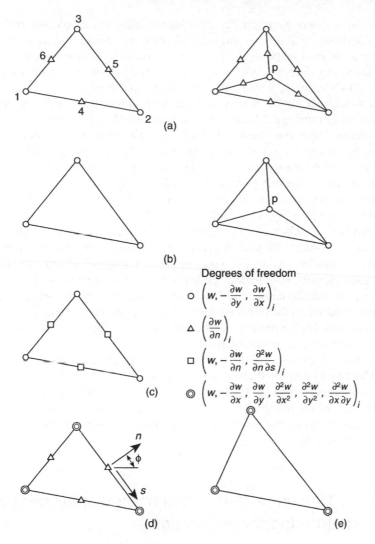

Degrees of freedom

$\circ \quad \left(w, -\dfrac{\partial w}{\partial y}, \dfrac{\partial w}{\partial x} \right)_i$

$\triangle \quad \left(\dfrac{\partial w}{\partial n} \right)_i$

$\square \quad \left(w, -\dfrac{\partial w}{\partial n}, \dfrac{\partial^2 w}{\partial n \, \partial s} \right)_i$

$\circledcirc \quad \left(w, -\dfrac{\partial w}{\partial x}, \dfrac{\partial w}{\partial y}, \dfrac{\partial^2 w}{\partial x^2}, \dfrac{\partial^2 w}{\partial y^2}, \dfrac{\partial^2 w}{\partial x \, \partial y} \right)_i$

Fig. 11.19 Various conforming triangular elements.

On combining Eq. (11.45) and the last three relations we have

$$\bar{\mathbf{Z}}\tilde{\mathbf{u}}^e = \mathbf{Z}\tilde{\mathbf{u}}^e + \gamma \tag{11.63}$$

from which it immediately follows on finding γ that

$$w = \mathbf{N}^0\tilde{\mathbf{u}}^e + \begin{bmatrix} \varepsilon_{23}, & \varepsilon_{31}, & \varepsilon_{12} \end{bmatrix} \left(\bar{\mathbf{Z}} - \mathbf{Z} \right) \tilde{\mathbf{u}}^e \tag{11.64}$$

in which $\mathbf{N}^0$ are the non-conforming shape functions defined in Eq. (11.45). Thus, new shape functions are now available from Eq. (11.64).

An alternative way of generating compatible triangles was developed by Clough and Tocher.[11] As shown in Fig. 11.19(a) each element triangle is first divided into three parts based on an internal point p. For each abp triangle a complete cubic expansion is written involving 10 terms which may be expressed in terms of the displacement and slopes at each vertex and the mid-side slope along the ab edge.

Matching the values at the vertices for the three subtriangles produces an element with 15 degrees of freedom: 12 conventional degrees of freedom at nodes 1, 2, 3 and p; and three normal slopes at nodes 4, 5, 6. Full C_1 continuity in the interior of the element is achieved by *constraining* the three parameters at the p node to satisfy continuous normal slope at each internal mid-side. Thus, we achieve an element with 12 degrees of freedom similar to the one previously outlined using the singular shape functions. Constraining the normal slopes on the exterior mid-sides leads to an element with 9 degrees of freedom [see Fig. 11.19(b)].

These elements are achieved at the expense of providing non-unique values of second derivatives at the corners. We note, however, that strains are in general also non-unique in elements surrounding a node (e.g. constant strain triangles in elasticity have different strains in each element surrounding each node). In the previously developed shape functions ε_{bc} an infinite number of values to the second derivatives are obtained at each node depending on the direction the corner is approached. Indeed, the derivation of the Clough and Tocher triangle can be obtained by defining an alternative set of ε functions, as has been shown in reference 12.

As both types of elements lead to almost identical numerical results the preferable one is that leading to simplified computation. If numerical integration is used (as indeed is always strongly recommended for such elements) the form of functions continuously defined over the whole triangle as given by Eqs (11.45) and (11.64) is advantageous, although a fairly high order of numerical integration is necessary because of the singular nature of the functions.

11.11 An 18 degree-of-freedom triangular element with conforming shape functions

An element that presents a considerable improvement over the type illustrated in Fig. 11.19(a) is shown in Fig. 11.19(c). Here, the 12 degrees of freedom are increased to 18 by considering both the values of w and its cross derivative $\partial^2 w/\partial s\,\partial n$, in addition to the normal slope $\partial w/\partial n$, at element mid-sides.[*]

Thus an equal number of degrees of freedom is presented at each node. Imposition of the continuity of cross derivatives at *mid-sides* does not involve any additional constraint as this indeed must be continuous in physical situations.

The derivation of this element is given by Irons[14] and it will suffice here to say that in addition to the modes already discussed, fourth-order terms of the type illustrated in Fig. 11.10(d) and 'twist' functions of Fig. 11.18(b) are used. Indeed, it can be simply verified that the element contains *all* 15 terms of the quartic expansion in addition to the 'singularity' functions.

---

[*] This is, in fact, identical to specifying both $\partial w/\partial n$ and $\partial w/\partial s$ at the mid-side.

11.12 Compatible quadrilateral elements

Any of the previous triangles can be combined to produce 'composite' compatible quadrilateral elements with or without internal degrees of freedom. Three such quadrilaterals are illustrated in Fig. 11.20 and, in all, no mid-side nodes exist on the external boundaries. This avoids the difficulties of defining a unique parameter and of assembly already mentioned.

In the first, no internal degrees of freedom are present and indeed no improvement on the comparable triangles is expected. In the following two, 3 and 7 internal degrees of freedom exist, respectively. Here, normal slope continuity imposed in the last one does not interfere with the assembly, as internal degrees of freedom are in all cases eliminated by static condensation.[67] Much improved accuracy with these elements has been demonstrated by Clough and Felippa.[15]

An alternative direct derivation of a quadrilateral element was proposed by Sander[68] and Fraeijs de Veubeke.[13,16] This is along the following lines. Within a quadrilateral of Fig. 11.21(a) a complete cubic with 10 constants is taken, giving the first component of the displacement which is defined by three functions. Thus,

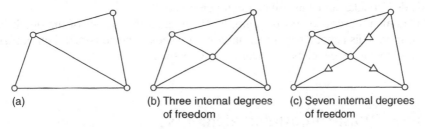

| (a) | (b) Three internal degrees of freedom | (c) Seven internal degrees of freedom |

Fig. 11.20 Some composite quadrilateral elements.

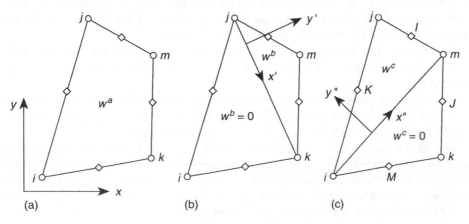

(a)   (b)   (c)

Fig. 11.21 The compatible functions of Fraeijs de Veubeke.[13,16]

$$w = w^{\mathrm{a}} + w^{\mathrm{b}} + w^{\mathrm{c}}$$
$$w^{\mathrm{a}} = \alpha_1 + \alpha_2 x + \cdots + \alpha_{10} y^3 \tag{11.65}$$

The second function w^{b} is defined in a piecewise manner. In the lower triangle of Fig. 11.21(b) it is taken as zero; in the upper triangle a cubic expression with three constants merges with slope discontinuity into the field of the lower triangle. Thus, in jkm,

$$w^{\mathrm{b}} = \alpha_{11} y'^2 + \alpha_{12} y'^3 + \alpha_{13} x' y'^2 \tag{11.66}$$

in terms of the locally specified coordinates x' and y'. Similarly, for the third function, Fig. 11.21(c), $w^{\mathrm{c}} = 0$ in the lower triangle, and in imj we define

$$w^{\mathrm{c}} = \alpha_{14} y''^2 + \alpha_{15} y''^3 + \alpha_{16} x'' y''^2 \tag{11.67}$$

The 16 external degrees of freedom are provided by 12 usual corner variables and four normal mid-side slopes and allow the 16 constants α_1 to α_{16} to be found by inversion. Compatibility is assured and once again non-unique second derivatives arise at corners.

Again it is possible to constrain the mid-side nodes if desired and thus obtain a 12 degree-of-freedom element. The expansion can be found explicitly, as shown by Fraeijs de Veubeke, and a useful element generated.[16]

The element described above cannot be formulated if a corner of the quadrilateral is re-entrant. This is not a serious limitation but needs to be considered on occasion if such an element degenerates to a near triangular shape.

11.13 Quasi-conforming elements

The performance of some of the conforming elements discussed in Secs 11.10–11.12 is shown in the comparison graphs of Fig. 11.16. It should be noted that although monotonic convergence in energy norm is now guaranteed, by subdividing each mesh to obtain the next one, the conforming triangular elements of references 11 and 12 perform almost identically but are considerably stiffer and hence less accurate than many of the non-conforming elements previously cited.

To overcome this inaccuracy a *quasi-conforming* or *smoothed* element was derived by Razzaque and Irons.[33,34] For the derivation of this element *substitute shape functions are used*.

The substitute functions are cubic functions (in area coordinates) so designed as to approximate in a least-square sense the singular functions ε and their derivatives used to enforce continuity [see Eqs (11.58)–(11.64)], as shown in Fig. 11.22.

The algebra involved is complex but a full subprogram for stiffness computations is available in reference 33. It is noted that this element performs very similarly to the simpler, non-conforming element previously derived for the triangle. It is interesting to observe that here the non-conforming element is developed by choice and not to avoid difficulties. Its validity, however, is established by patch tests.

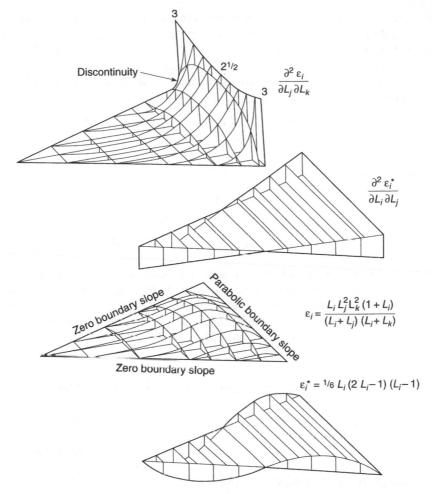

$$\frac{\partial^2 \varepsilon_i}{\partial L_j \partial L_k}$$

$$\frac{\partial^2 \varepsilon_i^*}{\partial L_i \partial L_j}$$

$$\varepsilon_i = \frac{L_i\,L_j^2\,L_k^2\,(1+L_i)}{(L_i+L_j)\,(L_i+L_k)}$$

$$\varepsilon_i^* = {}^{1}\!/_6\, L_i\,(2\,L_i-1)\,(L_i-1)$$

Fig. 11.22 Least-square substitute cubic shape function ε^* in place of rational function ε for plate bending triangles.

Conforming shape functions with additional degrees of freedom

11.14 Hermitian rectangle shape function

With the rectangular element of Fig. 11.7 the specification of $\partial^2 w/\partial x \partial y$ as a nodal parameter is always permissible as it does not involve 'excessive continuity'. It is easy to show that for such an element polynomial shape functions giving compatibility can be easily determined.

A polynomial expansion involving 16 constants [equal to the number of nodal parameters w_a, $(\partial w/\partial x)_a$, $(\partial w/\partial y)_a$ and $(\partial^2 w/\partial x \partial y)_a$] could, for instance, be written retaining terms that do not produce a higher-order variation of w or its normal slope along the sides. Many alternatives will be present here and some may not produce invertible **C** matrices [see Eq. (11.33)].

An alternative derivation uses Hermitian polynomials used for shape functions of the Euler–Bernoulli beam (see Sec. 10.4.1) which permit the writing down of suitable functions directly. It is easy to verify that the following shape functions

$$\mathbf{N}_a = \left[H_a^{(0)}(x)H_a^{(0)}(y), \quad H_a^{(1)}(x)H_a^{(0)}(y), \quad H_a^{(0)}(x)H_a^{(1)}(y), \quad H_a^{(1)}(x)H_a^{(1)}(y) \right]$$

$$(11.68)$$

correspond to the values of

$$w, \quad \frac{\partial w}{\partial x}, \quad \frac{\partial w}{\partial y}, \quad \frac{\partial^2 w}{\partial x \partial y},$$

specified at the corner nodes, taking successively unit values at node a and zero at other nodes.

An element based on these shape functions has been developed by Bogner *et al.*[17] and used with success. Indeed it is the most accurate rectangular element available as indicated by results in Fig. 11.16. A development of this type of element to include continuity of higher derivatives is simple and outlined in reference 18. In their undistorted form the above elements are, as for all rectangles, of very limited applicability.

11.15 The 21 and 18 degree-of-freedom triangle

If continuity of higher derivatives than first is accepted at nodes (thus imposing a certain constraint on non-homogeneous material and discontinuous thickness situations as explained in Sec. 11.2.4), the generation of slope and deflection compatible elements presents less difficulty.

Considering as nodal degrees of freedom

$$w, \quad \frac{\partial w}{\partial x}, \quad \frac{\partial w}{\partial y}, \quad \frac{\partial^2 w}{\partial x^2}, \quad \frac{\partial^2 w}{\partial x \partial y}, \quad \frac{\partial^2 w}{\partial y^2},$$

a triangular element will involve at least 18 degrees of freedom. However, a complete fifth-order polynomial contains 21 terms. If, therefore, we add three normal slopes at the mid-side as additional degrees of freedom a sufficient number of equations appear to exist for which the shape functions can be found with a complete quintic polynomial.

Along any edge we have six quantities determining the variation of w (displacement, slopes, and curvature at corner nodes), that is, specifying a fifth-order variation. Thus, this is uniquely defined and therefore w is continuous between elements. Similarly, $\partial w/\partial n$ is prescribed by five quantities and varies as a fourth-order polynomial. Again this is as required by the slope continuity between elements.

If we write the complete quintic polynomial as[*]

$$w = \alpha_1 + \alpha_2 x + \cdots + \alpha_{21} y^5$$

$$(11.69)$$

[*] For this derivation use of simple Cartesian coordinates is recommended in preference to area coordinates. Symmetry is assured as the polynomial is complete.

proceed along the lines of the argument used to develop the rectangle in Sec. 11.3 and write

$$w_a = \alpha_1 + \alpha_2 x_a + \cdots + \alpha_{21} y_a^5$$

$$\left. \frac{\partial w}{\partial x} \right|_a = \alpha_2 + \cdots + \alpha_{20} y_a^4$$

$$\left. \frac{\partial w}{\partial y} \right|_a = \alpha_3 + \cdots + 5\alpha_{21} y_a^4$$

$$\left. \frac{\partial^2 w}{\partial x^2} \right|_i = 2\alpha_4 + \cdots + 2\alpha_{19} y_a^3$$

and so on, and finally obtain an expression

$$\tilde{\mathbf{u}}^e = \mathbf{C}\boldsymbol{\alpha} \tag{11.70}$$

in which $\mathbf{C}$ is a 21×21 matrix.

The only apparent difficulty in the process that the reader may experience in forming this is that of the definition of the normal slopes at the mid-side nodes. However, if one notes that

$$\frac{\partial w}{\partial n} = \cos\phi \frac{\partial w}{\partial x} + \sin\phi \frac{\partial w}{\partial y} \tag{11.71}$$

in which ϕ is the angle of a particular side to the x axis, the manner of formulation becomes simple. It is not easy to determine an explicit inverse of $\mathbf{C}$, and the stiffness expressions, etc., are evaluated as in Eqs (11.22)–(11.25) by a numerical inversion.

The existence of the mid-side nodes with their single degree of freedom is an inconvenience. It is possible, however, to constrain these by allowing only a cubic variation of the normal slope along each triangle side. Now, explicitly, the matrix $\mathbf{C}$ and the degrees of freedom can be reduced to 18, giving an element illustrated in Fig. 11.19(e) with three corner nodes and 6 degrees of freedom at each node.

Both of these elements were described in several independently derived publications appearing during 1968 and 1969. The 21 degree-of-freedom element was described independently by Argyris et al.,[23] Bell,[19] Bosshard,[22] and Visser,[24] listing the authors alphabetically. The reduced 18 degree-of-freedom version was developed by Argyris et al.,[23] Bell,[19] Cowper et al.,[21] and Irons.[14] An essentially similar, but more complicated, formulation has been developed by Butlin and Ford,[20] and mention of the element shape functions was made earlier by Withum[69] and Felippa.[70]

It is clear that many more elements of this type could be developed and indeed some are suggested in the above references. A very inclusive study is found in the work of Zenisek,[71] Peano,[72] and others.[73–75] However, it should always be borne in mind that all the elements discussed in this section involve an inconsistency when discontinuous variation of material properties occurs. Further, the existence of higher-order derivatives makes it more difficult to impose boundary conditions and indeed the simple interpretation of energy conjugates as 'nodal forces' is more complex. Thus, the engineer may still feel a justified preference for the more intuitive formulation involving displacements and slopes only, despite the fact that very good accuracy is demonstrated in the references cited for the quartic and quintic elements.

Avoidance of continuity difficulties in mixed and constrained elements

11.16 Mixed formulations – general remarks

Equations (11.10a)–(11.11b) of this chapter provide for many possibilities to approximate both thick and thin plates by using mixed (i.e. reducible) forms. In these, more than one set of variables is approximated directly, and generally continuity requirements for such approximations can be of either C_1 or C_0 type. The options open are large and indeed so is the number of publications proposing various alternatives. We shall therefore limit the discussion to those that appear most useful.

To avoid constant reference to the beginning of this chapter, the four governing equations (11.10a)–(11.11b) are rewritten below in their abbreviated form with dependent variable sets $\mathbf{M}$, ϕ, $\mathbf{S}$, and w:

$$\mathbf{M} - \mathbf{D}\mathcal{L}\phi = \mathbf{0}$$
$$\frac{1}{\alpha}\mathbf{S} - \phi - \nabla w = \mathbf{0}$$
$$\mathcal{L}^{\mathrm{T}}\mathbf{M} + \mathbf{S} = \mathbf{0} \tag{11.72}$$
$$\nabla^{\mathrm{T}}\mathbf{S} + q = 0$$

in which $\alpha = \kappa\, Gt$. To these, of course, the appropriate boundary conditions can be added. For details of the operators, etc., the fuller forms previously quoted need to be consulted.

Mixed forms that utilize direct approximations to all the four variables are not common. The most obvious set arises from elimination of the moments $\mathbf{M}$, that is

$$\mathcal{L}^{\mathrm{T}}\mathbf{D}\mathcal{L}\phi + \mathbf{S} = \mathbf{0}$$
$$\frac{1}{\alpha}\mathbf{S} - \phi - \nabla w = \mathbf{0} \tag{11.73}$$
$$\nabla^{\mathrm{T}}\mathbf{S} + q = 0$$

and is the basis of a formulation directly related to the three-dimensional elasticity consideration. This is so important that we shall devote Chapter 12 entirely to it, and, of course, there it can be used for both thick and thin plates. We shall, however, return to one of its derivations in Sec. 11.18.

One of the earliest mixed approaches leaves the variables $\mathbf{M}$ and w to be approximated and eliminates $\mathbf{S}$ and ϕ. The form given is restricted to thin plates and thus $\alpha = \infty$ is taken.

We now can write for the first two of Eqs (11.72),

$$\mathbf{D}^{-1}\mathbf{M} + \mathcal{L}\nabla w = \mathbf{0} \tag{11.74}$$

and for the last two of Eqs (11.72),

$$\nabla^{\mathrm{T}} \mathcal{L}^{\mathrm{T}} \mathbf{M} - q = 0 \tag{11.75}$$

The approximation can now be made directly putting

$$\mathbf{M} = \mathbf{N}_M \tilde{\mathbf{M}} \quad \text{and} \quad w = \mathbf{N}_w \tilde{w} \tag{11.76}$$

where $\tilde{\mathbf{M}}$ and $\tilde{w}$ list the nodal (or other) parameters of the expansions, and $\mathbf{N}_M$ and $\mathbf{N}_w$ are appropriate shape functions.

The approximation equations can, as is well known, be made either via a suitable variational principle or directly in a weighted residual, Galerkin form, both leading to identical results. We choose here the latter, although the first presentations of this approximation by Herrmann[76] and others[52,77–84] all use the Hellinger–Reissner principle.

A weak form from which the plate approximation may be deduced is given by

$$\delta \Pi = \int_\Omega \delta \mathbf{M} \left(\mathbf{D}^{-1} \mathbf{M} + \mathcal{L} \nabla w \right) \mathrm{d}\Omega + \int_\Omega \delta w \left(\nabla^{\mathrm{T}} \mathcal{L}^{\mathrm{T}} \mathbf{M} - q \right) \mathrm{d}\Omega + \delta \Pi_{\mathrm{bt}} = 0 \tag{11.77}$$

where $\delta \Pi_{\mathrm{bt}}$ describes appropriate boundary condition terms. Using the Galerkin weighting approximations

$$\delta \mathbf{M} = \mathbf{N}_M \delta \tilde{\mathbf{M}} \quad \text{and} \quad \delta w = \mathbf{N}_w \delta \tilde{w} \tag{11.78}$$

gives on integration by parts the following equation set

$$\begin{bmatrix} \mathbf{A} & \mathbf{C} \\ \mathbf{C}^{\mathrm{T}} & \mathbf{0} \end{bmatrix} \begin{Bmatrix} \tilde{\mathbf{M}} \\ \tilde{w} \end{Bmatrix} = \begin{Bmatrix} \mathbf{f}_1 \\ \mathbf{f}_2 \end{Bmatrix} \tag{11.79}$$

where

$$\mathbf{A} = \int_\Omega \mathbf{N}_M^{\mathrm{T}} \mathbf{D}^{-1} \mathbf{N}_M \, \mathrm{d}\Omega \qquad \mathbf{f}_1 = \int_{\Gamma_t} (\nabla \mathbf{N}_w)^{\mathrm{T}} \begin{Bmatrix} \bar{M}_n \\ \bar{M}_{ns} \end{Bmatrix} \mathrm{d}\Gamma$$

$$\mathbf{C} = \int_\Omega (\mathcal{L} \mathbf{N}_M)^{\mathrm{T}} \nabla \mathbf{N}_w \, \mathrm{d}\Omega \qquad \mathbf{f}_2 = \int_\Omega \mathbf{N}_w^{\mathrm{T}} q \, \mathrm{d}\Omega + \int_{\Gamma_t} \mathbf{N}_w^{\mathrm{T}} S_n \, \mathrm{d}\Gamma \tag{11.80}$$

where $\bar{M}_n$ and $\bar{M}_{ns}$ are the prescribed boundary moments, and S_n is the prescribed boundary shear force.

Immediately, it is evident that only C_0 continuity is required for both $\mathbf{M}$ and w interpolation,[*] and many forms of elements are therefore applicable. Of course, appropriate patch tests for the mixed formulation must be enforced[43] and this requires a necessary condition that

$$n_m \geq n_w \tag{11.81}$$

where n_m stands for the number of parameters describing the moment field and n_w the number in the displacement field.

---

[*] It should be observed that, if C_0 continuity to the whole $\mathbf{M}$ field is taken, excessive continuity will arise and it is usual to ensure the continuity of M_n and M_{ns} at interfaces only.

Many excellent elements have been developed by using this type of approximation, though their application is limited because of the difficulty of interconnection with other structures as well as the fact that the coefficient matrix in Eq. (11.79) is indefinite with many zero diagonal terms.

Indeed, a similar fate is encountered in numerous 'equilibrium element' forms in which the moment (stress) field is chosen *a priori* in a manner satisfying Eq. (11.75). Here the research of Fraeijs de Veubeke[83] and others[30,68] has to be noted. It must, however, be observed that the second of these elements[30] is in fact identical to the mixed element developed by Herrmann[77] and Hellan[85] (see also reference 52).

11.17 Hybrid plate elements

Hybrid elements are essentially mixed elements in which the field inside the element is defined by one set of parameters and the one on the element frame by another, as shown in Fig. 11.23. The latter are generally chosen to be of a type identical to other displacement models and thus can be readily incorporated in a general program and indeed used in conjunction with the standard displacement types we have already discussed. The internal parameters can be readily eliminated (being confined to a single element) and thus the difference from displacement forms are confined to the element subprogram. The original concept is attributable to Pian[86,87] who pioneered this approach, and today many variants of the procedures exist in the context of thin plate theory.[65,88–97]

In the majority of approximations, an equilibrating stress field is assumed to be given by a number of suitable shape functions and unknown parameters. In others, a mixed stress field is taken in the interior. A more refined procedure, introduced by Jirousek,[65,97] assumes in the interior a series solution exactly satisfying all the differential equations involved for a homogeneous field.

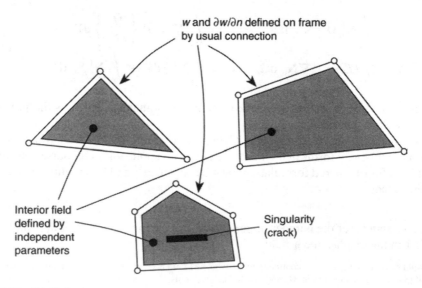

Fig. 11.23   Hybrid elements.

All procedures use a suitable linking of the interior parameters with those defined on the boundary by the 'frame parameters'. The procedures for doing this are described in Chapter 12 of reference 7 in the context of elasticity equations, and only a small change of variables is needed to adapt these to the present case. We leave this extension to the reader who can also consult appropriate references for details.

Some remarks need to be made in the context of hybrid elements.

Remark: The first is that the number of internal parameters, n_I, must be at least as large as the number of frame parameters, n_F, which describe the displacements, less the number of rigid body modes if singularity of the final (stiffness) matrix is to be avoided. Thus, we require that

$$n_I \geq n_F - 3 \tag{11.82}$$

for plates.

Remark: The second remark is a simple statement that it is possible, but counterproductive, to introduce an excessive number of internal parameters that simply give a more exact solution to a 'wrong' problem in which the frame is constraining the interior of an element. Thus additional accuracy is not achieved overall.

Remark: Most of the formulations are available for non-homogeneous plates (and hence non-linear problems). However, this is not true for the Trefftz-hybrid elements[65,97] where an exact solution to the differential equation needs to be available for the element interior. Such solutions are not known for arbitrary non-homogeneous interiors and hence the procedure fails. However, for homogeneous problems the elements can be made much more accurate than any of the others and indeed allow a general polygonal element with singularities and/or internal boundaries to be developed by the use of special functions (see Fig. 11.23). Obviously, this advantage needs to be borne in mind.

A number of elements matching (or duplicating) the displacement method have been developed and the performance of some of the simpler ones is shown in Fig. 11.16. Indeed, it can be shown that many hybrid-type elements duplicate precisely the various incompatible elements that pass the convergence requirement. Thus, it is interesting to note that the triangle of Allman[96] gives precisely the same results as the 'smoothed' Razzaque element of references 33 and 34 or, indeed, the element of Sec. 11.5.

11.18 Discrete Kirchhoff constraints

Another procedure for achieving excellent element performance is achieved as a constrained (mixed) element. Here it is convenient (though by no means essential) to use a variational principle to describe the first and third of Eqs (11.73). This can be written simply as the minimization of the functional

$$\Pi = \frac{1}{2} \int_\Omega (\mathcal{L}\theta)^T \mathbf{D}(\mathcal{L}\theta) \, d\Omega + \frac{1}{2} \int_\Omega \mathbf{S}^T \frac{1}{\alpha} \mathbf{S} \, d\Omega - \int_\Omega wq \, d\Omega + \Pi_{bt} = \text{minimum} \tag{11.83}$$

subject to the constraint that the second of Eqs (11.73) be satisfied, that is,

$$\frac{1}{\alpha}\mathbf{S} - \boldsymbol{\theta} - \nabla w = 0 \tag{11.84}$$

We shall use this form for general thick plates in Chapter 12, but in the case of thin plates with which this chapter is concerned, we can specialize by putting $\alpha = \infty$ and rewrite the above as

$$\Pi = \frac{1}{2}\int_{\Omega} (\mathcal{L}\boldsymbol{\theta})^{\mathsf{T}}\mathbf{D}(\mathcal{L}\boldsymbol{\theta})\,\mathrm{d}\Omega - \int_{\Omega} wq\,\mathrm{d}\Omega + \Pi_{\text{bt}} = \text{minimum} \tag{11.85}$$

subject to

$$\boldsymbol{\theta} + \nabla w = 0 \tag{11.86}$$

and we note that the explicit mention of shear forces $\mathbf{S}$ is no longer necessary.

To solve the problem posed by Eqs (11.85) and (11.86) we can

1. approximate w and $\boldsymbol{\theta}$ by independent interpolations of C_0 continuity as

$$w = \mathbf{N}_w\tilde{\mathbf{w}} \quad \text{and} \quad \boldsymbol{\theta} = \mathbf{N}_\theta\tilde{\boldsymbol{\theta}} \tag{11.87}$$

2. impose a discrete approximation to the constraint of Eq. (11.86) and solve the minimization problem resulting from substitution of Eq. (11.87) into Eq. (11.85) by either discrete elimination, use of suitable Lagrangian multipliers, or penalty procedures.

In the application of the so-called *discrete Kirchhoff constraints*, Eq. (11.86) is approximated by point (or subdomain) *collocation* and direct elimination is used to *reduce the number of nodal parameters*. Of course, the other means of imposing the constraints could be used with *identical* effect and we shall return to these in the next chapter. However, direct elimination is advantageous in reducing the final total number of variables and can be used effectively.

The procedure for constructing the discrete Kirchhoff relations was presented in Sec. 10.5.3 and applied to beams. For two-dimensional plate elements the situation is a little more complex, but if we imagine x to coincide with the direction tangent to an element side, precisely identical elimination to that presented in Sec. 10.5.3 enforces *complete compatibility* along an element side when both gradients of w are specified at the ends. However, with discrete imposition of the constraints it is not clear *a priori* that convergence will always occur – though, of course, one can argue heuristically that collocation applied in numerous directions should result in an acceptable element. Indeed, patch tests turn out to be satisfied by most elements in which the w interpolation (and hence the $\partial w/\partial s$ interpolation) have C_0 continuity.

The constraints frequently applied in practice involve the use of line or subdomain collocation to increase their number (which must, of course, always be less than the number of remaining variables) and such additional constraint equations as

$$I_\Gamma \equiv \int_{\Gamma_e} \left(\frac{\partial w}{\partial s} + \phi_s \right) ds = 0$$

$$I_{\Omega x} \equiv \int_{\Omega_e} \left(\frac{\partial w}{\partial x} + \phi_x \right) d\Omega = 0 \qquad (11.88)$$

$$I_{\Omega y} \equiv \int_{\Omega_e} \left(\frac{\partial w}{\partial y} + \phi_y \right) d\Omega = 0$$

are frequently used. The algebra involved in the elimination is not always easy and the reader is referred to original references for details pertaining to each particular element.

The concept of discrete Kirchhoff constraints was first introduced by Wempner et al.,[98] Stricklin et al.,[61] and Dhatt[62] in 1968–69, but it has been applied extensively since.[99–110]

In particular, the 9 degree-of-freedom triangle[99,100] and the complex semi-loof element of Irons[102] are elements which have been successfully used.

Figure 11.24 illustrates some of the possible types of quadrilateral elements achieved in these references.

11.19 Rotation-free elements

It is possible to construct elements for thin plates in terms of transverse displacement parameters alone. Nay and Utku used quadratic displacement approximation and minimum potential energy to construct a least-square fit for an element configuration shown in Fig. 11.25(a).[111] The element is non-conforming but passes the patch test and therefore is an admissible form. An alternative, mixed field, construction is given by Oñate and Zárate for a composite element constructed from linear interpolation on each triangle.[112,113] In this work a mixed variational principle is used together with a special approximation for the curvature. We summarize here the steps in the better approach.

A three-field mixed variational form for a thin plate problem based on the Hu–Washizu functional may be written as

$$\Pi = \frac{1}{2} \int_A \chi^T \mathbf{D} \chi \, dA - \int_A \mathbf{M}^T \left[(\boldsymbol{\mathcal{L}} \nabla) w - \chi \right] dA - \int_A wq \, dA + \Pi_{bt} \qquad (11.89)$$

where now χ and $\mathbf{M}$ are mixed variables to be approximated, $(\boldsymbol{\mathcal{L}} \nabla) w$ are again second derivatives of displacement w given in Eq. (11.13) and integration is over the area of the plate middle surface. Variation of Eq. (11.89) with respect to χ gives the discrete constitutive equation

$$\delta \Pi_\chi = \int_{A_e} \delta \chi^T \left[\mathbf{D} \chi - \mathbf{M} \right] dA = 0 \qquad (11.90)$$

where A_e is the domain of the patch for the element. Two alternatives for A_e are considered in reference 112 and named BPT and BPN as shown in Figs 11.25(a) and 11.25(b), respectively. For the BPT form the integration is taken over the area of the element 'abc' with area A_p and boundary Γ_p. For the type BPN integration is over

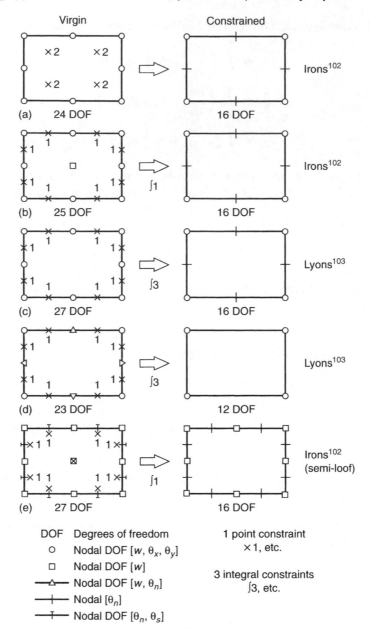

Fig. 11.24 A series of discrete Kirchhoff theory (DKT)-type elements of quadrilateral type.

the more complex area A_a with boundary Γ_a. Each, however, is simple to construct. Similarly, variation of Eq. (11.89) with respect to moment gives the discrete curvature relation

$$\delta\Pi_M = \int_{A_e} \delta\mathbf{M}^{\mathrm{T}}\,[(\mathcal{L}\nabla)w - \chi]\,\mathrm{d}A = 0 \tag{11.91}$$

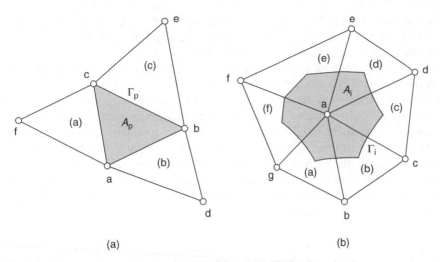

(a)   (b)

Fig. 11.25 Elements for rotation-free thin plates: (a) patch for Nay and Utku procedure[111] BPT triangle; and (b) patch for BPN triangle.[112]

Finally, the equilibrium equations are obtained from the variation with respect to the displacement, and are expressed as

$$\delta\Pi_w = \int_{A_e} [(\mathcal{L}\nabla)\delta w]^{\mathrm{T}} \mathbf{M}\,\mathrm{d}A - \int_{A_e} \delta wq\,\mathrm{d}A + \delta\Pi_{bt} = 0 \tag{11.92}$$

A finite element approximation may be constructed in the standard manner by writing

$$\mathbf{M} = N_a^m \tilde{\mathbf{M}}_a, \quad \chi = N_a^\chi \tilde{\chi}_a \quad \text{and} \quad w = N_a^w \tilde{w}_a \tag{11.93}$$

The simplest approximations are for $N_a^m = N_a^\chi = 1$ and linear interpolation over each triangle for N_a^w. Equation (11.90) is easily evaluated; however, the other two integrals have apparent difficulty since a linear interpolation yields zero derivatives within each triangle. Indeed the curvature is now concentrated in the 'kinks' which occur between contiguous triangles. To obtain discrete approximations to the curvature changes an integration by parts (i.e. application of Green's theorem) is used to rewrite Eq. (11.91) as

$$\int_{A_e} \delta\mathbf{M}^{\mathrm{T}}\chi\,\mathrm{d}A + \int_{\Gamma_e} \delta\mathbf{M}^{\mathrm{T}}\mathbf{g}\nabla w\,\mathrm{d}\Gamma = 0 \tag{11.94}$$

where

$$\mathbf{g} = \begin{bmatrix} n_x & 0 \\ 0 & n_y \\ n_y & n_x \end{bmatrix} \tag{11.95}$$

is a matrix of the direction cosines for an outward pointing normal vector $\mathbf{n}$ to the boundary Γ_e and

$$\nabla w = \begin{Bmatrix} w_{,x} \\ w_{,y} \end{Bmatrix} \tag{11.96}$$

In these expressions Γ_e is the part of the boundary within the area of integration A_e. Thus, for the element type BPT it is just the contour Γ_p as shown in Fig. 11.25(a).

For the element type BPN no slope discontinuity occurs on the boundary Γ_a shown in Fig. 11.25(b); however, it is necessary to integrate along the half sides of each triangle within the patch bounded by Γ_a. The remainder of the derivation is now straightforward and the reader is referred to reference 112 for additional details and results. In this paper results are also presented for thin shells.

We note that the type of element discussed in this section is quite different from those presented previously in that nodes exist outside the boundary of the element. Thus, the definition of an element and the assembly process are somewhat different. In addition, boundary conditions need some special treatments to include in a general manner.[112] Because of these differences we do not consider additional members in this family. We do note, however, that for explicit dynamic programs some advantages occur since no rotation parameters need be integrated. Results for thin shells subjected to impulsive loading are particularly noteworthy.[112]

11.20 Inelastic material behaviour

The preceding discussion has assumed the plate to be a linear elastic material. In many situations it is necessary to consider a more general constitutive behaviour in order to represent the physical problem correctly. For thin plates, only the bending and twisting moment are associated with deformations and are related to the local stresses through

$$\mathbf{M} = \left\{ \begin{array}{c} M_x \\ M_y \\ M_{xy} \end{array} \right\} = \int_{-t/2}^{t/2} \left\{ \begin{array}{c} \sigma_x \\ \sigma_y \\ \tau_{xy} \end{array} \right\} z \, dz \tag{11.97}$$

Any of the material models discussed in Chapter 4 *which have symmetric stress behaviour with respect to strains* may be used in plate analysis provided an appropriate plane stress form is available (either analytically or through iterative reduction of the three-dimensional equations). The symmetry is necessary to avoid the generation of in-plane force resultants – which are assumed to decouple from the bending behaviour. If such conditions do not exist it is necessary to use a shell formulation as described in Chapters 13–16.

In practice two approaches are considered – one dealing with the individual lamina using local stress components σ_x, σ_y and τ_{xy} and the other using plate resultant forces M_x, M_y and M_{xy} directly.

Numerical integration through thickness

The most direct approach is to use a *plane stress* form of the stress–strain relation and perform the through-thickness integration numerically. In order to capture the maximum stresses at the top and bottom of the plate it is best to use Gauss–Lobatto-type quadrature formulae[114] where integrals are approximated by

$$\int_{-1}^{1} f(\xi) \, d\xi \approx f(-1) W_1 + \sum_{n=2}^{N-1} f(\xi_n) W_n + f(1) W_N \tag{11.98}$$

These formula differ from the typical Gaussian quadrature considered previously and use the end points on the interval directly. This allows computation of first yield to be

Table 11.4 Gauss–Lobatto quadrature points and weights

N	ξ_n	W_n
3	± 1.0	1/3
	0.0	4/3
4	± 1.0	1/6
	$\pm \sqrt{0.2}$	5/6
5	± 1.0	0.1
	$\pm \sqrt{3/7}$	49/90
	0.0	64/90
6	± 1.0	1/15
	$\pm \sqrt{t_1}$	$0.6/[t_1(1-t_0)^2]$
	$\pm \sqrt{t_2}$	$0.6/[t_2(1-t_0)^2]$
	$t_0 = \sqrt{7}$	$t_1 = (7 + 2t_0)/21$
		$t_2 = (7 - 2t_0)/21$

more accurate. The values of the parameters ξ_n and W_n are given in Table 11.4 up to the six-point formula. Parameters for higher-order formulae may be found in reference 114.

Noting that the strain components in plates [see Eq. (11.8a)] are asymmetric with respect to the middle surface of the plate and that the z coordinate is also asymmetric we can compute the plate resultants by evaluating only half the integral. Accordingly, we may use

$$\mathbf{M} = \begin{Bmatrix} M_x \\ M_y \\ M_{xy} \end{Bmatrix} = 2 \int_0^{t/2} \begin{Bmatrix} \sigma_x \\ \sigma_y \\ \tau_{xy} \end{Bmatrix} z \, dz \qquad (11.99)$$

and here a six-point formula or less will generally be sufficient to compute integrals.

Equation (11.21) is replaced by the non-linear equation given as

$$\Psi(\tilde{\mathbf{w}}) = \mathbf{f} - \int_\Omega \mathbf{B}^{\mathrm{T}} \mathbf{M} \, d\Omega = \mathbf{0} \qquad (11.100)$$

The solution process (for a static case) may now proceed by using, for instance, a Newton scheme in which the *tangent moduli* for the plate are obtained by using the tangent moduli for the stress components as

$$d\mathbf{M} = \begin{Bmatrix} dM_x \\ dM_y \\ dM_{xy} \end{Bmatrix} = \left[2 \int_0^{t/2} \mathbf{D}_{\mathrm{T}}^{(\mathrm{ps})} z^2 \, dz \right] \mathcal{L} \nabla dw = \mathbf{D}_{\mathrm{T}} (\mathcal{L} \nabla) dw \qquad (11.101)$$

where $\mathbf{D}_{\mathrm{T}}^{(\mathrm{ps})}(z)$ is the tangent modulus matrix of a plane stress material model at each lamina level z, and $\mathbf{D}_{\mathrm{T}}$ is the resulting bending tangent stiffness matrix of the plate.

The Newton iteration for the displacement increment is computed as

$$\mathbf{K}_{\mathrm{T}}^{(k)} d\tilde{\mathbf{w}}^{(k)} = \Psi^{(k)} \qquad (11.102)$$

with iterative updates

$$\tilde{\mathbf{w}}^{(k+1)} = \tilde{\mathbf{w}}^{(k)} + d\tilde{\mathbf{w}}^{(k)} \qquad (11.103)$$

until a suitable convergence criterion is satisfied. This follows precisely methods previously defined for solids.

Resultant constitutive models

A resultant yield function for plates with Huber–von Mises-type material is given by[115]

$$F(\mathbf{M}) = \left(M_x^2 + M_y^2 - M_x M_y + 3M_{xy}^2 \right) - M_u^2(\kappa) \leq 0 \tag{11.104}$$

where κ is an 'isotropic' hardening parameter and M_u denotes a uniaxial yield moment and which for homogeneous plates is generally given by

$$M_u = \tfrac{1}{4}t^2 \sigma_y(\kappa) \tag{11.105}$$

in which σ_y is the material uniaxial yield stress in tension (and compression). We observe that, in the absence of hardening, M_u is the moment that exists when the entire cross-section is at a yield stress.

11.21 Concluding remarks – which elements?

The extensive bibliography of this chapter outlining the numerous approaches capable of solving the problems of thin, Kirchhoff, plate flexure shows both the importance of the subject in structural engineering – particularly as a preliminary to shell analysis – and the wide variety of possible approaches. Indeed, only part of the story is outlined here, as the next chapter, dealing with thick plate formulation, presents many practical alternatives of dealing with the same problem.

We hope that the presentation, in addition to providing a guide to a particular problem area, is useful in its direct extension to other fields where governing equations lead to C_1 continuity requirements.

Users of practical computer programs will be faced with a problem of 'which element' is to be used to satisfy their needs. We have listed in Table 11.3 some of the more widely known simple elements and compared their performance in Fig. 11.16. The choice is not always unique, and much more will depend on preferences and indeed extensions desired. As will be seen in Chapter 13 for general shell problems, triangular elements are an optimal choice for many applications and configurations. Further, such elements are most easily incorporated if adaptive mesh generation is to be used for achieving errors of predetermined magnitude.

References

1. G. Kirchhoff. Über das Gleichgewicht und die Bewegung einer elastichen Scheibe. *J. Reine und Angewandte Mathematik*, 40:51–88, 1850.
2. S.P. Timoshenko and S. Woinowski-Krieger. *Theory of Plates and Shells*. McGraw-Hill, New York, 2nd edition, 1959.
3. L. Bucciarelly and N. Dworsky. *Sophie Germain, an Essay on the History of Elasticity*. Reidel, New York, 1980.

4. E. Reissner. Reflections on the theory of elastic plates. *Appl. Mech. Rev.*, 38:1453–1464, 1985.
5. E. Reissner. The effect of transverse shear deformation on the bending of elastic plates. *J. Appl. Mech.*, 12:69–76, 1945.
6. R.D. Mindlin. Influence of rotatory inertia and shear in flexural motions of isotropic elastic plates. *J. Appl. Mech.*, 18:31–38, 1951.
7. O.C. Zienkiewicz, R.L. Taylor and J.Z. Zhu. *The Finite Element Method: Its Basis and Fundamentals*. Butterworth-Heinemann, Oxford, 6th edition, 2005.
8. I. Babuška and T. Scapolla. Benchmark computation and performance evaluation for a rhombic plate bending problem. *International Journal for Numerical Methods in Engineering*, 28:155–180, 1989.
9. I. Babuška. The stability of domains and the question of formulation of plate problems. *Appl. Math.*, pages 463–467, 1962.
10. B.M. Irons and J.K. Draper. Inadequacy of nodal connections in a stiffness solution for plate bending. *Journal of AIAA*, 3:5, 1965.
11. R.W. Clough and J.L. Tocher. Finite element stiffness matrices for analysis of plate bending. In *Proc. 1st Conf. Matrix Methods in Structural Mechanics*, volume AFFDL-TR-66-80, pages 515–545, Wright Patterson Air Force Base, Ohio, October 1966.
12. G.P. Bazeley, Y.K. Cheung, B.M. Irons and O.C. Zienkiewicz. Triangular elements in bending – conforming and non-conforming solutions. In *Proc. 1st Conf. Matrix Methods in Structural Mechanics*, volume AFFDL-TR-66-80, pages 547–576, Wright Patterson Air Force Base, Ohio, October 1966.
13. B. Fraeijs de Veubeke. Bending and stretching of plates. Special models for upper and lower bounds. In *Proc. 1st Conf. Matrix Methods in Structural Mechanics*, volume AFFDL-TR-66-80, pages 863–886, Wright Patterson Air Force Base, Ohio, October 1966.
14. B.M. Irons. A conforming quartic triangular element for plate bending. *International Journal for Numerical Methods in Engineering*, 1:29–46, 1969.
15. R.W. Clough and C.A. Felippa. A refined quadrilateral element for analysis of plate bending. In *Proc. 2nd Conf. Matrix Methods in Structural Mechanics*, volume AFFDL-TR-68-150, pages 399–440, Wright Patterson Air Force Base, Ohio, October 1968.
16. B. Fraeijs de Veubeke. A conforming finite element for plate bending. *International Journal of Solids and Structures*, 4:95–108, 1968.
17. F.K. Bogner, R.L. Fox and L.A. Schmit. The generation of interelement-compatible stiffness and mass matrices by the use of interpolation formulae. In *Proc. 1st Conf. Matrix Methods in Structural Mechanics*, volume AFFDL-TR-66-80, pages 397–443, Wright Patterson Air Force Base, Ohio, October 1966.
18. I.M. Smith and W. Duncan. The effectiveness of nodal continuities in finite element analysis of thin rectangular and skew plates in bending. *International Journal for Numerical Methods in Engineering*, 2:253–258, 1970.
19. K. Bell. A refined triangular plate bending element. *International Journal for Numerical Methods in Engineering*, 1:101–122, 1969.
20. G.A. Butlin and R. Ford. A compatible plate bending element. Technical Report 68–15, University of Leicester Engineering Department, 1968.
21. G.R. Cowper, E. Kosko, G.M. Lindberg and M.D. Olson. Formulation of a new triangular plate bending element. *Trans. Canad. Aero-Space Inst.*, 1:86–90, 1968 (see also NRC Aero Report LR514, 1968).
22. W. Bosshard. Ein neues vollverträgliches endliches Element für Plattenbiegung. *Mt. Ass. Bridge Struct. Eng. Bull.*, 28:27–40, 1968.
23. J.H. Argyris, I. Fried and D.W. Scharpf. The TUBA family of plate elements for the matrix displacement method. *The Aeronaut. J., Roy. Aeronaut. Soc.*, 72:701–709, 1968.
24. W. Visser. *The finite element method in deformation and heat conduction problems*. Dr. Wiss. dissertation, Technische Hochschule, Delft, 1968.

25. B.M. Irons, J.G. Ergatoudis and O.C. Zienkiewicz. Comments on 'complete polynomial displacement fields for finite element method' (by P.C. Dunne). *Trans. Roy. Aeronaut. Soc.*, 72:709, 1968.

26. O.C. Zienkiewicz and Y.K. Cheung. The finite element method for analysis of elastic isotropic and orthotropic slabs. *Proc. Inst. Civ. Eng.*, 28:471–488, 1964.

27. R.W. Clough. The finite element method in structural mechanics. In O.C. Zienkiewicz and G.S. Holister, editors, *Stress Analysis*, Chapter 7. John Wiley & Sons, Chichester, 1965.

28. D.J. Dawe. Parallelogram element in the solution of rhombic cantilever plate problems. *J. Strain Anal.*, 1:223–230, 1966.

29. J.H. Argyris. Continua and discontinua. In *Proc. 1st Conf. Matrix Methods in Structural Mechanics*, volume AFFDL-TR-66-80, pages 11–189, Wright Patterson Air Force Base, Ohio, October 1966.

30. L.S.D. Morley. On the constant moment plate bending element. *J. Strain Anal.*, 6:20–24, 1971.

31. R.D. Wood. A shape function routine for the constant moment triangular plate bending element. *Engineering Computations*, 1:189–198, 1984.

32. R. Narayanaswami. New triangular plate bending element with transverse shear flexibility. *Journal of AIAA*, 12:1761–1763, 1974.

33. A. Razzaque. Program for triangular bending element with derivative smoothing. *International Journal for Numerical Methods in Engineering*, 5:588–589, 1973.

34. B.M. Irons and A. Razzaque. Shape function formulation for elements other than displacement models. In C.A. Brebbia and H. Tottenham, editors, *Proc. of the International Conference on Variational Methods in Engineering*, volume II, pages 4/59–4/72. Southampton University Press, 1973.

35. J.E. Walz, R.E. Fulton and N.J. Cyrus. Accuracy and convergence of finite element approximations. In *Proc. 2nd Conf. Matrix Methods in Structural Mechanics*, volume AFFDL-TR-68-150, pages 995–1027, Wright Patterson Air Force Base, Ohio, October 1968.

36. R.J. Melosh. Structural analysis of solids. *J. Structural Engineering, ASCE*, 4:205–223, August 1963.

37. A. Adini and R.W. Clough. Analysis of plate bending by the finite element method. Technical Report G-7337, Report to National Science Foundation USA, 1961.

38. Y.K. Cheung, I.P. King and O.C. Zienkiewicz. Slab bridges with arbitrary shape and support conditions. *Proc. Inst. Civ. Eng.*, 40:9–36, 1968.

39. J.L. Tocher and K.K. Kapur. Comment on basis of derivation of matrices for direct stiffness method (by R. Melosh). *Journal of AIAA*, 3:1215–1216, 1965.

40. R.D. Henshell, D. Walters and G.B. Warburton. A new family of curvilinear plate bending elements for vibration and stability. *J. Sound Vibr.*, 20:327–343, 1972.

41. R.L. Taylor, O.C. Zienkiewicz, J.C. Simo and A.H.C. Chan. The patch test – a condition for assessing FEM convergence. *International Journal for Numerical Methods in Engineering*, 22:39–62, 1986.

42. O.C. Zienkiewicz, S. Qu, R.L. Taylor and S. Nakazawa. The patch test for mixed formulations. *International Journal for Numerical Methods in Engineering*, 23:1873–1883, 1986.

43. O.C. Zienkiewicz and D. Lefebvre. Three field mixed approximation and the plate bending problem. *Comm. Appl. Num. Meth.*, 3:301–309, 1987.

44. P.G. Bergen and L. Hanssen. A new approach for deriving 'good' element stiffness matrices. In J.R. Whiteman, editor, *The Mathematics of Finite Elements and Applications*, pages 483–497. Academic Press, London, 1977.

45. R.V. Southwell. *Relaxation Methods in Theoretical Physics*. Clarendon Press, Oxford, 1st edition, 1946.

46. P.G. Bergan and M.K. Nygard. Finite elements with increased freedom in choosing shape functions. *International Journal for Numerical Methods in Engineering*, 20:643–663, 1984.

47. C.A. Felippa and P.G. Bergan. A triangular plate bending element based on energy orthogonal free formulation. *Computer Methods in Applied Mechanics and Engineering*, 61:129–160, 1987.

48. A. Samuelsson. The global constant strain condition and the patch test. In R. Glowinski, E.Y. Rodin and O.C. Zienkiewicz, editors, *Energy Methods in Finite Element Analysis*, Chapter 3, pages 49–68. John Wiley & Sons, Chichester, 1979.

49. B. Specht. Modified shape functions for the three node plate bending element passing the patch test. *International Journal for Numerical Methods in Engineering*, 26:705–715, 1988.

50. T.J.R. Hughes. *The Finite Element Method: Linear Static and Dynamic Analysis*. Prentice-Hall, Englewood Cliffs, NJ, 1987.

51. F. Kikuchi and Y. Ando. A new variational functional for the finite element method and its application to plate and shell problems. *Nuclear Engineering and Design*, 21(1):95–113, 1972.

52. L.S.D. Morley. The triangular equilibrium element in the solution of plate bending problems. *Aero. Q.*, 19:149–169, 1968.

53. W. Ritz. Über eine neue Methode zur Lösung gewisser variationsproblem der mathematischen physik. *Journal für die reine und angewandte Mathematik*, 135:1–61, 1908.

54. B.G. Galerkin. Series solution of some problems in elastic equilibrium of rods and plates. *Vestn. Inzh. Tech.*, 19:897–908, 1915.

55. H. Hencky. Der Spannungszustand in rechteckigen Platten. Technical Report VI, 94 S, München, 1913.

56. H. Hencky. *Der Spannungszustand in rechteckigen Platten*. PhD thesis, Darmstadt, Published by R. Oldenbourg, Munich and Berlin, Germany, 1913.

57. I.A. Wojtaszak. The calculation of maximum deflection, moment, and shear for uniformly loaded rectangular plate with clamped edges. *J. Applied Mechanics, ASME*, 59:A173–A176, 1937.

58. R.L. Taylor and S. Govindjee. Solution of clamped rectangular plate problems. *Communications in Numerical Methods of Engineering*, 20:757–765, 2004.

59. H. Marcus. *Die Theorie elastisher Geweve und ihre Anwendung auf die Berechnung biegsamer Platten*. Springer, Berlin, 1932.

60. P. Ballesteros and S.L. Lee. Uniformly loaded rectangular plate supported at the corners. *Int. J. Mech. Sci.*, 2:206–211, 1960.

61. J.A. Stricklin, W. Haisler, P. Tisdale and K. Gunderson. A rapidly converging triangle plate element. *Journal of AIAA*, 7:180–181, 1969.

62. G.S. Dhatt. Numerical analysis of thin shells by curved triangular elements based on discrete Kirchhoff hypotheses. In W.R. Rowan and R.M. Hackett, editors, *Proc. Symp. on Applications of FEM in Civil Engineering*, Vanderbilt University, Nashville, Tennessee, 1969. ASCE.

63. J.L. Batoz, K.-J. Bathe and L.W. Ho. A study of three-node triangular plate bending elements. *International Journal for Numerical Methods in Engineering*, 15:1771–1812, 1980.

64. M.M. Hrabok and T.M. Hrudey. A review and catalogue of plate bending finite elements. *Computers and Structures*, 19:479–495, 1984.

65. J. Jirousek and L. Guex. The hybrid-Trefftz finite element model and its application to plate bending. *Int. J. Num. Meth. Eng.*, 23:651–693, 1986.

66. A. Razzaque. *Finite element analysis of plates and shells*. PhD thesis, Civil Engineering Department, University of Wales, Swansea, 1972.

67. E.L. Wilson. The static condensation algorithm. *International Journal for Numerical Methods in Engineering*, 8:1974, 199–203.

68. G. Sander. Bournes supérieures et inférieures dans l'analyse matricielle des plates en flexion-torsion. *Bull. Soc. Royale des Sci. de Liège*, 33:456–494, 1974.

69. D. Withum. Berechnung von Platten nach dem Ritzchen Verfahren mit Hilfe dreieckförmiger Meshnetze. Technical report, Mittl. Inst. Statik, Technische Hochschule, Hanover, 1966.

70. C.A. Felippa. *Refined finite element analysis of linear and non-linear two-dimensional structures*. PhD dissertation, Department of Civil Engineering, SEMM, University of California, Berkeley, 1966. Also: SEL Report 66–22, Structures Materials Research Laboratory.

71. A. Zanisek. Interpolation polynomials on the triangle. *International Journal for Numerical Methods in Engineering*, 10:283–296, 1976.

72. A.G. Peano. Conforming approximation for Kirchhoff plates and shells. *International Journal for Numerical Methods in Engineering*, 14:1273–1291, 1979.

73. J.J. Göel. Construction of basic functions for numerical utilization of Ritz's method. *Numerische Math.*, 12:435–447, 1968.

74. G. Birkhoff and L. Mansfield. Compatible triangular finite elements. *J. Math. Anal. Appl.*, 47:531–553, 1974.

75. C.L. Lawson. C^1-compatible interpolation over a triangle. Technical Report RM 33–770, NASA Jet Propulsion Laboratory, Pasadena, California, 1976.

76. L.R. Herrmann. Finite element bending analysis of plates. In *Proc. 1st Conf. Matrix Methods in Structural Mechanics*, AFFDL-TR-66-80, pages 577–602, Wright Patterson Air Force Base, Ohio, 1965.

77. L.R. Herrmann. Finite element bending analysis of plates. *J. Engineering Mechanics, ASCE*, 94(EM5):13–25, 1968.

78. W. Visser. A refined mixed type plate bending element. *Journal of AIAA*, 7:1801–1803, 1969.

79. J.C. Boot. On a problem arising from the derivation of finite element matrices using Reissner's principle. *International Journal for Numerical Methods in Engineering*, 12:1879–1882, 1978.

80. A. Chaterjee and A.V. Setlur. A mixed finite element formulation for plate problems. *International Journal for Numerical Methods in Engineering*, 4:67–84, 1972.

81. J.W. Harvey and S. Kelsey. Triangular plate bending elements with enforced compatibility. *Journal of AIAA*, 9:1023–1026, 1971.

82. B. Fraeijs de Veubeke and O.C. Zienkiewicz. Strain energy bounds in finite element analysis by slab analogy. *J. Strain Anal.*, 2:265–271, 1967.

83. B. Fraeijs de Veubeke. An equilibrium model for plate bending. *International Journal of Solids and Structures*, 4:447–468, 1968.

84. J. Bron and G. Dhatt. Mixed quadrilateral elements for plate bending. *Journal of AIAA*, 10: 1359–1361, 1972.

85. K. Hellan. Analysis of elastic plates in flexure by a simplified finite element method. Technical Report Civ. Eng. Series 46, Acta Polytechnica Scandinavia, Trondheim, 1967.

86. T.H.H. Pian. Derivation of element stiffness matrices by assumed stress distribution. *Journal of AIAA*, 2:1332–1336, 1964.

87. T.H.H. Pian and P. Tong. Basis of finite element methods for solid continua. *International Journal for Numerical Methods in Engineering*, 1:3–28, 1969.

88. R.J. Allwood and G.M.M. Cornes. A polygonal finite element for plate bending problems using the assumed stress approach. *International Journal for Numerical Methods in Engineering*, 1:135–160, 1969.

89. B.E. Greene, R.E. Jones, R.M. McLay and D.R. Strome. Generalized variational principles in the finite element method. *Journal of AIAA*, 7:1254–1260, 1969.

90. P. Tong. New displacement hybrid models for solid continua. *International Journal for Numerical Methods in Engineering*, 2:73–83, 1970.

91. B.K. Neale, R.D. Henshell and G. Edwards. Hybrid plate bending elements. *J. Sound Vibr.*, 22:101–112, 1972.

92. R.D. Cook. Two hybrid elements for analysis of thick, thin and sandwich plates. *International Journal for Numerical Methods in Engineering*, 5:277–299, 1972.

93. C. Johnson. On the convergence of a mixed finite element method for plate bending problems. *Num. Math.*, 21:43–62, 1973.

94. R.D. Cook and S.G. Ladkany. Observations regarding assumed-stress hybrid plate elements. *International Journal for Numerical Methods in Engineering*, 8:513–520, 1974.

95. I. Torbe and K. Church. A general quadrilateral plate element. *International Journal for Numerical Methods in Engineering*, 9:855–868, 1975.

96. D.J. Allman. A simple cubic displacement model for plate bending. *International Journal for Numerical Methods in Engineering*, 10:263–281, 1976.

97. J. Jirousek. Improvement of computational efficiency of the 9 dof triangular hybrid-Trefftz plate bending element. *International Journal for Numerical Methods in Engineering*, 23:2167–2168, 1986. (Letter to editor.)

98. G.A. Wempner, J.T. Oden and D.K. Cross. Finite element analysis of thin shells. *Proc. Am. Soc. Civ. Eng.*, 94(EM6):1273–1294, 1968.

99. G. Dhatt. An efficient triangular shell element. *Journal of AIAA*, 8:2100–2102, 1970.

100. J.L. Batoz and G. Dhatt. Development of two simple shell elements. *Journal of AIAA*, 10: 237–238, 1972.

101. J.T. Baldwin, A. Razzaque and B.M. Irons. Shape function subroutine for an isoparametric thin plate element. *International Journal for Numerical Methods in Engineering*, 7:431–440, 1973.

102. B.M. Irons. The semi-Loof shell element. In D.G. Ashwell and R.H. Gallagher, editors, *Finite Elements for Thin Shells and Curved Members*, Chapter 11, pages 197–222. John Wiley & Sons, Chichester, 1976.

103. L.P.R. Lyons. *A general finite element system with special analysis of cellular structures*. PhD thesis, Imperial College of Science and Technology, London, 1977.

104. R.A.F. Martins and D.R.J. Owen. Thin plate semi-Loof element for structural analysis including stability and structural vibration. *International Journal for Numerical Methods in Engineering*, 12:1667–1676, 1978.

105. J.L. Batoz and M.B. Tahar. Evaluation of a new quadrilateral thin plate bending element. *International Journal for Numerical Methods in Engineering*, 18:1655–1677, 1982.

106. J.L. Batoz. An explicit formulation for an efficient triangular plate bending element. *International Journal for Numerical Methods in Engineering*, 18:1077–1089, 1982.

107. M.A. Crisfield. A new model thin plate bending element using shear constraints: A modified version of Lyons' element. *Computer Methods in Applied Mechanics and Engineering*, 38: 93–120, 1983.

108. M.A. Crisfield. A qualitative Mindlin element using shear constraints. *Computers and Structures*, 18:833–852, 1984.

109. M.A. Crisfield. *Finite Elements and Solution Procedures for Structural Analysis, Vol. 1, Linear Analysis*. Pineridge Press, Swansea, 1986.

110. G. Dhatt, L. Marcotte and Y. Matte. A new triangular discrete Kirchhoff plate-shell element. *International Journal for Numerical Methods in Engineering*, 23:453–470, 1986.

111. R.A. Nay and S. Utku. An alternative for the finite element method. *Variational Methods in Engineering*, 1, 1972.

112. E. Oñate and F. Zárate. Rotation-free triangular plate and shell elements. *International Journal for Numerical Methods in Engineering*, 47:557–603, 2000.

113. E. Oñate and G. Bugeda. A study of mesh optimality criteria in adaptive finite element analysis. *Engineering Computations*, 10:307–321, 1993.

114. M. Abramowitz and I.A. Stegun, editors. *Handbook of Mathematical Functions*. Dover Publications, New York, 1965.

115. G.S. Shapiro. On yield surfaces for ideal plastic shells. In *Problems of Continuum Mechanics*, pages 414–418. SIAM, Philadelphia, 1961.

12

'Thick' Reissner–Mindlin plates – irreducible and mixed formulations

12.1 Introduction

We have already introduced in Chapter 11 the full theory of thick plates from which the thin plate, Kirchhoff, theory arises as the limiting case. In this chapter we shall show how the numerical solution of thick plates can easily be achieved and how, in the limit, an alternative procedure for solving all problems of Chapter 11 appears. The development of approximations to the Timoshenko beam form discussed in Chapter 10 plays an important role in the development of viable solutions of the thick, Reissner–Mindlin, theory that work in 'thin' plate applications.

To ensure continuity we repeat below the governing equations [Eqs (11.10a)–(11.11b), or Eqs (11.72)]. Referring to Fig. 11.3 of Chapter 11 and the text for definitions, we remark that all the equations could equally well be derived from full three-dimensional analysis of a flat and relatively thin portion of an elastic continuum illustrated in Fig. 12.1. All that it is now necessary to do is to assume that whatever form of the approximating shape functions in the xy plane those in the z direction are only linear. Further, it is assumed that σ_z stress is zero, thus eliminating the effect of vertical strain.[*] The first approximations of this type were introduced quite early[1,2] and the elements then derived are exactly of the Reissner–Mindlin type discussed in Chapter 11.

The equations from which we shall start and on which we shall base all subsequent discussion are the moment constitutive equation [see Eqs (11.10a) and (11.72)],

$$\mathbf{M} - \mathbf{D}\mathcal{L}\phi = \mathbf{0} \qquad (12.1a)$$

where $\mathbf{D}$ is the matrix of bending rigidities, the shear constitutive equation [see Eqs (11.10c) and (11.72)]

$$\frac{1}{\alpha}\mathbf{S} - \phi - \nabla w = \mathbf{0} \qquad (12.1b)$$

where $\alpha = \kappa G t$ is the shear rigidity, the moment equilibrium (angular momentum) equation [see Eqs (11.11b) and (11.72)]

$$\mathcal{L}^{\mathrm{T}}\mathbf{M} + \mathbf{S} = \mathbf{0} \qquad (12.1c)$$

---

[*] Reissner includes the effect of σ_z in bending but, for simplicity, this is disregarded here.

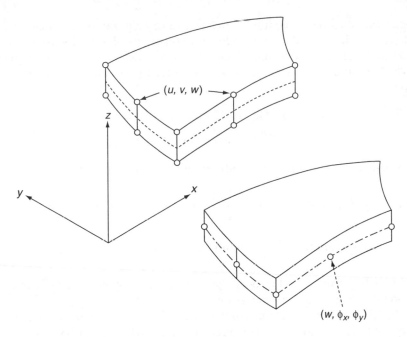

Fig. 12.1 An isoparametric three-dimensional element with linear interpolation in the transverse (thickness) direction and the 'thick' plate element.

and the shear equilibrium equation [see Eqs (11.11a) and (11.72)]

$$\nabla^T \mathbf{S} + q = 0 \qquad (12.1d)$$

In the above the moments $\mathbf{M}$, the transverse shear forces $\mathbf{S}$, and the elastic matrices $\mathbf{D}$ are as defined in Chapter 11, and

$$
\mathcal{L} =
\begin{bmatrix}
\dfrac{\partial}{\partial x} & 0 \\[2mm]
0 & \dfrac{\partial}{\partial y} \\[2mm]
\dfrac{\partial}{\partial y} & \dfrac{\partial}{\partial x}
\end{bmatrix}
\qquad (12.2)
$$

defines the strain–displacement operator on rotations ϕ, and its transpose the equilibrium operator on moments, $\mathbf{M}$. Boundary conditions are of course imposed on w and ϕ or the corresponding plate forces S_n, M_n, M_{ns} in the manner discussed in Sec. 11.2.2.

It is convenient to eliminate $\mathbf{M}$ from Eqs (12.1a)–(12.1d) and write the system of three equations [Eqs (11.73)] as

$$\mathcal{L}^T \mathbf{D} \mathcal{L} \phi + \mathbf{S} = 0$$

$$\frac{1}{\alpha} \mathbf{S} - \phi - \nabla w = 0 \qquad (12.3)$$

$$\nabla^T \mathbf{S} + q = 0$$

This equation system can serve as the basis on which a mixed discretization is built – or alternatively can be reduced further to yield an irreducible form. In Chapter 11 we have dealt with the irreducible form which is given by a fourth-order equation in terms of w alone and which could only serve for solution of *thin plate* problems, that is, when $\alpha = \infty$ [Eq. (11.14a)]. On the other hand, it is easy to derive an alternative irreducible form which is valid only if $\alpha \neq \infty$. Thus, the shear forces can be eliminated yielding two equations:

$$
\mathcal{L}^{\mathrm{T}} \mathbf{D} \mathcal{L} \phi + \alpha \left(\nabla w + \phi \right) = \mathbf{0}
$$
$$
\nabla^{\mathrm{T}} \left[\alpha \left(\nabla w + \phi \right) \right] + q = 0
$$
(12.4)

This is an *irreducible* system corresponding to minimization of the total potential energy

$$
\Pi = \frac{1}{2} \int_{\Omega} \left(\mathcal{L} \phi \right)^{\mathrm{T}} \mathbf{D} \mathcal{L} \phi \, \mathrm{d}\Omega + \frac{1}{2} \int_{\Omega} \left(\nabla w + \phi \right)^{\mathrm{T}} \alpha \left(\nabla w + \phi \right) \mathrm{d}\Omega
$$
$$
- \int_{\Omega} w \, q \, \mathrm{d}\Omega + \Pi_{bt} = \text{minimum}
$$
(12.5)

as can easily be verified. In the above the first term is simply the bending energy and the second the shear distortion energy [see Eq. (11.83)].

Clearly, this irreducible system is only possible when $\alpha \neq \infty$, but it can, obviously, be interpreted as a solution of the potential energy given by Eq. (11.83) for 'thin' plates with the constraint of Eq. (11.84) being imposed in a *penalty manner* with α being now a penalty parameter. Thus, as indeed is physically evident, the thin plate formulation is simply a limiting case of such analysis.

We shall see that the penalty form can yield a satisfactory solution only when discretization of the corresponding mixed formulation satisfies the necessary convergence criteria.

The thick plate form now permits independent specification of three conditions at each point of the boundary. The options which exist are:

$$
\begin{array}{ccc}
w & \text{or} & S_n \\
\phi_n & \text{or} & M_n \\
\phi_s & \text{or} & M_{ns}
\end{array}
$$

in which the subscript n refers to a normal direction to the boundary and s a tangential direction. Clearly, now there are many combinations of possible boundary conditions.

A 'fixed' or 'clamped' situation exists when all three conditions are given by displacement components, which are generally zero, as

$$
w = \phi_n = \phi_s = 0
$$

and a free boundary when all conditions are the 'resultant' components

$$
S_n = M_n = M_{ns} = 0
$$

When we discuss the so-called simply supported conditions (see Sec. 11.2.2), we shall usually refer to the specification

$$
w = 0 \quad \text{and} \quad M_n = M_{ns} = 0
$$

as a 'soft' support, SS1 (and indeed the most realistic support), and to

$$w = 0 \qquad M_n = 0 \qquad \text{and} \qquad \phi_s = 0$$

as a 'hard' support, SS2. The latter in fact replicates the thin plate assumptions and, incidentally, leads to some of the difficulties associated with it.

Finally, there is an important difference between thin and thick plates when 'point' loads are involved. In the thin plate case the displacement w remains finite at locations where a point load is applied; however, for thick plates the presence of shearing deformation leads to an infinite displacement (as indeed three-dimensional elasticity theory also predicts). In finite element approximations one always predicts a finite displacement at point locations with the magnitude increasing without limit as a mesh is refined near the loads. Thus, it is meaningless to compare the deflections at point load locations for different element formulations and we will not do so in this chapter. It is, however, possible to compare the total strain energy for such situations and here we immediately observe that for cases in which a single point load is involved the displacement provides a direct measure for this quantity.

12.2 The irreducible formulation – reduced integration

The procedures for discretizing Eq. (12.4) appear to be straightforward. However, we will find that the process is very sensitive. First, we consider standard isoparametric interpolation in which the two displacement variables are approximated by shape functions and parameters as

$$\phi = \mathbf{N}_\phi \tilde{\phi} \qquad \text{and} \qquad w = \mathbf{N}_w \tilde{w} \tag{12.6}$$

We recall that the rotation parameters ϕ may be transformed into physical rotations about the coordinate axes, θ, using Eq. (11.7). The parameters θ are often more convenient for calculations and are essential in shell developments. The approximation equations are now obtained directly by the use of the total potential energy principle [Eq. (12.5)], the Galerkin process on the weak form, or by the use of virtual work expressions. Here we note that the appropriate generalized strain components, corresponding to the moments $\mathbf{M}$ and shear forces $\mathbf{S}$, are

$$\varepsilon_m = \mathcal{L}\phi = (\mathcal{L}\mathbf{N}_\phi)\tilde{\phi} \tag{12.7a}$$

and

$$\varepsilon_s = \nabla w + \phi = \nabla \mathbf{N}_w \tilde{w} + \mathbf{N}_\phi \tilde{\phi} \tag{12.7b}$$

We thus obtain the discretized problem

$$\left(\int_\Omega (\mathcal{L}\mathbf{N}_\phi)^{\mathrm{T}} \mathbf{D} \mathcal{L}\mathbf{N}_\phi \, d\Omega + \int_\Omega \mathbf{N}_\phi^{\mathrm{T}} \alpha \mathbf{N}_\phi \, d\Omega \right) \tilde{\phi} - \left(\int_\Omega \mathbf{N}_\phi^{\mathrm{T}} \alpha \nabla \mathbf{N}_w \, d\Omega \right) \tilde{w} = \mathbf{f}_\phi$$

and

$$\left(\int_\Omega (\nabla \mathbf{N}_w)^{\mathrm{T}} \alpha \mathbf{N}_\phi \, d\Omega \right) \tilde{\phi} + \left(\int_\Omega (\nabla \mathbf{N}_w)^{\mathrm{T}} \alpha \nabla \mathbf{N}_w \, d\Omega \right) \tilde{w} = \mathbf{f}_w$$

or simply

$$\begin{bmatrix} \mathbf{K}_{ww} & \mathbf{K}_{w\phi} \\ \mathbf{K}_{\phi w} & \mathbf{K}_{\phi\phi} \end{bmatrix} \begin{Bmatrix} \tilde{w} \\ \tilde{\phi} \end{Bmatrix} = \mathbf{K}\tilde{\mathbf{u}} = (\mathbf{K}_b + \mathbf{K}_s)\,\tilde{\mathbf{u}} = \begin{Bmatrix} \mathbf{f}_w \\ \mathbf{f}_\phi \end{Bmatrix} = \mathbf{f} \qquad (12.8a)$$

with

$$\tilde{\mathbf{u}} = \begin{Bmatrix} \tilde{w} \\ \tilde{\phi} \end{Bmatrix} \qquad \tilde{\phi} = \begin{Bmatrix} \tilde{\phi}_x \\ \tilde{\phi}_y, \end{Bmatrix}$$

$$\mathbf{K}_b = \begin{bmatrix} 0 & 0 \\ 0 & \mathbf{K}_{\phi\phi}^b \end{bmatrix} \qquad \mathbf{K}_s = \begin{bmatrix} \mathbf{K}_{ww}^s & \mathbf{K}_{w\phi}^s \\ \mathbf{K}_{\phi w}^s & \mathbf{K}_{\phi\phi}^s \end{bmatrix} \qquad (12.8b)$$

where the arrays are defined by

$$\begin{aligned} \mathbf{K}_{\phi\phi}^b &= \int_\Omega (\mathcal{L}\mathbf{N}_\phi)^{\mathrm{T}}\mathbf{D}\mathcal{L}\mathbf{N}_\phi\,\mathrm{d}\Omega & \mathbf{K}_{ww}^s &= \int_\Omega (\nabla\mathbf{N}_w)^{\mathrm{T}}\alpha\nabla\mathbf{N}_w\,\mathrm{d}\Omega \\ \mathbf{K}_{\phi\phi}^s &= \int_\Omega \mathbf{N}_\phi^{\mathrm{T}}\alpha\mathbf{N}_\phi\,\mathrm{d}\Omega & \mathbf{K}_{\phi w}^s &= \int_\Omega \mathbf{N}_\phi^{\mathrm{T}}\alpha\nabla\mathbf{N}_w\,\mathrm{d}\Omega = (\mathbf{K}_{w\phi}^s)^{\mathrm{T}} \end{aligned} \qquad (12.8c)$$

and forces are given by

$$\mathbf{f}_w = \int_\Omega \mathbf{N}_w^{\mathrm{T}} q\,\mathrm{d}\Omega + \int_{\Gamma_s} \mathbf{N}_w^{\mathrm{T}}\bar{S}_n\,\mathrm{d}\Gamma$$

$$\mathbf{f}_\phi = \int_{\Gamma_m} \mathbf{N}_\phi^{\mathrm{T}}\bar{\mathbf{M}}\,\mathrm{d}\Gamma \qquad (12.8d)$$

where $\bar{S}_n$ is the prescribed shear on boundary Γ_s, and $\bar{\mathbf{M}}$ is the prescribed moment on boundary Γ_m.

The formulation is straightforward and there is little to be said about it *a priori*. Since the form contains only first derivatives apparently any C_0 shape functions of a two-dimensional kind can be used to interpolate the two rotations and the lateral displacement. Figure 12.2 shows some rectangular (or with isoparametric distortion,

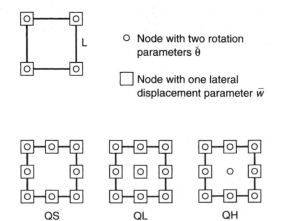

Fig. 12.2 Some early thick plate elements.

quadrilateral) elements used in the early work.[1-3] All should, in principle, be convergent as C_0 continuity exists and constant strain states are available. In Fig. 12.3 we show what in fact happens with a fairly fine subdivision of quadratic serendipity and Lagrangian rectangles as the ratio of span to thickness, L/t, varies. Here L is a characteristic length of the plate and may be a side length, a loading length or a normal mode characteristic. We note that the magnitude of the coefficient α is best measured by the ratio of the bending to shear rigidities and we could assess its value in a non-dimensional form. Thus, for an isotropic material with $\alpha = Gt$ this ratio becomes

$$\frac{12(1-\nu^2)GtL^2}{Et^3} \propto \left(\frac{L}{t}\right)^2 \qquad (12.9)$$

Obviously, 'thick' and 'thin' behaviour therefore depends on the L/t ratio.

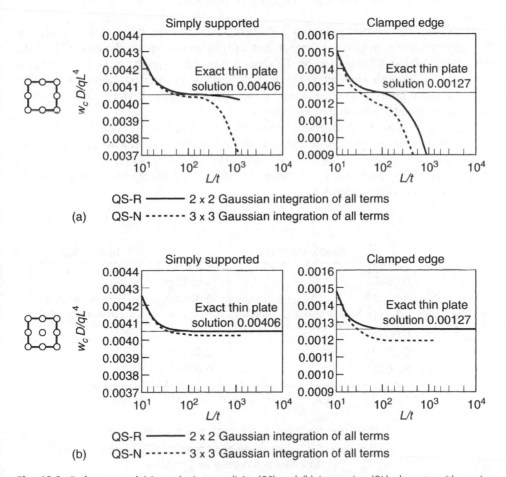

Fig. 12.3 Performance of (a) quadratic serendipity (QS) and (b) Lagrangian (QL) elements with varying span-to-thickness L/t, ratios, uniform load on a square plate with 4×4 normal subdivisions in a quarter. R is reduced 2×2 quadrature and N is normal 3×3 quadrature.

It is immediately evident from Fig. 12.3 that, while the answers are quite good for smaller L/t ratios, the serendipity quadratic fully integrated elements (QS) rapidly depart from the thin plate solution, and in fact tend to zero results (locking) when this ratio becomes large. For Lagrangian quadratics (QL) the answers are better, but again as the plate tends to be thin they are on the small side.

The reason for this 'locking' performance is similar to those considered for the nearly incompressible problem[4] and the Timoshenko beam problem (viz. Chapter 10). In the case of plates the shear constraint implied by the third of Eq. (12.3), and used to eliminate the shear resultant, is too strong if the terms in which this is involved are fully integrated. Indeed, we see that the effect is more pronounced in the serendipity element than in the Lagrangian one. In early work the problem was thus mitigated by using a *reduced* quadrature, either on all terms, which we label R in the figure,[5,6] or only on the offending shear terms selectively[7,8] (labelled S). The dramatic improvement in results is immediately noted. The use of reduced quadrature to develop a four-node plate element has been revisited recently by Gruttmann and Wagner using a stabilized form.[9]

The same improvement in results is observed for linear quadrilaterals in which the full (exact) integration gives results that are totally unacceptable (as shown in Fig. 12.4), but where a reduced integration on the shear terms (single point) gives excellent performance,[10] although a careful assessment of the element stiffness shows it to be rank deficient in an 'hourglass' mode in transverse displacements. (Reduced integration on all terms gives additional matrix singularity.)

A remedy thus has been suggested; however, it is not universal. We note in Fig. 12.3 that even without reduction of integration order, Lagrangian elements perform better in the quadratic expansion. In cubic elements (Fig. 12.5), however, we note that (a) almost no change occurs when integration is 'reduced' and (b), again, Lagrangian-type elements perform very much better.

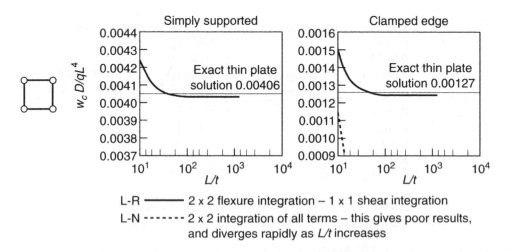

Fig. 12.4 Performance of bilinear elements with varying span-to-thickness, L/t, values.

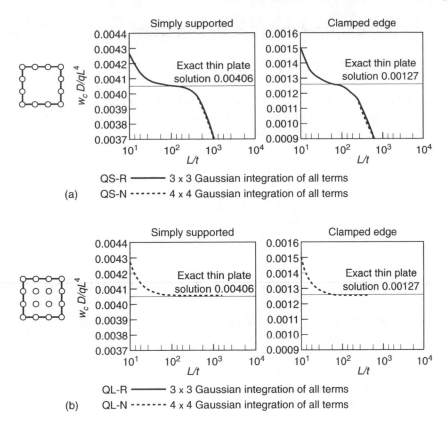

Fig. 12.5 Performance of cubic quadrilaterals: (a) serendipity (QS) and (b) Lagrangian (QL) with varying span-to-thickness, L/t, values.

In the late 1970s many heuristic arguments were advanced for devising better elements,[11–14] all making use of reduced integration concepts. Some of these perform quite well, for example the so-called 'heterosis' element of Hughes and Cohen[11] illustrated in Fig. 12.3 (in which the serendipity-type interpolation is used on w and a Lagrangian one on ϕ), but all of the elements suggested in that era fail on some occasions, either locking or exhibiting singular behaviour. Thus such elements are not 'robust' and should not be used universally.

A better explanation of their failure is needed and hence an understanding of how such elements could be designed. In the next section we shall address this problem by considering a mixed formulation. The reader will recognize here arguments used in the nearly incompressible problem in Chapter 2 and in reference 4 which led to a better understanding of the failure of some straightforward elasticity elements as incompressible behaviour was approached. The situation is completely parallel here.

12.3 Mixed formulation for thick plates

12.3.1 The approximation

The problem of thick plates can, of course, be solved as a mixed one starting from Eq. (12.3) and approximating directly each of the variables w, ϕ and $\mathbf{S}$ independently. Using Eq. (12.3), we construct a weak form as

$$\int_\Omega \delta w \left[\nabla^{\mathrm{T}}\mathbf{S} + q\right] d\Omega = 0$$

$$\int_\Omega \delta\phi^{\mathrm{T}} \left[\mathcal{L}^{\mathrm{T}}\mathbf{D}\mathcal{L}\phi + \mathbf{S}\right] d\Omega = 0 \tag{12.10}$$

$$\int_\Omega \delta\mathbf{S}^{\mathrm{T}} \left[-\frac{1}{\alpha}\mathbf{S} + \phi + \nabla w\right] d\Omega = 0$$

We now write the independent approximations, using the standard Galerkin procedure, as

$$\begin{aligned}
\mathbf{w} &= \mathbf{N}_w\tilde{\mathbf{w}} & \phi &= \mathbf{N}_\phi\tilde{\phi} & \text{and} & & \mathbf{S} &= \mathbf{N}_s\tilde{\mathbf{S}} \\
\delta\mathbf{w} &= \mathbf{N}_w\delta\tilde{\mathbf{w}} & \delta\phi &= \mathbf{N}_\phi\delta\tilde{\phi} & \text{and} & & \delta\mathbf{S} &= \mathbf{N}_s\delta\tilde{\mathbf{S}}
\end{aligned} \tag{12.11}$$

though, of course, other interpolation forms can be used, as we shall note later.

After appropriate integrations by parts of Eq. (12.10), we obtain the discrete symmetric equation system

$$\begin{bmatrix} \mathbf{0} & \mathbf{0} & \mathbf{E}^T \\ \mathbf{0} & \mathbf{K}_b & \mathbf{C}^T \\ \mathbf{E} & \mathbf{C} & \mathbf{H} \end{bmatrix} \begin{Bmatrix} \tilde{\mathbf{w}} \\ \tilde{\phi} \\ \tilde{\mathbf{S}} \end{Bmatrix} = \begin{Bmatrix} \mathbf{f}_w \\ \mathbf{f}_\phi \\ \mathbf{0} \end{Bmatrix} \tag{12.12}$$

where

$$\mathbf{K}_b = \int_\Omega \left(\mathcal{L}\mathbf{N}_\phi\right)^T \mathbf{D}\left(\mathcal{L}\mathbf{N}_\phi\right) d\Omega$$

$$\mathbf{E} = \int_\Omega \mathbf{N}_s^T \nabla\mathbf{N}_w \, d\Omega$$

$$\mathbf{C} = \int_\Omega \mathbf{N}_s^T \mathbf{N}_\phi \, d\Omega \tag{12.13}$$

$$\mathbf{H} = -\int_\Omega \mathbf{N}_s^T \frac{1}{\alpha}\mathbf{N}_s \, d\Omega$$

and where $\mathbf{f}_w$ and $\mathbf{f}_\phi$ are as defined in Eq. (12.8d).

The above represents a typical three-field mixed problem of the type discussed in Sec. 10.5.1 of reference 4, which has to satisfy certain criteria for stability of approximation as the thin plate limit (which can now be solved exactly) is approached. For this limit we have

$$\alpha = \infty \quad \text{and} \quad \mathbf{H} = \mathbf{0} \tag{12.14}$$

In this limiting case it can readily be shown that *necessary* criteria of solution stability for any element assembly and boundary conditions are that

$$n_\phi + n_w \geq n_s \quad \text{or} \quad \alpha_P \equiv \frac{n_\phi + n_w}{n_s} \geq 1 \qquad (12.15a)$$

and

$$n_s \geq n_w \quad \text{or} \quad \beta_P \equiv \frac{n_s}{n_w} \geq 1 \qquad (12.15b)$$

where n_ϕ, n_s and n_w are the number of $\tilde{\phi}$, $\tilde{S}$ and $\tilde{w}$ parameters in Eqs (12.11).

When the count condition is not satisfied then the *equation system either will be singular or will lock*. Equations (12.15a) and (12.15b) must be satisfied for the whole system but, in addition, they need to be satisfied for element patches if local instabilities and oscillations are to be avoided.[15–17] We remind the reader that Eqs (12.15a) are (12.15b) arc *necessary conditions*; however, they are not *sufficient conditions*. It is always necessary to conduct consistency and stability tests to ensure the proposed element passes the complete mixed patch test.[4]

The above criteria will, as we shall see later, help us to design suitable thick plate elements which show convergence to correct thin plate solutions.

12.3.2 Continuity requirements

The approximation of the form given in Eqs (12.12) and (12.13) implies certain continuities. It is immediately evident that C_0 continuity is needed for rotation shape functions $\mathbf{N}_\phi$ (as products of first derivatives are present in the approximation), but that either $\mathbf{N}_w$ or $\mathbf{N}_s$ can be discontinuous. In the form given in Eq. (12.13) a C_0 approximation for w is implied; however, after integration by parts a form for C_0 approximation of $\mathbf{S}$ results. Of course, physically only the component of $\mathbf{S}$ normal to boundaries should be continuous, as we noted also previously for moments in the mixed form discussed in Sec. 11.16.

In all the early approximations discussed in the previous section, C_0 continuity was assumed for both ϕ and w variables, this being very easy to impose. We note that such continuity cannot be described as *excessive* (as no physical conditions are violated), but we shall show later that successful elements also can be generated with discontinuous w interpolation (which is indeed not motivated by physical considerations).

For $\mathbf{S}$ it is obviously more convenient to use a completely discontinuous interpolation as then the shear can be eliminated at the element level and the final stiffness matrices written simply in standard $\tilde{\phi}$, $\tilde{w}$ terms for element boundary nodes. We shall show later that some formulations permit a limited case where α^{-1} is identically zero while others require it to be non-zero.

The continuous interpolation of the normal component of $\mathbf{S}$ is, as stated above, physically correct in the absence of line or point loads. However, with such interpolation, elimination of $\tilde{S}$ is not possible and the retention of such additional system variables is usually too costly to be used in practice and has so far not been adopted. However, we should note that an iterative solution process applicable to mixed forms can reduce substantially the cost of such additional variables.[18]

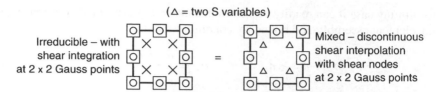

Fig. 12.6 Equivalence of mixed form and reduced shear integration in quadratic serendipity rectangle.

12.3.3 Equivalence of mixed forms with discontinuous S interpolation and reduced (selective) integration

An equivalence of penalized mixed forms with discontinuous interpolation of the constraint variable and of the corresponding irreducible forms with the same penalty variable may be demonstrated following work of Malkus and Hughes for incompressible problems.[19] Indeed, an exactly analogous proof can be used for the present case, and we leave the details of this to the reader; however, below we summarize some equivalencies that result.

Thus, for instance, we consider a serendipity quadrilateral, shown in Fig. 12.6 (a), in which integration of shear terms (involving α) is made at four Gauss points (i.e. 2×2 reduced quadrature) in an irreducible formulation [see Eqs (12.8a)–(12.8d)], we find that the answers are *identical* to a mixed form in which the **S** variables are given by a bilinear interpolation from nodes placed at the same Gauss points.

This result can also be argued from the limitation principle first given by Fraeijs de Veubeke.[20] This states that if the mixed form in which the stress is independently interpolated is precisely capable of reproducing the stress variation which is given in a corresponding irreducible form then the analysis results will be identical. It is clear that the four Gauss points at which the shear stress is sampled can only define a bilinear variation and thus the identity applies here.

The equivalence of reduced integration with the mixed discontinuous interpolation of **S** will be useful in our discussion to point out reasons why many elements mentioned in the previous section failed. However, in practice, it will be found equally convenient (and often more effective) to use the mixed interpolation explicitly and eliminate the **S** variables by element-level condensation rather than to use special integration rules. Moreover, in problems where the material properties lead to coupling between bending and shear response (e.g. elastic–plastic behaviour) use of selective reduced integration is not convenient. It must also be pointed out that the equivalence fails if α varies within an element or indeed if the isoparametric mapping implies different interpolations. In such cases the mixed procedures are generally more accurate.

12.4 The patch test for plate bending elements

12.4.1 Why elements fail

The nature and application of the patch test have changed considerably since its early introduction. As shown in references 15–17 and 21–25 this test can prove, in addition to *consistency* requirements (which were initially the only item tested), the *stability* of the approximation by requiring that for a patch consisting of an assembly of one or more elements the stiffness matrices are non-singular whatever the boundary conditions imposed.

To be absolutely sure of such non-singularity the test must, at the final stage, be performed numerically. However, we find that the 'count' conditions given in Eqs (12.15a) and (12.15b) are *necessary* for avoiding such non-singularity. Frequently, they also prove *sufficient* and make the numerical test only a final confirmation.[16,17] We shall demonstrate how the simple application of such counts immediately indicates *which elements fail* and which have a chance of success. Indeed, it is easy to show why the original quadratic serendipity element with reduced integration (QS-R) is not robust.

In Fig. 12.7 we consider this element in a single-element and four-element patch subjected to so-called *constrained* boundary conditions, in which all displacements on the external boundary of the patch are prescribed and a *relaxed* boundary condition in which only three displacements (conveniently two ϕs and one w) eliminate the rigid body modes. To ease the presentation of this figure, as well as in subsequent tests, we shall simply quote the values of α_P and β_P parameters as defined in Eqs (12.15a) and (12.15b) with subscript replaced by C or R to denote the constrained or relaxed tests, respectively. The symbol F will be given to any failure to satisfy the *necessary* condition. In the tests of Fig. 12.7 we note that both patch tests fail with the parameter α_C being less than 1, and hence the elements will lock under certain circumstances (or show a singularity in the evaluation of $\mathbf{S}$). A failure in the relaxed tests generally

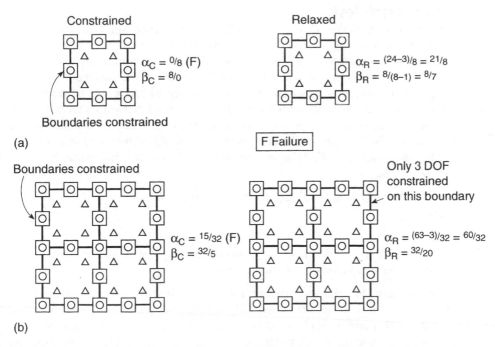

(a)

(b)

Fig. 12.7 'Constrained' and 'relaxed' patch test/count for serendipity (quadrilateral). (In the C test all boundary displacements are fixed. In the R test only three boundary displacements are fixed, eliminating rigid body modes.) (a) Single-element test; (b) four-element test.

predicts a singularity in the final stiffness matrix of the assembly, and this is also where frequently computational failures have been observed.

As the mixed and reduced integration elements are identical in this case we see immediately why the element fails in the problem of Fig. 12.3 (more severely under clamped conditions). Indeed, it is clear why in general the performance of Lagrangian-type elements is better as it adds further degrees of freedom to increase n_ϕ (and also n_w unless heterosis-type interpolation is used).[11]

In Table 12.1 we show a list of the α_P and β_P values for single- and four-element patches of various rectangles, and again we note that none of these satisfies completely the *necessary* requirements, and therefore none can be considered robust. However, it is interesting to note that the elements closest to satisfaction of the count perform best, and this explains why the heterosis elements[26] are quite successful and indeed why the Lagrangian cubic is nearly robust and often is used with success.[27]

Of course, similar approximation and counts can be made for various triangular elements. We list some typical and obvious ones, together with patch test counts, in the first part of Table 12.2. Again, none perform adequately and all will result in locking and spurious modes in finite element applications.

We should note again that the failure of the patch test (with regard to stability) means that under some circumstances the element will fail. However, in many problems a reasonable performance can still be obtained and non-singularity observed in its performance, providing consistency is, of course, also satisfied.

Numerical patch test

While the 'count' condition of Eqs (12.15a) and (12.15b) is a necessary one for stability of patches, on occasion singularity (and hence instability) can still arise even with its satisfaction. For this reason numerical tests should always be conducted ascertaining the rank sufficiency of the stiffness matrices and also testing the consistency condition.

In Chapter 9 of reference 4 we discussed in detail the consistency test for irreducible forms in which a single variable set **u** occurred. It was found that with a second-order operator the discrete equations should satisfy *at least* the solution corresponding to a linear field **u** exactly, thus giving constant strains (or first derivatives) and stresses. For the mixed equation set [Eq. (12.3)] again the lowest-order exact solution that has to be satisfied corresponds to:

1. constant values of moments or $\mathcal{L}\phi$ and hence a linear ϕ field;
2. linear w field;
3. constant **S** field.

The exact solutions for which plate elements commonly are tested and where full satisfaction of nodal equations is required consist of:

1. arbitrary constant **M** fields and arbitrary linear ϕ fields with zero shear forces (**S** = **0**); here a quadratic w form is assumed still yielding an exact finite element solution;
2. constant **S** and linear w fields yielding a constant ϕ field. The solution requires a distributed couple on the right-hand side of the first of Eq. (12.3) and this was not included in the original formulation. A simple procedure is to disregard the satisfaction of the moment equilibrium in this test. This may be done simply by inserting a very large value of the bending rigidity **D**.

Table 12.1 Quadrilateral mixed elements: patch count

Element	Reference	Single-element patch				Four-element patch			
		α_C	β_C	α_R	β_R	α_C	β_C	α_R	β_R
I									
Q4S1[5,8,10]		$\frac{0}{2}$ (F)	$\frac{2}{0}$	$\frac{9}{2}$	$\frac{2}{3}$ (F)	$\frac{3}{8}$ (F)	$\frac{8}{1}$	$\frac{24}{8}$	$\frac{8}{8}$
Q8S4[5]		$\frac{0}{8}$ (F)	$\frac{8}{0}$	$\frac{21}{8}$	$\frac{8}{7}$	$\frac{15}{32}$ (F)	$\frac{32}{5}$	$\frac{60}{32}$	$\frac{32}{20}$
Q9S4[7,8]		$\frac{3}{8}$ (F)	$\frac{8}{1}$	$\frac{16}{8}$	$\frac{8}{8}$	$\frac{27}{32}$ (F)	$\frac{32}{9}$	$\frac{72}{32}$	$\frac{32}{20}$
Q9HS4[11]		$\frac{2}{8}$ (F)	$\frac{8}{0}$	$\frac{15}{8}$	$\frac{8}{7}$	$\frac{23}{32}$ (F)	$\frac{32}{5}$	$\frac{68}{32}$	$\frac{32}{20}$
Q12S9[5]		$\frac{0}{18}$ (F)	$\frac{18}{0}$	$\frac{23}{18}$	$\frac{18}{11}$	$\frac{27}{72}$ (F)	$\frac{72}{9}$	$\frac{96}{72}$	$\frac{72}{32}$
Q16S9[27]		$\frac{12}{18}$ (F)	$\frac{18}{4}$	$\frac{45}{18}$	$\frac{18}{15}$	$\frac{75}{72}$	$\frac{72}{25}$	$\frac{150}{72}$	$\frac{72}{50}$
II									
Q4S1B1[28] Q4S1B1L[29,30]		$\frac{2}{2}$	$\frac{2}{0}$	$\frac{11}{2}$	$\frac{2}{3}$ (F)	$\frac{11}{8}$	$\frac{8}{1}$	$\frac{32}{8}$	$\frac{8}{8}$
Q4S2B2L[31]		$\frac{4}{4}$	$\frac{4}{0}$	$\frac{13}{4}$	$\frac{4}{4}$	$\frac{19}{16}$	$\frac{16}{1}$	$\frac{40}{16}$	$\frac{16}{8}$

[F] Failure to satisfy necessary conditions.

12.4.2 Design of some useful elements

The simple patch count test indicates how elements could be designed to pass it, and thus avoid the singularity (instability). Equation (12.15b) is always trivial to satisfy for elements in which **S** is interpolated independently in each element. In a single-element test it will be necessary to restrain at least one $\tilde{w}$ degree of freedom to prevent rigid body translations. Thus, the *minimum* number of terms which can be included in **S** for each element is always one less than the number of $\tilde{w}$ parameters in each element. As patches with more than one element are constructed the number of w parameters will increase proportionally with the number of nodes and the number of shear constraints increase by the number of elements. For both quadrilateral and triangular elements the requirement that $n_s \geq n_w - 1$ *for no boundary restraints* ensures that Eq. (12.15b) is satisfied on all patches for both constrained and relaxed boundary conditions. Failure to satisfy

Table 12.2 Triangular mixed elements: patch count

Element	Reference	Single-element patch				Six-element patch			
		α_C	β_C	α_R	β_R	α_C	β_C	α_R	β_R
I									
T3S1	T3S1	$\frac{0}{2}$ (F)	$\frac{2}{0}$	$\frac{6}{2}$	$\frac{2}{2}$	$\frac{3}{12}$ (F)	$\frac{12}{1}$	$\frac{18}{12}$	$\frac{12}{6}$
T6S3	T6S3	$\frac{0}{6}$ (F)	$\frac{6}{0}$	$\frac{15}{6}$	$\frac{6}{5}$	$\frac{21}{36}$ (F)	$\frac{36}{7}$	$\frac{54}{36}$	$\frac{36}{18}$
T10S3	T10S3	$\frac{3}{6}$ (F)	$\frac{6}{6}$	$\frac{27}{6}$	$\frac{6}{9}$	$\frac{57}{36}$ (F)	$\frac{36}{19}$	$\frac{108}{36}$	$\frac{36}{36}$
II									
T3S1B1L	T3S1B1L[32,33]	$\frac{2}{2}$	$\frac{2}{0}$	$\frac{8}{2}$	$\frac{2}{2}$	$\frac{20}{12}$	$\frac{12}{6}$	$\frac{35}{12}$	$\frac{12}{11}$
T3S1B1A	T3S1B1A[34]	$\frac{2}{2}$	$\frac{2}{0}$	$\frac{8}{2}$	$\frac{2}{2}$	$\frac{15}{12}$	$\frac{12}{1}$	$\frac{30}{12}$	$\frac{12}{6}$
T6S1B1	T6S1B1	$\frac{2}{2}$	$\frac{2}{0}$	$\frac{17}{2}$	$\frac{2}{5}$ (F)	$\frac{33}{12}$	$\frac{12}{7}$	$\frac{66}{12}$	$\frac{12}{18}$ (F)
T6S3B3	T6S3B3[17]	$\frac{6}{6}$	$\frac{6}{0}$	$\frac{21}{6}$	$\frac{6}{5}$	$\frac{75}{36}$	$\frac{36}{7}$	$\frac{108}{36}$	$\frac{36}{18}$

F Failure to satisfy necessary conditions.

this simple requirement explains clearly why certain of the elements in Tables 12.1 and 12.2 failed the single-element patch test for the relaxed boundary condition case.

Thus, a successful satisfaction of the count condition requires now only the consideration of Eq. (12.15a). In the remainder of this chapter we will discuss two approaches which can successfully satisfy Eq. (12.15a). The first is the use of discrete collocation constraints in which the third of Eq. (12.3) is enforced at preselected points on the boundary and occasionally in the interior of elements. Boundary constraints are often 'shared' between two elements and thus reduce the rate at which n_s increases. The other approach is to introduce bubble or enhanced modes for the rotation parameters in the interior of elements. Here, for convenience, we refer to both as a 'bubble mode' approach. The inclusion of at least as many bubble modes as shear modes will automatically satisfy Eq. (12.15a). This latter approach is similar to that used in Sec. 12.7 of Volume 1 to stabilize elements for solving the (nearly) incompressible problem and is a clear violation of 'intuition' since for the thin plate problem the rotations appear as derivatives of w. Its use in this case is justified by patch counts and performance.

12.5 Elements with discrete collocation constraints

12.5.1 General possibilities of discrete collocation constraints – quadrilaterals

The possibility of using conventional interpolation to achieve satisfactory performance of mixed-type elements is limited, as is apparent from the preceding discussion.

One feasible alternative is that of increasing the element order, and we have already observed that the cubic Lagrangian interpolation nearly satisfies the stability require-ment and often performs well.[3,8,27] However, the complexity of the formulation is formidable and this direction is not often used.

A different approach uses collocation constraints for the shear approximation on the element boundaries, thus limiting the number of **S** parameters and making the patch count more easily satisfied. This direction is indicated in the work of Hughes and Tezduyar,[35] Bathe and co-workers,[36,37] and Hinton and Huang,[38,39] as well as in generalizations by Zienkiewicz *et al.*,[40] and others.[41–48]

The procedure bears a close relationship to the so-called DKT (discrete Kirchhoff theory) developed in Chapter 11 (see Sec. 11.18) and indeed explains why these, essen-tially thin plate, approximations are successful.

The key to the discrete formulation is evident if we consider Fig. 12.8, where a simple bilinear element is illustrated. We observe that with a C_0 interpolation of ϕ and w, the shear strain

$$\gamma_x = \frac{\partial w}{\partial x} + \phi_x \tag{12.16}$$

is uniquely determined at any point of the side 1–2 (such as point I, for instance) and

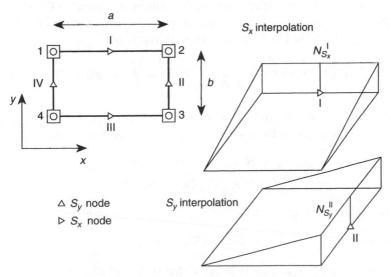

Fig. 12.8 Collocation constraints on a bilinear element: independent interpolation of S_x and S_y.

that hence [by Eq. (12.1b)]

$$S_x = \alpha \gamma_x \tag{12.17}$$

is also uniquely determined there.

Thus, if a node specifying the shear resultant distribution were placed at that point and if the constraints [or satisfaction of Eq. (12.1b)] were only imposed there, then

1. the nodal value of S_x would be shared by adjacent elements (assuming continuity of α);
2. the nodal values of S_x would be prescribed if the $\tilde{\phi}$ and $\tilde{w}$ values were constrained as they are in the constrained patch test.

Indeed if α, the shear rigidity, were to vary between adjacent elements the values of S_x would only differ by a multiplying constant and arguments remain essentially the same.

The prescription of the shear field in terms of such boundary values is simple. In the case illustrated in Fig. 12.8 we interpolate independently

$$S_x = \mathbf{N}_{sx}\tilde{\mathbf{S}}_x \quad \text{and} \quad S_y = \mathbf{N}_{sy}\tilde{\mathbf{S}}_y \tag{12.18}$$

using the shape functions

$$\begin{aligned}
\mathbf{N}_{sx} &= \frac{1}{y_I - y_{III}} \left[y - y_{III}, \quad y_I - y \right] \\
\mathbf{N}_{sy} &= \frac{1}{x_{II} - x_{IV}} \left[x - x_{IV}, \quad x_{II} - x \right]
\end{aligned} \tag{12.19}$$

as illustrated. Such an interpolation, of course, defines $\mathbf{N}_s$ of Eq. (12.11).

The introduction of the discrete constraint into the analysis is a little more involved. We can proceed by using different (Petrov–Galerkin) weighting functions, and in particular applying a Dirac delta weighting or point collocation to Eq. (12.1b) in the approximate form. However, it is advantageous here to return to the constrained variational principle [see Eq. (11.83)] and seek stationarity of

$$\Pi = \frac{1}{2} \int_{\Omega} (\mathcal{L}\phi)^{\mathrm{T}} \mathbf{D} \mathcal{L}\phi \, d\Omega + \frac{1}{2} \int_{\Omega} \mathbf{S}^{\mathrm{T}} \frac{1}{\alpha} \mathbf{S} \, d\Omega - \int_{\Omega} w\, q \, d\Omega + \Pi_{bt} = \text{stationary} \tag{12.20}$$

where the first term on the right-hand side denotes the *bending* and the second the *transverse shear* energy. In the above we again use the approximations

$$\begin{aligned}
\phi &= \mathbf{N}_{\phi}\tilde{\phi} \qquad & w &= \mathbf{N}_w \tilde{w} \\
\mathbf{S} &= \mathbf{N}_s \tilde{\mathbf{S}} \qquad & \mathbf{N}_s &= [\mathbf{N}_{sx}, \mathbf{N}_{sy}]
\end{aligned} \tag{12.21}$$

subject to the constraint Eq. (12.1b):

$$\mathbf{S} = \alpha \, (\nabla w + \phi) \tag{12.22}$$

being applied directly in a discrete manner, that is, by collocation at such points as I to IV in Fig. 12.8 and appropriate direction selection. We shall eliminate $\mathbf{S}$ from the computation but before proceeding with any details of the algebra it is interesting to

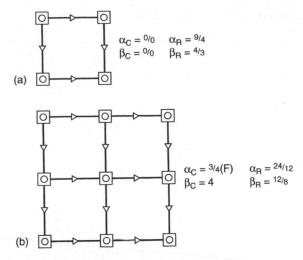

$$\alpha_C = 0/0 \quad \alpha_R = 9/4$$
$$\beta_C = 0/0 \quad \beta_R = 4/3$$

$$\alpha_C = 3/4(F) \quad \alpha_R = 24/12$$
$$\beta_C = 4 \quad \beta_R = 12/8$$

Fig. 12.9 Patch test on (a) one and (b) four elements of the type given in Fig. 12.8. (Observe that in a constrained test boundary values of **S** are prescribed.)

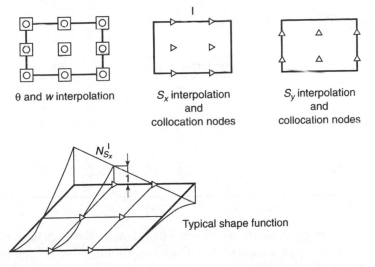

θ and w interpolation

S_x interpolation
and
collocation nodes

S_y interpolation
and
collocation nodes

N_{S_x}

Typical shape function

Fig. 12.10 The quadratic Lagrangian element with collocation constraints on boundaries and in the internal domain.[38,39]

observe the relation of the element of Fig. 12.8 to the patch test, noting that we still have a mixed problem requiring the count conditions to be satisfied. (This indeed is the element of references 36 and 37.) We show the counts on Fig. 12.9 and observe that although they fail in the four-element assembly the margin is small here (and for larger patches, counts are satisfactory).* The results given by this element are quite good, as will be shown in Sec. 12.9.

The discrete constraints and the boundary-type interpolation can of course be used in other forms. In Fig. 12.10 we illustrate the quadratic element of Huang and Hinton.[38,39] Here two points on each side of the quadrilateral define the shears S_x and S_y but in

---

* Reference 37 reports a mathematical study of stability for this element.

Table 12.3 Elements with collocation constraints: patch count. Degrees of freedom: $\square, w - 1; \bigcirc, \phi - 2; \triangle, S - 1; \cap, \phi_n - 1$

Element	Reference	Single-element patch				Four-element patch			
		α_C	β_C	α_R	β_R	α_C	β_C	α_R	β_R
	Q9D12[38,39]	$\frac{3}{4}$	$\frac{4}{1}$	$\frac{24}{12}$	$\frac{12}{8}$	$\frac{27}{24}$	$\frac{24}{9}$	$\frac{72}{40}$	$\frac{40}{24}$
	Q9D10	$\frac{3}{2}$ (F)	$\frac{2}{1}$	$\frac{24}{10}$	$\frac{10}{8}$	$\frac{27}{16}$	$\frac{16}{9}$	$\frac{72}{32}$	$\frac{32}{24}$
	Q8D8	$\frac{0}{0}$	$\frac{0}{0}$	$\frac{21}{8}$	$\frac{8}{7}$	$\frac{15}{8}$	$\frac{8}{5}$	$\frac{60}{24}$	$\frac{24}{21}$
	Q5D6	$\frac{3}{2}$	$\frac{2}{1}$	$\frac{12}{6}$	$\frac{6}{4}$	$\frac{15}{12}$	$\frac{12}{5}$	$\frac{36}{20}$	$\frac{20}{12}$
	Q4D4[36,37]	$\frac{0}{0}$	$\frac{0}{0}$	$\frac{8}{4}$	$\frac{4}{3}$	$\frac{3}{4}$ (F)	$\frac{4}{1}$	$\frac{24}{12}$	$\frac{12}{8}$

Element	Reference	Single-element patch				Six-element patch			
		α_C	β_C	α_R	β_R	α_C	β_C	α_R	β_R
	T6D6[40]	$\frac{0}{0}$	$\frac{0}{0}$	$\frac{15}{6}$	$\frac{6}{5}$	$\frac{21}{12}$	$\frac{12}{7}$	$\frac{43}{24}$	$\frac{24}{23}$
	T3D3[40,44]	$\frac{0}{0}$	$\frac{0}{0}$	$\frac{9}{3}$	$\frac{3}{2}$	$\frac{9}{6}$	$\frac{6}{1}$	$\frac{45}{12}$	$\frac{12}{6}$

[F] Failure to satisfy necessary conditions.

addition four internal parameters are introduced as shown. Now both the boundary and internal 'nodes' are again used as collocation points for imposing the constraints.

The count for single-element and four-element patches is given in Table 12.3. This element only fails in a single-element patch under constrained conditions, and again numerical verification shows generally good performance. Details of numerical examples will be given later.

It is clear that with discrete constraints many more alternatives for design of satisfactory elements that pass the patch test are present. In Table 12.3 several quadrilaterals and triangles that satisfy the count conditions are illustrated. In the first a modification of the Hinton–Huang element with reduced internal shear constraints is shown (second element). Here biquadratic 'bubble functions' are used in the interior shear component interpolation, as shown in Fig. 12.11. Similar improvements in the count can be achieved by using a serendipity-type interpolation, but now, of course, the distorted performance of the element may be impaired (for reasons we discussed in Volume 1, Sec. 9.7). Addition of bubble functions on all the w and ϕ parameters can, as shown, make the Bathe–Dvorkin fully satisfy the count condition. We shall pursue this further in Sec. 12.6.

All quadrilateral elements can, of course, be mapped isoparametrically, remembering of course that components of shear S_ξ and S_η parallel to the ξ, η coordinates have to be used to ensure the preservation of the desirable constrained properties previously discussed. Such 'directional' shear interpolation is also essential when considering triangular elements, to which the next section is devoted. Before doing this, however, we shall complete the algebraic derivation of element properties.

12.5.2 Element matrices for discrete collocation constraints

The starting point here will be to use the variational principle given by Eq. (12.20) with the shear variables eliminated directly.

The application of the discrete constraints of Eq. (12.22) allows the 'nodal' parameters $\tilde{\mathbf{S}}$ defining the shear force distribution to be determined explicitly in terms of the $\tilde{\mathbf{w}}$ and $\tilde{\boldsymbol{\phi}}$ parameters. This gives in general terms

$$\tilde{\mathbf{S}} = \alpha \left[\mathbf{Q}_w \tilde{\mathbf{w}} + \mathbf{Q}_\phi \tilde{\boldsymbol{\phi}} \right] \tag{12.23}$$

in each element. For instance, for the rectangular element of Fig. 12.8 we can write

$$S_x^I = \alpha \left[\frac{\tilde{w}_2 - \tilde{w}_1}{a} + \frac{\tilde{\phi}_{x1} + \tilde{\phi}_{x2}}{2} \right]$$

$$S_y^{II} = \alpha \left[\frac{\tilde{w}_2 - \tilde{w}_3}{b} + \frac{\tilde{\phi}_{y2} + \tilde{\phi}_{y3}}{2} \right]$$

$$S_x^{III} = \alpha \left[\frac{\tilde{w}_3 - \tilde{w}_4}{a} + \frac{\tilde{\phi}_{x3} + \tilde{\phi}_{x4}}{2} \right] \tag{12.24}$$

$$S_y^{IV} = \alpha \left[\frac{\tilde{w}_1 - \tilde{w}_4}{b} + \frac{\tilde{\phi}_{y1} + \tilde{\phi}_{y4}}{2} \right]$$

which can readily be rearranged into the form of Eq. (12.23) as

$$\mathbf{Q}_w = \frac{1}{ab} \begin{bmatrix} -b & b & 0 & 0 \\ 0 & a & -a & 0 \\ 0 & 0 & b & -b \\ a & 0 & 0 & -a \end{bmatrix} \quad \text{and} \quad \mathbf{Q}_\phi = \frac{1}{2} \begin{bmatrix} 1 & 0 & 1 & 0 & 0 & 0 & 0 & 0 \\ 0 & 0 & 0 & 1 & 0 & 1 & 0 & 0 \\ 0 & 0 & 0 & 0 & 1 & 0 & 1 & 0 \\ 0 & 1 & 0 & 0 & 0 & 0 & 0 & 1 \end{bmatrix}$$

where

$$\tilde{\mathbf{S}} = \begin{Bmatrix} S_x^I \\ S_x^{II} \\ S_x^{III} \\ S_y^{IV} \end{Bmatrix} ; \quad \tilde{\mathbf{w}} = \begin{Bmatrix} \tilde{w}_1 \\ \tilde{w}_2 \\ \tilde{w}_3 \\ \tilde{w}_4 \end{Bmatrix} ; \quad \tilde{\boldsymbol{\phi}} = \begin{Bmatrix} \tilde{\phi}_1 \\ \tilde{\phi}_2 \\ \tilde{\phi}_3 \\ \tilde{\phi}_4 \end{Bmatrix} ; \quad \tilde{\boldsymbol{\phi}}_a = \begin{Bmatrix} \tilde{\phi}_{xa} \\ \tilde{\phi}_{ya} \end{Bmatrix}$$

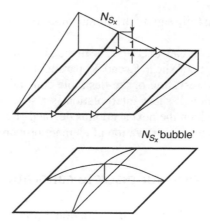

Fig. 12.11 A biquadratic hierarchical bubble for S_x.

Including the above discrete constraint conditions in the variational principle of Eq. (12.20) we obtain

$$
\Pi = \frac{1}{2} \int_\Omega (\mathcal{L} \mathbf{N}_\phi \tilde{\phi})^{\mathrm{T}} \mathbf{D} \mathcal{L} \mathbf{N}_\phi \tilde{\phi} \, d\Omega
$$
$$
+ \frac{1}{2} \int_\Omega [\mathbf{N}_s (\mathbf{Q}_\phi \tilde{\phi} + \mathbf{Q}_w \tilde{\mathbf{w}})]^{\mathrm{T}} \alpha [\mathbf{N}_s (\mathbf{Q}_\phi \tilde{\phi} + \mathbf{Q}_w \tilde{\mathbf{w}})] \, d\Omega \qquad (12.25)
$$
$$
- \int_\Omega wq \, d\Omega + \Pi_{bt} = \text{minimum}
$$

This is a constrained potential energy principle from which on minimization we obtain the system of equations

$$
\begin{bmatrix} \mathbf{K}_{ww} & \mathbf{K}_{w\phi} \\ \mathbf{K}_{\phi w} & \mathbf{K}_{\phi\phi} \end{bmatrix} \begin{Bmatrix} \tilde{\mathbf{w}} \\ \tilde{\phi} \end{Bmatrix} = \begin{Bmatrix} \mathbf{f}_w \\ \mathbf{f}_\phi \end{Bmatrix} \qquad (12.26)
$$

The element contributions are

$$
\mathbf{K}_{ww} = \mathbf{Q}_w^{\mathrm{T}} \mathbf{K}_{ss} \mathbf{Q}_w
$$
$$
\mathbf{K}_{w\phi} = \mathbf{Q}_w^{\mathrm{T}} \mathbf{K}_{ss} \mathbf{Q}_\phi = \mathbf{K}_{\phi w}^{\mathrm{T}}
$$
$$
\mathbf{K}_{\phi\phi} = \int_\Omega (\mathcal{L} \mathbf{N}_\phi)^{\mathrm{T}} \mathbf{D} (\mathcal{L} \mathbf{N}_\phi) \, d\Omega + \mathbf{Q}_\phi^{\mathrm{T}} \mathbf{K}_{ss} \mathbf{Q}_\phi \qquad (12.27)
$$
$$
\mathbf{K}_{ss} = \int_\Omega \mathbf{N}_s^{\mathrm{T}} \alpha \mathbf{N}_s \, d\Omega
$$

with the force terms identical to those defined in Eq. (12.8d).

These general expressions derived above can be used for any form of discrete constraint elements described and present no computational difficulties.

In the preceding we have imposed the constraints by point collocation of nodes placed on external boundaries or indeed the interior of the element. Other integrals could be used without introducing any difficulties in the final construction of the stiffness matrix. One could, for instance, require integrals such as

$$
\int_\Gamma W \left[S_s - \alpha \left(\frac{\partial w}{\partial s} + \phi \right) \right] \, d\Gamma = 0
$$

on segments of the boundary, or

$$\int_\Omega W \left[S_s - \alpha \left(\frac{\partial w}{\partial s} + \phi \right) \right] d\Omega = 0$$

in the interior of an element. All would achieve the same objective, providing elimination of the S_s parameters is still possible.

The use of discrete constraints can easily be shown to be equivalent to use of *substitute shear strain matrices* in the irreducible formulation of Eq. (12.8a). This makes introduction of such forms easy in standard computer programs. Details of such an approach are given by Oñate *et al.*[47,48]

12.5.3 Relation to the discrete Kirchhoff formulation

In Chapter 11, Sec. 11.18, we have discussed the so-called discrete Kirchhoff theory (DKT) formulation for beams in which the Kirchhoff constraints [i.e. Eq. (12.22) with $\alpha = \infty$] were applied in a discrete manner. The reason for the success of such discrete constraints was not obvious previously, but we believe that the formulation presented here in terms of the mixed form fully explains its basis. It is well known that the study of mixed forms frequently reveals the robustness or otherwise of irreducible approaches.

In Chapter 11 of reference 4 we explained why certain elements of irreducible form perform well as the limit of incompressibility is approached and why others fail. Here an analogous situation is illustrated.

It is clear that every one of the elements so far discussed has its analogue in the DKT form. Indeed, the thick plate approach we have adopted here with $\alpha \neq \infty$ is simply a penalty approach to the DKT constraints in which direct elimination of variables was used. Many opportunities for development of interesting and perhaps effective plate elements are thus available for both the thick and thin range.

We shall show in the next section some triangular elements and their DKT counterparts. Perhaps the whole range of the present elements should be termed 'QnDc' and 'TnDc' (discrete Reissner–Mindlin) elements in order to ease the classification. Here 'n' is the number of displacement nodes and 'c' the number of shear constraints.

12.5.4 Collocation constraints for triangular elements

Figure 12.12 illustrates a triangle in which a straightforward quadratic interpolation of ϕ and w is used. In this we shall take the shear forces to be given as a complete linear field defined by six shear force values on the element boundaries in directions parallel to these. The shear 'nodes' are located at Gauss points along each edge and the constraint collocation is made at the same position.

Writing the interpolation in standard area coordinates[4] we have

$$\mathbf{S} = \sum_{a=1}^{3} L_a \tilde{\mathbf{S}}_a \tag{12.28}$$

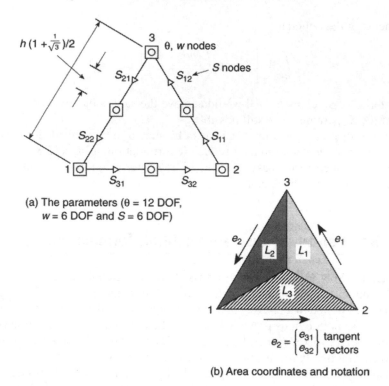

(a) The parameters (θ = 12 DOF, w = 6 DOF and S = 6 DOF)

(b) Area coordinates and notation

Fig. 12.12 The T6D6 triangular plate element.

where $\tilde{\mathbf{S}}_a$ are six parameters which can be determined by writing the tangential shear at the six constraint nodes. Introducing the tangent vector to each edge of the element as

$$\mathbf{e}_b = \left\{ \begin{array}{c} e_{bx} \\ e_{by} \end{array} \right\} \tag{12.29}$$

a tangential component of shear on the b-edge (for which $L_b = 0$) is obtained from

$$S_b = \mathbf{S} \cdot \mathbf{e}_b \tag{12.30}$$

Evaluation of Eq. (12.30) at the two Gauss points (defined on interval 0 to 1)

$$g_1 = \frac{1}{2\sqrt{3}}(\sqrt{3} - 1) \quad \text{and} \quad g_2 = \frac{1}{2\sqrt{3}}(\sqrt{3} + 1) \tag{12.31}$$

yields a set of six equations which can be used to determine the six parameters **a** in terms of the tangential edge shears $\tilde{S}_{b1}$ and $\tilde{S}_{b2}$. The final solution for the shear interpolation then becomes

$$\mathbf{S} = \sum_{a=1}^{3} \frac{L_a}{\Delta_a} \begin{bmatrix} e_{cy}, & -e_{by} \\ -e_{cx}, & e_{bx} \end{bmatrix} \left\{ \begin{array}{c} g_1\tilde{S}_{b1} + g_2\tilde{S}_{b2} \\ g_1\tilde{S}_{c1} + g_2\tilde{S}_{c2} \end{array} \right\} \tag{12.32}$$

in which a, b, c is a cyclic permutation of $1, 2, 3$. This defines uniquely the shape functions of Eq. (12.11) and, on application of constraints, expresses finally the shear

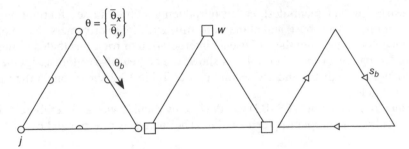

Fig. 12.13 The T3D3 (discrete Reissner-Mindlin) triangle of reference 44 with w linear, ϕ constrained quadratic, and S_b constant and parallel to sides.

field of nodal displacements $\tilde{w}$ and rotations $\tilde{\phi}$ in the manner of Eq. (12.23). The full derivation of the above expression is given in reference 40, and the final derivation of element matrices follows the procedures of Eqs (12.25)–(12.27).

The element derived satisfies fully the patch test count conditions as shown in Table 12.3 as the T6D6 element. This element yields results which are generally 'too flexible'. An alternative triangular element which shows considerable improvement in performance is indicated in Fig. 12.13. Here the w displacement is interpolated linearly and ϕ is initially a quadratic but is constrained to have linear behaviour along the tangent to each element edge.[44] Only a single shear constraint is introduced on each element side with the shear interpolation obtained from Eq. (12.32) by setting

$$\tilde{S}_{b1} = \tilde{S}_{b2} = \tilde{S}_b \tag{12.33}$$

The 'count' conditions are again fully satisfied for single and multiple element patches as shown in Table 12.3.

This element is of particular interest as it turns out to be the exact equivalent of the DKT triangle with 9 degrees of freedom which gave a satisfactory solution for thin plates.[49–51] Indeed, in the limit the two elements have an identical performance, though of course the T3D3 element is applicable also to plates with shear deformation. We note that the original DKT element can be modified in a different manner to achieve shear deformability[52] and obtain similar results. However, this element as introduced in reference 52 is not fully convergent.

12.6 Elements with rotational bubble or enhanced modes

As a starting point for this class of elements we may consider a standard functional of Reissner type given by

$$\Pi = \frac{1}{2} \int_\Omega (\mathcal{L}\phi)^{\mathrm{T}} \mathbf{D}\mathcal{L}\phi \, d\Omega - \frac{1}{2} \int_\Omega \mathbf{S}^{\mathrm{T}} \alpha^{-1} \mathbf{S} \, d\Omega + \int_\Omega \mathbf{S}^{\mathrm{T}} (\nabla w + \phi) \, d\Omega$$
$$- \int_\Omega w \, q \, d\Omega + \Pi_{bt} = \text{stationary} \tag{12.34}$$

in which approximations for w, ϕ and $\mathbf{S}$ are required.

Three triangular elements designed by introducing 'bubble modes' for rotation parameters are found to be robust, and at the same time excellent performers. None of these elements is 'obvious', and they all use an interpolation of rotations that is of higher or equal order than that of w. Figure 12.14 shows the degree-of-freedom assignments for these triangular elements and the second part of Table 12.2 shows again their performance in patches.

The quadratic element (T6S3B3) was devised by Zienkiewicz and Lefebvre[17] starting from a quadratic interpolation for w and ϕ. The shear **S** is interpolated by a complete

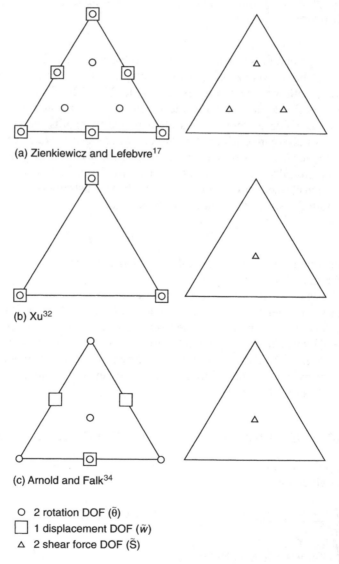

(a) Zienkiewicz and Lefebvre[17]

(b) Xu[32]

(c) Arnold and Falk[34]

○ 2 rotation DOF ($\bar{\theta}$)
□ 1 displacement DOF ($\bar{w}$)
△ 2 shear force DOF ($\tilde{S}$)

Fig. 12.14 Three robust triangular elements: (a) the T6S3B3 element of Zienkiewicz and Lefebvre;[17] (b) the T6S1B1 element of Xu;[32] (c) the T3S1B1A element of Arnold and Falk.[34]

linear polynomial in each element, giving here six parameters, $\tilde{\mathbf{S}}$. Three hierarchical quartic bubbles are added to the rotations giving the interpolation

$$\boldsymbol{\phi} = \sum_{a=1}^{6} N_a(L_k)\tilde{\boldsymbol{\phi}}_a + \sum_{b=1}^{3} \Delta N_b(L_k)\Delta\tilde{\boldsymbol{\phi}}_b \tag{12.35}$$

where $N_a(L_k)$ are conventional quadratic isoparametric interpolations on the six-node triangle[4] and shape functions for the quartic bubble modes are given as

$$\Delta N_b(L_k) = L_b(L_1 L_2 L_3)$$

Thus, we have introduced six additional rotation parameters but have left the number of w parameters unaltered from those given by a quadratic interpolation. This element has very desirable properties and good performance when the integral for $\mathbf{K}_b$ in Eq. (12.13) is computed by using seven point quadrature[4] and the other integrals are computed by using four points. Some improvement is gained using a mixed form in which the bending moments are approximated as discontinuous quadratic interpolations for each component and the first integral in Eq. (12.34) is replaced as

$$\frac{1}{2}\int_{\Omega} (\mathcal{L}\boldsymbol{\phi})^{\mathrm{T}}\mathbf{D}\mathcal{L}\boldsymbol{\phi}\,\mathrm{d}\Omega \rightarrow \int_{\Omega} (\mathcal{L}\boldsymbol{\phi})^{\mathrm{T}}\mathbf{M}\,\mathrm{d}\Omega - \frac{1}{2}\int_{\Omega}\mathbf{M}^{\mathrm{T}}\mathbf{D}^{-1}\mathbf{M}\,\mathrm{d}\Omega \tag{12.36}$$

All other terms in Eq. (12.34) remain the same. The mixed element computed in this way is denoted as T6S3B3M in subsequent results. We shall show later that optimal performance can be attained using 'linked' interpolation on this element.

Since the T6S3B3 type elements use a complete quadratic to describe the displacement and rotation field, an isoparametric mapping may be used to produce curved-sided elements and, indeed, curved-shell elements. Furthermore, by design this type passes the count test and by numerical testing is proved to be quite robust when used to analyse both thick and thin plate problems.[17,53] Since the w displacement interpolation is a standard quadratic interpolation the element may be joined compatibly to tetrahedral or prism solid elements which have six-node faces. We shall show later that optimal performance can be attained for this element by using 'linked' interpolation for w with additional enhanced strain modes – however, using this form the compatability between any attached tetrahedral elements is lost.

A linear triangular element [T3S1B1 – Fig. 12.14(b)] with a total of 9 nodal degrees of freedom adds a single cubic bubble to the linear rotational interpolation and uses linear interpolation for w with constant discontinuous shear. This element satisfies all count conditions for solution (see Table 12.2); however, without further enhancements it locks as the thin plate limit is approached.[32] As we have stated previously the count condition is necessary but not sufficient to define successful elements and numerical testing is always needed. In a later section we discuss a 'linked interpolation' modification which also makes this element robust.

A third element employing bubble modes [T3S1B1A – Fig. 12.14(c)] was devised by Arnold and Falk.[34] It is of interest to note that this element uses a discontinuous (non-conforming) w interpolation with parameters located at the mid-side of each triangle. The rotation interpolation is a standard linear interpolation with an added cubic bubble. The shear interpolation is constant on each element. This element is a direct opposite of the triangular element of Morley discussed in Chapter 11 in that location of the displacement and rotation parameters is reversed. The location of the displacement parameters, however, precludes its use in combination with standard solid elements. Thus, this element is of little general interest.

The introduction of successful bubble modes in quadrilaterals is more difficult. The first condition examined was the linear quadrilateral with a single bubble mode (Q4S1B1). For this element the patch count test fails when only a single element is considered but for assemblies above four elements it is passed and much hope was placed on this condition.[29,30] Unfortunately, one singular mode with a zero eigenvalue persists in all assemblies when the completely relaxed support conditions are considered. Despite this singularity the element does not lock and usually gives an excellent performance.[30]

To avoid, however, any singularity it is necessary to have at least three shear stress components and a similar number of rotation components of bubble form. No simple way of achieving a three-term interpolation exists but a successful four-component form was obtained by Auricchio and Taylor.[31] This four-term interpolation for shear is given by

$$
\mathbf{S} = \begin{bmatrix} J_{11}^0 & J_{21}^0 & J_{11}^0\eta & J_{21}^0\xi \\ J_{12}^0 & J_{22}^0 & J_{12}^0\eta & J_{22}^0\xi \end{bmatrix} \begin{Bmatrix} \tilde{S}_1 \\ \tilde{S}_2 \\ \tilde{S}_3 \\ \tilde{S}_4 \end{Bmatrix}
\tag{12.37}
$$

The Jacobian transformation J_{ij}^0 is identical to that introduced when describing the Pian–Sumihara element[4,54] and is computed as

$$
\begin{aligned}
J_{11}^0 &= x_{,\xi}\big|_{\xi=\eta=0}, & J_{12}^0 &= x_{,\eta}\big|_{\xi=\eta=0} \\
J_{21}^0 &= y_{,\xi}\big|_{\xi=\eta=0}, & J_{22}^0 &= y_{,\eta}\big|_{\xi=\eta=0}
\end{aligned}
\tag{12.38}
$$

To satisfy Eq. (12.15a) it is necessary to construct a set of four bubble modes. An appropriate form is found to be

$$
\Delta \mathbf{N}_{\text{bub}} = \frac{1}{j} N_b(\xi, \eta) \begin{bmatrix} J_{22}^0 & -J_{12}^0 & J_{22}^0\eta & -J_{12}^0\xi \\ -J_{21}^0 & J_{11}^0 & -J_{21}^0\eta & J_{11}^0\xi \end{bmatrix}
\tag{12.39}
$$

in which j is the determinant of the Jacobian transformation $\mathbf{J}$ (i.e. not the determinant of $\mathbf{J}^0$) and $N_{\text{bub}} = (1 - \xi^2)(1 - \eta^2)$ is a bubble mode. Thus, the rotation parameters are interpolated by using

$$
\phi = \sum_{a=1}^{4} N_a \tilde{\phi}_a + \Delta \mathbf{N}_{\text{bub}} \Delta \tilde{\phi}_{\text{bub}}
\tag{12.40}
$$

where N_a are the standard bilinear interpolations for the four-node quadrilateral. The element so achieved (Q4S2B2) is stable.

12.7 Linked interpolation – an improvement of accuracy

In the previous section we outlined various procedures which are effective in ensuring the necessary count conditions and which are, therefore, essential to make the elements 'work' without locking or singularity. In this section we shall try to improve the interpolation used to increase the accuracy without involving additional parameters.

The reader will here observe that in the primary interpolation we have used equal-order polynomials to interpolate both the displacement (w) and the rotations (ϕ). Clearly, if we consider the limit of thin plate theory

$$\phi = \left\{ \begin{matrix} \phi_x \\ \phi_y \end{matrix} \right\} = -\nabla w \tag{12.41}$$

and hence one order lower interpolation for ϕ appears necessary. To achieve this we introduce here the concept of *linked interpolation* in which the primary expression is written as

$$\phi = \mathbf{N}_\phi \tilde{\phi} \tag{12.42a}$$

and

$$w = \mathbf{N}_w \tilde{w} + \mathbf{N}_{w\phi} \tilde{\phi} \tag{12.42b}$$

This is precisely the form we obtained when seeking interpolations for the Timoshenko beam that gave exact interelement nodal results and forms the basis from which we now develop an interpolation for the Reissner–Mindlin plate theory. Such an expression ensures that a higher-order polynomial can be introduced for the representation of w without adding additional element parameters. This procedure can, of course, be applied to any of the elements we have listed in Tables 12.1 and 12.2 to improve the accuracy attainable. We shall here develop such linking for two types of elements in which the essential interpolations are linear on each edge and one element where they are quadratic.

We thus improve the triangular T3S1B1 and by linking L to its formulation we arrive at T3S1B1L. The same procedure can, of course, be applied to the quadrilaterals Q4S1B1 and Q4S2B2 of which only the second is unconditionally stable and add the letter L.

In the context of thick plates linked interpolation on a three-node triangle was intro-duced by Lynn and co-workers[55,56] and first extended to permit also thin plate analysis by Xu.[32,57] Similar interpolations have also been used by Tessler and Hughes[58,59] and termed 'anisoparametric'. Additional presentations dealing with the simple triangular element with 9 degrees of freedom in its reduced form [see Eq. (12.66)] have been given by Taylor and co-workers.[33,44,60]

Quadrilateral elements employing linked interpolation have been developed by Crisfield,[61] Zienkiewicz and co-workers,[28,30] and Auricchio and Taylor.[31]

12.7.1 Linking function for linear triangles and quadrilaterals

Based on the linked form for beams given in Sec. 10.5.1 the function $\mathbf{N}_{w\phi}$ for the linear triangle T3 and the linear quadrilaterals may be expressed as

$$w = N_a \tilde{w}_a - \frac{1}{8} \sum_{a=1}^{n_{el}} (N_{w\phi})_a l_{bc} (\phi_{sb} - \phi_{sc}) \tag{12.43}$$

where n_{el} is the number of vertex nodes on the element (i.e. 3 or 4), l_{bc} is the length of the b–c side, ϕ_{sb} is the rotation at node b in the tangent direction of the ath side, and $(N_{w\phi})_a$ are shape functions defining the quadratic w along the side but still maintaining zero w at corner nodes. For the triangular element these are the shape functions identical to those arising in the plane six-node element at mid-side nodes and are given by[4]

$$\mathbf{N}_{w\phi} = 4 \begin{bmatrix} L_1L_2, & L_2L_3, & L_3L_1 \end{bmatrix} \tag{12.44}$$

and for the quadrilateral element these are the shape functions for the eight-node serendipity functions given by

$$\mathbf{N}_{w\phi} = \tfrac{1}{2} \begin{bmatrix} (1-\xi^2)(1-\eta), & (1+\xi)(1-\eta^2), & (1-\xi^2)(1+\eta), & (1-\xi)(1-\eta^2) \end{bmatrix}$$

$$\tag{12.45}$$

Example 12.1 Derivation of linked function for three-node triangle

The development of a shape function for the three-node triangular element with a total of 9 degrees of freedom is developed using the fact that both shear and moment in the equivalent Timoshenko beam form were constant. A process to develop a linked interpolation for the plate transverse displacement, w, can start from a full quadratic expansion written in hierarchical form. Thus, for a triangle we have the interpolation in area coordinates

$$w = L_1\tilde{w}_1 + L_2\tilde{w}_2 + L_3\tilde{w}_3 + 4L_1L_2\tilde{\alpha}_1 + 4L_2L_3\tilde{\alpha}_2 + 4L_3L_1\tilde{\alpha}_3 \tag{12.46}$$

where $\tilde{w}_a$ are the nodal displacements and $\tilde{\alpha}_a$ are hierarchical parameters. The hierarchical parameters are then expressed in terms of rotational parameters. Along any edge, say the 1–2 edge (where $L_3 = 0$), the displacement is given by

$$w = L_1\tilde{w}_1 + L_2\tilde{w}_2 + 4L_1L_2\alpha_1 \tag{12.47}$$

The expression used to eliminate α_1 is deduced by constraining the transverse edge shear to be a constant. Along the edge the transverse shear is given by

$$\gamma_{12} = \frac{\partial w}{\partial s} + \phi_s \tag{12.48}$$

where s is the coordinate tangential to the edge and ϕ_s is the component of the rotation along the edge. The derivative of Eq. (12.47) along the edge is given by

$$\frac{\partial w}{\partial s} = \frac{\tilde{w}_2 - \tilde{w}_1}{l_{12}} + 4\frac{L_2 - L_1}{l_{12}}\tilde{\alpha}_1 \tag{12.49}$$

where l_{12} is the length of the 1–2 side. Assuming a linear interpolation for ϕ_s along the edge we have

$$\phi_s = L_1\tilde{\phi}_{s1} + L_2\tilde{\phi}_{s2} \tag{12.50}$$

which, after noting that $L_1 + L_2 = 1$ along the edge, may also be expressed as

$$\phi_s = \tfrac{1}{2}(\tilde{\phi}_{s1} + \tilde{\phi}_{s2}) + \tfrac{1}{2}(\tilde{\phi}_{s1} - \tilde{\phi}_{s2})(L_2 - L_1) \tag{12.51}$$

The transverse shear may now be given as

$$\gamma_{12} = \frac{\tilde{w}_2 - \tilde{w}_1}{l_{12}} + \frac{1}{2}(\tilde{\phi}_{s1} + \tilde{\phi}_{s2}) + \left[\frac{4}{l_{12}}\tilde{\alpha}_1 + \frac{1}{2}(\tilde{\phi}_{s1} - \tilde{\phi}_{s2})\right](L_2 - L_1) \qquad (12.52)$$

Constraining the strain to be constant gives

$$\tilde{\alpha}_1 = -\frac{l_{12}}{8}(\tilde{\phi}_{s1} - \tilde{\phi}_{s2}) \qquad (12.53)$$

yielding the 'linked' edge interpolation

$$w = L_1\tilde{w}_1 + L_2\tilde{w}_2 + \tfrac{1}{2}l_{12}L_1L_2(\tilde{\phi}_{s1} - \tilde{\phi}_{s2}) \qquad (12.54)$$

The normal rotations may now be expressed in terms of the nodal Cartesian components by using

$$\phi_s = \cos\psi_{12}\phi_x + \sin\psi_{12}\phi_y \qquad (12.55)$$

where ψ_{12} is the angle that the normal to the edge makes with the x axis. Repeating this process for the other two edges gives the final interpolation for the transverse displacement.

A similar process can be followed to develop the linked interpolations for the quadrilateral element. The reader can verify that the use of the constant $1/8$ ensures that constant shear strain on the element side occurs. Further, a rigid body rotation with $\tilde{\phi}_a = \tilde{\phi}_b$ in the element causes no straining. Finally, with rotation $\tilde{\phi}_a$ being the same for adjacent elements C_0 continuity is ensured.

12.7.2 Linked interpolation for quadratic elements

The interpolation for the quadratic element shown in Fig. 12.15 proceeds in a similar way to the linear form just presented. The development of the quadratic element, which

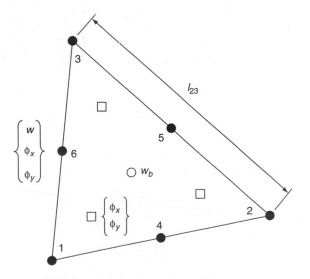

Fig. 12.15 Quadratic linked element.

is similar to the Zienkiewicz–Lefebvre element described above in Sec. 12.6, was presented by Auricchio and Lovadina.[62] The element was extended to include anisotropic materials and coupling with in-plane displacements by Taylor and Govindjee.[63] The development starts with quadratic interpolation for the displacement w and rotation ϕ. The transverse displacement is then increased to a complete cubic using the linked concept in which w is written as

$$
w = \sum_{a=1}^{6} N_a(L_k)\tilde{w}_a + N_{\text{bub}}\tilde{w}_{\text{bub}}
$$

$$
+ \frac{1}{3}\sum_{a=1}^{3} l_{ab}L_aL_b(L_b - L_a)\mathbf{t}_{ab}^T(\tilde{\phi}_a + \tilde{\phi}_b - 2\,\tilde{\phi}_{a+3}); \quad a = 1, 2, 3; \ b = 2, 3, 1 \tag{12.56}
$$

with

$$
N_{\text{bub}} = L_1 L_2 L_3 \tag{12.57}
$$

Here w_{bub} is an hierarchical parameter. We note that the bubble mode can be retained as a displacement degree of freedom or its gradient added as an enhanced mode to shearing strains. As noted above the rotation interpolation is given by the quadratic form

$$
\phi = \sum_{a=1}^{6} N_a(L_k)\tilde{\phi}_a \tag{12.58}
$$

To satisfy the mixed patch test the strains for the linked element need additional enhanced terms which are computed from the derivatives and values of additional displacement functions. Accordingly, we consider the enhanced curvature given by

$$
\chi = \left\{ \begin{array}{c} \chi_x \\ \chi_y \\ \chi_{xy} \end{array} \right\} = \sum_{a=1}^{6} \left[\begin{array}{cc} \dfrac{\partial N_a}{\partial x} & 0 \\[2ex] 0 & \dfrac{\partial N_a}{\partial y} \\[2ex] \dfrac{\partial N_a}{\partial y} & \dfrac{\partial N_a}{\partial x} \end{array} \right] \left\{ \begin{array}{c} \tilde{\phi}_x \\ \tilde{\phi}_y \end{array} \right\} + \left\{ \begin{array}{c} \dfrac{\partial \phi_x^{en}}{\partial x} \\[2ex] \dfrac{\partial \phi_y^{en}}{\partial x} \\[2ex] \dfrac{\partial \phi_x^{en}}{\partial y} + \dfrac{\partial \phi_y^{en}}{\partial x} \end{array} \right\} \tag{12.59}
$$

where

$$
\phi^{en} = \sum_{b=1}^{3}(L_b - \tfrac{1}{3})N_{\text{bub}}\,\tilde{\beta} + (\nabla N_{\text{bub}})N_{\text{bub}}\,\tilde{\phi}_{\text{bub}} \tag{12.60}
$$

in which $\tilde{\beta}$ and $\tilde{\phi}_{\text{bub}}$ add seven parameters.

The enhanced transverse shear strain is given by

$$
\gamma = \left\{ \begin{array}{c} \gamma_x \\ \gamma_y \end{array} \right\} = \left\{ \begin{array}{c} \dfrac{\partial w}{\partial x} + \phi_x \\[2ex] \dfrac{\partial w}{\partial y} + \phi_y \end{array} \right\} + \phi^{en} \tag{12.61}
$$

where we assume that w_{bub} is treated as a displacement parameter.

Finally, the shear interpolation is given by

$$S + \sum_{a=1} 3L_a\tilde{S}_a + \nabla N_{bub}\tilde{S}_{bub} \tag{12.62}$$

For the form given above each node has 3 degrees of freedom ($\tilde{w}_a$, ϕ_{xa} and ϕ_{ya}), one internal n_w, seven internal n_ϕ and seven internal n_x. The element satisfies the count condition of the mixed patch test for all element configurations including a single element with all external degrees of freedom restrained.

Example 12.2 Clamped plate by linked elements

For a clamped plate a solution based on the Reissner–Mindlin plate theory has been given by Chinosi and Lovadina.[64]   Using an inverse method we may write the solution as

$$w = w_b + w_s \quad \text{with} \quad \phi = -\nabla w_b$$

$$w_b = \frac{1}{3}x^3(x-1)^3y^3(y-1)^3$$

$$w_s = -\frac{t^2}{6\kappa(1-\nu)}\nabla^2 w_b$$

which gives the load

$$q = D\left[12y(y-1)(5x(x-1)+1)(2y^2(y-1)^2 + x(x-1)(5y(y-1)+1) \right. $$
$$\left. + 12x(x-1)(5y(y-1)+1)(2x^2(x-1)^2 + y(y-1)(5x(x-1)+1)\right]$$

This solution is used to illustrate the convergence of the energy error for a clamped plate as shown in Fig. 12.16.

12.8 Discrete 'exact' thin plate limit

Discretization of Eq. (12.34) using interpolations of the form[*]

$$w = N_w\tilde{w} + N_{w\phi}\tilde{\phi}, \quad \phi = N_\phi\tilde{\phi} + N_b\Delta\tilde{\phi}_b, \quad S = N_s\tilde{S} \tag{12.63}$$

where here we use the abbreviated notation b (instead of bub) for bubble modes, leads to the algebraic system of equations

$$\begin{bmatrix} 0 & 0 & 0 & K_{sw}^T \\ 0 & K_{\phi\phi} & K_{b\phi}^T & K_{s\phi}^T \\ 0 & K_{b\phi} & K_{bb} & K_{sb}^T \\ K_{sw} & K_{s\phi} & K_{sb} & K_{ss} \end{bmatrix} \begin{Bmatrix} \tilde{w} \\ \tilde{\phi} \\ \Delta\tilde{\phi}_b \\ \tilde{S} \end{Bmatrix} = \begin{Bmatrix} f_w \\ f_\phi \\ 0 \\ 0 \end{Bmatrix} \tag{12.64}$$

where, for simplicity, only the forces f_w and f_ϕ due to transverse load q and boundary conditions are included [see Eq. (12.8d)]. The arrays appearing in Eq. (12.64) are given by

[*] The term $N_{w\phi}$ will be exploited in the next part of this section and thus is included for completeness.

$$\mathbf{K}_{\phi\phi} = \int_{\Omega} (\mathcal{L}\mathbf{N}_{\phi})^{\mathrm{T}}\mathbf{D}(\mathcal{L}\mathbf{N}_{\phi})\,\mathrm{d}\Omega, \qquad \mathbf{K}_{ss} = -\int_{\Omega} \mathbf{N}_s \alpha^{-1}\mathbf{N}_s\,\mathrm{d}\Omega,$$

$$\mathbf{K}_{b\phi} = \int_{\Omega} (\mathcal{L}\mathbf{N}_b)^{\mathrm{T}}\mathbf{D}(\mathcal{L}\mathbf{N}_{\phi})\,\mathrm{d}\Omega, \qquad \mathbf{K}_{s\phi} = \int_{\Omega} \mathbf{N}_s^{\mathrm{T}}\left[\nabla\mathbf{N}_{w\phi} - \mathbf{N}_{\phi}\right]\mathrm{d}\Omega,$$

$$\mathbf{K}_{bb} = \int_{\Omega} (\mathcal{L}\mathbf{N}_b)^{\mathrm{T}}\mathbf{D}(\mathcal{L}\mathbf{N}_b)\,\mathrm{d}\Omega, \qquad \mathbf{K}_{sb} = -\int_{\Omega} \mathbf{N}_s^{\mathrm{T}}\mathbf{N}_b\,\mathrm{d}\Omega, \tag{12.65}$$

$$\mathbf{K}_{sw} = \int_{\Omega} \mathbf{N}_s^{\mathrm{T}}\nabla\mathbf{N}_w\,\mathrm{d}\Omega$$

Adopting a static condensation process at the element level[65] in which the internal rotational parameters are eliminated first, followed by the shear parameters, yields the element stiffness matrix in terms of the element $\tilde{\mathbf{w}}$ and $\tilde{\phi}$ parameters given by

$$\begin{bmatrix} \mathbf{K}_{sw}^{\mathrm{T}}\mathbf{A}_{ss}^{-1}\mathbf{K}_{sw} & -\mathbf{K}_{sw}^{\mathrm{T}}\mathbf{A}_{ss}^{-1}\mathbf{A}_{s\phi} \\ -\mathbf{A}_{s\phi}^{\mathrm{T}}\mathbf{A}_{ss}^{-1}\mathbf{K}_{sw} & -\mathbf{A}_{\phi\phi} + \mathbf{A}_{s\phi}^{\mathrm{T}}\mathbf{A}_{ss}^{-1}\mathbf{A}_{s\phi} \end{bmatrix} \begin{Bmatrix} \tilde{\mathbf{w}} \\ \tilde{\phi} \end{Bmatrix} = \begin{Bmatrix} \mathbf{f}_w \\ \mathbf{f}_\phi \end{Bmatrix} \tag{12.66}$$

in which

$$\mathbf{A}_{ss} = \mathbf{K}_{sb}\mathbf{K}_{bb}^{-1}\mathbf{K}_{sb}^{\mathrm{T}} - \mathbf{K}_{ss}, \qquad \mathbf{A}_{s\phi} = \mathbf{K}_{sb}\mathbf{K}_{bb}^{-1}\mathbf{K}_{b\phi} - \mathbf{K}_{s\phi}$$

$$\mathbf{A}_{\phi\phi} = \mathbf{K}_{b\phi}^{\mathrm{T}}\mathbf{K}_{bb}^{-1}\mathbf{K}_{b\phi} - \mathbf{K}_{\phi\phi} \tag{12.67}$$

This solution strategy requires the inverse of $\mathbf{K}_{bb}$ and $\mathbf{A}_{ss}$ only. In particular, we note that the inverse of $\mathbf{A}_{ss}$ can be performed even if $\mathbf{K}_{ss}$ is zero (provided the other term is non-singular). The vanishing of $\mathbf{K}_{ss}$ defines the *thin plate limit*. Thus, the above

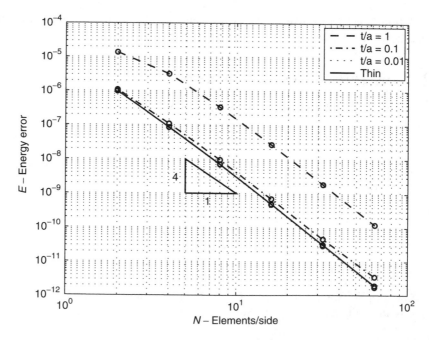

Fig. 12.16 Energy error for quadratic linked element. Clamped plate.

strategy leads to a solution process in which the thin plate limit is defined without recourse to a penalty method. Indeed, all terms in the process generally are not subject to ill-conditioning due to differences in large and small numbers. In the context of thick and thin plate analysis this solution strategy has been exploited with success in references 31 and 33.

12.9 Performance of various 'thick' plate elements – limitations of thin plate theory

The performance of both 'thick' and 'thin' elements is frequently compared for the examples of clamped and simply supported square plates, though, of course, more stringent tests can and should be devised. Figure 12.17(a)–12.17(d) illustrates the behaviour of various elements we have discussed in the case of a span-to-thickness ratio (L/t) of 100, which generally is considered within the scope of thin plate theory. The results are indeed directly comparable to those of Fig. 11.16 of Chapter 11, and it is evident that here the thick plate elements perform as well as the best of the thin plate forms.

It is of interest to note that in Fig. 12.17 we have included some elements that do not fully pass the patch test and hence are not robust. Many such elements are still used as their failure occurs only occasionally – although new developments should always strive to use an element which is robust.

All 'robust' elements of the thick plate kind can be easily mapped isoparametrically and their performance remains excellent and convergent. Figure 12.18 shows isoparametric mapping used on a curved-sided mesh in the solution of a circular plate for two elements previously discussed. Obviously, such a lack of sensitivity to distortion will be of considerable advantage when shells are considered, as we shall show in Chapter 15.

Of course, when the span-to-thickness ratio decreases and thus shear deformation importance increases, the thick plate elements are capable of yielding results not obtainable with thin plate theory. In Table 12.4 we show some results for a simply supported, uniformly loaded plate for two L/t ratios and in Table 12.5 we show results for the clamped uniformly loaded plate for the same ratios. In this example we show also the effect of the *hard* and *soft* simple support conditions. In the hard support we assume just as in thin plates that the rotation along the support (ϕ_s) is zero. In the soft support case we take, more rationally, a zero twisting moment along the support (see Chapter 11, Sec. 11.2.2).

It is immediately evident that:

1. the thick plate $(L/t = 10)$ shows deflections converging to very different values depending on the support conditions, both being considerably larger than those given by thin plate theory;
2. for the thin plate $(L/t = 1000)$ the deflections converge uniformly to the thin (Kirchhoff) plate results for *hard* support conditions, but for *soft* support conditions give answers about 0.2% higher in centre deflection.

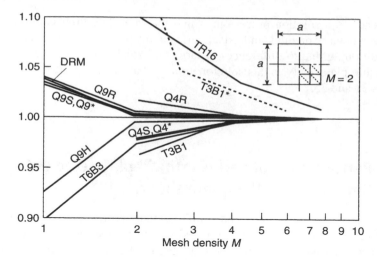

(a) Centre displacement normalized with respect to thin plate theory
for simply supported, uniformly loaded square plate

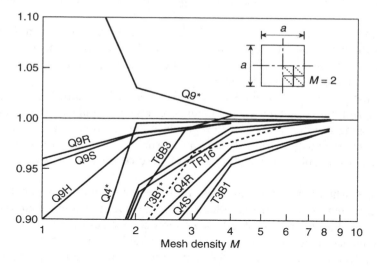

(b) Moment at Gauss point nearest centre (or central point)
normalized by centre moment of thin plate theory for
simply supported, uniformly loaded square plate

Fig. 12.17   Convergence study for relatively thin plate ($L/t = 100$): (a) centre displacement (simply supported, uniform load, square plate); (b) moment at Gauss point nearest centre (simply supported, uniform load, square plate). Tables 12.1 and 12.2 give keys to elements used.

It is perhaps an insignificant difference that occurs in this example between the support conditions but this can be more pronounced in different plate configurations.

In Fig. 12.19 we show the results of a study of a simply supported rhombic plate with $L/t = 100$ and 1000. For this problem an accurate Kirchhoff plate theory solution

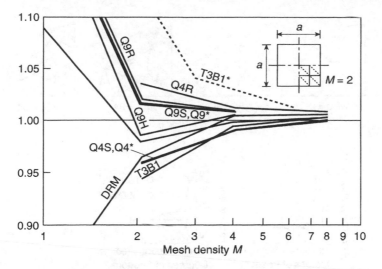

(c) Centre displacement normalized with respect to thin plate theory
   for clamped, uniformly loaded square plate

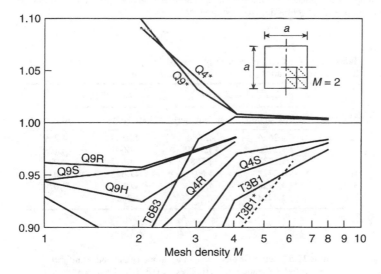

(d) Moment at Gauss point nearest centre (or central point)
   normalized by centre moments of thin plate theory for
   clamped, uniformly loaded square plate

Fig. 12.17 *Cont.*   (c) Centre displacement (clamped, uniform load, square plate); (d) moment at Gauss point nearest centre (clamped, uniform load, square plate).

is available,[66] but as will be noticed the thick plate results converge uniformly to a displacement nearly 4% in excess of the thin plate solution for all the $L/t = 100$ cases.

This problem is illustrative of the substantial difference that can on occasion arise in situations that fall well within the limits assumed by conventional thin plate theory

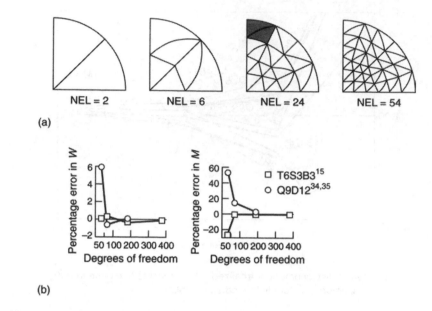

(a)

(b)

Fig. 12.18   Mapped curvilinear elements in solution of a circular clamped plate under uniform load: (a) meshes used; (b) percentage error in centre displacement and moment.

Table 12.4   Centre displacement of a simply supported plate under uniform load for two L/t ratios; $E = 10.92$, $\nu = 0.3$, $L = 10$, $q = 1$

Mesh, M	$L/t = 10$; $w \times 10^{-1}$		$L/t = 1000$; $w \times 10^{-7}$	
	hard support	soft support	hard support	soft support
2	4.2626	4.6085	4.0389	4.2397
4	4.2720	4.5629	4.0607	4.1297
8	4.2727	4.5883	4.0637	4.0928
16	4.2728	4.6077	4.0643	4.0773
32	4.2728	4.6144	4.0644	4.0700
Series	4.2728		4.0624	

Table 12.5   Centre displacement of a clamped plate under uniform load for two L/t ratios; $E = 10.92$, $\nu = 0.3$, $L = 10$, $q = 1$

Mesh, M	$L/t = 10$; $w \times 10^{-1}$	$L/t = 1000$; $w \times 10^{-7}$
2	1.4211	1.1469
4	1.4858	1.2362
8	1.4997	1.2583
16	1.5034	1.2637
32	1.5043	1.2646
Series	1.499[29]	1.2653

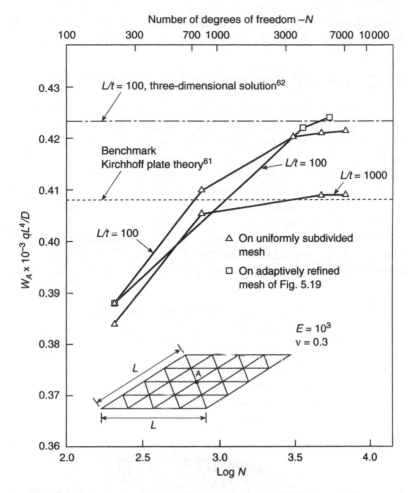

Fig. 12.19 Skew 30° simply supported plate (soft support); maximum deflection at A, the centre for various degrees of freedom N. The triangular element of reference 17 is used.

($L/t = 100$), and for this reason the problem has been thoroughly investigated by Babuška and Scapolla,[67] who solve it as a fully three-dimensional elasticity problem using support conditions of the 'soft' type which appear to be the closest to physical reality. Their three-dimensional result is very close to the thick plate solution, and confirms its validity and, indeed, superiority over the thin plate forms. However, we note that for very thin plates, even with soft support, convergence to the thin plate results occurs.

12.10 Inelastic material behaviour

We have discussed in some detail the problem of inelastic behaviour in Sec. 11.19 of Chapter 11. The procedures of dealing with the same situation when using the

Reissner–Mindlin theory are nearly identical and here we will simply refer the reader to the literature on the subject[68,69] and to the previous chapter.

12.11 Concluding remarks – adaptive refinement

The simplicity of deriving and using elements in which independent interpolation of rotations and displacements is postulated and shear deformations are included assures popularity of the approach. The final degrees of freedom used are exactly the same type as those used in the direct approach to thin plate forms in Chapter 14, and at no additional complexity shear deformability is included in the analysis.

If care is used to ensure robustness, elements of the type discussed in this chapter are generally applicable and indeed could be used with similar restrictions to other finite element approximations requiring C_1 continuity in the limit.

The ease of element distortion will make elements of the type discussed here the first choice for curved element solutions and they can easily be adapted to non-linear material behaviour. Extension to geometric non-linearity is also possible; however, in

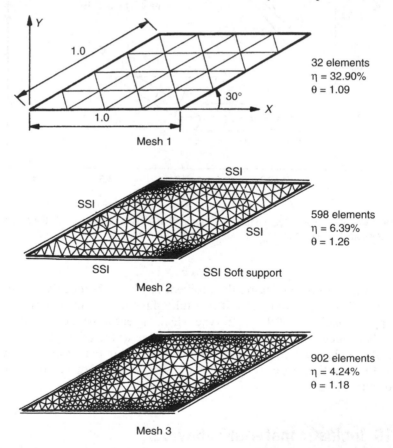

Fig. 12.20 Simply supported 30° skew plate with uniform load (problem of Fig. 12.19); adaptive analysis to achieve 5% accuracy; $L/t = 100$, $\nu = 0.3$, six-node element T6S3B3;[17] ϕ = effectivity index, η = percentage error in energy norm of estimator.

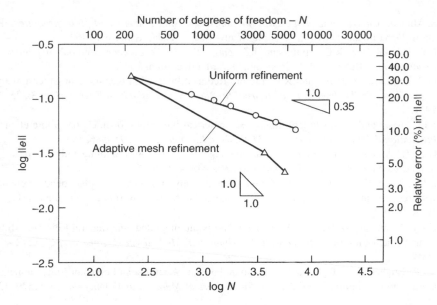

Fig. 12.21 Energy norm rate of convergence for the 30° skew plate of Fig. 12.19 for uniform and adaptive refinement; adaptive analysis to achieve 5% accuracy.

this case the effects of in-plane forces must be included and this renders the problem identical to shell theory. We shall discuss this more in Chapter 11.

In Chapters 14 and 15 of Volume 1 we discussed the need for an adaptive approach in which error estimation is used in conjunction with mesh generation to obtain answers of specified accuracy. Such adaptive procedures are easily used in plate bending problems with an almost identical form of error estimation.[70]

In Figs 12.20 and 12.21 we show a sequence of automatically generated meshes for the problem of the skew plate. It is of particular interest to note:

1. the initial refinement in the vicinity of corner singularity;
2. the final refinement near the simply support boundary conditions where the effects of transverse shear will lead to a 'boundary layer'.

Indeed, such boundary layers can occur near all boundaries of shear deformable plates and it is usually found that the shear error represents a very large fraction of the total error when approximations are made.

References

1. S. Ahmad, B.M. Irons and O.C. Zienkiewicz. Curved thick shell and membrane elements with particular reference to axi-symmetric problems. In *Proc. 2nd Conf. Matrix Methods in Structural Mechanics*, volume AFFDL-TR-68-150, pages 539–572, Wright Patterson Air Force Base, Ohio, October 1968.
2. S. Ahmad, B.M. Irons and O.C. Zienkiewicz. Analysis of thick and thin shell structures by curved finite elements. *International Journal for Numerical Methods in Engineering*, 2:419–451, 1970.

3. S. Ahmad. *Curved finite elements in the analysis of solids, shells and plate structures.* PhD thesis, Department of Civil Engineering, University of Wales, Swansea, 1969.

4. O.C. Zienkiewicz, R.L. Taylor and J.Z. Zhu. *The Finite Element Method: Its Basis and Fundamentals.* Butterworth-Heinemann, Oxford, 6th edition, 2005.

5. O.C. Zienkiewicz, J. Too and R.L. Taylor. Reduced integration technique in general analysis of plates and shells. *International Journal for Numerical Methods in Engineering*, 3:275–290, 1971.

6. S.F. Pawsey and R.W. Clough. Improved numerical integration of thick slab finite elements. *International Journal for Numerical Methods in Engineering*, 3:575–586, 1971.

7. O.C. Zienkiewicz and E. Hinton. Reduced integration, function smoothing and non-conformity in finite element analysis. *J. Franklin Inst.*, 302:443–461, 1976.

8. E.D.L. Pugh, E. Hinton and O.C. Zienkiewicz. A study of quadrilateral plate bending elements with reduced integration. *International Journal for Numerical Methods in Engineering*, 12: 1059–1079, 1978.

9. F. Gruttmann and W. Wagner. A stabilized one-point integrated quadrilateral Reissner–Mindlin plate element. *International Journal for Numerical Methods in Engineering*, 61:2273–2295, 2004.

10. T.J.R. Hughes, R.L. Taylor and W. Kanoknukulchai. A simple and efficient finite element for plate bending. *International Journal for Numerical Methods in Engineering*, 11:1529–1543, 1977.

11. T.J.R. Hughes and M. Cohen. The 'heterosis' finite element for plate bending. *Computers and Structures*, 9:445–450, 1978.

12. T.J.R. Hughes, M. Cohen and M. Harou. Reduced and selective integration techniques in the finite element analysis of plates. *Nuclear Engineering and Design*, 46:203–222, 1978.

13. E. Hinton and N. Bićanić. A comparison of Lagrangian and serendipity Mindlin plate elements for free vibration analysis. *Computers and Structures*, 10:483–493, 1979.

14. R.D. Cook. *Concepts and Applications of Finite Element Analysis.* John Wiley & Sons, Chichester, 1982.

15. R.L. Taylor, O.C. Zienkiewicz, J.C. Simo and A.H.C. Chan. The patch test – a condition for assessing FEM convergence. *International Journal for Numerical Methods in Engineering*, 22:39–62, 1986.

16. O.C. Zienkiewicz, S. Qu, R.L. Taylor and S. Nakazawa. The patch test for mixed formulations. *International Journal for Numerical Methods in Engineering*, 23:1873–1883, 1986.

17. O.C. Zienkiewicz and D. Lefebvre. A robust triangular plate bending element of the Reissner–Mindlin type. *International Journal for Numerical Methods in Engineering*, 26:1169–1184, 1988.

18. O.C. Zienkiewicz, J.P. Vilotte, S. Toyoshima and S. Nakazawa. Iterative method for constrained and mixed approximation. An inexpensive improvement of FEM performance. *Computer Methods in Applied Mechanics and Engineering*, 51:3–29, 1985.

19. D.S. Malkus and T.J.R. Hughes. Mixed finite element methods in reduced and selective integration techniques: a unification of concepts. *Computer Methods in Applied Mechanics and Engineering*, 15:63–81, 1978.

20. B. Fraeijs de Veubeke. Displacement and equilibrium models in finite element method. In O.C. Zienkiewicz and G.S. Holister, editors, *Stress Analysis*, Chapter 9, pages 145–197. John Wiley & Sons, Chichester, 1965.

21. K.-J. Bathe. *Finite Element Procedures.* Prentice-Hall, Englewood Cliffs, NJ, 1996.

22. W.X. Zhong. FEM patch test and its convergence. Technical Report 97–3001, Research Institute Engineering Mechanics, Dalian University of Technology, 1997 (in Chinese).

23. W.X. Zhong. Convergence of FEM and the conditions of patch test. Technical Report 97–3002, Research Institute Engineering Mechanics, Dalian University of Technology, 1997 (in Chinese).

24. I. Babuška and R. Narasimhan. The Babuška–Brezzi condition and the patch test: an example. *Computer Methods in Applied Mechanics and Engineering*, 140:183–199, 1997.

25. O.C. Zienkiewicz and R.L. Taylor. The finite element patch test revisited: a computer test for convergence, validation and error estimates. *Computer Methods in Applied Mechanics and Engineering*, 149:523–544, 1997.

26. T.J.R. Hughes and W.K. Liu. Implicit–explicit finite elements in transient analyses. Part I and Part II. *J. Appl. Mech.*, 45:371–378, 1978.

27. K.-J. Bathe and L.W. Ho. Some results in the analysis of thin shell structures. In W. Wunderlich *et al.*, editors, *Nonlinear Finite Element Analysis in Structural Mechanics*, pages 122–156. Springer-Verlag, Berlin, 1981.

28. O.C. Zienkiewicz, M. Huang, J. Wu and S. Wu. A new algorithm for the coupled soil-pore fluid problem. *Shock and Vibration*, 1:3–14, 1993.

29. O.C. Zienkiewicz, Z. Xu, L.F. Zeng, A. Samuelsson and N.-E. Wiberg. Linked interpolation for Reissner–Mindlin plate elements. Part I – a simple quadrilateral element. *International Journal for Numerical Methods in Engineering*, 36:3043–3056, 1993.

30. Z. Xu, O.C. Zienkiewicz and L.F. Zeng. Linked interpolation for Reissner–Mindlin plate elements. Part III – an alternative quadrilateral. *International Journal for Numerical Methods in Engineering*, 36:3043–3056, 1993.

31. F. Auricchio and R.L. Taylor. A shear deformable plate element with an exact thin limit. *Computer Methods in Applied Mechanics and Engineering*, 118:393–412, 1994.

32. Z. Xu. A simple and efficient triangular finite element for plate bending. *Acta Mechanica Sinica*, 2:185–192, 1986.

33. F. Auricchio and R.L. Taylor. A triangular thick plate finite element with an exact thin limit. *Finite Elements in Analysis and Design*, 19:57–68, 1995.

34. D.N. Arnold and R.S. Falk. A uniformly accurate finite element method for Mindlin–Reissner plate. Technical Report IMA Preprint Series No. 307, Institute for Mathematics and its Application, University of Maryland, 1987.

35. T.J.R. Hughes and T. Tezduyar. Finite elements based upon Mindlin plate theory with particular reference to the four node bilinear isoparametric element. *J. Appl. Mech.*, 46:587–596, 1981.

36. E.N. Dvorkin and K.-J. Bathe. A continuum mechanics based four node shell element for general non-linear analysis. *Engineering Computations*, 1:77–88, 1984.

37. K.-J. Bathe and A.B. Chaudhary. A solution method for planar and axisymmetric contact problems. *International Journal for Numerical Methods in Engineering*, 21:65–88, 1985.

38. H.C. Huang and E. Hinton. A nine node Lagrangian Mindlin element with enhanced shear interpolation. *Engineering Computations*, 1:369–380, 1984.

39. E. Hinton and H.C. Huang. A family of quadrilateral Mindlin plate elements with substitute shear strain fields. *Computers and Structures*, 23:409–431, 1986.

40. O.C. Zienkiewicz, R.L. Taylor, P. Papadopoulos and E. Oñate. Plate bending elements with discrete constraints; new triangular elements. *Computers and Structures*, 35:505–522, 1990.

41. K.-J. Bathe and F. Brezzi. On the convergence of a four node plate bending element based on Mindlin–Reissner plate theory and a mixed interpolation. In J.R. Whiteman, editor, *The Mathematics of Finite Elements and Applications*, volume V, pages 491–503. Academic Press, London, 1985.

42. H.K. Stolarski and M.Y.M. Chiang. On a definition of the assumed shear strains in formulation of the C^0 plate elements. *European Journal of Mechanics, A/Solids*, 8:53–72, 1989.

43. H.K. Stolarski and M.Y.M. Chiang. Thin plate elements with relaxed Kirchhoff constraints. *International Journal for Numerical Methods in Engineering*, 26:913–933, 1988.

44. P. Papadopoulos and R.L. Taylor. A triangular element based on Reissner–Mindlin plate theory. *International Journal for Numerical Methods in Engineering*, 30:1029–1049, 1990.

45. R.S. Rao and H.K. Stolarski. Finite element analysis of composite plates using a weak form of the Kirchhoff constraints. *Finite Elements in Analysis and Design*, 13:191–208, 1993.

46. K.-J. Bathe, M.L. Bucalem and F. Brezzi. Displacement and stress convergence of four MITC plate bending elements. *Engineering Computations*, 7:291–302, 1990.

47. E. Oñate, R.L. Taylor and O.C. Zienkiewicz. Consistent formulation of shear constrained Reissner–Mindlin plate elements. In C. Kuhn and H. Mang, editors, *Discretization Methods in Structural Mechanics*, pages 169–180. Springer-Verlag, Berlin, 1990.

48. E. Oñate, O.C. Zienkiewicz, B. Suárez and R.L. Taylor. A general methodology for deriving shear constrained Reissner–Mindlin plate elements. *International Journal for Numerical Methods in Engineering*, 33:345–367, 1992.

49. G.S. Dhatt. Numerical analysis of thin shells by curved triangular elements based on discrete Kirchhoff hypotheses. In W.R. Rowan and R.M. Hackett, editors, *Proc. Symp. on Applications of FEM in Civil Engineering*, Vanderbilt University, Nashville, Tennessee, 1969. ASCE.

50. J.L. Batoz, K.-J. Bathe and L.W. Ho. A study of three-node triangular plate bending elements. *International Journal for Numerical Methods in Engineering*, 15:1771–1812, 1980.

51. J.L. Batoz. An explicit formulation for an efficient triangular plate bending element. *International Journal for Numerical Methods in Engineering*, 18:1077–1089, 1982.

52. J.L. Batoz and P. Lardeur. A discrete shear triangular nine d.o.f. element for the analysis of thick to very thin plates. *International Journal for Numerical Methods in Engineering*, 28:533–560, 1989.

53. T. Tu. *Performance of Reissner–Mindlin elements*. PhD thesis, Rutgers University, Department of Mathematics, 1998.

54. T.H.H. Pian and K. Sumihara. Rational approach for assumed stress finite elements. *International Journal for Numerical Methods in Engineering*, 20:1685–1695, 1985.

55. L.F. Greimann and P.P. Lynn. Finite element analysis of plate bending with transverse shear deformation. *Nuclear Engineering and Design*, 14:223–230, 1970.

56. P.P. Lynn and B.S. Dhillon. Triangular thick plate bending elements. In *Proceedings 1st International Conference on Structural Mechanics in Reactor Technology*, page M 6/5, Berlin, 1971.

57. Z. Xu. A thick–thin triangular plate element. *International Journal for Numerical Methods in Engineering*, 33:963–973, 1992.

58. A. Tessler and T.J.R. Hughes. A three node Mindlin plate element with improved transverse shear. *Computer Methods in Applied Mechanics and Engineering*, 50:71–101, 1985.

59. A. Tessler. A C^0 anisoparametric three node shallow shell element. *Computer Methods in Applied Mechanics and Engineering*, 78:89–103.

60. R.L. Taylor and F. Auricchio. Linked interpolation for Reissner–Mindlin plate elements: Part II – a simple triangle. *International Journal for Numerical Methods in Engineering*, 36:3057–3066, 1993.

61. M.A. Crisfield. *Non-linear Finite Element Analysis of Solids and Structures*, volume 1. John Wiley & Sons, Chichester, 1991.

62. F. Auricchio and C. Lovadina. Analysis of kinematic linked interpolation methods for Reissner–Mindlin plate problems. *Computer Methods in Applied Mechanics and Engineering*, 190: 2465–2482, 2001.

63. R.L. Taylor and S. Govindjee. A quadratic linked plate element with an exact thin plate limit. Technical Report UCB/SEMM-02/10, University of California, Berkeley, November 2002.

64. C. Chinosi and C. Lovadina. Numerical analysis of some mixed finite element methods for Reissner–Mindlin plates. *Computational Mechanics*, 16:36–44, 1995.

65. E.L. Wilson. The static condensation algorithm. *International Journal for Numerical Methods in Engineering*, 8:1974, 199–203.

66. L.S.D. Morley. *Skew Plates and Structures*. Macmillan, New York, 1963. International Series of Monographs in Aeronautics and Astronautics.

67. I. Babuška and T. Scapolla. Benchmark computation and performance evaluation for a rhombic plate bending problem. *International Journal for Numerical Methods in Engineering*, 28: 155–180, 1989.

68. P. Papadopoulos and R.L. Taylor. Elasto-plastic analysis of Reissner–Mindlin plates. *Appl. Mech. Rev.*, 43(5):S40–S50, 1990.
69. P. Papadopoulos. *On the Finite Element Solution of General Contact Problems*. PhD dissertation, Department of Civil Engineering, University of California at Berkeley, Berkeley, USA, 1991.
70. O.C. Zienkiewicz and J.Z. Zhu. Error estimation and adaptive refinement for plate bending problems. *International Journal for Numerical Methods in Engineering*, 28:2839–2853, 1989.

13

Shells as an assembly of flat elements

13.1 Introduction

A shell is, in essence, a structure that can be derived from a plate by initially forming the middle surface as a singly (or doubly) curved surface. The same assumptions as used in thin plates regarding the transverse distribution of strains and stresses are again valid. However, the way in which the shell supports external loads is quite different from that of a flat plate. The stress resultants acting on the middle surface of the shell now have both tangential and normal components which carry a major part of the load, a fact that explains the economy of shells as load-carrying structures and their well-deserved popularity.

The derivation of detailed governing equations for a curved shell problem presents many difficulties and, in fact, leads to many alternative formulations, each depending on the approximations introduced. For details of classical shell treatment the reader is referred to standard texts on the subject, for example the well-known treatise by Flügge[1] or the classical book by Timoshenko and Woinowski-Krieger.[2]

In the finite element treatment of shell problems to be described in this chapter the difficulties referred to above are eliminated, at the expense of introducing a further approximation. This approximation is of a physical, rather than mathematical, nature. In this it is assumed that the behaviour of a continuously curved surface can be adequately represented by the behaviour of a surface built up of small flat elements. Intuitively, as the size of the subdivision decreases it would seem that convergence must occur and indeed experience indicates such a convergence.

It will be stated by many shell experts that when we compare the *exact* solution of a shell approximated by flat facets to the exact solution of a truly curved shell, considerable differences in the distribution of bending moments, etc., occur. It is arguable if this is true, but for simple elements the discretization error is approximately of the same order and excellent results can be obtained with flat shell element approximation. The mathematics of this problem is discussed in detail by Ciarlet.[3]

In a shell, the element generally will be subject both to bending and to 'in-plane' force resultants. For a flat element these cause independent deformations, provided the local deformations are small, and therefore the ingredients for obtaining the necessary stiffness matrices are available in the material already covered in the preceding chapters.

In the division of an arbitrary shell into flat elements only triangular elements can be used for doubly curved surfaces. Although the concept of the use of such elements in the analysis was suggested as early as 1961 by Greene et al.,[4] the success of such analysis was hampered by the lack of a good stiffness matrix for triangular plate elements in bending.[5-8] The developments described in Chapters 11 and 12 open the way to adequate models for representing the behaviour of shells with such a division.

Some shells, for example those with general cylindrical shapes, can be well represented by flat elements of rectangular or quadrilateral shape provided the mesh subdivision does not lead to 'warped' elements. With good stiffness matrices available for such elements the progress here has been more satisfactory. Practical problems of arch dam design and others for cylindrical shape roofs have been solved quite early with such subdivisions.[9,10]

Clearly, the possibilities of analysis of shell structures by the finite element method are enormous. Problems presented by openings, variation of thickness, or anisotropy are no longer of consequence.

A special case is presented by axisymmetrical shells. Although it is obviously possible to deal with these in the way described in this chapter, a simpler approach can be used. This will be presented in Chapters 14–16.

As an alternative to the type of analysis described here, curved shell elements could be used. Here curvilinear coordinates are essential and general procedures in Chapter 5 of reference 11 can be extended to define these. The physical approximation involved in flat elements is now avoided at the expense of reintroducing an arbitrariness of various shell theories. Several approaches using a direct displacement approximation are given in references 12–32 and the use of 'mixed variational principles' is given in references 33–36.

A very simple and effective way of deriving curved shell elements is to use the so-called 'shallow' shell theory approach.[14,15,37,38] Here the variables u, v, w define the *tangential* and *normal* components of displacement to the curved surface. If all the elements are assumed to be tangential to each other, no need arises to transfer these from local to global values. The element is assumed to be 'shallow' with respect to a local coordinate system representing its projection on a plane defined by nodal points, and its strain energy is defined by appropriate equations that include derivatives with respect to *coordinates in the plane of projection*. Thus, precisely the same shape functions can be used as in flat elements discussed in this chapter and all integrations are in fact carried out in the 'plane' as before.

Such shallow shell elements, by coupling the effects of membrane and bending strain in the energy expression, are slightly more efficient than flat ones where such coupling occurs on the interelement boundary only. For simple, small elements the gains are marginal, but with few higher-order large elements advantages appear. A good discussion of such a formulation is given in reference 23.

For many practical purposes the flat element approximation gives very adequate answers and also permits an easy coupling with edge beam and rib members, a facility sometimes not present in a curved element formulation. Indeed, in many practical problems the structure is in fact composed of flat surfaces, at least in part, and these can be simply reproduced. For these reasons curved, general, thin shell forms will not be discussed here and instead a general formulation of thick curved shells (based directly on three-dimensional behaviour and avoiding the shell equation ambiguities) will be presented in Chapter 15. The development of curved elements for general shell theories can also be effected in a direct manner; however, additional transformations over those discussed in this chapter are involved. The interested reader is referred to references 39 and 40 for additional discussion on this approach. In many respects the differences in the two approaches are quite similar, as shown by Bischoff and Ramm.[41]

In most arbitrary-shaped, curved shell elements the coordinates used are such that complete smoothness of the surface between elements is not guaranteed. The shape discontinuity occurring there, and indeed on any shell where 'branching' occurs, is precisely of the same type as that encountered in this chapter and therefore the methodology of assembly discussed here is perfectly general.

13.2 Stiffness of a plane element in local coordinates

Consider a typical polygonal flat element in a *local* coordinate system $\bar{x}\,\bar{y}\,\bar{z}$ subject simultaneously to 'in-plane' and 'bending' actions (Fig. 13.1).

Taking first the *in-plane* (plane stress) action, we know that the state of strain is uniquely described in terms of the $\bar{u}$ and $\bar{v}$ displacement of each typical node a. The minimization of the total potential energy led to the stiffness matrices described there and gives 'nodal' forces due to displacement parameters $\bar{\mathbf{u}}^{\mathrm{p}}$ as

$$(\bar{\mathbf{f}}^e)^{\mathrm{p}} = (\bar{\mathbf{K}}^e)^{\mathrm{p}}\bar{\mathbf{u}}^{\mathrm{p}} \quad \text{with} \quad \bar{\mathbf{u}}_a^{\mathrm{p}} = \begin{Bmatrix} \bar{u}_a \\ \bar{v}_a \end{Bmatrix} \quad \bar{\mathbf{f}}_a^{\mathrm{p}} = \begin{Bmatrix} F_{\bar{x}a} \\ F_{\bar{y}a} \end{Bmatrix} \tag{13.1}$$

Similarly, when bending was considered in Chapters 11 and 12, the state of strain was given uniquely by the nodal displacement in the $\bar{z}$ direction ($\bar{w}$) and the two rotations $\bar{\theta}_{\bar{x}}$ and $\bar{\theta}_{\bar{y}}$.[*] This resulted in stiffness matrices of the type

$$(\bar{\mathbf{f}}^e)^{\mathrm{b}} = (\bar{\mathbf{K}}^e)^{\mathrm{b}}\bar{\mathbf{u}}^{\mathrm{b}} \quad \text{with} \quad \bar{\mathbf{u}}_a^{\mathrm{b}} = \begin{Bmatrix} \bar{w}_a \\ \bar{\theta}_{\bar{x}a} \\ \bar{\theta}_{\bar{y}a} \end{Bmatrix} \quad \bar{\mathbf{f}}_a^{\mathrm{b}} = \begin{Bmatrix} F_{\bar{z}a} \\ M_{\bar{x}a} \\ M_{\bar{y}a} \end{Bmatrix} \tag{13.2}$$

Before combining these stiffnesses it is important to note two facts. The first is that the displacements prescribed for 'in-plane' forces do not affect the bending deformations and vice versa. The second is that the rotation $\theta_{\bar{z}}$ does not enter as a parameter into the definition of deformations in either mode. While one could neglect this entirely at the present stage it is convenient, for reasons which will be apparent later when assembly is considered, to take this rotation into account and associate with it a fictitious couple $M_{\bar{z}}$. The fact that it does not enter into the minimization procedure can be accounted for simply by inserting an appropriate number of zeros into the stiffness matrix.

Redefining the combined nodal displacement as

$$\bar{\mathbf{u}}_a = \begin{bmatrix} \bar{u}_a & \bar{v}_a & \bar{w}_a & \bar{\theta}_{\bar{x}a} & \bar{\theta}_{\bar{y}a} & \bar{\theta}_{\bar{z}a} \end{bmatrix}^T \tag{13.3}$$

and the appropriate nodal 'forces' as

$$\bar{\mathbf{f}}_a^e = \begin{bmatrix} F_{\bar{x}a} & F_{\bar{y}a} & F_{\bar{z}a} & M_{\bar{x}a} & M_{\bar{y}a} & M_{\bar{z}a} \end{bmatrix}^T \tag{13.4}$$

---

[*] In dealing with shells it is now convenient to use the rotation parameters θ instead of the director angle ϕ. This is accomplished using Eq. (11.7).

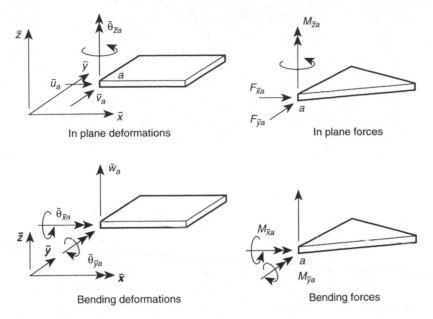

In plane deformations

In plane forces

Bending deformations

Bending forces

Fig. 13.1 A flat element subject to 'in-plane' and 'bending' actions.

we can write

$$\bar{\mathbf{K}}^e \bar{\mathbf{u}} = \bar{\mathbf{f}}^e \tag{13.5}$$

The stiffness matrix is now made up from the following submatrices

$$
\bar{\mathbf{K}}_{ab} =
\left[
\begin{array}{cc:ccc:c}
 & & 0 & 0 & 0 & 0 \\
\bar{\mathbf{K}}_{ab}^p & & 0 & 0 & 0 & 0 \\
\hdashline
0 & 0 & & & & 0 \\
0 & 0 & & \bar{\mathbf{K}}_{ab}^b & & 0 \\
0 & 0 & & & & 0 \\
\hdashline
0 & 0 & 0 & 0 & 0 & 0
\end{array}
\right]
\tag{13.6}
$$

if we note that

$$\bar{\mathbf{u}}_a = \begin{bmatrix} \bar{\mathbf{u}}_a^p & \bar{\mathbf{u}}_a^b & \bar{\theta}_{\bar{z}a} \end{bmatrix}^{\mathrm{T}} \tag{13.7}$$

The above formulation is valid for any shape of polygonal element and, in particular, for the two important types illustrated in Fig. 13.1.

13.3 Transformation to global coordinates and assembly of elements

The stiffness matrix derived in the previous section used a system of local coordinates as the 'reference plane', and forces and bending components also are originally derived for this system.

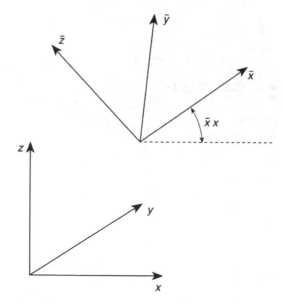

Fig. 13.2 Local and global coordinates.

Transformation of coordinates to a common global system (which will be denoted by xyz with the local system still $\bar{x}\,\bar{y}\,\bar{z}$) will be necessary to assemble the elements and to write the appropriate equilibrium equations.

In addition it will be initially more convenient to specify the element nodes by their global coordinates and to establish from these the local coordinates, thus requiring an inverse transformation. All the transformations are accomplished by a simple process.

The two systems of coordinates are shown in Fig. 13.2. The forces and displacements of a node transform from the global to the local system by a matrix $\mathbf{T}$ giving

$$\bar{\mathbf{u}}_a = \mathbf{T}\,\mathbf{u}_a \qquad \bar{\mathbf{f}}_a = \mathbf{T}\mathbf{f}_a \tag{13.8}$$

in which

$$\mathbf{T} = \begin{bmatrix} \mathbf{\Lambda} & \mathbf{0} \\ \mathbf{0} & \mathbf{\Lambda} \end{bmatrix} \tag{13.9}$$

with $\mathbf{\Lambda}$ being a 3×3 matrix of direction cosines between the two sets of axes,[42,43] that is,

$$\mathbf{\Lambda} = \begin{bmatrix} \cos(\bar{x}, x) & \cos(\bar{x}, y) & \cos(\bar{x}, z) \\ \cos(\bar{y}, x) & \cos(\bar{y}, y) & \cos(\bar{y}, z) \\ \cos(\bar{z}, x) & \cos(\bar{z}, y) & \cos(\bar{z}, z) \end{bmatrix} = \begin{bmatrix} \Lambda_{\bar{x}x} & \Lambda_{\bar{x}y} & \Lambda_{\bar{x}z} \\ \Lambda_{\bar{y}x} & \Lambda_{\bar{y}y} & \Lambda_{\bar{y}z} \\ \Lambda_{\bar{z}x} & \Lambda_{\bar{z}y} & \Lambda_{\bar{z}z} \end{bmatrix} \tag{13.10}$$

where $\cos(\bar{x}, x)$ is the cosine of the angle between the $\bar{x}$ axis and the x axis, and so on.

By the rules of orthogonal transformation the inverse of $\mathbf{T}$ is given by its transpose; thus we have

$$\mathbf{u}_a = \mathbf{T}^\mathrm{T}\bar{\mathbf{u}}_a \qquad \mathbf{f}_a = \mathbf{T}^\mathrm{T}\bar{\mathbf{f}}_a \tag{13.11}$$

which permits the stiffness matrix of an element in the global coordinates to be computed as

$$\mathbf{K}_{ab}^e = \mathbf{T}^T \bar{\mathbf{K}}_{ab}^e \mathbf{T} \qquad (13.12)$$

in which $\bar{\mathbf{K}}_{ab}^e$ is determined by Eq. (13.6) in the local coordinates.

The determination of the local coordinates follows a similar pattern. The relationship between global and local systems is given by

$$\left\{ \begin{array}{c} \bar{x} \\ \bar{y} \\ \bar{z} \end{array} \right\} = \mathbf{\Lambda} \left\{ \begin{array}{c} x - x_0 \\ y - y_0 \\ z - z_0 \end{array} \right\} \qquad (13.13)$$

where x_0, y_0, z_0 is the distance from the origin of the global coordinates to the origin of the local coordinates. As in the computation of stiffness matrices for flat plane and bending elements the position of the origin is immaterial, this transformation will always suffice for determination of the local coordinates in the plane (or a plane parallel to the element).

Once the stiffness matrices of all the elements have been determined in a common global coordinate system, the assembly of the elements and forces follow the standard solution pattern. The resulting displacements calculated are referred to the global system, and before the stresses can be computed it is necessary to change these to the local system for each element. The usual stress calculations for 'in-plane' and 'bending' components can then be used.

13.4 Local direction cosines

The determination of the direction cosine matrix $\mathbf{\Lambda}$ gives rise to some algebraic difficulties and, indeed, is not unique since the direction of one of the local axes is arbitrary, provided it lies in the plane of the element. We shall first deal with the assembly of rectangular elements in which this problem is particularly simple; later we shall consider the case for triangular elements arbitrarily orientated in space.

13.4.1 Rectangular elements

Such elements are limited in use to representing a cylindrical or box type of surface. It is convenient to take one side of each element and the corresponding $\bar{x}$ axis parallel to the global x axis. For a typical element 1234, illustrated in Fig. 13.3, it is now easy to calculate all the relevant direction cosines. Direction cosines of $\bar{x}$ are, obviously,

$$\Lambda_{\bar{x}x} = 1 \qquad \Lambda_{\bar{x}y} = \Lambda_{\bar{x}z} = 0 \qquad (13.14)$$

The direction cosines of the $\bar{y}$ axis have to be obtained by consideration of the coordinates of the various nodal points. Thus,

$$\Lambda_{\bar{y}x} = 0$$

$$\Lambda_{\bar{y}y} = \frac{y_4 - y_1}{\sqrt{(y_4 - y_1)^2 + (z_4 - z_1)^2}} \qquad (13.15)$$

$$\Lambda_{\bar{y}z} = \frac{z_4 - z_1}{\sqrt{(y_4 - y_1)^2 + (z_4 - z_1)^2}}$$

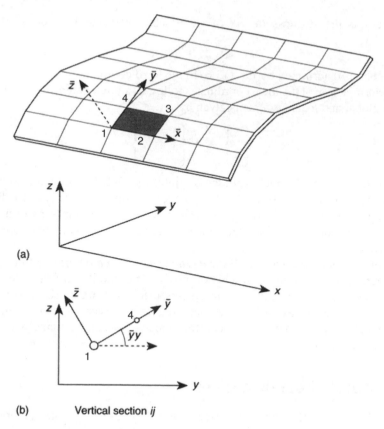

(a)

(b)   Vertical section ij

Fig. 13.3 A cylindrical shell as an assembly of rectangular elements: local and global coordinates.

are simple geometrical relations which can be obtained by consideration of the sectional plane passing vertically through 1–4 in the z direction. Similarly, from the same section, we have for the $\bar{z}$ axis

$$\Lambda_{\bar{z}x} = 0$$

$$\Lambda_{\bar{z}y} = \frac{z_1 - z_4}{\sqrt{(y_4 - y_1)^2 + (z_4 - z_1)^2}} = -\Lambda_{\bar{y}z}$$

$$\Lambda_{\bar{z}z} = \frac{y_4 - y_1}{\sqrt{(y_4 - y_1)^2 + (z_4 - z_1)^2}} = \Lambda_{\bar{y}y}$$

(13.16)

Clearly, the numbering of points in a consistent fashion is important to preserve the correct signs of the expression.

13.4.2 Triangular elements arbitrarily orientated in space

An arbitrary shell divided into triangular elements is shown in Fig. 13.4(a). Each element has an orientation in which the angles with the coordinate planes are arbitrary. The problem of defining local axes and their direction cosines is therefore more complex

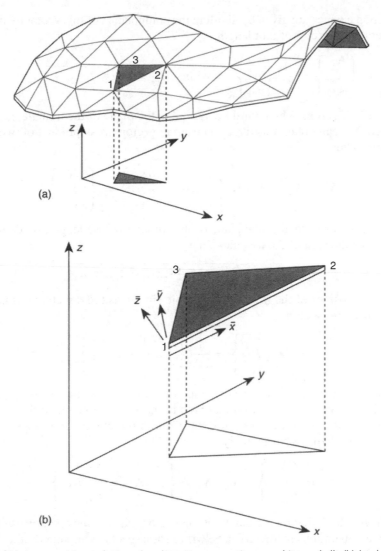

Fig. 13.4 (a) An assemblage of triangular elements representing an arbitrary shell; (b) local and global coordinates for a triangular element.

than in the previous simple example. The most convenient way of dealing with the problem is to use some properties of geometrical vector algebra.

One arbitrary but convenient choice of local axis direction is given here. We shall specify that the $\bar{x}$ axis is to be directed along the side 1–2 of the triangle, as shown in Fig. 13.4(b).

The vector $\mathbf{V}_{21}$ defines this side and in terms of global coordinates we have

$$\mathbf{V}_{21} = \begin{Bmatrix} x_2 - x_1 \\ y_2 - y_1 \\ z_2 - z_1 \end{Bmatrix} = \begin{Bmatrix} x_{21} \\ y_{21} \\ z_{21} \end{Bmatrix} \qquad (13.17)$$

The direction cosines are given by dividing the components of this vector by its length, that is, defining a vector of unit length

$$
\mathbf{v}_{\bar{x}} = \begin{Bmatrix} \Lambda_{\bar{x}x} \\ \Lambda_{\bar{x}y} \\ \Lambda_{\bar{x}z} \end{Bmatrix} = \frac{1}{l_{21}} \begin{Bmatrix} x_{21} \\ y_{21} \\ z_{21} \end{Bmatrix} \quad \text{with} \quad l_{21} = \sqrt{x_{21}^2 + y_{21}^2 + z_{21}^2} \tag{13.18}
$$

Now, the $\bar{z}$ direction, which must be normal to the plane of the triangle, needs to be established. We can obtain this direction from a 'vector' cross-product of two sides of the triangle. Thus,

$$
\mathbf{V}_{\bar{z}} = \mathbf{V}_{21} \times \mathbf{V}_{31} = \begin{Bmatrix} y_{21}z_{31} - z_{21}y_{31} \\ z_{21}x_{31} - x_{21}z_{31} \\ x_{21}y_{31} - y_{21}x_{31} \end{Bmatrix} = \begin{Bmatrix} yz_{123} \\ zx_{123} \\ xy_{123} \end{Bmatrix} \tag{13.19}
$$

represents a vector normal to the plane of the triangle whose length, by definition, is equal to twice the area of the triangle. Thus,

$$
l_{\bar{z}} = 2\Delta = \sqrt{(yz_{123})^2 + (zx_{123})^2 + (xy_{123})^2}
$$

The direction cosines of the $\bar{z}$ axis are available simply as the direction cosines of $\mathbf{V}_{\bar{z}}$, and we have a unit vector

$$
\mathbf{v}_{\bar{z}} = \begin{Bmatrix} \Lambda_{\bar{z}x} \\ \Lambda_{\bar{z}y} \\ \Lambda_{\bar{z}z} \end{Bmatrix} = \frac{1}{2\Delta} \begin{Bmatrix} y_{21}z_{31} - z_{21}y_{31} \\ z_{21}x_{31} - x_{21}z_{31} \\ x_{21}y_{31} - y_{21}x_{31} \end{Bmatrix} \tag{13.20}
$$

Finally, the direction cosines of the $\bar{y}$ axis are established in a similar manner as the direction cosines of a vector normal both to the $\bar{x}$ direction and to the $\bar{z}$ direction. If vectors of unit length are taken in each of these directions [as in fact defined by Eqs (13.18)–(13.20)] we have simply

$$
\mathbf{v}_{\bar{y}} = \begin{Bmatrix} \Lambda_{\bar{y}x} \\ \Lambda_{\bar{y}y} \\ \Lambda_{\bar{y}z} \end{Bmatrix} = \mathbf{v}_{\bar{z}} \times \mathbf{v}_{\bar{x}} = \begin{Bmatrix} \Lambda_{\bar{z}y}\Lambda_{\bar{x}z} - \Lambda_{\bar{z}z}\Lambda_{\bar{x}y} \\ \Lambda_{\bar{z}z}\Lambda_{\bar{x}x} - \Lambda_{\bar{z}x}\Lambda_{\bar{x}z} \\ \Lambda_{\bar{z}x}\Lambda_{\bar{x}y} - \Lambda_{\bar{z}y}\Lambda_{\bar{x}x} \end{Bmatrix} \tag{13.21}
$$

without having to divide by the length of the vector, which is now simply unity.

The vector operations involved can be written as a special computer routine in which vector products, normalizing (i.e. division by length), etc., are automatically carried out[44] and there is no need to specify in detail the various operations given above.

In the preceding outline the direction of the $\bar{x}$ axis was taken as lying along one side of the element. A useful alternative is to specify this by the section of the triangle plane with a plane parallel to one of the coordinate planes. Thus, for instance, if we desire to erect the $\bar{x}$ axis along a horizontal contour of the triangle (i.e. a section parallel to the xy plane) we can proceed as follows.

First, the normal direction cosines $\mathbf{v}_{\bar{z}}$ are defined as in Eq. (13.20). Now, the matrix of direction cosines of $\bar{x}$ has to have a zero component in the z direction and thus we have

$$
\mathbf{v}_{\bar{x}} = \begin{Bmatrix} \Lambda_{\bar{x}x} \\ \Lambda_{\bar{x}y} \\ 0 \end{Bmatrix} \tag{13.22}
$$

As the length of the vector is unity

$$\Lambda_{\bar{x}x}^2 + \Lambda_{\bar{x}y}^2 = 1 \qquad (13.23)$$

and as further the *scalar* product of the $\mathbf{v}_{\bar{x}}$ and $\mathbf{v}_{\bar{z}}$ must be zero, we can write

$$\Lambda_{\bar{x}x}\Lambda_{\bar{z}x} + \Lambda_{\bar{x}y}\Lambda_{\bar{z}y} = 0 \qquad (13.24)$$

and from these two equations $\mathbf{v}_{\bar{x}}$ can be uniquely determined. Finally, as before

$$\mathbf{v}_{\bar{y}} = \mathbf{v}_{\bar{z}} \times \mathbf{v}_{\bar{x}} \qquad (13.25)$$

It should be noted that this transformation will be singular if there is no line in the plane of the element which is parallel to the xy plane, and some other orientation must then be selected. Yet another alternative of a specification of the $\bar{x}$ axis is given in Chapter 15 where we discuss the development of 'shell' elements directly from the three-dimensional equations of solids.

13.5 'Drilling' rotational stiffness – 6 degree-of-freedom assembly

In the formulation described above a difficulty arises if all the elements meeting at a node are co-planar. This situation will happen for flat (folded) shell segments and at straight boundaries of developable surfaces (e.g. cylinders or cones). The difficulty is due to the assignment of a zero stiffness in the $\theta_{\bar{z}i}$ direction of Fig. 13.1 and the fact that classical shell equations do not produce equations associated with this rotational parameter. Inclusion of the third rotation and the associated 'force' $F_{\bar{z}i}$ has obvious benefits for a finite element model in that both rotations and displacements at nodes may be treated in a very simple manner using the transformations just presented.

If the set of assembled equilibrium equations in *local coordinates* is considered at such a point we have six equations of which the last (corresponding to the $\theta_{\bar{z}}$ direction) is simply

$$0 \, \theta_{\bar{z}} = 0 \qquad (13.26)$$

As such, an equation of this type presents no special difficulties (solution programs usually detect the problem and issue a warning). However, if the global coordinate directions differ from the local ones and a transformation is accomplished, the six equations mask the fact that the equations are singular. Detection of this singularity is somewhat more difficult and depends on round-off in each computer system.

A number of alternatives have been presented that avoid the presence of this singular behaviour. Two simple ones are:

1. assemble the equations (or just the rotational parts) at points where elements are co-planar in local coordinates (and delete the $0\theta_{\bar{z}} = 0$ equation); and/or
2. insert an arbitrary stiffness coefficient $\bar{k}_{\theta_{\bar{z}}}$ at such points only.

This leads in the local coordinates to replacing Eq. (13.26) by

$$\bar{k}_{\theta_{\bar{z}}}\theta_{\bar{z}_b} = 0 \tag{13.27}$$

which, on transformation, leads to a perfectly well-behaved set of equations from which, by usual processes, all displacements now including $\theta_{\bar{z}i}$ are obtained. As $\theta_{\bar{z}i}$ does not affect the stresses and indeed is uncoupled from all equilibrium equations any value of $\bar{k}_{\theta_{\bar{z}}}$ similar to values already in Eq. (13.6) can be inserted as an external stiffness without affecting the result.

These two approaches lead to programming complexity (as a decision on the co-planar nature is necessary) and an alternative is to modify the formulation so that the rotational parameters arise more naturally and have a real physical significance. This has been a topic of much study[45-57] and the $\theta_{\bar{z}}$ parameter introduced in this way is commonly called a *drilling* degree of freedom, on account of its action to the surface of the shell. An early application considering the rotation as an additional degree of freedom in plane analysis is contained in reference 15. In reference 8 a set of rotational stiffness coefficients was used in a general shell program for all elements whether co-planar or not. These were defined such that in local coordinates overall equilibrium is not disturbed. This may be accomplished by adding to the formulation for each element the term

$$\Pi^* = \Pi + \int_\Omega \alpha_n E t^n \left(\theta_{\bar{z}} - \bar{\theta}_{\bar{z}}\right)^2 d\Omega \tag{13.28}$$

in which the parameter α_n is a fictitious elastic parameter and $\bar{\theta}_{\bar{z}}$ is a mean rotation of each element which permits the element to satisfy local equilibrium in a weak sense. The above is a generalization of that proposed in reference 8 where the value of n is unity in the scaling value t^n. Since the term will lead to a stiffness that will be in terms of rotation parameters the scaling indicated above permits values proportional to those generated by the bending rotations – namely, proportional to t cubed. In numerical experiments this scaling leads to less sensitivity in the choice of α_n. For a triangular element in which a linear interpolation is used for $\theta_{\bar{z}}$ minimization with respect to $\theta_{\bar{z}a}$ leads to the form

$$\begin{Bmatrix} M_{\bar{z}1} \\ M_{\bar{z}2} \\ M_{\bar{z}3} \end{Bmatrix} = \frac{1}{36}\alpha_n E t^n \Delta \begin{bmatrix} 1 & -0.5 & -0.5 \\ -0.5 & 1 & -0.5 \\ -0.5 & -0.5 & 1 \end{bmatrix} \begin{Bmatrix} \theta_{\bar{z}1} \\ \theta_{\bar{z}2} \\ \theta_{\bar{z}3} \end{Bmatrix} \tag{13.29}$$

where α_n is yet to be specified. This additional stiffness does in fact affect the results where nodes are not co-planar and indeed represents an approximation; however, effects of varying α_n over fairly wide limits are quite small in many applications. For instance in Table 13.1 a set of displacements of an arch dam analysed in reference 4 is given for various values of α_1. For practical purposes extremely small values of α_n are possible, providing a large computer word length is used.[58]

Table 13.1   Nodal rotation coefficient in dam analysis[4]

α_1	1.00	0.50	0.10	0.03	0.00
Radial displacement (mm)	61.13	63.551	64.52	64.78	65.28

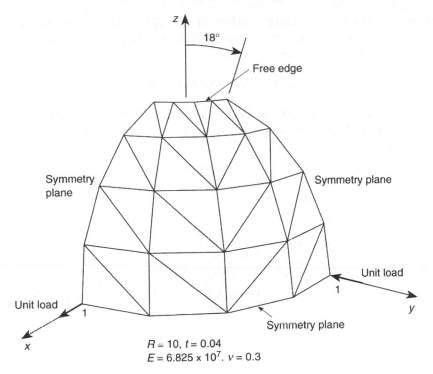

$R = 10, t = 0.04$
$E = 6.825 \times 10^7. \, v = 0.3$

Fig. 13.5 Spherical shell test problem.[59]

The analysis of the spherical test problem proposed by MacNeal and Harter as a standard test[59] is indicated in Fig. 13.5. For this test problem a constant strain triangular membrane together with the discrete Kirchhoff triangular plate bending element is combined with the rotational treatment. The results for regular meshes are shown in Table 13.2 for several values of α_3 and mesh subdivisions.

The above development, while quite easy to implement, retains the original form of the membrane interpolations. For triangular elements with corner nodes only, the membrane form utilizes linear displacement fields that yield only constant strain terms. Most bending elements discussed in Chapters 11 and 12 have bending strains with higher than constant terms. Consequently, the membrane error terms will dominate the behaviour of many shell problem solutions. In order to improve the situation it is desirable to increase the order of interpolation. Using conventional interpolations this implies the introduction of additional nodes on each element; however, by utilizing a

Table 13.2 Sphere problem: radial displacement at load

Mesh	α_3 value					
	10.0	1.00	0.100	0.010	0.001	0.000
4 × 4	0.0639	0.0919	0.0972	0.0979	0.0980	0.0980
8 × 8	0.0897	0.0940	0.0945	0.0946	0.0946	0.0946
16 × 16	0.0926	0.0929	0.0929	0.0929	0.0930	0.0930

drill parameter these interpolations can be transformed to a form that permits a 6 degree-of-freedom assembly at each vertex node. Quadratic interpolations along the edge of an element can be expressed as

$$\bar{\mathbf{u}}(\xi) = N_1(\xi)\bar{\mathbf{u}}_1 + N_2(\xi)\bar{\mathbf{u}}_2 + N_3(\xi)\Delta\bar{\mathbf{u}}_3 \tag{13.30}$$

where $\bar{\mathbf{u}}_1$ are nodal displacements $(\bar{u}_1, \bar{v}_1)$ at an end of the edge (vertex), similarly $\bar{\mathbf{u}}_2$ is the other end, and $\Delta\bar{\mathbf{u}}_3$ are hierarchical displacements at the centre of the edge (Fig. 13.6).

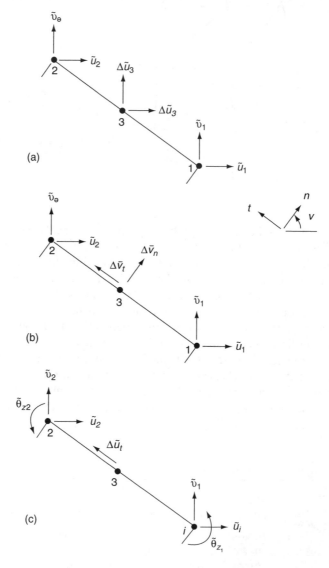

Fig. 13.6 Construction of in-plane interpolations with drilling parameters.

The centre displacement parameters may be expressed in terms of *normal* ($\Delta\bar{u}_n$) and *tangential* ($\Delta\bar{u}_t$) components as

$$\Delta\bar{\mathbf{u}}_3 = \Delta\bar{u}_n\mathbf{n} + \Delta\bar{u}_t\mathbf{t} \qquad (13.31)$$

where **n** is a unit outward normal and **t** is a unit tangential vector to the edge:

$$\mathbf{n} = \begin{Bmatrix} \cos\nu \\ \sin\nu \end{Bmatrix} \quad \text{and} \quad \mathbf{t} = \begin{Bmatrix} -\sin\nu \\ \cos\nu \end{Bmatrix} \qquad (13.32)$$

where ν is the angle that the normal makes with the $\bar{x}$ axis. The normal displacement component may be expressed in terms of drilling parameters at each end of the edge (assuming a quadratic expansion).[45,53] Accordingly,

$$\Delta u_n = \tfrac{1}{8}l_{12}(\theta_{\bar{z}2} - \theta_{\bar{z}1}) \qquad (13.33)$$

in which l_{12} is the length of the 1–2 side. This construction produces an interpolation on each edge given by

$$\bar{\mathbf{u}}(\xi) = N_1(\xi)\bar{\mathbf{u}}_1 + N_2(\xi)\bar{\mathbf{u}}_2 + N_3(\xi)\left[\tfrac{1}{8}l_{12}(\theta_{\bar{z}2} - \theta_{\bar{z}1}) + \Delta\bar{u}_t\mathbf{t}\right] \qquad (13.34)$$

The reader will undoubtedly observe the similarity here with the process used for linked interpolation for the bending element (see Sec. 12.7).

The above interpolation may be further simplified by constraining the $\Delta\bar{u}_t$ parameters to zero. We note, however, that these terms are beneficial in a three-node triangular element. If a common sign convention is used for the hierarchical tangential displacement at each edge, this tangential component maintains compatibility of displacement even in the presence of a kink between adjacent elements. For example, an appropriate sign convention can be accomplished by directing a positive component in the direction in which the end (vertex) node numbers increase. The above structure for the in-plane displacement interpolations may be used for either an irreducible or a mixed element model and generates stiffness coefficients that include terms for the $\theta_{\bar{z}}$ parameters as well as those for $\bar{u}$ and $\bar{v}$. It is apparent, however, that the element generated in this manner must be singular (i.e. has spurious zero-energy modes) since for equal values of the end rotation the interpolation is independent of the $\theta_{\bar{z}}$ parameters. Moreover, when used in non-flat shell applications the element is not free of local equilibrium errors. This later defect may be removed by using the procedure identified above in Eq. (13.28), and results for a quadrilateral element generated according to this scheme are given by Jetteur[54] and Taylor.[55]

A structure of the plane stress problem which includes the effects of a drill rotation field is given by Reissner[60] and is extended to finite element applications by Hughes and Brezzi.[61] A variational formulation for the in-plane problem may be stated as

$$\Pi_d(\bar{\mathbf{u}}, \theta_{\bar{z}}, \tau) = \frac{1}{2}\int_\Omega \boldsymbol{\varepsilon}^{\mathrm{T}}\mathbf{D}\boldsymbol{\varepsilon}\,\mathrm{d}\Omega + \int_\Omega \tau(\omega_{\bar{x}\bar{y}} - \theta_{\bar{z}})\,\mathrm{d}\Omega \qquad (13.35)$$

where τ is a *skew symmetric* stress component and $\omega_{\bar{x}\bar{y}}$ is the rotational part of the displacement gradient, which for the $\bar{x}\bar{y}$ plane is given by

$$\omega_{\bar{x}\bar{y}} = \frac{\partial\bar{v}}{\partial\bar{x}} - \frac{\partial\bar{u}}{\partial\bar{y}} \qquad (13.36)$$

In addition to the terms shown in Eq. (13.35), terms associated with initial stress and strain as well as boundary and body load must be appended for the general shell problem as discussed in Chapter 2 and also in reference 11.

A variation of Eq. (13.35) with respect to τ gives the constraint that the skew symmetric part of the displacement gradients is the rotation $\theta_{\bar{z}}$. Conversely, variation with respect to $\theta_{\bar{z}}$ gives the result that τ must vanish. Thus, the equations generated from Eq. (13.35) are those of the conventional membrane but include the rotation field. A penalty form of the above equations suitable for finite element applications may be constructed by modifying Eq. (13.35) to

$$\bar{\Pi}_d = \Pi_d - \int_\Omega \frac{1}{\alpha_\tau E t} \tau^2 \, d\Omega \qquad (13.37)$$

where α_τ is a penalty number.

It is important to use this mixed representation of the problem with the mixed patch test to construct viable finite element models. Use of constant τ and isoparametric interpolation of $\theta_{\bar{z}}$ in each element together with the interpolations for the displacement approximation given by Eq. (13.34) lead to good triangular and quadrilateral membrane elements. Applications to shell solutions using this form are given by Ibrahimbegovic et al.[57] Also the solution for a standard barrel vault problem is contained in Sec. 13.8.

13.6 Elements with mid-side slope connections only

Many of the difficulties encountered with the nodal assembly in global coordinates disappear if the element is so constructed as to require only the continuity of displacements u, v, and w at the corner nodes, with continuity of the normal slope being imposed along the element sides. Clearly, the corner assembly is now simple and the introduction of the sixth nodal variable is unnecessary. As the normal slope rotation along the sides is the same both in local and in global coordinates its transformation there is unnecessary – although again it is necessary to have a unique definition of parameters for the adjacent elements.

Elements of this type arise naturally in hybrid forms[11] and we have already referred to a plate bending element of a suitable type in Sec. 11.6. This element of the simplest possible kind has been used in shell problems by Dawe[26] with some success. A considerably more sophisticated and complex element of such type is derived by Irons and named 'semi-loof'.[30] This element is briefly mentioned in Chapter 9 and although its derivation is far from simple it performs well in many situations.

13.7 Choice of element

Numerous membrane and bending element formulations are now available, and, in both, conformity is achievable in flat assemblies. Clearly, if the elements are not coplanar conformity will, in general, be violated and only approached in the limit as smooth shell conditions are reached.

It would appear consistent to use expansions of similar accuracy in both the membrane and bending approximations but much depends on which action is dominant. For thin shells, the simplest triangular element would thus appear to be one with a linear in-plane displacement field and a quadratic bending displacement – thus approximating the stresses as constants in membrane and in bending actions. Such an element is used by Dawe[26] but gives rather marginal (though convergent) results.

In the examples shown we use the following elements which give quite adequate performance.

Element A: This is a mixed rectangular membrane with four corner nodes (Sec. 10.4.4 of reference 11) combined with the non-conforming bending rectangle with four corner nodes (Sec. 11.3). This was first used in references 9 and 10.

Element B: This is a constant strain triangle with three nodes combined with the incompatible bending triangle with 9 degrees of freedom (Sec. 11.5). Use of this in the shell context is given in references 8 and 62.

Element C: In this a more consistent linear strain triangle with six nodes is combined with a 12 degree-of-freedom bending triangle using shape function smoothing. This element has been introduced by Razzaque.[63]

Element D: This is a four-node quadrilateral with drilling degrees of freedom [Eq. (13.34) with $\Delta \bar{u}_t$ constrained to zero] combined with a discrete Kirchhoff quadrilateral.[55,64]

13.8 Practical examples

The first example given here is that for the solution of an arch dam shell. The simple geometrical configuration, shown in Fig. 13.7, was taken for this particular problem as results of model experiments and alternative numerical approaches were available.

A division based on rectangular elements (type A) was used as the simple cylindrical shape permitted this, although a rather crude approximation for the fixed foundation had to be used.

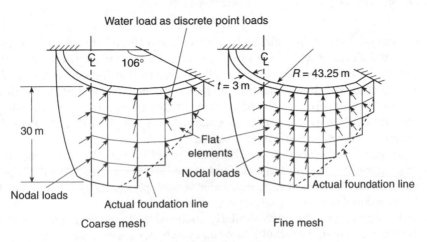

Fig. 13.7 An arch dam as an assembly of rectangular elements.

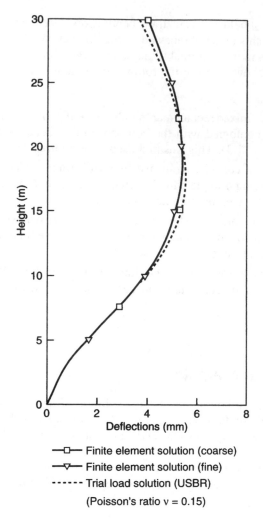

Fig. 13.8 Arch dam of Fig. 13.7: horizontal deflections on centre-line.

Two sizes of division into elements are used, and the results given in Figs 13.8 and 13.9 for deflections and stresses on the centre-line section show that little change occurred by the use of the finer mesh. This indicates that the convergence of both the physical approximation to the true shape by flat elements and of the mathematical approximation involved in the finite element formulation is more than adequate. For comparison, stresses and deflection obtained using the USBR trial load solution (another approximate method) are also shown.

A large number of examples have been computed by Parekh[62] using the triangular, non-conforming element (type B), and indeed show for equal division a general improvement over the conforming triangular version presented by Clough and Johnson.[7] Some examples of such analyses are now shown.

A doubly curved arch dam was similarly analysed using the triangular flat element (type B) representation. The results show an even better approximation.[8]

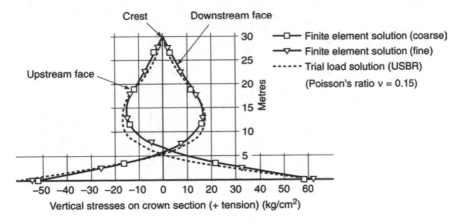

Fig. 13.9 Arch dam of Fig. 13.7: vertical stresses on centre-line.

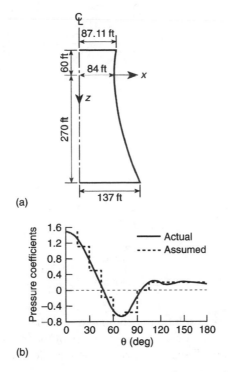

Fig. 13.10 Cooling tower: geometry and pressure load variation about circumference.

13.8.1 Cooling tower

This problem of a general axisymmetric shape could be more efficiently dealt with by the axisymmetric formulations to be presented in Chapters 11 and 12. However, here this example is used as a general illustration of the accuracy attainable. The answers

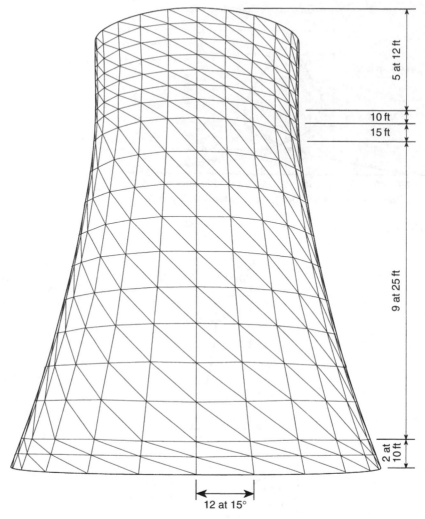

Fig. 13.11   Cooling tower of Fig. 13.10: mesh subdivisions.

against which the numerical solution is compared have been derived by Albasiny and Martin.[65] Figures 13.10 to 13.12 show the geometry of the mesh used and some results for a 5 inch and a 7 inch thick shell. Unsymmetric wind loading is used here.

13.8.2 Barrel vault

This typical shell used in many civil engineering applications is solved using analytical methods by Scordelis and Lo[66] and Scordelis.[67] The barrel is supported on rigid diaphragms and is loaded by its own weight. Figures 13.13 and 13.14 show some comparative answers, obtained by elements of type B, C and D. Elements of type C are obviously more accurate, involving more degrees of freedom, and with a mesh

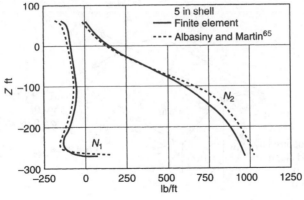

(a)

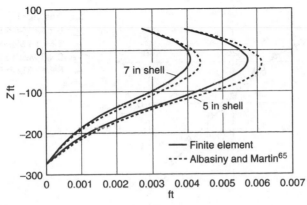

(b)

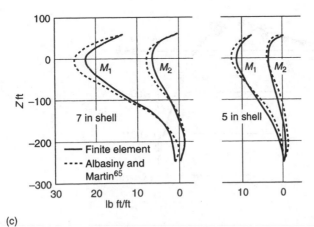

(c)

Fig. 13.12 Cooling tower of Fig. 13.10: (a) membrane forces at $\theta = 0°$; N_1, tangential; N_2, meridian; (b) radial displacements at $\theta = 0°$; (c) moments at $\theta = 0°$; M_1, tangential; M_2, meridian.

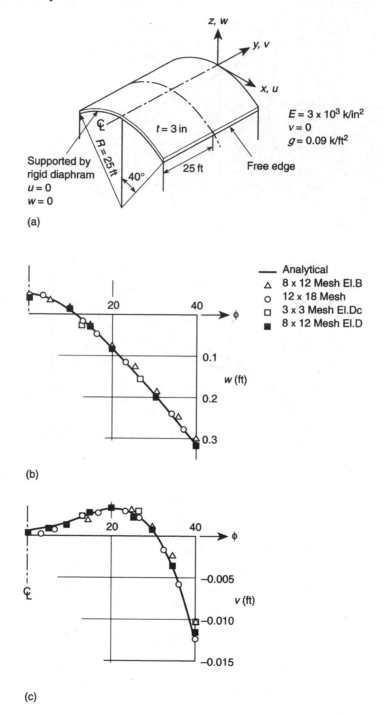

Fig. 13.13 Barrel (cylindrical) vault: flat element model results. (a) Barrel vault geometry and properties; (b) vertical displacement of centre section; (c) longitudinal displacement of support.

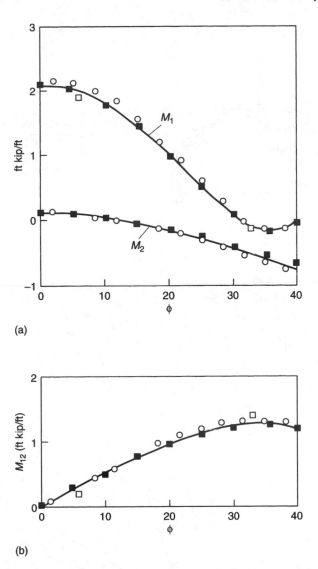

Fig. 13.14 Barrel vault of Fig. 13.13. (a) M_1, transverse; M_2, longitudinal; centre-line moments; (b) M_{12}, twisting moment at support.

of 6×6 elements the results are almost indistinguishable from analytical ones. This problem has become a classic on which various shell elements are compared and we shall return to it in Chapter 15. It is worthwhile remarking that only a few, second-order, curved elements give superior results to those presented here with a flat element approximation.

13.8.3 Folded plate structure

As no analytical solution of this problem is known, comparison is made with a set of experimental results obtained by Mark and Riesa.[68]

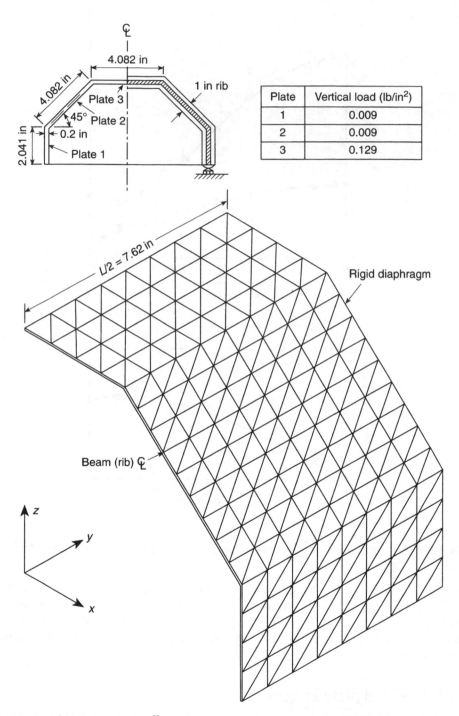

Plate	Vertical load (lb/in^2)
1	0.009
2	0.009
3	0.129

Fig. 13.15 A folded plate structure;[68] model geometry, loading and mesh, $E = 3560$ lb/in^2, $\nu = 0.43$.

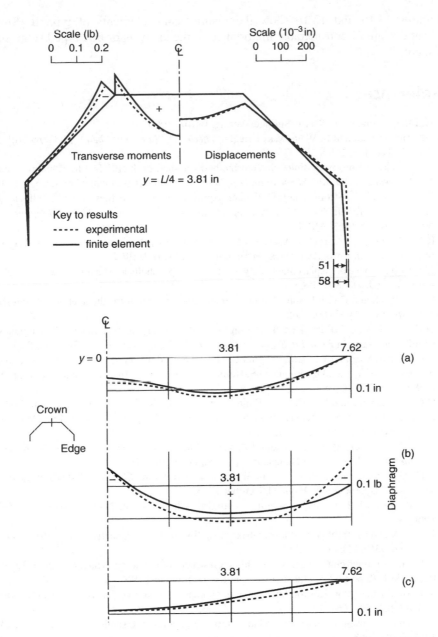

Scale (lb)
0   0.1   0.2

Scale (10^{-3} in)
0   100   200

Transverse moments | Displacements

$y = L/4 = 3.81$ in

Key to results
----- experimental
——— finite element

51
58

Crown

Edge

$y = 0$

3.81

7.62

0.1 in

(a)

3.81

0.1 lb

Diaphragm

(b)

3.81

7.62

0.1 in

(c)

Fig. 13.16 Folded plate of Fig. 13.15; moments and displacements on centre section. (a) Vertical displacements along the crown; (b) longitudinal moments along the crown; (c) horizontal displacements along edge.

This example presents a problem in which actual flat finite element representation is physically exact. Also a frame stiffness is included by suitable superposition of beam elements – thus illustrating also the versatility and ease by which different types of elements may be used in a single analysis.

Figures 13.15 and 13.16 show the results using elements of type B. Similar applications are of considerable importance in the analysis of box type bridge structures, etc.

References

1. W. Flügge. *Stresses in Shells*. Springer-Verlag, Berlin, 1960.
2. S.P. Timoshenko and S. Woinowski-Krieger. *Theory of Plates and Shells*. McGraw-Hill, New York, 2nd edition, 1959.
3. P.G. Ciarlet. Conforming finite element method for shell problem. In J.R. Whiteman, editor, *The Mathematics of Finite Elements and Application*, volume II. Academic Press, London, 1977.
4. B.E. Greene, D.R. Strome and R.C. Weikel. Application of the stiffness method to the analysis of shell structures. In *Proc. Aviation Conf. of American Society of Mechanical Engineers*, Los Angeles, March 1961. ASME.
5. R.W. Clough and J.L. Tocher. Analysis of thin arch dams by the finite element method. In *Proc. Int. Symp. on Theory of Arch Dams*. Pergamon Press, Oxford, 1965.
6. J.H. Argyris. Matrix displacement analysis of anisotropic shells by triangular elements. *J. Roy. Aero. Soc.*, 69:801–805, 1965.
7. R.W. Clough and C.P. Johnson. A finite element approximation for the analysis of thin shells. *J. Solids Struct.*, 4:43–60, 1968.
8. O.C. Zienkiewicz, C.J. Parekh and I.P. King. Arch dams analysed by a linear finite element shell solution program. In *Proc. Int. Symp. on Theory of Arch Dams*. Pergamon Press, Oxford, 1965.
9. O.C. Zienkiewicz and Y.K. Cheung. Finite element procedures in the solution of plate and shell problems. In O.C. Zienkiewicz and G.S. Holister, editors, *Stress Analysis*, Chapter 8. John Wiley & Sons, Chichester, 1965.
10. O.C. Zienkiewicz and Y.K. Cheung. Finite element methods of analysis for arch dam shells and comparison with finite difference procedures. In *Proc. Int. Symp. on Theory of Arch Dams*, pages 123–140. Pergamon Press, Oxford, 1965.
11. O.C. Zienkiewicz, R.L. Taylor and J.Z. Zhu. *The Finite Element Method: Its Basis and Fundamentals*. Butterworth-Heinemann, Oxford, 6th edition, 2005.
12. R.H. Gallagher. Shell elements. In *World Conf. on Finite Element Methods in Structural Mechanics*, Bournemouth, England, October 1975.
13. F.K. Bogner, R.L. Fox and L.A. Schmit. A cylindrical shell element. *Journal of AIAA*, 5:745–750, 1966.
14. J. Connor and C. Brebbia. Stiffness matrix for shallow rectangular shell element. *Proc. Am. Soc. Civ. Eng.*, 93(EM1):43–65, 1967.
15. A.J. Carr. A refined element analysis of thin shell structures including dynamic loading. Technical Report SEL Report 67–9, University of California, Berkeley, 1967.
16. S. Utku. Stiffness matrices for thin triangular elements of non-zero Gaussian curvature. *Journal of AIAA*, 5:1659–1667, 1967.
17. G. Cantin. Strain-displacement relationships for cylindrical shells. *Journal of AIAA*, 6:1787–1788, 1968.
18. G. Cantin and R.W. Clough. A curved, cylindrical shell finite element. *AIAA Journal*, 6(6):1057–1062, 1968.
19. G. Bonnew, G. Dhatt, Y.M. Giroux and L.P.A. Robichaud. Curved triangular elements for analysis of shells. In *Proc. 2nd Conf. Matrix Methods in Structural Mechanics*, volume AFFDL-TR-68-150, Wright Patterson Air Force Base, Ohio, October 1968.
20. G.E. Strickland and W.A. Loden. A doubly curved triangular shell element. In *Proc. 2nd Conf. Matrix Methods in Structural Mechanics*, volume AFFDL-TR-68-150, Wright Patterson Air Force Base, Ohio, October 1968.

21. B.E. Greene, R.E. Jones and D.R. Strome. Dynamic analysis of shells using doubly curved finite elements. In *Proc. 2nd Conf. Matrix Methods in Structural Mechanics*, volume AFFDL-TR-68-150, Wright Patterson Air Force Base, Ohio, October 1968.

22. S. Ahmad. *Curved finite elements in the analysis of solids, shells and plate structures*. PhD thesis, Department of Civil Engineering, University of Wales, Swansea, 1969.

23. G.R. Cowper, G.M. Lindberg and M.D. Olson. A shallow shell finite element of triangular shape. *International Journal of Solids and Structures*, 6:1133–1156, 1970.

24. G. Dupuis and J.J. Goël. A curved finite element for thin elastic shells. *International Journal of Solids and Structures*, 6:987–996, 1970.

25. S.W. Key and Z.E. Beisinger. The analysis of thin shells by the finite element method. In *High Speed Computing of Elastic Structures*, volume 1, pages 209–252. University of Liège Press, 1971.

26. D.J. Dawe. The analysis of thin shells using a facet element. Technical Report CEGB RD/B/N2038, Berkeley Nuclear Laboratory, England, 1971.

27. D.J. Dawe. Rigid-body motions and strain-displacement equations of curved shell finite elements. *Int. J. Mech. Sci.*, 14:569–578, 1972.

28. D.G. Ashwell and A. Sabir. A new cylindrical shell finite element based on simple independent strain functions. *Int. J. Mech. Sci.*, 4:37–47, 1973.

29. G.R. Thomas and R.H. Gallagher. A triangular thin shell finite element: linear analysis. Technical Report CR-2582, NASA, 1975.

30. B.M. Irons. The semi-Loof shell element. In D.G. Ashwell and R.H. Gallagher, editors, *Finite Elements for Thin Shells and Curved Members*, Chapter 11, pages 197–222. John Wiley & Sons, Chichester, 1976.

31. D.G. Ashwell. Strain elements with application to arches, rings, and cylindrical shells. In D.G. Ashwell and R.H. Gallagher, editors, *Finite Elements for Thin Shells and Curved Members*, Chapter 6, pages 91–111. John Wiley & Sons, Chichester, 1976.

32. N. Carpenter, H. Stolarski and T. Belytschko. A flat triangular shell element with improved membrane interpolation. *Communications in Applied Numerical Methods*, 1:161–168, 1985.

33. C. Pratt. Shell finite element via Reissner's principle. *International Journal of Solids and Structures*, 5:1119–1133, 1969.

34. J. Connor and G. Will. A mixed finite element shallow shell formulation. In R.H. Gallagher *et al.*, editors, *Advances in Matrix Methods of Structural Analysis and Design*, pages 105–137. University of Alabama Press, 1969.

35. L.R. Herrmann and W.E. Mason. Mixed formulations for finite element shell analysis. In *Conf. on Computer-Oriented Analysis of Shell Structures*, Paper AFFDL-TR-71-79, June 1971.

36. G. Edwards and J.J. Webster. Hybrid cylindrical shell elements. In D.G. Ashwell and R.H. Gallagher, editors, *Finite Elements for Thin Shells and Curved Members*, Chapter 10, pages 171–195. John Wiley & Sons, Chichester, 1976.

37. H. Stolarski and T. Belytschko. Membrane locking and reduced integration for curved elements. *J. Appl. Mech.*, 49:172–176, 1982.

38. Ph. Jetteur and F. Frey. A four node Marguerre element for non-linear shell analysis. *Engineering Computations*, 3:276–282, 1986.

39. J.C. Simo and D.D. Fox. On a stress resultant geometrically exact shell model. Part I: Formulation and optimal parametrization. *Computer Methods in Applied Mechanics and Engineering*, 72:267–304, 1989.

40. J.C. Simo, D.D. Fox and M.S. Rifai. On a stress resultant geometrically exact shell model. Part II: The linear theory; computational aspects. *Computer Methods in Applied Mechanics and Engineering*, 73:53–92, 1989.

41. M. Bischoff and E. Ramm. Solid-like shell or shell-like solid formulation? A personal view. In W. Wunderlich, editor, *Proc. Eur. Conf. on Comp. Mech. (ECCM'99 on CD-ROM)*, Munich, September 1999.

42. L.E. Malvern. *Introduction to the Mechanics of a Continuous Medium.* Prentice-Hall, Englewood Cliffs, NJ, 1969.

43. I.H. Shames and F.A. Cozzarelli. *Elastic and Inelastic Stress Analysis.* Taylor & Francis, Washington, DC, 1997. (Revised printing.)

44. S. Ahmad, B.M. Irons and O.C. Zienkiewicz. A simple matrix-vector handling scheme for three-dimensional and shell analysis. *International Journal for Numerical Methods in Engineering,* 2:509–522, 1970.

45. D.J. Allman. A compatible triangular element including vertex rotations for plane elasticity analysis. *Computers and Structures,* 19:1–8, 1984.

46. D.J. Allman. A quadrilateral finite element including vertex rotations for plane elasticity analysis. *International Journal for Numerical Methods in Engineering,* 26:717–730, 1988.

47. D.J. Allman. Evaluation of the constant strain triangle with drilling rotations. *International Journal for Numerical Methods in Engineering,* 26:2645–2655, 1988.

48. P.G. Bergan and C.A. Felippa. A triangular membrane element with rotational degrees of freedom. *Computer Methods in Applied Mechanics and Engineering,* 50:25–69, 1985.

49. P.G. Bergan and C.A. Felippa. Efficient implementation of a triangular membrane element with drilling freedoms. In T.J.R. Hughes and E. Hinton, editors, *Finite Element Methods for Plate and Shell Structures,* volume 1, pages 128–152. Pineridge Press, Swansea, 1986.

50. R.D. Cook. On the Allman triangle and a related quadrilateral element. *Computers and Structures,* 2:1065–1067, 1986.

51. R.D. Cook. A plane hybrid element with rotational d.o.f. and adjustable stiffness. *International Journal for Numerical Methods in Engineering,* 24:1499–1508, 1987.

52. T.J.R. Hughes, L.P. Franca and G.M. Hulbert. A new finite element formulation for computational fluid dynamics: VIII. The Galerkin/least-squares method for advective-diffusive equations. *Computer Methods in Applied Mechanics and Engineering,* 73:173–189, 1989.

53. R.L. Taylor and J.C. Simo. Bending and membrane elements for analysis of thick and thin shells. In G.N. Pande and J. Middleton, editors, *Proc. NUMETA 85 Conf.,* volume 1, pages 587–591. A.A. Balkema, Rotterdam, 1985.

54. Ph. Jetteur. Improvement of the quadrilateral JET shell element for a particular class of shell problems. Technical Report IREM 87/1, Ecole Polytechnique Federale de Lausanne, February 1987.

55. R.L. Taylor. Finite element analysis of linear shell problems. In J.R. Whiteman, editor, *The Mathematics of Finite Elements and Applications VI,* pages 191–203. Academic Press, London, 1988.

56. R.H. MacNeal and R.L Harter. A refined four-noded membrane element with rotational degrees of freedom. *Computers and Structures,* 28:75–88, 1988.

57. A. Ibrahimbegovic, R.L. Taylor and E.L. Wilson. A robust quadrilateral membrane finite element with drilling degrees of freedom. *International Journal for Numerical Methods in Engineering,* 30:445–457, 1990.

58. R.W. Clough and E.L. Wilson. Dynamic finite element analysis of arbitrary thin shells. *Computers and Structures,* 1:1971, 33–56.

59. R.H. MacNeal and R.L. Harter. A proposed standard set of problems to test finite element accuracy. *Journal of Finite Elements in Analysis and Design,* 1:3–20, 1985.

60. E. Reissner. A note on variational theorems in elasticity. *International Journal of Solids and Structures,* 1:93–95, 1965.

61. T.J.R. Hughes and F. Brezzi. On drilling degrees-of-freedom. *Computer Methods in Applied Mechanics and Engineering,* 72:105–121, 1989.

62. C.J. Parekh. *Finite element solution system.* PhD thesis, Department of Civil Engineering, University of Wales, Swansea, 1969.

63. A. Razzaque. *Finite element analysis of plates and shells.* PhD thesis, Civil Engineering Department, University of Wales, Swansea, 1972.

64. J.L. Batoz and M.B. Tahar. Evaluation of a new quadrilateral thin plate bending element. *International Journal for Numerical Methods in Engineering,* 18:1655–1677, 1982.

65. E.L. Albasiny and D.W. Martin. Bending and membrane equilibrium in cooling towers. *Proc. Am. Soc. Civ. Eng.*, 93(EM3):1–17, 1967.
66. A.C. Scordelis and K.S. Lo. Computer analysis of cylindrical shells. *J. Am. Concr. Inst.*, 61:539–561, 1964.
67. A.C. Scordelis. Analysis of cylindrical shells and folded plates. In *Concrete Thin Shells*, Report SP 28–N. American Concrete Institute, Farmington, Michigan, 1971.
68. R. Mark and J.D. Riesa. Photoelastic analysis of folded plate structures. *Proc. Am. Soc. Civ. Eng.*, 93(EM4):79–83, 1967.

14

Curved rods and axisymmetric shells

14.1 Introduction

The problem of axisymmetric shells is of sufficient practical importance to include in this chapter special methods dealing with their solution. While the general method described in the previous chapter is obviously applicable here, it will be found that considerable simplification can be achieved if account is taken of axial symmetry of the structure. In particular, if both the shell and the loading are axisymmetric it will be found that the elements become 'one dimensional'. This is the simplest type of element, to which little attention was given in earlier chapters.

The first approach to the finite element solution of axisymmetric shells was presented by Grafton and Strome.[1] In this, the elements are simple conical frustra and a direct approach via displacement functions is used. Refinements in the derivation of the element stiffness are presented in Popov *et al.*[2] and in Jones and Strome.[3] An extension to the case of unsymmetrical loads, which was suggested in Grafton and Strome, is elaborated in Percy *et al.*[4] and others.[5,6]

Later, much work was accomplished to extend the process to curved elements and indeed to refine the approximations involved. The literature on the subject is considerable, no doubt promoted by the interest in aerospace structures, and a complete bibliography is here impractical. References 7–15 show how curvilinear coordinates of various kinds can be introduced to the analysis, and references 9 and 14 discuss the use of additional nodeless degrees of freedom in improving accuracy. 'Mixed' formulations have found here some use.[16] Early work on the subject is reviewed comprehensively by Gallagher[17,18] and Stricklin.[19]

In axisymmetric shells, in common with all other shells, both bending and 'in-plane' or 'membrane' forces will occur. These will be specified uniquely in terms of the generalized 'strains', which now involve extensions and changes in curvatures of the middle surface. If the displacement of each point of the middle surface is specified, such 'strains' and the internal stress resultants, or simply 'stresses', can be determined by formulae available in standard references dealing with shell theory.[20–23]

14.2 Straight element

As a simple example of an axisymmetric shell subjected to axisymmetric loading we consider the case shown in Figs 14.1 and 14.2 in which the displacement of a point

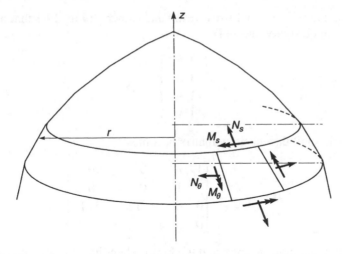

Fig. 14.1 Axisymmetric shell, loading, displacements, and stress resultants; shell represented as a stack of conical frustra.

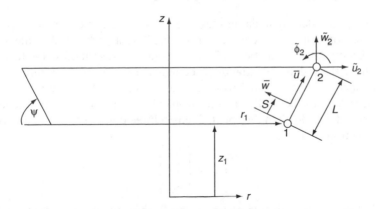

Fig. 14.2 An element of an axisymmetric shell.

on the middle surface of the meridian plane at an angle ψ measured positive from the x axis is uniquely determined by two components $\bar{u}$ and $\bar{w}$ in the tangential (s) and normal directions, respectively.

Using the Kirchhoff–Love assumption (which excludes transverse shear deformations) and assuming that the angle ψ does not vary (i.e. elements are straight), the four strain components are given by[20–23]

$$
\varepsilon = \begin{Bmatrix} \varepsilon_s \\ \varepsilon_\theta \\ \chi_s \\ \chi_\theta \end{Bmatrix} = \begin{Bmatrix} d\bar{u}/ds \\ [\bar{u}\cos\psi - \bar{w}\sin\psi]/r \\ -d^2\bar{w}/ds^2 \\ -(d\bar{w}/ds)\cos\psi/r \end{Bmatrix}
\tag{14.1}
$$

This results in the four internal stress resultants shown in Fig. 14.1 that are related to the strains by an elasticity matrix $\mathbf{D}$:

$$
\sigma = \begin{Bmatrix} N_s \\ N_\theta \\ M_s \\ M_\theta \end{Bmatrix} = \mathbf{D}\varepsilon \tag{14.2}
$$

For an isotropic shell the elasticity matrix becomes

$$
\mathbf{D} = \frac{Et}{1-\nu^2} \begin{bmatrix} 1 & \nu & 0 & 0 \\ \nu & 1 & 0 & 0 \\ 0 & 0 & t^2/12 & \nu t^2/12 \\ 0 & 0 & \nu t^2/12 & t^2/12 \end{bmatrix} \tag{14.3}
$$

the upper part being a plane stress and the lower a bending stiffness matrix with shear terms omitted as 'thin' conditions are assumed.

14.2.1 Element characteristics – axisymmetrical loads

Let the shell be divided by nodal circles into a series of conical frustra, as shown in Fig. 14.2. The nodal displacements at points 1 and 2 for a typical 1–2 element will have to define uniquely the deformations of the element via prescribed shape functions.

At each node the radial and axial displacements, u and w, and a rotation, ϕ, will be used as parameters. From virtual work by edge forces we find that all three components are necessary as the shell can carry in-plane forces and bending moments. The displacements of a node i can thus be defined by three components, the first two being in global directions r and z,

$$
\tilde{\mathbf{u}}_a = \begin{Bmatrix} \tilde{u}_a \\ \tilde{w}_a \\ \tilde{\phi}_a \end{Bmatrix} \tag{14.4}
$$

The simplest elements with two nodes, 1 and 2, thus possess 6 degrees of freedom, determined by the element displacements

$$
\tilde{\mathbf{u}}^e = \begin{Bmatrix} \tilde{\mathbf{u}}_1 \\ \tilde{\mathbf{u}}_2 \end{Bmatrix} \tag{14.5}
$$

The displacements within the element have to be uniquely determined by the nodal displacements $\mathbf{u}^e$ and the position s (as shown in Fig. 14.2) and maintain slope and displacement continuity.

Thus in local (s) coordinates we have

$$
\bar{\mathbf{u}} = \begin{Bmatrix} \bar{u} \\ \bar{w} \end{Bmatrix} = \mathbf{N}(s)\tilde{\mathbf{u}}^e \tag{14.6}
$$

Based on the strain–displacement relations (14.1) we observe that $\bar{u}$ can be of C_0 type while $\bar{w}$ must be of type C_1. The simplest approximation takes $\bar{u}$ varying linearly with

s and $\bar{w}$ as cubic in s. We shall then have six undetermined constants which can be determined from nodal values of u, w and ϕ.

At the node a,

$$\left\{ \begin{array}{c} \bar{u}_a \\ \bar{w}_a \\ (d\bar{w}/ds)_a \end{array} \right\} = \left[\begin{array}{ccc} \cos\psi & \sin\psi & 0 \\ -\sin\psi & \cos\psi & 0 \\ 0 & 0 & 1 \end{array} \right] \left\{ \begin{array}{c} \tilde{u}_a \\ \tilde{w}_a \\ \tilde{\phi}_a \end{array} \right\} = \mathbf{T}\tilde{\mathbf{u}}_a \tag{14.7}$$

Introducing the interpolations

$$\bar{u} = N_1(\xi)\bar{u}_1 + N_2(\xi)\bar{u}_2 \tag{14.8}$$

$$\bar{w} = H_1^{(0)}(\xi)\bar{w}_1 + H_2^{(0)}(\xi)\bar{w}_2 + \frac{L}{2}\left[H_1^{(1)}(\xi)\frac{d\bar{w}}{ds}\bigg|_1 + H_2^{(1)}(\xi)\frac{d\bar{w}}{ds}\bigg|_2 \right] \tag{14.9}$$

where N_a are the usual linear interpolations in ξ $(-1 \le \xi \le 1)$

$$N_1 = \tfrac{1}{2}(1 - \xi) \quad \text{and} \quad N_2 = \tfrac{1}{2}(1 + \xi)$$

and $H_a^{(0)}$ and $H_a^{(1)}$ are the Hermitian interpolations defined in Sec. 10.4 and repeated here for completeness

$$H_1^{(0)} = \frac{1}{4}\left(2 - 3\xi + \xi^3\right) \quad \text{and} \quad H_2^{(0)} = \frac{1}{4}\left(2 + 3\xi - \xi^3\right)$$

and

$$H_1^{(1)} = \tfrac{1}{4}\left(1 - \xi - \xi^2 + \xi^3\right) \quad \text{and} \quad H_2^{(1)} = \tfrac{1}{4}\left(-1 - \xi + \xi^2 + \xi^3\right)$$

in which, placing the origin of the meridian coordinate s at the 1 node,

$$s = N_2(\xi)L = \frac{1}{2}(1 + \xi)L$$

The global coordinates for the conical frustrum may also be expressed by using the N_a interpolations as

$$r = N_1(\xi)\tilde{r}_1 + N_2(\xi)\tilde{r}_2$$
$$z = N_1(\xi)\tilde{z}_1 + N_2(\xi)\tilde{z}_2 \tag{14.10}$$

and used to compute the length L as

$$L = \sqrt{(\tilde{r}_2 - \tilde{r}_1)^2 + (\tilde{z}_2 - \tilde{z}_1)^2}$$

Writing the interpolations as

$$\bar{\mathbf{u}} = \left[\begin{array}{ccc} N_a & 0 & 0 \\ 0 & H_a^{(0)} & H_a^{(1)} \end{array} \right] \left\{ \begin{array}{c} \bar{u}_a \\ \bar{w}_a \\ (d\bar{w}/ds)_a \end{array} \right\} = \bar{\mathbf{N}}_a\bar{\mathbf{u}}_a \tag{14.11}$$

we can now write the global interpolation as

$$\mathbf{u} = \begin{Bmatrix} u \\ w \end{Bmatrix} = \begin{bmatrix} \bar{\mathbf{N}}_1\mathbf{T} & \bar{\mathbf{N}}_2\mathbf{T} \end{bmatrix} \tilde{\mathbf{u}}^e = \mathbf{N}\tilde{\mathbf{u}}^e \tag{14.12}$$

From Eq. (14.12) it is a simple matter to obtain the strain matrix $\mathbf{B}$ by use of the definition (14.1). This gives

$$\varepsilon \equiv \mathbf{B}\tilde{\mathbf{u}}^e = \begin{bmatrix} \bar{\mathbf{B}}_1\mathbf{T} & \bar{\mathbf{B}}_2\mathbf{T} \end{bmatrix} \tilde{\mathbf{u}}^e \tag{14.13}$$

in which, noting from Eq. (14.7) that $u = \cos\psi\bar{u} - \sin\psi\bar{w}$, we have

$$\bar{\mathbf{B}}_a = \begin{bmatrix} \mathrm{d}N_a/\mathrm{d}s & 0 & 0 \\ N_a\cos\psi/r & -H_a^{(0)}\sin\psi/r & -H_a^{(1)}\sin\psi/r \\ 0 & -\mathrm{d}^2H_a^{(0)}/\mathrm{d}s^2 & -\mathrm{d}^2H_a^{(1)}/\mathrm{d}s^2 \\ 0 & -(\mathrm{d}H_a^{(0)}/\mathrm{d}s)\cos\psi/r & -(\mathrm{d}H_a^{(1)}/\mathrm{d}s)\cos\psi/r \end{bmatrix} \tag{14.14}$$

Derivatives are evaluated by using

$$\frac{\mathrm{d}N_a}{\mathrm{d}s} = \frac{2}{L}\frac{\mathrm{d}N_a}{\mathrm{d}\xi} \quad \text{and} \quad \frac{\mathrm{d}^2N_a}{\mathrm{d}s^2} = \frac{4}{L^2}\frac{\mathrm{d}^2N_a}{\mathrm{d}\xi^2}$$

with similar expressions for $H_a^{(0)}$ and $H_a^{(1)}$. Now all the 'ingredients' required for computing the stiffness matrix (or load, stress, and initial stress matrices) by standard formulae are known. The integrations required are carried out over the area, A, of each element, that is, with

$$\mathrm{d}A = 2\pi r\,\mathrm{d}s = \pi r L\,\mathrm{d}\xi \tag{14.15}$$

with ξ varying from -1 to 1.

Thus, the stiffness matrix $\mathbf{K}$ becomes, in local coordinates,

$$\bar{\mathbf{K}}_{ab} = \pi L \int_{-1}^{1} \bar{\mathbf{B}}_a^T \mathbf{D}\bar{\mathbf{B}}_b r\,\mathrm{d}\xi \tag{14.16}$$

On transformation, the stiffness $\mathbf{K}_{ab}$ of the global matrix is given by

$$\mathbf{K}_{ab} = \mathbf{T}^T\bar{\mathbf{K}}_{ab}\mathbf{T} \tag{14.17}$$

Once again it is convenient to evaluate the integrals numerically and the form above is written for Gaussian quadrature.[24] Grafton and Strome[1] give an explicit formula for the stiffness matrix based on a single average value of the integrand (one-point Gaussian quadrature) and using a $\mathbf{D}$ matrix corresponding to an orthotropic material. Percy et al.[4] and Klein[5] used a seven-point numerical integration; however, it is generally recommended to use only two points to obtain all arrays (especially if inertia forces are added, since one point then would yield a rank deficient mass matrix).

It should be remembered that if any external line loads or moments are present, their full circumferential value must be used in the analysis, just as was the case with axisymmetric solids discussed in Chapter 2.

14.2.2 Additional enhanced mode

A slight improvement to the above element may be achieved by adding an *enhanced strain* mode to the ε_s component. Here this is achieved by following the procedures outlined in Chapter 10 and in reference 24, and we can observe that the necessary condition not to affect a constant value of N_s is given by

$$2\pi \int_L \varepsilon_s^{en} r \, ds = \pi L \int_{-1}^{1} \varepsilon_s^{en} r \, d\xi = 0 \tag{14.18}$$

where ε_s^{en} denotes the enhanced strain component. A simple mode may thus be defined as

$$\varepsilon_s^{en} = \frac{\xi}{r} \tilde{\alpha}_{en} = B_{en} \tilde{\alpha}_{en} \tag{14.19}$$

in which $\tilde{\alpha}_{en}$ is a parameter to be determined. For the linear elastic case considered above the mode may be determined from

$$\begin{bmatrix} K_{en} & G^T \\ G & K \end{bmatrix} \begin{Bmatrix} \tilde{\alpha}_{en} \\ \tilde{u}^e \end{Bmatrix} = \begin{Bmatrix} 0 \\ f \end{Bmatrix} \tag{14.20}$$

where

$$K_{en} = 2\pi L \int_{-1}^{1} B_{en} D_{11} B_{en} \, r \, d\xi$$

$$G_a = 2\pi L \int_{-1}^{1} B_{en} DB_a \, r \, d\xi \tag{14.21}$$

Now a partial solution may be performed by means of static condensation[25] to obtain the stiffness for assembly

$$\bar{K} = K - G^T K_{en}^{-1} G \tag{14.22}$$

The effect of the added mode is most apparent in the force resultant N_s where solution oscillations are greatly reduced. This improvement is not needed for the purely elastic case but is more effective when the material properties are inelastic and the oscillations can cause errors in behaviour, such as erratic yielding in elasto-plastic solutions.

14.2.3 Examples and accuracy

In the treatment of axisymmetric shells described here, continuity between the shell elements is satisfied at all times. For an axisymmetric shell of polygonal meridian shape, therefore, convergence will always occur.

The problem of the physical approximation to a curved shell by a polygonal shape is similar to the one discussed in Chapter 13. Intuitively, convergence can be expected, and indeed numerous examples indicate this.

When the loading is such as to cause predominantly membrane stresses, discrepancies in bending moment values exist (even with reasonably fine subdivision). Again, however, these disappear as the size of the subdivisions decreases, particularly if correct sampling is used (see Chapter 10). This is necessary to eliminate the physical approximation involved in representing the shell as a series of conical frustra.

Figures 14.3 and 14.4 illustrate some typical examples taken from the Grafton and Strome paper which show quite remarkable accuracy. In each problem it should be noted that small elements are needed near free edges to capture the 'boundary layer' nature of shell solutions.

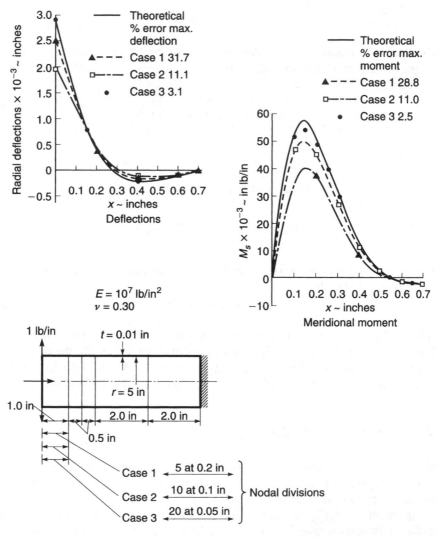

Fig. 14.3 A cylindrical shell solution by finite elements, from Grafton and Strome.[1]

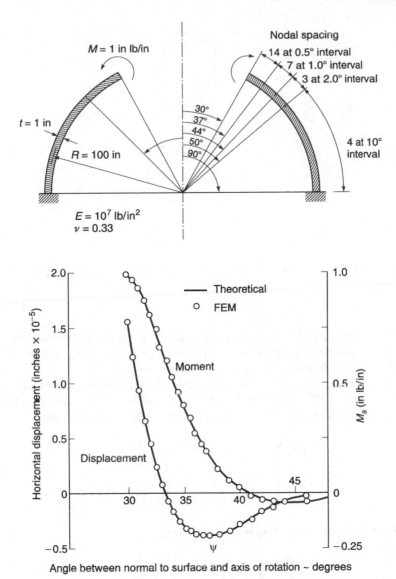

Fig. 14.4 A hemispherical shell solution by finite elements, from Grafton and Strome.[1]

14.3 Curved elements

Use of curved elements has already been described in the context of analyses that involved only first derivatives in the definition of strain. Here second derivatives exist [see Eq. (14.1)] and require additional effort to compute the strains.

It was previously mentioned that many possible definitions of curved elements have been proposed and used in the context of axisymmetric shells. The derivation used here is one due to Delpak[14] and is of the subparametric type.[24]

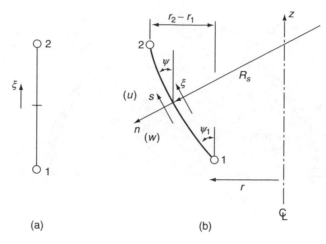

Fig. 14.5 Curved, isoparametric, shell element for axisymmetric problems: (a) parent element; (b) curvilinear coordinates.

The basis of curved element definition is one that gives a common tangent between adjacent elements (or alternatively, a specified tangent direction). This is physically necessary to avoid 'kinks' in the description of a smooth shell.

If a general curved form of a shell of revolution is considered, as shown in Fig. 14.5, the expressions for strain quoted in Eq. (14.1) have to be modified to take into account the curvature of the shell in the meridian plane.[20,21] These now become

$$
\varepsilon =
\begin{Bmatrix}
\varepsilon_s \\
\varepsilon_\theta \\
\chi_s \\
\chi_\theta
\end{Bmatrix}
=
\begin{Bmatrix}
d\bar{u}/ds + \bar{w}/R_s \\
[\bar{u}\cos\psi - \bar{w}\sin\psi]/r \\
-d^2\bar{w}/ds^2 - d(\bar{u}/R_s)/ds \\
-[(d\bar{w}/ds + \bar{u}/R_s)]\cos\psi/r
\end{Bmatrix}
\tag{14.23}
$$

In the above the angle ψ is a function of s, that is,

$$
\frac{dr}{ds} = \cos\psi \quad \text{and} \quad \frac{dz}{ds} = \sin\psi
$$

R_s is the principal radius in the meridian plane, and the second principal curvature radius R_θ is given by

$$
r = R_\theta \sin\psi
$$

The reader can verify that for $R_s = \infty$ Eq. (14.23) coincides with Eq. (14.1).

14.3.1 Shape functions for a curved element

We shall now consider the 1–2 element to be curved as shown in Fig. 14.5(b), where the coordinate is in 'parent' form ($-1 \le \xi \le 1$) as shown in Fig. 14.5(a). The coordinates and the unknowns are 'mapped' in the manner of Chapter 2. As we wish to interpolate a quantity with slope continuity we can write for a typical function $g(\xi)$

$$g(\xi) = \sum_{a=1}^{2} \left[H_a^{(0)}(\xi) g_a + H_a^{(1)} \frac{dg}{d\xi} \Big|_a \right] = \mathbf{N}\tilde{\mathbf{g}} \tag{14.24}$$

where again the Hermitian interpolations have been used. We can now simultaneously use these functions to describe variations of the global displacements u and w as[*]

$$u = \sum_{a=1}^{2} \left[H_a^{(0)}(\xi) \tilde{u}_a + H_a^{(1)} \frac{d\tilde{u}}{d\xi} \Big|_a \right]$$

$$w = \sum_{a=1}^{2} \left[H_a^{(0)}(\xi) \tilde{w}_a + H_a^{(1)} \frac{d\tilde{w}}{d\xi} \Big|_a \right] \tag{14.25}$$

and of the coordinates r and z which define the shell (mid-surface). Indeed, if the thickness of the element is also variable the same interpolation could be applied to it. Such an element would then be isoparametric. Accordingly, we can define the geometry as

$$r = \sum_{a=1}^{2} \left[H_a^{(0)}(\xi) \tilde{r}_a + H_a^{(1)} \frac{d\tilde{r}}{d\xi} \Big|_a \right]$$

$$z = \sum_{a=1}^{2} \left[H_a^{(0)}(\xi) \tilde{z}_a + H_a^{(1)} \frac{d\tilde{z}}{d\xi} \Big|_a \right] \tag{14.26}$$

and, provided the nodal values in the above can be specified, a one-to-one relation between ξ and the position on the curved element surface is defined [Fig. 14.5(b)].

While specification of $\tilde{r}_a$ and $\tilde{z}_a$ is obvious, at the ends only the slope

$$\cot \psi_a = -\frac{dr}{dz} \Big|_a \tag{14.27}$$

is defined. The specification to be adopted with regard to the derivatives occurring in Eq. (14.26) depends on the *scaling* of ξ along the tangent length s. Only the ratio

$$\frac{dr}{dz} \Big|_a = \frac{(dr/d\xi)_a}{(dz/d\xi)_a} \tag{14.28}$$

is unambiguously specified. Thus $(dr/d\xi)_a$ or $(dz/d\xi)_a$ can be given an arbitrary value. Here, however, practical considerations intervene as with the wrong choice a very uneven relationship between s and ξ will occur. Indeed, with an unsuitable choice the shape of the curve can depart from the smooth one illustrated and loop between the end values.

---

[*] One immediate difference will be observed from that of the previous formulation. Now both displacement components vary in a cubic manner along an element while previously a linear variation of the tangential displacement was permitted. This additional degree of freedom does not, however, introduce excessive constraints provided the shell thickness is itself continuous.

To achieve a reasonably uniform spacing it suffices for well-behaved surfaces to approximate

$$\frac{dr}{d\xi} \approx \frac{\Delta r}{\Delta \xi} = \frac{\tilde{r}_2 - \tilde{r}_1}{2} \quad \text{or} \quad \frac{dz}{d\xi} \approx \frac{\Delta z}{\Delta \xi} = \frac{\tilde{z}_2 - \tilde{z}_1}{2} \tag{14.29}$$

using whichever is largest and noting that the whole range of ξ is 2 between the nodal points.

14.3.2 Strain expressions and properties of curved elements

The variation of global displacements are specified by Eq. (14.25) while the strains are described in locally directed displacements in Eq. (14.23). Some transformations are therefore necessary before the strains can be determined.

We can express the locally directed displacements $\bar{u}$ and $\bar{w}$ in terms of the global displacements by using Eq. (14.7), that is,

$$\left\{ \begin{matrix} \bar{u} \\ \bar{w} \end{matrix} \right\} = \begin{bmatrix} \cos\psi & \sin\psi \\ -\sin\psi & \cos\psi \end{bmatrix} \left\{ \begin{matrix} u \\ w \end{matrix} \right\} = \bar{\mathbf{T}}\mathbf{u} \tag{14.30}$$

where ψ is the angle of the tangent to the curve and the r axis (Fig. 14.5). We note that this transformation may be expressed in terms of the ξ coordinate using Eqs (14.27) and (14.28) and the interpolations for r and z. With this transformation the continuity of displacement between adjacent elements is achieved by matching the global nodal displacements $\tilde{u}_a$ and $\tilde{w}_a$. However, in the development for the conical element we have specified continuity of *rotation* of the cross-section only. Here we shall allow usually the continuity of both s derivatives in displacements. Thus, the parameters du/ds and dw/ds will be given common values at nodes. As

$$\frac{du}{ds}\frac{ds}{d\xi} = \frac{du}{d\xi} \quad \text{and} \quad \frac{dw}{ds}\frac{ds}{d\xi} = \frac{dw}{d\xi} \tag{14.31}$$

where

$$\frac{ds}{d\xi} = \sqrt{\left(\frac{dr}{d\xi}\right)^2 + \left(\frac{dz}{d\xi}\right)^2}$$

No difficulty exists in substituting these new variables in Eqs (14.25) and (14.30) which now take the form

$$\bar{\mathbf{u}} = \mathbf{N}(\xi)\tilde{\mathbf{u}}^e \quad \text{with} \quad \tilde{\mathbf{u}}_a = \begin{bmatrix} \tilde{u}_a & \tilde{w}_a & (d\tilde{u}/ds)_a & (dw/ds)_a \end{bmatrix}^T \tag{14.32}$$

The form of the 2×4 shape function submatrices $\mathbf{N}_a$ can now be explicitly determined by using the above transformations in Eq. (14.25).[14] We note that the meridian radius of curvature R_s can be calculated explicitly from the mapped, parametric, form of the element by using

$$R_s = \frac{\left[(dr/d\xi)^2 + (dz/d\xi)^2\right]^{3/2}}{(dr/d\xi)(d^2z/d\xi^2) - (dz/d\xi)(d^2r/d\xi^2)} \tag{14.33}$$

in which all the derivatives are directly determined from expression (14.26).

If shells that branch or in which abrupt thickness changes occur are to be treated, the nodal parameters specified in Eq. (14.32) are not satisfactory. It is better to rewrite these as

$$\tilde{\mathbf{u}}_a = \begin{bmatrix} \tilde{u}_a & \tilde{w}_a & \tilde{\phi}_a & (d\tilde{u}/ds)_a \end{bmatrix}^T \tag{14.34}$$

where $\tilde{\phi}_a$, equal to $(d\tilde{w}/ds)_a$, is the nodal rotation, and to connect only the first three parameters. The fourth is now an unconnected element parameter with respect to which, however, the usual treatment is still carried out. Transformations needed in the above are implied in Eq. (14.7).

In the derivation of the **B** matrix expressions which define the strains, both first and second derivatives with respect to s occur, as seen in the definition of Eq. (14.23). If we observe that the derivatives can be obtained by the simple (chain) rules already implied in Eq. (14.31), for any function F we can write

$$\frac{dF}{d\xi} = \frac{dF}{ds}\frac{ds}{d\xi} \qquad \frac{d^2F}{d\xi^2} = \frac{d^2F}{ds^2}\left(\frac{ds}{d\xi}\right)^2 + \frac{dF}{ds}\left(\frac{d^2s}{d\xi^2}\right) \tag{14.35}$$

and all the expressions of **B** can be found.

Finally, the stiffness matrix is obtained in a similar way as in Eq. (14.16), changing the variable

$$ds = \frac{ds}{d\xi}\,d\xi \tag{14.36}$$

and integrating ξ within the limits -1 and $+1$. Once again the quantities contained in the integral expressions prohibit exact integration, and numerical quadrature must be used. As this is carried out in one coordinate only it is not very time consuming and an adequate number of Gauss points can be used to determine the stiffness (generally three points suffice). Initial stress and other load matrices are similarly obtained.

The particular isoparametric formulation presented in summary form here differs somewhat from the alternatives of references 15, 7, 8 and 13 and has the advantage that, because of its *isoparametric* form, rigid body displacement modes and indeed the states of constant first derivatives are available. Proof of this is similar to that contained in Sec. 5.5 of reference 24. The fact that the forms given in the alternative formulations have strain under rigid body nodal displacements may not be serious in some applications, as discussed by Haisler and Stricklin.[26] However, in some modes of non-axisymmetric loads (see Chapter 16) this incompleteness may be a serious drawback and may indeed lead to very wrong results.

Constant states of curvature cannot be obtained for a *finite* element of any kind described here and indeed are not physically possible. When the size of the element decreases it will be found that such arbitrary constant curvature states are available in the limit.

14.3.3 Additional nodeless variables

As in the straight frustrum element, addition of nodeless (enhanced) variables in the analysis of axisymmetric shells is particularly valuable when large curved elements are capable of reproducing with good accuracy the geometric shapes. Thus an addition of

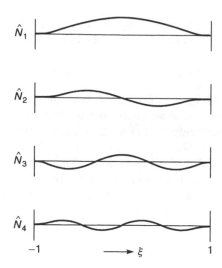

Fig. 14.6 Internal shape functions for a linear element.

a set of internal, hierarchical, element variables

$$\sum_{b=1}^{n} \hat{N}_b \Delta \tilde{u}_b \tag{14.37}$$

to the definition of the normal displacement defined in Eq. (14.6) or Eq. (14.25), in which $\Delta \tilde{u}_b$ is a set of internal parameters and $\hat{N}_b$ is a set of functions having zero values and zero first derivatives at the nodal points, allows considerable improvement in representation of the displacements to be achieved without violating any of the convergence requirements. For tangential displacements the requirement of zero first derivatives at nodes could be omitted. Webster also uses such additional functions in the context of straight elements.[9] In transient situations where these modes affect the mass matrix one can also use these functions as a basis for developing *enhanced strain modes* (see Sec. 14.2) since these by definition do not influence the assumed displacement field and, hence, the mass and surface loading terms.

Whether the element is in fact straight or curved does not matter and indeed we can supplement the definitions of displacements contained in Eq. (14.25) by Eq. (14.37) for each of the components. If this is done only in the displacement definition and *not* in the coordinate definition [Eq. (14.26)] the element now becomes of the category of subparametric.[*] As proved in Chapter 5 of reference 24, the same advantages are retained as in isoparametric forms.

The question as to the expression to be used for additional, internal shape functions is of some importance though the choice is wide. While it is no longer necessary to use polynomial representation, Delpak does so and uses a special form of Legendre polynomial (hierarchical functions). The general shapes are shown in Fig. 14.6.

---

[*] While it would obviously be possible to include the new shape function in the element coordinate definition, little practical advantage would be gained as a cubic represents realistic shapes adequately. Further, the development would then require 'fitting' the $\Delta \tilde{r}_b$ and $\Delta \tilde{z}_b$ for coordinates to the shape, complicating even further the development of derivatives.

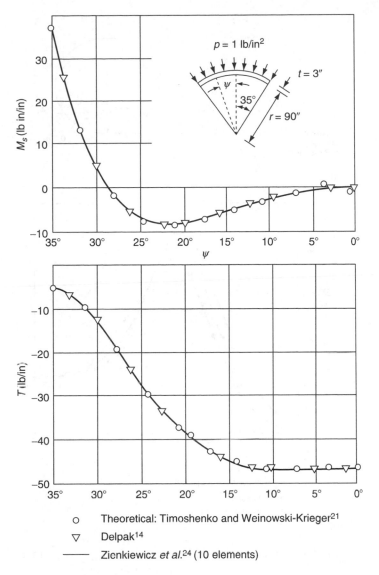

Fig. 14.7 Spherical dome under uniform pressure.

A series of examples shown in Figs 14.7–14.9 illustrate the applications of the isoparametric curvilinear element of the previous section with additional internal parameters.

In Fig. 14.7 a spherical dome with clamped edges is analysed and compared with analytical results of reference 21. Figures 14.8 and 14.9 show, respectively, more complex examples. In the first a torus analysis is made and compared with alternative finite element and analytical results.[12,13,27–29] The second case is one where branching occurs, and here alternative analytical results are given by Kraus.[30]

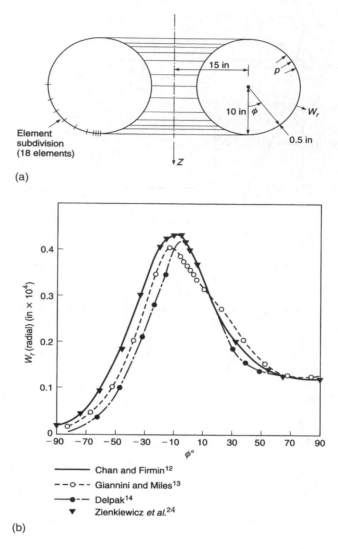

Fig. 14.8 Toroidal shell under internal pressure: (a) element subdivision; (b) radial displacements.

14.4 Independent slope–displacement interpolation with penalty functions (thick or thin shell formulations)

In Chapters 10 to 12 we discussed the use of independent slope and displacement interpolation in the context of beams and plates, respectively. Continuity was assured by the introduction of the shear force as an independent *mixed* variable which was defined within each element. The elimination of the shear variable led to a penalty type formulation in which the shear rigidity played the role of the penalty parameter. The

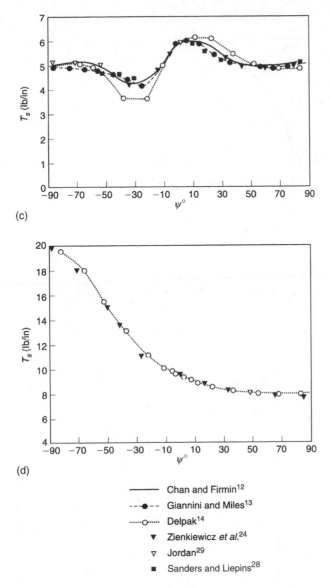

(c)

(d)

——— Chan and Firmin[12]

---●--- Giannini and Miles[13]

······○······ Delpak[14]

▼   Zienkiewicz *et al.*[24]

▽   Jordan[29]

■   Sanders and Liepins[28]

Fig. 14.8 *Cont.*   (c) in-plane stress resultants; (d) in-plane stress resultants.

equivalence of the number of parameters used in defining the shear variation and the number of integration points used in evaluating the penalty terms was demonstrated there in special cases, and this justified the success of *reduced* integration methods. This equivalence is not exact in the case of the axisymmetric problem in which the radius, r, enters the integrals, and hence slightly different results can be expected from the use of the mixed form and simple use of reduced integration. The differences become greatest near the axis of rotation and disappear completely when $r \rightarrow \infty$

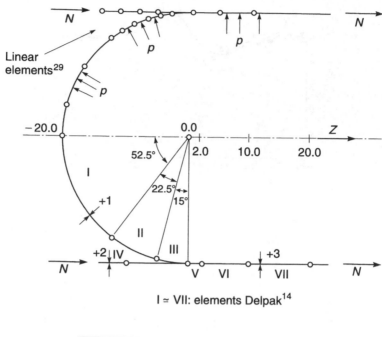

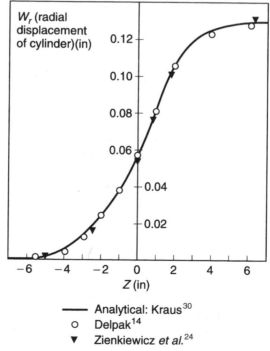

Fig. 14.9 Branching shell.

where the axisymmetric form results in an equivalent beam (or cylindrical bending plate) element.

Although in general the use of the mixed form yields a superior result, for simplicity we shall here derive only the reduced integration form, leaving the former to the reader as an exercise accomplished following the rules of Chapter 10.

In what follows we shall develop in detail the simplest possible element of this class. This is a direct descendant of the linear beam and plate elements.[27,31] (We note, however, that the plate element formulated in this way has singular modes and can on occasion give completely erroneous results; no such deficiency is present in the beam or the axisymmetric shell.)

Consider the strain expressions of Eq. (14.1) for a straight element. When using these the need for C_1 continuity was implied by the second derivative of w existing there. If now we use

$$\frac{d\bar{w}}{ds} = -\phi \tag{14.38}$$

the strain expression becomes

$$\varepsilon = \begin{Bmatrix} \varepsilon_s \\ \varepsilon_\theta \\ \chi_s \\ \chi_\theta \end{Bmatrix} = \begin{Bmatrix} d\bar{u}/ds \\ [\bar{u}\cos\psi - \bar{w}\sin\psi]/r \\ d\phi/ds \\ \phi\cos\psi/r \end{Bmatrix} \tag{14.39}$$

As ϕ can vary independently, a constraint has to be imposed:

$$C(\bar{w}, \phi) \equiv \frac{d\bar{w}}{ds} + \phi = 0 \tag{14.40}$$

This can be done by using the energy functional with a penalty multiplier α. We can thus write

$$\Pi = \pi \int_L \varepsilon^T \mathbf{D}\varepsilon\, r\, ds + \pi \int_L \alpha \left(\frac{dw}{ds} + \phi\right)^2 r\, ds + \Pi_{\text{ext}} \tag{14.41}$$

where Π_{ext} is a potential for boundary and loading terms and ε and $\mathbf{D}$ are defined as in Eq. (14.3). Immediately, α can be identified as the shear rigidity:

$$\alpha = \kappa Gt \quad \text{where for a homogeneous shell } \kappa = 5/6 \tag{14.42}$$

The penalty functional (14.41) can be identified on purely physical grounds. Washizu[22] quotes this on pages 199–201, and the general theory indeed follows that earlier suggested by Naghdi[23] for shells with shear deformation.

With first derivatives occurring in the energy expression only C_0 continuity is now required for the interpolation of u, w, and ϕ, and in place of Eqs (14.6)–(14.12) we can write directly

$$\bar{\mathbf{u}} = \begin{Bmatrix} \bar{u} \\ \bar{w} \\ \phi \end{Bmatrix} = \sum_{a=1}^{2} N_a(\xi)\mathbf{T}\tilde{\mathbf{u}}_a \tag{14.43}$$

$$\mathbf{u}_a^T = \begin{bmatrix} \tilde{u}_a & \tilde{w}_a & \tilde{\phi}_a \end{bmatrix}$$

where for $N_a(\xi)$ we can use any one-dimensional C_0 interpolation.[24] Once again, an isoparametric transformation could be used for curvilinear elements with strains defined by Eq. (14.23), and a formulation that we shall discuss in Chapter 15 is but an alternative to this process. If linear elements are used, we can write the expression without consequent use of isoparametric transformation. Indeed, we can replace the interpolations in Eq. (14.9) and now simply use

$$u = N_1(\xi)\tilde{u}_1 + N_2(\xi)\tilde{u}_2$$
$$w = N_1(\xi)\tilde{w}_1 + N_2(\xi)\tilde{w}_2 \qquad (14.44)$$
$$\phi = N_1(\xi)\tilde{\phi}_1 + N_2(\xi)\tilde{\phi}_2$$

and evaluate the integrals arising from expression (14.41) at one Gauss point, which is sufficient to maintain convergence and yet here does not give a singularity.

This extremely simple form will, of course, give very poor results with exact integration, even for thick shells, but now with reduced integration shows excellent performance. In Figs 14.7–14.9 we superpose results obtained with this simple, straight element, and the results speak for themselves.

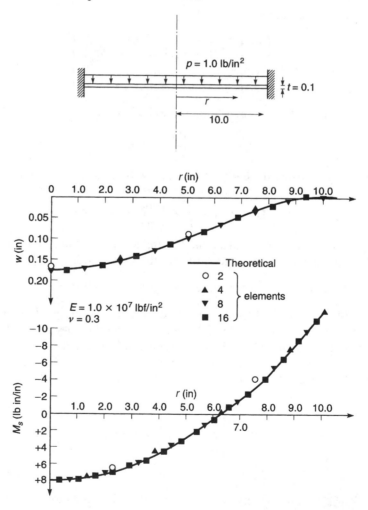

Fig. 14.10 Bending of a circular plate under uniform load; convergence study.

For other examples the reader can consult reference 27, but in Fig. 14.10 we show a very simple example of a bending of a circular plate with use of different numbers of equal elements. This purely bending problem shows the type of results and convergence attainable.

Interpreting the single integrating point as a single shear variable and applying the patch test count of Chapter 12, the reader can verify that this simple formulation passes the test in assemblies of two or more elements. In a similar way it can be verified that a quadratic interpolation of displacements and the use of two quadrature points (or a linear shear force) will also result in a robust element of excellent performance.

One final word of caution when the element is used in transient analyses is in order. Here it is necessary to compute a mass matrix which can be deduced from the term

$$\delta\Pi_{in} = 2\pi \int_L \left[\delta u \, \rho t \, \ddot{u} + \delta w \, \rho t \, \ddot{w} + \delta\phi \, \rho \frac{t^3}{12} \ddot{\phi} \right] r \, ds \qquad (14.45)$$

Evaluation of this integral with a single quadrature point will lead to a rank deficient mass matrix, which when used with any time stepping scheme can lead to large numerical errors (generally after many time steps have been computed). Accordingly, it is necessary to compute the mass matrix with at least two quadrature points (nodal quadrature giving immediately a diagonal 'lumped' mass).

References

1. P.E. Grafton and D.R. Strome. Analysis of axi-symmetric shells by the direct stiffness method. *Journal of AIAA*, 1:2342–2347, 1963.
2. E.P. Popov, J. Penzien and Z.A. Liu. Finite element solution for axisymmetric shells. *Proc. Am. Soc. Civ. Eng.*, EM5:119–145, 1964.
3. R.E. Jones and D.R. Strome. Direct stiffness method of analysis of shells of revolution utilizing curved elements. *Journal of AIAA*, 4:1519–1525, 1966.
4. J.H. Percy, T.H.H. Pian, S. Klein and D.R. Navaratna. Application of matrix displacement method to linear elastic analysis of shells of revolution. *Journal of AIAA*, 3:2138–2145, 1965.
5. S. Klein. A study of the matrix displacement method as applied to shells of revolution. In *Proc. 1st Conf. Matrix Methods in Structural Mechanics*, volume AFFDL-TR-66-80, Wright Patterson Air Force Base, Ohio, October 1966.
6. R.E. Jones and D.R. Strome. A survey of analysis of shells by the displacement method. In *Proc. 1st Conf. Matrix Methods in Structural Mechanics*, volume AFFDL-TR-66-80, Wright Patterson Air Force Base, Ohio, October 1966.
7. J.A. Stricklin, D.R. Navaratna and T.H.H. Pian. Improvements in the analysis of shells of revolution by matrix displacement method (curved elements). *Journal of AIAA*, 4:2069–2072, 1966.
8. M. Khojasteh-Bakht. Analysis of elastic–plastic shells of revolution under axi-symmetric loading by the finite element method. Technical Report SESM 67–8, University of California, Berkeley, 1967.
9. J.J. Webster. Free vibration of shells of revolution using ring elements. *J. Mech. Sci.*, 9:559–570, 1967.
10. S. Ahmad, B.M. Irons and O.C. Zienkiewicz. Curved thick shell and membrane elements with particular reference to axi-symmetric problems. In *Proc. 2nd Conf. Matrix Methods in Structural Mechanics*, volume AFFDL-TR-68-150, pages 539–572, Wright Patterson Air Force Base, Ohio, October 1968.

11. E.A. Witmer and J.J. Kotanchik. Progress report on discrete element elastic and elastic–plastic analysis of shells of revolution subjected to axisymmetric and asymmetric loading. In *Proc. 2nd Conf. Matrix Methods in Structural Mechanics*, volume AFFDL-TR-68-150, pages 1341–1453, Wright Patterson Air Force Base, Ohio, October 1968.

12. A.S.L. Chan and A. Firmin. The analysis of cooling towers by the matrix finite element method. *Aeronaut. J.*, 74:826–835, 1970.

13. M. Giannini and G.A. Miles. A curved element approximation in the analysis of axi-symmetric thin shells. *International Journal for Numerical Methods in Engineering*, 2:459–476, 1970.

14. R. Delpak. *Role of the curved parametric element in linear analysis of thin rotational shells.* PhD thesis, Department of Civil Engineering and Building, the Polytechnic of Wales, 1975.

15. P.L. Gould and S.K. Sen. Refined mixed method finite elements for shells of revolution. In *Proc. 3rd Conf. Matrix Methods in Structural Mechanics*, volume AFFDL-TR-71-160, Wright-Patterson Air Force Base, Ohio, 1972.

16. Z.M. Elias. Mixed finite element method for axisymmetric shells. *International Journal for Numerical Methods in Engineering*, 4:261–272, 1972.

17. R.H. Gallagher. Analysis of plate and shell structures. In W.R. Rowan and R.M. Hackett, editors, *Applications of Finite Element Method in Engineering*, pages 155–205. ASCE, Vanderbilt University, Nashville, Tennessee, 1969.

18. R.H. Gallagher. *Finite Element Analysis: Fundamentals.* Prentice-Hall, Englewood Cliffs, NJ, 1975.

19. J.A. Stricklin. Geometrically nonlinear static and dynamic analysis of shells of revolution. In *High Speed Computing of Elastic Structures*, pages 383–411. University of Liège, 1976.

20. V.V. Novozhilov. *Theory of Thin Shells.* Noordhoff, Dordrecht, 1959. (English translation.)

21. S.P. Timoshenko and S. Woinowski-Krieger. *Theory of Plates and Shells.* McGraw-Hill, New York, 2nd edition, 1959.

22. K. Washizu. *Variational Methods in Elasticity and Plasticity.* Pergamon Press, New York, 3rd edition, 1982.

23. P.M. Naghdi. Foundations of elastic shell theory. In I.N. Sneddon and R. Hill, editors, *Progress in Solid Mechanics*, volume IV, Chapter 1. North Holland, Amsterdam, 1963.

24. O.C. Zienkiewicz, R.L. Taylor and J.Z. Zhu. *The Finite Element Method: Its Basis and Fundamentals.* Butterworth-Heinemann, Oxford, 6th edition, 2005.

25. E.L. Wilson. The static condensation algorithm. *International Journal for Numerical Methods in Engineering*, 8:1974, 199–203.

26. W.E. Haisler and J.A. Stricklin. Rigid body displacements of curved elements in the analysis of shells by the matrix displacement method. *Journal of AIAA*, 5:1525–1527, 1967.

27. O.C. Zienkiewicz, J. Bauer, K. Morgan and E. Oñate. A simple element for axi-symmetric shells with shear deformation. *International Journal for Numerical Methods in Engineering*, 11:1545–1558, 1977.

28. J.L. Sanders Jr. and A. Liepins. Toroidal membrane under internal pressure. *Journal of AIAA*, 1:2105–2110, 1963.

29. F.F. Jordan. Stresses and deformations of the thin-walled pressurized torus. *J. Aero. Sci.*, 29: 213–225, 1962.

30. H. Kraus. *Thin Elastic Shells.* John Wiley & Sons, New York, 1967.

31. T.J.R. Hughes, R.L. Taylor and W. Kanoknukulchai. A simple and efficient finite element for plate bending. *International Journal for Numerical Methods in Engineering*, 11:1529–1543, 1977.

15

Shells as a special case of three-dimensional analysis – Reissner–Mindlin assumptions

15.1 Introduction

In the analysis of solids the use of isoparametric, curved, two- and three-dimensional elements is particularly effective, as illustrated in Chapters 2 and 4 and presented in reference 1. It seems obvious that use of such elements in the analysis of curved shells could be made directly simply by reducing their dimension in the thickness direction as shown in Fig. 15.1. Indeed, in an axisymmetric situation such an application is illustrated in the example of Fig. 9.25 of reference 1. With a straightforward use of the three-dimensional concept, however, certain difficulties will be encountered.

In the first place the retention of 3 displacement degrees of freedom at each node leads to large stiffness coefficients from strains in the shell thickness direction. This presents numerical problems and may lead to ill-conditioned equations when the shell thickness becomes small compared with other dimensions of the element.

The second factor is that of economy. The use of several nodes across the shell thickness ignores the well-known fact that even for thick shells the 'normals' to the mid-surface remain practically straight after deformation. Thus an unnecessarily high number of degrees of freedom has to be carried, involving penalties of computer time.

In this chapter we present specialized formulations which overcome both of these difficulties. The constraint of straight 'normals' is introduced to improve economy and the strain energy corresponding to the stress perpendicular to the mid-surface is ignored to improve numerical conditioning.[2-4] With these modifications an efficient tool for analysing curved thick shells becomes available. The accuracy and wide range of applicability of the approach is demonstrated in several examples.

15.2 Shell element with displacement and rotation parameters

The reader will note that the two constraints introduced correspond precisely to the so-called Reissner–Mindlin assumptions already discussed in Chapter 12 to describe

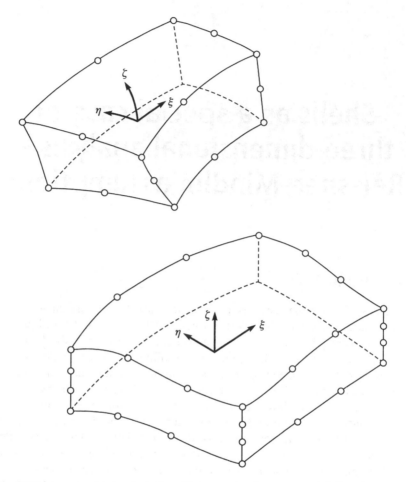

Fig. 15.1 Curved, isoparametric hexahedra in a direct approximation to a curved shell.

the behaviour of thick plates. The omission of the third constraint associated with the thin plate theory (normals remaining normal to the mid-surface after deformation) permits the shell to experience transverse shear deformations – an important feature of thick shell situations.

The formulation presented here leads to additional complications compared with the straightforward use of a three-dimensional element. The elements developed here are in essence an alternative to the processes discussed in Chapter 12, for which an independent interpolation of slopes and displacement are used with a penalty function imposition of the continuity requirements. The use of reduced integration is useful if thin shells are to be dealt with – and, indeed, it was in this context that this procedure was first discovered.[5-8] Again the same restrictions for robust behaviour as those discussed in Chapter 12 become applicable and generally elements that perform well in plate situations will do well in shells.

15.2.1 Geometric definition of an element

Consider a typical shell element illustrated in Fig. 15.2. The external faces of the element are curved, while the sections across the thickness are generated by straight lines. Pairs of points, a_{top} and a_{bottom}, each with given Cartesian coordinates, prescribe the shape of the element.

Let ξ, η be the two curvilinear coordinates in the mid-surface of the shell and let ζ be a linear coordinate in the thickness direction. If, further, we assume that ξ, η, ζ vary between -1 and 1 on the respective faces of the element we can write a relationship between the Cartesian coordinates of any point of the shell and the curvilinear coordinates in the form

$$
\begin{Bmatrix} x \\ y \\ z \end{Bmatrix} = \sum N_a(\xi, \eta) \left(\frac{1+\zeta}{2} \begin{Bmatrix} \tilde{x}_a \\ \tilde{y}_a \\ \tilde{z}_a \end{Bmatrix}_{\text{top}} + \frac{1-\zeta}{2} \begin{Bmatrix} \tilde{x}_a \\ \tilde{y}_a \\ \tilde{z}_a \end{Bmatrix}_{\text{bottom}} \right) \tag{15.1}
$$

Here $N_a(\xi, \eta)$ is a standard two-dimensional shape function taking a value of unity at the top and bottom nodes a and zero at all other nodes (see Appendix A). If the basic

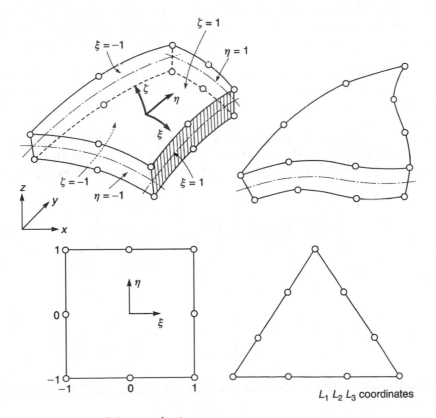

Fig. 15.2 Curved thick shell elements of various types.

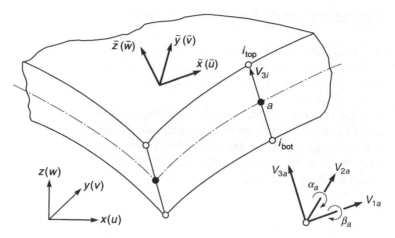

Fig. 15.3 Local and global coordinates.

functions N_a are derived as 'shape functions' of a 'parent', two-dimensional element, square or triangular* in plan, and are so 'designed' that compatibility is achieved at interfaces, then the curved space elements will fit into each other. Arbitrary curved shapes of the element can be achieved by using shape functions of higher order than linear. Indeed, any of the two-dimensional shape functions of Chapter 2 or Appendix A can be used here.

The relation between the Cartesian and curvilinear coordinates is now established and it will be found desirable to operate with the curvilinear coordinates as the basis. It should be noted that often the coordinate direction ζ is *only approximately normal* to the mid-surface.

It is convenient to rewrite the relationship, Eq. (15.1), in a form specified by the 'vector' connecting the upper and lower points (i.e. a vector of length equal to the shell thickness t) and the mid-surface coordinates. Thus we can rewrite Eq. (15.1) as (Fig. 15.3)

$$\begin{Bmatrix} x \\ y \\ z \end{Bmatrix} = \sum N_a(\xi, \eta) \left(\begin{Bmatrix} \tilde{x}_a \\ \tilde{y}_a \\ \tilde{z}_a \end{Bmatrix}_{\text{mid}} + \frac{1}{2}\zeta \mathbf{V}_{3a} \right) \qquad (15.2)$$

where

$$\begin{Bmatrix} \tilde{x}_a \\ \tilde{y}_a \\ \tilde{z}_a \end{Bmatrix} = \frac{1}{2} \left(\begin{Bmatrix} \tilde{x}_a \\ \tilde{y}_a \\ \tilde{z}_a \end{Bmatrix}_{\text{top}} + \begin{Bmatrix} \tilde{x}_a \\ \tilde{y}_a \\ \tilde{z}_a \end{Bmatrix}_{\text{bottom}} \right) \quad \text{and} \quad \mathbf{V}_{3a} = \begin{Bmatrix} \tilde{x}_a \\ \tilde{y}_a \\ \tilde{z}_a \end{Bmatrix}_{\text{top}} - \begin{Bmatrix} \tilde{x}_a \\ \tilde{y}_a \\ \tilde{z}_a \end{Bmatrix}_{\text{bottom}} \qquad (15.3)$$

with $\mathbf{V}_{3a}$ defining a vector whose length represents the *shell director* and has length of shell thickness.

---

* Area coordinates L_k would be used in this case in place of ξ, η.[1]

For relatively thin shells, it is convenient to replace the vector $\mathbf{V}_{3a}$ by a unit vector $\mathbf{v}_{3a}$ in the direction normal to the mid-surface. Now Eq. (15.2) is written simply as

$$\begin{Bmatrix} x \\ y \\ z \end{Bmatrix} = \sum N_a(\xi, \eta) \left(\begin{Bmatrix} \tilde{x}_a \\ \tilde{y}_a \\ \tilde{z}_a \end{Bmatrix}_{\text{mid}} + \frac{1}{2} \zeta t_a \mathbf{v}_{3a} \right)$$

where t_a is the shell thickness at the node a. Construction of a vector normal to the mid-surface is a simple process (see Sec. 13.4.2).

15.2.2 Displacement field

The displacement field is now specified for the element. As the strains in the direction normal to the mid-surface will be assumed to be negligible, the displacement throughout the element will be taken to be uniquely defined by the *three Cartesian components* of the mid-surface node displacement and *two rotations* about two orthogonal directions normal to the nodal vector $\mathbf{V}_{3a}$. If these two orthogonal directions are denoted by unit vectors $\mathbf{v}_{1a}$ and $\mathbf{v}_{2a}$ with corresponding rotations $\tilde{\alpha}_a$ and $\tilde{\beta}_a$ (see Fig. 15.3), we can write, similar to Eq. (15.2) but dropping the subscript 'mid' for simplicity,

$$\begin{Bmatrix} u \\ v \\ w \end{Bmatrix} = \sum N_a(\xi, \eta) \left(\begin{Bmatrix} \tilde{u}_a \\ \tilde{v}_a \\ \tilde{w}_a \end{Bmatrix} + \frac{1}{2} \zeta t_a \begin{bmatrix} \tilde{v}_{1a}, & -\tilde{v}_{2a} \end{bmatrix} \begin{Bmatrix} \tilde{\alpha}_a \\ \tilde{\beta}_a \end{Bmatrix} \right) \tag{15.4}$$

from which the usual form is readily obtained as

$$\begin{Bmatrix} u \\ v \\ w \end{Bmatrix} = \mathbf{N}\tilde{\mathbf{u}}^e; \quad \tilde{\mathbf{u}}^e = \begin{Bmatrix} \tilde{\mathbf{u}}_1^e \\ \vdots \\ \tilde{\mathbf{u}}_n^e \end{Bmatrix} \quad \text{with} \quad \tilde{\mathbf{u}}_a^e = \begin{Bmatrix} \tilde{u}_a \\ \tilde{v}_a \\ \tilde{w}_a \\ \tilde{\alpha}_a \\ \tilde{\beta}_a \end{Bmatrix} \tag{15.5}$$

where u, v and w are displacements in the directions of the global x, y and z axes.

As an infinity of vector directions normal to a given direction can be generated, a particular scheme has to be devised to ensure a *unique* definition. Some such schemes were discussed in Chapter 13. Here another unique alternative will be given,[3,5] but other possibilities are open.[8]

Here $\mathbf{V}_{3a}$ is the vector to which a normal direction is to be constructed. A coordinate vector in a Cartesian system may be defined by

$$\mathbf{x} = x\mathbf{e}_x + y\mathbf{e}_y + z\mathbf{e}_z \tag{15.6}$$

in which $\mathbf{e}_x$, $\mathbf{e}_y$ and $\mathbf{e}_z$ are three (orthogonal) base vectors. To find the first normal vector we find the minimum component of $\mathbf{V}_{3a}$ and construct a vector cross-product with the unit vector in this direction to define $\mathbf{V}_{1a}$. For example, if the x component of $\mathbf{V}_{3a}$ is the smallest one we construct

$$\mathbf{V}_{1a} = \mathbf{e}_x \times \mathbf{V}_{3a} \tag{15.7}$$

where

$$\mathbf{e}_x = \begin{bmatrix} 1 & 0 & 0 \end{bmatrix}^T$$

is the form of the unit vector in the x direction. Now

$$\mathbf{v}_{1a} = \frac{\mathbf{V}_{1a}}{|\mathbf{V}_{1a}|} \qquad \text{where} \qquad |\mathbf{V}_{1a}| = \sqrt{\mathbf{V}_{1a}^T \mathbf{V}_{1a}} \tag{15.8}$$

defines the first unit vector.

The second normal vector may now be computed from

$$\mathbf{V}_{2a} = \mathbf{V}_{3a} \times \mathbf{V}_{1a} \tag{15.9}$$

and normalized using the form in Eq. (15.8). We have thus three local, orthogonal axes defined by unit vectors

$$\mathbf{v}_{1a}, \mathbf{v}_{2a} \qquad \text{and} \qquad \mathbf{v}_{3a} \tag{15.10}$$

Once again if N_a are C_0 functions then displacement compatibility is maintained between adjacent elements.

The element coordinate definition is now given by the relation Eq. (15.2) and has more degrees of freedom than the definition of the displacements. The element is therefore of the 'superparametric' kind[1] and the constant strain criteria are not automatically satisfied. Nevertheless, it will be seen from the definition of strain components involved that both rigid body motions and constant strain conditions are available.

Physically it has been assumed in the definition of Eq. (15.4) that no strains occur in the 'thickness' direction ζ. While this direction is not always exactly normal to the mid-surface it still represents a good approximation of one of the usual shell assumptions.

At each mid-surface node a of Fig. 15.3 we now have the five basic degrees of freedom, and the connection of elements will follow precisely the patterns described in Chapter 13 (Secs 13.3 and 13.4).

15.2.3 Definition of strains and stresses

To derive the properties of a finite element the essential strains and stresses need first to be defined. The components in directions of *orthogonal axes* related to the surface ζ (constant) are essential if account is to be taken of the basic shell assumptions. Thus, if at any point in this surface we erect a normal $\bar{z}$ with two other orthogonal axes $\bar{x}$ and $\bar{y}$ tangential to it (Fig. 15.3), the strain components of interest are given simply by the three-dimensional relationships in Chapter 2:

$$\bar{\boldsymbol{\varepsilon}} = \begin{Bmatrix} \varepsilon_{\bar{x}} \\ \varepsilon_{\bar{y}} \\ \gamma_{\bar{x}\bar{y}} \\ \gamma_{\bar{y}\bar{z}} \\ \gamma_{\bar{z}\bar{x}} \end{Bmatrix} = \begin{Bmatrix} \bar{u}_{,\bar{x}} \\ \bar{v}_{,\bar{y}} \\ \bar{u}_{,\bar{y}} + \bar{v}_{,\bar{x}} \\ \bar{v}_{,\bar{z}} + \bar{w}_{,\bar{y}} \\ \bar{w}_{,\bar{x}} + \bar{u}_{,\bar{z}} \end{Bmatrix} \tag{15.11}$$

with the strain in direction $\bar{z}$ neglected so as to be consistent with the usual shell assumptions. It must be noted that in general none of these directions coincide with those of the curvilinear coordinates ξ, η, ζ, although $\bar{x}$, $\bar{y}$ are in the $\xi\eta$ plane ($\zeta =$ constant).[*]

The stresses corresponding to these strains are defined by a matrix $\bar{\sigma}$ and for elastic behaviour are related to the usual elasticity matrix $\bar{D}$. Thus

$$\bar{\sigma} = \begin{Bmatrix} \sigma_{\bar{x}} \\ \sigma_{\bar{y}} \\ \tau_{\bar{x}\bar{y}} \\ \tau_{\bar{y}\bar{z}} \\ \tau_{\bar{z}\bar{x}} \end{Bmatrix} = \bar{D}\,(\bar{\varepsilon} - \bar{\varepsilon}_0) + \bar{\sigma}_0 \tag{15.12}$$

where $\bar{\varepsilon}_0$ and $\bar{\sigma}_0$ represent any 'initial' strains and stresses, respectively.

The 5×5 matrix $\bar{D}$ can now include any anisotropic properties and indeed may be prescribed as a function of ζ if sandwich or laminated construction is used. For the present we shall define it only for an isotropic material. Here

$$\bar{D} = \frac{E}{1 - \nu^2} \begin{bmatrix} 1 & \nu & 0 & 0 & 0 \\ \nu & 1 & 0 & 0 & 0 \\ 0 & 0 & (1-\nu)/2 & 0 & 0 \\ 0 & 0 & 0 & \kappa(1-\nu)/2 & 0 \\ 0 & 0 & 0 & 0 & \kappa(1-\nu)/2 \end{bmatrix} \tag{15.13}$$

in which E and ν are Young's modulus and Poisson's ratio, respectively. The factor κ included in the last two shear terms is taken as $5/6$ and its purpose is to improve the shear displacement approximations (see Chapter 11). From the displacement definition it will be seen that the shear distribution is approximately constant through the thickness, whereas in reality the shear distribution for elastic behaviour is approximately parabolic. The value $\kappa = 5/6$ is the ratio of relevant strain energies.

It is important to note that this matrix is *not* defined by deleting appropriate terms from the equivalent three-dimensional stress matrix. It must be derived by substituting $\sigma_{\bar{z}} = 0$ into the three-dimensional constitutive equations and performing suitable elimination so that this important shell assumption is satisfied. This is similar to the procedure for deriving plane stress behaviour in two-dimensional analyses.

15.2.4 Element properties and necessary transformations

The stiffness matrix – and indeed all other 'element' property matrices – involves integrals over the volume of the element, which are quite generally of the form

$$\int_{\Omega^e} H \, dx \, dy \, dz \tag{15.14}$$

where the matrix H is a function of the coordinates. For instance, in the stiffness matrix

$$H = \bar{B}^T \bar{D} \bar{B} \tag{15.15}$$

---

[*] Indeed, these directions will only approximately agree with the nodal directions v_{1a}, v_{2a} previously derived, as in general the vector v_{3a} is only approximately normal to the mid-surface.

and with the usual definition

$$\bar{\varepsilon} = \bar{\mathbf{B}}\tilde{\mathbf{u}}^e \tag{15.16}$$

we have $\bar{\mathbf{B}}$ defined in terms of the displacement derivatives with respect to the local Cartesian coordinates $\bar{x}, \bar{y}, \bar{z}$ by Eq. (15.11). Now, therefore, *two sets of transformations* are necessary before the element can be integrated with respect to the curvilinear coordinates ξ, η, ζ.

First, by identically the same process as that used for isoparametric elements, the derivatives with respect to the x, y, z directions are obtained. As Eq. (15.4) relates the global displacements u, v, w to the curvilinear coordinates, the derivatives of these displacements with respect to the global x, y, z coordinates are given by a matrix relation:

$$\begin{bmatrix} \dfrac{\partial u}{\partial x} & \dfrac{\partial v}{\partial x} & \dfrac{\partial w}{\partial x} \\[2mm] \dfrac{\partial u}{\partial y} & \dfrac{\partial v}{\partial y} & \dfrac{\partial w}{\partial y} \\[2mm] \dfrac{\partial u}{\partial z} & \dfrac{\partial v}{\partial z} & \dfrac{\partial w}{\partial z} \end{bmatrix} = \mathbf{J}^{-1} \begin{bmatrix} \dfrac{\partial u}{\partial \xi} & \dfrac{\partial v}{\partial \xi} & \dfrac{\partial w}{\partial \xi} \\[2mm] \dfrac{\partial u}{\partial \eta} & \dfrac{\partial v}{\partial \eta} & \dfrac{\partial w}{\partial \eta} \\[2mm] \dfrac{\partial u}{\partial \zeta} & \dfrac{\partial v}{\partial \zeta} & \dfrac{\partial w}{\partial \zeta} \end{bmatrix} \tag{15.17}$$

In this, the Jacobian matrix is defined as

$$\mathbf{J} = \begin{bmatrix} \dfrac{\partial x}{\partial \xi} & \dfrac{\partial y}{\partial \xi} & \dfrac{\partial z}{\partial \xi} \\[2mm] \dfrac{\partial x}{\partial \eta} & \dfrac{\partial y}{\partial \eta} & \dfrac{\partial z}{\partial \eta} \\[2mm] \dfrac{\partial x}{\partial \zeta} & \dfrac{\partial y}{\partial \zeta} & \dfrac{\partial z}{\partial \zeta} \end{bmatrix} \tag{15.18}$$

and calculated from the coordinate definitions of Eq. (15.2). Now, for every set of curvilinear coordinates the global displacement derivatives can be obtained numerically.

A second transformation to the local displacements $\bar{x}, \bar{y}, \bar{z}$ will allow the strains, and hence the $\bar{\mathbf{B}}$ matrix, to be evaluated. The directions of the local axes can be established from a vector normal to the $\xi\eta$ mid-surface ($\zeta = 0$). This vector can be found from two vectors $\partial\mathbf{x}/\partial\xi$ and $\partial\mathbf{x}/\partial\eta$ that are tangential to the mid-surface. Thus

$$\mathbf{V}_3 = \begin{bmatrix} \dfrac{\partial x}{\partial \xi} \\[2mm] \dfrac{\partial y}{\partial \xi} \\[2mm] \dfrac{\partial z}{\partial \xi} \end{bmatrix} \times \begin{bmatrix} \dfrac{\partial x}{\partial \eta} \\[2mm] \dfrac{\partial y}{\partial \eta} \\[2mm] \dfrac{\partial z}{\partial \eta} \end{bmatrix} = \begin{bmatrix} \dfrac{\partial y}{\partial \xi}\dfrac{\partial z}{\partial \eta} - \dfrac{\partial y}{\partial \eta}\dfrac{\partial z}{\partial \xi} \\[2mm] \dfrac{\partial z}{\partial \xi}\dfrac{\partial x}{\partial \eta} - \dfrac{\partial z}{\partial \eta}\dfrac{\partial x}{\partial \xi} \\[2mm] \dfrac{\partial x}{\partial \xi}\dfrac{\partial y}{\partial \eta} - \dfrac{\partial x}{\partial \eta}\dfrac{\partial y}{\partial \xi} \end{bmatrix} \tag{15.19}$$

We can now construct two perpendicular vectors $\mathbf{V}_1$ and $\mathbf{V}_2$ following the process given previously to describe the $\bar{x}$ and $\bar{y}$ directions, respectively. The three orthogonal vectors can be reduced to unit magnitudes to obtain a matrix of vectors in the $\bar{x}$, $\bar{y}$, $\bar{z}$ directions (which is in fact the direction cosine matrix) given as

$$\boldsymbol{\theta} = \begin{bmatrix} \mathbf{v}_1, & \mathbf{v}_2, & \mathbf{v}_3 \end{bmatrix} \tag{15.20}$$

The global derivatives of displacement u, v and w are now transformed to the local derivatives of the local orthogonal displacements by a standard operation

$$\begin{bmatrix} \dfrac{\partial \bar{u}}{\partial \bar{x}} & \dfrac{\partial \bar{v}}{\partial \bar{x}} & \dfrac{\partial \bar{w}}{\partial \bar{x}} \\[2mm] \dfrac{\partial \bar{u}}{\partial \bar{y}} & \dfrac{\partial \bar{v}}{\partial \bar{y}} & \dfrac{\partial \bar{w}}{\partial \bar{y}} \\[2mm] \dfrac{\partial \bar{u}}{\partial \bar{z}} & \dfrac{\partial \bar{v}}{\partial \bar{z}} & \dfrac{\partial \bar{w}}{\partial \bar{z}} \end{bmatrix} = \boldsymbol{\theta}^T \begin{bmatrix} \dfrac{\partial u}{\partial x} & \dfrac{\partial v}{\partial x} & \dfrac{\partial w}{\partial x} \\[2mm] \dfrac{\partial u}{\partial y} & \dfrac{\partial v}{\partial y} & \dfrac{\partial w}{\partial y} \\[2mm] \dfrac{\partial u}{\partial z} & \dfrac{\partial v}{\partial z} & \dfrac{\partial w}{\partial z} \end{bmatrix} \boldsymbol{\theta} \tag{15.21}$$

From this the components of the $\bar{\mathbf{B}}$ matrix can now be found explicitly, noting that five degrees of freedom exist at each node:

$$\bar{\boldsymbol{\varepsilon}} = \bar{\mathbf{B}} \tilde{\mathbf{u}}^e \tag{15.22}$$

where the form of $\tilde{\mathbf{u}}^e$ is given in Eq. (15.5).

The infinitesimal volume is given in terms of the curvilinear coordinates as

$$\mathrm{d}x\,\mathrm{d}y\,\mathrm{d}z = \det \mathbf{J}\,\mathrm{d}\xi\,\mathrm{d}\eta\,\mathrm{d}\zeta = j\,\mathrm{d}\xi\,\mathrm{d}\eta\,\mathrm{d}\zeta \tag{15.23}$$

where $j = \det \mathbf{J}$. This standard expression completes the basic formulation.

Numerical integration within the appropriate limits is carried out in exactly the same way as for three-dimensional elements using Gaussian quadrature formulae.[1,9] An identical process serves to define all the other relevant element matrices arising from body and surface loading, inertia matrices, etc.

As the variation of the strain quantities in the thickness, or ζ direction, is linear, two Gauss points in that direction are sufficient for homogeneous elastic sections, while two to four in the ξ, η directions are needed for parabolic and cubic shape functions N_a.

It remarked here that, in fact, the integration with respect to ζ can be performed explicitly if desired, thus saving computation time.[2,5]

15.2.5 Some remarks on stress representation

The element properties are now defined, and the assembly and solution are in standard form. It remains to discuss the presentation of the stresses, and this problem is of some consequence. The strains being defined in local direction, $\bar{\sigma}$, are readily available. Such components are indeed directly of interest but as the directions of local axes are not easily visualized (and indeed may not be continuously defined between adjacent elements) it is sometimes convenient to transfer the components to the global system using the standard transformation

$$
\begin{bmatrix} \sigma_x & \tau_{xy} & \tau_{xz} \\ \tau_{yx} & \sigma_y & \tau_{yz} \\ \tau_{zx} & \tau_{zy} & \sigma_z \end{bmatrix} = \boldsymbol{\theta} \begin{bmatrix} \sigma_{\bar{x}} & \tau_{\bar{x}\bar{y}} & \tau_{\bar{x}\bar{z}} \\ \tau_{\bar{y}\bar{x}} & \sigma_{\bar{y}} & \tau_{\bar{y}\bar{z}} \\ \tau_{\bar{z}\bar{x}} & \tau_{\bar{z}\bar{y}} & \sigma_{\bar{z}} \end{bmatrix} \boldsymbol{\theta}^T \tag{15.24}
$$

This transformation should be performed only for elements which belong to the approximation for the same smooth surface and/or same material.

In a general shell structure, the stresses in a global system do not, however, give a clear picture of shell surface stresses. It is thus convenient always to compute the principal stresses (or invariants of stress) by a suitable transformation. Regarding the shell stresses more rationally, one may note that the shear components $\tau_{\bar{x}\bar{z}}$ and $\tau_{\bar{y}\bar{z}}$ are often zero on the top and bottom surfaces and this may be noted when making the transformation of Eq. (15.24) before converting to global components to ensure in this case that the principal stresses lie on the surface of the shell. The values obtained directly for these shear components are the average values across the section. The maximum transverse shear on a solid cross-section occurs on the mid-surface and is equal to about 1.5 times the average value.

15.3 Special case of axisymmetric, curved, thick shells

For axisymmetric shells the formulation is simplified. Now the element mid-surface is defined by only two coordinates ξ, η and a considerable saving in computer effort is obtained.[2]

The element now is derived in a similar manner by starting from a two-dimensional definition of Fig. 15.4.

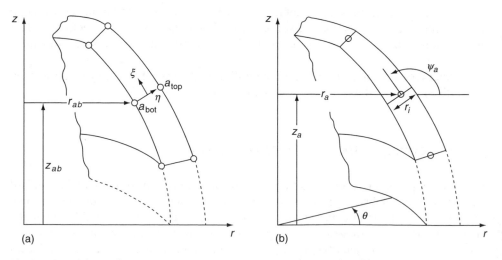

Fig. 15.4 Coordinates for an axisymmetric shell: (a) coordinate representation; (b) shell representation.

Equations (15.1) and (15.2) are now replaced by their two-dimensional equivalents defining the relation between coordinates as

$$\left\{ \begin{matrix} r \\ z \end{matrix} \right\} = \sum N_a(\xi) \left(\frac{1+\eta}{2} \left\{ \begin{matrix} \tilde{r}_a \\ \tilde{z}_a \end{matrix} \right\}_{\text{top}} + \frac{1-\eta}{2} \left\{ \begin{matrix} \tilde{r}_a \\ \tilde{z}_a \end{matrix} \right\}_{\text{bottom}} \right)$$

$$= \sum N_a(\xi) \left(\left\{ \begin{matrix} \tilde{r}_a \\ \tilde{z}_a \end{matrix} \right\}_{\text{mid}} + \frac{1}{2} \eta t_a \mathbf{v}_{3a} \right) \tag{15.25}$$

with

$$\mathbf{v}_{3a} = \left\{ \begin{matrix} \cos \psi_a \\ \sin \psi_a \end{matrix} \right\}$$

in which ψ_a is the angle defined in Fig. 15.4(b) and t_a is the shell thickness. Similarly, the displacement definition is specified by following the lines of Eq. (15.4).

Here we consider the case of axisymmetric loading only. Non-axisymmetric loading is addressed in Chapter 16 along with other schemes which permit treatment of problems in a reduced manner. Thus, we specify the two displacement components as

$$\left\{ \begin{matrix} u \\ w \end{matrix} \right\} = \sum N_a \left(\left\{ \begin{matrix} \tilde{u}_a \\ \tilde{w}_a \end{matrix} \right\} + \frac{\eta t_a}{2} \left\{ \begin{matrix} -\sin \tilde{\psi}_a \\ \cos \tilde{\psi}_a \end{matrix} \right\} \tilde{\phi}_a \right) \tag{15.26}$$

In this $\tilde{\phi}_a$ is the rotation illustrated in Fig. 15.5, and $\tilde{u}_a$, $\tilde{w}_a$ are the displacements of the middle surface node.

Global strains are conveniently defined by the relationship[10]

$$\boldsymbol{\varepsilon} = \left\{ \begin{matrix} \varepsilon_r \\ \varepsilon_z \\ \varepsilon_\theta \\ \gamma_{rz} \end{matrix} \right\} = \left\{ \begin{matrix} \dfrac{\partial u}{\partial r} \\[2mm] \dfrac{\partial w}{\partial z} \\[2mm] \dfrac{u}{r} \\[2mm] \dfrac{\partial u}{\partial z} + \dfrac{\partial w}{\partial r} \end{matrix} \right\} \tag{15.27}$$

These strains are transformed to the local coordinates and the component normal to η ($\eta = \text{constant}$) is neglected.

All the transformations follow the pattern described in previous sections and need not be further commented on except perhaps to remark that they are now carried out only between sets of directions ξ, η, r, z, and $\tilde{r}$, $\tilde{z}$, thus involving only two variables.

Similarly the integration of element properties is carried out numerically with respect to ξ and η only, noting, however, that the volume element is

$$dx \, dy \, dz = 2\pi r \, \det \mathbf{J} \, d\xi \, d\eta \, d\theta = 2\pi r \, j \, d\xi \, d\eta \, d\theta \tag{15.28}$$

By suitable choice of shape functions $N_a(\xi)$, straight, parabolic, or cubic shapes of variable thickness elements can be used as shown in Fig. 15.6.

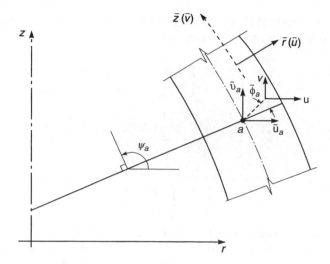

Fig. 15.5 Global displacements in an axisymmetric shell.

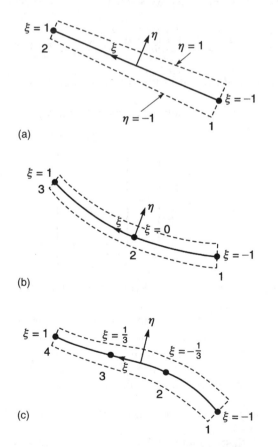

Fig. 15.6 Axisymmetric shell elements: (a) linear; (b) parabolic; (c) cubic.

15.4 Special case of thick plates

The transformations necessary in this chapter are somewhat involved and the programming steps are quite sophisticated. However, the application of the principle involved is available for thick plates and readers are advised to first test their comprehension on such a simple problem.

Here the following obvious simplifications arise.

1. $\zeta = 2z/t$ and unit vectors v_1, v_2 and v_3 can be taken as e_x, e_y, and e_z, respectively.
2. $\tilde{\alpha}_a$ and $\tilde{\beta}_a$ are simply the rotations $\tilde{\theta}_y$ and $\tilde{\theta}_x$, respectively (see Chapter 12).
3. It is no longer necessary to transform stress and strain components to a local system of axes $\bar{x}, \bar{y}, \bar{z}$ and global definitions x, y, z can be used throughout. For elements of this type, numerical thickness integration can be avoided and, as an exercise, readers are encouraged to derive the stiffness matrices, etc., for, say, linear, rectangular elements. Forms will be found which are identical to those derived in Chapter 12 with an independent displacement and rotation interpolation and using shear constraints. This demonstrates the essential identity of the alternative procedures.

15.5 Convergence

Whereas in three-dimensional analysis it is possible to talk about absolute convergence to the true exact solution of the elasticity problem, in equivalent plate and shell problems such a convergence cannot happen. As the element size decreases the so-called convergent solution of a plate bending problem approaches only to the exact solution of the approximate model implied in the formulation. Thus, here again convergence of the above formulation will only occur to the exact solution constrained by the requirement that straight 'normals' remain straight during deformation.

In elements of finite size it will be found that pure bending deformation modes are nearly always accompanied by some shear strains which in fact do not exist in the conventional thin plate or shell bending theory (although quite generally shear stresses may be deduced by equilibrium considerations on an element of the model, similar to the manner by which shear stresses in beams are deduced). Thus large elements deforming mainly under bending action (as would be the case of the shell element degenerated to a flat plate) tend to be appreciably too stiff. In such cases certain limits of the ratio of size of element to its thickness need to be imposed. However, it will be found that such restrictions often are relaxed by the simple expedient of *reducing the integration order*.[5]

Figure 15.7 shows, for instance, the application of the quadratic eight-node element to a square plate situation. Here results for integration with 3×3 and 2×2 Gauss points are given and results plotted for different thickness-to-span ratios. For reasonably thick situations, the results are similar and both give the additional shear deformation not available by thin plate theory. However, for thin plates the results with the more exact integration tend to diverge rapidly from the now correct thin plate results whereas the reduced integration still gives excellent results. The reasons for this improved performance are fully discussed in Chapter 12 and the reader is referred there for further plate examples using different types of shape functions.

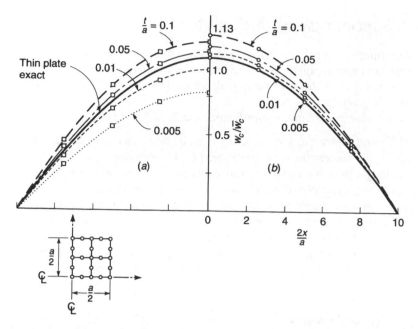

Fig. 15.7 A simply supported square plate under uniform load q_0: plot of central deflection w_c for eight-node elements with (a) 3×3 Gauss point integration and (b) with 2×2 (reduced) Gauss point integration. Central deflection is $\bar{w}_c$ for thin plate theory.

15.6 Inelastic behaviour

All the formulations presented in this chapter can of course be used for all non-linear materials. The procedures are similar to those mentioned in Chapters 11 and 12 dealing with plates. Now it is only necessary to replace Eqs (15.12) and (15.13) by the appropriate constitutive equation and tangent operator, respectively. In this case it is necessary always to perform the through thickness integration numerically since *a priori* knowledge of the behaviour will not be available. Any of the constitutive models described in Chapter 4 may be used for this purpose provided appropriate transformations are made to make $\sigma_{\bar{z}}$ zero (viz. Sec. 6.2.4).

15.7 Some shell examples

A limited number of examples which show the accuracy and range of application of the axisymmetric shell formulation presented in this chapter will be given. For a fuller selection the reader is referred to references 2–8.

15.7.1 Spherical dome under uniform pressure

The 'exact' solution of shell theory is known for this axisymmetric problem, illustrated in Fig. 15.8. Twenty-four cubic-type elements are used with graded size more closely spaced towards supports. Contrary to the 'exact' shell theory solution, the present

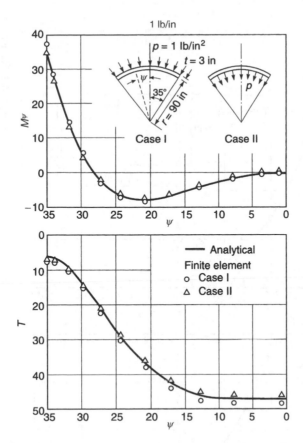

Fig. 15.8 Spherical dome under uniform pressure analysed with 24 cubic elements (first element subtends an angle of 0.1° from fixed end, others in arithmetic progression).

formulation can distinguish between the application of pressure on the inner and outer surfaces as shown in the figure.

15.7.2 Edge loaded cylinder

A further axisymmetric example is shown in Fig. 15.9 to study the effect of subdivision. Two, six, or 14 cubic elements of unequal length are used and the results for both of the finer subdivisions are almost coincident with the exact solution. Even the two-element solution gives reasonable results and departs only in the vicinity of the loaded edge.

Once again the solutions are basically identical to those derived with independent slope and displacement interpolation in the manner presented in Chapter 12.

15.7.3 Cylindrical vault

This is a test example of application of the full process to a shell in which bending action is dominant as a result of supports restraining deflection at the ends (see also Sec. 13.8.2).

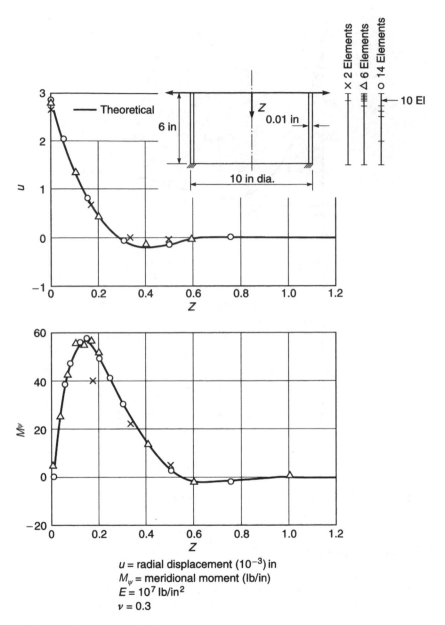

Fig. 15.9 Thin cylinder under a unit radial edge load.

In Fig. 15.10 the geometry, physical details of the problem, and subdivision are given, and in Fig. 15.11 the comparison of the effects of 3×3 and 2×2 integration using eight-node quadratic elements is shown on the displacements calculated. Both integrations result, as expected, in convergence. For the more exact integration, this is rather slow, but, with reduced integration order, very accurate results are obtained, even with one element. The improved convergence of displacements is matched by rapid convergence of stress components.

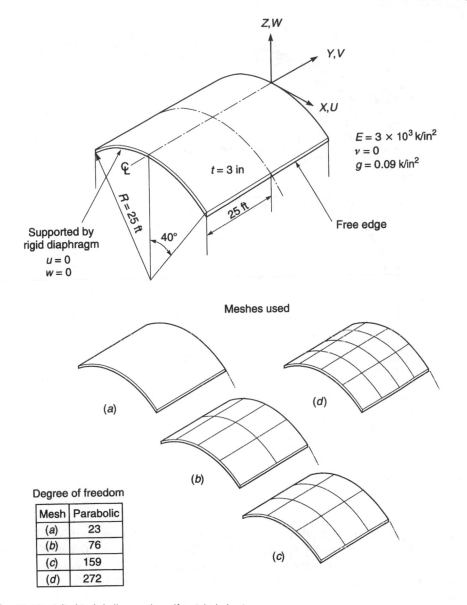

Meshes used

(a)

(b)

(d)

(c)

Degree of freedom

Mesh	Parabolic
(a)	23
(b)	76
(c)	159
(d)	272

Fig. 15.10 Cylindrical shell example: self-weight behaviour.

This example illustrates most dramatically the advantages of this simple expedient and is described more fully in references 5 and 7. The comparison solution for this problem is one derived along more conventional lines by Scordelis and Lo.[11]

15.7.4 Curved dams

All the previous examples were rather thin shells and indeed demonstrated the applicability of the process to these situations. At the other end of the scale, this formulation

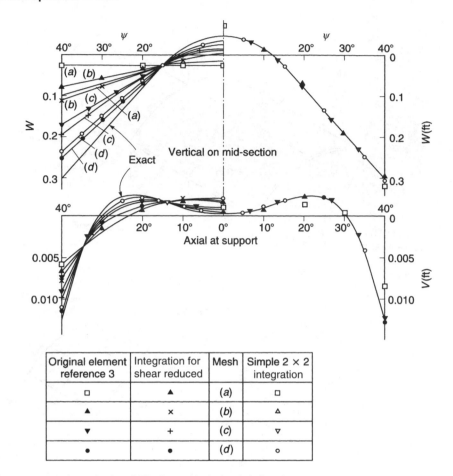

Original element reference 3	Integration for shear reduced	Mesh	Simple 2 × 2 integration
□	▲	(a)	□
▲	×	(b)	△
▼	+	(c)	▽
●	●	(d)	○

Fig. 15.11 Displacement (parabolic element), cylindrical shell roof.

has been applied to the doubly curved dams illustrated in Chapter 9 of Volume 1 (Fig. 9.28). Indeed, exactly the same subdivision is again used and *results repro-duce almost exactly those of the three-dimensional solution*.[4] This remarkable result is achieved at a very considerable saving in both degrees of freedom and computer solution time.

Clearly, the range of application of this type of element is very wide.

15.7.5 Pipe penetration and spherical cap

The last two examples, a pipe penetration[12] shown in Figs 15.12 and 15.13 and a spherical cap[8] shown in Fig. 15.14, illustrate applications in which the irregular shape of elements is used. Both illustrate practical problems of some interest and show that with reduced integration a useful and very general shell element is available, even when the elements are quite distorted.

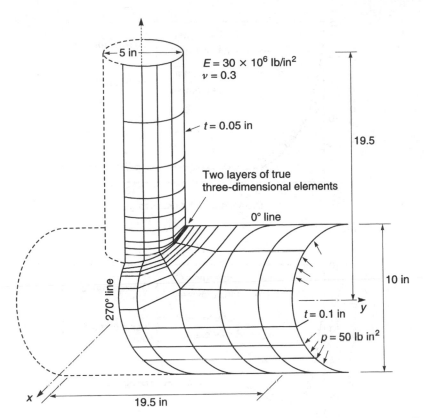

Fig. 15.12 An analysis of cylinder intersection by means of reduced integration shell-type elements.[12]

15.8 Concluding remarks

The elements described in this chapter using degeneration of solid elements are shown in plate and axisymmetric problems to be nearly identical to those described in Chapters 12 and 14 where an independent slope and displacement interpolation is directly used in the middle plane. For the general curved shell the analogy is less obvious but clearly still exists. We should therefore expect that the conditions established in Chapter 12 for robustness of plate elements to be still valid. Further, it appears possible that other additional conditions on the various interpolations may have to be imposed in curved element forms. Both statements are true. The eight- and nine-node elements which we have shown in the previous section to perform well will fail under certain circumstances and for this reason many of the more successful plate elements also have been adapted to the shell problem.

The introduction of additional degrees of freedom in the interior of the eight-node serendipity element was first suggested by Cook[13,14] and later by Hughes[15-17] without, however, achieving complete robustness. The full Lagrangian cubic interpolation as shown in Chapter 12 is quite effective and has been shown to perform well. However, the

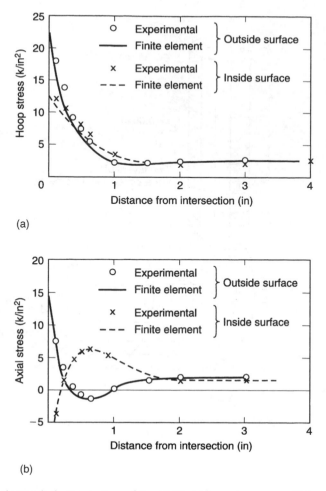

(a)

(b)

Fig. 15.13 Cylinder-to-cylinder intersections of Fig. 15.12: (a) hoop stresses near 0° line; (b) axial stresses near 0° line.

best results achieved to date appear to be those in which 'local constraints' are applied (see Sec. 12.5) and such elements as those due to Dvorkin and Bathe,[18] Huang and Hinton,[19] and Simo et al.[20,21] fall into this category.

While the importance of transverse shear strain constraints is now fully understood, the constraints introduced by the 'in-plane' (membrane) stress resultants are less amenable to analysis (although the elastic parameters Et associated with these are of the same order as those of shear Gt). It is well known that *membrane locking* can occur in situations that do not permit inextensional bending. Such locking has been thoroughly discussed[22–24] but to date the problem has not been rigorously solved and further developments are required.

Much effort is continuing to improve the formulation of the processes described in this chapter as they offer an excellent solution to the curved shell problem.[24–27]

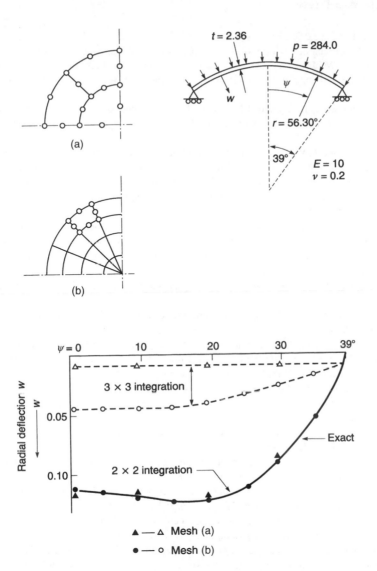

Fig. 15.14 A spherical cap analysis with irregular isoparametric shell elements using full 3 × 3 and reduced 2 × 2 integration.

References

1. O.C. Zienkiewicz, R.L. Taylor and J.Z. Zhu. *The Finite Element Method: Its Basis and Fundamentals*. Butterworth-Heinemann, Oxford, 6th edition, 2005.
2. S. Ahmad, B.M. Irons and O.C. Zienkiewicz. Curved thick shell and membrane elements with particular reference to axi-symmetric problems. In *Proc. 2nd Conf. Matrix Methods in Structural Mechanics*, volume AFFDL-TR-68-150, pages 539–572, Wright Patterson Air Force Base, Ohio, October 1968.

3. S. Ahmad. *Curved finite elements in the analysis of solids, shells and plate structures*. PhD thesis, Department of Civil Engineering, University of Wales, Swansea, 1969.

4. S. Ahmad, B.M. Irons and O.C. Zienkiewicz. Analysis of thick and thin shell structures by curved finite elements. *International Journal for Numerical Methods in Engineering*, 2:419–451, 1970.

5. O.C. Zienkiewicz, J. Too and R.L. Taylor. Reduced integration technique in general analysis of plates and shells. *International Journal for Numerical Methods in Engineering*, 3:275–290, 1971.

6. S.F. Pawsey and R.W. Clough. Improved numerical integration of thick slab finite elements. *International Journal for Numerical Methods in Engineering*, 3:575–586, 1971.

7. S.F. Pawsey. *The analysis of moderately thick to thin shells by the finite element method*. PhD dissertation, Department of Civil Engineering, University of California, Berkeley, 1970 (also SESM Report 70–12).

8. J.J.M. Too. *Two dimensional, plate, shell and finite prism isoparametric elements and their application*. PhD thesis, Department of Civil Engineering, University of Wales, Swansea, 1970.

9. M. Abramowitz and I.A. Stegun, editors. *Handbook of Mathematical Functions*. Dover Publications, New York, 1965.

10. I.S. Sokolnikoff. *The Mathematical Theory of Elasticity*. McGraw-Hill, New York, 2nd edition, 1956.

11. A.C. Scordelis and K.S. Lo. Computer analysis of cylindrical shells. *J. Am. Concr. Inst.*, 61: 539–561, 1964.

12. S.A. Bakhrebah and W.C. Schnobrich. Finite element analysis of intersecting cylinders. Technical Report UILU-ENG-73–2018, University of Illinois, Civil Engineering Studies, 1973.

13. R.D. Cook. More on reduced integration and isoparametric elements. *International Journal for Numerical Methods in Engineering*, 5:141–142, 1972.

14. R.D. Cook. *Concepts and Applications of Finite Element Analysis*. John Wiley & Sons, Chichester, 1982.

15. T.J.R. Hughes and M. Cohen. The 'heterosis' finite element for plate bending. *Computers and Structures*, 9:445–450, 1978.

16. T.J.R. Hughes and W.K. Liu. Non linear finite element analysis of shells. Part I. *Computer Methods in Applied Mechanics and Engineering*, 26:331–362, 1981.

17. T.J.R. Hughes and W.K. Liu. Non linear finite element analysis of shells. Part II. *Computer Methods in Applied Mechanics and Engineering*, 27:331–362, 1981.

18. E.N. Dvorkin and K.-J. Bathe. A continuum mechanics based four node shell element for general non-linear analysis. *Engineering Computations*, 1:77–88, 1984.

19. E.C. Huang and E. Hinton. Elastic, plastic and geometrically non-linear analysis of plates and shells using a new, nine-noded element. In P. Bergan *et al.*, editors, *Finite elements for Non Linear Problems*, pages 283–297. Springer-Verlag, Berlin, 1986.

20. J.C. Simo and D.D. Fox. On a stress resultant geometrically exact shell model. Part I: Formulation and optimal parametrization. *Computer Methods in Applied Mechanics and Engineering*, 72: 267–304, 1989.

21. J.C. Simo, D.D. Fox and M.S. Rifai. On a stress resultant geometrically exact shell model. Part II: The linear theory; computational aspects. *Computer Methods in Applied Mechanics and Engineering*, 73:53–92, 1989.

22. H. Stolarski and T. Belytschko. Membrane locking and reduced integration for curved elements. *J. Appl. Mech.*, 49:172–176, 1982.

23. H. Stolarski and T. Belytschko. Shear and membrane locking in curved C^0 elements. *Computer Methods in Applied Mechanics and Engineering*, 41:279–296, 1983.

24. R.V. Milford and W.C. Schnobrich. Degenerated isoparametric finite elements using explicit integration. *International Journal for Numerical Methods in Engineering*, 23:133–154, 1986.

25. D. Bushnell. Computerized analysis of shells – governing equations. *Computers and Structures*, 18:471–536, 1984.
26. M.A. Cilia and N.G. Gray. Improved coordinate transformations for finite elements: the Lagrange cubic case. *International Journal for Numerical Methods in Engineering*, 23:1529–1545, 1986.
27. S. Vlachoutsis. Explicit integration for three dimensional degenerated shell finite elements. *International Journal for Numerical Methods in Engineering*, 29:861–880, 1990.

16

Semi-analytical finite element processes – use of orthogonal functions and 'finite strip' methods

16.1 Introduction

Standard finite element methods have been shown to be capable, in principle, of dealing with any two- or three- (or even four-)[*] dimensional situations. Nevertheless, the cost of solutions increases greatly with each dimension added and indeed, on occasion, overtaxes the available computer capability. It is therefore always desirable to search for alternatives that may reduce computational effort. One such class of processes of quite wide applicability will be illustrated in this chapter.

In many physical problems the situation is such that the *geometry* and *material properties* do not vary along one coordinate direction. However, the 'load' terms may still exhibit a variation in that direction, preventing the use of such simplifying assumptions as those that, for instance, permitted a two-dimensional plane strain or axisymmetric analysis to be substituted for a full three-dimensional treatment. In such cases it is possible still to consider a 'substitute' problem, not involving the particular coordinate (along which the geometry and properties do not vary), and to synthesize the true answer from a series of such simplified solutions.

The method to be described is of quite general use and, obviously, is not limited to structural situations. It will be convenient, however, to use the nomenclature of structural mechanics and to use potential energy minimization as an example.

We shall confine our attention to problems of minimizing a quadratic functional for a linear elastic material. The interpretation of the process involved as the application of partial discretization leading to semi-discrete forms[1] followed (or preceded) by the use of a Fourier series expansion should be noted.

Let (x, y, z) be the coordinates describing the domain (in this context these do not necessarily have to be the Cartesian coordinates). The last one of these, z, is the coordinate along which the geometry and material properties do not change and which is limited to lie between two values

$$0 \leq z \leq a$$

The boundary values are thus specified at $z = 0$ and $z = a$.

---

[*] See finite elements in the time domain in reference 1.

We shall assume that the shape functions defining the variation of displacement **u** can be written in a separation of variables form as

$$
\mathbf{u} = \mathbf{N}(x, y, z)\tilde{\mathbf{u}}^e = \sum_{l=1}^{L} \left(\bar{\mathbf{N}}(x, y) \cos \frac{l\pi z}{a} + \bar{\bar{\mathbf{N}}}(x, y) \sin \frac{l\pi z}{a} \right) (\tilde{\mathbf{u}}^l)^e
$$
$$
= \sum_{l=1}^{L} \mathbf{N}^l(x, y, z)(\tilde{\mathbf{u}}^l)^e
\tag{16.1}
$$

In this type of representation completeness is preserved in view of the capability of Fourier series to represent any continuous function within a given region (naturally assuming that the shape functions $\bar{\mathbf{N}}$ and $\bar{\bar{\mathbf{N}}}$ in the domain x, y satisfy the same requirements).

The loading terms will similarly be given a form

$$
\mathbf{b} = \sum_{l=1}^{L} \left(\cos \frac{l\pi z}{a} \bar{\mathbf{b}}^l + \sin \frac{l\pi z}{a} \bar{\bar{\mathbf{b}}}^l \right) = \sum_{l=1}^{L} \mathbf{b}^l(x, y, z)
\tag{16.2}
$$

for body force, with similar form for any concentrated loads and boundary tractions. Indeed, initial strains and stresses, if present, would be expanded again in the above form.

Finally, we restrict our attention to linear elastic materials in which the constitutive equation is given by

$$
\boldsymbol{\sigma} = \mathbf{D}(\boldsymbol{\varepsilon} - \boldsymbol{\varepsilon}_0) + \boldsymbol{\sigma}_0
\tag{16.3}
$$

with stress and strain ordered as

$$
\boldsymbol{\sigma} = \begin{bmatrix} \sigma_x & \sigma_y & \sigma_z & \tau_{xy} & \tau_{yz} & \tau_{zx} \end{bmatrix}^T
$$
$$
\boldsymbol{\varepsilon} = \begin{bmatrix} \varepsilon_x & \varepsilon_y & \varepsilon_z & \gamma_{xy} & \gamma_{yz} & \gamma_{zx} \end{bmatrix}^T
$$

and the elastic material matrix has a z-symmetry direction such that

$$
\mathbf{D} = \begin{bmatrix}
D_{11} & D_{12} & D_{13} & D_{14} & 0 & 0 \\
D_{21} & D_{22} & D_{23} & D_{24} & 0 & 0 \\
D_{31} & D_{32} & D_{33} & D_{34} & 0 & 0 \\
D_{41} & D_{42} & D_{43} & D_{44} & 0 & 0 \\
0 & 0 & 0 & 0 & D_{55} & D_{56} \\
0 & 0 & 0 & 0 & D_{65} & D_{66}
\end{bmatrix}
\tag{16.4}
$$

Applying the standard process for the determination of the element contribution to the equation minimizing the potential energy Π, and limiting our attention to the contribution of body forces **b** only, we can write

$$
\frac{\partial \Pi}{\partial \tilde{\mathbf{u}}^e} = \mathbf{K}^e \begin{Bmatrix} \tilde{\mathbf{u}}^{1e} \\ \vdots \\ \tilde{\mathbf{u}}^{Le} \end{Bmatrix} + \begin{Bmatrix} \mathbf{f}^{1e} \\ \vdots \\ \mathbf{f}^{Le} \end{Bmatrix} = \mathbf{0}
\tag{16.5}
$$

In the above, to avoid summation signs, the vectors $\tilde{\mathbf{u}}^e$, etc., are expanded, listing the contribution of each value of l separately.

Now a typical submatrix of $\mathbf{K}^e$ is

$$(\mathbf{K}^{lm})^e = \int_\Omega (\mathbf{B}^l)^T \mathbf{D} \, \mathbf{B}^m \, d\Omega \, dz \tag{16.6}$$

and a typical term of the 'force' vector becomes

$$(\mathbf{f}^l)^e = \int_\Omega (\mathbf{N}^l)^T \mathbf{b}^l \, d\Omega \tag{16.7}$$

Without going into details, it is obvious that the matrix given by Eq. (16.6) will contain the following integrals as products of various submatrices:

$$
\begin{aligned}
I_1 &= \int_0^a \sin \frac{l\pi z}{a} \cos \frac{m\pi z}{a} \, dz \\
I_2 &= \int_0^a \sin \frac{l\pi z}{a} \sin \frac{m\pi z}{a} \, dz \\
I_3 &= \int_0^a \cos \frac{l\pi z}{a} \cos \frac{m\pi z}{a} \, dz
\end{aligned}
\tag{16.8}
$$

These integrals arise from products of the derivatives contained in the definition of $\mathbf{B}^l$ and, owing to the well-known orthogonality property, give

$$I_2 = I_3 = \begin{cases} \frac{1}{2}a, & \text{for } l = m \\ 0, & \text{for } l \neq m \end{cases} ; \qquad l, m = 1, 2, \ldots \tag{16.9}$$

The first integral I_1 is only zero when l and m are both even or odd numbers. The term involving I_1, however, vanishes in many applications because of the structure of $\mathbf{B}^l$. This means that $\mathbf{K}^e$ becomes a block diagonal matrix and that the assembled final equations of the system have the form

$$
\begin{bmatrix}
\mathbf{K}^{11} & & & \\
& \mathbf{K}^{22} & & \\
& & \ddots & \\
& & & \mathbf{K}^{LL}
\end{bmatrix}
\begin{Bmatrix}
\tilde{\mathbf{u}}^1 \\
\tilde{\mathbf{u}}^2 \\
\vdots \\
\tilde{\mathbf{u}}^L
\end{Bmatrix}
+
\begin{Bmatrix}
\mathbf{f}^1 \\
\mathbf{f}^2 \\
\vdots \\
\mathbf{f}^L
\end{Bmatrix}
\tag{16.10}
$$

and *the large system of equations splits into L separate problems*:

$$\mathbf{K}^{ll}\tilde{\mathbf{u}}^l + \mathbf{f}^l = \mathbf{0} \tag{16.11}$$

in which

$$\mathbf{K}^{ll}_{ab} = \int_\Omega (\mathbf{B}^l_a)^T \mathbf{D} \, \mathbf{B}^l_b \, d\Omega \tag{16.12}$$

Further, from Eqs (16.7) and (16.2) we observe that owing to the orthogonality property of the integrals given by Eqs (16.8), the typical load term becomes simply

$$\mathbf{f}^l_a = \int_\Omega (\mathbf{N}^l_i)^T \mathbf{b}^l \, d\Omega \tag{16.13}$$

This means that the force term of the lth harmonic only affects the lth system of Eq. (16.11) and contributes nothing to the other equations. This extremely important property is of considerable practical significance for, *if the expansion of the loading factors involves only one term, only one set of equations need be solved.* The solution of this will tend to the exact one with increasing subdivision in the xy domain only. Thus, what was originally a three-dimensional problem now has been reduced to a two-dimensional one with consequent reduction of computational effort.

The preceding derivation was illustrated on a three-dimensional, elastic situation. Clearly, the arguments could equally well be applied for reduction of two-dimensional problems to one-dimensional ones, etc., and the arguments are not restricted to problems of elasticity. Any physical problem governed by a minimization of a quadratic functional or by linear differential equations is amenable to the same treatment.

A word of warning should be added regarding the boundary conditions imposed on **u**. For a complete decoupling to be possible these must be satisfied separately by each and every term of the expansion given by Eq. (16.1). Insertion of a zero displacement in the final reduced problem implies in fact a zero displacement fixed throughout all terms in the z direction by definition. Care must be taken not to treat the final matrix therefore as a simple reduced problem. Indeed, this is one of the limitations of the process described.

When the loading is complicated and many Fourier components need to be considered the advantages of the approach outlined here reduce and the full three-dimensional solution sometimes becomes more efficient.

Other permutations of the basic definitions of the type given by Eq. (16.1) are obviously possible. For instance, two independent sets of parameters $\tilde{\mathbf{u}}^e$ may be specified with each of the trigonometric terms. Indeed, on occasion use of other orthogonal functions may be possible. The appropriate functions are often related to a reduction of the differential equation directly using separation of variables.[2]

16.2 Prismatic bar

Consider a prismatic bar, illustrated in Fig. 16.1, which is assumed to be held at $z = 0$ and $z = a$ in a manner preventing all displacements in the xy plane but permitting unrestricted motion in the z direction (traction $t_z = 0$). The problem is fully three dimensional and three components of displacement u, v, and w have to be considered.

Subdividing into finite elements in the xy plane we can prescribe the lth displacement components as

$$\mathbf{u}^l = \left\{ \begin{array}{c} u^l \\ v^l \\ w^l \end{array} \right\} = \sum_b N_b \begin{bmatrix} \sin \gamma_l z & 0 & 0 \\ 0 & \sin \gamma_l z & 0 \\ 0 & 0 & \cos \gamma_l z \end{bmatrix} \left\{ \begin{array}{c} \tilde{u}_b^l \\ \tilde{v}_b^l \\ \tilde{w}_b^l \end{array} \right\} \tag{16.14}$$

where $\gamma_l = l\pi/a$. In this, N_b are simply the (scalar) shape functions appropriate to the elements used in the xy plane and again $\gamma_l = l\pi/a$. If, as shown in Fig. 16.1, simple triangles are used then the shape functions are given in area coordinates by $N_b = L_b$, but any other isoparametric elements would be equally suitable. The displacement expansion ensures zero u and v displacements at the ends and the zero $\bar{t}_z$ traction condition can be imposed in a standard manner.

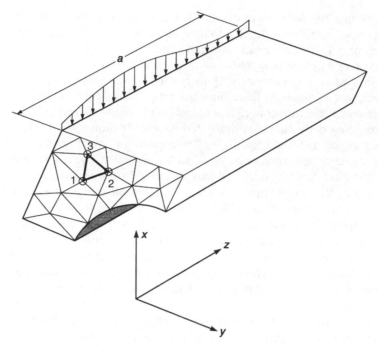

Fig. 16.1 A prismatic bar reduced to a series of two-dimensional finite element solutions.

As the problem is fully three dimensional, the appropriate expression for strain involving all six components needs to be considered. This expression is given in Eq. (2.9a) of Chapter 2. On substitution of the shape function given by Eq. (16.14) for a typical term of the **B** matrix we have

$$
\mathbf{B}_b^l =
\begin{bmatrix}
N_{b,x} \sin \gamma_l z & 0 & 0 \\
0 & N_{b,y} \sin \gamma_l z & 0 \\
0 & 0 & -N_b \gamma_l \sin \gamma_l z \\
N_{b,y} \sin \gamma_l z & N_{b,x} \sin \gamma_l z & 0 \\
0 & N_b \gamma_l \cos \gamma_l z & N_{b,y} \cos \gamma_l z \\
N_b \gamma_l \cos \gamma_l z & 0 & N_{b,x} \cos \gamma_l z
\end{bmatrix}
\tag{16.15}
$$

It is convenient to separate the above as

$$
\mathbf{B}_b^l = \bar{\mathbf{B}}_b^l \sin \gamma_l z + \bar{\bar{\mathbf{B}}}_b^l \cos \gamma_l z
\tag{16.16}
$$

where

$$
\bar{\mathbf{B}}_b^l =
\begin{bmatrix}
N_{b,x} & 0 & 0 \\
0 & N_{b,y} & 0 \\
0 & 0 & -N_b \gamma_l \\
N_{b,y} & N_{b,x} & 0 \\
0 & 0 & 0 \\
0 & 0 & 0
\end{bmatrix}
\quad \text{and} \quad
\bar{\bar{\mathbf{B}}}_b^l =
\begin{bmatrix}
0 & 0 & 0 \\
0 & 0 & 0 \\
0 & 0 & 0 \\
0 & 0 & 0 \\
0 & N_b \gamma_l & N_{b,y} \\
N_b \gamma_l & 0 & N_{b,x}
\end{bmatrix}
$$

In all of the above it is assumed that the parameters are listed in the usual order:

$$\tilde{\mathbf{u}}_a^l = \begin{bmatrix} \tilde{u}_a^l & \tilde{v}_a^l & \tilde{w}_a^l \end{bmatrix}^{\mathrm{T}} \tag{16.17}$$

and that the axes are as shown in Fig. 16.1.

The stiffness matrix can be computed in the usual manner, noting that

$$(\mathbf{K}_{ab}^{ll})^e = \int_{\Omega^e} \mathbf{B}_a^{lT} \mathbf{D} \mathbf{B}_b^l \, d\Omega \tag{16.18}$$

On substitution of Eq. (16.16), multiplying out, and noting the value of the integrals from Eq. (16.9), this reduces to

$$(\mathbf{K}_{ab}^{ll})^e = \frac{a}{2} \iint_{A^e} \left(\bar{\mathbf{B}}_a^{lT} \mathbf{D} \, \bar{\mathbf{B}}_b^l + \bar{\bar{\mathbf{B}}}_a^{lT} \mathbf{D} \, \bar{\bar{\mathbf{B}}}_b^l \right) dx \, dy \quad l = 1, 2, \ldots \tag{16.19}$$

The integration is now simply carried out over the element *area.*[*]

The contributions from distributed loads, initial stresses, etc., are found as the loading terms. To match the displacement expansions distributed body forces may be expanded in the Fourier series

$$\mathbf{b}_a^l = \int_0^a N_a \begin{bmatrix} \sin \gamma_l z & 0 & 0 \\ 0 & \sin \gamma_l z & 0 \\ 0 & 0 & \cos \gamma_l z \end{bmatrix} \begin{Bmatrix} b_x(x, y, z) \\ b_y(x, y, z) \\ b_z(x, y, z) \end{Bmatrix} dz \tag{16.20}$$

Similarly, concentrated line loads can be expressed directly as nodal forces

$$\mathbf{f}_a^l = \int_0^a N_a \begin{bmatrix} \sin \gamma_l z & 0 & 0 \\ 0 & \sin \gamma_l z & 0 \\ 0 & 0 & \cos \gamma_l z \end{bmatrix} \begin{Bmatrix} f_x(x, y, z) \\ f_y(x, y, z) \\ f_z(x, y, z) \end{Bmatrix} dz \tag{16.21}$$

in which $\mathbf{f}_a^l$ are intensities per unit length.

The boundary conditions used here have been of a type ensuring *simply supported* conditions for the prism. Other conditions can be inserted by suitable expansions.

The method of analysis outlined here can be applied to a range of practical problems – one of these being a popular type of box girder, concrete bridge, illustrated in Fig. 16.2. Here a particularly convenient type of element is the distorted, serendipity or Lagrangian quadratic or cubic element.[3] Finally, it should be mentioned that some restrictions placed on the general shapes defined by Eq. (16.1) or Eq. (16.14) can be removed by doubling the number of parameters and writing expansions in the form of two sums:

$$\mathbf{u} = \sum_{l=1}^{L} \bar{\mathbf{N}}(x, y) \cos \gamma_l z \, \tilde{\mathbf{u}}^{Al} + \sum_{l=1}^{L} \bar{\bar{\mathbf{N}}}(x, y) \sin \gamma_l z \, \tilde{\mathbf{u}}^{Bl} \tag{16.22}$$

Parameters $\tilde{\mathbf{u}}^{Al}$ and $\tilde{\mathbf{u}}^{Bl}$ are independent and for every component of displacement two values have to be found and two equations formed.

---

[*] It should be noted that now, even for a single triangle, the integration is not trivial as some linear terms from N_a will remain in $\bar{\mathbf{B}}$ and $\bar{\bar{\mathbf{B}}}$.

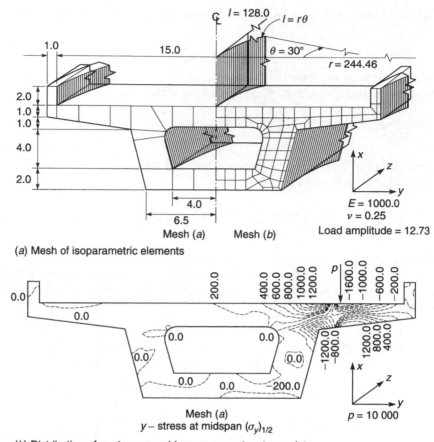

Fig. 16.2 A thick box bridge prism of straight or curved platform.

An alternative to the above process is to write the expansion as

$$\mathbf{u} = \sum \left[\mathbf{N}(x, y) \exp(i\gamma_l z) \right] \tilde{\mathbf{u}}^e$$

and to observe that both $\mathbf{N}$ and $\tilde{\mathbf{u}}$ are then complex quantities.

Complex algebra is available in standard programming languages and the identity of the above expression with Eq. (16.22) will be observed, noting that

$$\exp i\theta = \cos\theta + i\sin\theta$$

16.3 Thin membrane box structures

In the previous section a three-dimensional problem was reduced to that of two dimensions. Here we shall see how a somewhat similar problem can be reduced to one-dimensional elements (Fig. 16.3).

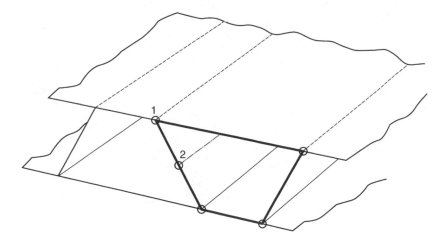

Fig. 16.3 A 'membrane' box with one-dimensional elements.

A box-type structure is made up of thin shell components capable of sustaining stresses only in its own plane. Now, just as in the previous case, three displacements have to be considered at every point and indeed similar variation can be prescribed for these. However, a typical element *ab* is 'one dimensional' in the sense that integrations have to be carried out only along the line *ab* and only stresses in that direction need be considered. Indeed, it will be found that the situation and solution are similar to that of a pin-jointed framework.

16.4 Plates and boxes with flexure

Consider now a rectangular plate simply supported at the ends and in which all strain energy is contained in flexure. Only one displacement, w, is needed to specify fully the state of strain (see Chapter 11).

For consistency of notation with the plate and shell chapters where z is the thickness direction, the direction in which geometry and material properties do not change is now taken as y (see Fig. 16.4). To preserve slope continuity the functions now need to include a 'rotation' parameter θ_a.

Use of simple beam functions (cubic Hermitian interpolations) is easy and for a typical element *ab* we can write the plate transverse displacement as (with $\gamma_l = l\pi/a$)

$$w^l = \bar{N}(x) \sin \gamma_l y (\tilde{u}^l)^e \tag{16.23}$$

ensuring *simply supported* end conditions. In this, the typical nodal parameters are

$$\tilde{u}_a^l = \begin{Bmatrix} \tilde{w}_a^l \\ \tilde{\theta}_a^l \end{Bmatrix} \tag{16.24}$$

The shape functions of the cubic type are easy to write and are in fact identical to the Hermitian polynomials given in Sec. 10.4.1 and also those used for the asymmetric thin shell problem [Chapter 14, Eq. (14.9)].

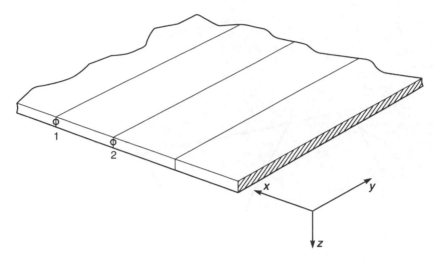

Fig. 16.4 The 'strip' method in slabs.

Using all definitions of Chapter 11 the strains (curvatures) are found and the **B** matrices determined; now with C_1 continuity satisfied in a trivial manner, the problem of a two-dimensional kind has here been reduced to that of one dimension.

This application has been developed by Cheung and others,[4-18] named the 'finite strip' method, and used to solve many rectangular plate problems, box girders, shells, and various folded plates.

It is illuminating to quote an example from the above papers here. This refers to a square, uniformly loaded plate with three sides simply supported and one clamped. Ten strips or elements in the x direction were used in the solution, and Table 16.1 gives the results corresponding to the first three harmonics.

Not only is an accurate solution of each l term a simple one involving only some 19 unknowns but the importance of higher terms in the series is seen to decrease rapidly.

Extension of the process to box structures in which both *membrane* and *bending effects* are present is almost obvious when this example is considered together with the one in the previous section.

In the examples just quoted a thin plate theory using the single displacement variable w and enforcing C_1 compatibility in the x direction was employed. Obviously, any of the independently interpolated slope and displacement elements of Chapter 12 could be used here, again employing either reduced integration or mixed methods.

Table 16.1   Square plate, uniform load q; three sides simply supported, one clamped (Poisson ratio $= 0.3$)

Term l	Central deflection	Central M_x	Maximum negative M_x
1	0.002832	0.0409	−0.0858
2	−0.000050	−0.0016	0.0041
3	0.002786	0.0396	−0.0007
Σ	0.002786	0.0396	−0.0824
Series	0.0028	0.039	−0.084
Multiplier	qa^4/D	qa^2	qa^2

Parabolic-type elements with reduced integration are employed in references 14 and 15, and linear interpolation with a single integration point is shown to be effective in reference 16.

Other applications for plate and box-type structures abound and additional information is given in the text of reference 18.

16.5 Axisymmetric solids with non-symmetrical load

One of the most natural and indeed earliest applications of Fourier expansion occurs in axisymmetric bodies subject to non-axisymmetric loads. Now, not only the radial (u) and axial (w) displacement will have to be considered but also a tangential component (v) associated with the tangential angular direction θ (Fig. 16.5). It is in this direction that the geometric and material properties do not vary and hence here that the elimination will be applied.

To simplify matters we shall consider first components of load which are symmetric about the $\theta = 0$ axis and later include those which are antisymmetric. Describing now only the nodal loads (with similar expansion holding for body forces, boundary conditions, initial strains, etc.) we specify forces per unit of circumference as

$$R = \sum_{l=1}^{L} \bar{R}^l \cos l\theta$$

$$T = \sum_{l=1}^{L} \bar{T}^l \sin l\theta \qquad (16.25)$$

$$Z = \sum_{l=1}^{L} \bar{Z}^l \cos l\theta$$

in the direction of the various coordinates for symmetric loads [Fig. 16.6(a)]. The apparently non-symmetric sine expansion is used for T, since to achieve symmetry the direction of T has to change for $\theta > \pi$.

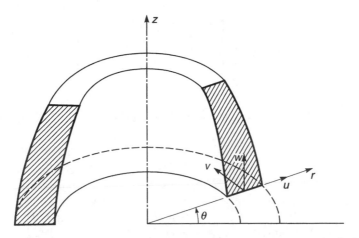

Fig. 16.5 An axisymmetric solid; coordinate displacement components in an axisymmetric body.

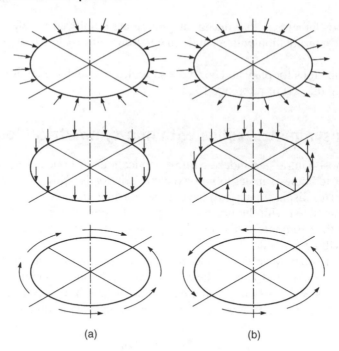

Fig. 16.6 Load and displacement components in an axisymmetric body: (a) symmetric; (b) antisymmetric.

The displacement components are described again in terms of the two-dimensional (r, z) shape functions appropriate to the element subdivision, and, observing symmetry, we write, as in Eq. (16.14),

$$
\mathbf{u}^l = \begin{Bmatrix} u^l \\ v^l \\ w^l \end{Bmatrix} = \sum_a N_a \begin{bmatrix} \cos l\theta & 0 & 0 \\ 0 & \sin l\theta & 0 \\ 0 & 0 & \cos l\theta \end{bmatrix} \begin{Bmatrix} u_a^l \\ v_a^l \\ w_a^l \end{Bmatrix} \tag{16.26}
$$

To proceed further it is necessary to specify the general, three-dimensional expression for strains in cylindrical coordinates. These are given by[19]

$$
\boldsymbol{\varepsilon} = \begin{Bmatrix} \varepsilon_r \\ \varepsilon_z \\ \varepsilon_\theta \\ \gamma_{rz} \\ \gamma_{z\theta} \\ \gamma_{\theta r} \end{Bmatrix} = \begin{Bmatrix} u_{,r} \\ w_{,z} \\ \left[u + v_{,\theta}\right]/r \\ u_{,z} + w_{,r} \\ v_{,z} + w_{,\theta}/r \\ u_{,\theta}/r + v_{,r} - v/r \end{Bmatrix} \tag{16.27}
$$

We have on substitution of Eq. (16.26) into Eq. (16.27), and grouping the variables as in Eq. (16.17):

$$
\mathbf{B}_a^l = \begin{bmatrix} N_{a,r}\cos l\theta & 0 & 0 \\ 0 & 0 & N_{a,z}\cos l\theta \\ N_a/r\cos l\theta & lN_a/r\cos l\theta & 0 \\ N_{a,z}\cos l\theta & 0 & N_{a,r}\cos l\theta \\ 0 & N_{a,z}\sin l\theta & -lN_a/r\sin l\theta \\ -lN_a/r\sin l\theta & \left(N_{a,r} - N_a/r\right)\sin l\theta & 0 \end{bmatrix} \tag{16.28}
$$

A purely axisymmetric problem may be described for the complete zero harmonic ($l = 0$) and a further simplification arises in that the strains split into two problems: the first involves the displacement components u and w which appear only in the first four components of strain; and the second involves only the v displacement component and appears only in the last two shearing strains. This second problem is associated with a *torsion* problem on the axisymmetric body – with the first problem sometimes referred to as a *torsionless* problem. For an isotropic elastic material the stiffness matrix for these two problems completely decouples as a result of the structure of the **D** matrix, and they can be treated separately. However, for inelastic problems a coupling occurs whenever both torsionless and torsional loading are both applied as loading conditions on the same problem. Thus, it is often expedient to form the axisymmetric case including all three displacement components (as is necessary also for the other harmonics).

For the elastic case the remaining steps of the formulation follow precisely the previous derivations and can be performed by the reader as an exercise.

For the antisymmetric loading, of Fig. 16.6(b), we shall simply replace the sine by cosine and vice versa in Eqs (16.25) and (16.26).

The load intensity terms in each harmonic are obtained by virtual work as

$$
\mathbf{\bar{f}}^l = \int_0^{2\pi} \left\{ \begin{array}{l} \bar{R}^l \cos^2 l\theta \\ \bar{T}^l \sin^2 l\theta \\ \bar{Z}^l \cos^2 l\theta \end{array} \right\} d\theta = \left\{ \begin{array}{l} \pi \left\{ \begin{array}{l} \bar{R}^l \\ \bar{T}^l \\ \bar{Z}^l \end{array} \right\} \quad \text{when } l = 1, 2, \dots \\ \\ 2\pi \left\{ \begin{array}{l} \bar{R}^l \\ 0 \\ \bar{Z}^l \end{array} \right\} \quad \text{when } l = 0 \end{array} \right.
\tag{16.29}
$$

for the symmetric case. Similarly, for the antisymmetric case

$$
\mathbf{\bar{f}}^l = \int_0^{2\pi} \left\{ \begin{array}{l} \bar{R}^l \sin^2 l\theta \\ \bar{T}^l \cos^2 l\theta \\ \bar{Z}^l \sin^2 l\theta \end{array} \right\} d\theta = \left\{ \begin{array}{l} \pi \left\{ \begin{array}{l} \bar{R}^l \\ \bar{T}^l \\ \bar{Z}^l \end{array} \right\} \quad \text{when } l = 1, 2, \dots \\ \\ 2\pi \left\{ \begin{array}{l} 0 \\ \bar{T}^l \\ 0 \end{array} \right\} \quad \text{when } l = 0 \end{array} \right.
\tag{16.30}
$$

The final nodal loads are then computed from the appropriate integrals

$$
\mathbf{f}_a^l = \int_{A_e} N_a \, \bar{f}^l \, dA
\tag{16.31}
$$

for body loads and

$$
\mathbf{f}_a^l = \int_{\Gamma_e} N_a \, \bar{f}^l \, d\Gamma
\tag{16.32}
$$

for surface loads.

We see from this and from an expansion of $\mathbf{K}^e$ that, as expected, for $l = 0$ the problem reduces to only two variables and the axisymmetric case is retrieved when

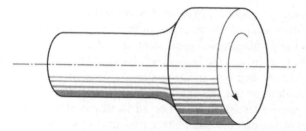

Fig. 16.7 Torsion of a variable section circular bar.

symmetric terms only are involved. Similarly, when $l = 0$ only one set of equations remains in the variable for v for the antisymmetric case.

This corresponds to constant tangential traction and solves simply the torsion problem of shafts subject to known torques (Fig. 16.7). This problem is classically treated by the use of a stress function[20] and indeed in this way has been solved by using a finite element formulation.[21] Here, an alternative, more physical, approach is available.

The first application of the above concepts to the analysis of axisymmetric solids was made by Wilson.[22] A simple example illustrating the effects of various harmonics is shown in Figs 16.8(a) and 16.8(b).

16.6 Axisymmetric shells with non-symmetrical load

16.6.1 Thin case – no shear deformation

The extension of analysis of axisymmetric thin shells as described in Chapter 14 to the case of non-axisymmetric loads is simple and will again follow the standard pattern.

It is, however, necessary to extend the definition of strains to include now all three displacements and force components (Fig. 16.9). Three membrane and three bending effects are now present and, extending Eq. (7.1) involving straight generators, we now define strains as[23,24] *

$$
\bar{\varepsilon} =
\begin{Bmatrix}
\varepsilon_s \\
\varepsilon_\theta \\
\gamma_{s\theta} \\
\chi_s \\
\chi_\theta \\
\chi_{s\theta}
\end{Bmatrix}
=
\begin{Bmatrix}
\bar{u}_{,s} \\
\bar{v}_{,\theta}/r + (\bar{u}\cos\psi - \bar{w}\sin\psi)/r \\
\bar{u}_{,\theta}/r + \bar{v}_{,s} - \bar{v}\cos\psi/r \\
-\bar{w}_{,ss} \\
-\bar{w}_{,\theta\theta}/r^2 - \bar{w}_{,s}\cos\psi/r + \bar{v}_{,\theta}\sin\psi/r \\
2\left(-\bar{w}_{,s\theta}/r + \bar{w}_{,\theta}\cos\psi/r^2 + \bar{v}_{,s}\sin\psi/r - \bar{v}\sin\psi\cos\psi/r^2\right)
\end{Bmatrix}
$$

(16.33)

---

* Various alternatives are available as a result of the multiplicity of shell theories. The one presented is quite commonly accepted.

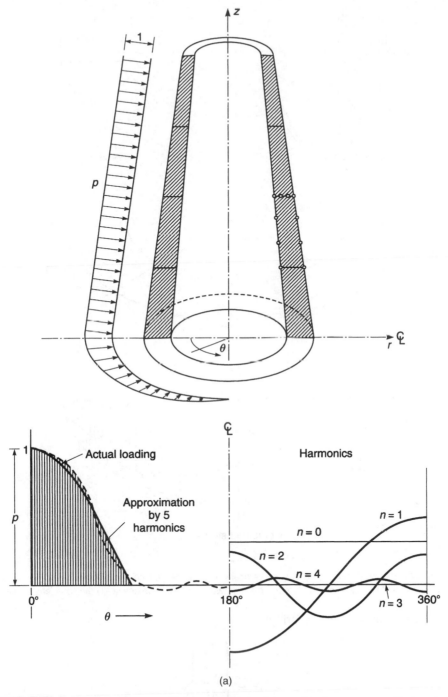

Fig. 16.8 (a) An axisymmetric tower under non-symmetric load; four cubic elements are used in the solution; the harmonics of load expansion used in the analysis are shown.

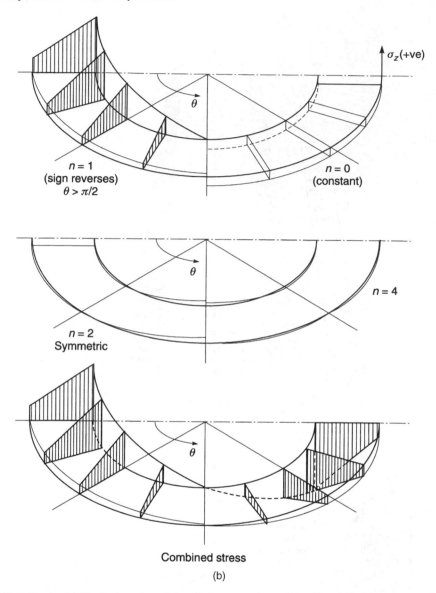

Fig. 16.8 *Cont.*   (b) Distribution of σ_z, the vertical stress on base arising from various harmonics and their combination (third harmonic identically zero), the first two harmonics give practically the complete answer.

The corresponding 'stress' matrix is

$$
\sigma = \begin{Bmatrix} N_s \\ N_\theta \\ N_{s\theta} \\ M_s \\ M_\theta \\ M_{s\theta} \end{Bmatrix}
\tag{16.34}
$$

with the three membrane and bending stresses defined as in Fig. 16.9.

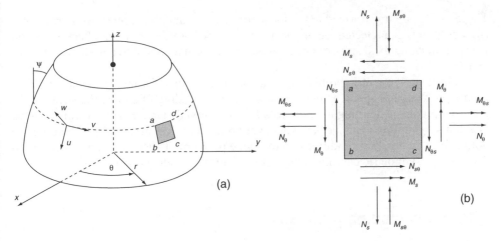

Fig. 16.9 Axisymmetric shell with non-symmetric load; (a) geometry and displacements, (b) stress resultants.

Once again, symmetric and antisymmetric variation of loads and displacements can be assumed, as in the previous section. As the processes involved in executing this extension of the application are now obvious, no further description is needed here, but note again should be made of the more elaborate form of equations necessary when curved elements are involved [see Chapter 14, Eq. (14.23)].

The reader is referred to the original paper by Grafton and Strome[24] in which this problem is first treated and to the many later papers on the subject listed in Chapter 14.

16.6.2 Thick case – with shear deformation

The displacement definition for a shell which includes the effects of transverse shearing deformation is specified using the forms given in Eqs (8.4) and (8.26). For a case of loading which is symmetric about $\theta = 0$, the decomposition into global trigonometric components involves the three displacement components of the nth harmonic as

$$\begin{Bmatrix} u^n \\ v^n \\ w^n \end{Bmatrix} = \sum N_a \begin{bmatrix} \cos n\theta & 0 & 0 \\ 0 & \sin n\theta & 0 \\ 0 & 0 & \cos n\theta \end{bmatrix} \left(\begin{Bmatrix} \tilde{u}_a^n \\ \tilde{v}_a^n \\ \tilde{w}_a^n \end{Bmatrix} + \frac{\eta t_a}{2} \begin{bmatrix} -\sin \psi_a & 0 \\ 0 & 1 \\ \cos \psi_a & 0 \end{bmatrix} \begin{Bmatrix} \tilde{\alpha}_a^n \\ \tilde{\beta}_a^n \end{Bmatrix} \right)$$

(16.35)

In this $\tilde{u}_a$, $\tilde{w}_a$, and $\tilde{\alpha}_a$ stand for the displacements and rotation illustrated in Fig. 15.5, $\tilde{v}_a$ is a displacement of the middle surface node in the tangential (θ) direction, and $\tilde{\beta}_a$ is a rotation about the vector tangential to the mid-surface.

Global strains are conveniently defined by the relationship[19]

$$\varepsilon = \begin{Bmatrix} \varepsilon_r \\ \varepsilon_z \\ \varepsilon_\theta \\ \gamma_{rz} \\ \gamma_{z\theta} \\ \gamma_{\theta r} \end{Bmatrix} = \begin{Bmatrix} u_{,r} \\ w_{,z} \\ [u + v_{,\theta}]/r \\ u_{,z} + w_{,r} \\ v_{,z} + w_{,\theta}/r \\ v_{,r} - v/r + u_{,\theta}/r \end{Bmatrix}$$

(16.36)

These strains are transformed to the local coordinates, and the component normal to η ($\eta = $ constant) is neglected. As in the axisymmetric case described in Chapter 15, the $\mathbf{D}$ matrix relating local stresses and strains takes a form identical to that defined by Eq. (8.13).

A purely axisymmetric problem may again be described for the complete zero harmonic problem and again, as in the non-symmetric loading of solids, the strains split into two problems defining a torsionless and a torsional state. However, for inelastic problems a coupling again occurs whenever both torsionless and torsional loading are both applied as loading conditions on the same problem. Thus, it is often expedient to form the axisymmetric case including all three displacement components.

16.7 Concluding remarks

A fairly general process combining some of the advantages of finite element analysis with the economy of expansion in terms of generally orthogonal functions has been illustrated in several applications. Certainly, these only touch on the possibilities offered, but it should be borne in mind that the economy is achieved only in certain geometrically constrained situations and those to which the number of terms requiring solution is limited.

Similarly, other 'prismatic' situations can be dealt with in which only a segment of a body of revolution is developed (Fig. 16.10). Clearly, the expansion must now be taken in terms of the angle $l\pi\theta/\alpha$, but otherwise the approach is identical to that described previously.[3]

In the methods of this chapter it was assumed that material properties remain constant with one coordinate direction. This restriction can on occasion be lifted with the

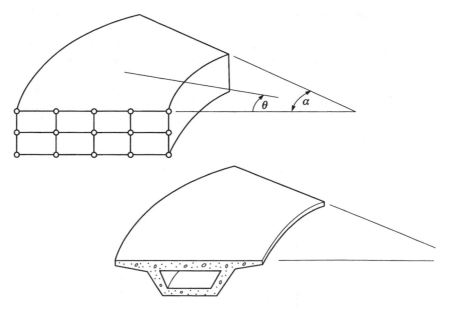

Fig. 16.10   Other segmental, prismatic situations.

same general process maintained. An early example of this type was presented by Stricklin and DeAndrade.[25] Inclusion of inelastic behaviour has also been successfully treated.[26-29]

All the problems we have described in this chapter could be derived in terms of semi-discretization in time. We would thus *first* semi-discretize, describing the problem in terms of an ordinary differential equation in z of the form

$$\mathbf{K}_1 \frac{d^2\mathbf{a}}{dz^2} + \mathbf{K}_2 \frac{d\mathbf{a}}{dz} + \mathbf{K}_3\mathbf{a} + \mathbf{f} = 0$$

Second, the above equation system would be solved in the domain $0 < z < a$ by means of orthogonal functions that *naturally* enter the problem as solutions of ordinary differential equations *with constant coefficients*. This second solution step is most easily found by using a diagonalization process described in dynamic applications.[1] Clearly, the final result of such computations would turn out to be identical with the procedures here described, but on occasion the above formulation is more self-evident.

References

1. O.C. Zienkiewicz, R.L. Taylor and J.Z. Zhu. *The Finite Element Method: Its Basis and Fundamentals*. Butterworth-Heinemann, Oxford, 6th edition, 2005.
2. P.M. Morse and H. Feshbach. *Methods of Theoretical Physics*. McGraw-Hill, New York, 1953.
3. O.C. Zienkiewicz and J.J.M. Too. The finite prism in analysis of thick simply supported bridge boxes. *Proc. Inst. Civ. Eng.*, 53:147–172, 1972.
4. Y.K. Cheung. The finite strip method in the analysis of elastic plates with two opposite simply supported ends. *Proc. Inst. Civ. Eng.*, 40:1–7, 1968.
5. Y.K. Cheung. Finite strip method of analysis of elastic slabs. *Proc. Am. Soc. Civ. Eng.*, 94(EM6):1365–1378, 1968.
6. Y.K. Cheung. Folded plates by the finite strip method. *Proc. Am. Soc. Civ. Eng.*, 95(ST2):963–979, 1969.
7. Y.K. Cheung. The analysis of cylindrical orthotropic curved bridge decks. *Publ. Int. Ass. Struct. Eng.*, 29–II:41–52, 1969.
8. Y.K. Cheung, M.S. Cheung and A. Ghali. Analysis of slab and girder bridges by the finite strip method. *Building Sci.*, 5:95–104, 1970.
9. Y.C. Loo and A.R. Cusens. Development of the finite strip method in the analysis of cellular bridge decks. In K. Rockey *et al.*, editors, *Conf. on Developments in Bridge Design and Construction*. Crosby Lockwood, London, 1971.
10. Y.K. Cheung and M.S. Cheung. Static and dynamic behaviour of rectangular plates using higher order finite strips. *Building Sci.*, 7:151–158, 1972.
11. G.S. Tadros and A. Ghali. Convergence of semi-analytical solution of plates. *Proc. Am. Soc. Civ. Eng.*, 99(EM5):1023–1035, 1973.
12. A.R. Cusens and Y.C. Loo. Application of the finite strip method in the analysis of concrete box bridges. *Proc. Inst. Civ. Eng.*, 57–II:251–273, 1974.
13. T.G. Brown and A. Ghali. Semi-analytic solution of skew plates in bending. *Proc. Inst. Civ. Eng.*, 57–II:165–175, 1974.
14. A.S. Mawenya and J.D. Davies. Finite strip analysis of plate bending including transverse shear effects. *Building Sci.*, 9:175–180, 1974.
15. P.R. Benson and E. Hinton. A thick finite strip solution for static, free vibration and stability problems. *International Journal for Numerical Methods in Engineering*, 10:665–678, 1976.

16. E. Hinton and O.C. Zienkiewicz. A note on a simple thick finite strip. *International Journal for Numerical Methods in Engineering*, 11:905–909, 1977.

17. H.C. Chan and O. Foo. Buckling of multilayer plates by the finite strip method. *Int. J. Mech. Sci.*, 19:447–456, 1977.

18. Y.K. Cheung. *Finite Strip Method in Structural Analysis*. Pergamon Press, Oxford, 1976.

19. I.S. Sokolnikoff. *The Mathematical Theory of Elasticity*. McGraw-Hill, New York, 2nd edition, 1956.

20. S.P. Timoshenko and J.N. Goodier. *Theory of Elasticity*. McGraw-Hill, New York, 3rd edition, 1969.

21. O.C. Zienkiewicz, P.L. Arlett and A.K. Bahrani. Solution of three-dimensional field problems by the finite element method. *The Engineer*, October 1967.

22. E.L. Wilson. Structural analysis of axi-symmetric solids. *Journal of AIAA*, 3:2269–2274, 1965.

23. V.V. Novozhilov. *Theory of Thin Shells*. Noordhoff, Dordrecht, 1959. (English translation).

24. P.E. Grafton and D.R. Strome. Analysis of axi-symmetric shells by the direct stiffness method. *Journal of AIAA*, 1:2342–2347, 1963.

25. J.A. Stricklin and J.C. de Andrade. Linear and non-linear analysis of shells of revolution with asymmetrical stiffness properties. In *Proc. 2nd Conf. Matrix Methods in Structural Mechanics*, volume AFFDL-TR-68-150, Wright Patterson Air Force Base, Ohio, October 1968.

26. L.A. Winnicki and O.C. Zienkiewicz. Plastic or visco-plastic behaviour of axisymmetric bodies subject to non-symmetric loading; semi-analytical finite element solution. *International Journal for Numerical Methods in Engineering*, 14:1399–1412, 1979.

27. W. Wunderlich, H. Cramer and H. Obrecht. Application of ring elements in the nonlinear analysis of shells of revolution under nonaxisymmetric loading. *Computer Methods in Applied Mechanics and Engineering*, 51:259–275, 1985.

28. H. Obrecht, F. Schnabel and W. Wunderlich. Elastic–plastic creep buckling of circular cylindrical shells under axial compression. *Z. Angew. Math. Mech.*, 67:T118–T120, 1987.

29. W. Wunderlich and C. Seiler. Nonlinear treatment of liquid-filled storage tanks under earthquake excitation by a quasistatic approach. In B.H.V. Topping, editor, *Advances in Computational Structural Mechanics, Proceedings 4th International Conference on Computational Structures*, pages 283–291, August 1998.

17
___

Non-linear structural problems –
large displacement and instability

17.1 Introduction

In the previous chapter the question of finite deformations and non-linear material behaviour was discussed and methods were developed to allow the standard linear forms to be used in an iterative way to obtain solutions. In the present chapter we consider the more specialized problem of large displacements but with strains restricted to be small. Generally, we shall assume that 'small strain' stress–strain relations are adequate but for accurate determination of the displacements geometric non-linearity needs to be considered. Here, for instance, stresses arising from membrane action, usually neglected in plate flexure, may cause a considerable decrease of displacements as compared with the linear solution discussed in Chapters 11 and 12, even though displacements remain quite small. Conversely, it may be found that a load is reached where indeed a state may be attained where load-carrying capacity *decreases* with continuing deformation. This classic problem is that of structural stability and obviously has many practical implications. The applications of such an analysis are clearly of considerable importance in aerospace and automotive engineering applications, design of telescopes, wind loading on cooling towers, box girder bridges with thin diaphragms and other relatively 'slender' structures.

In this chapter we consider the above class of problems applied to beam, plate, and shell systems by examining the basic non-linear equilibrium equations. Such considerations lead also to the formulation of classical initial stability problems. These concepts are illustrated in detail by formulating the large deflection and initial stability problems for beams and flat plates. A Lagrangian approach is adopted throughout in which displacements are referred to the original (reference) configuration.

17.2 Large displacement theory of beams

17.2.1 Geometrically exact formulation

In Chapter 10 we described the behaviour for the bending of a beam for the small strain theory. Here we present a form for cases in which large displacements with finite rotations occur. We shall, however, assume that the strains which result are small.

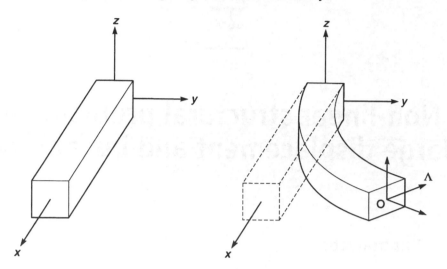

Fig. 17.1 Finite motion of three-dimensional beams.

A two-dimensional theory of beams (rods) was developed by Reissner[1] and was extended to a three-dimensional dynamic form by Simo.[2] In these developments the normal to the cross-section is followed, as contrasted to following the tangent to the beam axis, by an orthogonal frame.

Here we consider an initially straight beam for which the orthogonal triad of the beam cross-section is denoted by the vectors $\mathbf{a}_i$ (Fig. 17.1). The motion for the beam can then be written as

$$\varphi_i \equiv x_i = x_i^0 + \Lambda_{iI} Z_I \tag{17.1}$$

where the orthogonal matrix is related to the $\mathbf{a}_i$ vectors as

$$\Lambda = \begin{bmatrix} \mathbf{a}_1 & \mathbf{a}_2 & \mathbf{a}_3 \end{bmatrix} \tag{17.2}$$

If we assume that the reference coordinate X_1 (X) is the beam axis and X_2, X_3 (Y, Z) are the axes of the cross-section the above motion may be written in matrix form as

$$\begin{Bmatrix} x_1 \\ x_2 \\ x_3 \end{Bmatrix} = \begin{Bmatrix} x \\ y \\ z \end{Bmatrix} = \begin{Bmatrix} X \\ 0 \\ 0 \end{Bmatrix} + \begin{Bmatrix} u \\ v \\ w \end{Bmatrix} + \begin{bmatrix} \Lambda_{11} & \Lambda_{12} & \Lambda_{13} \\ \Lambda_{21} & \Lambda_{22} & \Lambda_{23} \\ \Lambda_{31} & \Lambda_{32} & \Lambda_{33} \end{bmatrix} \begin{Bmatrix} 0 \\ Y \\ Z \end{Bmatrix} \tag{17.3}$$

where $u(X)$, $v(X)$, and $w(X)$ are displacements of the beam reference axis and where $\Lambda(X)$ is the rotation of the beam cross-section which does not necessarily remain normal to the beam axis and thus admits the possibility of transverse shearing deformations.

The derivation of the deformation gradient for Eq. (17.3) requires computation of the derivatives of the displacements and the rotation matrix. The derivative of the rotation matrix is given by[2,3]

$$\Lambda_{,X} = \hat{\boldsymbol{\theta}}_{,X} \Lambda \tag{17.4}$$

where $\hat{\boldsymbol{\theta}}_{,X}$ denotes a skew symmetric matrix for the derivatives of a rotation vector $\boldsymbol{\theta}$ and is expressed by

$$\hat{\theta} = \begin{bmatrix} 0 & -\theta_{Z,X} & \theta_{Y,X} \\ \theta_{Z,X} & 0 & -\theta_{X,X} \\ -\theta_{Y,X} & \theta_{X,X} & 0 \end{bmatrix} \tag{17.5}$$

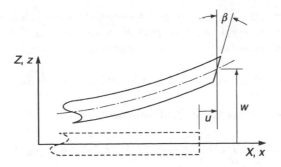

Fig. 17.2 Deformed beam configuration.

Here we consider in detail the two-dimensional case where the motion is restricted to the X–Z plane. The orthogonal matrix may then be represented as $(\theta_Y = \beta)$

$$\Lambda = \begin{bmatrix} \cos\beta & 0 & \sin\beta \\ 0 & 1 & 0 \\ -\sin\beta & 0 & \cos\beta \end{bmatrix} \tag{17.6}$$

Inserting this in Eq. (17.3) and expanding, the deformed position then is described compactly by

$$
\begin{aligned}
x &= X + u(X) + Z \sin\beta(X) \\
y &= Y \\
z &= w(X) + Z \cos\beta(X)
\end{aligned}
\tag{17.7}
$$

This results in the deformed configuration for a beam shown in Fig. 17.2. It is a two-dimensional specialization of the theory presented by Simo and co-workers[2,4,5] and is called *geometrically exact* since no small-angle approximations are involved. The deformation gradient for this displacement is given by the relation

$$F_{iI} = \begin{bmatrix} [1 + u_{,X} + Z\beta_{,X}\cos\beta] & 0 & \sin\beta \\ 0 & 1 & 0 \\ [w_{,X} - Z\beta_{,X}\sin\beta] & 0 & \cos\beta \end{bmatrix} \tag{17.8}$$

Using Eq.(10.15) and computing the Green–Lagrange strain tensor, two non-zero components are obtained which, ignoring a quadratic term in Z, are expressed by

$$
\begin{aligned}
E_{XX} &= u_{,X} + \tfrac{1}{2}(u_{,X}^2 + w_{,X}^2) + Z\Lambda\beta_{,X} = E^0 + ZK^b \\
2E_{XZ} &= (1 + u_{,X})\sin\beta + w_{,X}\cos\beta = \Gamma
\end{aligned}
\tag{17.9}
$$

where E^0 and Γ are strains which are constant on the cross-section and K^b measures change in rotation (curvature) of the cross-sections and

$$\Lambda = (1 + u_{,X})\cos\beta - w_{,X}\sin\beta \tag{17.10}$$

A variational equation for the beam can be written now by introducing second Piola–Kirchhoff stresses as described in Chapter 5 to obtain

$$\delta\Pi = \int_\Omega (\delta E_{XX} S_{XX} + 2\delta E_{XZ} S_{XZ})\, \mathrm{d}V - \delta\Pi_{\text{ext}} \tag{17.11}$$

where $\delta\Pi_{ext}$ denotes the terms from end forces and loading along the length. If we separate the volume integral into one along the length times an integral over the beam cross-sectional area A and define force resultants as

$$T^{p} = \int_{A} S_{XX} \, dA, \quad S^{p} = \int_{A} S_{XZ} \, dA \quad \text{and} \quad M^{b} = \int_{A} S_{XX} Z \, dA \tag{17.12}$$

the variational equation may be written compactly as

$$\delta\Pi = \int_{L} (\delta E^{0} T^{p} + \delta\Gamma S^{p} + \delta K^{b} M^{b}) \, dX - \delta\Pi_{ext} \tag{17.13}$$

where virtual strains for the beam are given by

$$\begin{aligned}
\delta E^{0} &= (1 + u_{,X})\delta u_{,X} + w_{,X}\delta w_{,X} \\
\delta\Gamma &= \sin\beta\delta u_{,X} + \cos\beta\delta w_{,X} + \Lambda\delta\beta \\
\delta K^{b} &= \Lambda\delta\beta_{,X} + \Gamma\delta\beta + \cos\beta\delta u_{,X} + \sin\beta\delta w_{,X}
\end{aligned} \tag{17.14}$$

A finite element approximation for the displacements may be introduced in a manner identical to that used in Sec. 17.4 for axisymmetric shells. Accordingly, we can write

$$\begin{Bmatrix} u \\ w \\ \beta \end{Bmatrix} = N_{a}(X) \begin{Bmatrix} \tilde{u}_{a} \\ \tilde{w}_{a} \\ \tilde{\beta}_{a} \end{Bmatrix} \tag{17.15}$$

where the shape functions for each variable are the same. Using this approximation the virtual work is computed as

$$\delta\Pi = \begin{bmatrix} \delta\tilde{u}_{a} & \delta\tilde{w}_{a} & \delta\tilde{\beta}_{a} \end{bmatrix} \int_{L} \mathbf{B}_{a}^{T} \begin{Bmatrix} T^{p} \\ S^{p} \\ M^{b} \end{Bmatrix} dX - \delta\Pi_{ext} \tag{17.16}$$

where

$$\mathbf{B}_{a} = \begin{bmatrix} (1 + u_{,X})N_{a,X} & w_{,X}N_{a,X} & 0 \\ \sin\beta N_{a,X} & \cos\beta N_{a,X} & \Lambda N_{a} \\ \beta_{,X}\cos\beta N_{a,X} & -\beta_{,X}\sin\beta N_{a,X} & (\Lambda N_{a,X} - \Gamma\beta_{,X}N_{a}) \end{bmatrix} \tag{17.17}$$

Just as for the axisymmetric shell described in Sec. 17.4 this interpolation will lead to 'shear locking' and it is necessary to compute the integrals for stresses by using a 'reduced quadrature'. For a two-node beam element this implies use of one quadrature point for each element. Alternatively, a mixed formulation where Γ and S^{p} are assumed constant in each element can be introduced as was done in Sec. 12.6 for the bending analysis of plates using the T6S3B3 element.

The non-linear equilibrium equation for a quasi-static problem that is solved at each load level (or time) is given by

$$\Psi_{n+1} = \mathbf{f}_{n+1} - \int_{L} \mathbf{B}_{a}^{T} \begin{Bmatrix} T^{p}_{n+1} \\ S^{p}_{n+1} \\ M^{b}_{n+1} \end{Bmatrix} dX = 0 \tag{17.18}$$

For a Newton-type solution the tangent stiffness matrix is deduced by a linearization of Eq. (17.18). To give a specific relation for the derivation we assume, for simplicity, the strains are small and the constitution may be expressed by a linear elastic relation between the Green–Lagrange strains and the second Piola–Kirchhoff stresses. Accordingly, we take

$$S_{XX} = E E_{XX} \quad \text{and} \quad S_{XZ} = 2G E_{XZ} \tag{17.19}$$

where E is a Young's modulus and G a shear modulus. Integrating Eq. (17.12) the elastic behaviour of the beam resultants becomes

$$T^{\mathrm{p}} = EAE^0, \quad S^{\mathrm{p}} = \kappa GA\Gamma \quad \text{and} \quad M^{\mathrm{b}} = EIK^{\mathrm{b}}$$

in which A is the cross-sectional area, I is the moment of inertia about the centroid, and κ is a shear correction factor to account for the fact that S_{XZ} is not constant on the cross-section. Using these relations the linearization of Eq. (17.18) gives the tangent stiffness

$$(\mathbf{K}_{\mathrm{T}})_{ab} = \int_L \mathbf{B}_a^T \mathbf{D}_{\mathrm{T}} \mathbf{B}_b \, dX + (\mathbf{K}_G)_{ab} \tag{17.20}$$

where for the simple elastic relation Eq. (17.20)

$$\mathbf{D}_{\mathrm{T}} = \begin{bmatrix} EA & & \\ & \kappa GA & \\ & & EI \end{bmatrix} \tag{17.21}$$

and $\mathbf{K}_G$ is the geometric stiffness resulting from linearization of the non-linear expression for $\mathbf{B}$. After some algebra the reader can verify that the geometric stiffness is given by

$$
(\mathbf{K}_G)_{ab} = \int_L \left(N_{a,X} \begin{bmatrix} T^{\mathrm{p}} & 0 & M^{\mathrm{b}}\cos\beta \\ 0 & T^{\mathrm{p}} & -M^{\mathrm{b}}\sin\beta \\ M^{\mathrm{b}}\cos\beta & -M^{\mathrm{b}}\sin\beta & 0 \end{bmatrix} N_{b,X} + N_a \begin{bmatrix} 0 & 0 & 0 \\ 0 & 0 & 0 \\ 0 & 0 & G_3 \end{bmatrix} N_b \right.
$$

$$
\left. + N_{a,X} \begin{bmatrix} 0 & 0 & G_1 \\ 0 & 0 & G_2 \\ 0 & 0 & -M^{\mathrm{b}}\Gamma \end{bmatrix} N_b + N_a \begin{bmatrix} 0 & 0 & 0 \\ 0 & 0 & 0 \\ G_1 & G_2 & -M^{\mathrm{b}}\Gamma \end{bmatrix} N_{b,X} \right) dX
\tag{17.22}
$$

where

$$G_1 = S^{\mathrm{p}}\cos\beta - M^{\mathrm{b}}\beta_{,X}\sin\beta, \quad G_2 = -S^{\mathrm{p}}\sin\beta - M^{\mathrm{b}}\beta_{,X}\cos\beta,$$

and

$$G_3 = -S^{\mathrm{p}}\Gamma - M^{\mathrm{b}}\beta_{,X}\Lambda$$

17.2.2 Large displacement formulation with small rotations

In many applications the full non-linear displacement field with finite rotations is not needed; however, the behaviour is such that limitations of the small displacement theory are not appropriate. In such cases we can assume that rotations are small so that the trigonometric functions may be approximated as

$$\sin \beta \approx \beta \quad \text{and} \quad \cos \beta \approx 1$$

In this case the displacement approximations become

$$
\begin{aligned}
x &= X + u(X) + Z\beta(X) \\
y &= Y \\
z &= w(X) + Z
\end{aligned}
\tag{17.23}
$$

which yield now the non-zero Green–Lagrange strain expressions

$$
\begin{aligned}
E_{XX} &= u_{,X} + \tfrac{1}{2}(u_{,X}^2 + w_{,X}^2) + Z\beta_{,X} = E^0 + ZK^{\mathrm{b}} \\
2E_{XZ} &= w_{,X} + \beta = \Gamma
\end{aligned}
\tag{17.24}
$$

where terms in Z^2 as well as products of β with derivatives of displacements are ignored. With this approximation and again using Eq. (17.15) for the finite element representation of the displacements in each element we obtain the set of non-linear equilibrium equations given by Eq. (17.18) in which now

$$
\mathbf{B}_a =
\begin{bmatrix}
(1 + u_{,X})N_{a,X} & w_{,X}N_{a,X} & 0 \\
0 & N_{a,X} & N_a \\
0 & 0 & N_{a,X}
\end{bmatrix}
\tag{17.25}
$$

This expression results in a much simpler geometric stiffness term in the tangent matrix given by Eq. (17.20) and may be written simply as

$$
(\mathbf{K}_{\mathrm{G}})_{ab} = \int_L N_{a,X}
\begin{bmatrix}
T^{\mathrm{p}} & 0 & 0 \\
0 & T^{\mathrm{p}} & 0 \\
0 & 0 & 0
\end{bmatrix}
N_{b,X}\, \mathrm{d}X
\tag{17.26}
$$

It is also possible to reduce the theory further by assuming shear deformations to be negligible so that from $\Gamma = 0$ we have

$$\beta = -w_{,X} \tag{17.27}$$

Taking the approximations now in the form

$$
\begin{aligned}
u &= N_a^u \tilde{u}_a \\
w &= N_a^w \tilde{w}_a + N_a^\beta \tilde{\beta}_a
\end{aligned}
\tag{17.28}
$$

in which $\tilde{\beta}_a \equiv -\tilde{w}_{a,X}$ at nodes.

The equilibrium equation is now given by

$$
\mathbf{\Psi}_{n+1} = \mathbf{f}_{n+1} - \int_L \mathbf{B}_a^T
\begin{Bmatrix}
T_{n+1}^{\mathrm{p}} \\
M_{n+1}^{\mathrm{b}}
\end{Bmatrix}
\mathrm{d}X = \mathbf{0}
\tag{17.29}
$$

where the strain–displacement matrix is expressed as

$$
\mathbf{B}_a = \begin{bmatrix} (1 + u_{,X})N^u_{a,X} & w_{,X}N^w_{a,X} & w_{,X}N^\beta_{a,X} \\ 0 & -N^w_{a,XX} & -N^\beta_{a,XX} \end{bmatrix}
\tag{17.30}
$$

The tangent matrix is given by Eq. (17.20) where the elastic tangent moduli involve only the terms from T^p and M^b as

$$
\mathbf{D}_T = \begin{bmatrix} EA & 0 \\ 0 & EI \end{bmatrix}
\tag{17.31}
$$

and the geometric tangent is given by

$$
(\mathbf{K}_G)_{ab} = \int_L N_{a,X} \begin{bmatrix} N^u_{a,X} T^p N^u_{b,X} & 0 & 0 \\ 0 & N^w_{a,X} T^p N^w_{b,X} & N^w_{a,X} T^p N^\beta_{b,X} \\ 0 & N^\beta_{a,X} T^p N^w_{b,X} & N^\beta_{a,X} T^p N^\beta_{b,X} \end{bmatrix} dX
\tag{17.32}
$$

Example 17.1   A clamped–hinged arch

To illustrate the performance and limitations of the above formulations we consider the behaviour of a circular arch with one boundary clamped, the other boundary hinged and loaded by a single point load, as shown in Fig. 17.3(a). Here it is necessary to introduce a transformation between the axes used to define each beam element and the global axes used to define the arch. This follows standard procedures as used many times previously. The cross-section of the beam is a unit square with other properties as shown in the figure. An analytical solution to this problem has been obtained by da Deppo and Schmidt[6] and an early finite element solution by Wood and Zienkiewicz.[7] Here a solution is obtained using 40 two-node elements of the types presented in this section. The problem produces a complex load displacement history with 'softening' behaviour that is traced using the arc-length method described in Sec. 3.2.6 [Fig. 17.3(b)]. It is observed from Fig. 17.3(b) that the assumption of small rotation produces an accurate trace of the behaviour only during the early parts of loading and also produces a limit state which is far from reality. This emphasizes clearly the type of discrepancies that can occur by misusing a formulation in which assumptions are involved.

Deformed configurations during the deformation history are shown for the load parameter $\beta = EI/PR^2$ in Fig. 17.4. In Fig. 17.4(a) we show the deformed configuration for five loading levels – three before the limit load is reached and two after passing the limit load. It will be observed that continued loading would not lead to correct solutions unless a contact state is used between the support and the arch member. This aspect was considered by Simo et al.[8] and loading was applied much further into the deformation process. In Fig. 17.4(b) we show a comparison of the deformed shapes for $\beta = 3.0$ where the small-angle assumption is still valid.

17.3 Elastic stability – energy interpretation

The energy expression given in Eq. (10.37) and the equilibrium behaviour deduced from the first variation given by Eq. (10.42) may also be used to assess the *stability* of

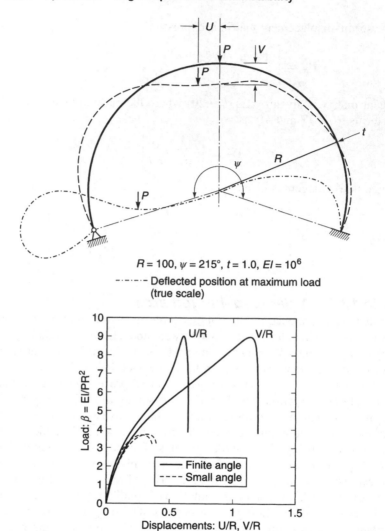

Fig. 17.3 Clamped–hinged arch: (a) problem definition; (b) load deflection.

equilibrium.[9] For an equilibrium state we always have

$$\delta\Pi = -\delta\tilde{\mathbf{u}}^T \mathbf{\Psi} = 0 \tag{17.33}$$

that is, *the total potential energy is stationary* [which, ignoring inertia effects, is equivalent to Eq. (10.65)].

The second variation of Π is

$$\delta^2\Pi = \delta(\delta\Pi) = -\delta\tilde{\mathbf{u}}^T \delta\mathbf{\Psi} = \delta\tilde{\mathbf{u}}^T \mathbf{K}_T \delta\tilde{\mathbf{u}} \tag{17.34}$$

The stability criterion is given by a positive value of this second variation and, conversely, instability by a negative value (as in the first case energy has to be added to the structure whereas in the second it contains surplus energy). In other words, *if* $\mathbf{K}_T$ *is positive definite, stability exists.* This criterion is well known[9] and of considerable

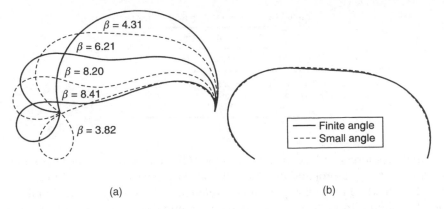

Fig. 17.4 Clamped–hinged arch: deformed shapes. (a) Finite-angle solution; (b) finite-angle form compared with small-angle form.

use when investigating stability during large deformation.[10,11] An alternative test is to investigate the sign of the determinant of $\mathbf{K}_T$, a positive sign denoting stability.[12]

A limit on stability exists when the second variation is zero. We note from Eq. (10.66) that the stability test can then be written as (assuming $\mathbf{K}_L$ is zero)

$$\delta\tilde{\mathbf{u}}^T\mathbf{K}_M\delta\tilde{\mathbf{u}} + \delta\tilde{\mathbf{u}}^T\mathbf{K}_G\delta\tilde{\mathbf{u}} = 0 \qquad (17.35)$$

This may be written in the *Rayleigh quotient form*[13]

$$\frac{\delta\tilde{\mathbf{u}}^T\mathbf{K}_M\delta\tilde{\mathbf{u}}}{\delta\tilde{\mathbf{u}}^T\mathbf{K}_G\delta\tilde{\mathbf{u}}} = -\lambda \qquad (17.36)$$

where we have

$$\lambda \begin{cases} < 1, & \text{stable} \\ = 1, & \text{stability limit} \\ > 1, & \text{unstable} \end{cases} \qquad (17.37)$$

The limit of stability is sometimes called *neutral equilibrium* since the configuration may be changed by a small amount without affecting the value of the second variation (i.e. equilibrium balance). Several options exist for implementing the above test and the simplest is to let $\lambda = 1 + \Delta\lambda$ and write the problem in the form of a generalized linear eigenproblem given by

$$\mathbf{K}_T\delta\mathbf{u} = \Delta\lambda\mathbf{K}_G\delta\mathbf{u} \qquad (17.38)$$

Here we seek the solution where $\Delta\lambda$ is zero to define a stability limit. This form uses the usual tangent matrix directly and requires only a separate implementation for the geometric term and availability of a general eigensolution routine. To maintain numerical conditioning in the eigenproblem near a buckling or limit state where $\mathbf{K}_T$ is singular a *shift* may be used as described for the vibration problem in Chapter 16 of reference 14.

Example 17.2   *Euler buckling – propped cantilever*
As an example of the stability test we consider the buckling of a straight beam with one end fixed and the other on a roller support. We can also use this example to show the usefulness of the small-angle beam theory.

Table 17.1 Linear buckling load estimates

Number of elements		
20	100	500
20.36	20.19	20.18
61.14	59.67	59.61
124.79	118.85	118.62

An axial compressive load is applied to the roller end and the Euler buckling load computed. This is a problem in which the displacement prior to buckling is purely axial. The buckling load may be estimated relative to the small deformation theory by using the solution from the first tangent matrix computed. Alternatively, the buckling load can be computed by increasing the load until the tangent matrix becomes singular. In the case of a structure where the distribution of the internal forces does not change with load level and material is linear elastic there is no difference in the results obtained. Table 17.1 shows the results obtained for the propped cantilever using different numbers of elements. Here it is observed that accurate results for higher modes require use of more elements; however, both the finite rotation and small rotation formulations given above yield identical answers since no rotation is present prior to buckling. The properties used in the analysis are $E = 12 \times 10^6$, $A = 1$, $I = 1/12$, and length $L = 100$. The classical Euler buckling load is given by

$$P_{cr} = \alpha \frac{EI}{L^2} \tag{17.39}$$

with the lowest buckling load given as $\alpha = 20.18$.[15]

17.4 Large displacement theory of thick plates

17.4.1 Definitions

The small rotation form for beams described in Sec. 17.2.2 may be used to consider problems associated with deformation of plates subject to 'in-plane' and 'lateral' forces, when displacements are not infinitesimal but also not excessively large (Fig. 17.5). In this situation the 'change-in-geometry' effect is less important than the relative magnitudes of the linear and non-linear strain–displacement terms, and in fact for 'stiffening' problems the non-linear displacements are always less than the corresponding linear ones (see Fig. 17.6). It is well known that in such situations the lateral displacements will be responsible for development of 'membrane'-type strains and now the two problems of 'in-plane' and 'lateral' deformation can no longer be dealt with separately but are *coupled*.

Generally, for plates the rotation angles remain small unless in-plane strains also become large. To develop the equations for small rotations in which plate bending is modelled using the formulations discussed in Chapter 12 we generalize the displacement field given in Eq. (4.9) to include the effects of in-plane displacements.

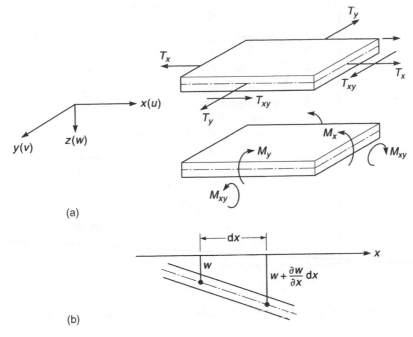

(a)

(b)

Fig. 17.5 (a) 'In-plane' and bending resultants for a flat plate; (b) increase of middle surface length owing to lateral displacement.

Accordingly, we write

$$\mathbf{u} = \begin{Bmatrix} u_1 \\ u_2 \\ u_3 \end{Bmatrix} = \begin{Bmatrix} u(X, Y) \\ v(X, Y) \\ w(X, Y) \end{Bmatrix} + Z \begin{Bmatrix} \phi_X(X, Y) \\ \phi_Y(X, Y) \\ 0 \end{Bmatrix} \qquad (17.40)$$

where ϕ are small rotations defined according to Fig. 11.3 and X, Y, Z denote positions in the reference configuration of the plate. Using these to compute the Green–Lagrange strains given by Eq. (5.15) we can write the non-zero terms as

$$\begin{Bmatrix} E_{XX} \\ E_{YY} \\ 2E_{XY} \\ 2E_{XZ} \\ 2E_{YZ} \end{Bmatrix} = \begin{Bmatrix} u_{,X} + \frac{1}{2}(w_{,X})^2 \\ v_{,Y} + \frac{1}{2}(w_{,Y})^2 \\ u_{,Y} + v_{,X} + w_{,X}w_{,Y} \\ \phi_X + w_{,X} \\ \phi_Y + w_{,Y} \end{Bmatrix} + Z \begin{Bmatrix} \phi_{X,X} \\ \phi_{Y,Y} \\ \phi_{X,Y} + \phi_{Y,X} \\ 0 \\ 0 \end{Bmatrix} \qquad (17.41)$$

In these expressions we have used classical results[16] that ignore all square terms involving ϕ and derivatives of u and v, as well as terms which contain quadratic powers of Z.

Generally, the position of the in-plane reference coordinates X and Y change very little during deformations and we can replace them with the current coordinates x and y just as is implicitly done for the small strain case considered in Chapter 11. Thus,

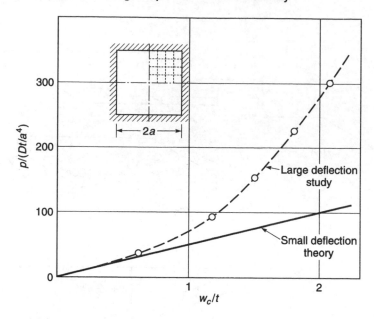

Fig. 17.6 Central deflection w_c of a clamped square plate under uniform load p;[12] $u = v = 0$ at edge.

we can represent the Green–Lagrange strains in terms of the middle surface strains and changes in curvature as

$$\mathbf{E} = \left\{ \begin{array}{c} u_{,x} + \frac{1}{2}(w_{,x})^2 \\ v_{,y} + \frac{1}{2}(w_{,y})^2 \\ u_{,y} + v_{,x} + w_{,x}w_{,y} \end{array} \right\} + Z \left\{ \begin{array}{c} \phi_{x,x} \\ \phi_{y,y} \\ \phi_{x,y} + \phi_{y,x} \end{array} \right\} = \mathbf{E}^{\mathrm{p}} + Z\mathbf{K}^{\mathrm{b}} \qquad (17.42)$$

where $\mathbf{E}^{\mathrm{p}}$ denotes the in-plane membrane strains and $\mathbf{K}^{\mathrm{b}}$ the change in curvatures owing to bending. In addition we have the transverse shearing strains given by

$$\mathbf{\Gamma}^{\mathrm{s}} = \left\{ \begin{array}{c} \phi_x + w_{,x} \\ \phi_y + w_{,y} \end{array} \right\} \qquad (17.43)$$

The variations of the strains are given by

$$\delta\mathbf{E}^{\mathrm{p}} = \left\{ \begin{array}{c} \delta u_{,x} \\ \delta v_{,y} \\ \delta u_{,y} + \delta v_{,x} \end{array} \right\} + \begin{bmatrix} w_{,x} & 0 \\ 0 & w_{,y} \\ w_{,y} & w_{,x} \end{bmatrix} \left\{ \begin{array}{c} \delta w_{,x} \\ \delta w_{,y} \end{array} \right\} \qquad (17.44)$$

$$\delta\mathbf{K}^{\mathrm{b}} = \left\{ \begin{array}{c} \delta\phi_{x,x} \\ \delta\phi_{y,y} \\ \delta\phi_{x,y} + \delta\phi_{y,x} \end{array} \right\} \quad \text{and} \quad \delta\mathbf{\Gamma}^{\mathrm{s}} = \left\{ \begin{array}{c} \delta\phi_x + \delta w_{,x} \\ \delta\phi_y + \delta w_{,y} \end{array} \right\} \qquad (17.45)$$

Using these expressions the variation of the plate equations may be expressed as

$$\delta\Pi = \int_{\Omega} (\delta\mathbf{E}^{\mathrm{p}})^T \mathbf{S} \, \mathrm{d}\Omega + \int_{\Omega} (\delta\mathbf{\Gamma}^{\mathrm{s}})^T \mathbf{S}^{\mathrm{s}} \, \mathrm{d}\Omega + \int_{\Omega} (\delta\mathbf{K}^{\mathrm{b}})^T \mathbf{S} \, Z \, \mathrm{d}\Omega - \delta\Pi_{\mathrm{ext}} \qquad (17.46)$$

Defining the integrals through the thickness in terms of the 'in-plane' membrane forces

$$\mathbf{T}^{\text{p}} = \begin{Bmatrix} T_x \\ T_y \\ T_{xy} \end{Bmatrix} = \int_{-t/2}^{t/2} \mathbf{S}\, \mathrm{d}Z \equiv \int_{-t/2}^{t/2} \begin{Bmatrix} S_{XX} \\ S_{YY} \\ S_{XY} \end{Bmatrix} \mathrm{d}Z \tag{17.47}$$

transverse shears

$$\mathbf{T}^{\text{s}} = \begin{Bmatrix} T_{xz} \\ T_{yz} \end{Bmatrix} = \int_{-t/2}^{t/2} \mathbf{S}^{\text{s}}\, \mathrm{d}Z \equiv \int_{-t/2}^{t/2} \begin{Bmatrix} S_{XZ} \\ S_{YZ} \end{Bmatrix} \mathrm{d}Z \tag{17.48}$$

and bending forces

$$\mathbf{M}^{\text{b}} = \begin{Bmatrix} M_{xx} \\ M_{yy} \\ M_{xy} \end{Bmatrix} = \int_{-t/2}^{t/2} \mathbf{S}Z\, \mathrm{d}Z \equiv \int_{-t/2}^{t/2} \begin{Bmatrix} S_{XX} \\ S_{YY} \\ S_{XY} \end{Bmatrix} Z\, \mathrm{d}Z \tag{17.49}$$

we obtain the virtual work expression for the plate, given as

$$\delta\Pi = \int_A [(\delta\mathbf{E}^{\text{p}})^T \mathbf{T}^{\text{p}} + \delta(\mathbf{\Gamma}^{\text{s}})^T \mathbf{T}^{\text{s}} + \delta(\mathbf{K}^{\text{b}})^T \mathbf{M}^{\text{b}}]\, \mathrm{d}A - \delta\Pi_{\text{ext}} \tag{17.50}$$

This may now be used to construct a finite element solution.

17.4.2 Finite element evaluation of strain–displacement matrices

For further evaluation it is necessary to establish expressions for the finite element $\mathbf{B}$ and $\mathbf{K}_T$ matrices. Introducing the finite element approximations, we have

$$\begin{Bmatrix} u \\ v \\ w \end{Bmatrix} = \begin{bmatrix} N_a & & \\ & N_a & \\ & & N_a^w \end{bmatrix} \begin{Bmatrix} \tilde{u}_a \\ \tilde{v}_a \\ \tilde{w}_a \end{Bmatrix} \tag{17.51}$$

and

$$\begin{Bmatrix} \phi_x \\ \phi_y \end{Bmatrix} = N_a^\phi \begin{Bmatrix} (\tilde{\phi}_x)_a \\ (\tilde{\phi}_y)_a \end{Bmatrix} \tag{17.52}$$

The expressions for the strain–displacement matrices are deduced from Eqs (17.44) and (17.45) as

$$\delta\mathbf{E}^{\text{p}} = \mathbf{B}_a^{\text{p}}\delta\tilde{\mathbf{a}}_a = \begin{bmatrix} N_{a,x} & 0 \\ 0 & N_{a,y} \\ N_{a,y} & N_{a,x} \end{bmatrix} \begin{Bmatrix} \delta\tilde{u}_a \\ \delta\tilde{v}_a \end{Bmatrix} + \begin{bmatrix} w_{,x} & 0 \\ 0 & w_{,y} \\ w_{,y} & w_{,x} \end{bmatrix} \mathbf{G}_a \begin{Bmatrix} \delta\tilde{w}_a \\ \delta(\tilde{\phi}_x)_a \\ \delta(\tilde{\phi}_y)_a \end{Bmatrix} \tag{17.53a}$$

$$= \mathbf{B}_a\delta\tilde{\mathbf{u}}_a + \mathbf{B}_a^{\text{L}}\delta\tilde{\mathbf{w}}_a$$

$$\delta\mathbf{\Gamma}^{\text{s}} = \mathbf{B}_a^{\text{s}}\delta\mathbf{w}_a = \begin{bmatrix} N_{a,x}^w & -N_a^\phi & 0 \\ N_{a,y}^w & 0 & -N_a^\phi \end{bmatrix} \begin{Bmatrix} \delta\tilde{w}_a \\ \delta(\tilde{\phi}_x)_a \\ \delta(\tilde{\phi}_y)_a \end{Bmatrix} \tag{17.53b}$$

and

$$\delta \mathbf{K}^b = \mathbf{B}_a^b \delta \tilde{\mathbf{w}}_a = \begin{bmatrix} 0 & N_{a,x}^\phi & 0 \\ 0 & 0 & N_{a,y}^\phi \\ 0 & N_{a,y}^\phi & N_{a,x}^\phi \end{bmatrix} \begin{Bmatrix} \delta \tilde{w}_a \\ \delta (\tilde{\phi}_x)_a \\ \delta (\tilde{\phi}_y)_a \end{Bmatrix} \tag{17.53c}$$

where

$$\mathbf{G}_a = \begin{bmatrix} N_{a,x}^w & 0 & 0 \\ N_{a,y}^w & 0 & 0 \end{bmatrix} \tag{17.53d}$$

with nodal parameters defined by

$$\tilde{\mathbf{a}}_a^T = \begin{bmatrix} \tilde{u}_a & \tilde{v}_a & \tilde{w}_a & (\tilde{\phi}_x)_a & (\tilde{\phi}_y)_a \end{bmatrix} = \begin{bmatrix} \tilde{\mathbf{u}}_a^T & \tilde{\mathbf{w}}_a^T \end{bmatrix}$$
$$\tilde{\mathbf{u}}_a^T = \begin{bmatrix} \tilde{u}_a & \tilde{v}_a \end{bmatrix} \quad \text{and} \quad \tilde{\mathbf{w}}_a^T = \begin{bmatrix} \tilde{w}_a & (\tilde{\phi}_x)_a & (\tilde{\phi}_y)_a \end{bmatrix}$$

We here immediately recognize an in-plane term which is identical to the small strain (linear) plane stress (membrane) form and a term which is identical to the small strain bending and transverse shear form. The added non-linear in-plane term results from the quadratic displacement terms in the membrane strains.

Using the above strain–displacement matrices we can now write Eq. (17.50) as

$$\delta \Pi = \delta \tilde{\mathbf{u}}_a^T \int_A (\mathbf{B}_a^p)^T \mathbf{T}^p \, dA + \delta \tilde{\mathbf{w}}_a^T \int_A (\mathbf{B}_a^s)^T \mathbf{T}^s \, dA + \delta \tilde{\mathbf{w}}_a^T \int_A (\mathbf{B}_a^b)^T \mathbf{M}^b \, dA - \delta \Pi_{\text{ext}} = 0 \tag{17.54}$$

Grouping the force terms as

$$\bar{\boldsymbol{\sigma}} = \begin{Bmatrix} \mathbf{T}^p \\ \mathbf{T}^s \\ \mathbf{M}^b \end{Bmatrix} \tag{17.55}$$

and the strain matrices as

$$\bar{\mathbf{B}}_a = \begin{bmatrix} \mathbf{B}_a & \mathbf{B}_a^L \\ 0 & \mathbf{B}_a^s \\ 0 & \mathbf{B}_a^b \end{bmatrix} \tag{17.56}$$

the virtual work expression may be written compactly as

$$\delta \Pi = \delta \tilde{\mathbf{a}}_a^T \int_A \bar{\mathbf{B}}_a^T \bar{\boldsymbol{\sigma}} \, dA - \delta \Pi_{\text{ext}} = 0 \tag{17.57}$$

The non-linear problem to be solved is thus expressed as

$$\boldsymbol{\Psi}_a = \mathbf{f}_a - \int_A \bar{\mathbf{B}}_a^T \bar{\boldsymbol{\sigma}} \, dA = \mathbf{0} \tag{17.58}$$

This may be solved by using a Newton process for which a tangent matrix is required.

17.4.3 Evaluation of tangent matrix

A tangent matrix for the non-linear plate formulation may be computed by a linearization of Eq. (17.57). Formally, this may be written as

$$d(\delta \Pi) = \delta \tilde{\mathbf{a}}_a^T \int_A \left[d(\bar{\mathbf{B}}_a^T) \bar{\boldsymbol{\sigma}} + \bar{\mathbf{B}}_a^T d(\bar{\boldsymbol{\sigma}}) \right] \, dA - d(\delta \Pi_{\text{ext}}) \tag{17.59}$$

We shall assume for simplicity that loading is conservative so that $d(\delta\Pi_{\text{ext}}) = 0$ and hence the only terms to be linearized are the strain–displacement matrix and the stress–strain relation. If we assume linear elastic behaviour, the relation between the plate forces and strains may be written as

$$
\begin{Bmatrix} \mathbf{T}^{\text{p}} \\ \mathbf{T}^{\text{s}} \\ \mathbf{M}^{\text{b}} \end{Bmatrix} = \begin{bmatrix} \mathbf{D}^{\text{p}} & \mathbf{0} & \mathbf{0} \\ \mathbf{0} & \mathbf{D}^{\text{s}} & \mathbf{0} \\ \mathbf{0} & \mathbf{0} & \mathbf{D}^{\text{b}} \end{bmatrix} \begin{Bmatrix} \mathbf{E}^{\text{p}} \\ \boldsymbol{\Gamma}^{\text{s}} \\ \mathbf{K}^{\text{b}} \end{Bmatrix}
\tag{17.60}
$$

where for an isotropic homogeneous plate

$$
\mathbf{D}^{\text{p}} = \frac{Et}{1-\nu^2} \begin{bmatrix} 1 & \nu & 0 \\ \nu & 1 & 0 \\ 0 & 0 & (1-\nu)/2 \end{bmatrix}, \quad \mathbf{D}^{\text{s}} = \frac{\kappa Et}{2(1+\nu)} \begin{bmatrix} 1 & 0 \\ 0 & 1 \end{bmatrix}, \quad \text{and} \quad \mathbf{D}^{\text{b}} = \frac{t^2}{12}\mathbf{D}^{\text{p}}
\tag{17.61}
$$

Again, κ is a shear correction factor which, for homogeneous plates, is usually taken as $5/6$. Thus, the linearization of the constitution becomes

$$
d(\bar{\sigma}) = \begin{Bmatrix} d(\mathbf{T}^{\text{p}}) \\ d(\mathbf{T}^{\text{s}}) \\ d(\mathbf{M}^{\text{b}}) \end{Bmatrix} = \begin{bmatrix} \mathbf{D}^{\text{p}} & \mathbf{0} & \mathbf{0} \\ \mathbf{0} & \mathbf{D}^{\text{s}} & \mathbf{0} \\ \mathbf{0} & \mathbf{0} & \mathbf{D}^{\text{b}} \end{bmatrix} \begin{Bmatrix} d(\mathbf{E}^{\text{p}}) \\ d(\boldsymbol{\Gamma}^{\text{s}}) \\ d(\mathbf{K}^{\text{b}}) \end{Bmatrix}
\tag{17.62}
$$

$$
= \begin{bmatrix} \mathbf{D}^{\text{p}} & \mathbf{0} & \mathbf{0} \\ \mathbf{0} & \mathbf{D}^{\text{s}} & \mathbf{0} \\ \mathbf{0} & \mathbf{0} & \mathbf{D}^{\text{b}} \end{bmatrix} \begin{bmatrix} \mathbf{B}_b & \mathbf{B}_b^{\text{L}} \\ \mathbf{0} & \mathbf{B}_b^{\text{s}} \\ \mathbf{0} & \mathbf{B}_b^{\text{b}} \end{bmatrix} \begin{Bmatrix} d(\tilde{\mathbf{u}}_b) \\ d(\tilde{\mathbf{w}}_b) \end{Bmatrix}
$$

Using this result the material part of the tangent matrix is expressed as

$$
(\mathbf{K}_{\text{M}})_{ab} = \int_A \begin{bmatrix} (\mathbf{B}_a^{\text{p}})^T & \mathbf{0} & \mathbf{0} \\ (\mathbf{B}_a^{\text{L}})^T & (\mathbf{B}_a^{\text{s}})^T & (\mathbf{B}_a^{\text{b}})^T \end{bmatrix} \begin{bmatrix} \mathbf{D}^{\text{p}} & \mathbf{0} & \mathbf{0} \\ \mathbf{0} & \mathbf{D}^{\text{s}} & \mathbf{0} \\ \mathbf{0} & \mathbf{0} & \mathbf{D}^{\text{b}} \end{bmatrix} \begin{bmatrix} \mathbf{B}_b^{\text{p}} & \mathbf{B}_b^{\text{L}} \\ \mathbf{0} & \mathbf{B}_b^{\text{s}} \\ \mathbf{0} & \mathbf{B}_b^{\text{b}} \end{bmatrix} \, dA
\tag{17.63}
$$

$$
= \int_A \bar{\mathbf{B}}_a^T \mathbf{D}_{\text{T}} \bar{\mathbf{B}}_b \, dA = \begin{bmatrix} (\mathbf{K}_{\text{M}}^{\text{p}})_{ab} & (\mathbf{K}_{\text{M}}^{\text{L}})_{ab} \\ (\mathbf{K}_{\text{M}}^{\text{L}})_{ab}^T & (\mathbf{K}_{\text{M}}^{\text{b}})_{ab} \end{bmatrix}
$$

where $\mathbf{D}_{\text{T}}$ is the coefficient matrix from Eq. (17.60), and the individual parts of the tangent matrix are

$$
(\mathbf{K}_{\text{M}}^{\text{p}})_{ab} = \int_A (\mathbf{B}_a^{\text{p}})^T \mathbf{D}^{\text{p}} \mathbf{B}_b \, dA
$$

$$
(\mathbf{K}_{\text{M}}^{\text{L}})_{ab} = \int_A (\mathbf{B}_a^{\text{p}})^T \mathbf{D}^{\text{p}} \mathbf{B}_b^{\text{L}} \, dA
\tag{17.64}
$$

$$
(\mathbf{K}_{\text{M}}^{\text{b}})_{ab} = \int_A \left[(\mathbf{B}_a^{\text{s}})^T \mathbf{D}^{\text{s}} \mathbf{B}_b^{\text{s}} + (\mathbf{B}_a^{\text{b}})^T \mathbf{D}^{\text{b}} \mathbf{B}_b^{\text{b}} \right] dA
$$

We immediately recognize that the material part of the tangent matrix consists of the same result as that of the small displacement analysis except for the added term $\mathbf{K}_{\text{M}}^{\text{L}}$ which establishes coupling between membrane and bending behaviour.

The remainder of the computation for the tangent involves the linearization of the non-linear part of the strain–displacement matrix, $\mathbf{B}_a^{\mathrm{L}}$. As in the continuum problem discussed in Chapter 5 it is easiest to rewrite this term as

$$d(\mathbf{B}_a^{\mathrm{L}})^T \mathbf{T}^{\mathrm{b}} = \mathbf{G}_a^T \begin{bmatrix} d(w_{,x}) & 0 & d(w_{,y}) \\ 0 & d(w_{,y}) & d(w_{,x}) \end{bmatrix} \begin{Bmatrix} T_x^{\mathrm{p}} \\ T_y^{\mathrm{p}} \\ T_{xy}^{\mathrm{p}} \end{Bmatrix}$$

$$= \mathbf{G}_a^T \begin{bmatrix} T_x^{\mathrm{p}} & T_{xy}^{\mathrm{p}} \\ T_{xy}^{\mathrm{p}} & T_y^{\mathrm{p}} \end{bmatrix} \begin{Bmatrix} d(w_{,x}) \\ d(w_{,y}) \end{Bmatrix}$$

(17.65)

This may now be expressed in terms of finite element interpolations to obtain the geometric part of the tangent as

$$(\mathbf{K}_G^{\mathrm{L}})_{ab} = \int_A \mathbf{G}_a^T \begin{bmatrix} T_x^{\mathrm{p}} & T_{xy}^{\mathrm{p}} \\ T_{xy}^{\mathrm{p}} & T_y^{\mathrm{p}} \end{bmatrix} \mathbf{G}_b \, dA$$

(17.66)

which is inserted into the total geometric tangent as

$$(\mathbf{K}_G)_{ab} = \begin{bmatrix} \mathbf{0} & \mathbf{0} \\ \mathbf{0} & (\mathbf{K}_G^{\mathrm{L}})_{ab} \end{bmatrix}$$

(17.67)

This geometric matrix is also referred to in the literature as the *initial stress matrix* for plate bending.

17.5 Large displacement theory of thin plates

The above theory may be specialized to the thin plate formulation by neglecting the effects of transverse shearing strains as discussed in Chapter 11. Thus setting $E_{XZ} = E_{YZ} = 0$ in Eq. (17.41), this yields the result

$$\phi_X = -w_{,X} \quad \text{and} \quad \phi_Y = -w_{,Y}$$

(17.68)

The displacements of the plate middle surface may then be approximated as

$$\mathbf{u} = \begin{Bmatrix} u_1 \\ u_2 \\ u_3 \end{Bmatrix} = \begin{Bmatrix} u(X, Y) \\ v(X, Y) \\ w(X, Y) \end{Bmatrix} - Z \begin{Bmatrix} w_{,X}(X, Y) \\ w_{,Y}(X, Y) \\ 0 \end{Bmatrix}$$

(17.69)

Once again we can note that in-plane positions X and Y do not change significantly, thus permitting substitution of x and y in the strain expressions to obtain Green–Lagrange strains as

$$\mathbf{E} = \begin{Bmatrix} u_{,x} + \frac{1}{2}(w_{,x})^2 \\ v_{,y} + \frac{1}{2}(w_{,y})^2 \\ u_{,y} + v_{,x} + w_{,x} w_{,y} \end{Bmatrix} - Z \begin{Bmatrix} w_{,xx} \\ w_{,yy} \\ 2w_{,xy} \end{Bmatrix} = \mathbf{E}^{\mathrm{p}} + Z\mathbf{K}^{\mathrm{b}}$$

(17.70)

where we have once again neglected square terms involving derivatives of the in-plane displacements and terms in Z^2. We note now that introduction of Eq. (17.68) modifies the expression for change in curvature to the same form as that used for thin plates in Chapter 11.

17.5.1 Evaluation of strain–displacement matrices

For further formulation it is again necessary to establish expressions for the $\bar{\mathbf{B}}$ and $\mathbf{K}_T$ matrices. The finite element approximations to the displacements now involve only u, v, and w. Here we assume these to be expressed in the form

$$\begin{Bmatrix} u \\ v \end{Bmatrix} = N_a \begin{Bmatrix} \tilde{u}_a \\ \tilde{v}_a \end{Bmatrix} = N_a \tilde{\mathbf{u}}_a \tag{17.71}$$

and

$$w = N_a^w \tilde{w}_a + \mathbf{N}_a^\phi \tilde{\phi}_a \tag{17.72}$$

where now the rotation parameters are defined as

$$\tilde{\phi}_a^T = \left[(\tilde{\phi}_x)_a \quad (\tilde{\phi}_y)_a \right] = - \left[(\tilde{w}_{,x})_a \quad (\tilde{w}_{,y})_a \right] \tag{17.73}$$

The expressions for $\mathbf{B}^p$ and $\mathbf{B}^L$ are identical to those given previously except for the definition of $\mathbf{G}$. Owing to the form of the interpolation for w, we now obtain

$$\mathbf{G}_a = \begin{bmatrix} N_{a,x}^w & N_{a,x}^{\phi x} & N_{a,x}^{\phi y} \\ N_{a,y}^w & N_{a,y}^{\phi x} & N_{a,y}^{\phi y} \end{bmatrix} \tag{17.74}$$

The variation in curvature for the thin plate is given by

$$\delta \mathbf{K}^b = - \begin{bmatrix} N_{a,xx}^w & N_{a,xx}^{\phi x} & N_{a,xx}^{\phi y} \\ N_{a,yy}^w & N_{a,yy}^{\phi x} & N_{a,yy}^{\phi y} \\ 2N_{a,xy}^w & 2N_{a,xy}^{\phi x} & 2N_{a,xy}^{\phi y} \end{bmatrix} \begin{Bmatrix} \delta \tilde{w}_a \\ (\delta \tilde{\phi}_x)_a \\ (\delta \tilde{\phi}_y)_a \end{Bmatrix} \tag{17.75}$$

$$= \mathbf{B}_a^b \delta \tilde{\mathbf{w}}_a \tag{17.76}$$

Grouping the force terms, now without the shears $\mathbf{T}^s$, as

$$\bar{\sigma} = \begin{Bmatrix} \mathbf{T}^p \\ \mathbf{M}^b \end{Bmatrix} \tag{17.77}$$

and the strain matrices as

$$\bar{\mathbf{B}}_a = \begin{bmatrix} \mathbf{B}_a & \mathbf{B}_a^L \\ \mathbf{0} & \mathbf{B}_a^b \end{bmatrix} \tag{17.78}$$

the virtual work expression may be written in matrix form as

$$\delta \Pi = \delta \tilde{\mathbf{u}}_a^T \int_A \bar{\mathbf{B}}_a^T \bar{\sigma} \, dA - \delta \Pi_{\text{ext}} = 0 \tag{17.79}$$

and once again a non-linear problem in the form of Eq. (17.58) is obtained.

17.5.2 Evaluation of tangent matrix

A tangent matrix for the non-linear plate formulation may be computed by a linearization of Eq. (17.57). If we again assume linear elastic behaviour, the relation between the plate forces and strains may be written as

$$\begin{Bmatrix} \mathbf{T}^p \\ \mathbf{M}^b \end{Bmatrix} = \begin{bmatrix} \mathbf{D}^p & \mathbf{0} \\ \mathbf{0} & \mathbf{D}^b \end{bmatrix} \begin{Bmatrix} \mathbf{E}^p \\ \mathbf{K}^b \end{Bmatrix} \tag{17.80}$$

where the elastic constants are given in Eq. (17.61). Thus, the linearization of the constitution becomes

$$d(\sigma) = \begin{bmatrix} \mathbf{D}^p & \mathbf{0} \\ \mathbf{0} & \mathbf{D}^b \end{bmatrix} \begin{bmatrix} \mathbf{B}^p_b & \mathbf{0} \\ \mathbf{B}^L_b & \mathbf{B}^b_b \end{bmatrix} \begin{Bmatrix} d(\tilde{\mathbf{u}}_b) \\ d(\tilde{\mathbf{w}}_b) \end{Bmatrix} \tag{17.81}$$

Using this result the material part of the tangent matrix is expressed as

$$\begin{aligned} (\mathbf{K}_M)_{ab} &= \int_A \begin{bmatrix} (\mathbf{B}^p_a)^T & \mathbf{0} \\ (\mathbf{B}^L_a)^T & (\mathbf{B}^b_a)^T \end{bmatrix} \begin{bmatrix} \mathbf{D}^p & \mathbf{0} \\ \mathbf{0} & \mathbf{D}^b \end{bmatrix} \begin{bmatrix} \mathbf{B}^p_b & \mathbf{B}^L_b \\ \mathbf{0} & \mathbf{B}^b_b \end{bmatrix} dA \\ &= \begin{bmatrix} (\mathbf{K}^p_M)_{ab} & (\mathbf{K}^L_M)_{ab} \\ (\mathbf{K}^L_M)^T_{ab} & (\mathbf{K}^b_M)_{ab} \end{bmatrix} \end{aligned} \tag{17.82}$$

where $\mathbf{K}^p_M$ and $\mathbf{K}^L_M$ are given as in Eq. (17.64), and $\mathbf{K}^b_M$ simplifies to

$$(\mathbf{K}^b_M)_{ab} = \int_A (\mathbf{B}^b_a)^T \mathbf{D}^b \mathbf{B}^b_b \, dA \tag{17.83}$$

and now $\mathbf{B}^b_a$ is given by Eq. (17.76). Using Eq. (17.74) the geometric matrix has identical form to Eqs (17.66) and (17.67).

17.6 Solution of large deflection problems

All the ingredients necessary for computing the 'large deflection' plate problem are now available. Here we may use results from either the thick or thin plate formulations described above. Below we describe the process for the thin plate formulation.

As a first step displacements $\tilde{\mathbf{a}}^0$ are found according to the small displacement uncoupled solution. This is used to determine the actual strains by considering the non-linear relations for $\mathbf{E}^p$ and the linear curvature relations for $\mathbf{K}^b$ defined in Eq. (17.70). Corresponding stresses can be found by the elastic relations and a Newton iteration process set up to solve Eq. (17.58) [which is obtained from Eq. (17.79)].

A typical solution which shows the stiffening of the plate with increasing deformation arising from the development of 'membrane' stresses was shown in Fig. 17.6.[12] The results show excellent agreement with an alternative analytical solution. The element properties were derived using for the in-plane deformation the simplest bilinear rectangle and for the bending deformation the non-conforming shape function for a rectangle (Sec. 11.3).

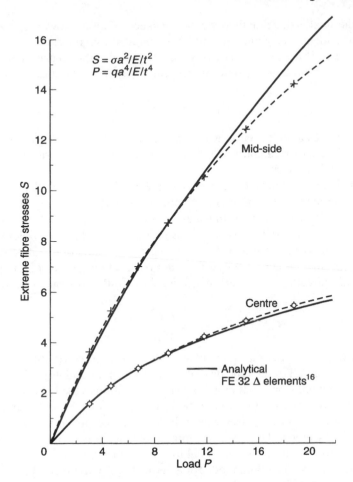

Fig. 17.7 Clamped square plate: stresses.

Example 17.3   Clamped plate subjected to uniform load

An example of the stress variation with loads for a clamped square plate under uniform dead load is shown in Fig. 17.7.[17] A quarter of the plate is analysed as above with 32 triangular elements, using the 'in-plane' triangular three-node linear element together with a modified version of the non-conforming plate bending element of Sec. 11.5.[18] Many other examples of large plate deformation obtained by finite element methods are available in the literature.[17,19–24]

17.6.1 Bifurcation instability

In a few practical cases, as in the classical Euler problem, a bifurcation instability is possible similar to the case considered for straight beams in Sec. 17.3. Consider the situation of a plate loaded purely in its own plane. As lateral deflections, w, are not produced, the small deflection theory gives an exact solution. However, even with zero lateral displacements, the geometric stiffness (initial stress) matrix can be found while $\mathbf{B}^L$ remains zero. If the in-plane stresses are compressive this matrix will be such that real eigenvalues of the bending deformation can be found by solving the eigenproblem

$$\mathbf{K}_M^b \, d\tilde{\mathbf{w}} = -\lambda \mathbf{K}_G^L \, d\tilde{\mathbf{w}} \tag{17.84}$$

in which λ denotes a multiplying factor on the in-plane stresses necessary to achieve neutral equilibrium (limit stability), and $\delta\tilde{\mathbf{w}}$ is the eigenvector describing the shape that a 'buckling' mode may take.

At such an increased load incipient buckling occurs and lateral deflections can occur without any lateral load. The problem is simply formulated by writing only the bending equations with $\mathbf{K}_M^b$ determined as in Chapter 11 and with $\mathbf{K}_G^L$ found from Eq. (17.66).

Points of such incipient stability (buckling) for a variety of plate problems have been determined using various element formulations.[25–30] Some comparative results for a simple problem of a square, simply supported plate under a uniform compression T_x applied in one direction are given in Table 17.2. In this the buckling parameter is defined as

$$C = \frac{T_x a^2}{\pi^2 D}$$

where a is the side length of a square plate and D the bending rigidity.

The elements are all of the type described in Chapter 11 and it is of interest to note that all those that are slope compatible always overestimate the buckling factor. This result is obtained only for cases where the in-plane stresses $\mathbf{T}^p$ are exact solutions to the differential equations; in cases where these are approximate solutions this bound property is not assured. The non-conforming elements in this case underestimate the load, although there is now no theoretical lower bound available.

Figure 17.8 shows a buckling mode for a geometrically more complex case.[28] Here again the non-conforming triangle was used.

Such incipient stability problems in plates are of limited practical importance. As soon as lateral deflection occurs a stiffening of the plate follows and additional loads can be carried. This stiffening was noted in the example of Fig. 17.6. Postbuckling behaviour thus should be studied by the large deformation process described in previous sections.[31,32]

Table 17.2   Values of C for a simply supported square plate and compressed axially by T_x

Elements in quarter plate	Non-compatible		Compatible	
	rectangle[27] 12 d.o.f.	triangle[28] 9 d.o.f.	rectangle[29] 16 d.o.f.	quadrilateral[30] 16 d.o.f.
2×2		3.22		
4×4	3.77	3.72	4.015	4.029
8×8	3.93	3.90	4.001	4.002

Exact $C = 4.00$.[15]
d.o.f. = degrees of freedom.

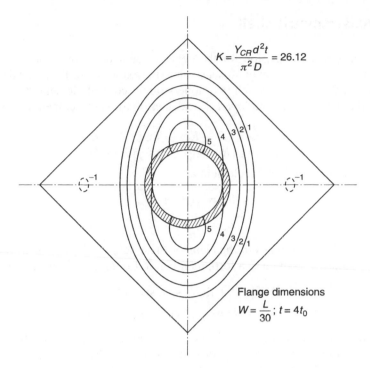

$$K = \frac{Y_{CR} d^2 t}{\pi^2 D} = 26.12$$

Flange dimensions

$$W = \frac{L}{30}; \; t = 4t_0$$

Fig. 17.8 Buckling mode of a square plate under shear: clamped edges, central hole stiffened by flange.[20]

17.7 Shells

In shells, non-linear response and stability problems are much more relevant than in plates. Here, in general, the problem is one in which the tangential stiffness matrix $\mathbf{K}_T$ should always be determined taking the actual displacements into account, as now the special case of uncoupled membrane and bending effects does not occur under load except in the most trivial cases. If the *initial stability* matrix $\mathbf{K}_G$ is determined for the elastic stresses it is, however, sometimes possible to obtain useful results concerning the stability factor λ, and indeed in the classical work on the subject of shell buckling this initial stability often has been considered. The true collapse load may, however, be well *below* the initial stability load and it is important to determine at least approximately the deformation effects.

   If the shell is assumed to be built up of flat plate elements, the same transformations as given in Chapter 13 can be followed with the plate tangential stiffness matrix.[33] If curved shell elements are used it is important to revert to the equations of shell theory and to include in these the non-linear terms.[12,34–36] Alternatively, one may approach the problem from a degeneration of solids, as described in Chapter 15 for the small deformation case, suitably extended to the large deformation form. This approach was introduced by several authors and extensively developed in recent years.[37–48] A key to successful implementation of this approach is the treatment of finite rotations. For details on the complete formulation the reader is referred to the cited references.

17.7.1 Axisymmetric shells

Here we consider the extension for the beam presented above in Sec. 17.2 to treat axisymmetric shells. We limit our discussion to the extension of the small deformation case treated in Sec. 17.4 in which two-node straight conical elements (see Fig. 14.2) and reduced quadrature are employed. Local axes on the shell segment may be defined by

$$
\begin{aligned}
\bar{R} &= \cos \psi (R - R_0) - \sin \psi (Z - Z_0) \\
\bar{z} &= \sin \psi (R - R_0) + \cos \psi (Z - Z_0)
\end{aligned}
\tag{17.85}
$$

where R_0, Z_0 are centred on the element as

$$
\begin{aligned}
R_0 &= \tfrac{1}{2}(R_1 + R_2) \\
Z_0 &= \tfrac{1}{2}(Z_1 + Z_2)
\end{aligned}
\tag{17.86}
$$

with R_I, Z_I nodal coordinates of the element. The deformed position with respect to the local axes may be written in a form identical to Eq. (17.7). Accordingly, we have

$$
\begin{aligned}
\bar{r} &= \bar{R} + \bar{u}(\bar{R}) + \bar{Z} \sin \beta(\bar{R}) \\
\bar{z} &= \bar{w}(\bar{R}) + \bar{Z} \cos \beta(\bar{R})
\end{aligned}
\tag{17.87}
$$

To consider the axisymmetric shell it is necessary to integrate over the volume of the shell and to include the axisymmetric hoop strain effects. Accordingly, we now consider a segment of shell in the R–Z plane (i.e. X is replaced by the radius R). The volume of the shell in the reference configuration is obtained by multiplying the beam volume element by the factor $2\pi R$. In axisymmetry the deformation gradient in the tangential (hoop) direction must be included. Accordingly, in the local coordinate frame the deformation gradient is given by

$$
F_{iI} =
\begin{bmatrix}
[1 + \bar{u}_{,\bar{R}} + \bar{Z}(\cos \beta)\beta_{,\bar{R}}] & 0 & \sin \beta \\
0 & r/R & 0 \\
[\bar{w}_{,\bar{R}} - \bar{Z}(\sin \beta)\beta_{,\bar{R}}] & 0 & \cos \beta
\end{bmatrix}
\tag{17.88}
$$

Following the same procedures as indicated for the beam we obtain the expressions for Green–Lagrange strains as

$$
\begin{aligned}
E_{\bar{R}\bar{R}} &= \bar{u}_{,\bar{R}} + \tfrac{1}{2}(\bar{u}^2_{,\bar{R}} + \bar{w}^2_{,\bar{R}}) + \bar{Z}\Lambda\beta_{,\bar{R}} = E^0_{\bar{R}\bar{R}} + \bar{Z}K^b_{\bar{R}\bar{R}} \\
E_{TT} &= \frac{u}{R} + \frac{1}{2}\left(\frac{u}{R}\right)^2 + \bar{Z}\left(1 + \frac{u}{R}\right)\frac{\sin \beta}{R} = E^0_{TT} + ZK^b_{TT} \\
2E_{\bar{R}\bar{Z}} &= (1 + \bar{u}_{,\bar{R}})\sin \beta + \bar{w}_{,\bar{R}}\cos \beta = \Gamma
\end{aligned}
\tag{17.89}
$$

where $\Lambda = (1 + \bar{u}_{,\bar{R}})\cos \beta - \bar{w}_{,\bar{R}}\sin \beta$, and the additional hoop strain results in two additional strain components, E^0_{TT} and K^b_{TT}.

With the above modifications, the virtual work expression for the shell now becomes

$$
\delta\Pi = \int_\Omega (\delta E_{\bar{R}\bar{R}} S_{\bar{R}\bar{R}} + \delta E_{TT} S_{TT} + 2\delta E_{\bar{R}\bar{Z}} S_{\bar{R}\bar{Z}})\, \mathrm{d}V - \delta\Pi_{\text{ext}} = 0
\tag{17.90}
$$

in which S_{TT} is the hoop stress in the cylindrical direction. The remainder of the development follows the procedures presented in Sec. 17.2.1 and is left as an exercise for the reader. It is also possible to develop a small rotation theory following the methods described in Sec. 17.2.2.

Here we demonstrate the use of the axisymmetric shell theory by considering a shallow spherical cap subjected to an axisymmetric vertical ring load (Fig. 17.9). The case where the ring load is concentrated at the crown has been examined analytically by Biezeno[49] and Reissner.[50] Solutions using finite difference methods on the equations of Reissner are presented by Mescall.[51] Solutions by finite elements have been presented earlier by Zienkiewicz and co-workers.[7,52] Owing to the shallow nature of the shell, rotations remain small, and excellent agreement exists between the finite rotation and small rotation forms.

17.7.2 Shallow shells – co-rotational forms

In the case of shallow shells the transformations of Chapter 13 may conveniently be avoided by adopting a formulation based on Marguerre shallow shell theory.[24,53,54] A simple extension to a shallow shell theory for the formulation presented for thin plates may be obtained by replacing the displacements by

$$\begin{Bmatrix} u \\ v \\ w \end{Bmatrix} \rightarrow \begin{Bmatrix} u_0 + u \\ v_0 + v \\ w_0 + w \end{Bmatrix} \tag{17.91}$$

in which u_0, v_0, and w_0 describe the position of the shell reference configuration from the X–Y plane. Now the current configuration of the shell (where, often, u_0 and v_0 are taken as zero) may be described by

$$x_1(t) = X + u_0(X, Y) + u(X, Y, t) - Z\,[w_{0,X}(X, Y) + w_{,X}(X, Y, t)]$$
$$x_2(t) = Y + v_0(X, Y) + v(X, Y, t) - Z\,[w_{0,Y}(X, Y) + w_{,Y}(X, Y, t)] \tag{17.92}$$
$$x_3(t) = w_0(X, Y) + w(X, Y, t)$$

where a time t is introduced to remind the reader that at time zero the reference configuration is described by

$$x_1(0) = X + u_0(X, Y) - Z\,w_{0,X}(X, Y)$$
$$x_2(0) = Y + v_0(X, Y) - Z\,w_{0,Y}(X, Y) \tag{17.93}$$
$$x_3(0) = w_0(X, Y)$$

where u, v, w vanish. Using these expressions we can compute the deformation gradient for the deformed configuration and for the reference configuration. Denoting these by F_{iI} and F_{iI}^0, respectively, we can deduce the Green–Lagrange strains from

$$E_{IJ} = \tfrac{1}{2}[F_{iI}F_{iJ} - F_{iI}^0 F_{iJ}^0] \tag{17.94}$$

The remainder of the derivations are straightforward and left as an exercise for the reader. This approach may be generalized and also used to deduce the equations for deep shells.[44] Alternatively, we can note that as finite elements become small they are essentially shallow shells relative to a rotated plane. This observation led to the development of many general shells based on a concept named 'co-rotational'. Here the reader is referred to the literature for additional details.[55–69]

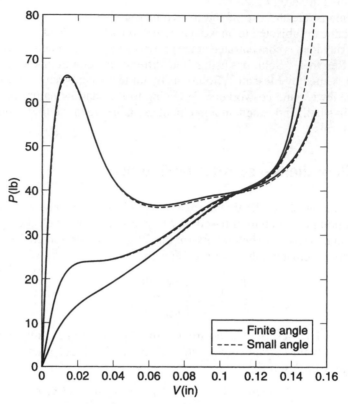

Load–deflection curves for various ring loads

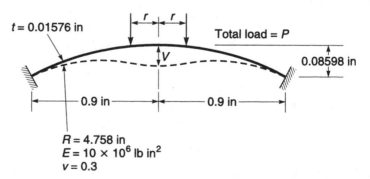

Fig. 17.9 Spherical cap under vertical ring load: (a) load–deflection curves for various ring loads. Spherical cap under vertical ring load; (b) geometry definition and deflected shape.

17.7.3 Stability of shells

It is extremely important to emphasize again that instability calculations are meaningful only in special cases and that they often overestimate the collapse loads considerably. For correct answers a full non-linear process has to be invoked. A progressive 'softening' of a shell under load is shown in Fig. 17.10 and the result is well below the one given by linearized buckling.[12] Figure 17.11 shows the progressive collapse of an arch at a load much below that given by the linear stability value. The solution from the finite rotation beam formulation is compared with an early solution obtained by Marcal[70] who employed small-angle approximations. Here again it is evident that use of finite angles is important.

The determination of the actual collapse load of a shell or other slender structure presents obvious difficulties (of a kind already discussed in Chapter 3 and encountered above for beams), as convergence of displacements cannot be obtained when load is 'increased' near the peak carrying capacity. In such cases one can proceed by prescribing displacement increments and computing the corresponding reactions if only one concentrated load is considered. By such processes, Argyris[71] and others[36,52] succeeded in following a complete snap-through behaviour of a shallow arch.

Pian and Tong[72] show how the process can be generalized simply when a system of proportional loads is considered. This and other 'arc-length' methods are considered in Sec. 3.2.6.

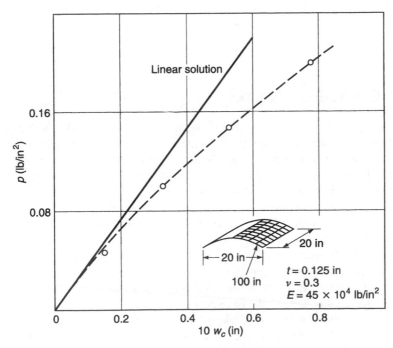

Fig. 17.10 Deflection of cylindrical shell at centre: all edges clamped.[12]

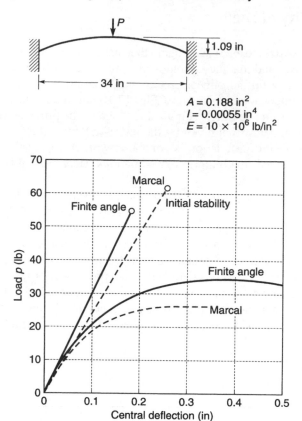

Fig. 17.11 'Initial stability' and incremental solution for large deformation of an arch under central load p.[70]

17.8 Concluding remarks

This chapter presents a summary of approaches that can be used to solve problems in structures composed of beams (rods), plates, and shells. The various procedures follow the general theory presented in Chapter 5 combined with solution methods for non-linear algebraic systems as presented in Chapter 3. Again we find that solution of a non-linear large displacement problem is efficiently approached by using a Newton-type approach in which a residual and a tangent matrix are used. We remind the reader, however, that use of modified approaches, such as use of a constant tangent matrix, is often as, or even more, economical than use of the full Newton process.

If a full load deformation study is required it has been common practice to proceed with small load increments and treat, for each such increment, the problem by a form of the Newton process. It is recommended that each solution step be accurately solved so as not to accumulate errors. We have observed that for problems which have a limit load, beyond which the system is stable, a full solution can be achieved only by use of an 'arc-length' method (except in the trivial case of one point load as noted above).

Extension of the problem to dynamic situations is readily accomplished by adding the inertial terms. In the geometrically exact approach in three dimensions one may encounter quite complex forms for these terms and here the reader should consult literature on the subject before proceeding with detailed developments.[2-5] For the small-angle assumptions the treatment of rotations is identical to the small deformation problem and no such difficulties arise.

References

1. E. Reissner. On one-dimensional finite strain beam theory: the plane problem. *J. Appl. Math. Phys.*, 23:795–804, 1972.
2. J.C. Simo. A finite strain beam formulation. The three-dimensional dynamic problem: Part I. *Computer Methods in Applied Mechanics and Engineering*, 49:55–70, 1985.
3. A. Ibrahimbegovic and M. Al Mikdad. Finite rotations in dynamics of beams and implicit time-stepping schemes. *International Journal for Numerical Methods in Engineering*, 41:781–814, 1998.
4. J.C. Simo and L. Vu-Quoc. A three-dimensional finite strain rod model. Part II: Geometric and computational aspects. *Computer Methods in Applied Mechanics and Engineering*, 58:79–116, 1986.
5. J.C. Simo, N. Tarnow and M. Doblare. Non-linear dynamics of three-dimensional rods: exact energy and momentum conserving algorithms. *International Journal for Numerical Methods in Engineering*, 38:1431–1473, 1995.
6. D.A. da Deppo and R. Schmidt. Instability of clamped-hinged circular arches subjected to a point load. *Trans. Am. Soc. Mech. Eng.*, pages 894–896, 1975.
7. R.D. Wood and O.C. Zienkiewicz. Geometrically non-linear finite element analysis of beams–frames–circles and axisymmetric shells. *Computers and Structures*, 7:725–735, 1977.
8. J.C. Simo, P. Wriggers, K.H. Schweizerhof and R.L. Taylor. Finite deformation post-buckling analysis involving inelasticity and contact constraints. *International Journal for Numerical Methods in Engineering*, 23:779–800, 1986.
9. H.L. Langhaar. *Energy Methods in Applied Mechanics*. John Wiley & Sons, New York, 1962.
10. K. Marguerre. Über die Anwendung der energetishen Methode auf Stabilitätsprobleme. *Hohrb.*, DVL, pages 252–262, 1938.
11. B. Fraeijs de Veubeke. The second variation test with algebraic and differential constraints. In *Advanced Problems and Methods for Space Flight Optimization*. Pergamon Press, Oxford, 1969.
12. C.A. Brebbia and J. Connor. Geometrically non-linear finite element analysis. *Proc. Am. Soc. Civ. Eng.*, 95(EM2):463–483, 1969.
13. B.N. Parlett. *The Symmetric Eigenvalue Problem*. Prentice-Hall, Englewood Cliffs, NJ, 1980.
14. O.C. Zienkiewicz, R.L. Taylor and J.Z. Zhu. *The Finite Element Method: Its Basis and Fundamentals*. Butterworth-Heinemann, Oxford, 6th edition, 2005.
15. S.P. Timoshenko and J.M. Gere. *Theory of Elastic Stability*. McGraw-Hill, New York, 1961.
16. R. Szilard. *Theory and Analysis of Plates*. Prentice-Hall, Englewood Cliffs, NJ, 1974.
17. R.D. Wood. *The application of finite element methods to geometrically non-linear analysis*. PhD thesis, Department of Civil Engineering, University of Wales, Swansea, 1973.
18. A. Razzaque. Program for triangular bending element with derivative smoothing. *International Journal for Numerical Methods in Engineering*, 5:588–589, 1973.
19. M.J. Turner, E.H. Dill, H.C. Martin and R.J. Melosh. Large deflection of structures subjected to heating and external loads. *J. Aero. Sci.*, 27:97–106, 1960.
20. L.A. Schmit, F.K. Bogner and R.L. Fox. Finite deflection structural analysis using plate and cylindrical shell discrete elements. *Journal of AIAA*, 5:1525–1527, 1968.
21. R.H. Mallett and P.V. Marçal. Finite element analysis of non-linear structures. *Proc. Am. Soc. Civ. Eng.*, 94(ST9):2081–2105, 1968.

22. D.W. Murray and E.L. Wilson. Finite element large deflection analysis of plates. *Proc. Am. Soc. Civ. Eng.*, 94(EM1):143–165, 1968.

23. T. Kawai and N. Yoshimura. Analysis of large deflection of plates by finite element method. *International Journal for Numerical Methods in Engineering*, 1:123–133, 1969.

24. P.G. Bergan and R.W. Clough. Large deflection analysis of plates and shallow shells using the finite element method. *International Journal for Numerical Methods in Engineering*, 5:543–556, 1973.

25. R.H. Gallagher and J. Padlog. Discrete element approach to structural instability analysis. *Journal of AIAA*, 1:1537–1539, 1963.

26. H.C. Martin. On the derivation of stiffness matrices for the analysis of large deflection and stability problems. In *Proc. 1st Conf. Matrix Methods in Structural Mechanics*, volume AFFDL-TR-66-80, Wright Patterson Air Force Base, Ohio, October 1966.

27. K.K. Kapur and B.J. Hartz. Stability of thin plates using the finite element method. *Proc. Am. Soc. Civ. Eng.*, 92(EM2):177–195, 1966.

28. R.G. Anderson, B.M. Irons and O.C. Zienkiewicz. Vibration and stability of plates using finite elements. *International Journal of Solids and Structures*, 4:1033–1055, 1968.

29. W.G. Carson and R.E. Newton. Plate buckling analysis using a fully compatible finite element. *Journal of AIAA*, 8:527–529, 1969.

30. Y.K. Chan and A.P. Kabaila. A conforming quadrilateral element for analysis of stiffened plates. Technical Report UNICIV Report R-121, University of New South Wales, 1973.

31. D.W. Murray and E.L. Wilson. Finite element post buckling analysis of thin plates. In *Proc. 2nd Conf. Matrix Methods in Structural Mechanics*, volume AFFDL-TR-68-150, Wright Patterson Air Force Base, Ohio, October 1968.

32. K.C. Rockey and D.K. Bagchi. Buckling of plate girder webs under partial edge loadings. *Int. J. Mech. Sci.*, 12:61–76, 1970.

33. R.H. Gallagher, R.A. Gellately, R.H. Mallett and J. Padlog. A discrete element procedure for thin shell instability analysis. *Journal of AIAA*, 5:138–145, 1967.

34. R.H. Gallagher and H.T.Y. Yang. Elastic instability predictions for doubly curved shells. In *Proc. 2nd Conf. Matrix Methods in Structural Mechanics*, volume AFFDL-TR-68-150, Wright Patterson Air Force Base, Ohio, October 1968.

35. J.L. Batoz, A. Chattapadhyay and G. Dhatt. Finite element large deflection analysis of shallow shells. *International Journal for Numerical Methods in Engineering*, 10:35–38, 1976.

36. T. Matsui and O. Matsuoka. A new finite element scheme for instability analysis of thin shells. *International Journal for Numerical Methods in Engineering*, 10:145–170, 1976.

37. E. Ramm. Geometrishe nichtlineare Elastostatik und Finite elemente. Technical Report Vol. 76–2, Institut für Baustatik, Universität Stuttgart, 1976.

38. H. Parisch. Efficient non-linear finite element shell formulation involving large strains. *Engineering Computations*, 3:121–128, 1986.

39. J.C. Simo and D.D. Fox. On a stress resultant geometrically exact shell model. Part I: Formulation and optimal parametrization. *Computer Methods in Applied Mechanics and Engineering*, 72:267–304, 1989.

40. J.C. Simo, D.D. Fox and M.S. Rifai. On a stress resultant geometrically exact shell model. Part II: The linear theory; computational aspects. *Computer Methods in Applied Mechanics and Engineering*, 73:53–92, 1989.

41. J.C. Simo, M.S. Rifai and D.D. Fox. On a stress resultant geometrically exact shell model. Part IV: Variable thickness shells with through-the-thickness stretching. *Computer Methods in Applied Mechanics and Engineering*, 81:91–126, 1990.

42. J.C. Simo and N. Tarnow. On a stress resultant geometrically exact shell model. Part VI: 5/6 dof treatments. *International Journal for Numerical Methods in Engineering*, 34:117–164, 1992.

43. H. Parisch. A continuum-based shell theory for nonlinear applications. *International Journal for Numerical Methods in Engineering*, 38:1855–1883, 1993.

44. K.-J. Bathe. *Finite Element Procedures*. Prentice Hall, Englewood Cliffs, NJ, 1996.

45. P. Betsch, F. Gruttmann and E. Stein. A 4-node finite shell element for the implementation of general hyperelastic 3d-elasticity at finite strains. *Computer Methods in Applied Mechanics and Engineering*, 130:57–79, 1996.

46. M. Bischoff and E. Ramm. Shear deformable shell elements for large strains and rotations. *International Journal for Numerical Methods in Engineering*, 40:4427–4449, 1997.

47. E. Ramm. From Reissner plate theory to three dimensions in large deformation shell analysis. *Z. Angew. Math. Mech.*, 79:1–8, 1999.

48. M. Bischoff, E. Ramm and D. Braess. A class of equivalent enhanced assumed strain and hybrid stress finite elements. *Computational Mechanics*, 22:443–449, 1999.

49. C.B. Biezeno. Über die Bestimmung der Durchschlagkraft einer schwach-gekrummten kneinformigen Platte. *Zeitschrift für Mathematik und Mechanik*, 15:10, 1935.

50. E. Reissner. On axisymmetric deformation of thin shells of revolution. In *Proc. Symp. in Appl. Math.*, page 32, 1950.

51. J.F. Mescall. Large deflections of spherical caps under concentrated loads. *Trans. ASME, J. Appl. Mech.*, 32:936–938, 1965.

52. O.C. Zienkiewicz and G.C. Nayak. A general approach to problems of plasticity and large deformation using isoparametric elements. In *Proc. 3rd Conf. Matrix Methods in Structural Mechanics*, volume AFFDL-TR-71-160, Wright-Patterson Air Force Base, Ohio, 1972.

53. T.Y. Yang. A finite element procedure for the large deflection analysis of plates with initial imperfections. *Journal of AIAA*, 9:1468–1473, 1971.

54. T.M. Roberts and D.G. Ashwell. The use of finite element mid-increment stiffness matrices in the post-buckling analysis of imperfect structures. *International Journal of Solids and Structures*, 7:805–823, 1971.

55. T. Belytschko and R. Mullen. Stability of explicit-implicit time domain solution. *International Journal for Numerical Methods in Engineering*, 12:1575–1586, 1978.

56. T. Belytschko, J.I. Lin and C.-S. Tsay. Explicit algorithms for the nonlinear dynamics of shells. *Computer Methods in Applied Mechanics and Engineering*, 42:225–251, 1984.

57. C.C. Rankin and F.A. Brogan. An element independent co-rotational procedure for the treatment of large rotations. *ASME, J. Press. Vessel Tech.*, 108:165–174, 1986.

58. Kuo-Mo Hsiao and Hung-Chan Hung. Large-deflection analysis of shell structure by using co-rotational total Lagrangian formulation. *Computer Methods in Applied Mechanics and Engineering*, 73:209–225, 1989.

59. H. Stolarski, T. Belytschko and S.-H. Lee. A review of shell finite elements and co-rotational theories. *Computational Mechanics Advances*, 2:125–212, 1995.

60. E. Madenci and A. Barut. A free-formulation-based flat shell element for non-linear analysis of thin composite structures. *International Journal for Numerical Methods in Engineering*, 30:3825–3842, 1994.

61. E. Madenci and A. Barut. Dynamic response of thin composite shells experiencing non-linear elastic deformations coupled with large and rapid overall motions. *International Journal for Numerical Methods in Engineering*, 39:2695–2723, 1996.

62. T.M. Wasfy and A.K. Noor. Modeling and sensitivity analysis of multibody systems using new solid, shell and beam elements. *Computer Methods in Applied Mechanics and Engineering*, 138:187–211, 1996.

63. A. Barut, E. Madenci and A. Tessler. Nonlinear elastic deformations of moderately thick laminated shells subjected to large and rapid rigid-body motion. *Finite Elements in Analysis and Design*, 22:41–57, 1996.

64. M.A. Crisfield and G.F. Moita. A unified co-rotational framework for solids, shells and beams. *International Journal of Solids and Structures*, 33:2969–2992, 1996.

65. M.A. Crisfield and J. Shi. An energy conserving co-rotational procedure for non-linear dynamics with finite elements. *Nonlinear Dynamics*, 9:37–52, 1996.

66. A.A. Barut, E. Madenci and A. Tessler. Nonlinear analysis of laminates through a Mindlin-type shear deformable shallow shell element. *Computer Methods in Applied Mechanics and Engineering*, 143:155–173, 1997.

67. J.L. Meek and S. Ristic. Large displacement analysis of thin plates and shells using a flat facet finite element formulation. *Computer Methods in Applied Mechanics and Engineering*, 145 (3–4):285–299, 1997.

68. H.G. Zhong and M.A. Crisfield. An energy-conserving co-rotational procedure for the dynamics of shell structures. *Engineering Computations*, 15:552–576, 1998.

69. C. Pacoste. Co-rotational flat facet triangular elements for shell instability analyses. *Computer Methods in Applied Mechanics and Engineering*, 156:75–110, 1998.

70. P.V. Marçal. Effect of initial displacement on problem of large deflection and stability. Technical Report ARPA E54, Brown University, 1967.

71. J.H. Argyris. Continua and discontinua. In *Proc. 1st Conf. Matrix Methods in Structural Mechanics*, volume AFFDL-TR-66-80, pages 11–189, Wright Patterson Air Force Base, Ohio, October 1966.

72. T.H.H. Pian and P. Tong. Variational formulation of finite displacement analysis. In *Symp. on High Speed Electronic Computation of Structures*, Liège, 1970.

18

Multiscale modelling*

18.1 Introduction

In the previous chapters we mainly have used single-scale models to study material and structural behaviour at a *macroscopic level*. However, many natural and man-made materials exhibit an internal structure at more than one length scale. These internal structures may be of a translational nature, where the structure is more or less invariant with respect to a translation corresponding to the smallest length scale. Materials with internal structure may show also *multiscale* features, i.e. they may be invariant with respect to scaling. Such materials can be considered to be fractal-like, but they are not true fractals since the exponent n remains finite and the volume fraction does not go to zero even for large n. These examples of scalable structures are by no means exhaustive and many other possibilities exist. Thus, in many cases the microstructure may not be scalable and may be different at each structural level.[1] Examples of different scales and their domain of application are shown in Table 18.1.

Materials with internal structure have in common that each structural level plays its own role in the global response: the material behaviour is controlled by the physical phenomena which take place at the various scales and by the interaction of these phenomena across scales. Single-scale models, usually at a macro scale, make use of constitutive equations which should reflect the behaviour of the underlying finer scales. These constitutive equations are generally of a phenomenological type. An alternative to the use of constitutive equations at a single (macro) scale is provided by multiscale modelling, in which the relevant physics is explicitly captured on multiple spatial and temporal scales.[2]

This does not exclude the use of phenomenological constitutive equations at lower scale, except when the lowest scale is pushed to the electronic structure and *ab-initio* calculations are performed. However, multiscale modelling permits one to reduce the uncertainty incorporated in many single-scale constitutive models. The output of multiscale material modelling are usually effective material properties which are used at the higher scale. This analysis may also span over several scales where the output from the preceding, lower level is used as input for the next higher level.

Examples of materials with internal structure are: dense hierarchical materials, composites and polycrystals, polymers, hierarchical cellular materials, cellular solids, hierarchical honeycombs of second order, hierarchical composites comprising continuous

---

* This chapter was contributed by Professor Bernhard A. Schrefler, University of Padova, Italy.

carbon nanotube fibres in a nanotube-reinforced matrix; natural materials with structural hierarchy are biological materials such as bone or soft tissue, wood, etc.; further we have granular materials such as soils, clays and foams.

The reason why man-made materials with structural hierarchy are common in many engineering applications is that they permit one to obtain special desired properties due to the behaviour of the individual components, their geometric structure and the interaction between the components. Examples of such properties are: materials with both high stiffness and high damping, improved strength and toughness, improved thermal and electrical conductivity, permeability, unusual physical properties such as negative Poisson ratio, negative stiffness inclusions, etc.

Beyond material behaviour the overall structural behaviour is also of interest. It is at this level that external loads are usually specified and the structural response has to be determined. At the structural level a direct simulation of multiscale systems is usually rather complex and time consuming because the discretization has to be reduced to the lowest scale at which information is needed. This would allow one to obtain at the same time the macroscopic behaviour of the entire structure as well as all the information at the lower scales. But this is not necessary and in a case where we start from an atomic or a nano-scale level is not yet possible. Multiscale modelling which links the different single-scale models in a hierarchical way is an answer to the problem. In the case of material multiscale modelling it is usually of interest to proceed from the lower scales upward in order to obtain homogenized material properties. Alternatively, in the case of structural modelling it is important to be able to step down through the scales until the desired scale of the real, not homogenized, material is reached. This technique is often known as unsmearing or localization. Usually in a global analysis both aspects need to be pursued.

As can be seen from Table 18.1, there are many possible scales and the techniques for scale bridging are different. At the lower end we have a linkage between the electronic

Table 18.1 Different length scales and domain of interest; reconstructed from NASA web site

Level	Length scale (m)	Scientific domain	Subject of manipulation	Nature of prediction
Quantum	10^{-12}	Computational chemistry	Molecular assembly, nuclei	Qualitative predictions
Nano	10^{-9}	Computational material mechanics	Molecular fragments, molecular interactions	Qualitative predictions
Micro	10^{-6}	Computational material mechanics	Surface interactions, orientation, anisotropy, crystals, molecular weight, free volumes	Qualitative predictions– Quantitative predictions
Meso	10^{-3}	Computational mechanics– computational material mechanics	Different constituents, different phases, damage	Quantitative predictions
Macro	10^{0}	Computational mechanics– structural mechanics	Composed structures	Quantitative predictions

structure and the atomistic one and there a discrete-to-discrete linkage is commonly used. At the upper end usually we have a continuum-to-continuum linkage. In between these scales there is a transition from discrete to continuum. The scale at which this transition is made, depends largely on the problem. For instance, continuum elasticity may be used down to the atomic scale, whereas classical plasticity breaks down at the micrometre scale, at which dislocation self-organization takes place.[2,3] Discrete-to-continuum linkage may also take place between the micro- or mesoscopic and the macroscopic level when particle mechanics or discrete elements are used at the lower scale. Discrete-to-discrete linkage at the lower end of Table 18.1 requires the definition of atomic potentials and is still in the research stage, e.g. see references 2,4.

We deal here more extensively with continuum-to-continuum linkages and with discrete-to-continuum linkages at the upper end of Table 18.1. However, one example of multiple-scale bridging will go down to the micro level. For linkages at the lowest scale the reader is referred to the references mentioned above,[2,4] the references therein and to specialized literature.

The most common methods for scale bridging are *bounds, self-consistent methods* and *asymptotic analysis*. Voigt[5] was the first to give an evaluation of effective mechanical properties for heterogeneous solids, followed later by Reuss.[6] Voigt and Reuss provide lower and upper bounds for equivalent material properties. Self consistent methods and the Mori–Tanaka[7] approach give instead an estimate of overall mechanical properties for composite materials. Further work on bounds, self consistent and other estimation methods can be found in Eshelby,[8] Hashin and Strikmann,[9] Kröner[10,11] and Willis.[12]

An alternative approach to scale bridging is asymptotic analysis of media with periodic (or quasi-periodic) structure at the micro level, also called *homogenization theory*.

18.2 Asymptotic analysis

Asymptotic analysis not only permits one to obtain equivalent material properties but also allows one to solve the full structural problem down to stresses in the constituent materials at a micro (or local) scale. It is mostly applied to linear two-scale problems, but it can be extended to non-linear analysis and to several scales as will be shown later in Sec. 18.9. We do not intend to give a full account of the underlying theory in this chapter. In references 13 and 14 the interested reader will find a rigorous formulation of the method, its application in many fields and further references. We will show in detail, however, its finite element solution methodology that is a basic ingredient of many multiscale analyses. For the moment we consider just two levels, the micro (or local) level and the macro (or global) level. These levels are shown in Fig. 18.1 for a structure assumed to be periodic and, thus, asymptotic analysis can be successfully applied. Periodicity means that if we consider a body Ω with periodic structure and a generic mechanical or geometric property f (for example, the constitutive tensor), we have

$$\text{if } \mathbf{x} \in \Omega \quad and \quad (\mathbf{x} + \mathbf{Y}) \in \Omega \Rightarrow f(\mathbf{x} + \mathbf{Y}) = f(\mathbf{x}) \tag{18.1}$$

where $\mathbf{Y}$ is the (geometric) period of the structure. Hence the elements of f are $\mathbf{Y}$-periodic functions of the position vector $\mathbf{x}$. The characteristic size of a single cell

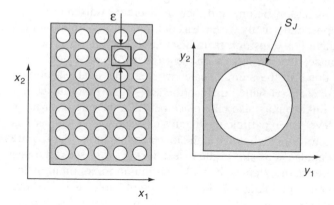

Fig. 18.1 Example of a periodic structure with two levels: global on the left and local on the right.

of periodicity is assumed to be much smaller than the geometric dimensions of the structure under analysis which means that a clear scale separation is possible.

18.3 Statement of the problem and assumptions

One important assumption for asymptotic analysis is that it must be possible to distinguish two length scales associated with the macroscopic and microscopic phenomena. The ratio of these scales defines a small parameter ε (Fig. 18.1). Two sets of coordinates related by

$$\mathbf{y} = \frac{1}{\varepsilon}\mathbf{x} \tag{18.2}$$

formally express this separation of scales between macro and micro phenomena. The global coordinate vector $\mathbf{x}$ refers to the whole body Ω and the stretched local coordinate vector $\mathbf{y}$ is related to the single, repetitive cell of periodicity. In this way the single cell is mapped into the unitary domain Y [here and in the remainder of this chapter Y indicates the unitary domain occupied by the cell of periodicity and not the period of the composite material as in Eq. (18.1)].

In an asymptotic analysis the normalized cell of periodicity is mapped onto a sequence of finer and finer structures as ε tends to 0. If, as defined below, equivalent material properties are employed, the fields considered (temperature, displacement, etc.) converge toward the homogeneous macroscopic solution as the micro-structural parameter ε tends to 0. In this sense problems for the heterogeneous body and the homogenized one are equivalent. (For more details concerning the mathematical meaning see references 13 and 14.)

We consider now a small strain problem of thermoelasticity in a heterogeneous body, such as that depicted in Fig. 18.1, defined by the following equations:

Balance equations

$$\sigma^{\varepsilon}_{ji,j}(\mathbf{x}) + f_i(\mathbf{x}) = 0; \quad \sigma_{ij} = \sigma_{ji}$$
$$q^{\varepsilon}_{i,i} - r = 0 \tag{18.3}$$

Constitutive equations

$$\sigma_{ij}^{\varepsilon}(\mathbf{x}) = C_{ijkl}^{\varepsilon}(\mathbf{x}) e_{kl} (\mathbf{u}^{\varepsilon}(\mathbf{x})) - \alpha_{ij}^{\varepsilon}\theta$$
$$q_i^{\varepsilon} = -k_{ij}^{\varepsilon}\theta_{,j} \tag{18.4}$$

Strain definition

$$e_{ij}(\mathbf{u}^{\varepsilon}(\mathbf{x})) = \tfrac{1}{2}\left(u_{i,j}^{\varepsilon}(\mathbf{x}) + u_{j,i}^{\varepsilon}(\mathbf{x})\right) \equiv u_{(i,j)}^{\varepsilon} \tag{18.5}$$

Boundary and discontinuity conditions

$$t_i = \sigma_{ij}^{\varepsilon}(\mathbf{x})n_j = 0 \ \text{on} \ \Omega_t \quad \text{and} \quad u_i^{\varepsilon}(\mathbf{x}) = 0 \ \text{on} \ \Omega_u$$
$$q_i^{\varepsilon}(\mathbf{x})n_i = 0 \ \text{on} \ \Omega_q \quad \text{and} \quad \theta^{\varepsilon}(\mathbf{x}) = 0 \ \text{on} \ \Omega_\theta \tag{18.6a}$$

$$\left[u_i^{\varepsilon}(\mathbf{x})\right] = 0 \quad \text{and} \quad \left[\sigma_{ij}^{\varepsilon}(\mathbf{x})n_j\right] = 0 \ \text{on} \ S_J$$
$$\left[\theta^{\varepsilon}(\mathbf{x})\right] = 0 \quad \text{and} \quad \left[q_i^{\varepsilon}(\mathbf{x})n_i\right] = 0 \ \text{on} \ S_J \tag{18.6b}$$

The superscript ε is used to indicate that the variables of the problem depend on the cell dimensions related to the global length. Square parentheses denote the jump of the enclosed value. The other symbols have the usual meaning: $\mathbf{u}$ is the displacement vector, $\mathbf{e}(\mathbf{x})$ denotes the linearized strain tensor, $\sigma_{ij}(\mathbf{x})$ the stress tensor, $C_{ijkl}(\mathbf{x})$ the tensor of elastic moduli, $k_{ij}(\mathbf{x})$ the tensor of thermal conductivity, $\alpha_{ij}(\mathbf{x}) = C_{ijkl}(\mathbf{x})\beta_{kl}$ where $\beta_{kl}(\mathbf{x})$ the tensor of thermal expansion coefficients, $\theta(\mathbf{x})$, $q_i(\mathbf{x})$ temperature and heat flux respectively, and $r(\mathbf{x})$, $f_i(\mathbf{x})$ stand for thermal sources and body forces, respectively.

Since the components of the elasticity and thermal conductivity tensors are discontinuous, differentiation [in the above equations and in Eqs (18.10a) and (18.10b) below] should be understood in the weak (variational) sense. This is the main reason why most of the problems posed in the sequel will be presented in a variational formulation.

We introduce now the second hypothesis of homogenization theory: we assume that the periodicity of the material characteristics imposes an analogous periodical perturbation on quantities describing the mechanical behaviour of the body; hence we will use the following representation for displacements and temperatures

$$\mathbf{u}^{\varepsilon}(\mathbf{x}) = \mathbf{u}^0(\mathbf{x}) + \varepsilon^1 \mathbf{u}^1(\mathbf{x}, \mathbf{y}) + \varepsilon^2 \mathbf{u}^2(\mathbf{x}, \mathbf{y}) + \cdots + \varepsilon^k \mathbf{u}^k(\mathbf{x}, \mathbf{y})$$
$$\theta^{\varepsilon}(\mathbf{x}) = \theta^0(\mathbf{x}) + \varepsilon^1 \theta^1(\mathbf{x}, \mathbf{y}) + \varepsilon^2 \theta^2(\mathbf{x}, \mathbf{y}) + \cdots + \varepsilon^k \theta^k(\mathbf{x}, \mathbf{y}) \tag{18.7}$$

Using Eq. (18.7), similar expansion with respect to powers of ε give expressions for stresses, strains and heat fluxes as

$$\sigma^{\varepsilon}(\mathbf{x}) = \mathbf{u}^0(\mathbf{x}) + \varepsilon^1 \sigma^1(\mathbf{x}, \mathbf{y}) + \varepsilon^2 \sigma^2(\mathbf{x}, \mathbf{y}) + \cdots + \varepsilon^k \sigma^k(\mathbf{x}, \mathbf{y})$$
$$e^{\varepsilon}(\mathbf{x}) = \mathbf{u}^0(\mathbf{x}) + \varepsilon^1 e^1(\mathbf{x}, \mathbf{y}) + \varepsilon^2 e^2(\mathbf{x}, \mathbf{y}) + \cdots + \varepsilon^k e^k(\mathbf{x}, \mathbf{y}) \tag{18.8}$$
$$\mathbf{q}^{\varepsilon}(\mathbf{x}) = \mathbf{u}^0(\mathbf{x}) + \varepsilon^1 \mathbf{q}^1(\mathbf{x}, \mathbf{y}) + \varepsilon^2 \mathbf{q}^2(\mathbf{x}, \mathbf{y}) + \cdots + \varepsilon^k \mathbf{q}^k(\mathbf{x}, \mathbf{y})$$

where $\mathbf{u}^k$, σ^k, e^k, θ^k, $\mathbf{q}^k$, for $k > 0$ are Y-periodic, i.e. take the same values on the opposite sides of the cell of periodicity. The term scaled with the nth power of ε in Eqs (18.7) and (18.8) is called a term of order n.

18.4 Formalism of the homogenization procedure

The necessary mathematical tools are the chain rule of differentiation with respect to the micro variable and averaging over a cell of periodicity. We introduce the assumption (18.7) and (18.8) into equations of the heterogeneous problem (18.3)–(18.6b) and use the differential calculus rule (see reference 14)

$$\frac{\mathrm{d}f}{\mathrm{d}x_i} = \left(\frac{\partial f}{\partial x_i} + \frac{1}{\varepsilon} \frac{\partial f}{\partial y_i} \right) = f_{,i(x)} + \frac{1}{\varepsilon} f_{,i(y)} \tag{18.9}$$

This equation also defines the notation used in the sequel for differentiation with respect to local and global independent variables.

Because of Eq. (18.9) the equilibrium equations and the heat balance equation split into terms of different orders [the terms of the same power of ε are equated to zero separately: e.g. Eqs (18.10a) and (18.10b) are of order $1/\varepsilon$].

For the equilibrium equation we have

$$\sigma^0_{ij,j(y)}(\mathbf{x}, \mathbf{y}) = 0$$
$$\sigma^0_{ij,j(x)}(\mathbf{x}, \mathbf{y}) + \sigma^1_{ij,j(y)}(\mathbf{x}, \mathbf{y}) + f_i(\mathbf{x}) = 0$$
$$\sigma^1_{ij,j(x)}(\mathbf{x}, \mathbf{y}) + \sigma^2_{ij,j(y)}(\mathbf{x}, \mathbf{y}) = 0 \tag{18.10a}$$
$$\cdots$$

We have a similar expression for the heat balance equation:

$$q^0_{i,i(y)}(\mathbf{x}, \mathbf{y}) = 0$$
$$q^0_{i,i(x)}(\mathbf{x}, \mathbf{y}) + q^1_{i,i(y)}(\mathbf{x}, \mathbf{y}) - r(\mathbf{x}) = 0$$
$$q^1_{i,i(x)}(\mathbf{x}, \mathbf{y}) + q^2_{i,i(y)}(\mathbf{x}, \mathbf{y}) = 0 \tag{18.10b}$$
$$\cdots$$

From Eqs (18.5) and (18.9) it follows that the main term of e_{ij} in expansions (18.8$_2$) depends not only on $\mathbf{u}^0$, but also on $\mathbf{u}^1$

$$e^0_{ij}(\mathbf{x}, \mathbf{y}) = u^0_{(i,j)(x)} + u^1_{(i,j)(y)} \equiv e_{ij(x)}(\mathbf{u}^0) + e_{ij(y)}(\mathbf{u}^1) \tag{18.11}$$

The constitutive relationships (18.4) now assume the form

$$\sigma^0_{ij}(\mathbf{x}, \mathbf{y}) = C_{ijkl}(\mathbf{y}) \left(e_{kl(x)}(\mathbf{u}^0) + e_{kl(y)}(\mathbf{u}^1) \right) - \alpha_{ij}(\mathbf{y}) \theta^0$$
$$\sigma^1_{ij}(\mathbf{x}, \mathbf{y}) = C_{ijkl}(\mathbf{y}) \left(e_{kl(x)}(\mathbf{u}^1) + e_{kl(y)}(\mathbf{u}^2) \right) - \alpha_{ij}(\mathbf{y}) \theta^1 \tag{18.12a}$$
$$\cdots$$

$$q^0_k(\mathbf{x}, \mathbf{y}) = -k_{kl}(\mathbf{y}) \left(\theta^0_{,l(x)} + \theta^1_{,l(y)} \right)$$
$$q^1_k(\mathbf{x}, \mathbf{y}) = -k_{kl}(\mathbf{y}) \left(\theta^1_{,l(x)} + \theta^2_{,l(y)} \right) \tag{18.12b}$$
$$\cdots$$

It can be seen that the terms of order n in the asymptotic expansions for stresses (18.12a) and heat flux (18.12b) depend respectively on the displacement and temperature terms of order n and $n + 1$. In this way the influence of the local perturbation on the global quantities is accounted for. This is the reason why for instance we need $\mathbf{u}^1(\mathbf{x}, \mathbf{y})$ to define via the constitutive relationship the main term in expansion (18.8) for stresses [and $\mathbf{u}^2(\mathbf{x}, \mathbf{y})$ for the term of order 1, if needed, see below].

18.5 Global solution

Referring separately to the terms of the same powers of ε leads to the following variational formulations for unknowns of successive order of the problem. Starting with the first order, using separation of variables it can be formally shown that $\mathbf{u}^1(\mathbf{x}, \mathbf{y})$ and $\theta^1(\mathbf{x}, \mathbf{y})$ can be represented by[14,15]

$$
\begin{aligned}
u_i^1(\mathbf{x}, \mathbf{y}) &= e_{pq(x)}(\mathbf{u}^O(\mathbf{x}))\chi_i^{pq}(\mathbf{y}) + C_i(\mathbf{x}) \\
\theta^1(\mathbf{x}, \mathbf{y}) &= \theta^0_{,p(x)}(\mathbf{x})\vartheta^P(\mathbf{y}) + C(\mathbf{x})
\end{aligned}
\tag{18.13}
$$

We call $\chi^{pq}(\mathbf{y})$ and $\vartheta^P(\mathbf{y})$ the homogenization functions for displacements and temperature, respectively.

The zero-order (often also referred to as first-order) component of the equation of equilibrium (18.10a) and of heat balance (18.10b) in the light of Eq. (18.13) yields the following boundary value problems for functions of homogenization:

find $\chi^{pq} \in V_Y$ such that: $\forall \nu_i \in V_Y$

$$
\int_Y C_{ijkl}(\mathbf{y}) \left(\delta_{ip}\delta_{jq} + \chi_{i,j(y)}^{pq}(\mathbf{y})\right) \nu_{k,l(y)}(\mathbf{y}) \, d\Omega = 0
\tag{18.14a}
$$

find $\vartheta^P \in V_Y$ such that: $\forall \phi \in V_Y$

$$
\int_Y k_{ij}(\mathbf{y}) \left(\delta_{ip} + \vartheta_{,i(y)}^{pq}(\mathbf{y})\right) \phi_{,j(y)}(\mathbf{y}) \, d\Omega = 0
\tag{18.14b}
$$

In the above equations V_Y is the subset of the space of kinematically admissible functions that contains the functions with equal values on the opposite sides of the cell of periodicity Y. The tensor χ_{pq} and the vector ϑ^P are functions that depend only on the geometry of the cell of periodicity and on the values of the jumps of material coefficients across S_J. Functions $\nu_i(\mathbf{y})$ and $\phi(\mathbf{y})$ are the usual arbitrary test functions having the meaning of Y-periodic displacements and temperature fields, respectively. They are used here to write explicitly the counterparts of the expressions (18.10a) and (18.10b), in which the prescribed differentiations again are understood in a weak (variational) sense.

The solutions χ^{pq} and ϑ^P of the 'local' (i.e. defined for a single cell of periodicity) boundary value problems with periodic boundary condition (18.14a) and (18.14b) can be interpreted as obtained for the cell subject to a unitary average strain e^{pq} and unitary average temperature gradient $\vartheta_{,p(y)}$, respectively. The true value of perturbations are obtained after by scaling χ^{pq} and ϑ^P with true global strains (gradient of global temperature), as prescribed by Eq. (18.13).

In the asymptotic expansion for displacements and temperature given by Eq. (18.7) the dependence on $\mathbf{x}$ alone occurs only in the first term. The independence on $\mathbf{y}$ of these functions can be proved (see, for example, reference 14). The functions depending only on $\mathbf{x}$ define the macro behaviour of the structure and we will call these the *global terms*. To obtain the global behaviour of stresses and heat flux the following mean values over the cell of periodicity are defined[14]

$$
\tilde{\sigma}_{ij}^0(\mathbf{x}) = |Y|^{-1} \int_Y \sigma_{ij}^0(\mathbf{x}, \mathbf{y}) \, dY \quad \text{and} \quad \tilde{q}_i^0(\mathbf{x}) = |Y|^{-1} \int_Y q_i^0(\mathbf{x}, \mathbf{y}) \, dY \tag{18.15}
$$

Averaging of Eqs (18.12a$_1$) and (18.12b$_1$) results in the following, effective constitutive relationships

$$\tilde{\sigma}^0_{ij}(\mathbf{x}) = C^h_{ijkl}\, e_{kl}(\mathbf{u}^0) - \alpha^h_{ij}\, \theta^0 \quad \text{and} \quad \tilde{q}^0_i = -k^h_{ij}\, \theta^0_{,j} \qquad (18.16)$$

where the effective material coefficients are computed according to

$$C^h_{ijkl} = |Y|^{-1} \int_Y C_{ijpq}(\mathbf{y})\left(\delta_{kp}\delta_{lq} + \chi^{pq}_{k,l(y)}(\mathbf{y})\right)\,\mathrm{d}Y$$

$$k^h_{ij} = |Y|^{-1} \int_Y k_{ip}(\mathbf{y})\left(\delta_{jp} + \vartheta^P_{,j(y)}(\mathbf{y})\right)\,\mathrm{d}Y \qquad (18.17)$$

$$\alpha^h_{ij} = |Y|^{-1} \int_Y \alpha_{ij}(\mathbf{y})\,\mathrm{d}Y$$

The macro behaviour can be defined now by averaging first-order terms in the equilibrium equations (18.10a$_2$), flux balance equations (18.10b$_2$) and boundary conditions (18.6a) and then substituting the averaged counterparts of stress and heat flux (18.15) [first-order perturbations vanish in averaging of Eqs (18.10a$_2$), (18.10b$_2$) because of periodicity]. Equation (18.4) is replaced by Eq. (18.16) and Eq. (18.6b) is no longer needed since we deal now with homogeneous uncoupled thermoelasticity.

The heterogeneous structure can now be studied as a homogeneous one with effective material coefficients given by Eq. (18.17) from which the global displacements, strains and average stresses and heat fluxes can be computed. We then go back to Eq. (18.12a) to recover a local approximation of stresses. This last step corresponds to the unsmearing or localization mentioned in the introduction.

18.6 Local approximation of the stress vector

We note that the homogenization approach results in two different kinds of stress tensors. The first one is the average stress field defined by Eq. (18.16). It represents the stress tensor for the homogenized, equivalent but unreal macro body. Once the effective material coefficients are known, the stress field and the heat flux may be obtained from a standard finite element structural analysis and heat transfer program as described in the previous section.

The other stress field is associated with a family of uniform states of strains $e_{pq(x)}(\mathbf{u}^0)$ over each cell of periodicity Y. This local stress is obtained by introducing Eq. (18.11) into Eq. (18.12a) and results in

$$\sigma^0_{ij}(\mathbf{x}, \mathbf{y}) = C_{ijkl}(\mathbf{y})\left(\delta_{kp}\delta_{iq} - \chi^{pq}_{k,l(y)}\right) e_{pq(x)}(\mathbf{u}^0) - \alpha_{ij}(\mathbf{y})\,\theta^0 \qquad (18.18)$$

Because of Eqs (18.10a) and (18.14a) this tensor satisfies the equations of equilibrium everywhere in Y. If needed, the stress description can be completed with a higher-order term in Eq. (18.8). This approach is presented in references 16 and 17.

Finally the local approximation of heat flux is as follows:

$$q^0_j(\mathbf{y}) = -k_{ij}(\mathbf{y})[\delta_{ip} + \vartheta^P_{,i(y)}(\mathbf{y})]\,\theta^0_{,p(x)} \qquad (18.19)$$

18.7 Finite element analysis applied to the local problem

For the finite element formulation it is convenient to introduce matrix notation for the quantities introduced above. Accordingly, the homogenization functions are ordered as defined by Eq. (18.20) [the numbers in the superscripts in Eq. (18.20) and subscripts in Eq. (18.21), refer to the reference coordinate axes 1, 2, 3]:

$$
\mathbf{X}^T(\mathbf{y}) = \left[\chi^{11}(\mathbf{y}) \quad \chi^{22}(\mathbf{y}) \quad \chi^{33}(\mathbf{y}) \quad \chi^{12}(\mathbf{y}) \quad \chi^{23}(\mathbf{y}) \quad \chi^{31}(\mathbf{y}) \right]_{3 \times 6}
$$
$$
\mathbf{T}^T(\mathbf{y}) = \left[\vartheta^1(\mathbf{y}) \quad \vartheta^2(\mathbf{y}) \quad \vartheta^3(\mathbf{y}) \right]_{1 \times 3}
$$
(18.20)

This is in accordance with the ordering of strains and fluxes

$$
\mathbf{c} = \left[e_{11} \quad e_{22} \quad e_{33} \quad e_{12} \quad e_{23} \quad e_{31} \right]^T = \left\{ e_{pq} \right\}_{6 \times 1}
$$
$$
\mathbf{q} = \left[q_1 \quad q_2 \quad q_3 \right]^T = \left\{ q_p \right\}_{3 \times 1}
$$
(18.21)

In the following a tilde again denotes a nodal value in the finite element mesh. We have the usual representations for each element:

$$
\mathbf{X}(\mathbf{y}) = \mathbf{N}(\mathbf{y})\tilde{\mathbf{X}}; \qquad T(\mathbf{y}) = \mathbf{N}(\mathbf{y})\tilde{\mathbf{T}}
$$
(18.22)

where $\mathbf{N}$ contains the values of standard shape functions.

It is easy to show that the variational formulation (18.14a) can be rewritten as follows:

$$
\text{find } \mathbf{X} \in V_Y \text{ such that: } \forall \mathbf{v} \in V_Y
$$
$$
\int_Y \mathbf{e}^T(\mathbf{v}(\mathbf{y}))\mathbf{D}(\mathbf{y}) \left(1 + \mathcal{L}\mathbf{X}(\mathbf{y}) \right) \, dY = 0
$$
(18.23)

In the above $\mathcal{L}$ denotes the matrix of differential operators, and $\mathbf{D}$ contains the material coefficients a_{ijkl} in the repetitive domain. Matrix $\tilde{\mathbf{X}}$ which contains the values of homogenization functions at the nodes of the mesh is obtained as a finite element solution of Eq. (18.23). The equation to solve is the following:

$$
\mathbf{K}_X \tilde{\mathbf{X}} - \mathbf{f} = 0
$$
(18.24)

where $\tilde{\mathbf{X}}$ is Y-periodic, with zero mean value over the cell, and

$$
\mathbf{f} = \int_Y \mathbf{B}^T \mathbf{D}(\mathbf{y}) \, dY; \quad \mathbf{K}_X = \int_Y \mathbf{B}^T \mathbf{D}(\mathbf{y}) \mathbf{B} \, dY; \quad \mathbf{B} = \mathcal{L}\mathbf{N}(\mathbf{y})
$$
(18.25)

$\mathbf{D}$ contains the material coefficients a_{ijkl}.

It can be shown that $\mathbf{X}$ in Eq. (18.23) and thus Eq. (18.24) is a solution of a boundary value problem, for which the loading consists of unitary average strains over the cell. This is seen in the form of the right-hand side of Eq. (18.24) which forms a matrix. We solve thus six equations for six functions of homogenization.

The variational formulation (18.14b) can be represented in a form similar to Eq. (18.24), $\tilde{\mathbf{T}}$ being Y-periodic, with given mean zero value over the cell

$$
\mathbf{K}_T \tilde{\mathbf{T}} + \mathbf{f} = 0
$$
(18.26)

where

$$\mathbf{f} = \int_Y \mathbf{B}_\theta^T \mathbf{k}_\theta(\mathbf{y}) \, dY; \quad \mathbf{K}_T = \int_Y \mathbf{B}_\theta^T \mathbf{k}_\theta(\mathbf{y}) \mathbf{B}_\theta \, dY; \quad \mathbf{B}_\theta = \mathcal{L}_\theta \mathbf{N}(\mathbf{y}) \tag{18.27}$$

$\mathbf{k}_\theta$ contains the conductivities k_{ij} of materials in the repetitive domain. Differential operators in $\mathcal{L}_\theta$ are ordered suitably for the thermal problem.

The periodicity conditions can be taken into account using Lagrange multipliers in the construction of a finite element code. The requirements of the zero mean value also has to be included in the program.

Having computed $\tilde{\mathbf{X}}$, $\tilde{\mathbf{T}}$ and by consequence $\mathbf{u}^1$ and θ^1 one can derive the effective material coefficients, according to:

$$\mathbf{D}^h = |Y|^{-1} \int_Y \mathbf{D}(\mathbf{y}) \left(\mathbf{1} + \mathbf{B}\tilde{\mathbf{X}}\right) dY$$

$$\mathbf{k}^h = |Y|^{-1} \int_Y \mathbf{k}_\theta(\mathbf{y}) \left(\mathbf{1} + \mathbf{B}_\theta \tilde{\mathbf{T}}\right) dY \tag{18.28}$$

$$\boldsymbol{\alpha}^h = |Y|^{-1} \int_Y \boldsymbol{\alpha}(\mathbf{y}) \, dY$$

With the homogenized material coefficients (18.28) any thermoelastic finite element program can be used to obtain global displacements and temperatures. For the unsmearing procedure we need the gradients of temperatures and strains in the regions of interest, see Eqs (18.18) and (18.19). Strains and temperature gradients are directly obtained from the finite element interpolations.

To present the plots of stresses and heat fluxes over the single cell nodal projection can be used. To assure continuity of tangential stresses, this projection should be extended to patches of cells, for example, using SPR.[18]

18.7.1 Example

As an example of the procedure described above, the temperature effects on a super-conducting coil are analysed. The structure has an overall shape of a large D, and is constructed by winding of a superconducting cable with a rectangular cross-section (Fig. 18.2). The whole D-shaped frame is immersed in liquid helium at a temperature of about 4K to assure superconductivity. The supports are designed to eliminate only the rigid motion of the structure. This is a typical example where asymptotic analysis can be successfully applied and the structure is clearly periodic: the macro scale is defined by a typical dimension of the coil cross-section, while the micro scale is given by the height of the section of a single cable (cell of periodicity).

In the following a beam-like kinematics with deformable cross-section is used, instead of a full three-dimensional analysis. Homogenization is carried out in the cross-section plane (1,2) only, while in the axial direction (3) there is no periodic structure.[16,17]

First effective elastic coefficients and six vectors of homogenization functions for unsmearing are computed. Then the thermal effective characteristics, i.e. effective conductivity and effective thermal expansion are calculated. Three scalar functions of

Coil cross-section   Cable cross-section

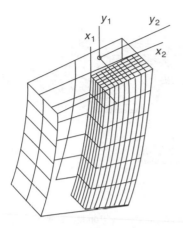

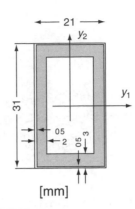

[mm]

Fig. 18.2 A superconducting coil cross-section and the single cell of periodicity (cable). The material properties are discontinuous, with discontinuities along regular surfaces S_J.

homogenization are further obtained for computations of the heat flux. These functions are shown in Figs 18.3–18.5.

The macro analysis with the homogenized material (a standard thermoelastic finite element program) yields the global displacements and the temperature field. Unsmearing as described in the previous section then yields the local distribution of stresses in the region of interest in the coil. Stresses in each different material of a cell of periodicity can be recovered. In Fig. 18.6 the graphs of stress in the cross-section

Plane homogenization functions

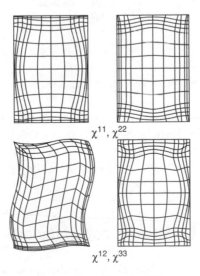

χ^{11}, χ^{22}

χ^{12}, χ^{33}

Fig. 18.3 Plane functions of homogenization χ^{11}, χ^{12}, χ^{22}, χ^{33}.

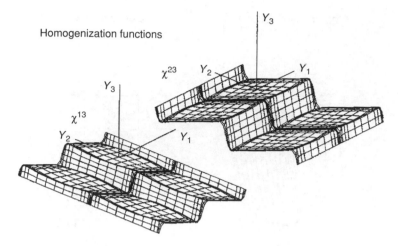

Fig. 18.4 Plots of the anti-plane functions of homogenization χ^{13}, χ^{23} on four strands.

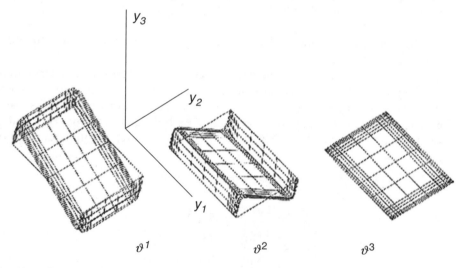

Fig. 18.5 Three scalar functions of homogenization for temperature.

perpendicular to the fibres are shown for uniform cooling. Figure 18.7 shows the full complex state of stress in the neighbourhood of a single conductor for non-uniform cooling. Uniform cooling means here that the whole coil is immersed in liquid helium while non-uniform cooling indicates that the coil is only partially immersed in liquid helium.

18.7.2 Corrections for stresses and boundary effects

A refined stress description over the cell of periodicity can be obtained by considering the second of Eq. (18.10a) (for simplicity, in the following we omit temperature effects).

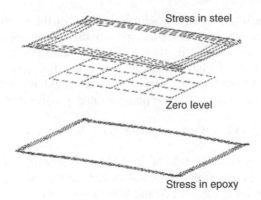

Stress in steel

Zero level

Stress in epoxy

Fig. 18.6 Graphs of stress σ_{33} on the cross-section perpendicular to the fibres for uniform cooling (one cell of periodicity only).

This requires taking into account the second of Eq. (18.12a) and the solution for $\mathbf{u}^2$ in the expansion for displacements (18.7). The rather lengthy procedure is fully described in reference 19 and has been applied to the example problem in reference 17. It also has to be extended to account for localization after the solution of the global problem. An advantage is that, over a single cell of periodicity, the equilibrium conditions are also fully satisfied in the presence of a non-uniform global state of strain. An alternative is to make the localization over a patch of neighbouring cells using the first-order homogenization described in Sec. 18.5.[19]

There is still another open problem: the periodicity condition used to find the local perturbation is strictly applicable only inside the body. We have hence the solution of a (thermo-)elasticity problem based on an assumed stress or displacement field which is valid nearly everywhere in the region occupied by the body under investigation, except

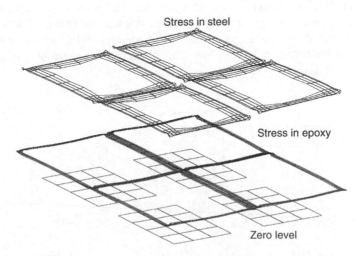

Stress in steel

Stress in epoxy

Zero level

Fig. 18.7 Graphs of stress σ_{33} on the cross-section perpendicular to fibres for non-uniform cooling (patch of four cells of periodicity).

on the boundary. The use of material coefficients based on the assumption of periodicity in the global solution (where the real boundary conditions are imposed) may implicitly impose some unrealistic constraints close to the boundary. This problem can be solved by some corrections which change the solution in the domain close to the boundary while away from the boundary the correction field should asymptotically decrease.[20] Such a correction can be obtained by replacing the expansion (18.7) by

$$\mathbf{u}^{\varepsilon}(\mathbf{x}) = \mathbf{u}^{0}(\mathbf{x}) + \varepsilon \left[\mathbf{u}^{1}(\mathbf{x}, \mathbf{y}) + \mathbf{b}^{1}(\mathbf{x}, \mathbf{z}) \right] + \cdots \tag{18.29}$$

where $\mathbf{z}$ represents the coordinates of an additional system with the origin on the external surface, the axis z_3 directed normal to the boundary surface with the other two axes oriented tangential to the surface and $\mathbf{b}$ is the boundary layer correction. Vector $\mathbf{b}$ should vanish exponentially when z_3 goes to infinity. The procedure is described in detail in reference 21 and has been implemented by making use of one cell of periodicity and infinite elements with homogenized material properties around it. The use of infinite elements assures the desired exponential decay. Although the procedure is theoretically necessary near free edges, it is seldom applied in practical engineering problems.

18.8 The non-linear case and bridging over several scales

If applied iteratively, asymptotic theory of homogenization may also be used for non-linear situations. Furthermore, it can obviously be used to bridge several scales. Here we deal with the case where three scales are bridged by applying in sequential manner the two-scale asymptotic analysis. The behaviour of the components is physically non-linear. Again we refer to thermomechanical behaviour and introduce a micro, meso and macro level as shown in Fig. 18.8.

At the stage of micro or meso modelling, some main features of the local structure can be extracted and used later in the macro analysis. The behaviour of the components, even if elastic–plastic, is assumed to be piecewise linear, so that the homogenization we perform is piecewise linear. Only monotonic loading and/or temperature increase (decrease) is considered, otherwise we need to store the whole history and use an incremental analysis.

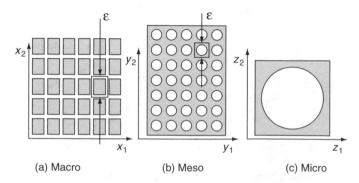

(a) Macro   (b) Meso   (c) Micro

Fig. 18.8 Example of a periodic structure with three levels: macro, meso and micro.

Because of the assumed form of material properties we deal with a sequence of problems of linear elasticity written for a non-homogeneous material domain and with coefficients that are functions of both temperature and stress level.

At the top level of the hierarchy we consider an elastic body contained in the domain Ω with a smooth boundary $\partial\Omega$ where on the part $\partial\Omega_t$ of the boundary tractions are given and on the remaining part $\partial\Omega_u$ displacements are prescribed. The domain Ω is filled with repetitive cells of periodicity Y, shown in Fig. 18.8, where the material of the body is supposed to be piecewise homogeneous inside Y, as defined in Eq. (18.1). The governing equations are given by Eqs (18.3) to (18.5).

For the lowest level all the formulations are formally the same with one difference: the boundary conditions are those of an infinite body. It is worth mentioning that all the macro fields at the micro level become the micro fields at the higher structural level. Effective material coefficients and mean fields obtained with the homogenization procedure at the lowest level enter as local perturbations at the next higher step.

Before explaining the application of the homogenization procedure in sequential form to multi-level non-linear material behaviour we mention the solution by Terada and Kikuchi[22] who write a two-scale variational statement within the theory of homogenization. The solution of the microscopic problem at each Gauss point of the finite element mesh for the overall structure, and the deformation histories at time t_{n-1} must be stored until the macroscopic equilibrium state at current time t_n is obtained. This procedure has not been applied to bridging of more than two scales. A three-scale asymptotic analysis is used by Fish and Yu[23] to analyse damage phenomena occurring at micro, meso, and macro scales in brittle composite materials (woven composites). These authors also retain the second-order term in the displacement expansion (18.7) and introduce a similar form for the expansion of the damage variable. We recall further that also stochastic aspects can be introduced in the homogenization procedure.[24]

18.9 Asymptotic homogenization at three levels: micro, meso and macro

The two usual tools of homogenization of the previous section are used, i.e. volume averaging and total differentiation with respect to the global variable $\mathbf{x}$ that involves the local variable $\mathbf{y}$. The homogenization functions are obtained similarly to Eqs (18.14a) and (18.14b). Only a factor λ is introduced in Eq. (18.14a) to adapt the solution to the real strain level as explained below.

find $\chi_i^{pq} \in V_Y$ such that: $\forall \nu_i \in V_Y$

$$\int_{Y(\lambda)} C_{ijkl}(\mathbf{y}, \lambda, \theta^O)[\delta_{ip}\delta_{jq} + \chi_{i,j(y)}^{pq}]\,\nu_{k,l(y)}\,\mathrm{d}\Omega = 0, \quad \sigma(\lambda, \chi_i^{pq}) \in P \tag{18.30}$$

Material properties are assumed to depend on temperature and a set of representative temperatures is considered for the material input data where linear interpolation is used between the given values. P is the domain inside the surface of plasticity. The requirement that the stress belongs to the admissible region P [introduced in Eq. (18.30)] is verified using the classical unsmearing procedure described in the preceding section.

Table 18.2 Updating yield surface algorithm scheme

1. Compute effective coefficients at micro level.
2. Compute effective coefficients at meso level.
3. Apply increment of forces and/or temperature at the macro level, solve global homogeneous problem.
4. Compute global strain E_{ij}: $E_{ij} = e_{ij}(\mathbf{u}^0)$ reminding that $E_{ij} = \tilde{e}_{ij}^{\varepsilon}(\mathbf{x})$ [see Eq. (18.31)].
5. Apply E_{ij} to meso-level cell by equivalent kinematic loading (displacement on the border).
6. Solve the kinematic problem at the meso level for $\mathbf{w}(\mathbf{y})$, compute stress (unsmearing for meso level) and strain E_{ij}; now $E_{ij} = e_{ij}(\mathbf{w}^0)$ and $E_{ij} = \tilde{e}_{ij}^{\varepsilon}(\mathbf{y})$.
7. Apply E_{ij} from meso to micro level cell by equivalent kinematic loading (displacements on the border).
8. Solve the kinematic problem at the micro level for $\mathbf{w}^1(\mathbf{z})$, compute stress (unsmearing for micro level).
9. Verify yielding of the material **in the physically true situation at micro level.** If 'yes' change mechanical parameter of the material and go to 1; else exit.

It is to be noted that for 'solving the boundary value problem' mentioned in points 6 and 8 it is not always necessary to use the true finite element solution. If the cell of periodicity has not been changed before, this solution can be composed according to Eq. (18.13) (suitably rewritten).

The modification of the algorithm required by the non-linearity starts with the composite cell of periodicity with given elastic components. The uniform strain is increased step by step. Effective material coefficients are constant until the stress reaches the yield surface at some points of the cell. The yield surface in the space of stresses is different for each material component, being thus a function of position. The region where the material yield is of finite volume at the end of the step, thus it is easy to replace the material with the yielded one, with the elastic modulus equal to the hardening modulus of the elastic–plastic material and with the Poisson ratio tending to 0.5.

The cell of periodicity is hence transformed into a form with one more material and the usual analysis procedure is restarted with a uniform strain, a new homogenization function and a new stress map over the cell. We identify each new region where further local yielding occurs, then redefine the cell and perform the analysis. The loop is repeated as many times as needed. In Eq. (18.30) the history of this replacement of materials at the micro level is marked by λ, the level of the average stress, for which the micro yielding occurs each time. The algorithm is summarized in Table 18.2 where $w(y)$ indicates displacement at meso level.

At the end of each step we can compute also the mean stress over the cell having (generalized) homogenization functions [see Eq. (18.16)] and the effective coefficients can be computed using Eq. (18.17).

18.10 Recovery of the micro description of the variables of the problem

An important part of multiscale modelling is the recovery of stress and heat flux as well as strain, temperature and displacements at the level of the microstructure. This is obtained by Eqs (18.18) and (18.19) with the following procedure: first global (mean) fields are obtained from the homogeneous analysis where the material is characterized by the effective coefficients (18.17), then we go back to the original problem formulation, using homogenization functions. Thus we recover the main parts of the stress and heat flux. Because of the specified three-level hierarchical structure we are dealing with, the recovery process must be applied twice, also since material characteristics are temperature dependent and non-linear the procedure must be applied for each representative temperature within the context of the correct stress state. We recall that the recovery process starts at the highest structural level while the homogenization begins at the lowest part of the structural hierarchy.

18.10.1 Example: the VAC strand analysis

As an example of application, we consider a superconducting strand which is used to build the cable of Fig. 18.2. The structures and the three scales are shown in Figs 18.9 and 18.10, where the single filament (micro scale), groups of filaments (meso scale) and the superconducting strand (macro scale) are shown.

The homogeneous effective properties will be defined for the inner part of the strand, shown on the left of Fig. 18.10. The diameter of the strand is about 0.80 mm. The application of the theory of homogenization is justified by the scale separation as clearly shown in Figs 18.9 and 18.10.

As already indicated, periodic homogenization is applicable to structures obtained by a multiple translation of a representative volume element (RVE), called in this case the cell of periodicity. The considered strand shows two different levels of such a translational structure. On the meso level we have the repetitive pattern of the superconducting filament in the bronze matrix (micro scale RVE), filling the hexagonal region as illustrated in Fig. 18.11. The second translational structure is the net of the hexagonal filament groups (meso scale RVE) in the body of the single strand shown in Fig. 18.12. The homogenization splits thus into two steps, each one dealing with rather similar geometry and a comparable scale separation factor. Boundary conditions for the macro problem will be given in terms of interaction of the strand with the other strands in the cable[25] and will be of the type of Eq. (18.6).

To form the superconducting alloy Nb3Sn the coil is kept for 175 hours at 923K. Afterward, to reach an operating temperature, it is cooled from 923K to 4.2K. In this example we analyse the effects of such cool-down, using the homogenization procedure to define the strain state of the strand at 4.2K which is caused by the different thermal

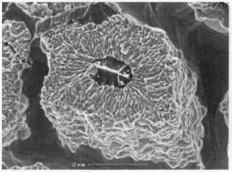

Fig. 18.9 A single Nb3Sn filament (left) and Nb3Sn filaments groups (right); the respective scales are also evidenced. Each filament group is made of 85 filaments.

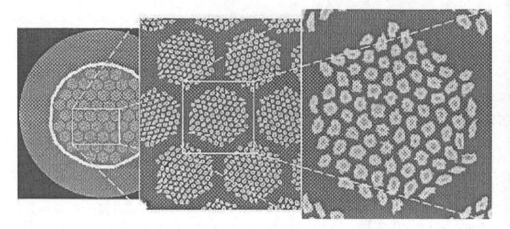

Fig. 18.10 Three-level hierarchy in the VAC strand. The central part of the strand itself (left) consists of 55 groups of 85 filaments, embedded in tin-rich bronze matrix, while the outer region is made of high conductivity copper.

contractions of the materials. This strain state is the initial condition for successive operations of the cable. It is recalled that superconductivity of the employed material is strain sensitive and hence a precise knowledge of these strains is of paramount importance. At the end of the cool-down in a reacted strand the filaments are in a compressive strain state while the bronze and copper matrices are in a tensile state. We assume that the strand components are stress free at 923K, since the strands remained at that temperature for 175 hours.

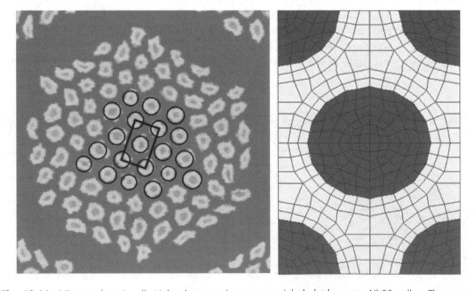

Fig. 18.11 Micro scale unit cell. Light elements: bronze material, dark elements: Nb3Sn alloy. The area of the cell is 90×5.6 μm.

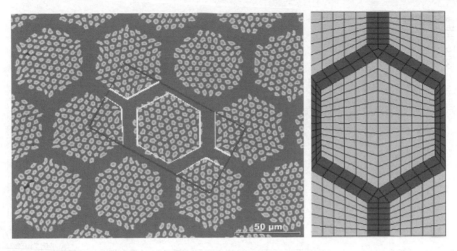

Fig. 18.12 Meso scale unit cell. Dark element: bronze material, light elements: homogenized material at micro level. The area of the cell is 100.0 × 60.1 μm.

18.11 Material characteristics and homogenization results

The Nb3Sn compound has a low thermal contraction but a relatively high elastic modulus and a very high yield strength. The bronze and copper reach their yield limits as soon as the temperature starts decreasing, so that they are plastically flowing for nearly the whole thermomechanical analysis.

Material thermal characteristics are taken from the conductor database.[26] There are very few measurements of elastic–plastic properties of the strand components over the whole temperature range 4−923K, see references 27–30. Variation of the different materials' elastic moduli versus temperature is shown in Fig. 18.13. Due to their high yield limit the Nb3Sn filaments can be assumed to remain elastic over the whole temperature range, with a constant elastic modulus of 160 GPa.[31] Bronze and Nb3Sn thermal conductivity (W/mK) and thermal expansion (%) as a function of temperature (K) are illustrated in Figs 18.14 and 18.15, respectively.

After the homogenization procedure, the resulting equivalent material has an orthotropic behaviour, depending upon the material characteristics and the geometrical configuration of the unit cell.

Concerning the thermal behaviour, the Nb3Sn compound has a very low thermal conductivity, and is practically zero when compared to that of bronze. The bronze material has a thermal conductivity of less than 500 W/mK from reaction temperature until to about 100K, then, with decreasing temperature, thermal conductivity undergoes a strong increase of almost 2500 W/mK and then decreases again (Fig. 18.14). It is quite obvious that the resulting material will be guided by the bronze behaviour, both at meso level (Fig. 18.14, grey continuous lines) and on macro level (Fig. 18.14, grey dashed lines): k_{11}, k_{22}, k_{33} stand for the value of the conductivity referred to a Cartesian system of coordinates, where the third axis is parallel to the longitudinal axis of the strand.

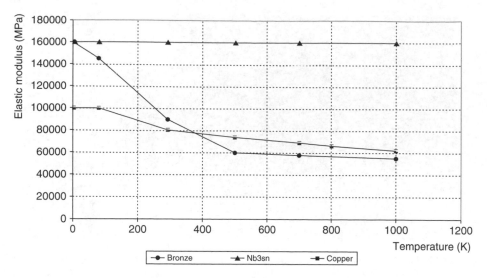

Fig. 18.13 Variation of bronze (circles), Nb3Sn (triangles) and copper (squares) elastic modulus versus temperature.

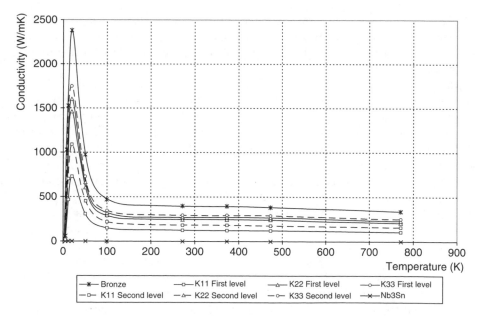

Fig. 18.14 Thermal conductivity (W/mK) of bronze (stars), Nb3Sn (crosses), meso and macro level homogenization results (grey continuous and grey dashed lines, respectively).

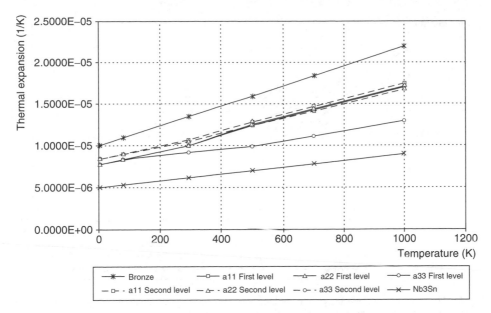

Fig. 18.15 Thermal expansion [1/K] of bronze (stars), Nb3Sn (crosses), meso and macro level homogenization results (grey continuous and grey dashed lines, respectively).

Thermal expansion is almost linear with temperature, that of bronze being higher than that of Nb3Sn (Fig. 18.15). Resulting effective coefficients are illustrated in Fig. 18.15 for the meso level (grey continuous lines) and for the macro one (grey dashed lines): α_{11}, α_{22}, α_{33} stand for the value of the expansion coefficients referred to the same Cartesian system of coordinates as for the conductivity values described above.

Mechanical characteristics of the individual materials and homogeneous results are compared in Fig. 18.16, showing the diagonal terms of the elasticity tensors as a function of temperature. The peculiar disposition of the superconducting filaments gathered into groups results in an almost isotropic behaviour in the strand cross-section, while along the longitudinal direction of the strand the material behaviour is strongly influenced by the superconducting material. The procedure has been validated by comparing results of a homogenized group of filaments and those of a three-dimensional discretization – involving about 8000 elements – subjected to the cool-down analysis of a strand.[59]

18.12 Multilevel procedures which use homogenization as an ingredient

There exist several procedures for multilevel approaches which use asymptotic homogenization as an ingredient. We will consider here two of them. Feyel[32–35] introduced an integrated multiscale analysis, also called the FE² method which consists of the following ingredients typical for multiscale techniques:

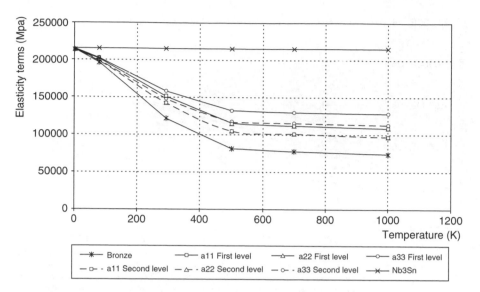

Fig. 18.16 Main diagonal elasticity terms for bronze (stars), Nb3Sn (crosses), meso and macro level homogenization results (grey continuous and grey dashed lines, respectively).

1. The mechanical behaviour is modelled on the lower scale, i.e. on the representative volume element (RVE).
2. A localization rule is chosen to determine the local solution inside RVE for any given overall strain.
3. A homogenization rule is selected which yields the macroscopic stress tensor, knowing the micro-mechanical stress state.
4. At each integration point of the macroscopic level an RVE is located where the above operations are carried out.
5. The overall strain and displacements are obtained at the macroscopic level.

In principle any localization/homogenization rule can be used for steps 1 to 3, as indicated in the next section, but the most common method is the periodic homogenization described in the previous sections which is considered further in this section.

Also in the integrated analysis two finite element meshes are needed as in the sequential application of the homogenization method described above, one for the cell of periodicity and the other for the macroscopic structure. The difference from that considered previously is that the computation is carried out simultaneously on both scales.

In the cell of periodicity located at each Gauss point of the macroscopic structure the method computes the stress tensor at time t using the current strain, strain rate and mechanical history since the beginning of the analysis. In classical phenomenological models at macroscopic scale the mechanical history is taken into account through a set of internal variables. Here the internal variable set is constructed by assembling all microscopic data required by the finite element computation at a lower level. This includes of course microscopic internal variables needed to describe dissipative phenomena.

The local analysis yields the macroscopic (average) stresses and strains, defined as

$$\Sigma_{ij} = \bar{\sigma}_{ij} = \frac{1}{V} \int_V \sigma_{ij} \, dV; \qquad E_{ij} = \bar{e}_{ij} = \frac{1}{V} \int_V e_{ij} \, dV \qquad (18.31)$$

also known as Bishop–Hill relations.[36]

The integrated approach can be easily implemented in a classical finite element program based on a Newton algorithm to handle all non-linearities and in fact consists in a sequence of Newton algorithms at the global and local levels.

For optimum performance the tangent stiffness matrix at the macroscopic level has to be computed. For this calculation we need the macroscopic (algorithmic) tangent matrix in the Gauss points which can be computed from the variation of the average strains and stresses of Eq. (18.31) as

$$\mathbf{D}^{\text{macro}} = \frac{\partial \Delta \Sigma}{\partial \Delta \mathbf{E}} \qquad (18.32)$$

where Δ denotes the increment of the quantity between time t and time $t + \Delta t$. This computation depends of course on the homogenization theory used and on its finite element implementation.

In the case of the periodic homogenization theory it is convenient to add at the local level of each element in the cell of periodicity mesh some degrees of freedom corresponding to the average or macroscopic strain $\mathbf{E}$. This involves at most 6 degrees of freedom shared by every element in the mesh. It is recalled that at a local level the unknown displacements are the periodic part $\mathbf{u}^l$ of the total displacement $\mathbf{u}$ on the cell where $\mathbf{u}^l$ is obtained using Eqs (18.24) and (18.13). These, together with the first of Eq. (18.31), are contained in the condensation procedure outlined below, proposed in reference 35.

The deformation tensor is computed by Eq. (18.11) and may be written as

$$\mathbf{e}^0(\mathbf{x}, \mathbf{y}) = \mathbf{e}_{(x)}(\mathbf{u}^0) + \mathbf{e}_{(y)}(\mathbf{u}^l) = \mathbf{E} + \mathbf{e}_{(y)}(\mathbf{u}^l) \qquad (18.33)$$

The $\mathbf{B}$ matrix needed in this case, called $\hat{\mathbf{B}}$, is similar to the usual one, except that a new part comes from degrees of freedom associated with $\mathbf{E}$ (these degrees of freedom may be inserted at the end of the finite element degree of freedom list):

$$\hat{\mathbf{B}} = \begin{bmatrix} \mathbf{B} & \mathbf{1} \end{bmatrix} \qquad (18.34)$$

where $\mathbf{B}$ denotes as usual the standard symmetric gradient of the shape functions. According to our choice, $\mathbf{E}$ has associated degrees of freedom and hence the associated reactions give the mean stress Σ, to be multiplied by the volume of the cell.

The assembled tangent stiffness matrix at the cell scale can be written as

$$\mathbf{K} = \int_{\text{cell}} \hat{\mathbf{B}}^T \mathbf{D} \hat{\mathbf{B}} \, d\Omega \qquad (18.35)$$

Taking into account Eq. (18.34) this leads to

$$\mathbf{K} = \int_{\Omega_e} \begin{bmatrix} \mathbf{B}^T \mathbf{D} \mathbf{B} & \mathbf{B}^T \mathbf{D} \\ \mathbf{D} \mathbf{B} & \mathbf{D} \end{bmatrix} d\Omega = \begin{bmatrix} \mathbf{k} & \mathbf{G}^T \\ \mathbf{G} & \mathbf{H} \end{bmatrix} \qquad (18.36)$$

where **D** represents the tangent matrix given by all microscopic phenomenological constitutive equations.

Recalling the meaning of the additional degrees of freedom and the associated reactions, the macroscopic tangent matrix $\mathbf{D}^{\text{macro}}$ of Eq. (18.32) is then nothing but a condensation of the previous matrix onto the degrees of freedom associated with **E**. That is

$$\mathbf{D}^{\text{macro}} = \frac{1}{V}[\mathbf{H} - \mathbf{G}\,\mathbf{k}^{-1}\mathbf{G}^T] \qquad (18.37)$$

Conceptually this condensed matrix is then very easy to compute. However, for large structures the method requires a great computational effort to perform the reductions involving **k** and parallel computing is usually employed in this step.

Another multiscale computational strategy which makes use of the homogenization theory is proposed by Ladevèze and co-workers.[37–39] In that case the structure is considered as an assembly of substructures and interfaces. The junction between the macro and micro scales takes place only at the interfaces.

18.13 General first-order and second-order procedures

The main characteristics of the methods of the previous section are:

1. The constitutive response at the macro scale is undetermined *a priori* and ensues from the solution of the micro scale boundary value problem.
2. The method can deal with large displacements in a straightforward way, provided that the micro structural constituents are modelled adequately.
3. The different constituents in the microstructure can be modelled with any desired non-linear constitutive model.
4. The micro scale problem is a classical boundary value problem; therefore also methods different from that of asymptotic homogenization can be used, i.e. any appropriate solution strategy is applicable.
5. The macroscopic constituent tangent operator can be obtained from the overall microscopic stiffness tensor by static condensation, as indicated above. This scale transition preserves consistency.[40]

The procedure is called a first-order computational homogenization since it includes only first-order gradients of the macroscopic displacement field. Other methods have been published by Ghosh,[41,42] Miehe,[43,44] Kouznetsova,[45] Suquet[46] and Wriggers and co-workers.[47,48] As already mentioned, they also make use of solution methods of the micro scale problem which differ from the asymptotic homogenization. First-order methods require, however, that the characteristic length of the spatial variation of the macroscopic loading must be very large with respect to the size of the micro structure and the macroscopic gradients must remain very small with respect to the micro scale. Possible size effects which may arise from the micro scale and localization problems cannot be dealt with properly. For this purpose the method has been extended by Geers *et al.*[40] and Kouznetsova *et al.*[45] to higher-order continua.

18.13.1 Example

As an example, we show micro scale computations including damage at interfaces on a single RVE under two different boundary conditions.[47] The investigation of a single RVE without embedding in a concurrent method as above may also be used for numerical material testing.[47] The considered unit cube of Fig. 18.17 contains 10 particles. Around the particles there is a cohesive zone as in reference 49 in which damage can occur. Since meshing of a complex three-dimensional structure is difficult and often leads to badly shaped element geometries, here the geometry is discretized and approximated with cube-shaped elements and subsequently refined at the boundaries of the particles and the cohesive zone to obtain a better geometry approximation, see Fig. 18.17. The matrix material, the particle material and the cohesive zone material are considered to be a simple neo-Hookean material with the strain energy function

$$\Psi(\mathbf{C}) = \tfrac{1}{2}\mu\left(J^{-2/3}\mathrm{tr}(\mathbf{C}) - 3\right) + K\left(J - 1 - \log(J)\right); \quad \mathrm{tr}(\mathbf{C}) = C_{II}$$

where $\mathbf{C} = \mathbf{F}^T\mathbf{F}$ and $J^2 = \det(\mathbf{C})$. The material parameters of the particle material and the matrix material are chosen to be $\mu_{\text{Matrix}} = 3$, $K_{\text{Matrix}} = 7$, $\mu_{\text{particle}} = 30$ and $K_{\text{particle}} = 70$. The damage law used for the cohesive zone is the one proposed by Zohdi and Wriggers.[48] The initial material parameters for the cohesive zone material are chosen to be the same as the parameters for the matrix material. When damage occurs they are weakened by a factor α.

$$\begin{aligned} \mu^{(cz)} &= \alpha\,\mu_0^{(cz)} \\ K^{(cz)} &= \alpha\,K_0^{(cz)} \end{aligned} \quad 0 < \alpha \le 1$$

where $(\cdot)^{(cz)}$ means *cohesive zone*.

Fig. 18.17 Cut through the initial mesh.

The local constraint condition from which α can be computed is

$$\Psi(\alpha) = M(\alpha) - K(\alpha) \leq 0$$

where $M(\alpha)$ is a scalar valued term representing the stress state of the material point

$$M(\alpha) = \left(\mathbf{g}(\sigma_{\text{deg}}(\alpha)) : \mathbf{g}(\sigma_{\text{deg}}(\alpha)) \right)^{1/2}$$

$$\mathbf{g}(\sigma_{\text{deg}}(\alpha)) = \frac{1}{3} \eta_1 \, \text{tr}(\sigma_{\text{deg}}) \mathbf{1} + \eta_2 \left(\sigma_{\text{deg}} - \frac{1}{3} \text{tr}(\sigma_{\text{deg}}) \mathbf{1} \right)$$

and η_1 and η_2 are parameters scaling the isochoric and deviatoric parts of $\mathbf{g}(\sigma_{\text{deg}}(\alpha))$. $K(\alpha)$ is a threshold value which depends on the damage variable itself.

$$K(\alpha) = \Phi_{\text{lim}} + (\Phi_{\text{crit}} - \Phi_{\text{lim}}) \alpha^P$$

where Φ_{crit} is the initial threshold value, and Φ_{lim} is the threshold value in the limiting case that the material point has degraded completely ($\alpha = 0$). Finally P is an exponent which controls the rate of degradation.

The parameters chosen here are $\eta_1 = 1$, $\eta_2 = 1$, $\Phi_{\text{lim}} = 0.1$, $\Phi_{\text{crit}} = 1$ and $P = 0.1$. The tests done here are a comparison between the results for pure Dirichlet boundary conditions and pure Neumann boundary conditions. For pure displacement boundary conditions a displacement gradient

$$\mathbf{H} = \gamma \begin{bmatrix} 1 & 1 & 1 \\ 1 & 1 & 1 \\ 1 & 1 & 1 \end{bmatrix}, \quad \gamma = 0.1$$

has been enforced on the entire boundary of the RVE, γ is a load factor. The resulting damage distribution and the first principal stresses are shown in Figs 18.18 and 18.19. For pure traction boundary conditions the Cauchy stress tensor

$$\sigma = \gamma \begin{bmatrix} 3.5 & 1 & 1 \\ 1 & 3.5 & 1 \\ 1 & 1 & 3.5 \end{bmatrix}, \quad \gamma = 0.48$$

has been prescribed. The same quantities as above are shown in Figs 18.20 and 18.21.

More details about the homogenization procedure and the statistical testing method chosen can be found in reference 47.

As for the boundary conditions, it is recalled that linear deformations and uniform traction constraints yield upper and lower bounds of the sought material characteristics, respectively. On the contrary, in classical homogenization theory as shown in the previous sections periodicity conditions reflect exact results of the global stiffness for periodic structures. Hence, where applicable, periodicity conditions should be applied, see also the next section.

18.14 Discrete-to-continuum linkage

The methods described in the previous section in principle also can be applied to a discrete-to-continuum linkage in which the discrete body is contained in the RVE

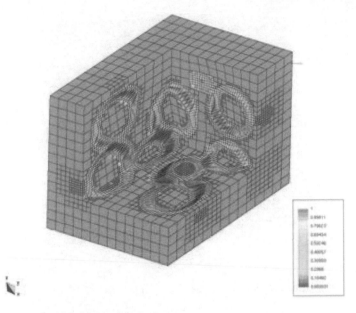

Fig. 18.18 Displacement boundary conditions: damage.

while the macroscopic body represents the continuum, see, for example, Miehe and Dettmar,[50] where a package of irregular discs is studied. In this case a periodic RVE is assumed but its size is rather large because of the irregularity of the disc size.

As an alternative we show here a sequential method that can be used with any localization/homogenization rule. A numerical homogenized constitutive relation for the global behaviour is defined first by constructing subsequent yield surfaces by means

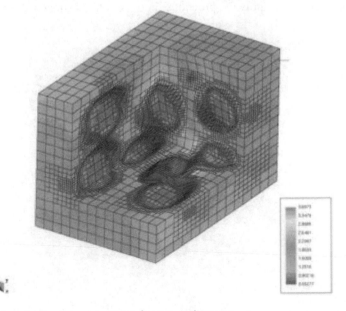

Fig. 18.19 Displacement boundary conditions: first principal stresses.

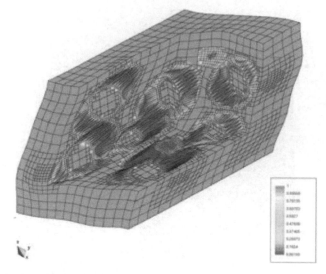

Fig. 18.20 Traction boundary conditions: damage.

of the solution of a series of local problems on a RVE. These permit the definition of so-called interpolation points in the global stress space, which are assumed to belong to the global yield surfaces. At a macroscopic level the method uses an elastic–plastic algorithm, where the necessary information (hardening law, flow rule) is extracted numerically from the interpolation points.

This method is advantageous if several macroscopic loading conditions have to be considered, since the numerical constitutive relationship is established and stored once for all cases. In the following we select asymptotic homogenization for the solution of the local problems. We apply the procedure and the related stress recovery for the case

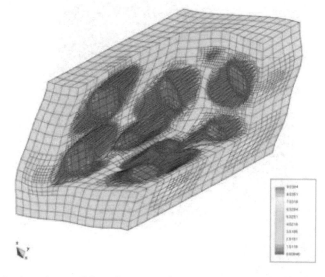

Fig. 18.21 Traction boundary conditions: first principal stresses.

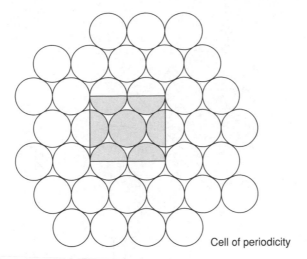

Fig. 18.22 Periodic structure composed of discs in contact.

where in the RVE there are elastic–plastic discs in contact, as depicted in Fig. 18.22. This represents clearly a discrete structure, while the overall macroscopic behaviour is dealt with as a continuum. Plane situations, monotonic proportional loading and small strains are assumed. With enhanced interpolation procedures the method can be extended to more generic situations, such as the discrete element methods described in Chapter 9, if they are used to obtain macroscopic constitutive behaviour.[51] For an alternative approach see references 52 and 53.

18.14.1 Representative volume element and boundary conditions

We take here advantage of the periodicity of the structure and hence the choice of the RVE is straightforward. The size of the RVE takes advantage of the regularity of the packing. As in the previous sections, microscopic quantities are indicated with lower case letters while capital letters are used to indicate macroscopic quantities. To derive the constitutive equations from micro mechanics and homogenization, it is necessary to relate the fields in the interior of the RVE to their volume and boundary averages. This can be obtained through the macroscopic stress and strain defined in Eq. (18.31) which in our case can also be expressed as

$$
\Sigma_{ij} = \frac{1}{V} \int_{V^*} \sigma_{ij} \, dV
$$

$$
E_{ij} = \frac{1}{2V} \int_{S_{RVE}} (u_i n_j + u_j n_i) \, dS - \frac{1}{2V} \int_{S_c} \left([u_i]n_j + [u_j]n_i \right) \, dS_c
$$

(18.38)

where V^* is the portion of RVE occupied by the solid phase, S_{RVE} is the external surface of the RVE, S_c is the common surface of the discs in contact, $[u_j]$ is the jump of the displacement field across the surface S_c, and n_i is a vector normal to the relevant surface, see Figs 18.23 and 18.26.[25]

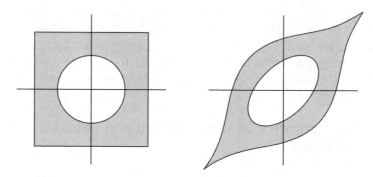

Fig. 18.23 Bodies in elastic or elastic–plastic contact.

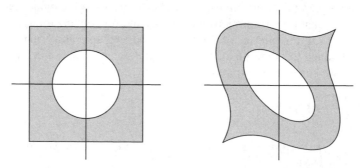

Fig. 18.24 Periodic boundary conditions on a quadratic cell with circular inclusion: deformed configuration due to $\mathbf{u}^1$ (on the right).

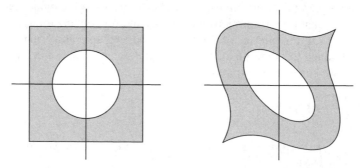

Fig. 18.25 Periodic boundary conditions on a quadratic cell with circular inclusion: periodic component of the deformation due to $\mathbf{u}^2$ (on the right).

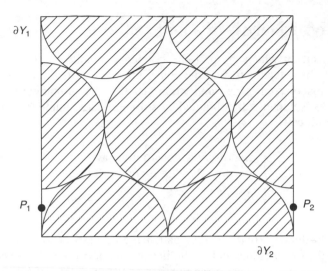

Fig. 18.26 Single cell of periodicity.

We simulate a large number of different radial loading paths on the unit cell in such a way that any generic monotonic proportional loading can be approximated by an interpolation between some paths previously simulated. One of the main problems in this approach is that the unit cell on which the different loading paths are simulated is constrained in a manner which cannot represent all the possible *in situ* conditions. We adopt here the following periodic boundary conditions on ∂V

$$u_i = E_{ij}x_j + u_i^*; \ u_i^* \text{ periodic on } \partial V$$
$$\sigma_{ij}n_j; \text{ anti-periodic on } \partial V \qquad (18.39)$$

where $U_i = (u_1, u_2)$, for planar problems, is the displacement applied to the boundary of the cell and is the sum of two contributions: a linear part given by $E_{ij}y_j$ and a periodic part u_j^* which gives no contribution to the global problem. u_i^* can contain terms of the first order or also higher order of Eq. (18.7). As an example, in Fig. 18.24 a typical deformed configuration of the cell of periodicity shown in Fig. 18.1 is presented and in Fig. 18.25 we show the periodic component due to $\mathbf{u}^2$.[54] In the example only the linear part $\mathbf{u}^1$ has been used.

As for the anti-periodicity condition (18.39$_2$), the boundary of the unit cell is decomposed into two parts

$$\partial V = \partial V_1 + \partial V_2 \qquad (18.40)$$

where each point $P_1 := P_1 \in \partial V_1$ has a corresponding point $P_2 := P_2 \in \partial V_2$, see Fig. 18.26, such that

$$(\sigma_{ij}n_j)_{P_2} = -(\sigma_{ij}n_j)_{P_1} \qquad (18.41)$$

From a numerical point of view periodic boundary conditions can be easily implemented in general-purpose finite element codes (see reference 18 for a discussion on periodic boundary conditions).

18.15 Local analysis of a unit cell

Given a unit cell on which the periodic boundary conditions (18.39) are imposed, the problem to be solved is to find the constitutive law of the homogenized material described by macroscopic stress and strain (18.38). The unknown relation between these quantities is numerically obtained by solving a large number of local problems given on the unit cell

$$\begin{cases} \text{microscopic constitutive laws} \\ \operatorname{div}\boldsymbol{\sigma} = \mathbf{0}; \quad \text{micro-equilibrium} \\ E_{ij} = \left(\alpha_0 + \frac{1}{c}\,\alpha_0\,m\right) E_{ij}^0; \quad \text{given} \end{cases} \tag{18.42}$$

E_{ij}^0 indicates the direction of the loading path in the strain space, α_0 is a strain multiplier which allows the first yielding and/or slip in some points of the RVE to be reached and $m = 0, 1, 2, \ldots, m_{\max}$ indicates the number of loading steps which follow the first elastic step; in the numerical examples shown below we choose $m_{\max} = 25$. The factor $1/c$ is the multiplier for plastic strain. For a description of contact formulations, see Chapter 5.

A global strain tensor E_{ij}^0 is hence imposed to the cell and it is monotonically increased to generate a kinematic loading path: in particular, an elastic step is applied so as to reach the global elastic frontier, which is defined as the set of points in stress space corresponding to the first yielding of any point of the unit cell or the first activation of relative displacements in the contact points for a fixed loading direction E_{ij}^0. Then $m_{\max}$ 'small' load increments are applied to induce plastic deformations in the unit cell and/or slip in the contact points. Each 'small' load increment is equal to $1/c$ times the first (elastic) increment.

The homogenized stress tensor Σ_{ij} is computed, by means of Eq. (18.38), for each step of the load history. Therefore we have one point in stress space for each loading step; all the points characterized by the same E_{ij}^0 form a loading path in stress space. These points are called interpolation points: there the behaviour of the homogenized material (and precisely the value of the homogenized strains E_{ij} and stresses Σ_{ij}) is known.

Repeating the procedure for several different given tensors E_{ij}^0 we know the behaviour of the homogenized material at a discrete number of points and for a discrete variety of load situations. Interpolation points and typical loading paths are shown in Fig. 18.27.[55]

18.16 Homogenization procedure – definition of successive yield surfaces

Equation (18.38) defines macroscopic stresses and strains, which are assumed to be linked by a macroscopic constitutive law. Such a law is constructed starting from the constitutive relations of the single components and the geometry of the unit cell; this step is referred to as homogenization procedure. The macroscopic behaviour is assumed

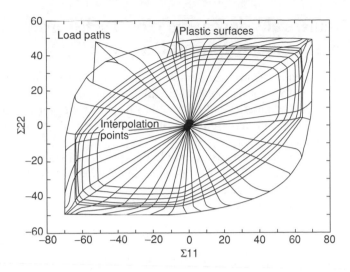

Fig. 18.27   Grid of interpolation points and plastic surfaces.

to be of elastic–plastic type, even if at microscopic level the behaviour may be of a different type.[54,56] There the behaviour should, however, follow the complementary rule as it does in the present case. We have hence to determine the following macroscopic quantities: elastic frontier, flow rule, hardening law and extremal surface of the global homogenized materials. In other words our aim is to determine the global elastic–plastic constitutive law

$$\Sigma_{ij} = \Sigma_{ij}(E_{ij}) \tag{18.43}$$

Given the nature of the components, the microscopic stress field is constrained by the usual relation

$$\sigma(\mathbf{y}) \in P(\mathbf{y}) \tag{18.44}$$

where $P(\mathbf{y})$ is the set of stress states that the material can admit in the space of stresses. $P(\mathbf{y})$ depends on the single material and on the contact conditions, and hence on the position $\mathbf{y}$ in the RVE. $P(\mathbf{y})$ can be defined by means of a yield function $f(\mathbf{y}, \sigma)$

$$P(\mathbf{y}) = \{\sigma \,|\, f(\mathbf{y}, \sigma, \varepsilon^p) \leq 0\} \tag{18.45}$$

Since local stress states σ must lie within the set given in Eq. (18.44), it seems reasonable that all physical global states Σ are contained in a macroscopic (or effective) domain P^{eff} whose frontier is called global (or effective) extremal yield surface

$$\Sigma \in P^{\text{eff}} \tag{18.46}$$

The global constitutive behaviour of the homogenized material usually has an initial range with a linear relation between global strains and global stresses followed by a non-linear range, see Fig. 18.27. The limit of the initial range, the successive yield surfaces for different values of $\mathbf{m}$ (loading steps) and the extremal yield surfaces can be constructed with the numerical experiments on local problems described above. They are all shown in Fig. 18.27 for the case of only E_{11} and E_{22}. An example of a three-dimensional surface for the case of local damage behaviour is presented in reference 56.

18.17 Numerically developed global self-consistent elastic–plastic constitutive law

From the results obtained from the solution of many local problems we extract now the information necessary for a global elastic–plastic analysis.

The behaviour of the material is linear within the most interior closed curve, see Fig. 18.27; a flow rule and hardening law have to be defined for the homogenized material outside the elastic zone.

We consider here the simplest case where in the space of global stresses each interpolation point is characterized by the values of three variables: the ratios E_{11}^0/E_{22}^0, E_{12}^0/E_{22}^0 and m (number of steps).

The constitutive law at the macro level can be written as

$$\Sigma \in P^{\text{eff}} = \{\Sigma \mid f(\Sigma, \mathbf{E}^p) \le 0\} \tag{18.47}$$

The consistency condition of Eq. (18.47) can be expressed as

$$\partial f(\Sigma, \mathbf{E}^p) = \frac{\partial f}{\partial \Sigma} d\Sigma + \frac{\partial f}{\partial \mathbf{E}^p} d\mathbf{E}^p = 0 \tag{18.48}$$

The gradient of the yield function with respect to stress $\partial f(\Sigma)/\partial \Sigma$, for a generic point lying within interpolation points, can be obtained easily by an interpolation technique which is the same as that generally used in post-processing to obtain contour lines or surfaces, see Fig. 18.27. The plastic strain increment $d\mathbf{E}^p$ can be obtained from the local analysis with the following additive decomposition and relationship

$$E_{ij} = E_{ij}^E + E_{ij}^P$$
$$\Sigma_{ij} = D_{ijkl}^E E_{kl}^E \tag{18.49}$$

where E_{ij}^E, the macroscopic purely elastic strain, is defined as that part of the macro strain that is non-zero if and only if the macro stress is non-zero, and D_{ijkl}^E is the elastic constitutive tensor obtained with elastic homogenization.

If we assume an associative plastic flow for the macro level constitutive model, once the vector $\partial f(\Sigma)/\partial \Sigma$ has been obtained the plastic flow direction is known at all points in the stress space for the constitutive law at the macro level. However, to implement the algorithm systematically, a hardening modulus has to be determined in advance. This requires an iterative procedure due to the unknown final state for one incremental step. We assume here that the mean hardening modulus can be expressed as

$$H = \frac{\partial f(\Sigma, \mathbf{E}^P)}{\partial \mathbf{E}^P} \tag{18.50}$$

Hence, from Eq. (18.48), the following iterative procedure can be established

$$\mathbf{H}^{(i)} = \begin{cases} -\dfrac{1}{\Delta E^P}\left((1-\eta)\dfrac{\partial f^{(i)}}{\partial \Sigma} + \eta \dfrac{\partial f^{(i-1)}}{\partial \Sigma}\right)\Delta \Sigma^{(i)}, & \text{when } \Delta \mathbf{E}^P \neq 0 \\ 0 & , \quad \text{when } \Delta \mathbf{E}^P = 0 \end{cases} \tag{18.51}$$

where $0 \le \eta \le 1$ is an integration parameter such as that used in a mid-point algorithm.

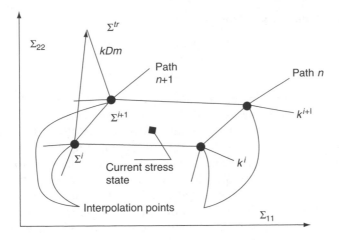

Fig. 18.28 Interpolation of the flow direction.

When a non-associative plastic flow is used, the flow rule can be extracted from the radial return mapping algorithm written in a standard form as

$$\mathbf{R} := -\boldsymbol{\Sigma}^{tr} + \boldsymbol{\Sigma}^{(i)} + k\,\mathbf{D}\,\mathbf{m} = \mathbf{0}, \qquad k \geq 0 \tag{18.52}$$

where $\mathbf{D}$ is the elastic constitutive tensor obtained with the elastic homogenization, k is the increment of the plastic flow, $\mathbf{m}$ is the flow direction and $\boldsymbol{\Sigma}^{tr}$, $\boldsymbol{\Sigma}^{(i)}$ are the global stresses.

The flow direction (which refers to an unknown plastic potential) can be calculated, at each interpolation point, in the following manner, see Fig. 18.28. Starting from an interpolation point corresponding to step i one multiplies the known strain increment $\mathbf{E}$ by the elastic effective tensor $\mathbf{A}$, obtaining the trial stress $\boldsymbol{\Sigma}^{tr}$. Since there is plastic deformation or slip in some part of the unit cell, the strain $\Delta\mathbf{E}$ actually generates the new stress $\boldsymbol{\Sigma}^{(i+l)}$, different from $\boldsymbol{\Sigma}^{tr}$. The flow direction at the interpolation point for the level $i + 1$ can be computed as

$$\mathbf{m} = c\,\mathbf{A}^{-1}\left[\boldsymbol{\Sigma}^{tr} + \boldsymbol{\Sigma}^{(i)}\right] = c\left[\Delta\mathbf{E} - \mathbf{A}^{-1}\boldsymbol{\Sigma}^{(i)}\right] \tag{18.53}$$

The value of k has been arbitrarily chosen to be $1/c$ [see local problem (18.42)], hence the quantity $\mathbf{m}$ not only gives information about the flow direction but also about the amount of the plastic flow. For stress points inside a patch, the flow direction is obtained again by interpolation.

Once a consistency condition and a flow rule have been determined, the global constitutive law is fully defined and can be assumed as constitutive law of the homogenized material.

18.18 Global solution and stress-recovery procedure

At this point a global problem can be solved. Given a periodic structure subjected to assigned external loads $\mathbf{f}$ and to assigned boundary conditions, the global displacements can be found solving the problem

$$
\begin{cases}
\text{macroscopic constitutive law} \\
\text{div } \Sigma = \mathbf{f}; \quad \text{macro-equilibrium} \\
\text{global boundary conditions}
\end{cases}
\tag{18.54}
$$

The solution of the global problem (18.54) gives a reasonably good estimate of the displacements of the structure. Nevertheless, very often stresses are the most relevant mechanical quantities but they cannot be easily derived from global displacements. The global solution is used to evaluate the local distribution of micro stresses by solving a local problem (18.42). This is the third step of the homogenization procedure.

If the stress distribution in a specific region is required, we take into consideration the integration points used in the global solution which are close to the examined region of the real structure. If the number of unit cells is high a single integration point corresponds to a group of cells, but usually one RVE corresponds to one integration point. The strains computed in the global solution are assumed as global strains of the investigated cell (or group of cells); in general the finite element computation gives a sequence of n values of strains if the load history is composed of n load steps. Such a sequence of values is taken as load history on a unit cell with periodic boundary conditions and the following local problem is solved:

$$
\begin{cases}
\text{microscopic constitutive laws} \\
\text{div } \sigma = \mathbf{0}; \quad \text{micro-equilibrium} \\
E_{ij} = E_{ij}(t) \text{ given by the global problem}
\end{cases}
\tag{18.55}
$$

The solution of such a local problem gives a stress distribution which usually is a good approximation of the real one, as shown in the next example.

Numerical examples

We consider an assembly of discs shown in Fig. 18.22, under assumption of plane stress behaviour. The discs have the following mechanical properties

$$
E = 128\,000 \text{ MPa}, \quad v = 0.34, \quad \sigma_Y = 80 \text{ MPa}
$$

The elastic–plastic constitutive law of the discs is of the von Mises type. The overall property of the structure will clearly be controlled by the behaviour of the discs around their contact points and by the elastic–plastic material behaviour. This character is captured by the analysis of three cases of different interface parameters.

(a) Elastic–plastic material with purely frictional contact

$$
\mu = 0.25, \quad p_{no} = p_{to} = 0, \quad \delta^\star \neq 0
$$

(b) Elastic–plastic material with cohesive frictional contact:

$$
\mu = 0.25, \quad p_{no}0, \quad p_{to} = \frac{1}{\sqrt{3}} 80 \text{ MPa}, \quad \delta^\star \neq 0
$$

(c) Elastic–plastic material with cohesive frictional contact:

$$
\mu = 0.10, \quad p_{no}0, \quad p_{to} = \frac{1}{\sqrt{3}} 80 \text{ MPa}, \quad \delta^\star \neq 0
$$

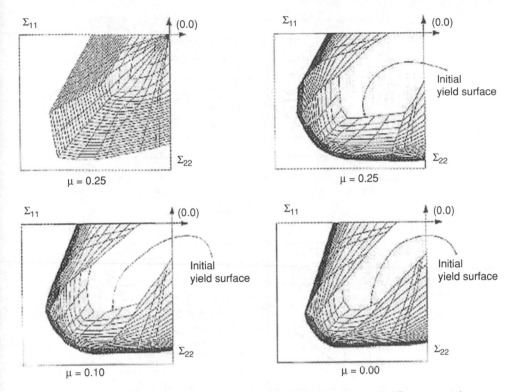

Fig. 18.29 Local numerical constitutive law for elastic–plastic bodies in contact with different material parameters: (a) purely frictional contact; (b) cohesive–frictional contact with $\mu = 0.25$; (c) cohesive–frictional contact with $\mu = 0.10$; (d) purely cohesive contact.

(d) Elastic–plastic material with purely cohesive contact:

$$\mu = 0.00, \quad p_{no}0, \quad p_{to} = \frac{1}{\sqrt{3}}80 \text{ MPa}, \quad \delta^\star \neq 0$$

p_n, p_t are the limits of the normal and tangential contact forces respectively, and $\delta^\star$ is the gap, see Fig. 18.23.

The contact algorithm used in the example is that of Zhang *et al.*[57] The cell of periodicity, indicated in Fig. 18.22, is discretized with four-node plane stress elements. Our interest is focused on the macro stress domain

$$\left\{ \Sigma_{11}, \ \Sigma_{22} \ \middle| \ \Sigma_{11} \leq 0, \ \Sigma_{22} \leq 0 \right\}$$

and the obtained interpolation points (yield surfaces) corresponding to the above four cases are presented in Figs 18.29 (a)–(d) in that quadrant. For the elastic case see reference 55.

To verify the capability of the procedure a group of 140 discs arranged in a rectangle (10 × 14 discs) is analysed under uniform compression. For this purpose a uniform vertical displacement is assigned to the top nodes. The vertical displacements of the

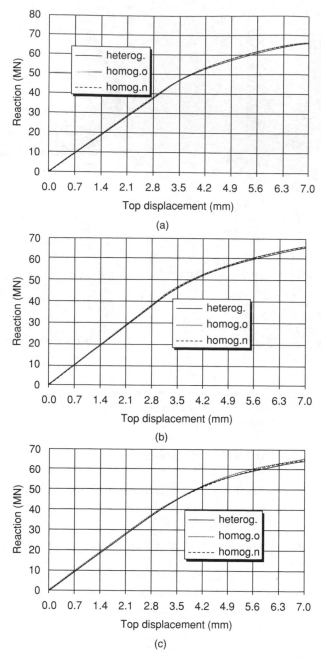

Fig. 18.30 Comparison between the reactions in the case of uniform top displacement. Homog.n indicates the results with the self-consistent procedure of this chapter,[25] homog.o the results of reference 58.

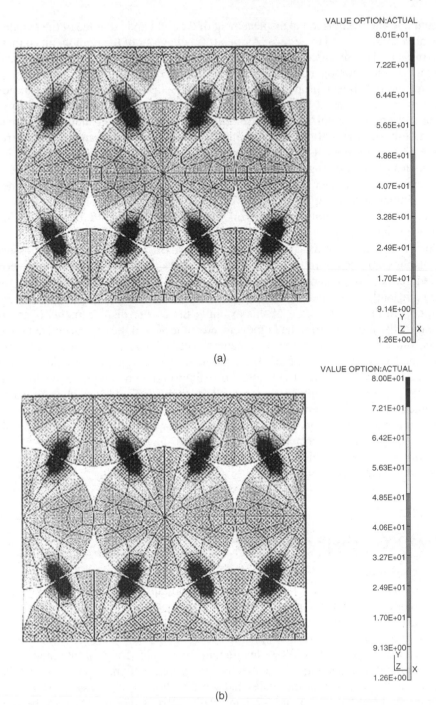

(a)

(b)

Fig. 18.31 Comparison between the stress distributions obtained for elastic–plastic bodies in contact: (a) stresses recovered from the homogeneous computation; (b) stresses recovered from the heterogeneous computation.

bottom line and the horizontal displacement of the right and left edge of the rectangular assembly are restrained to be zero. Associative behaviour of the homogenized material is assumed. The problem is solved using a rough discretization (43 nodes and 35 elements, called homogeneous) with the numerical constitutive law which takes into account the non-linear property due to the stick–slip behaviour between the discs. The vertical reactions at the middle point of the bottom line are compared to those of the equivalent model with a finite element discretization (called heterogeneous) which describes the real material distribution and the real mechanical characteristics of the single components. The comparisons between the vertical reactions in the homogeneous and heterogeneous models are shown in Figs 18.30(a)–(c) for the three different cases of contact parameters b, c, and d. The results obtained in reference 58 with a similar procedure, which is, however, not self-consistent, are also shown in the figure. The results obtained by the proposed method are in good agreement with the heterogeneous results.

The case of a non-uniform displacement distribution on the top of the assembly was also investigated in reference 25. We show here the results of the stress recovery procedure, i.e. we consider a unit cell from a local point of view and impose to it a kinematic load represented by the displacement field computed with the homogenized solution. In the specific case of this example the corner displacements of the RVE were directly obtained from the homogenized solution and the displacements imposed on the other boundary nodes were obtained with a linear interpolation between the corner nodes. The solution of this local problem gives a stress distribution in the unit cell which can be compared to the stress distribution obtained with the heterogeneous (discrete) model. The results for one unit cell are shown in Fig. 18.31 where an excellent agreement can be observed.

In conclusion, the approach requires the preliminary solution of many local problems and proceeds then with a macroscopic numerical analysis of the overall structure. Solution of local problems for obtaining the local stress distribution is needed then only at the Gauss points of interest.

18.19 Concluding remarks

Discrete–continuum linkage becomes particularly important for nanoscale mechanics and materials. Continuum-based approaches are not applicable down to the nanoscale as non-continuum behaviour is observed at that scale, such as for instance in large deformation of carbon nanotubes or ion deposition processes.[4] Further, nanoscale components are generally used in conjunction with components that are larger and have a mechanical response at different length and time scale. Single-scale methods such as molecular dynamics or quantum mechanics are generally not applicable in this last case due to the disparity of the scales and scale bridging is a necessity. This is a new and rapidly developing area and the interested reader is referred, e.g., to the already mentioned special issue devoted to that topic.[4]

Finally, the contributions to this chapter by D.P. Boso, M. Lefik, S. Loehnert and V. Salomoni are gratefully acknowledged.

References

1. R. Lakes. Materials with structural hierarchy. *Nature*, 361:511–515, 1993.
2. R.C. Picu. Foreword to special issue on linking discrete and continuum models. *Int. J. Multiscale Computational Engng.*, 1(1):vii–viii, 2003.
3. P. Ladevéze and J. Fish. Preface to special issue on multiscale computational mechanics for materials and structure. *Computer Methods in Applied Mechanics and Engineering*, 192:28–30, 2003.
4. W.K. Liu, D. Qian and M.F. Horstemeyer. Preface to special issue on multiple scale methods for nanoscale mechanics and materials. *Computer Methods in Applied Mechanics and Engineering*, 193:17–20, 2004.
5. W. Voigt. *Lehrbuch der Kristallphysik*. Teubner, Leipzig-Berlin, 1910.
6. A. Reuss. Berechnung der Fliessgrenze von Mischkristallen auf Grund der Plastizitatsdingung für Einkristale. *Z. Angew. Math. Mech.*, 9:49, 1926.
7. T. Mori and K. Tanaka. Average stress in matrix and average elastic energy of materials with misfitting inclusions. *Acta Metallurgica*, 21:571–574, 1973.
8. J.D. Eshelby. The determination of the elastic field of an ellipsoidal inclusion and related problems. *Proc. Roy. Soc.*, A241:376–396, 1957.
9. Z. Hashin and S. Shtrikman. A variational approach to the theory of the elastic behaviour of multiphase materials. *J. Mech. Phys. Sol.*, 11(2):127–141, 1964.
10. E. Kröner. Bounds for effective elastic moduli of disordered materials. *J. Mech. Phys. Sol.*, 25(2):137–155, 1977.
11. E. Kröner. Self-consistent scheme and graded disorder in polycristal elasticity. *J. Phys.*, 8:2261–2267, 1978.
12. J.R. Willis. Bounds and self-consistent estimates for the overall properties of anisotropic composites. *J. Mech. Phys. Sol.*, 25:185–202, 1977.
13. A. Bensoussan, J.L. Lions and G. Papanicolau. *Asymptotic Analysis for Periodic Structures*. North-Holland, Amsterdam, 1976.
14. E. Sanchez-Palencia. *Non-Homogeneous Media and Vibration Theory*. Springer Verlag, Berlin, 1980.
15. G.A. Francfort. Homogenisation and fast oscillations in linear thermoelasticity. In R. Lewis *et al.*, editors, *Numerical Methods for Transient and Coupled Problems*, pages 382–392. Pineridge Press, Swansea, 1984.
16. M. Lefik and B.A. Schrefler. 3d finite element analysis of composite beams with parallel fibres based on the homogenisation theory. *Computational Mechanics*, 14:2–15, 1994.
17. M. Lefik and B.A. Schrefler. Application of the homogenisation method to the analysis of superconducting coils. *Fusion Engineering and Design*, 24:231–255, 1994.
18. O.C. Zienkiewicz, R.L. Taylor and J.Z. Zhu. *The Finite Element Method: Its Basis and Fundamentals*. Butterworth-Heinemann, Oxford, 6th edition, 2005.
19. B.A. Schrefler, M. Lefik and U. Galvanetto. Correctors in a beam model for unidirectional composites. *Mechanics of Composite Materials and Structures*, 4:159–190, 1997.
20. P. Ladevéze, editor. *Local Effects in the Analysis of Structures*, page 342. Elsevier, New York, 1985.
21. M. Lefik and B.A. Schrefler. Fe modelling of boundary layer correctors for composites using the homogenisation theory. *Engineering Computations*, 13:31–42, 1996.
22. K. Terada and N. Kikuchi. A class of general algorithms for multi-scale analyses of heterogeneous media. *Computer Methods in Applied Mechanics and Engineering*, 190:5427–5464, 2001.
23. J. Fish and Q. Yu. Multiscale damage modeling for composite materials: theory and computational framework. *International Journal for Numerical Methods in Engineering*, 52:161–192, 2001.

24. M. Kaminski and B.A. Schrefler. Probabilistic effective characteristics of cables for supercon-ducting coils. *Computer Methods in Applied Mechanics and Engineering*, 188:1–16, 2000.

25. H.W. Zhang, D.P. Boso and B.A. Schrefler. Homogeneous analysis of periodic assemblies of elastoplastic disks in contact. *Int. J. of Multiscale Computational Engineering*, 1:359–380, 2003.

26. Thermal, electrical and mechanical properties of materials at cryogenic temperatures. Conductor Database, Appendix C, Annex ll, 24 August 2000.

27. J. Ekin. Mechanical properties and strain effects in superconductors. In S. Foner and B. Schwartz, editors, *Superconductor Materials Science: Metallurgy, Fabrication and Applications*. Plenum Press, NATO Advanced Study Institute Series, 1980.

28. R.P. Reed and A.F. Clark, editors. *Materials at Low Temperature*. American Society for Metals, Metals Park, Ohio, 1983.

29. S. Ochiai and K. Osamura. Prediction of variation of upper critical magnetic field of nb_3sn superconducting composites as a function of applied stress at room temperature. *Acta Metall.*, 37(9):2539–2549, 1989.

30. G. Rupp. The importance of being prestressed (nb_3sn composite superconductor). In M. Suenaga and A.F. Clark, editors, *Filamentary A15 Superconductors*, pages 155–170. Plenum Press, New York, 1980.

31. N. Mitchell. Analysis of the effect of nb_3sn strand bending on CICC superconductor performance. *Cryogenics*, 42:311–325, 2002.

32. F. Feyel. Multiscale fe^2 elastoviscoplastic analysis of composite structures. *Comp. Mat. Sci.*, 16:344–354, 1999.

33. F. Feyel and J.L. Chaboche. fe^2 multiscale approach for modelling the elastoviscoplastic behaviour of long fibre SiC/Ti composite materials. *Computer Methods in Applied Mechanics and Engineering*, 183:309–330, 2000.

34. F. Feyel. Multiscale non-linear fe^2 analysis of composite structures: Fiber size effects. *J. Phys. IV, France*, 11:195–202, 2001.

35. F. Feyel and J.L. Chaboche. Multi-scale non-linear fe^2 analysis of composite structures: dam-age and fiber size effects. In Khèmais Saanouni, editor, *Revue Européenne des Eléments Finis, 'Numerical Modelling in Damage Mechanics' – NUMEDAM'OO*, volume 10, pages 449–472, 2001.

36. J.F. Bishop and R. Hill. A theory of the plastic distorsion of polycrystalline aggregate under combined stresses. *Philos. Mag.*, 42:414–427, 1951.

37. P. Ladevéze, O. Loiseau and D. Dureisseix. A micro–macro and parallel computational strategy for highly heterogeneous structures. *International Journal for Numerical Methods in Engineer-ing*, 52:121–138, 2001.

38. P. Ladevéze, A. Nouy and O. Loiseau. A multiscale computational approach for contact problems. *Computer Methods in Applied Mechanics and Engineering*, 191:4869–4891, 2002.

39. P. Ladevéze and A. Nouy. On a multiscale computational strategy with time and space homogeni-sation for structural mechanics. *Computer Methods in Applied Mechanics and Engineering*, 192:3061–3087, 2003.

40. M.G.D. Geers, V. Kouznetsova and W.A.M. Brekelmans. Gradient-enhanced computational homogenisation for the micro–macro scale transition. *Journal de Physique IV*, 11(5):5145–5152, 2001.

41. S. Ghosh, K. Lee and S. Moorthy. Two scales analysis of heterogeneous elastic–plastic materials with asymptotic homogenisation and voronoi cell finite element model. *Computer Methods in Applied Mechanics and Engineering*, 132:63–116, 1996.

42. S. Ghosh, K. Lee and P. Raghavan. A multilevel computational model for multiscale damage analysis in composite and porous materials. *International Journal of Solids and Structures*, 38:2335–2385, 2001.

43. C. Miehe, J. Schroder and J. Schotte. Computational homogenisation analysis in finite plasticity. Simulation of texture development in polycrystalline materials. *Computer Methods in Applied Mechanics and Engineering*, 171:387–418, 1999.

44. C. Miehe, J. Schotte and J. Schroder. Computational micro–macro transitions and overall moduli in the analysis of polycrystals at large strains. *Computational Materials Science*, 16:372–382, 1999.

45. V. Kouznetsova, W.A.M. Brekelmans and F.P.T. Baaijens. An approach to micro–macro modelling of heterogeneous materials. *Computational Mechanics*, 27:37–48, 2001.

46. P.M. Suquet. Plasticity today: modelling, methods and applications. In *Local and Global Aspects in the Mathematical Theory of Plasticity*, pages 279–310. Elsevier Applied Science Publishers, London, 1985.

47. S. Loehnert and P. Wriggers. Homogenisation of microheterogeneous materials considering interfacial delamination at finite strains. *Technische Mechanik*, 23(2-3):167–177, 2003.

48. T.I. Zohdi and P. Wriggers. Computational micro–macro material testing. *Archives of Computational Methods in Engineering*, 8(2):131–228, 2001.

49. A. Needleman. A continuum model for void nucleation by inclusion debonding. *J. Applied Mechanics, ASME*, 54:525–531, 1987.

50. C. Miehe and J. Dettmar. A framework for micro–macro transitions in periodic particle aggregates of granular materials. *Computer Methods in Applied Mechanics and Engineering*, 193:225–256, 2004.

51. A. Nardin, G. Zavarise and B.A. Schrefler. Modelling of cutting tool–soil interaction – Part I: Contact behaviour. *Computational Mechanics*, 31:327–339, 2003.

52. J.R. Wren and R.I. Borja. Micromechanics of granular media. Part I: Generation of overall constitutive equation for assemblies of circular disks. *Computer Methods in Applied Mechanics and Engineering*, 127:13–36, 1995.

53. J.R. Wren and R.I. Borja. Micromechanics of granular media Part II: Overall tangential moduli and localization model for periodic assemblies of circular disks. *Computer Methods in Applied Mechanics and Engineering*, 141:221–246, 1995.

54. C. Pellegrino, U. Galvanetto and B.A. Schrefler. Numerical homogenisation of periodic composite materials with non-linear material components. *International Journal for Numerical Methods in Engineering*, 46:1609–1637, 1999.

55. H.W. Zhang, U. Galvanetto and B.A. Schrefler. Local analysis and global non-linear behaviour of periodic assemblies of bodies in elastic contact. *Computational Mechanics*, 24:217–229, 1999.

56. D.P. Boso, C. Pellegrino, U. Galvanetto and B.A. Schrefler. Macroscopic damage in periodic composite materials. *Comm. Numer. Meth. Engng*, 16(9):615–623, 2000.

57. H.W. Zhang, W.X. Zhong and Y.X. Gu. A combined programming and iteration algorithm for finite element analysis of three-dimensional contact problems. *Acta Mech. Sinica*, 11:318–326, 1995.

58. H.W. Zhang and B.A. Schrefler. Global constitutive behaviour of periodic assemblies of inelastic bodies in contact. *Mechanics of Composite Materials and Structures*, 7:355–382, 2000.

59. D.P. Boso, M. Lefik and B.A. Schrefler. A multilevel homogenised model for superconducting strands thermomechanics. *Cryogenics*, 45:259–271, 2005.

Computer procedures for finite element analysis

19.1 Introduction

In this chapter we describe some features of the companion computer program which may be used to perform numerical studies for many of the topics discussed in this book. The source program and manuals are available at no cost from the publisher's web page:

http://books.elsevier.com/companions

or from the authors' web page:

http://www.ce.berkeley.edu/~rlt

The computer program described in this volume is intended for use by those who are undertaking a study of the finite element method and wish to implement and test specific elements or specific solution steps. The program also includes a library of simple elements to permit solution to many of the topics discussed in this book. The program is called *FEAPpv* to emphasize the fact that it may be used as a *personal version* system. With very few exceptions, the program is written using standard Fortran, hence it may be implemented on any personal computer, engineering workstation, or main frame computer which has access to a Fortran 77 or Fortran 90/95 compiler.

It still may be necessary to modify some routines to avoid system-dependent difficulties. Non-standard routines are restricted to the graphical interfaces and file handling for temporary data storage. Users should consult their compiler manuals on alternative options when such problems arise.

Users may also wish to add new features to the program. In order to accommodate a wide range of changes several options exist for users to write new modules without difficulty. There are options to add new mesh input routines through addition of routines named UMESHn or to include solution options through additions of routines named UMACRn. Finally, the addition of a user developed element module is accommodated by adding a single subprogram named ELMTnn. In adding new options the use of established algorithms as described in references 1–5 can be very helpful.

The current chapter presents a brief discussion to describe aspects of the program that are related to solution of non-linear problems. The program *FEAPpv* includes capabilities to solve general non-linear finite element models for transient and steady-state (static) problems. The transient problem types include solution algorithms for both first-order (diffusion-type) and second-order (vibration/wave-type) ordinary differential equations in time. In addition an eigensolution system is included to compute eigenpairs of typical problems. A simultaneous vector iteration algorithm (subspace method) is used to extract the eigenpairs nearest to a specified shift of a *symmetric tangent* matrix. Hence, the eigensystem may be used with either linear or non-linear problems. Non-linear problems are often difficult to solve and time consuming in computer resources. In many applications the complete analysis may not be performed during one execution of the program; hence, techniques to stop the program at key points in the analysis for a later restart to continue the solution are available.

The program described in this chapter has been developed and used in an educational and research environment over a period of nearly 35 years. The concept of the command language solution algorithm has permitted several studies that cover problems that differ widely in scope and concept, to be undertaken at the same time without need for different program systems. Unique features for each study may be provided as new solution commands. The ability to treat problems whose coefficient matrix may be either symmetric or unsymmetric often proves useful for testing the performance of algorithms that advocate substitution of a symmetric tangent matrix in place of an unsymmetric matrix resulting from a consistent linearization process. The element interface is quite straightforward and, once understood, permits users to test rapidly new types of finite elements.

We believe that the program in this book provides a very powerful solution system to assist the interested reader in performing finite element analyses. The program *FEAPpv* is by no means a complete software system that can be used to solve any finite element problem, and readers are encouraged to modify the program in any way necessary to solve their particular problem. While the program has been tested on several sample problems, it is likely that errors and mistakes still exist within the program modules. The authors need to be informed about these errors so that the available system can be continuously updated. We also welcome readers' comments and suggestions concerning possible future improvements.

19.2 Solution of non-linear problems

The general methods described in this volume are directed toward the solution of non-linear problems in solid and structural mechanics. The application of the finite element method to these problems leads to a set of non-linear algebraic equations. A solution to the non-linear algebraic problem by a Newton method, as described in Chapter 3, is given by[6]

$$\Psi(\mathbf{u}^{(k)}) = \mathbf{P}(\mathbf{u}^{(k)}) + \mathbf{f}^{(k)}$$

$$\mathbf{K}_T^{(k)} d\mathbf{u}^{(k)} = \Psi^{(k)} \text{ where } \mathbf{K}_T^{(k)} = -\left.\frac{\partial \Psi}{\partial \mathbf{u}}\right|^{(k)} \tag{19.1}$$

$$\mathbf{u}^{(k+1)} = \mathbf{u}^{(k)} + d\mathbf{u}^{(k)}$$

where $\mathbf{f}$ is a vector of applied loads and $\mathbf{P}$ is the non-linear *internal* force vector which is indicated as a function of the nodal parameters $\mathbf{u}$. The vector Ψ is the residual of the problem, $\mathbf{K}_T$ is the tangent matrix and a solution is defined as any set of nodal displacements, $\mathbf{u}$, for which the residual is zero. In general, there may be more than one set of displacements which define a solution and it is the responsibility of a user to ensure that a proper solution is obtained. This may be achieved by starting from a state which satisfies physical arguments for a solution and then applying small increments to the loading vector, $\mathbf{f}$. By taking small enough steps, a solution path may usually be traced.

FEAPpv uses the basic Newton strategy defined in Chapter 3 to perform all solution steps. A key feature of the program is a command language which permits users to construct many different linear and non-linear solution algorithms as data statements of the analysis process. For example, a simple set of commands given by

```
LOOP newton 10
    TANGent
    FORM residual
    SOLVe
NEXT newton
```

performs the necessary computations for a Newton algorithm. In the above `LOOP-NEXT` defines the necessary commands to form 10 iterations of the Newton scheme. The command `TANGent` constructs the tangent matrix, `FORM` constructs the residual and `SOLVe` performs a solution of the linear equations. In *FEAPpv* only the first four characters of command words are processed and the remainder can be used to provide additional clarity, thus for emphasis the required data is shown above in upper case letters. The user manual describes the available commands which may be used to construct general linear and non-linear steady state or transient algorithms.

Non-linear problems require use of many schemes to improve the convergence of the solution. Use of the BFGS algorithm described in Chapter 3 can lead to improved solution performance and/or reduced solution cost (see the necking example in Chapter 6). It is also particularly effective when no exact tangent matrix can be computed. In addition a linear line search is useful to limit the magnitude of $d\mathbf{u}^{(k)}$ during early iterations[7] and its use often allows use of larger load increments and still obtain rapid converge. A line search requires repeated computations of $\mathbf{\Psi}(\mathbf{u})$ which may increase solution times. Thus, some assessment of the need of a line search should be made before proceeding with large numbers of solution steps.

A modified Newton method also may be performed by removing the tangent computation from the loop structure given above (i.e. placing `TANG` before `LOOP`). This points out the power of the command language scheme to efficiently include many solution algorithms.

The solution of transient problems defined by the algorithms given in Chapter 2 may also be performed using *FEAPpv*. The program includes options to solve transient finite element problems which generate first- and second-order ordinary differential equations using the GN11 and Newmark (GN22) algorithms.[8] Options also exist to use an explicit version of the GN22 algorithm.

19.3 Eigensolutions

The solution of a general linear eigenproblem is a useful feature included in the *FEAPpv* program. The program can compute a set of the smallest eigenvalues (in absolute value) and their associated eigenvectors for the problem

$$\mathbf{K_T V} = \mathbf{M V \Lambda} \tag{19.2}$$

In the above, $\mathbf{K_T}$ is any symmetric tangent matrix which has been computed by using a TANG command statement; $\mathbf{M}$ is a mass or identity matrix computed using a MASS or IDEN command statement, respectively; the columns of $\mathbf{V}$ are the set of eigenvectors to be computed; and $\mathbf{\Lambda}$ is a diagonal matrix which contains the set of eigenvalues to be computed. For the second-order equations from solid or structural mechanics problems the eigenvalues λ are the frequencies squared, ω^2.

The tangent matrix can have zero eigenvalues and, for this case, the algorithm used requires the problem to be transformed to

$$(\mathbf{K_T} - \alpha\mathbf{M})\mathbf{V} = \mathbf{M V \Lambda}_\alpha \tag{19.3a}$$

where α is a parameter called the *shift*, which can be selected to make the coefficient matrix on the left-hand side of Eq. (19.3a) nonsingular. $\mathbf{\Lambda}_\alpha$ are the eigenvalues of the shift which are related to the desired values by

$$\mathbf{\Lambda} = \mathbf{\Lambda}_\alpha + \alpha\mathbf{I} \tag{19.3b}$$

The shift may also be used to compute the eigenpairs nearest to some specified value (e.g., a buckling load). The components of $\mathbf{\Lambda}$ are output as part of the eigenproblem solution. In addition, the vectors may be output as numerical values or presented graphically.

The program uses a subspace algorithm[9–11] to compute a small general eigenproblem defined as

$$\mathbf{K^\star x} = \mathbf{M^\star x}\lambda \tag{19.4a}$$

where

$$\mathbf{V} = \mathbf{Q x} \tag{19.4b}$$

and

$$\begin{aligned}\mathbf{K^\star} &= \mathbf{Q^T M^T} (\mathbf{K_T} - \alpha\mathbf{M})^{-1}\mathbf{M Q}\\ \mathbf{M^\star} &= \mathbf{Q^T M Q}\end{aligned} \tag{19.4c}$$

Accordingly, after the projection, the λ are reciprocals of $\mathbf{\Lambda}_\alpha$ (i.e. $\mathbf{\Lambda}_\alpha^{-1}$). An eigensolution of the small problem may be used to generate a sequence of iterates for $\mathbf{Q}$ which converge to the solution for the original problem (e.g. see reference 10). The solution of the projected small general problem is solved here using a transformation to a standard linear eigenproblem combined with a QL algorithm.[6]

The transformation is performed by computing the Choleski factors of $\mathbf{M^\star}$ to define the standard linear eigenproblem

$$\mathbf{H y} = \mathbf{y}\lambda \tag{19.5a}$$

where

$$\begin{aligned}\mathbf{M^\star} &= \mathbf{L L^T}\\ \mathbf{y} &= \mathbf{L^T x}\\ \mathbf{H} &= \mathbf{L^{-1} K^\star L^{-T}}\end{aligned} \tag{19.5b}$$

In the implementation described here scaling is introduced, which causes $\mathbf{M}^*$ to converge to an identity matrix; hence the above transformation is numerically stable. Furthermore, use of a standard eigenproblem solution permits calculation of positive and negative eigenvalues. The subspace algorithm implemented provides a means to compute a few eigenpairs for problems with many degrees of freedom or all of the eigenpairs of small problems. A subspace algorithm is based upon a power method to compute the dominant eigenvalues. Thus, the effectiveness of the solution strategy depends on the ratio of the absolute value of the largest eigenvalue sought in the subspace to that of the first eigenvalue not contained in the subspace. This ratio may be reduced by adding additional vectors to the subspace. That is, if p pairs are sought, the subspace is taken as q vectors so that

$$\left| \frac{\lambda_p}{\lambda_{q+1}} \right| < 1 \tag{19.6}$$

Of course, the magnitude of this ratio is unknown before the problem is solved and some analysis is necessary to estimate its value. The program tracks the magnitude of the shifted reciprocal eigenvalues Λ and computes the change in values between successive iterations. If the subspace is too small, convergence will be extremely slow owing to Eq. (19.6) having a ratio near unity. It may be desirable to increase the subspace size to speed the convergence. In some problems, characteristics of the eigenvalue magnitudes may be available to assist in the process. It should be especially noted that when p is specified as the total number of degrees of freedom in the problem (or q becomes this value), then λ_{q+1} is infinitely larger and the ratio given in Eq. (19.6) is zero. In this case subspace iteration converges in a single iteration, a fact which is noted by the program to limit the iterations to 1. Accordingly, it is usually more efficient to compute all the eigenpairs if q is very near the number of degrees of freedom.

19.4 Restart option

The program *FEAPpv* permits a user to save a solution state and subsequently use it later to continue the analysis. This is called a *restart* option. To use the restart feature, the file names given at initiation of the program must be appropriately specified.

The file name for the set of problem restart data files is specified at the time execution of *FEAPpv* is initiated. During a solution a restart file may be saved by using the command statement

        SAVE <extender>

This saves the current solution data in a file that has the restart file name with an extension `extender`. For example, if the restart write file has the name 'Rprob', issuing the command

        SAVE ti0

saves the data on a file named `Pprob.ti0`. Alternatively, issuing the command as

        SAVE

saves the data on the file named `Pprob`. For large problems the restart file can be quite large (especially if the elements use several history variables at each integration point) thus one should be cautious about use of too many files in these situations.

To restore a file the command

        RESTart

is given to load the file without an extender, and the command

        RESTart <extender>

to load the file with an extender.

19.5 Concluding remarks

In the discussion above we have presented a few of the ways the program *FEAPpv* may be used to solve non-linear finite element problems. The classes of non-linear problems which may be solved using this system is extensive and we cannot give a comprehensive summary here. The reader is encouraged to obtain a copy of the program source statements and companion documents from the publisher's website (http://books.elsevier.com/companions).

As noted in the introduction to this chapter the computer programs will undoubtedly contain some errors. We welcome being informed of these as well as comments and suggestions on how the programs may be improved. Although the programs available are written in Fortran it is quite easy to adapt these to permit program modules to be constructed in other languages. For example, an interface for element routines written in C has been developed by Govindjee.[12]

The program system *FEAPpv* contains only basic commands to generate structured meshes as blocks of elements. For problems where graded meshes are needed (e.g. adaptive mesh refinements) more sophisticated mesh generation techniques are needed. There are many locations where generators may be obtained and two are given in references 13 and 14. The program GiD offers two- and three-dimensional options for fluid and structure applications.

References

1. W.H. Press *et al.*, editors. *Numerical Recipes in Fortran: The Art of Scientific Computing.* Cambridge University Press, Cambridge, 2nd edition, 1992.
2. W.H. Press *et al.*, editors. *Numerical Recipes in Fortran 77 and 90: The Art of Scientific and Parallel Computing (Software).* Cambridge University Press, Cambridge, 1997.
3. W.H. Press *et al.*, editors. *Numerical Recipes in Fortran 90: The Art of Parallel Scientific Computing*, volume 2. Cambridge University Press, Cambridge, 1996.
4. G.H. Golub and C.F. Van Loan. *Matrix Computations.* The Johns Hopkins University Press, Baltimore MD, 3rd edition, 1996.
5. J. Demmel. *Applied Numerical Linear Algebra.* Society for Industrial and Applied Mathematics, Philadelphia, PA, 1997.
6. L. Collatz. *The Numerical Treatment of Differential Equations.* Springer, Berlin, 1966.

7. H. Matthies and G. Strang. The solution of nonlinear finite element equations. *International Journal for Numerical Methods in Engineering*, 14:1613–1626, 1979.

8. O.C. Zienkiewicz, R.L. Taylor and J.Z. Zhu. *The Finite Element Method: Its Basis and Fundamentals*. Butterworth-Heinemann, Oxford, 6th edition, 2005.

9. J.H. Wilkinson and C. Reinsch. *Linear Algebra. Handbook for Automatic Computation*, volume II. Springer-Verlag, Berlin, 1971.

10. K.-J. Bathe and E.L. Wilson. *Numerical Methods in Finite Element Analysis*. Prentice-Hall, Englewood Cliffs, NJ, 1976.

11. K.-J. Bathe. *Finite Element Procedures*. Prentice-Hall, Englewood Cliffs, NJ, 1996.

12. S. Govindjee. Interface for c-language routines for feap programs. Private communication (see also at internet address: http://www.ce.berkeley.edu/~sanjay), 2000.

13. J. Shewchuk. Triangle. http://www.cs.berkeley.edu/~jrs).

14. GiD – The Personal Pre/Postprocesor. www.gidhome.com, 2004.

Appendix A

Isoparametric finite element approximations

A.1 Introduction

An isoparametric formulation may be used for any problem in which the approximations are C^0 continuous. In an isoparametric formulation a parent element is defined in terms of a set of natural coordinates. The shape functions are constructed on a parent element and used to compute the coordinates within each element using

$$\mathbf{x} = N_a(\boldsymbol{\xi})\tilde{\mathbf{x}}_a \tag{A.1}$$

where N_a denotes the shape function, $\boldsymbol{\xi}$ are a set of natural coordinates and $\tilde{\mathbf{x}}_a$ are nodal coordinates. A dependent variable $\mathbf{u}$ is then approximated as

$$\mathbf{u} \approx \mathbf{u}^h = N_a(\boldsymbol{\xi})\tilde{\mathbf{u}}_a \tag{A.2}$$

The construction of shape functions requires the selection of an appropriate set of natural coordinates. Here we first summarize the form for quadrilateral and brick elements in two and three dimensions, respectively. We then consider triangular and tetrahedral elements.

A.2 Quadrilateral elements

The natural coordinates for a quadrilateral element are given by

$$\boldsymbol{\xi} = (\xi, \eta); \quad -1 \leq \xi, \eta \leq 1$$

as shown in Fig. A.1.

The simplest group of elements construct the shape functions from products of one dimensional Lagrangian interpolation functions given by

$$l_a^n(\xi) = \frac{(\xi-\xi_0)(\xi-\xi_1)\cdots(\xi-\xi_{a-1})(\xi-\xi_{a+1})\cdots(\xi-\xi_n)}{(\xi_a-\xi_0)(\xi_a-\xi_1)\cdots(\xi_a-\xi_{a-1})(\xi_a-\xi_{a+1})\cdots(\xi_a-\xi_n)}, \quad a=1, 2, \cdots, n-1$$

$$= \prod_{\substack{b=0 \\ b \neq a}}^{n} \frac{\xi - \xi_b}{\xi_a - \xi_b}$$

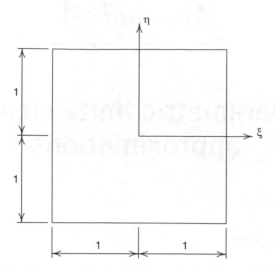

Fig. A.1 Natural coordinates for a quadrilateral.

and

$$l_a^n(\xi) = \frac{(\xi - \xi_1) \cdots (\xi - \xi_n)}{(\xi_a - \xi_1) \cdots (\xi_a - \xi_n)}, \quad a = 0$$

$$l_a^n(\xi) = \frac{(\xi - \xi_0) \cdots (\xi - \xi_{n-1})}{(\xi_a - \xi_0) \cdots (\xi_a - \xi_{n-1})}, \quad a = n$$

which gives a unit value at ξ_a and passes through zero at n points specified by the points $\xi_k, k = 0, 2, \cdots, n$. Using this form of interpolation we can construct the *Lagrangian family* of elements expressed as products of the one-dimensional functions given by

$$N_a(\xi, \eta) = l_a^m(\xi)\, l_a^n(\eta) \tag{A.3}$$

where m and n may be different orders in the natural coordinate directions. The simplest element uses linear interpolation in which $\xi_0, \eta_0 = -1$ and $\xi_1, \eta_1 = 1$ giving the four functions

$$\begin{aligned}
N_1 &= \frac{(\xi - 1)}{(-2)} \frac{(\eta - 1)}{(-2)} = \tfrac{1}{4}(1 - \xi)(1 - \eta) \\
N_2 &= \frac{(\xi + 1)}{(2)} \frac{(\eta - 1)}{(-2)} = \tfrac{1}{4}(1 + \xi)(1 - \eta) \\
N_3 &= \frac{(\xi + 1)}{(2)} \frac{(\eta + 1)}{(2)} = \tfrac{1}{4}(1 + \xi)(1 + \eta) \\
N_4 &= \frac{(\xi - 1)}{(-2)} \frac{(\eta + 1)}{(2)} = \tfrac{1}{4}(1 - \xi)(1 + \eta)
\end{aligned} \tag{A.4}$$

where nodes are numbered as shown in Fig. A.2(a). We shall also often use elements constructed from quadratic functions in which $\xi_0, \eta_0 = -1$, $\xi_1, \eta_1 = 0$, and $\xi_2, \eta_2 = 1$ giving a nine-node element. For an element with nodes numbered as shown in Fig. A.2(b) the shape functions are given by

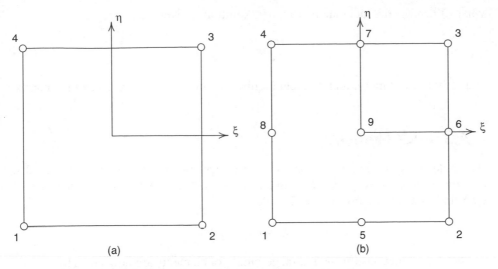

Fig. A.2 Node numbering for four-node and nine-node quadrilateral elements.

$$N_1 = \tfrac{1}{4}\xi(\xi - 1)\eta(\eta - 1)$$
$$N_2 = \tfrac{1}{4}\xi(\xi + 1)\eta(\eta - 1)$$
$$N_3 = \tfrac{1}{4}\xi(\xi + 1)\eta(\eta + 1)$$
$$N_4 = \tfrac{1}{4}\xi(\xi - 1)\eta(\eta + 1)$$
$$N_5 = \tfrac{1}{2}(1 - \xi^2)\eta(\eta - 1) \qquad\qquad \text{(A.5)}$$
$$N_6 = \tfrac{1}{2}\xi(\xi + 1)(1 - \eta^2)$$
$$N_7 = \tfrac{1}{2}(1 - \xi^2)\eta(\eta + 1)$$
$$N_8 = \tfrac{1}{2}\xi(\xi - 1)(1 - \eta^2)$$
$$N_9 = (1 - \xi^2)(1 - \eta^2)$$

In problems using C^0 shape functions it is necessary to construct the derivatives of variables with respect to the global coordinates. For an isoparametric formulation these derivatives are computed using the chain rule given by

$$\frac{\partial \mathbf{u}}{\partial \xi} = \frac{\partial \mathbf{u}}{\partial x_i}\frac{\partial x_i}{\partial \xi}$$
$$\frac{\partial \mathbf{u}}{\partial \eta} = \frac{\partial \mathbf{u}}{\partial x_i}\frac{\partial x_i}{\partial \eta}$$

which may be written in matrix form as

$$\begin{Bmatrix} \dfrac{\partial \mathbf{u}}{\partial \xi} \\[2mm] \dfrac{\partial \mathbf{u}}{\partial \eta} \end{Bmatrix} = \begin{bmatrix} \dfrac{\partial x_1}{\partial \xi} & \dfrac{\partial x_2}{\partial \xi} \\[2mm] \dfrac{\partial x_1}{\partial \eta} & \dfrac{\partial x_2}{\partial \eta} \end{bmatrix} \begin{Bmatrix} \dfrac{\partial \mathbf{u}}{\partial x_1} \\[2mm] \dfrac{\partial \mathbf{u}}{\partial x_2} \end{Bmatrix}$$

$$\frac{\partial \mathbf{u}}{\partial \xi} = \mathbf{J}\frac{\partial \mathbf{u}}{\partial \mathbf{x}}$$

where $\mathbf{J}$ denotes the Jacobian matrix. The solution is given by

$$\frac{\partial \mathbf{u}}{\partial \mathbf{x}} = \mathbf{J}^{-1} \frac{\partial \mathbf{u}}{\partial \boldsymbol{\xi}} \tag{A.6}$$

and for the two-dimensional problem requires the inverse of a 2×2 Jacobian matrix.

A.3 Brick elements

Brick elements are generalization to three dimensions of the quadrilateral used in two dimensions. The global coordinates $x_i, i = 1, 2, 3$ now require three natural coordinates which we give as (see Fig. A.3)

$$\boldsymbol{\xi} = (\xi, \eta, \zeta)$$

The shape functions for the Lagrange family of elements are now given by

$$N_a(\xi, \eta, \zeta) = l_a^m(\xi)\, l_a^n(\eta)\, l_a^p(\zeta) \tag{A.7}$$

with the Jacobian transformation given by the 3×3 matrix

$$\mathbf{J} = \begin{bmatrix} \dfrac{\partial x_1}{\partial \xi} & \dfrac{\partial x_2}{\partial \xi} & \dfrac{\partial x_3}{\partial \xi} \\[2ex] \dfrac{\partial x_1}{\partial \eta} & \dfrac{\partial x_2}{\partial \eta} & \dfrac{\partial x_3}{\partial \eta} \\[2ex] \dfrac{\partial x_1}{\partial \zeta} & \dfrac{\partial x_2}{\partial \zeta} & \dfrac{\partial x_3}{\partial \zeta} \end{bmatrix} \tag{A.8}$$

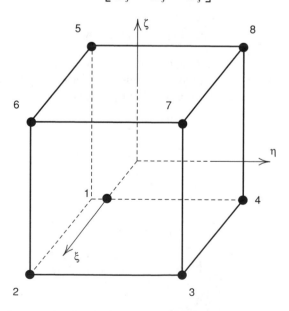

Fig. A.3 Natural coordinates and node order for an eight-node brick element.

with the gradients now given by

$$\frac{\partial \mathbf{u}}{\partial \mathbf{x}} = \left(\frac{\partial \mathbf{u}}{\partial x_1} \quad \frac{\partial \mathbf{u}}{\partial x_2} \quad \frac{\partial \mathbf{u}}{\partial x_3} \right)^T$$

$$\frac{\partial \mathbf{u}}{\partial \boldsymbol{\xi}} = \left(\frac{\partial \mathbf{u}}{\partial \xi} \quad \frac{\partial \mathbf{u}}{\partial \eta} \quad \frac{\partial \mathbf{u}}{\partial \zeta} \right)^T$$

(A.9)

The simplest member of the three-dimensional family is an eight-node brick with nodal order as shown in Fig. A.3. The shape functions are given by

$$N_1 = \tfrac{1}{8}(1 - \xi)(1 - \eta)(1 - \zeta)$$
$$N_2 = \tfrac{1}{8}(1 + \xi)(1 - \eta)(1 - \zeta)$$
$$N_3 = \tfrac{1}{8}(1 + \xi)(1 + \eta)(1 - \zeta)$$
$$N_4 = \tfrac{1}{8}(1 - \xi)(1 + \eta)(1 - \zeta)$$
$$N_5 = \tfrac{1}{8}(1 - \xi)(1 - \eta)(1 + \zeta)$$
$$N_6 = \tfrac{1}{8}(1 + \xi)(1 - \eta)(1 + \zeta)$$
$$N_7 = \tfrac{1}{8}(1 + \xi)(1 + \eta)(1 + \zeta)$$
$$N_8 = \tfrac{1}{8}(1 - \xi)(1 + \eta)(1 + \zeta)$$

(A.10)

Higher-order elements may be constructed similar to that given for the two-dimensional form above.

A.4 Triangular elements

For two-dimensional problems in which the elements are of triangular shape the natural coordinates are taken as the *area coordinates*, L_a, $a = 1, 2, 3$ as shown in Fig. A.4(a), in which the constraint

$$L_1 + L_2 + L_3 = 1$$

is used to limit the number of independent values to the same number as in the global x_1, x_2 Cartesian coordinates.

The shape functions for the three-node triangular element are given by

$$N_a = L_a, \quad a = 1, 2, 3$$

Derivatives with respect to x_i for shape functions expressed in area coordinates may be computed using the chain rule

$$\frac{\partial N_a}{\partial x_i} = \frac{\partial N_a}{\partial L_b} \frac{\partial L_b}{\partial x_i}$$

The derivatives $\partial L_b / \partial x_i$ may be computed by writing three equations

$$r_1 = x_1 - \sum_a N_a \tilde{x}_{1a} = 0$$

$$r_2 = x_2 - \sum_a N_a \tilde{x}_{2a} = 0$$

$$r_3 = 1 - \sum_a L_a = 0$$

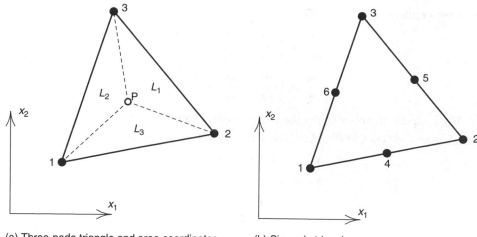

(a) Three-node triangle and area coordinates   (b) Six-node triangle

Fig. A.4 Natural coordinates and node order for triangular elements.

and using the chain rule

$$\frac{\partial r_j}{\partial x_i} = \frac{\partial r_j}{\partial L_b} \frac{\partial L_b}{\partial x_i}$$

This gives

$$
\begin{bmatrix} 1 & 0 \\ 0 & 1 \\ 0 & 0 \end{bmatrix} =
\begin{bmatrix}
\sum_a \dfrac{\partial N_a}{\partial L_1}\tilde{x}_{1a} & \sum_a \dfrac{\partial N_a}{\partial L_2}\tilde{x}_{1a} & \sum_a \dfrac{\partial N_a}{\partial L_3}\tilde{x}_{1a} \\
\sum_a \dfrac{\partial N_a}{\partial L_1}\tilde{x}_{2a} & \sum_a \dfrac{\partial N_a}{\partial L_2}\tilde{x}_{2a} & \sum_a \dfrac{\partial N_a}{\partial L_3}\tilde{x}_{2a} \\
1 & 1 & 1
\end{bmatrix}
\begin{bmatrix}
\dfrac{\partial L_1}{\partial x_1} & \dfrac{\partial L_1}{\partial x_2} \\
\dfrac{\partial L_2}{\partial x_1} & \dfrac{\partial L_2}{\partial x_2} \\
\dfrac{\partial L_3}{\partial x_1} & \dfrac{\partial L_3}{\partial x_2}
\end{bmatrix}
$$

which may be solved for the derivatives $\partial L_a/\partial x_j$.

Higher-order triangles may also be constructed. We leave it to the reader to show that the shape functions for the six-node triangle shown in Fig. A.4(b) may be written as

$$\begin{aligned} N_a &= \tfrac{1}{2}L_a(2L_a - 1) \\ N_{a+3} &= 4\,L_a L_b \end{aligned} \quad a = 1, 2, 3; \; b = 2, 3, 1$$

A.5 Tetrahedral elements

Tetrahedral elements may be constructed in a manner similar to that used for triangular elements. In the case of tetrahedral elements we use *volume coordinates* L_a, $a = 1, 2, 3, 4$ with a constraint

$$L_1 + L_2 + L_3 + L_4 = 1$$

The shape functions for a four-node tetrahedron are given by

$$N_a = L_a; \quad a = 1, 2, 3, 4$$

Derivatives are computed using the chain rule in the same way as for triangles, the difference being that arrays are of size 4 and 3 instead of 3 and 2. We leave this as an exercise for the reader to write appropriate relations for these steps.

Appendix B

Invariants of second-order tensors

B.1 Principal invariants

Given any second-order Cartesian tensor $\mathbf{a}$ with components expressed as

$$\mathbf{a} = \begin{bmatrix} a_{11} & a_{12} & a_{13} \\ a_{21} & a_{22} & a_{23} \\ a_{31} & a_{32} & a_{33} \end{bmatrix} \tag{B.1}$$

the *principal values* of $\mathbf{a}$, denoted as a_1, a_2, and a_3, may be computed from the solution of the eigenproblem

$$\mathbf{a}\,\mathbf{q}^{(m)} = a_m \mathbf{q}^{(m)} \tag{B.2}$$

in which the (right) eigenvectors $\mathbf{q}^{(m)}$ denote *principal directions* for the associated eigenvalue a_m. Non-trivial solutions of Eq. (B.2) require

$$\det \begin{vmatrix} (a_{11} - a) & a_{12} & a_{13} \\ a_{21} & (a_{22} - a) & a_{23} \\ a_{31} & a_{32} & (a_{33} - a) \end{vmatrix} = 0 \tag{B.3}$$

Expanding the determinant results in the cubic equation

$$a_m^3 - \mathrm{I}_a a_m^2 + \mathrm{II}_a a_m - \mathrm{III}_a = 0 \tag{B.4}$$

where:

$$
\begin{aligned}
\mathrm{I}_a &= a_{11} + a_{22} + a_{33} \\
\mathrm{II}_a &= a_{11}a_{22} + a_{22}a_{33} + a_{33}a_{11} - a_{12}a_{21} - a_{23}a_{32} - a_{31}a_{13} \\
\mathrm{III}_a &= a_{11}a_{22}a_{33} - a_{11}a_{23}a_{32} - a_{22}a_{31}a_{13} - a_{33}a_{12}a_{21} + a_{12}a_{23}a_{31} + a_{21}a_{32}a_{13} \\
&= \det \mathbf{a}
\end{aligned} \tag{B.5}
$$

The quantities I_a, II_a, and III_a are called the *principal invariants of* $\mathbf{a}$. The roots of Eq. (B.4) give the principal values a_m.

The invariants for the deviator of $\mathbf{a}$ may be obtained by using

$$\mathbf{a}' = \mathbf{a} - \bar{a}\mathbf{1} \tag{B.6}$$

where $\bar{a}$ is the mean defined as

$$\bar{a} = \tfrac{1}{3}(a_{11} + a_{22} + a_{33}) = \tfrac{1}{3}\mathrm{I}_a \tag{B.7}$$

Substitution of Eq. (B.6) into Eq. (B.2) gives

$$\left[\mathbf{a}' + \bar{a}\mathbf{I}\right]\mathbf{q}^{(m)} = a_m\mathbf{q}^{(m)} \tag{B.8}$$

or

$$\mathbf{a}'\mathbf{q}^{(m)} = (a_m - \bar{a})\mathbf{q}^{(m)} = a_m'\mathbf{q}^{(m)} \tag{B.9}$$

which yields a cubic equation for principal values of the deviator given as

$$(a_m')^3 + \mathrm{II}_a'a_m' - \mathrm{III}_a' = 0 \tag{B.10}$$

where invariants of $\mathbf{a}'$ are denoted as I_a', II_a', and III_a'.

Since the deviators $\mathbf{a}'$ differ from the total $\mathbf{a}$ by a mean term only, we observe from Eq. (B.9) that the directions of their principal values coincide, and the three principal values are related through

$$a_i = a_i' + \bar{a}; \quad i = 1, 2, 3 \tag{B.11}$$

Moreover Eq. (B.10) generally has a closed-form solution which may be constructed by using the Cardon formula.[1,2]

The definition of $\mathbf{a}'$ given by Eq. (B.6) yields

$$\mathrm{I}_a' = a_{11}' + a_{22}' + a_{33}' = 0 \tag{B.12}$$

Using this result, the second invariant of the deviator may be shown to have the indicial form[3]

$$\mathrm{II}_a' = -\tfrac{1}{2}a_{ij}'a_{ji}' \tag{B.13}$$

The third invariant is again given by

$$\mathrm{III}_a' = \det \mathbf{a}' \tag{B.14}$$

however, we show in Sec. B.2 that this invariant may be written in a form which is easier to use in many applications (e.g. yield functions for elasto-plastic materials).

B.2 Moment invariants

It is also possible to write the invariants in a form known as *moment invariants*.[4] The moment invariants are denoted as $\bar{\mathrm{I}}_a$, $\bar{\mathrm{II}}_a$, $\bar{\mathrm{III}}_a$, and are defined by the indicial forms

$$\bar{\mathrm{I}}_a = a_{ii}, \qquad \bar{\mathrm{II}}_a = \tfrac{1}{2}a_{ij}a_{ji}, \qquad \bar{\mathrm{III}}_a = \tfrac{1}{3}a_{ij}a_{jk}a_{ki} \tag{B.15}$$

We observe that moment invariants are directly related to the *trace* of products of $\mathbf{a}$. The trace (tr) of a matrix is defined as the sum of its diagonal elements. Thus, the first three moment invariants may be written in matrix form (using a square matrix for $\mathbf{a}$) as

$$\bar{\mathrm{I}}_a = \mathrm{tr}\,(\mathbf{a}), \qquad \bar{\mathrm{II}}_a = \tfrac{1}{2}\mathrm{tr}\,(\mathbf{aa}), \qquad \bar{\mathrm{III}}_a = \tfrac{1}{3}\mathrm{tr}\,(\mathbf{aaa}) \tag{B.16}$$

The moment invariants may be related to the principal invariants as[4]

$$\begin{aligned}
\bar{\mathrm{I}}_a &= \mathrm{I}_a, & \bar{\mathrm{II}}_a &= \tfrac{1}{2}\mathrm{I}_a^2 - \mathrm{II}_a, & \bar{\mathrm{III}}_a &= \mathrm{III}_a - \tfrac{1}{3}\mathrm{I}_a^3 + \mathrm{I}_a\mathrm{II}_a \\
\mathrm{I}_a &= \bar{\mathrm{I}}_a, & \mathrm{II}_a &= \tfrac{1}{2}\bar{\mathrm{I}}_a^2 - \bar{\mathrm{II}}_a, & \mathrm{III}_a &= \bar{\mathrm{III}}_a + \tfrac{1}{6}\bar{\mathrm{I}}_a^3 - \bar{\mathrm{I}}_a\bar{\mathrm{II}}_a
\end{aligned} \tag{B.17}$$

Using Eq. (B.12) and the identities given in Eq. (B.17) we can immediately observe that the principal invariants and the moment invariants for a deviatoric second-order tensor are related through

$$\mathrm{II}'_a = -\bar{\mathrm{II}}'_a \quad \text{and} \quad \mathrm{III}'_a = \bar{\mathrm{III}}'_a = \det \mathbf{a}' \tag{B.18}$$

B.3 Derivatives of invariants

We often also need to compute the derivative of the invariants with respect to their components and this is only possible when all components are treated independently – that is, we do not use any symmetry, if present. From the definitions of the principal and moment invariants given above, it is evident that derivatives of the moment invariants are the easiest to compute since they are given in concise indicial form. Derivatives of principal invariants can be computed from these by using the identities given in Eqs (B.17) and (B.18).

The first derivatives of the principal invariants for symmetric second-order tensors may be expressed in a matrix form directly, as shown by Nayak and Zienkiewicz;[5,6] however, second derivatives from these are not easy to construct and we now prefer the methods given here.

B.3.1 First derivatives of invariants

The first derivative of each moment invariant may be computed by using Eq. (B.15). For the first invariant we obtain

$$\frac{\partial \bar{\mathrm{I}}_a}{\partial a_{ij}} = \delta_{ij} \tag{B.19}$$

Similarly, for the second moment invariant we get

$$\frac{\partial \bar{\mathrm{II}}_a}{\partial a_{ij}} = a_{ji} \tag{B.20}$$

and for the third moment invariant

$$\frac{\partial \bar{\mathrm{III}}_a}{\partial a_{ij}} = a_{jk}a_{ki} \tag{B.21}$$

Using the identities, the derivative of the principal invariants may be written in indicial form as

$$\frac{\partial \mathrm{I}_a}{\partial a_{ij}} = \delta_{ij}, \quad \frac{\partial \mathrm{II}_a}{\partial a_{ij}} = \mathrm{I}_a\delta_{ij} - a_{ji}, \quad \frac{\partial \mathrm{III}_a}{\partial a_{ij}} = \mathrm{II}_a\delta_{ij} - \mathrm{I}_a a_{ji} + a_{jk}a_{ki}\mathrm{III}_a a_{ji}^{-1} \tag{B.22}$$

The third invariant may also be shown to have the representation[3]

$$\frac{\partial \mathrm{III}_a}{\partial a_{ij}} = \mathrm{III}_a a_{ji}^{-1} \tag{B.23}$$

where a_{ji}^{-1} is the inverse (transposed) of the a_{ij} tensor. Thus, in matrix form we may write the derivatives as

$$\frac{\partial I_a}{\partial \mathbf{a}} = 1, \qquad \frac{\partial II_a}{\partial \mathbf{a}} = I_a 1 - \mathbf{a}^{\mathrm{T}}, \qquad \frac{\partial III_a}{\partial \mathbf{a}} = III_a \mathbf{a}^{-\mathrm{T}} \qquad \text{(B.24)}$$

where here 1 denotes a 3×3 identity matrix.

The expression for the derivative of the determinant of a second-order tensor is of particular use as we shall encounter this in dealing with volume change in finite deformation problems and in plasticity yield functions and flow rules.

Performing the same steps for the invariants of the deviator yields

$$\frac{\partial I'}{\partial a'_{ij}} = \frac{\partial \bar{I}'}{\partial a'_{ij}} = 0, \qquad \frac{\partial II'}{\partial a'_{ij}} = -\frac{\partial \bar{II}'}{\partial a'_{ij}} = -a'_{ji}, \qquad \frac{\partial III'}{\partial a'_{ij}} = \frac{\partial \bar{III}'}{\partial a'_{ij}} = a'_{jk}a'_{ki} \qquad \text{(B.25)}$$

with only a sign change occurring in the second invariant to obtain the derivative of principal invariants from derivatives of moment invariants.

Often the derivatives of the invariants of a deviator tensor are needed with respect to the tensor itself, and these may be computed as

$$\frac{\partial (\cdot)'}{\partial a_{mn}} = \frac{\partial (\cdot)'}{\partial a'_{ij}} \frac{\partial a'_{ij}}{\partial a_{mn}} \qquad \text{(B.26)}$$

where

$$\frac{\partial a'_{ij}}{\partial a_{mn}} = \delta_{im}\delta_{jn} - \frac{1}{3}\delta_{ij}\delta_{mn} \qquad \text{(B.27)}$$

Combining the two expressions yields

$$\frac{\partial (\cdot)'}{\partial a_{mn}} = \frac{\partial (\cdot)'}{\partial a'_{mn}} - \frac{1}{3}\delta_{mn}\left[\delta_{ij}\frac{\partial (\cdot)'}{\partial a'_{ij}}\right] \qquad \text{(B.28)}$$

B.3.2 Second derivatives

In developments of tangent tensors we need second derivatives of the invariants. These may be computed directly from Eqs (B.19)–(B.21) by standard operations. The second derivatives of $\bar{I}_a, \bar{II}_a, \bar{III}_a$ yield

$$\frac{\partial^2 \bar{I}_a}{\partial a_{ij}\partial a_{kl}} = 0, \qquad \frac{\partial^2 \bar{II}_a}{\partial a_{ij}\partial a_{kl}} = \delta_{jk}\delta_{il}, \qquad \frac{\partial^2 \bar{III}_a}{\partial a_{ij}\partial a_{kl}} = \delta_{jk}a_{il} + a_{jk}\delta_{il} \qquad \text{(B.29)}$$

The computations for principal invariants follow directly from the above using the identities given in Eqs (B.17) and (B.18). Also, all results may be transformed to the vector form used extensively in this volume for the finite element constructions. These steps are by now a standard process and are left as an exercise for the reader.

References

1. H.M. Westergaard. *Theory of Elasticity and Plasticity*. Harvard University Press, Cambridge, 1952.
2. W.H. Press *et al.*, editors. *Numerical Recipes in Fortran: The Art of Scientific Computing*. Cambridge University Press, Cambridge, 2nd edition, 1992.
3. I.H. Shames and F.A. Cozzarelli. *Elastic and Inelastic Stress Analysis*. Taylor & Francis, Washington, DC, 1997. (Revised printing.)
4. J.L. Ericksen. Tensor fields. In S. Flügge, editor, *Encyclopedia of Physics*, volume III/1. Springer-Verlag, Berlin, 1960.
5. G.C. Nayak and O.C. Zienkiewicz. Convenient forms of stress invariants for plasticity. *Proc. Am. Soc. Civ. Eng.*, 98(ST4):949–953, 1972.
6. O.C. Zienkiewicz and R.L. Taylor. *The Finite Element Method*, volume 2. McGraw-Hill, London, 4th edition, 1991.

Author index

Page numbers in **bold** are for pages at the end of chapters with names of author references.

Subject index